**Progress in Mathematics**
Volume 95

# Representation Theory of Finite Groups and Finite-Dimensional Algebras

**Proceedings of the Conference at the University of Bielefeld from May 15–17, 1991, and 7 Survey Articles on Topics of Representation Theory**

**Edited by**
**G. O. Michler**
**C. M. Ringel**

**Springer Basel AG**

Editor's addresses:

Prof. Dr. G. O. Michler
Institut für Experimentelle
Mathematik
Universität GHS Essen
Ellernstrasse 29
D–4300 Essen 12

Prof. Dr. C. M. Ringel
Universität Bielefeld
Fakultät für Mathematik
Universitätsstrasse
D–4800 Bielefeld

**Library of Congress Cataloging-in-Publication Data**

**Representation theory of finite groups and finite-dimensional algebras**
/ edited by G. O. Michler, C. M. Ringel.
p. cm. – (Progress in mathematics ; v. 95)
Includes bibliographical references and index.
ISBN 978-3-0348-9720-4 ISBN 978-3-0348-8658-1 (eBook)
DOI 10.1007/978-3-0348-8658-1
1. Finite groups. 2. Representations of groups.
3. Representations of algebras. I. Michler, G. (Gerhard), 1938–
. II. Ringel, Claus Michael. III. Title: Finite-dimensional
algebras. IV. Series: Progress in mathematics (Boston, Mass.) ;
vol. 95
QA171.R429 1991
512'.2 – dc20

**Deutsche Bibliothek Cataloging-in-Publication Data**

**Representation theory of finite groups and finite-dimensional algebras** / ed. by G. O. Michler ; C. M. Ringel. – Basel ;
Boston ; Berlin : Birkhäuser, 1991
(Progress in mathematics ; Vol. 95)
ISBN 978-3-0348-9720-4
NE: Michler, Gerhard O. [Hrsg.]; GT


Originally published by Birkhäuser Verlag Basel in 1991
Softcover reprint of the hardcover 1st edition 1991
ISBN 978-3-0348-9720-4

# CONTENTS

# PREFACE

From April 1, 1984 until March 31, 1991 the Deutsche Forschungsgemeinschaft has sponsored the project "Representation Theory of Finite Groups and Finite Dimensional Algebras". The proposal for this project was submitted by B. Huppert (Mainz), B. Fischer (Bielefeld), G. Michler (Essen), H. Pahlings (Aachen) and C.M. Ringel (Bielefeld) in order to strengthen the interaction between the different research areas in representation theory.

The Deutsche Forschungsgemeinschaft has given many research positions and fellowships for young algebraists enabling them to do research at their own universities or as visitors at well known research institutions in America, Australia, England and France. The whole project benefitted very much from an extensive exchange programme between German and American scientists sponsored by the Deutsche Forschungsgemeinschaft and by the National Science Foundation of the United States.

This volume presents lectures given in a final conference and reports by members of the project. It is divided into two parts. The first part contains seven survey articles describing recent advances in different areas of representation theory. These articles do not only concentrate on the work done by the German research groups, but also inform on major developments of the subject at all. The volume omits those topics already treated in book form. In particular, it does not contain a survey on K. Erdmann's work on the structure of tame blocks of finite groups; her recent book explains in detail the fruitful interaction between the homological and combinatorial methods of the representation theory of finite dimensional algebras, and the group and character theoretical methods of group representation theory. D. Happel's book on triangulated categories and the representation theory of algebras and the book on Brauer trees written by G. Hiss and K. Lux present important recent developments. These books have already become standard references.

The second part contains 17 articles representing the contents of the lectures which have be delivered at the conference on "Representation Theory of Finite Groups and Finite Dimensional Algebras" at the University of Bielefeld from 15 until 17 May 1991. Invited lectures have been given by M. Auslander, J.A. Green, M. Isaacs and J.C. Jantzen. These mathematicians have influenced the whole project very much either as visitors at German universities or as hosts of German visitors at their home institutions. Their articles are followed by 13 research papers of German algebraists having benefitted from the support of the project.

The editors thank the Deutsche Forschungsgemeinschaft for its support.

G. Michler, C.M. Ringel

Bielefeld and Essen, January 1991

## CONTRIBUTORS


| | |
|---|---|
| **M. Auslander** | Department of Mathematics, Brandeis University, Waltham, Massachusetts 02254, USA |
| **C. Bessenrodt** | Institut für Experimentelle Mathematik, Universität GHS Essen, 4300 Essen 12, Germany |
| **E. Dieterich** | Mathematisches Institut, Universität Zürich, 8001 Zürich, Switzerland |
| **R. Dipper** | Department of Mathematics, University of Oklahoma, Norman, Oklahoma 73019, USA |
| **K. Erdmann** | Somerville College, University of Oxford, Oxford OX2 6HD, England |
| **B. Fischer** | Fakultät für Mathematik, Universität Bielefeld, 4800 Bielefeld 1, Germany |
| **J.A. Green** | Mathematics Institute, University of Warwick, Coventry CV47AL, England |
| **D. Happel** | Fakultät für Mathematik, Universität Bielefeld, 4800 Bielefeld 1, Germany |
| **G. Hiss** | Interdisziplinäres Zentrum für Wissenschaftliches Rechnen, Universität Heidelberg, 6900 Heidelberg, Germany |
| **B. Huppert** | Fachbereich Mathematik, Universität Mainz, 6500 Mainz, Germany |
| **M. Isaacs** | Department of Mathematics, University of Wisconsin, Madison, Wisconsin 537706, USA |
| **J.C. Jantzen** | Department of Mathematics, University of Oregon, Eugene, Oregon 97403-1222, USA |
| **R. Knörr** | Fachbereich Mathematik, Universität Mainz, 6500 Mainz, Germany |
| **B. Külshammer** | Institut für Mathematik, Universität Augsburg, 8900 Augsburg, Germany |
| **K. Lux** | Lehrstuhl D für Mathematik, Technische Universität, 5100 Aachen, Germany |
| **G. Malle** | Interdisziplinäres Zentrum für Wissenschaftliches Rechnen, Universität Heidelberg, 6900 Heidelberg, Germany |
| **O. Manz** | Interdisziplinäres Zentrum für Wissenschaftliches Rechnen, Universität Heidelberg, 6900 Heidelberg, Germany |

| | |
|---|---|
| **B.H. Matzat** | Interdisziplinäres Zentrum für Wissenschaftliches Rechnen, Universität Heidelberg, 6900 Heidelberg, Germany |
| **G.O. Michler** | Institut für Experimentelle Mathematik, Universität GHS Essen, 4300 Essen 12, Germany |
| **H. Pahlings** | Lehrstuhl D für Mathematik, Technische Universität, 5100 Aachen, Germany |
| **W. Plesken** | Lehrstuhl B für Mathematik, Technische Universität, 5100 Aachen, Germany |
| **I. Reiten** | Institutt for Matematikk og Statistikk, Universitetet i Trondheim, 7055 Dragvoll, Norway |
| **C.M. Ringel** | Fakultät für Mathematik, Universität Bielefeld, 4800 Bielefeld 1, Germany |
| **K.W. Roggenkamp** | Mathematisches Institut B, Universität Stuttgart, 7000 Stuttgart 80, Germany |
| **A. Wiedemann** | Mathematisches Institut B, Universität Stuttgart, 7000 Stuttgart 80, Germany |
| **W. Willems** | Institut für Experimentelle Mathematik, Universität GHS Essen, 4300 Essen 12, Germany |

Progress in Mathematics, Vol. 95, © 1991 Birkhäuser Verlag Basel

# Clifford–Matrices

**BERND FISCHER**

## Introduction

If $G$ is a finite group and $N$ a normal $p$–subgroup of $G$ then the irreducible complex characters of $G$ can be described in terms of irreducible characters $\varphi$ of $N$ and induced characters $\eta^G$ for irreducible characters $\eta$ of inertia group $T_G(\varphi)$ satisfying $(\eta_{N'}, \varphi)_N \neq 0$ by Clifford's theory. In complicated cases it is often easier to compute the characters of $G$ rather than the relevant characters of $T_G(\varphi)$.

In this paper we intend to describe the characters $\eta^G$ in terms of information we have for $T_G(\varphi)/N$. These inertia quotient groups are often very natural subgroups of $G/N$ and have much simpler structure then $T_G(\varphi)$. For each coset $Nx$ in $G/N$ we describe a certain matrix, which allows us to write $(\eta^G(x_1), \ldots, \eta^G(x_t))$ for $x_i \in Nx$ as a linear combination of rows of this matrix and coefficients which are values of certain "projective" characters $\hat{\eta}$ of $T_G(\varphi)/N$.

This method has the advantage that character–tables of pretty complicated groups (e.g. maximal local subgroups of the "monster") can be computed because the arithmetical properties of the matrices involved are very powerful.

The examples given in this paper are illustrations of the method. Our group–theoretic and character–theoretic notations are standard and we need only very elementary theorems on characters.

### 1. Basic definitions

Let $G$ be a finite group. Let $G^\vee$ be the set of complex irreducible characters of $G$. We denote classfunctions, generalized characters and characters of $G$ by $\mathbb{C}[G], \mathbb{Z}[G], \mathbb{Z}^+[G]$. If $U \leq G$ and $\varphi \in \mathbb{C}[U]$, $\chi \in \mathbb{C}[G]$ then $\varphi^G \in \mathbb{C}[G], \chi_U \in \mathbb{C}[U]$ denote the induced classfunction, resp. the restriction. $(\ ,\ )_G$ denotes the standard hermitian form on $\mathbb{C}[G]$.

Let $X \subseteq G$ and $f : X \longrightarrow \mathbb{C}$. For $g, h \in G$ let

$$ {}^g f(h) = \begin{cases} f(ghg^{-1}), & \text{if } ghg^{-1} \in X, \\ 0, & \text{if } ghg^{-1} \notin X, \end{cases} $$

and

$$T_G(f) = \{t \in G \mid {}^t f = {}^1 f\},$$

the inertia group of $f$ in $G$; if $X \leq G$ and $f \in \mathbb{C}[X]$ we use the same notations.

Let $N$ be a normal $p$–subgroup of $G$ for a fixed prime $p$. For $\varphi \in \mathbb{C}[N]$ let

$$N_\varphi = \langle x \in N \mid \varphi(x) \neq 0 \rangle \trianglelefteq N.$$

We fix $\varphi \in N^\vee$. Then we have

**(1.1)** There is a subgroup $L \leq N$ and $\lambda \in L^\vee$ such that $\lambda^N = \varphi$.

**(1.2)** $N_\varphi = N$ iff $\varphi(1) = 1$.

(1.1) holds since $\varphi$ is monomial. If $\varphi(1) = 1$ then $\varphi(x) \in \mathbf{C}^*$ for $x \in N$. If $\varphi(1) > 1$ the subgroup $L$ in (1.1) is contained in a proper normal subgroup of $N$.

**(1.3)** Let $N_1 \trianglerighteq \ldots N_m \trianglerighteq N_{m+1}$ and $\lambda_i \in N_i^\vee$ be recursively chosen by $N_1 = N, \lambda_1 = \varphi$ and

$$N_{i+1} = N_{i\lambda_i}, (\lambda_{i+1}, \lambda_i|_{N_{i+1}})_{N_{i+1}} \neq 0.$$

The choice of $\lambda_{i+1}$ is unique up to conjugation in $N_i$ by Clifford, and $\lambda_i^N \in \mathbf{N}\varphi$ by induction. By (1.2) $N_i = N_{i+1}$ iff $\lambda_i(1) = 1$.

We choose $N(\varphi) = N_m$ for $N_m = N_{m+1}$. Then $N(\varphi)$ is unique up to conjugation in $N$. Let $\lambda_\varphi = \lambda_m$ and $K(\varphi) = \ker(\lambda_\varphi)$. Then $K(\varphi) \leq \ker\varphi$. Let $L(\varphi) = L_G(\varphi) = T_G({}^1\lambda_\varphi) \trianglerighteq N(\varphi)$. Then $K(\varphi) \trianglelefteq L(\varphi)$. Furthermore

**(1.4)** $$T_G(\varphi) = NL(\varphi) = NT_G({}^1\lambda_i).$$

**(1.5)** $$N(\varphi)/K(\varphi) \text{ is cyclic.}$$

**(1.6)** $$N(\varphi)/K(\varphi) \leq \mathbf{Z}(L(\varphi)/K(\varphi)).$$

**Proof:** If $\varphi(1) = 1$ then $L(\varphi) = T_G(\varphi) \geq N$, and the statements are trivial. Since $\lambda_\varphi(1) = 1$ we have (1.5) and (1.6).

Let $\varphi(1) > 1$ but $N(\varphi) = N_\varphi$. Then $(\lambda_\varphi^{L(\varphi)})^N{}_N$ is a multiple of $\varphi$. Hence $T_G(\varphi) \geq NL(\varphi)$. Let $g \in T_G(\varphi)$. Then $N(\varphi)^g = N(\varphi)$ and ${}^g\lambda_{\varphi_N}$ is conjugate to $\lambda_\varphi$ in $N$. Hence (1.4) holds in this case.

(1.4) follows by induction on $m$ with $N_m = N_{m+1}$ if we replace $N$ by $N_2$.

Let $\eta \in T_G(\varphi)^\vee$. Then $\eta$ is $\varphi$–<u>relevant</u>, if $\eta_N \in \mathbf{N}\varphi$.

**(1.7) Theorem (Clifford).** *If $\eta \in T_G(\varphi)^\vee$ is $\varphi$–relevant then $\eta^G \in G^\vee$ and each $\chi \in G^\vee$ with $(\chi_N, \varphi)_N \neq 0$ is of this form. If $\eta, \tilde\eta$ are different $\varphi$–relevant characters then $\eta^G \neq \tilde\eta^G$.*

**(1.8)** Let $\eta \in L(\varphi)^\vee$ be $\lambda_\varphi$–relevant. Then $\eta^{T_G(\varphi)}$ is $\varphi$–relevant and each $\varphi$–relevant character is of this form.

**Proof:** The groups $N_i$ in (1.3) are normalized by $L(\varphi)$. Let $N_{m+1} = N_m < N_{m-1}$. Let $\theta_i = \eta^{N_iL(\varphi)}$, $i < m$. Then $\theta_{iN_i} \in \mathbf{N}\lambda_i$. By (1.4) $T_{N_{i-1}L(\varphi)}({}^1\lambda_i) = N_iL(\varphi)$.

Hence (1.7) implies that $\theta_{i-1}$ is irreducible and $\lambda_{i-1}$ relevant in $N_{i-2}L(\varphi)$.

**(1.9) Lemma.** *Let $x \in T_G(\varphi)$ such that $\eta(x) \neq 0$ for some $\varphi$–relevant character $\eta \in T_G(\varphi)^\vee$. Then there is a map $\zeta_x : Nx \longrightarrow \mathbf{C}$ such that*

$$\eta(y) = \zeta_x(y) \cdot \eta(x), \qquad y \in Nx,$$

*and each $\varphi$–relevant character $\eta \in T_G(\varphi)^\vee$.*

**Proof:** Let $\eta$ be $\varphi$–relevant and $\sigma(\eta)$ the $\lambda_\varphi$–relevant character of $L(\varphi)$ with $\sigma(\eta)^{T_G(\varphi)} = \eta$. Then $x^N \cap L(\varphi) \neq \emptyset$, and we assume $x \in L(\varphi)$.

Let $y \in Nx$. If $\eta'(y) \neq 0$ for some $\varphi$–relevant character $\eta'$ of $L(\varphi)$ then $y^N \cap L(\varphi) \neq \emptyset$, and we assume $y \in L(\varphi)$, $\sigma(\eta')(y) \neq 0$. Restrictions $\eta_{N\langle x\rangle}$ have only $\varphi$–relevant constituents in $N\langle x\rangle$. Thus we may assume $G = N\langle x\rangle$ in order to prove $y^N \cap N(\varphi)x \neq \emptyset$ : by (1.6) $N(\varphi)\langle x\rangle/K(\varphi)$ is abelian, and we get for $\lambda_\varphi$–relevant characters $\sigma$ of $L(\varphi)$

$$(*) \qquad \sigma(xz) = \lambda_\varphi(z)\sigma(x), \qquad \text{for } z \in N(\varphi);$$

and there are exactly $|\langle xN/N\rangle|$ characters of this type, since $\varphi$ is extendable to $N\langle x\rangle$; inducing $\lambda_\varphi$–relevant characters of $N(\varphi)\langle x\rangle$ to $N\langle x\rangle$, each $\varphi$–relevant character of $N\langle x\rangle$ appears as constituent (and no others).

Hence $y^N \cap N(\varphi)x \neq \emptyset$. The final statement follows from $(*)$.

**Remark.** In general, the choice of $x \in Nx$ in (1.9) depends on $\varphi$ : if $x \in T_G(\varphi) \cap T_G(\varphi')$ it may be that $\eta(x) \neq 0$ for some $\varphi$–relevant character but $\eta'(x) = 0$ for each $\varphi'$–relevant character.

**(1.10) Definition.** Let $\varphi_1 = 1_N, \varphi_2, \dots, \varphi_t$ be representatives of the $G$–conjugacy classes of $N^\vee$. Let $T_i = T_G(\varphi_i)/N$. Let $X \in G/N$.

**(1.10.1)** Let $C(X) = \{x_1, -, x_{c(X)}\}$ be a set of representatives of $G$–conjugacy-classes of elements in $X$.

**(1.10.2)** Let $R(X) = \{\beta_1, -, \beta_{r(X)}\}$ be a set of pairs $\beta_j = (i, y)$ where $i \in \{1, -, t\}$ and $y \in X^G$ ranges over representatives of conjugacy–classes of $T_i$ such that there is a $\varphi_i$–relevant character $\eta$ with $\eta(w) \neq 0$ for some $w \in y$. Set $\beta_1 = (1, 1)$.

**(1.10.3)** Let $x_j^{(i,\sigma)}$ for $\sigma \in B(i,j) = \{1, -, s_{ij}\}$ be representatives $z$ of conjugacy–classes of $T_G(\varphi_i)$ such that $z^G = x_j^G$.

**(1.10.4)** For $i \in \{1, -, t\}$ let $\pi_i : T_i \longrightarrow T_G(\varphi_i)$ such that $\pi_i(Y) \in Y$ and $\eta\pi_i(Y) \neq 0$ for some $\varphi_i$–relevant character $\eta$ provided that there is at least one such character (such a map $\pi_i$ exists by (1.9)).

**(1.10.5)** Let $\mathrm{Cl}(X) = (X^\beta_{c_\alpha})$ be the matrix with coefficients

$$c_j^{(i,y)} = {}^X c_j^{(i,y)} = \sum_\sigma{}' |\mathrm{C}_G(x_j)||\mathrm{C}_{T_G(\varphi_i)}(x_j^{(i,\sigma)})|^{-1} \zeta_{X_j}(i,\sigma)({}^{\pi_i(y)})$$

where the sum is taken over those $\sigma \in B(i,j)$ satisfying

$$(x_j^{(i,\sigma)}.N)^{T_G(\varphi_i)} = y^{T_i}.$$

We call $\mathrm{Cl}(X)$ the Clifford–matrix of $X$ (in $G$).

**(1.10.6)** If $\eta$ is a $\varphi_i$–relevant character, set $\hat{\eta}(w) = \eta(\pi_i(w))$ for $w \in T_i$ and

$$T_i^\wedge = \{\hat{\eta} | \eta \text{ is } \varphi_i\text{–relevant}\}.$$

**(1.10.7)** If we have to distinguish notations for subroups $U \leq G$, $\tilde{N} \trianglelefteq U$ we add symbols like $\mathrm{Cl}_U(X), Cl_{U,\tilde{N}}(\tilde{X})$ etc.

**(1.11) Lemma.** *If $\eta$ is $\varphi_i$–relevant then*

$$\eta^G(x_j) = \sum_{y|(i,y)\in R(X)} X_{c_j}^{(i,y)}\hat{\eta}(y).$$

**Proof:** The concepts in this section are introduced in order to be able to rewrite the standard induction formula in this way.

**Remark.** Cl$(x)$ depends on various conventions. We usually order $R(x)$, the row–indices, "lecico–graphically"; the maps $\pi_i$ can be chosen in many ways. If $G = NK, K \cap N = 1$, we prefer the choice $\pi_i(x) = K \cap T_G(\varphi_i)$ provided it is allowed. For arithmetical properties the following weights are important:

**(1.12)** $$b_{(i,y)} = |\mathrm{C}_{T_i}(y)|, \qquad (i,y) \in R(X).$$

**(1.13)** $$c_j = |\mathrm{C}_G(x_j)|, \qquad x_j \in C(X).$$

**(1.14)** $$m_j = |\mathrm{N}_G(X) : \mathrm{C}_G(x_j)|, \qquad x_j \in C(X).$$

In section 4 and 5 we will prove

**(1.15) Theorem.** Cl$(x)$ *satisfies* ($^-$: *complex conjugation*):

**(1.15.1)** $|C(x)| = |R(x)|$.

**(1.15.2)** The coefficients of Cl$(x)$ are sums of $|N(\varphi_i)/K(\varphi_i)|$–th roots of unity.

**(1.15.3)** $$\sum_{j\in C(X)} m_j c_j^{(1,y)}\overline{c_j^{(i',y')}} = \delta_{(i,y),(i',y')} b_{(1,1)} b_{(i,y)}^{-1} |N|.$$

**(1.15.4)** $$\sum_{(i,y)} \in R(X) c_j^{(i,y)}\overline{c_{j'}^{(i,y)}} \cdot b_{(i,y)} = \delta_{j,j'}, c_j.$$

**(1.15.5)** $$C_j^{(1,1)} = 1.$$

**(1.15.6)** If $N$ is elementary abelian and has a complement $K$ then $\pi_i$ can be chosen by $\pi_i(x) = K \cap T_G(\varphi_i) = x_1$. Furthermore,

$$c_1^{(i,y)} \geq |c_j^{(i,y)}|, c_1^{(i,y)} \cdot b_{(i,y)} = b_{(1,1)}.$$

## 2. Three examples

We describe in some easy cases the computation of the matrices and the resulting character–tables.

**I)** $G = \Sigma_4, N = E_4, G/N \simeq \Sigma_3$.

$G$ acts transitively on involutions in $N$ and hence on the dual space. Thus $T_1 \simeq \Sigma_3$ and $T_2 \simeq \Sigma_2$.

| | | | | | |
|---|---|---|---|---|---|
| $c_j$: | 24 | 8 | 3 | 4 | 4 |
| order | $1A$ | $2A$ | $3A$ | $2B$ | $4A$ |
| $\chi_1$ | 1 | 1 | 1 | 1 | 1 |
| $\chi_2$ | 1 | 1 | 1 | −1 | −1 |
| $\chi_3$ | 2 | 2 | −1 | . | . |
| $\chi_4$ | 3 | −1 | . | 1 | −1 |
| $\chi_5$ | 3 | −1 | . | −1 | −1 |

| | | | | | | |
|---|---|---|---|---|---|---|
| $Cl$: | 6 | 1 | 1 | 2 | 1 | 1 |
| | 2 | 3 | −1 | 2 | 1 | −1 |
| | | 1 | 3 | | 2 | 2 |

| | |
|---|---|
| 3 | 1 |
| | 4 |

We use the conventions: on the left of each Clifford–matrix we write the $b_{(i,y)}$'s; under each Clifford–matrix we write the $m_j$'s. The inertia groups are separated by lines.

In order to prove this result we determine the fusion of elements of $T_2$ into $T_1$. In this case, $G$ contains $K \simeq \Sigma_3$, and we have a trivial choice for $\pi_1, \pi_2$. Hence the shape of the matrices and their first line follow from (1.15.1) and (1.15.5). The $m_j$'s are consequences of (1.15.3) and the fact that they are obvious orbit–lengths. Now the absolute values of the remaining entries follow from (1.15.4). The maps $\pi_1, \pi_2$ force their signs.

**II)** $N \simeq E_8, G/N \simeq GL_3(2)$, $G$ acts naturally on $N$.

In this case, $T_1 \simeq GL_3(2), T_2 \simeq \Sigma_4$. Let $T = T_G(\varphi_2)$, and $K = \ker\varphi_2$. If $N/K$ is not a direct factor in $\mathrm{O}_2(T/K)$ the $\varphi_2$–relevant characters would vanish on elements of $T/K \setminus N/K$ of order 2 or 4; the Clifford–matrix for the involutory coset would be trivial and the elements in this coset would be conjugate in $G$, an obvious contradiction. Hence there are exactly 5 $\varphi_2$–relevant characters (in the split and non–split case).

| | $c_j$: 64.21 | 64.3 | 32 | 32 | 16 | 6 | 6 | 8 | 8 | 7 | 7 | |
|---|---|---|---|---|---|---|---|---|---|---|---|---|
| orders: | $1A$ | $2A$ | $2B$ | $2C$ | $4A$ | $3A$ | $6A$ | $4C$ | $4D$ | $7A$ | $7B$ | split |
| | $1A$ | $2A$ | $2B$ | $2C$ | $4A$ | $3A$ | $6A$ | $4C$ | $4D$ | $7A$ | $7B$ | non split |
| $\chi_1$ | 1 | 1 | 1 | 1 | 1 | 1 | 1 | 1 | 1 | 1 | 1 | |
| $\chi_2$ | 6 | 6 | 2 | 2 | 2 | · | · | · | · | −1 | −1 | |
| $\chi_3$ | 7 | −1 | 3 | −1 | −1 | 1 | −1 | 1 | −1 | · | · | |
| $\chi_4$ | 7 | −1 | −1 | 3 | −1 | 1 | −1 | −1 | 1 | · | · | |
| $\chi_5$ | 14 | | 2 | 2 | −2 | −1 | 1 | · | · | · | · | |
| $\chi_6$ | 21 | −3 | 1 | −3 | 1 | · | · | −1 | 1 | · | · | |
| $\chi_7$ | 21 | −3 | −3 | 1 | 1 | · | · | 1 | −1 | · | · | |

| | | |
|---|---|---|
| 168 | 1 | 1 |
| 24 | 7 | −1 |
| | 1 | 7 |

| | | |
|---|---|---|
| 3 | 1 | 1 |
| 3 | 1 | −1 |
| | 4 | 4 |

| | |
|---|---|
| 7 | 1 |
| | 8 |

| | |
|---|---|
| 7 | 1 |
| | 8 |

| | | | |
|---|---|---|---|
| 8 | 1 | 1 | 1 |
| 8 | 1 | 1 | −1 |
| 4 | 2 | −2 | · |
| | 2 | 2 | 4 |

| | | |
|---|---|---|
| 4 | 1 | 1 |
| 4 | 1 | −1 |
| | 4 | 4 |

The shape of these matrices and the $m_j$'s follow from (1.15) as well as the absolute values of the entries. The signs for entries under elements of odd order follow by the proper choice of $\pi_2$. Since columns can be interchanged we could have

| | | | |
|---|---|---|---|
| | 1 | 1 | 1 |
| | −1 | −1 | 1 |
| | 2 | −2 | · |
| $\chi_3$ | 1 | −3 | 1 |

for the involutory coset, but the resulting "character–table" for $G$ leads to a contradiction: we have $\chi_3(x) - \chi_3(1) \equiv 2(4)$ for $x$ in the involutory coset and hence these elements have order 4; this implies that $\chi_3^- : \chi_3^-(y) = 2^{-1}(\chi_3(y)^2 - \chi_3(y))$ is not a character.

Given the matrices the character–table follows. (4 non–faithful characters are not written). The orders of the elements follow from

$$\chi_3^- = \chi_6 \text{ resp. } \chi_3^- = \chi_3 + \chi_5$$

in the split and the non–split case.

**Remark.** In the "character–table" mentioned above, orthogonality relations hold; this is always true if we multiply some rows in Clifford–matrices for a proper table by $p$–th root of unity if $N$ is a $p$–group. For that reason the determination of these signs is in many cases the hardest problem in getting the matrices.

**III)** $N = E_{16}, G/N \simeq GL_4(2), G$ acts naturally on $N$.

In this case, $T_1 \simeq GL_3(2), T_2 \simeq N.K$ for $K \simeq GL_3(2), N \simeq E_8$. If $G$ splits, $G = N \cdot W, W \simeq GL_4(2)$ we get for the 2–cosets (indicated by their Jordan–form):

| $[1^4]$ | $1_A$ | $2_A$ | |
|---|---|---|---|
| | 1 | 2 | |
| | 15 | −1 | $1_A$ |
| | 1 | 15 | |

| $[2^1 1^2]$ | $4_A$ | $4_B$ | $2_B$ | |
|---|---|---|---|---|
| | 1 | 1 | 1 | |
| | 1 | 1 | −1 | $2_A$ |
| − | 6 | −2 | | $2_B$ |
| | 2 | 6 | 8 | |

| $[2^2]$ | $2_A$ | $4_B$ | |
|---|---|---|---|
| | 1 | 1 | |
| | 3 | −1 | $2_C$ |
| | 4 | 12 | |

| $[3^1 1]$ | $8_A$ | $8_B$ | $8_C$ | |
|---|---|---|---|---|
| | 1 | 1 | 1 | |
| − | 1 | 1 | −1 | $4_A$ |
| | 2 | −2 | · | $4_C$ |
| | 4 | 4 | 8 | |

| $[4^1]$ | $4_B$ | $8_C$ | |
|---|---|---|---|
| | 1 | 1 | |
| | 1 | −1 | $4_D$ |
| | 4 | 4 | |

Let $G$ be a subgroup of Co3 [1]. Then $G$ has a faithful character $\mu, \mu(1) = 23$. Since faithful characters of $G$ have multiples of 15 as degrees we get $\mu = 1_G + \kappa + \lambda$ with $\lambda(1) = 15$ and $\kappa(1) = 7$, and $\lambda$ is irreducible. Hence $T_G(\varphi_2)$ has a linear character which is $\varphi_2$–relevant. This forces the shape of all the matrices and the values of the entries up to sign–changes for the rows by (1.15). Since $\mu(7_A) = 2$, $\kappa$ is irreducible. For 2–elements in Co3 we have

| Co 3 : | $2A$ | $2B$ | $4A$ | $4B$ | $8A$ | $8B$ | $8C$ |
|---|---|---|---|---|---|---|---|
| $\mu$ | 7 | −1 | −5 | 3 | 1 | −3 | 1. |

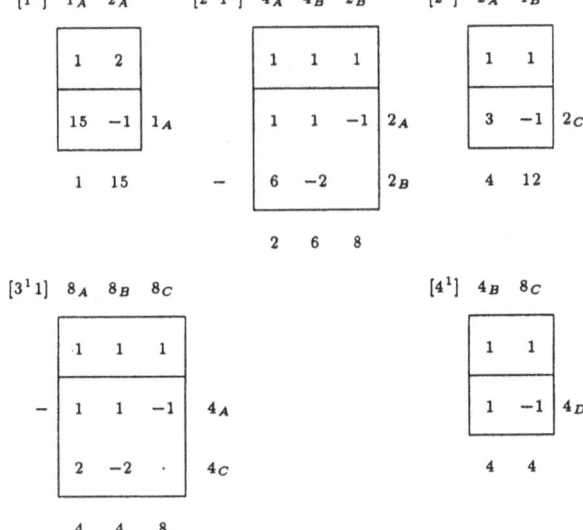

Hence the $[2^2]$–coset contains $2A, 4B$. The $[2^1 1^2]$–coset must contain 8 involutions and 8 elements of order 4; hence it contains $4A, 4B, 2B$, and the signs in the last row

have to be changed. The element $8B$ can only fuse into the $[3^1 1]$–coset (and has to, since $G$ contains a Sylow–2–subgroup of Co3). Therefore, the signs in the second row have to be changed. The remaining sign–problems can be used for definitions. Apart from the fusion of the $[4^1]$–coset the indicated fusions are now trivial: here, the element of order 8 is of type $8A$ or $8C$. Let $\tilde{\mu}$ be the smallest character distinguishing $8A, 8C$ :

| Co3 : | $1A$ | $2A$ | $4B$ | $8A$ | $8B$ | $8C$ |
|---|---|---|---|---|---|---|
| $\tilde{\mu}$ | 253 | 29 | 5 | −3 | −3 | 1 |

Then $\tilde{\mu}_G = \alpha+\beta$, where the character $\alpha$ contains $N$ in its kernel, and $\beta$ is faithful. We get $\alpha(1) = 43, \beta(1) = 210$. Let $\gamma$ be the character of $T_2$ defining $\beta$. Then $\gamma(1) = 14$, $\gamma(4C) = 0$, $\varphi(4A) = 2$. Hence $\gamma = \chi_1 + \chi_2 + \chi_4$ and $\gamma(4D) = 2$. Thus the element of order 8 in question fuses to $8C$.

### 3. The map $\pi$ and projective characters

The map $\pi_i$ in (1.10.4) is needed only to describe the set $T_i^\wedge \subseteq \mathbb{C}[T_i]$. In view of (1.9) it is sufficient to describe $T_i^\wedge$ in $L(\varphi_i)$. Let $H = N \cap L(\varphi_i)$. Since $N(\varphi_i)/K(\varphi_i)$ is central in $L(\varphi_i)$ each representation $S$ of $L = L(\varphi_i)/K(\varphi_i)$ which is $\varphi_i$–relevant is of the form $S = Y \otimes X$ where $Y$ and $X$ are projective representations of $L$, and the degree of $X$ is 1; $Y$ can be viewed as a projective representation of $L/N(\varphi_i)$ **[2,(51.7)]**. We have to consider only such representations $Y$ which have $H$ in their kernel.

For arithmetical properties of $T_i^\wedge$ we need only to consider the case

**(3.1)** $$N = \langle g \rangle \leq Z(G), \quad |N| = p^n, \quad \varphi \in N^\vee.$$

In this case, a factor system and the map $\pi$ can be written explicitly. We get $G_\varphi^\wedge = \{\hat{\zeta} | \zeta \ \varphi\text{–relevant}\}$.

For $X \in G/N$ set $\delta_\varphi(X) = 0$ if $\hat{\zeta}(X) = 0$ for each $\hat{\zeta} \in G\hat{\varphi}$ and $\delta_\varphi(X) = 1$ otherwise. Then

**(3.1.1)** $$\sum_{X \in G/N} \hat{\zeta}(x)\overline{\hat{\zeta}'(x)} = |G/N|.\delta_{\hat{\zeta},\hat{\zeta}'}.$$

**(3.1.2)** $$\sum_{\hat{\zeta} \in G\hat{\varphi}} \hat{\zeta}(x)\overline{\zeta'(x)} = |C_{G/N}(x)|.\delta\varphi(x)\delta_{x^G, x'^G}.$$

**Remark.** The conventions we use here for "projective" characters coincide with **[1]**.

For $X \in G/N$ there is a subgroup $N_X \leq N$ such that $x, y \in X$ are conjugate in $G$ iff $xy^{-1} \in N_X$. Let $p^{n(x)} = |N/N_X|$. The Clifford–matrices coincide in this case with the character–table of a cyclic group of order $p^{n(x)}$ (with the obvious notations for rows and columns).

| $Cl(1)$ | $l_1$ | $l_2$ | $l_3$ | $l_4$ |
|---|---|---|---|---|
| $Cl(2)$ | $a_1$ | $a_2$ | $a_3$ | $a_4$ |
| $4_1$ | 1 | 1 | 1 | 1 |
| $4_2$ | 1 | 1 | $-1$ | $-1$ |
| $4_3$ | 1 | $-1$ | $i$ | $-i$ |
| $4_4$ | 1 | $-1$ | $-i$ | $i$ |

It is always possible to choose $\pi_i$ in such a way that we get

(3.1.3.) $$\sum_{t=1,(t,0(x))=1}^{0(x)} \hat{\zeta}(x^t) \in \mathbf{Z},\ X \in G/N,\ \hat{\zeta} \in G_\varphi^\wedge.$$

Let

$$\chi_X^\varphi(z) = \delta_{X^G,Z^G} \cdot |\mathrm{C}_{G/N}(x)| \cdot \delta_\varphi(x),$$
$$\chi_0^\varphi(x) = |G/N|\delta_\varphi(X) \quad X, Z \in G/N.$$

With these notations we get

**(3.1.4)** $$\chi_X^\varphi = \sum_{\hat{\zeta} \in G_\varphi^\wedge} \hat{\zeta}(X)\hat{\zeta}.$$

**(3.1.5)** $$\chi_0^\varphi \in \mathbf{Z}G_\varphi^\wedge.$$

**Proof:** (3.1.4) follows from (3.1.1), (3.1.2).

$$\chi_0^\varphi = \sum_{X \in G/N} |G/N||\mathrm{C}_{G/N}(X)|^{-1}\chi_X^\varphi$$
$$= \sum_{\hat{\zeta} \in G_\varphi^\wedge} \left( \sum_{X \in G/N} |G/N||\mathrm{C}_{G/N}(X)|^{-1}\overline{\hat{\zeta}(X)} \right) \hat{\zeta},$$

thus (3.1.3) implies (3.1.5).

We conclude this section with a trivial example: Let $G$ be cyclic of order 8 and $N \leq G$ of order 4, $G = \{1, a\}$. Fix $i, j \in \mathbf{C}$, $i^2 = -1$, $j^2 = i$. Let

| $(G/N)^\vee$ | 1 | $a$ |
|---|---|---|
| | 1 | 1 |
| | 1 | $-1$ |

| $(G/N)^\wedge$ | 1 | $a$ |
|---|---|---|
| | 1 | $i$ |
| | 1 | $-i$ |

| $(G/N)^\sim$ | 1 | $a$ |
|---|---|---|
| | 1 | $j$ |
| | 1 | $-j$ |

If we denote the elements of $G$ by $1_i, a_i$ we get

and $G^{\wedge}_{\varphi_1} = (G/N)^{\vee}, G^{\wedge}_{\varphi_2} = (G/N)^{\sim}$, $G^{\wedge}_{\varphi_3} = G^{\wedge}_{\varphi_4} = (G/N)^{\sim}$. The character corresponding to $(1,-j) = \hat{\eta}$ and $\varphi_4$ then reads

| $1_1$ | $1_2$ | $1_3$ | $1_4$ | $1_1$ | $a_2$ | $a_3$ | $a_4$ |
|---|---|---|---|---|---|---|---|
| 1 | $-1$ | $-i$ | $i$ | $-j$ | $j$ | $ij$ | $-ij$ |

Obviously, we have better methods to compute characters of cyclic groups.

## 4. Basic properties of $\mathrm{Cl}(x)$

In this section $N$ is a normal $p$–subgroup of $G$ and $X$ is a fixed coset of $G/N$. Let $M = \mathrm{N}_G(X)$. Then $|M| = |N|.b_{(1,1)}$.

By (1.7) and (1.11) we have $rk\ Cl(X) = |C(X)| \leq |R(X)|$. For each pair $(i,y) \in R(X)$ the characteristic function $\chi^{\varphi_i}_{(i,y)}$ is non–trivial by the choice of $R(X)$. Because of (3.1.4) and (1.7) the class–functions

$$\sum_{z(i,z)\in R(x)} C_j^{(i,z)} \chi^{\varphi_i}_{(i,y)} | (i,y) \in R(x)$$

are linearly independent and involve only irreducible characters of $G$ associated with $\varphi_i$. Thus $|R(x)| \leq rk\ Cl(x)$ and (1.15.1) is proven.

We now prove the orthogonality–relations (1.15.3), (1.15.4): By (3.1.4) we have

$$\chi^{\varphi_i}_{(i,y)} = \sum_{\hat{\eta}\in G^{\wedge}_{\varphi}} z_\eta \hat{\eta},\ z_\eta \in \mathbb{C}.$$

Let

(*) $$\tilde{\chi}^{\varphi_i}_{(i,y)} = \sum_{\hat{\eta}\in G^{\wedge}_{\varphi}} z_\eta\ \eta \in \mathbb{C}[T_G(\varphi_i)].$$

By (1.11) we get

(4.1) $$\left(\tilde{\chi}^{\varphi_i}_{(i,y)}\right)^G (x_j) = c_j^{(i,y)} \cdot \sum_\eta z_\eta \hat{\eta}(y) = c_j^{(i,y)} b_{(i,y)}.$$

Because of (1.7) the map $\eta \mapsto \eta^G$ is an isometry for relevant characters. Hence (3.1.1) implies (1.15.3) in the case $(i,y) \neq (i',y')$.

The orthogonality–relations for $G^{\vee}$ together with (1.7) imply

$$\sum_\eta \eta^G(x_j)\eta^G(x_{j1}^{-1}) = \delta_{ij} c_j,$$

where we sum over relevant characters.

Thus, (1.15.4) follows from (1.11), (3.1.4).

Let $\chi = \tilde{\chi}^{\varphi}_{(i,y)}$ in (*). Then

$$(\chi,\chi)_{T_G(\varphi_i)} = b_{(i,y)} = (\chi^G,\chi^G)_G$$

by the isometry for relevant characters. Applying (1.11) and (3.1.1) we get (1.15.3).

**(4.2) Lemma.** *Let* $\tilde{\chi}_{(i,y)} = (\tilde{\chi}^{\varphi_i}_{(i,y)})G$ *in* (4.1) *and* $\chi = \eta^G$ *for a* $\varphi_i$*–relevant character* $\eta$. *Then*

$$(\tilde{\chi}_{(i,y)}, \chi)_G = \hat{\eta}(y)$$

*and*

$$(\tilde{\chi}_{(i,y)}, \psi)_G = 0$$

*for* $\psi \in G^\vee, \psi = \eta'^G$ *and* $\eta'$ *not* $\varphi_i$*–relevant.*

**Proof:** $(\tilde{\chi}_{(i,y)}, \chi)_G = \left(\chi^{\varphi_i}_{(i,y)}, \eta\right)_{L_G(\varphi_i)}$, and (3.1.4) implies the statement.

**(4.3) Lemma.** *Notations can be chosen such that the Clifford–matrix of* $X$ *in* $\mathsf{N}_G(X)$ *is equal to* $\mathrm{Cl}(X)$.

**Proof:** $X$ is a $TI$–set in $G$. Hence we may use the $C(X)$ as column–indices in both cases. Let $\psi_1, -, \psi_w$ be representatives of $M$–orbits of $N^\vee M = \mathsf{N}_G(X)$ and let $\psi_1, -, \psi_r$ be chosen such that they appear in (1.10.2), i.e. $R_M(X)$ contains some $(i, y)$ for each $1 \leq i \leq r$. Then there is a $\psi_i$–relevant character $\theta \in T_M(\psi_i)$ such that $\theta(w) \neq 0$ for some $w \in y$. In our special situation, $y = X$. Let $\tilde{\theta} = \theta^{T_G(\psi_i)}$. Then $\tilde{\theta}$ has only $\psi_i$–relevant constituents in $T_G(\psi_i)$; therefore there is some constituent $\eta$ with $\eta(w) \neq 0$. Suppose $\psi_i^G \ni \varphi_k$ and $\psi_i^{g_i} = \varphi_k$. Then $(g, y^{g_i})$ can be used in $R_G(X)$, and we have a bijection $R_M(X) \longrightarrow R_G(X)$, which we use to define the maps $\pi_{i,M}(1 \leq i \leq r = |Cl_G(X)|)$. Obviously

$$(**) \qquad b_{(i,y),M} = b_{(k,y^{g_i}),G}.$$

If $\eta$ is $\psi_i$–relevant in $T_G(\psi_i)$ then $(\eta, \psi_s)_N = 0$ for $s \neq i$. This property implies that the restriction of functions in (4.1) to $X \leq M$ is the $c$–fold of the corresponding row. By (**) and (1.10.4), $|c| = 1$; the choice of $\pi_{i,M}$ yields $c = 1$.

**(4.4) Lemma.** *Let* $G = \mathsf{N}_G(X)$ *and* $\alpha \in \mathrm{Aut}\, G$ *such that* $X^\alpha = X$. *Then notations can be chosen in such a way that* $\alpha$ *induces a permutation of the rows of* $\mathrm{Cl}(X)$.

**Proof:** We choose $C(X)$ such that $C(X) = C(X)^\alpha$. We choose $\{\varphi_1, -, \varphi_t\}$ such that $\varphi_i^\alpha = \varphi_{i\alpha} \in \{\varphi_1, -, \varphi_t\}$. If $(i, y) \in R(X)$ then $(i\alpha, y^\alpha)$ satisfies (1.10.2), and we can choose $R(X)$ such that $R(X) = R(X)\alpha$. Finally, we adjust the maps $\pi_i$ by $\pi_{i\alpha}(y) := \pi_i(y^\alpha)$.

**(4.5) Lemma.** *Let* $N \leq U \leq G$ *and* $X \subseteq U$. *Then notations can be chosen in such a way that rows of* $\mathrm{Cl}_G(X)$ *are sums of rows of* $\mathrm{Cl}_U(X)$.

**Proof:** Let $C_U(X) = \{x_{\alpha\beta}\}$ where $x_{\alpha\beta} \in x_\alpha^G$. We adjust the notations in $\mathrm{Cl}_\eta(X)$ in order to identify $\mathrm{Cl}_U(X)$ with $\mathrm{Cl}_{\mathsf{N}_U(X)}(X)$ using (4.3). Now we adjust the notations in such a way that $\mathsf{N}_G(X)$ induces row–permutations of $\mathrm{Cl}_{\mathsf{N}_U(X)}(X)$.

Now we may assume $G = \mathsf{N}_G(X)$. Using functions of type (4.1) it follows

$$c_\alpha^{(i,y)} = \sum_{(\gamma,\delta)} u_{\alpha\beta}^{(\gamma,\delta)}$$

where $\left(u_{\alpha\beta}^{(\gamma,\delta)}\right) = \mathrm{Cl}_U(X)$, and the sum is taken over the $G$–orbit of $R_U(X)$ corresponding to $(i, y)$.

**Remark.** There may be $G$–orbits of $R_U(X)$ which do not correspond to some $(i, y) \in R_G(X)$; in this case their sum is trivial.

For practical purposes it is important to observe that in the case $P \leq N$, $P \trianglelefteq G$ there are relations between sections of the corresponding matrices.

In (4.5) we do not state that each row of $\mathrm{Cl}_U(X)$ appears as a sum and for a row of $\mathrm{Cl}_G(X)$.

## 5. Vector spaces

In this section we assume

**(5.1)** $N$ is an elementary abelian $p$–group.

Set $N_X = [z, N]$ for $z \in X$, let $M = \mathbf{N}_G(X)$. Then $\langle X\rangle/N_X$ is abelian and $X/N_X$ is a coset of $N/N_X$.

**(5.2) Lemma.** *The rows of* $\mathrm{Cl}(X)$ *can be identified with restrictions of* $M$*–invariant characters of* $\langle X\rangle/N_X$ *to* $X/N_X$.

*The coset* $X$ *is a* <u>*split coset*</u>*, if* $X$ *contains an element* $x_1$ *such that* $M = N.\mathrm{C}_G(x_1)$. *(We do not require* $\langle x_1\rangle \cap N = \langle 1\rangle$*).*

**(5.3) Lemma.** *If* $X$ *is a split coset the rows of* $\mathrm{Cl}(X)$ *can be identified with* $M$*–invariant characters of* $N/N_X$ *multiplied by a* $p$*–th root of unity.*

**Remark.** In (5.3) these characters are exactly the orbit–sums. The scalar can be avoided by proper choices of the maps $\pi_i$ which we did in the case of the cyclic group $C_8$. However, if we want to use the characters of $T_G(\varphi_i)/N$ whenever possible we have to allow for scalars as shown in (2. III).

In (5.2) the rows will be an independent set of orbit–sums.

**Proof of (5.2), (5.3):** Let $\eta$ be an irreducible character of $\langle X\rangle$ and $N_X \leq \ker\eta$. Then $\eta(1) = 1$ and $\eta|_N = \varphi_i$. Let $T = T_G(\varphi_i)$ and $S = T_M(\eta)$. Set

$$\tilde{\eta} = [M : S]^{-1}\eta^M = \sum_{\eta^m, m\in M} \eta^m.$$

Choose $C(X) = \{x_1, \ldots, x_r\}$, $|\mathrm{C}_G(x_1)| \geq |\mathrm{C}_G(x_j)|$.

Let $\psi \in M^\vee$ with $\psi(x_j) \neq 0$ for some $j$. Since the conjugacy classes of $M/N_X \subseteq X/N_X$ correspond in a natural way, $N_X \leq \ker\psi$.

Let $\mu \in T^\vee$ be $\varphi_i$–relevant. Since $N/N \cap \ker\mu$ is cyclic and central in $T/N \cap \ker\mu$, $\mu(x) = 0$ for each element $x \in X$ or $\mu(a)\mu(b)^{-1}$ is a $p$–th root of unity. Therefore there is no restriction in (1.10.4) for the choice of $\pi_i$ apart from $\pi_i(Y) \in Y$.

The Clifford–matrix of $X/N_X$ in $M/N_X$ can be identified with $\mathrm{Cl}(X)$. Since $\langle x\rangle/N_X$ is abelian (5.2) is trivial in the case $M = \langle x\rangle$. By (4.3) and (4.5) we can choose the notation such that the rows are of type $\tilde{\eta}|_{\{x_1,-,x_r\}}$ and (5.2) follows.

In case of (5.3) the map

$$x \longrightarrow x^{-1}x_1, \quad x \in X$$

has the property that two elements $z, y$ of $x$ are conjugate in $M$ iff $z^{-1}x_1$ and $y^{-1}x_1$ are conjugate under $M$ modulo $N_X$. Hence (5.3) follows.

**(5.4) Lemma.** *Let $X$ be a split coset, $\mathsf{N}_G(X) = N\mathsf{C}_G(x_1)$ for $x_1 \in C(X)$. Then*

**(5.4.1)** $$|c_1^{(i,y)}| = |b_{(1,1)} \cdot b_{(i,j)}^{-1}|.$$

**(5.4.2)** $$|c_j^{(i,y)}| \leq |c_1^{(i,y)}|.$$

**(5.4.3)** $$\sum_{(i,y)\in R(X)} c_j^{(i,y)} \cdot c_1^{(i,y)} |\beta_{(1,1)} \cdot b_{i,j}^{-1}|^{-1} = \delta_{1,j}|N_X|.$$

**(5.4.4)** If $G = NK$, $N \cap K = 1$, $\pi(Y) = K \cap Y$, then $c_1^{(i,y)} \in \mathsf{N}$.

**Proof:** These statements are trivial consequences of (5.3) and (1.15.4); the last statement follows by splitting $1_K^G$.

**Example.** Let $G_0 = V \cdot S, S \simeq \Sigma_6, V = E_{64}$ and let $S$ act on a basis of $V$ in the natural way. Let $G = G_0/\mathbf{Z}(G_0)$, $N = V/\mathbf{Z}(G_0)$. Then $\mathrm{Cl}(N)$ reads

| | | | |
|---|---|---|---|
| 1 | 1 | 1 | 1 |
| 1 | 1 | −1 | −1 |
| 15 | −1 | 5 | −3 |
| 15 | −1 | −5 | 3 |

Let $|N| = 16, N \trianglelefteq G, X = 0_2(G)/N$. Then $\mathrm{Cl}(X)$ reads

| | |
|---|---|
| 1 | 1 |
| 5 | −3 |

or

| | |
|---|---|
| 1 | 1 |
| −5 | 3 |

## 6. $GL_n(q)$ on its natural module

**Notation.** Let $F = GF(q), N = F^n, T_1 \simeq GL_n(q), q = p^m$ and

$$N \leq G = \left\{ \begin{pmatrix} 1 & n \\ 0 & k \end{pmatrix} \middle| \, n \in N, k \in T_1 \right\} \leq GL_{n+1}(q)$$

$$T_2 = \left\{ \begin{pmatrix} 1 & 0 & 0 \\ 0 & 1 & w \\ 0 & 0 & k \end{pmatrix} \middle| \, w \in GF(q)^{n-1}, k \in GL_{n-1}(q) \right\} \leq T_1$$

Let $\varphi_2(n_1, -n_n) = \theta(n_1) \in \mathbb{C}$ for an irreducible representation $\theta$ of $F^+$. Then $\varphi_2$ is an irreducible representation of $N, T_G(\varphi_2) = NT_2$. For $Y \in G/N$ let $\pi_1(Y) = \pi_2(Y) = Y \cap T_1$.

If $k \in T_1$ is not a $p$–element then $k = ab = rs$ where $a$ and $r$ are $p$–elements and $S$ is a $p'$–element; $a$ and $b$ are uniquely determined by

$$[N, s] \leq \mathrm{C}_N(a),\ [N, a] \leq \mathrm{C}_N(s),\ [N, b] = [N, s].$$

Now, the Clifford–matrix for $Nk$ is identical with the Clifford–matrix of $\mathrm{C}_N(s) \cdot a$ in $\mathrm{C}_N(s) \cdot GL(\mathrm{C}_N(s), q)$; the $b_{(i,y)}$ have to be adjusted.

For this reason, we describe only the Clifford–matrices for $Nk$, where $k$ is a fixed $p$–element.

Jordan normal forms for $p$–elements $a \in T_1$ will be written

(*) $$[a] = [1, s_1^{r_1}, \ldots, s_t^{r_t}],\ s_i \cdot r_i \geq 1, \Sigma\, r_i s_i = n$$

and we mean by this the obvious upper–triangular matrix.

Two cases are trivial:

| $Cl([1^n])$ | | |
|---|---|---|
| $n>1$ | 1 | 1 |
| | $q^{n-1}$ | $-1$ |

| $Cl([2^1])$ | | |
|---|---|---|
| | 1 | 1 |
| | $q^{n-1}$ | $-1$ |

Let $c(k) = \Sigma r_i(s_i - 1)$ and $d(k) = n - c(k)$. Then $d(k) = \dim_F N/[N, k]$. Let $E_{i,j}$ be $K$–subspaces of $N$ such that $s_i < s_{i+1}, k = [1, s^{r_1}, \ldots, s_r^{r_{t(k)}}]$ $\dim_F E_{i,j} = S_i$ for $1 \leq j \leq r_i$, $N = \oplus E_{i,j}$, $E_{i,j}^k = E_{ij}$, $\dim_F \mathrm{C}_{E_{i,j}}(k) = 1$.

Set $x_1 = k$, $x_{i+1} \in E_{i,1}k$ such that

$$[x_{i+1}] = \left[s_1^{r_1}, -, s_{i-1}^{r_{i-1}}, (s_i + 1)^*, s_i^{r_i - 1}, -, s_{t(k)}^{r_{t(k)}}\right],$$

where $^*$ symbolizes an exceptional position. Then $\{x_1, -, x_{t(k)+1}\}$ have pairwise different Jordan forms in $GL_{n+1}(q)$.

Let 1 be the element

$$[1] = \left[1, s_1, -, s_{t(k)}, s_1^{r_1 - 1}, \ldots, s_{t(k)}^{r_{t(k)} - 1}\right] \in T_2$$

where $1 \leq s_i < s_{i+1}$. Let $\zeta$ be a $t(k)$–cycle and set

$$(1, 1) = [1] = (2, 1) \text{ for } n > 1$$

$$(2, i + 1) = \left[1, s_{1\zeta} - i, \ldots, s_{t(k)\zeta}^{-i}, s_1^{r_1 - 1}, -, s_{t(k)}^{r_{t(k)} - 1}\right].$$

| | $x_1$ | $x_{j-1}$ | $x_j$ | $x_{t+1}$ | |
|---|---|---|---|---|---|
| | 1 | 1 | 1 | 1 | (1,1) |
| | $(\prod_{\sigma>2} q^{r_\sigma})(q^{r_2-1})$ , | . | . | . | (2,1) |
| | | | . | . | (2,j−1) |
| | $(\prod_{\sigma>j} q^{r_\sigma})(q^{r_j-1})=W$ , | ,W, | $-\prod_{\sigma>j} q^{r_\sigma}$ , | . | (2,j) |
| | $(q^{r_t-1})$, | $(q^{r_t-1})$ , | $(q^{r_t-1})$ , | ,−1 | (2,t) |

Then the elements in $\{(2,1),-,(z,t(k))\}$ are conjugate in $T_1$ but are pairwise non-conjugate in $T$, since their Jordan forms modulo $\mathbf{O}_p(T_2)$ are different.

**(6.1) Lemma.** *We may choose $C(Nk) = \{x_1, -, x_{t(k)+1}\}$ and $R(Nk) = \{(1,1), (2,1), -(2,t(k))\}$ if $n > 1$. We write $k$ in the form (*) with $s_i < s_{i+1}$. Then $\mathrm{Cl}(N(k)$ has the form $(t = t(k))$.*

**Remark.** This matrix implies that $b_{(2,i)} < b_{(2,i+1)}$ and $c_j > c_{j+1}$.

For the proof it satisfies to show that $N/[N,k]$ is a uniserial $C_{T_2}(k)$-module with chief-factors of the indicated dimensions and only one orbit of non-trivial elements. We illustrate one example and leave the proof to the reader.

Let $n = 6$ and

$$k = \begin{pmatrix} 1 & & & & & \\ & 1 & 1 & & & \\ & & 1 & & & \\ & & & 1 & 1 & \\ & & & & 1 & 1 \\ & & & & & 1 \end{pmatrix} = x_1 = (1,1) = (2,1)$$

$$x_2 = \begin{pmatrix} 1 & 1 & & & & \\ & 1 & 1 & & & \\ & & 1 & & & \\ & & & 1 & 1 & \\ & & & & 1 & 1 \\ & & & & & 1 \end{pmatrix}, \quad x_3 = \begin{pmatrix} 1 & & & 1 & & \\ & 1 & 1 & & & \\ & & 1 & & & \\ & & & 1 & 1 & \\ & & & & 1 & 1 \\ & & & & & 1 \end{pmatrix},$$

$$(2,2) = \begin{pmatrix} 1 & & & & & \\ & 1 & 1 & & & \\ & & 1 & 1 & & \\ & & & 1 & & \\ & & & & 1 & 1 \\ & & & & & 1 \end{pmatrix}.$$

Let $(2,2) = (2,1)^x$ for $x \in T_1$. Then $\varphi_2$ and $\varphi_2^x$ are not conjugate in $T_2$ and $k^{-1}x_2 \in \ker\varphi_2^{xy}$ for $y \in C_{T_1}(k) = M_k$ but $k^{-1}x_3 \notin \ker\varphi_2^{xy}$. Since $M_k$ acts transitively on the sets

$$\{(1\ a\ b\ 0\ c\ d) \mid a,b,c,d \in F\} \leq N,$$
$$\{(1\ a\ b\ e\ c\ d) \mid a,b,c,d \in F,\ e \in F^x\} \subseteq N,$$

| | | | | | |
|---|---|---|---|---|---|
| c: | $d\cdot q^2$ | $d\cdot(q-1)^{-1}$ | $d\cdot(q-1)^{-1}q^{-1}$ | | |
| $Cl[2,3]$ | $x_1$ | $x_2$ | $x_3$ | | $b$ |
| $\varphi_1$ | 1 | 1 | 1 | (1.1) | $d$ |
| $\varphi_2$ | $q(q-1)$ | $-q$ | $\cdot$ | (2.1) | $d\cdot q^{-1}(q-1)^{-1}$ |
| $\varphi_2^x$ | $q-1$ | $q-1$ | $-1$ | (2.2) | $d\cdot(q-1)^{-1}$ |
| | $1\cdot m$ | $(q-1)m$ | $q(q-1)\cdot m$ | | |

we get
where $m = q^3 = |[N,k]|d = |C_{T_1}(k)|$. The identification $\varphi_2 \sim (2.1)$ follows from the order of $C_{T_2}(2.1)$.

## REFERENCES

[1] J.H. Conway et al, Atlas of Finite Groups, Oxford 1985.

[2] Ch.W. Curtis, I. Reiner, Representation Theory of Finite Groups and Associative Algebras, New York – London, 1962.

Progress in Mathematics, Vol. 95, © 1991 Birkhäuser Verlag Basel

# Research in Representation Theory at Mainz (1984 - 1990)

BERTRAM HUPPERT

This is a report on research conducted at Mainz from 1984 to 1990, made possible by the DFG-project.

Several special results, known since some time, made it clear that the structure of a finite group $G$ is controlled to a large extent by the type of the prime-number-decomposition of the degrees of the irreducible characters of $G$ over $\mathbf{C}$. A large part of the activities at Mainz was devoted to a systematic study of this question.

In §1 we describe mainly the relevant earlier results (theorems 2-6) and introduce convenient notations.

§2 describes the recent results in classical representation theory, proved since 1984 by mathematicians at Mainz and visitors. The special results (theorems 8-12) culminate in two very general questions, the $\rho - \sigma$-conjecture (theorem 13) and the study of the degree-graph (theorem 14).

All questions studied in §2 still make sense in the context of modular representation theory. But at present the corresponding theorems have been studied only in a small number of cases (theorems 16-24). One question of a new type is particularly interesting: To what extent does the knowledge of the degrees of the irreducible $p$-modular represention of a solvable group $G$ restrict the ordinary degrees of $G/O_p(G)$? (Theorems 19 and 20 give some answers.)

Rather late it was observed that there are surprising similarities between theorems in §2 and the influence of the lengths of the conjugacy-classes of $G$ on the structure of $G$. There are analogues of most of the theorems of §2, and nearly always the conditions on the class-lengths restrict the group much more than the corresponding restrictions on the degrees of the irreducible representations. Most striking seems to us the fact that C.Casola could recently prove an exact analogue of the $\rho - \sigma$-conjecture for class-lengths in a large class of groups including the perfect ones (theorem 13′). We do not know any universal reason for the surprising similarities of the results in §2 and §4.

Questions similar to §2 have also been studied for a "reduced degree" (see §5). This requires the control not only of the character-degrees but also the Schur-indices. As the determination of Schur-indices is often very difficult, there are limits to this approach. A general impression finally came up: The structure of a group $G$ is already to some extent controlled by the knowledge of the degrees of the characters with Schur-index 1 over $\mathbf{Q}$ (theorem 28).

In §6 we report on some study of the Loewy-series of a $p$-group $P$ in characteristic $p$. A conjecture, that there are no "dents", was disproved for $p = 2, 3, 5$, but there is some strong experimental evidence that it might be true if $p \geq 7$.

Finally in §7 we report on some applications of modular representation theory to coding-theory.

## §1 Introduction

**1.1.** Let $G$ be a finite group. The group-algebra $\mathbf{C}G$ of $G$ over the field $\mathbf{C}$ of complex numbers is semisimple, hence has a Wedderburn-decomposition

$$\mathbf{C}G \cong \bigoplus_{i=1}^{h} (\mathbf{C})_{n_i} .$$

The $n_i$ are the degrees of the irreducible representations of $G$ over $\mathbf{C}$. Two natural questions come to mind immediately:

**Question 1.** Which semisimple algebras $\bigoplus_{i=1}^{h} (\mathbf{C})_{n_i}$ are goup-algebras of a finite group?

**Question 2.** Suppose $G$ and $H$ are groups with $\mathbf{C}G \cong \mathbf{C}H$. How "similar" are the groups $G$ and $H$?

**1.2** On Question **1**.

Several conditions for the degrees $n_i$ of a group have been known since the beginning of representation theory.

1) If $n_i \leq ... \leq n_h$, then $n_1 = 1$ and the number of $n_i's$ equal to 1 is $|G/G'|$, hence is a divisor of $|G| = \dim_{\mathbf{C}} \mathbf{C}G$.

2) (G. Frobenius) The $n_i$ divide $|G| = \dim_{\mathbf{C}} \mathbf{C}G$.

Unfortunately these conditions are not at all sufficient to ensure that a semisimple $\mathbf{C}$-algebra $\bigoplus_{i=1}^{h} (\mathbf{C})_{n_i}$ is the group-algebra of a finite group. We give two examples:

**Examples.** a) Let $A = (\mathbf{C}) \oplus \,..9..\, \oplus (\mathbf{C}) \oplus (\mathbf{C})_3$.

If $A \cong \mathbf{C}G$, we would have $|G| = 18$ and $|G/G'| = 9$, hence $|G'| = 2$. But then $G = G' \times P$, where $P$ is a Sylow-3-subgroup of $G$. Hence $G$ is abelian and has only the degrees 1,...,1.

b) Let

$$A = (\mathbf{C}) \oplus (\mathbf{C}) \oplus (\mathbf{C}) \oplus (\mathbf{C}) \oplus (\mathbf{C})_2 \oplus (\mathbf{C})_2 \oplus (\mathbf{C})_2 \oplus (\mathbf{C})_4 .$$

If $A \cong \mathbf{C}G$, then $|G| = 2^5$ and $|G/G'| = 4$. But then by a theorem of O. Taussky (see Huppert $I, p.$ 339) $G$ is dihedral, semidihedral or generalized quaternion, and hence $G$ has only the degrees 1 and 2, but not the degree 4.

c) If

$$\mathbf{C}G \cong \bigoplus_{i=1}^{h} (\mathbf{C})_{n_i} ,$$

then $h = h(G)$ is the number of conjugacy-classes of $G$. Not much is known about $h(G)$. A theorem of W. Burnside says that for odd $|G|$ always

$$h(G) \equiv |G| \pmod{16}$$

(cf. Huppert I, p. 549). For odd dimensions, this provides an additional necessary condition for a semisimple $\mathbf{C}$-algebra to be a group-algebra. But again, these conditions are not sufficient: Let

$$A = (\mathbf{C}) \oplus \,..7..\, \oplus (\mathbf{C}) \oplus (\mathbf{C})_7 \oplus (\mathbf{C})_7 .$$

If $A \cong \mathbf{C}G$, then $|G| = 105$ and $|G/G'| = 7$. But then $G'$ is cyclic of order 15 and has no automorphism of order 7. This forces $G$ to be abelian, and then $\mathbf{C}G \cong A$.

Unfortunately, nothing is known or even conjectured about conditions sufficient to make

$\sum_{i=1}^{h}(\mathbf{C})_{n_i}$ the group-algebra of a group.

**1.3 On Question 2.**

From $\mathbf{C}G \cong \mathbf{C}H$ one cannot conclude that $G \cong H$. For instance, all abelian groups of the same order have isomorphic group-algebras. Roggenkamp and Scott have recently shown that $\mathbf{Z}G \cong \mathbf{Z}H$ for large classes of groups implies $G \cong H$. Over $\mathbf{C}$ we know of only one remarkable result:

**Theorem 1** (I.M. Isaacs [20]). Suppose $\mathbf{C}G \cong \mathbf{C}H$.

a) If $G$ is $p$-nilpotent, so is $H$. If $G_{p'}$ and $H_{p'}$ are the normal $p$-complements of $G$ resp. $H$, then $\mathbf{C}G_{p'} \cong \mathbf{C}H_{p'}$.

b) If $G$ is nilpotent, so is $H$.

c) If $G$ has a Sylow-tower of some type, then $H$ has a Sylow-tower of the same type.

Statement c) suggests the question, whether $\mathbf{C}G \cong \mathbf{C}H$ and $G$ supersolvable implies that $H$ is supersolvable. But this is not true. T. Hawkes has given an example where

$|G| = |H| = 2^2 \cdot 3^2 \cdot 13^2$, and

$(\mathbf{C}) \oplus \cdot\cdot{}_{12}\cdot\cdot \oplus (\mathbf{C}) \oplus (\mathbf{C})_2 \oplus \cdot\cdot{}_{6}\cdot\cdot \oplus (\mathbf{C})_2 (\mathbf{C})_{12} \oplus \cdot\cdot{}_{42}\cdot\cdot \oplus (\mathbf{C})_{12}$,

but where $G$ is supersolvable and $H$ is not. Even much more: T. Hawkes has constructed recently examples $G$ and $H$, where $\mathbf{C}G \cong \mathbf{C}H$, $G$ is supersolvable and metabelian, but $H$ has arbitrary high nilpotent length (Hawkes [13]). Hence theorem 1 seems to be a very isolated result.

It might be mentioned that the much more detailed information on $G$ contained in the character-table of $G$ determines the lattice of normal subgroups of $G$ including indices, hence determines solvability and supersolvability of $G$. Does it determine the derived length of $G$? This seems to be unknown.

One question is most natural:

If $\mathbf{C}G \cong \mathbf{C}H$ and $G$ is solvable, is then $H$ solvable? There does not seem to exist a reduction to the simple case.

**1.4.** In 1968 I.M. Isaacs and D. Passman made a detailed study of connections between the structure of a group $G$ and the degrees of the irreducible representations of $G$. Here we mention only one of their results, which was an important starting point for many of our own investigations:

**Theorem 2** (I.M. Isaacs, D. Passman [25]). Suppose that all the degrees of irreducible representations of $G$ are 1 or primes.

a) $G$ is solvable and $G''' = E$.

(The proof does not depend on the classification of finite simple groups!)

b) Among the degrees of $G$ are at most two different primes.

(Isaacs and Passman give even a complete description of these groups.)

We would also like to mention here a result of W. Burnside:

**Theorem 3** (W. Burnside [5]). Let $G$ be an irreducible solvable subgroup of $GL(p, \mathbf{C})$, where $p$ is a prime.

a) If $G$ is monomial, then $G$ has an abelian normal subgroup A such that $|G/A|$ divides $p(p-1)$. In particular, $G''' = E$.

b) If $G$ is primitive, then $F(G)/Z(G)$ is of type $(p,p)$ and $G/F(G)$ is isomorphic to a solvable subgroup of $SL(2,p)$. In particular $G^{(6)} = E$.

This theorem indicates that for irreducible $G$ not the size of the degree but rather its arithmetical nature is important.

To formulate arithmetical properties of the degree-configuration of a finite group in a convenient way, we introduce several invariants:

**Definition** a) Let $n = \prod_{i=1}^{k} p_i{}^{a_i}$ be the prime-factor-decomposition of the natural number $n$ (the $p_i$ being different primes and $a_i > 0$).

We define

$$\omega(n) = \sum_{i=1}^{n} a_i,$$

$$\sigma(n) = k,$$

$$\tau(n) = \max_i \ a_i$$

b) If $G$ is a finite group and

$$\mathbf{C}G \cong \bigoplus_{i\,=\,1}^{h} (\mathbf{C}\ )_{n_i}\ .$$

we consider the *set*

$$cd\ G = \ \{n_i | i = 1, ..., h\}$$

(neglecting multiplicities) and call this the set of degrees of $G$. We also define

$$\omega(G) = \max_i \ \omega(n_i),$$

$$\sigma(G) = \max_i \ \sigma(n_i),$$

$$\tau(G) = \max_i \ \tau(n_i),$$

Finally, we denote by $\rho(G)$ the set of primes which divide at least one of the degrees $n_i$ of $G$.

There is a good reason to work with the set $cd\ G$ rather than with the stronger information of all the $n_i$ 's including multiplicities: Arithmetical conditions on $cd\ G$ are usually inherited by factor-groups and normal subgroups of $G$ (by Clifford-theory). Hence induction often is possible. But no such inductive behaviour holds for the structure of $\mathbf{C}G$. (We do not know which direct summands of $\mathbf{C}G$ correspond to factor-groups of $G$.)

As theorem 2 already shows, by no means every finite set of natural numbers is the set of degrees of a group. Only one positive result is known to us:

**Theorem 4** (I.M. Isaacs [21]). Let $p$ be a prime and let

$$1 = p^{a_1} < p^{a_2} < \ ... \ < p^{a_k}$$

be any set of powers of $p$. Then there exists a $p$-group of class 2 such that

$$cd\ G = \{p^{a_i} \mid i = 1, ..., k\}.$$

As by a famous result of G. Higman and C. Sims "most" $p$-groups are of class 2, it is not very surprising that all possible degree-sets for a $p$-group can already be realized with groups of class 2. But this theorem does by no means characterize the group-algebras of $p$-groups (see example b) in 1.2).

Now we come to a fundamental theorem describing $\rho(G)$ :

**Theorem 5** (N. Ito, [26], [27], G. Michler [41]). Equivalent are:
a) $p \notin \rho(G)$, which means that $p$ does not divide any degree of $G$.
b) $G$ has a normal abelian Sylow$-p$-subgroup.

The implication b) $\Rightarrow$ a) follows from a theorem of Ito (1951), stating that every degree of $G$ divides the index of any abelian normal subgroup of $G$. The conclusion a) $\Rightarrow$ b) was proved for $p$-solvable $G$ also by Ito [27] in 1951. Ito's proof really gives a reduction to the case of a simple group $G$. Using the classification of finite simple groups and deep information about their representations, G. Michler proved in 1986 in full generality the implication a) $\Rightarrow$ b). The crucial step is to show that a simple Chevalley-group $G$ has for every prime $p$ dividing $|G|$ blocks of different defects. (Indeed, there are always blocks of defect 0; Michler [41], Willems [48].)

Next, we want to mention a rather isolated result which is sometimes useful.

**Theorem 6** (J.G. Thompson [47]). Suppose $p$ is a prime such that $p|n_i$ for all degrees $n_i > 1$ of $G$. Then $G$ is $p$-nilpotent.

The direct converse is not true; for some kind of converse see Pahlings [44].

Finally, we report on two rather exotic results:

**Theorem 7** (Huppert [14]). Suppose that

$$cd\ G = \{1, 2, ..., k\}.$$

a) Then $k \leq 6$ and $k \neq 5$.
b) $k = 6$ if and only if $G = G'Z(G)$ and $G' \cong SL(2,5)$.
c) $k = 4$ if and only if $G$ has the following properties:

There exists $M \lhd G$ such that $|G/M| = 2$ and $G/Z(M) \cong S_4$. If $N \lhd G$ and $|N/Z(M)| = 4$, then $|N'| = 2$ and $Z(M)/N'$ is central under $G$.
(The smallest example is $G = GL(2,3)$, where $Z(M) = N' = Z(G)$.)

The cases $k \leq 3$ are described by Isaacs-Passman (see theorem 2).

**Remark.** If $G = Z_p \wr S_4$ and $p \geq 5$, then the character-theory of wreath-products shows that cdG is the set of all divisors of 24. This case is not quite as exotic as it seems. For J. Pense observed the following: Let $H$ be any solvable group. Then there exists an abelian group $A$ such that in $H \times A$ the converse of Lagrange's theorem holds (McLain; Huppert I, p. 663). Then $G = Z_p \wr (H \times A)$ (regular wreath-product) has for $p > |H \times A|$ the property that $cd\ G$ is the set of all divisors of $|H \times A|$.

**Question.** If $G$ is a nilpotent group, then $cd\ G$ is obviously closed under taking greatest common divisors and least common multiples. Also, if $n = n_1 n_2 \in cd\ G$ with $(n_1, n_2) = 1$, then $n_i \in cd\ G$. These properties obviously do not characterize nilpotent groups, they also hold for $G_1 \times G_2$ if cd $G_1 = \{1, p\}$ and cd $G_2 = \{1, q\}$ with primes $p \neq q$. How big is the class of groups whose degree-set has these properties? Unfortunately, these properties are not inherited by normal subgroups and factorgroups.

## §2 Degree problems in ordinary representation theory

Many theorems considered in this section are rather special. But they are sometimes the induction-basis for more general results. Historically they led to the general theorems and conjectures in theorems 13 and 14.

**2.1.** If $\omega(G) = 1$, we have the complete description of $G$ by Isaacs-Passman in theorem 2. Also the case $\omega(G) = 2$ is rather well known:
**Theorem 8** (B. Huppert, O. Manz [17]).
Suppose $\omega(G) = 2$.
a) If $G$ is not solvable, then *either* $G$ is a direct product of the alternating group $A_7$ with an abelian group it or $G$ has only one nonsolvable composition factor, and this is isomorphic to $A_5 \cong SL(2,4)$. All such groups with $\omega(G) = 2$ have been described. Typical examples are

$$G = S_5,\ cd\ G = \{1,4,5,6\};$$

$$G = SL(2,5),\ cd\ G = \{1,2,3,4,5,6,\};$$

$G$ is the natural extension of the vector space of order $2^4$ by $SL(2,4)$, here $cd\ G = \{1,3,4,5,15\}$.
b) If $G$ is solvable, then the derived length $dl\ G$ of $G$ is bounded by 4. This bound is best-possible. There are many groups with $dlG = 4$ and $\omega(G) = 2$, all these have been determined by $A$. Feyzioglu (unpublished). There are 5 different series of such groups. Neglecting central direct factors, two series consist of (2,3)-groups, one of $(2,p,q)$-groups, where $p = 2q - 1 \geq 5$, one of $(3,p,q)$-groups, where $p = q^2 + q + 1$ and $q > 3$, and finally a series of $(2,p,q)$ groups, where $q$ divides $p^2 - 1$. Hence the prime 2 or 3 is always involved.

A more general result is
**Theorem 9** (Huppert [15]). Let $G$ be solvable.
a) If $\omega(G) > 1$, then $dl\ G \leq 2\omega(G)$.
(Observe that $\omega(G) = 1$ implies only $dl\ G \leq 3$; see theorem 2).
b)$|\rho(G)| \leq 2\omega(G)^2$.

b) is a very weak result, we will see in theorem 13 much better estimates. a) was improved by A. Feyzioglu [7] as follows:
c) If $\omega(G) = 3$, then $dl\ G \leq 5$.
If $\omega(G) \geq 4$, then

$$dl\ G \leq \left[\tfrac{3}{2}\omega(G) + 2\right].$$

(Using more precise results about the case $\omega(G) = 3$ and dl $G = 5$, Feyzioglu showed that

$$dl\ G \leq \left[\tfrac{3}{2}\omega(G) + \tfrac{3}{2}\right]$$

in general.)
Is a bound linear in $\omega(G)$ for dl $G$ asymptotically best-possible? This seems likely, for Feyzioglu proved the following:

Let $G$ be an irreducible solvable subgroup of $GL(n, \mathbf{C})$. Then dl $G$ is bounded by a linear function in $\omega(n)$, and this bound is indeed asymptotically best-possible. (The case $\omega(n) = 1$, hence $n$ prime, is mentioned in theorem 3.)

The most convenient construction of solvable groups of large derived length is by iterated wreath-products of small groups. Unfortunately, in this construction the degrees grow exponentially, while the derived length grows only linearly with the number of factors.

**2.2 Theorem 10** (O. Manz [34] und [35]).
Suppose $\sigma(G) = 1$, which means that all degrees of $G$ are powers of primes.
a) If $G$ is not solvable, then $G$ is a direct product of $PSL(2,4)$ ($\cong A_5$) or $PSL(2,8)$ with an abelian group.
Observe *cd* $PSL(2,4) = \{1,3,4,5\}$ and cd $PSL(2,8) = \{1,7,8,9\}$.
(The proof needs a consequence of the classification of simple groups.)
b) $G$ is solvable if and only if $|\rho(G)| \leq 2$.
c) If $\rho(G) = \{p\}$, then $G$ has (by theorem 6 or the much deeper theorem 5) an abelian normal $p$-complememt. Now the nilpotent length of $G$ is bounded by 2, but there is obviously no universal bound for the derived length of $G$.
d) But if $\rho(G) = \{p,q\}$, then dl $G \leq 5$. If dl $G = 5$, then $\{p,q\} = \{2,3\}$ and up to central factors $G$ is the (necessarily splitting) extension of an abelian group of type (3,3) by GL(2,3). In this case

$$\text{cd } G = \{1,2,3,4,8,16\}.$$

**2.3.** Finally we study the influence of the invariant $\tau(G)$ on the structure of $G$.
**Theorem 11** (B. Huppert, O. Manz [18]).
Suppose $\tau(G) = 1$, which means that all degrees of $G$ are squarefree.
a) If $G$ is not solvable, then $G = A_7 \times H$, where $H$ is solvable with only squarefree degrees and $2,3,5,7 \notin \rho(H)$.
(Observe cd $A_7 = \{1,6,10,14,15,21,35\}$.)
*b*) If $G$ is solvable, then *dl* $G \leq 4$ and the nilpotent length of $G$ is at most 3. If $F_i(G)$ denotes the members of the ascending Fitting series, then $F_2(G)/F_1(G)$ and $G/F_2(G)$ are cyclic of squarefree orders.

A general result is

**Theorem 12** (U. Leisering, O. Manz [33]).
If $G$ is solvable, then

$$dl\ G \leq 4(\tau(G)+1).$$

(Even *dl* $G \leq 4\ \tau(G)$ is true; Leisering, unpublished.)

**2.4 Remarks.** We want to make two short remarks about the proofs of these theorems.
a) In all the cases where we made a statement about insolvable groups, the classification of finite simple groups was used. As an example we sketch how to prove that $A_7$ is the only simple group with only squarefree degrees:

The 26 sporadic groups are excluded, using the character-tables of Aachen or the Cambridge-Atlas. For the alternating groups degree- formulas are known, and it is easy to find non-squarefree degrees, except for $A_7$. There remain the Chevalley-groups. But if $p$ is the characteristic of the defining field and $p^a || G|$, then the Chevalley-group $G$ has a Steinberg-character of degree $p^a$. This forces $a = 1$ and then the list of the orders of the simple Chevalley-groups shows $G \cong PSL(2,p)$. But then $G$ has the degrees $p-1$ and $p+1$, and one of these is divisible by 4. (In many other theorems, the Chevalley-groups offer a hard problem.)
b) In the solvable cases the degrees are determined by Clifford-theory. The typical situation is as follows: We have $N \triangleleft G$ and know a character $\psi$ of N. Let $T_G(\psi)$ be the inertia-group of $\psi$ in G. In many cases $\psi$ has an extension $\psi_o$ to $T_G(\psi)$. Then all characters of $G$ above $\psi$ are of the form

$$(\psi_o \tau)^G, \text{ where } \tau \in Irr\ T_G(\psi)/N.$$

This gives particularly large degrees if $T_G(\psi) = N$. Theorems about regular orbits of $p$-groups acting on vector-spaces are several times very efficient in producing this situation. Also, if the structure of $T_G(\psi)/N$ allows, there are $\tau \in Irr\ T_G(\psi)/N$ of large degree. Together, conditions on the degrees of $G$ restrict $|G : T_G(\psi)|$ and $T_G(\psi)/N$ substantially, hence restrict $G/N$.

Two general problems arose from the special results of this section.

**2.5 The $\rho$ - $\sigma$ - conjecture**
a) If $G$ is solvable, then $|\rho(G)| \leq 2\ \sigma(G)$.
b) In general, $|\rho(G)| \leq 3\ \sigma(G)$.
As O. Manz will report in his contribution in detail on the $\rho - \sigma$-conjecture, we just give the best results at present.

**Theorem 13**. Let $G$ be solvable.
a) (O. Manz [34]) If $\sigma(G) = 1$, then $|\rho(G)| \leq 2$.
b) (D. Gluck [8]) If $\sigma(G) = 2$, then $|\rho(G)| \leq 4$.
c) (D. Gluck, O. Manz, T. Wolf [10])

$$|\rho(G)| \leq 3\ \sigma(G) + 2.$$

If every normal Sylow-subgroup of $G$ is abelian, then even

$$|\rho(G)| \leq 2\ \sigma(G) + 32.$$

d) $|\rho(G)| \leq 3\ \sigma(G)$ is true for all simple $G$.

Easy examples show that $|\rho(G)| \leq 2\ \sigma(G)$ is certainly best-possible for solvable $G$.

Another interesting attempt to formulate general results about the degrees is by using the degree graph $\Gamma(G)$.

**Definition** The vertices of $\Gamma(G)$ are the primes in $\rho(G)$. We connect two different primes $p$ and $q$ in $\rho(G)$ by an edge, if pq $|\chi(1)$ for some $\chi \in Irr\ G$.

**Theorem 14** (O. Manz, P. Palfy, R. Staszewski, W. Willems, T. Wolf [38], [39]).
a) $\Gamma(G)$ has at most 3 components.
b) If $G$ is solvable, $\Gamma(G)$ has at most 2 components.
c) Suppose $G$ is solvable and $\Gamma(G)$ has 2 components. Then both of them are complete graphs. If $k$ is the cardinality of the vertices in the smaller component, the other has cardinality at least $2^k - 1$. Also $G$ is rather special, the nilpotent length of $G$ is at most 3, except the possibility $G/F(G) \cong GL(2,3)$.
*d*) If $G$ is solvable and $\Gamma(G)$ is connected, the diameter of $G$ is at most 3. (Is it always at most 2?)

The possible graphs $\Gamma(G)$ for solvable groups $G$ with at most 4 vertices are:

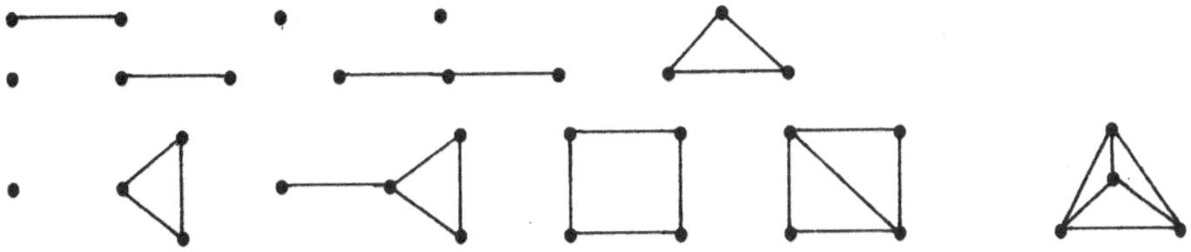

which do really exist and possibly (but most doubtfully) 

**2.7** We finally mention here a question, to the classical version of which no recent contribution has been made but several of whose analogues will be considered in §3-5.

**Theorem 15** Suppose that cd $G$ contains exactly $k$ different degrees (including 1).
a) (Isaacs-Passman [25]) If $k = 2$, then $G$ is solvable and $dl\ G \leq 2$.
b) (Isaacs [24]) If $k = 3$, then $G$ is solvable and $dl\ G \leq 3$.
c) (Garrison, unpublished) If $G$ is solvable and $k = 4$, then $dl\ G \leq 4$.
d) (Taketa [46]) If $G$ is a monomial group, then $dl\ G \leq k$. (This includes nilpotent and supersolvable groups.)
e) (D. Gluck [8]) If $G$ is solvable, then $dl\ G \leq 2k$.
f) (T. Berger [1] If $|G|$ is odd, then $dl\ G \leq k$.

The conjecture, that for solvable $G$ always $dl\ G \leq k$ holds, is still not proved in full generality.

The assumption of theorem 15 is (trivially) inherited by factor-groups of $G$, also by normal Hall-subgroups, but in general not by normal subgroups. Hence an approach by Clifford-induction is not obvious.

## §3 Modular degree problems

In this section let $K$ always be an algebraically closed field of characteristic $p > 0$. It is not true in general that the dimensions of the simple KG-modules divide $|G|$. (The simple group $PSL(2,7)$ of order 168 has simple modules of dimensions 1,3,5,7 in characteristic 7.) But if $G$ is $p$-solvable, by a theorem of P. Fong and Swan the dimensions of the simple $KG$-modules do divide $|G|$. Hence all the questions considered in §1 and §2 at least make sense. Also the methods of proof based on Clifford-theory are available, but they became weaker. This is mainly due to the fact that the structure of a $p$-group $P$ is largely controled by the knowledge of its degrees in characteristic 0, but in characteristic $p$ there is only one simple KG-module, the trivial module.

We use the notations of §1, but add an index 0 if we work over **C**, an index $p$ if we work over an algebraically closed field $K$ of characteristic $p$.

The theorem of P. Fong quoted above says more precisely

$$cd_p G \subseteq cd_o G$$

for $p$-solvable $G$.

This follows from the fact that all irreducible representations of $G$ over $K$ are obtained from integral $p$-adic representations by reduction mod $p$, if $G$ is $p$-solvable.

As the largest normal $p$-subgroup $O_p(G)$ of $G$ is the intersection of the kernels of all simple KG-modules, information about $cd_pG$ can only say something about $G/O_p(G)$.

We begin with the modular analogues of theorem 5:

**Theorem 16** (G. Michler [41], Okuyama [43]).
Equivalent are:
a) $G$ has a normal Sylow$-p$-subgroup.
b) Every modular degree in $cd_pG$ is prime to $p$.
For $p > 2$ this is a consequence of Michler's result about $p$-blocks of different defects for Chevalley-groups; for $p = 2$ there exists a beautiful proof by Okuyama, independent of the classification of simple groups.

But with theorem 16 the story is not complete in the modular case. What happens for primes $q$ different from the characteristic $p$?

**Theorem 17** (O. Manz, T. Wolf [40]).
Let $G$ be $p$-solvable and let $q$ be a prime, $q \neq p$, such that $q|n$ for all $n \in cd_pG$.
a) Then $G$ is $q$-solvable, the $q$-length of $G/O_{p,q}(G)$ is at most one and the $q$-sections in the ascending $q$-series of $G$ are abelian. Hence the Sylow$-q$-subgroups of $G$ are metabelian.
b) If $q \mid p-1$ and $(p,q) \neq (2,3)$, then $G/O_p(G)$ has a normal abelian Sylow$-q$-subgroup. (This is the exact analogue of theorem 5.)

Is it true, that without $p$-solvability the assumptions in theorem 17 force the Sy*low* $- q$-subgroups of $G$ to be metabelian?

Next we mention the extensions of theorems 1 and 6 to the modular case:

**Theorem 18** (G. Navarro [42]). Let $q$ be a prime and $q \neq p = Char\ K$.
a) If $KG \cong KH$ and $G$ is $q$-nilpotent, so is $H$.
b) Let $G$ be $p$-solvable. If the dimension of every simple $KG$-module is 1 or a multiple of $q$, then $G$ is $q$-nilpotent.
( b) was proved earlier for solvable $G$ by R. Gow.)
We remark that $cd_2(SL(2,2^f))$, $cd_2(Suz(2^f))$ and $cd_2(Sp(4,2^f))$ contain only powers of 2. Are there similar cases for odd primes $p$?

Next we turn to the modular analogue of theorem 2:

**Theorem 19** (B. Huppert [16]).
Let $G$ be $p$-solvable and $O_p(G) = E$. Suppose that $cd_pG$ contains only 1 and primes. Then
*either* $cd_oG$ contains only 1 and primes (and then $G$ is described by Isaacs-Passman)
*or* $p = 3$, $G' \cong SL(2,3)$, $|G/G'Z(G)| = 2$.
Then

$$cd_3G = \{1,2,3\} \text{ and } cd_oG = \{1,2,3,4\}.$$

(The smallest $G$ with these properties is $GL(2,3)$.)

Unlike theorem 2, $p$-solvability in theorem 19 is needed; this is shown by the examples

$$cd_5A_5 = \{1,3,5\},\ cd_7PSL(2,7) = \{1,3,5,7\}.$$

Theorem 19 leads to the following general question:

**Question.** Suppose $G$ is $p$-solvable and $O_p(G) = E$. If we know some arithmetic restriction for all the $p$-modular degrees in $cd_pG$, do there follow any arithmetic restrictions for the ordinary degrees in $cd_oG$?

Using the invariant $\tau_p(G)$, a generalization of theorem 19 has been proved.

**Theorem 20** (U. Leisering [32]).
Let $G$ be solvable and $O_p(G) = E$.
a) If $\tau_p(G) = 1$ (which means that all $p$-modular degrees are squarefree), then $\tau(G) \leq 2$. Further dl $G \leq 4$ and nil $G \leq 3$.

nil $G \leq 2\ \tau_p(G) + 4$,
dl $G \leq 3\ \tau_p(G) + \log_2 \tau_p(G) + 5$,
$\tau_o(G) \leq \tau_p(G)^2(8\ \log_2 p + 4) + \tau_p(G)(8\ \log_2 p + 7)$.

It would be nice to have a stronger statement like $\tau_o(G) \leq 2\ \tau_p(G)$.
Are similar theorems true, connecting $\omega_p(G)$ with $\omega_o(G)$ or $\sigma_p(G)$ with $\sigma_o(G)$?

Modular analogues of theorem 15 have been studied by F. Bernhardt.

**Theorem 21** (F. Bernhardt [2]). Let $G$ be solvable and $O_p(G) = E$.
Suppose $cd_pG = \{1, m\}$ with $m > 1$.
a) If $p > 2$, then $dl\ G \leq 3$.
b) If $p = 2$, then $dl\ G \leq 4$.

Both estimates are best-possible. This is shown by the examples
$cd_{13}A(3^3) = \{1, 26\}$ (semilinear group),.
$cd_2G = \{1, 8\}$, if $G$ is the extension of an elementary abelian group of type (3,3) by SL(2,3).

It is surprising, that solvability is not forced in this case, for

$$cd_2PGL(2, q) = \{1, q-1\}$$

if $q = 2^m + 1$ is a Fermat prime or $q = 9$. (Are there nonsolvable cases for odd $p$?)

**Theorem 22** (F. Bernhardt [3]). Let $G$ be solvable. Then

nil $G \leq 3|cd_pG| - 2$

and

$\ell_p(G) \leq |cd_pG| - 1$.

(The proof is an adaptation of the proof of Taketa's theorem 15 d). But this method does not allow a control of $dl\ G$.)
Much more general results have been obtained very recently by Isaacs.

**Theorem 23** (I. M. Isaacs [22]).
Suppose that $O_p(G) = E$ and $c = |cd_pG|$.
a) If $G$ is $p$-solvable, then $\ell_p(G) \leq c - 1$ (improving theorem 22).
b) If $G$ is solvable, then

$$\text{dl } G \leq c^2 + 3c - 5.$$

**Remark** Let $G = VP$ be a semidirect product of an elementary abelian $q$-group $V (q \neq p)$ with a $p$-group $P$, which operates faithfully on V. If $\infty$ is a character of $V (\alpha \neq 1)$ and $T_G(\infty)$ its inertia-subgroup, then $\alpha$ has exactly one extension $\alpha_o$ to $T_G(\alpha)$, and $\alpha_o^G$ is irreducible of degree $|G : T_G(\alpha)|$. Hence $cd_pG$ is the set of orbit-lengths of $P$ on the dual of V. To obtain a modular analogue of theorem 15 in this special case, one has to answer the following question:

If the $p$-group $P$ operates faithfully on $V$ and if there appear $b$ different orbit-lengths, does there exist some bound for dl $P$ by a function of $b$?

In this special case better bounds then in theorem 23 are available:

**Theorem 24** (I.M. Isaacs [22]). Suppose the $p$-group $P$ acts faithfully on the abelian $p'$-group $V$ with $b$ different orbit-lengths of $P$ on V. Then

$$dl\ P \leq \begin{cases} b-1 & \text{if } p>2 \\ b & \text{if } p=2 \end{cases}.$$

The proof uses earlier results by F. Bernhardt [3] for the case $b = 2$.

The existence of regular orbits is of some interest in this context. This question has been studied by several authors; a short and rather complete account is in Huppert-Manz [19].

Also the degree-graph of theorem 14 has been studied in the modular case. We refer to the report by O. Manz.

## §4 Lengths of conjugacy-classes

If $g \in G$, then the length of the conjugacy-class of $g$ in $G$ is $|G : C_G(g)|$, hence is a divisor of $|G|$. It therefore makes sense, to classify groups by the arithmetic properties of their class-lengths. We define in exact analogue of §1 :

$$\omega'(G) = \max_{g \in G} \omega(|G : C_G(g)|),$$

$$\sigma'(G) = \max_{g \in G} \sigma(|G : C_G(g)|),$$

$$\tau'(G) = \max_{g \in G} \tau(|G : C_G(g)|),$$

$cl\,G = \{|G : C_G(g)| \mid g \in G\}$ (as set)
$\rho'(G) = \{p \mid p$ divides some element from $cl\,G\}$.

A most astonishing fact is, that arithmetic restrictions on the class-lengths of $G$ lead also to restrictions on the structure of $G$, similar to those in §2 and in most cases much stronger. We begin with the analogue of theorem 5.

**Theorem 5'**. Suppose no class-length of $G$ is divisible by the prime $p$. Then $G$ has a central Sylow$-p$-subgroup.

**Proof**. Let $P$ be a Sylow $p$-subgroup of $G$ and $g \in G$. Then
$p \mid |G : C_G(g)|$, hence
$P^h \leq C_G(g)$ for some $h \in G$. Therefore $g \in C_G(P)^h$ and so

$$G \subseteq \bigcup_{h \in G} GC_G(P)^h.$$

It is a well-known elementary fact that this implies $C_G(P) = G$.

We gave the proof of this theorem to show how trivial it is, while the proof of theorem 5 needed the classification of simple groups!

Let us look at other parallels.

**Theorem 1'** (J. Cossey, T. Hawkes). Let $G$ and $H$ be finite groups. Suppose that for every integer $m$ the number of conjugacy-classes of length $m$ in $G$ and $H$ are the same. If $G$ is nilpotent, so is $H$.
(More general : If $G = O_p(G) \times O_{p'}(G)$, then also $H = O_p(H) \times O_{p'}(H)$.)

**Theorem 2'** . Let $G$ be solvable.
a) (D. Chillag, M. Herzog [6], S. Dolfi).
If $\omega'(G) = 1$, then *either* $G$ is nilpotent of class 2 and $|G'|$ is a prime *or* $G/Z(G)$ is a Frobenius-group of order pq ($p$ and $q$ primes), then $F(G)$ is abelian of index $q$ (say) and $|G'| = p$. In particular, $G$ is supersolvable and $G'' = E$.
b) (S. Dolfi). In general

$$dl\ G \leq 2|'(G) \text{ and nil } G \leq |'(G) + 1.$$

**Theorem 10'** (D. Chillag, M. Herzog [6], S. Dolfi). Suppose $\sigma'(G) = 1$. Then $G$ is solvable of nilpotent length at most 2 and $\rho'(G)$ contains at most 2 different primes. More precisely: (S. Dolfi) Up to abelian direct factors $G$ is a $p$-group *or* $G$ has the following structure:

$G = PQ$ with abelian Sylow-subgroups $P$ and $Q$ for for primes $p$ and $q$. Further $F(G) = O_p(G) \times Q$, $O_p(G) = Z(G)$ and $G/Z(G)$ is a Frobenius-group with Frobenius - kernel $F(G)/Z(G)$. In particular $G'' = E$

(Observe that these groups are much more restricted than the groups in theorem 10, where $dl\ G \leq 5$ was best-possible!)

S. Dolfi could also describe all groups, where every class-length is either a $\pi$-number or a $\pi'$-number, and both cases occur. The groups are only slightly more general than groups with $\sigma'(G) = 1$. (Observe that the corresponding condition for character-degrees allows much more complicated groups; cf. theorem 10 and the cases with a non-connected graph $\Gamma(G)$ in Theorem 14).

**Theorem 11'** (D. Chillag, M. Herzog [6]). Suppose all class-lengths of $G$ are square-free. Then $G$ is solvable, $G/F(G)$ is cyclic and dl $G \leq 3$.

In all cases, theorems $9'$ to $11'$ give stronger results than the corresponding degree-theorems, in particular solvability is enforced in all cases.

Even more surprising is that the exact analogue of the $\rho - \sigma$-injecture seems to hold.

**Theorem 13'** (C. Casola).
Let $G$ be a finite group which is $p$-nilpotent with abelian Sylow $p$-subgroup for at most one prime p dividing the order of $G$.
Then $|\rho'(G)| \leq 2\ |\sigma'(G)|$
( Note that perfect groups obviously satisfy the hypothesis in Theorem 13'. That $\sigma'(G) = 1$ implies $|\rho'(G)| \leq 2$ is stated in theorem 10'. )

Also here, $|\rho'(G)| \leq 2\ \sigma'(G)$ seems to be true in general, it again would be best-possible.

For $p$-groups much stronger results do exist:

**Theorem 25** (Vaughan-Lee, Leedham-Green, P. Neumann, J. Wiegold).
Let $G$ be a $p$-group. Then

$$|G'| \leq p^{\frac{\omega'(G)(\omega'(G)+1}{2}}$$

and

$$c(G) < \frac{p}{p-1}\omega'(G) + 1.$$

(See Blackburn-Huppert II, p.341-347.)

Obviously, no universal bound for the nilpotency class $c(G)$ of a $p$-group $G$ in terms of $\omega(G)$ is possible. There are $p$-groups with an abelian normal subgroup of index $p$ (hence with $\omega(G) = 1$), but of arbitrarily high nilpotency-class (dihedral-groups, wreath-products $Z_{p^m} \wr Z_p$.) Also here, class-length conditions are much more restrictive than degree conditions.

Simple relations between cd $G$ and $c\ l\ G$ do not seem to exist, even not for $p$-groups. If $G$ is extraspecial and $|G| = p^{2m+1}$, then cd $G = \{1, p^m\}$ and $c\ l\ G = \{1, p\}$. If $G = A \wr Z_p$, where A is abelian and $|A| = p^n$, then cd $G = \{1, p\}$ and $c\ l\ G = \{1, p, p^{n(p-1)}\}$.

We do not know of any general reason for the similarities listed above. There might even not be a universal reason.

Thompson's theorem 6 has no analogue for class-lengths. For the class-lengths of SL(2,3) are 1,4 and 6, but SL(2,3) is not 2-nilpotent.

Is it possible to bound the derived length of a solvable group by the number of it's different class-lengths, in strict analogy of theorem 15? Induction works even less here than for theorem 15. Some facts have been established by N. Ito long ago:

**Theorem 15'** (N. Ito [26], [27]).

a) If $c\,l\,G = \{1, m\}$, then $m = p^a$ is a power of a prime and $G$ is nilpotent. More precisely, $G = P \times A$, where A is an abelian $p'$-group, $P$ is a $p$ group with $P/Z(P)$ of exponent $p$. (But is it true in this case that $G'' = E$? Is there at least a universal bound for $dl\ G$?)

b) If $|c\,l\,G| = 3$, then $G$ is solvable. (This is an analogue of b) in theorem 15. But what about $dl\ G$?)

Ito proved also the following result on class-lengths: Suppose $p$ and $q$ are different primes and there is no conjugacy-class of $G$ of a length divisible by pq. Then $G$ is $p$-nilpotent or $q$-nilpotent. The direct analogue does not hold for degrees, as already the symmetric group $S_4$ and many groups with $\sigma(G) = 1$ show. It would be very interesting to have any analogue. For it would tell us something about primes $p, q$ in $\rho(G)$, not connected in $\Gamma(G)$. The general impression that $\Gamma(G)$ is a very rich graph could be made more explicit by such a theorem. (see O. Manz).

If $G$ has two conjugacy classes $g^G$ and $h^G$ of coprime lengths $> 1$, then obviously $G = C_G(g)C_G(h)$. By a conjecture of Szep (proved by Arad and Fisman, using the classification of simple groups) it follows that $G$ is not simple. The direct analogue of this deep statement for degrees is obviously false. Is there any analogue? (I doubt it.)

A theorem by Isaacs [22] states the following (see O. Manz, 5,5):

Let $H \neq E$ be acting faithfully and coprimely on G. If $H$ fixes every non-linear $\chi \in Irr\ G$, then $G'$ is nilpotent. (Isaacs gives some more detailed information about $[G, H]$.)

A quite trivial class-analogue of this theorem is:

Let $H \neq E$ be acting faithfully and coprimely on G. If $H$ fixes every conjugacy-class of $G$ outside of $Z(G)$, then $G$ is abelian. (If $H$ fixes all classes of $G$, then it is well-known that $H = E$; Speiser, Satz 108.)

## §5 Reduced degree problems

**5.1 Definition** Let $K$ be an arbitrary field of characteristic 0 and $L$ *it's* algebraic closure. If $G$ is a finite group, then

$$KG = \bigoplus_{j=1}^{k} (D_j)_{n_j},$$

where the $D_j$ are skew-fields. Each $D_j$ corresponds to a set of absolutely irreducible characters of $G$, Galois-conjugate over $K$. Let $\chi_j$ be a character corresponding to $D_j$.

Then

$$Z(D_j) = K(\chi_j(g) \mid g \in G)$$

and $dim_{Z(D_j)} D_j = m_K(\chi_j)^2$, where $m_K(\chi_j)$ is the Schur-index of $\chi_j$ over $K$.

We call $n_j$ the reduced degree of $\chi_j$ and write $n_j = t_K(\chi_j)$. We also put

$$t_K(G) = \{t_K(\chi_j) | \chi_j \in Irr_L G\}.$$

As $\chi_j(1) = m_K(\chi_j) t_K(\chi_j)$, the control of $G$ by the $t_K(\chi_j)$ requires knowledge of $\chi_j(1)$ *and* $m_K(\chi_j)$. The determination of $m_K(\chi_j)$ is often a very difficult task of a number-theoretic nature, even for very "small" groups it can be non-trivial. Nevertheless, some analogues of the theorems in §1 and §2 have been proved for reduced degrees.

**Theorem 26** (R. Gow, B. Huppert [11]). Suppose $p$ is a prime, not dividing any $t_K(\chi)$, where $\chi \in Irr_L G$.
a) $G$ has a normal Sylow–$p$-subgroup $P$.
b) Either $P$ is abelian or $p = 2$ and $G = Q \times H$, where $Q$ is a quaternion group of order 8 and $H$ has a normal elementary abelian Sylow-2-subgroup. If $q$ is any prime divisor of $|H|$, then $-1 = a^2 + b^2$ has no solutions $a, b$ in the cyclotomic field $K(\epsilon_q)$ of primitive $q$-th roots of unity over $K$.

If $K = \mathbf{Q}$, then conversely all these groups do have only reduced degrees prime to $p$.

This is an analogue of the Ito-Michler-theorem 5. The exact analogue of Thompson's theorem 6 holds also.

**Theorem 27** (R. Gow, B. Huppert [11]). If the prime $p$ divides $t_K(\chi)$ for all $\chi \in Irr_L G$ with $t_K(\chi) > 1$, then $G$ is $p$-nilpotent.

R. Gow showed recently, that it is even sufficient to assume in theorems 26 and 27 only that for every $\chi \in Irr_L G$ with $m_K(\chi) = 1$ (hence with $t_K(\chi) = \chi(1)$) that $p | \chi(1)$ resp. that $p | \chi(1)$ for all such $\chi$ with $\chi(1) > 1$. (This seems to show that there are "sufficiently many" characters with Schur-index 1.)

The general impression that the characters with Schur-index 1 already determine largely the group, is made more precise by the following very recent generalization of theorem 10. :

**Theorem 28** (R. Gow). Suppose $|G|$ is odd and all $\chi \in Irr\ G$ with $m_Q(\chi) = 1$ have prime-power degrees. Then $|\rho(G)| \leq 2$.

Are there more general theorems about the existence of "sufficiently many" characters with Schur-index 1 of a solvable group?

Also the analogue of theorem 15 a) has been studied in the context of reduced degrees.

**Theorem 29** (R. Gow, B. Huppert [12]). Suppose 1 and $k > 1$ are the only reduced degrees of $G$ over K. Let $\pi$ be the set of all prime-divisors of $k$.
a) $G$ has a normal $\pi$-complement $N$, and $G/N$ is nilpotent. $N$ has only reduced degree 1, hence $N$ is abelian or $N = Q \times A$ with a quaternion-group $Q$ of order 8 and abelian A.
b) If $k$ is odd, then $G'' = E$.
c) If $k$ is even, but not a power of a prime, then $G''' = E$; if $8 \nmid k$ then even $G'' = E$ (There are cases with $G'' \neq E$ if $8 | k$.)
d) If $k = 2^e$, then $G^{(4)} = E$.

(If 2-groups with only two reduced degrees are always metabelian, then even $G''' = E$ in this case. We do not know a 2-group $G$ with only two reduced degrees and $G'' \neq E$.)

There are further results by Gow and Huppert (unpublished):
**Theorem 30** a) Let $G$ be solvable and suppose that all reduced degrees of $G$ over $K$ are powers of primes. Then there appear among these primes at most two odd primes. *If* $-1 = a^2 + b^2$ with $a, b \in K$, there appear at most two primes.
b) If all reduced degrees over $K$ are 1 or primes, then $G$ is solvable. (The proof uses the classification of finite simple groups, quite different from theorem 2 by Isaacs-Passman!)
c) If $G$ is solvable and all reduced degrees of $G$ over $K$ are squarefree, then $G^{(4)} = E$. (The covering-group $\hat{S}_4$ of $S_4$ with only one involution has over $Q$ the reduced degrees 1,2,3, but has derived length 4.)

## §6 Loewy-series

Let $P$ be a $p$-group and char $K$ $p$. We define the Loewy-series of $P$ by

$$KP > J(KP) > \dots > J(KP)^s > J(KP)^{s+1} = 0$$

and put

$$c_j(P) = \dim_K J(KP)^j / J(KP)^{j+1}.$$

Jennings gave already in 1941 a formula to calculate $c_j(P)$:
**Theorem** . We define a central-series

$$P = \kappa_1(P) \geq \kappa_2(P) \geq \dots \geq \kappa_\ell(P) > \kappa_{\ell+1}(P) = E$$

by

$$\kappa_j(P) = [\kappa_{j-1}(P), P] \cdot \kappa_m(P)^p,$$

where $m$ is the least integer with $pm \geq j (j > 1)$. If $|\kappa_j(P)/\kappa_{j+1}(P)| = p^{d_j}$, then Jenning's formula says

$$\sum_{j=o}^{s} c_j t^j = \prod_{i=1}^{l} (1 + t^i + \dots + t^{i(p-1)})^{d_i}.$$

(See Blackburn-Huppert II, p. 252-262). In particular, $c_j = c_{s-j}$ $(j = 0, \dots, s)$

This formula is not easy to use, due to the rather irregular behaviour of the $d_j$. Namely very often is $\kappa_j(P) = \kappa_{j+1}(P)$, hence $d_j = 0$.

Based only on a small number of examples, the following question was asked:
(**Q**) Is it true that

$$c_{j-1}(P) \leq c_j(P) \text{ for } 0 \leq j \leq \frac{s}{2}?$$

The positive answers concern rather special groups.
**Theorem 31** (O. Manz, R. Staszewski [37], [36]).
a) ($Q$) is true if $P$ is regular (in the sense of P. Hall) and $P' \leq P^p$ or if $P$ is extraspecial. (This includes abelian and metacyclic $p$-groups.)
b) Let $d$ be the minimal number of generators of P. Suppose $d > 1$. We put $c_j(P) = c_j$,

$$a = \left[\frac{(p-1)(d-1)}{2}\right] \text{ and } b = \left[\frac{(p-1)d}{2}\right].$$

Then $c_o < c_1 < \ldots < c_b$ and $c_a < c_i$ for $a < i < s$-a.

In particular, $d \leq c_i$ for $1 \leq i \leq s-1$.

Counterexamples to $(Q)$ could be found rather easily for $p = 2, 3, 5$ (see Manz-Staszewski [37] , Stammbach-Stricker [45] and Neubüser). Most of the counterexamples were provided by $p$-groups of maximal class, where the calculation of $\kappa_j(P)$ is easy. But all attempts to find counterexamples for $p \geq 7$ were until now without sucess. Leedham-Green and Staszewski [31] have shown that there are no such examples if $|P|$ divides $7^{15}$, $11^{12}$, $13^{12}$. This result is based on several general properties of the numbers $d_j$ and a computer-aided calculation of all possible cases for the $P$ as described above.

Hence there seems to be quite strong evidence that $(Q)$ is true if $p \geq 7$.

If $M$ is a KG-module, we define the Loewy-series of $M$ by

$$M = M_o > M_1 > \ldots > M_\ell = E,$$

where $M_j = M\ J(KG)^j$. One is naturally interested in the composition factors of the semisimple modules $M_j/M_{j+1}$. Much less than in the case of $p$-groups is known. But for groups of $p$-length 1 a precise result can be proved:

**Theorem 32** (O. Manz, U. Stammbach, R. Staszewski [36]). Suppose $G$ is $p$-solvable of $p$-length 1 and $K$ is algebraically closed of characteristic $p$. Let $V$ be a simple KG-module in a block $B$ with defect group $\delta(B)$. Then the multiplicity of $V$ in $B\ J(KG)^n/B\ J(KG)^{n+1}$ is dim $V \cdot c_n$, where

$$c_n = \dim\ J(K\delta(B))^n/J(K\delta(B))^{n+1}.$$

By considering the semilinear groups $\Gamma(p^p)$ of $p$-length 2 in [36], it became clear that theorem 32 does not hold in general if the $p$-length is larger than 1. The situation seems to become very complex then.

**Theorem 33** (W. Willems [49]). Let $G$ be $p$-solvable and let $P$ be a projective indecomposable KG-module (Char $K = p$). If $\ell(P)$ is the Loewy-length of $P$, then

$$\ell(P) \leq Max\{|V|\ |V \text{ is the vertex of a composition factor of } P\}.$$

Equality holds if and only if the block of $P$ has cyclic defect groups. In particular, $\ell(P) \leq |D|$, if $D$ is a defect group for the block of $P$ (This last remark is an earlier result by Koshitani, Okuyama and Tsushima.)

## §7 Coding-theory

Some linear codes can be described conveniently as ideals in group-rings.

**7.1** (F. Bernhardt, P. Landrock, O. Manz [4]). Let $g_{24}$ resp. $g_{12}$ be the extended binary resp. ternary Golay-codes. It is well-known that

Aut $g_{24} \cong M_{24}$ (Mathieu-group $M_{24}$)

Aut $g_{12} \cong \hat{M}_{12}$ (Schur-covering group of the Mathieu group $M_{12}$).

$M_{24}$ contains a subgroup $S_4$ which operates fixed-point-freely. $\hat{M}_{12}$ contains a group $SL(2,3)$, and

$A_4 \cong SL(2,3)/\ Z \leq M_{12}$,

where $A_4$ operates fixed-point-freely. This observation leads to the following conclusion:

$g_{24}$ is an ideal in $GF(2)[S_4]$,

$g_{12}$ is an ideal in the twisted group ring $GF(3)[A_4]$.

The module-structure (direct decompositions, Loewy-series) of $g_{24}$ and $g_{12}$ have been determined. $g_{12}$ is projective, $g_{24}$ not. Using a computer, idempotents in $g_{12}$ and $g_{24}$ have been found, which can be used for decoding.

**7.2** (P. Landrock, O. Manz [30]).
Let $p$ be a prime, $F = GF(p)$, $K = GF(p^m)$, $E = K^+$ the additive group of $K$ and $Z = K^*$ the multiplicative group of K. Let $L$ be a field with $F \leq L \leq K$. We consider the element

$\bar{E} = \sum_{e \in E} e$

in the group ring $L[E]$. There is an $F$-linear mapping

$\Gamma_L : L[Z] \to L[E]/L\bar{E}$,

defined by,

$z_L = z + L\bar{E}$ (for $z \in Z$).

For $F = L$, this shows that the Reed-Muller-codes over $F$ are exactly the powers of $J(F[E])$.

(Proved first by S. Berman if $p = 2$, by P. Charpin if $p > 2$.)

For arbitrary $L$ this shows that the Reed-Muller-Codes over $L$ still are ideals in $L[E]$, but do not correspond to the powers of the radical.

In the case $F = L = GF(2)$, the structure of $F[E]$ is used to describe a new decoding algorithm for the Reed-Muller-codes.

**7.3** (R. Knörr, W. Willems [28]).
If $GRM(r,m)(0 \leq r \leq m(p-1))$ is the generalized Reed-Muller code over $GF(p)$, then

$Aut(GRM(r,m)) = GF(p)^* \times AGL(m,p)$,

where $AGL(m,p)$ is the affine group, with the obvious exceptions for
$r = 0,\ m(p-1)-1,\ m(p-1)$.

(This was known for $p = 2$ before. The proof uses the classification of doubly transitive groups.)

**7.4** (R. Knörr, W. Willems [29]).
Let $C$ be a linear code over $GF(p)$ of length $p^m$. Then $C$ is a GRM-code if and only if Aut $C$ contains a subgroup isomorphic to $AGL(m,p)$.

## REFERENCES

[1] T. Berger, "Characters and derived lengths in groups of odd order", Journal of Algebra 39 (1976), 199-207.

[2] F. Bernhardt, "Modular character degrees and the structure of solvable groups", to appear Arch. Math .

[3] F. Bernhardt, "On $p$-groups with three orbit sizes", to appear.

[4] F. Bernhardt, P. Landrock, O. Manz, "The extended Golay codes considered as ideals", Aarhus Preprint Series 1988/89 No.3.

[ 5] W. Burnside, "On soluble irreducible groups of linear substitutions substitutions in a prime number of variables", Acta mathematica 27 (1903), 217-224.
[ 6] D. Chillag, M. Herzog, "On the lengths of the conjugacy classes of finite groups", J. of Algebra 131 (1990), 110-125.
[ 7] A.K. Feyzioglu, "Charaktergrade und die Kommutatorlänge in auflösbaren Gruppen", J. Algebra 126, no.1 (1989), 225-251.
[ 8] D. Gluck, "Bounding the number of character degrees of a solvable group", Journal of the London Mathematical Society. Vol. 31 (1984), 457-462.
[ 9] D. Gluck, "A conjecture about character degrees of solvable groups".
[10] D. Gluck, O. Manz, "Prime factors of character degrees and group structure", Proc. of Symposia in Pure Math. 47, Vol. II (1987), 65-69.
[11] R. Gow, B. Huppert, "Degree problems of representation theory over arbitrary fields of characteristic 0, Part 0: On theorems of N. *Itô* and J.G. Thompson, J. reine angew. Math. 381 (1987), 136-147.
[12] R. Gow, B. Huppert, "Degree problems of representation theory over arbitrary fields of characteristic 0, Part 2: Groups which have only two reduced degrees, *J.* reine angew. Math. 389 (1988), 122-132.
[13] T. Hawkes, "Groups with isomorphic complex group algebras", to appear.
[14] B. Huppert, "A characterization of GL(2,3) and SL(2,5) by the degrees of their representations", Forum Mathematicum 1 (1989), 167-183.
[15] B. Huppert, "Inequalities for character degrees of solvable groups", Arch. Math., Vol.46 (1986), 387-392.
[16] B. Huppert,"Solvable groups, all of whose irreducible representations in characteristic $p$ have prime degrees", J. of Algebra, Vol. 104, No.1, 23-36.
[17] B. Huppert, O. Manz, "Non solvable groups, whose character degrees are degrees are products of at most two prime numbers", Osaka J. Math. 23 (1986), 491-502.
[18] B. Huppert, O. Manz, "Degree-problems I. Square free character degrees", Arch. Math., Vol. 45 (1985), 125-132.
[19] B. Huppert, O. Manz, "Orbit sizes of $p$-groups", Arch. Math. 54 (1990), 105-110.
[20] I.M. Isaacs, "Recovering information about a group from its complex algebra", Arch. Math. Vol. 47 (1986), 293-295.
[21] I.M. Isaacs, "Sets of $p$-powers as irreducible character degrees", Proc. Am. Math. Soc. Vol. 96, no.4.
[22] I.M. Isaacs, "Lengths and numbers of modular character degrees for solvable groups", to appear.
[23] I.M. Isaacs, "Coprime group actions fixing all nonlinear irreducible characters", Can. J. Math. Vol. XL, No.1 (1989), 68-82.
[24] I.M. Isaacs, "Groups having at most three irreducible character degrees", Proc. Amer. Math. Soc. 21 (1969), 185-188.
[25] I.M. Isaacs, D. Passman, "A characterization of groups in terms of the degrees of their characters II", Pacific Journal of Mathematics Vol. 24, No.3, 1968.
[26] N. Itô "On the degrees of irreducible representations of a finite group", Nagoya Math. J. 3 (1951), 5-6.
[27] N. Itô, "Some studies of group characters", Nagoya Math. J. 2 (1951), 17-28.

[28] R. Knörr, W. Willems, "The automorphism groups of generalized Reed Muller codes", to appear in Asterisque.
[29] R. Knörr, W. Willems, "A characterization of generalized Reed Muller codes", Preprint.
[30] P. Landrock, O. Manz, "Classical codes as ideals in group algebras", Aarhus Preprint Series 1986/87, No. 18.
[31] C.R. Leedham-Green, R. Staszewski, "On the Loewy and Jennings series of a finite $p$-group", to appear.
[32] U. Leisering, "Ordinary and $p$-modular character degrees of prime power", to appear in Arch. Math., Vol. 53.
[33] U. Leisering, O. Manz, "A note on character degrees of solvable groups", Arch. Math., Vol. 48 (1987), 32-35.
[34] O. Manz, "Endliche auflösbare Gruppen, deren sämtliche Charaktergrade Primzahlpotenzen sind", J. of Algebra, Vol. 94 No. 1, (1985).
[35] O. Manz, "Endliche nicht-auflösbare Gruppen, deren sämtliche Charaktergrade Primzahlpotenzen sind", J. of Algebra, Vol. 96, No.1, (1985), 114-119.
[36] O. Manz, U. Stammbach, R. Staszewski, "On the Loewy-series of the group algebra of groups of small $p$-length", Comm. in Algebra 17 (1989), 1249-1274.
[37] O. Manz, R. Staszewski, "On the Loewy-series of the modular group algebra of a finite $p$-group", J. reine angewandte Math. 368 (1986), 108-118.
[38] O. Manz, R. Staszewski, W. Willems, "On the number of components of a graph related to character degrees", Proc. Amer. Math. Soc. 103 (1988), 31-37.
[39] O. Manz, W. Willems, T. Wolf, "The diameter of the character degree graph", J. reine angew. Math. 402 (1989), 181-198.
[40] O. Manz, T. Wolf, "Brauer characters of $q'$-degrees in $p$-solvable groups", J. of Algebra, Vol. 115, No.1 (1988), 75-91.
[41] G. Michler, "A finite simple group of Lie-type has $p$-block with different defects, $p \neq 2$", J. of Algebra 104 (1986), 220-230.
[42] G. Navarro, "Two groups with isomorphic group algebras", to appear in Arch. Math.
[43] T. Okuyama, "On a problem of Wallace", to appear.
[44] H. Pahlings, "Character degrees and normal $p$-complements", Comm. in Algebra 3 (1) (1975), 75-80.
[45] U. Stammbach, M. Stricker, "A remark on the Loewy-layers of the modular group ring of a finite $p$-group", Topology Appl. 25 (1987), 121-124.
[46] K. Taketa, "Über die Gruppen, deren Darstellungen sich sämtlich auf monomiale Gestalt transformieren lassen", Proc. Acad. Tokyo 6 (1930), 31-33.
[47] J.G. Thompson, "Normal $p$-complements and irreducible characters", J. of Algebra 14 (1970), 129-134.
[48] W. Willems, "Blocks of defect zero in finite simple groups", J. of Algebra 113 (1988), 511-522.
[49] W. Willems, "An upper bound for Loewy-lengths of projective modules in $p$-solvable groups", Osaka J. Math. 24 (1987), 471-478.

Progress in Mathematics, Vol. 95, © 1991 Birkhäuser Verlag Basel

# Computational aspects of representation theory of finite groups

K. LUX AND H. PAHLINGS

In many applications of representation theory of finite groups numerical computations for particular groups are called for. Although there are cases where one has to construct matrices for representations, in the majority of cases it is sufficient to work with characters, in fact this seems to be the only way to deal with many problems for larger groups.
In order to deal effectively with characters of finite groups on computers, two computer systems CAS (Character Algorithm System) and MOC (Modular Characters) have been developed in Aachen. CAS deals mainly with ordinary characters and had been designed by J. Neubüser, H. Pahlings and W. Plesken and described in [NPP 84]. Since then it has been substantially revised and extended, see section 1. MOC on the other hand deals with modular characters and was developed by G. Hiß, K. Lux and R. Parker mainly to compute decomposition numbers for finite groups.
The usefulness of the computer systems mentioned is greatly enhanced by the fact, that they contain large libraries of character tables, ordinary and modular ones. CAS now contains all the character tables published in the Atlas of Finite Groups [Con 85] and many others, in particular the character tables of many maximal subgroups of the sporadic simple groups (although completeness has not been achieved so far). The latter have proved particularly useful for many applications and for solving some open questions. We mention just one, raised independently by several people, cf. e.g. [Jan66].
**Question:** Is there for any irreducible character $\chi$ of a finite group $G$ a maximal subgroup $H$ and an irreducible character $\varphi$ of $H$, such that $(\chi, \varphi^G) = 1$?
The answer is "No", as O. Bonten first showed by examining all the maximal subgroups of the sporadic simple group $J_4$. The irreducible character of degree 1579061136 has at least multiplicity 2 in any character induced from a proper subgroup of $J_4$. This appears to be the only known example of this kind so far.

One of the applications of CAS in recent years, for which the character tables of maximal subgroups have also been of great use, concerns the inverse problem of Galois Theory; in fact using a Theorem of Belyi, Fried, Matzat and Thompson (see e.g. [Mat 87] and [Mat 88]), the second author was able to prove for several sporadic simple groups that they are Galois groups over the field $\mathbb{Q}$ of rational numbers, see [Pah 88], [Pah 89]. At present the Mathieu group $M_{23}$ is the only sporadic simple group, for which it is not known whether or not it is a Galois group over $\mathbb{Q}$ (see e.g. [Pah 90]).
MOC on the other hand contains a large library of Brauer character tables, all those of the sporadic simple groups which are known at present but also samples of Brauer

character tables of some symmetric and linear groups.
It should be mentioned that CAS and MOC are now closely linked together, so that in particular both libraries of character tables can be used by either system.
The paper is organized as follows: in the first section the main new features and most important new algorithms of CAS concerning ordinary characters are described. Section 2 contains applications for analyzing the subgroup structure of a group, in particular an algorithm for computing the table of marks is presented. Also the Möbius functions of the lattice of subgroups and of the poset of conjugacy classes of subgroups are studied. In section 3 the main ideas and algorithms of MOC are outlined and an overview is given on the results achieved with it concerning the computation of decomposition numbers.

**Acknowledgement:** We wish to thank the Deutsche Forschungsgemeinschaft for financial support.

## 1. Ordinary characters of finite groups

The construction of the character table of a finite group and also the use of a character table for applications in group theory or elsewhere usually requires a great amount of numerical computations. To facilitate these, more than 10 years ago a computer system CAS (Character Algorithm System) has been developed in Aachen. This has been described in a joint paper by Neubüser, Pahlings and Plesken [NPP 84]. Since then the system has undergone major revisions and extensions and has found many new applications. The revisions mentioned were partly necessitated by the fast development in computer technology and software techniques. Whereas CAS has first been developed on a (CDC)-CYBER 175 mainframe it turned out in 1984 that it was advisable to switch to microcomputers with UNIX-operating system. Also the present version of CAS is the last one which is written in Fortran. Work is already under way towards a completely new system written in the GAP-language (see [NNS 88]) with some basic routines such as the cyclotomic arithmetic written in C. In fact some of the newer algorithms as indicated below are already written in GAP and the experience obtained this way has convinced us, not to make any further changes to the present Fortran version except for those necessary to correct any bugs which might still be found and to do any further development in the new GAP-system.

For a computer system for the handling of characters it is an absolute requirement to be able to compute with cyclotomic integers in an exact algebraic form and, of course, also with (almost) arbitrary large integers. Whereas the basic data structures have not been changed in CAS since [NPP 84], the arithmetic for cyclotomic fields has been completely revised. Instead of using an arithmetic for the residue class rings $\mathbb{Q}_n = \mathbb{Z}[X]/(\Phi_n(X))$, where $\Phi_n(X)$ is the n-th cyclotomic polynomial, a new integral basis of the n-th cyclotomic field is used (cf [Zum 89]), which consists of the $\varphi(n)$ different products

$$\xi_n^m \prod_{p|n} \xi_n^{n_{p'} \cdot j_p} \quad j_p = 0, 1, \ldots, \quad \varphi(n_p) - 1,$$

where $\xi_n$ is a fixed primitive n-th root of unity, $n = n_p \cdot n_{p'}$ with $n_p$ being the highest power of the prime $p$, which divides $n$ and $m$ is a number with $(m, n) = 1$. In fact it turns out that a convenient choice for $m$ is

$$m = \sum_{p|n} n_{p'} \, (\frac{n_p/p+1}{2})^{\epsilon_p} \quad with \quad \epsilon_p = \{ \begin{matrix} 0 & for & p = 2 \\ 1 & for & p \neq 2. \end{matrix}$$

If $n$ is square-free then the basis chosen is a normal basis for $\mathbb{Q}(\epsilon_n)$, which is certainly useful, since Galois conjugation is a very frequent operation. One of the main advantages of this new basis is, that it is easy to decide, whether a number of $\mathbb{Q}_n$ written in this basis is in fact contained in a smaller cyclotomic field $\mathbb{Q}_r$ with $r|n$ and also to determine its representation in the corresponding integral basis of this field. The arithmetic is organized in such a way that the result of a computation is always expressed in the basis of the smallest possible cyclotomic field. Thus a test of equality of two numbers is very easy. Also for computing in cyclotomic fields there is no longer the necessity to store the cyclotomic polynomials. Furthermore it is easy to find out, whether a cyclotomic integer is contained in a quadratic number field and if so, to find its representation in the usual form $a + b\sqrt{d}$. This form is presented in addition to the representation as a sum of roots of unity, when numbers are displayed.

One of the most important new algorithms which have been implemented in CAS is the so-called LLL-algorithm due to A. K. Lenstra, H. W. Lenstra and L. Lovasz [LLL 82] for finding short vectors in lattices. The algorithm proceeds roughly as follows: Assume $b_1, \ldots, b_n$ is a basis for a $\mathbb{Z}$-lattice L. To initialize the algorithm the Gram-Schmidt orthogonalization procedure is used to compute

$$b_i^* = b_i \; - \; \sum_{j=1}^{i-1} \mu_{ij} b_j^* \quad 1 \leq i \leq n$$

with

$$\mu_{ij} = \frac{(b_i, b_j^*)}{(b_j^*, b_j^*)} \quad 1 \leq j \leq i-1.$$

At each level $m$ $(2 \leq m \leq n)$ the vector $b_m$ is reduced modulo the previous $b_j (1 \leq j \leq m-1)$ to obtain

$$| \, \mu_{mj} \, | \; \leq \; \frac{1}{2} \quad \text{for} \quad 1 \leq j \leq m-1.$$

In case one has $m \geq 2$ and

$$|b_m^* + \mu_{m\, m-1} b_{m-1}^* \, |^{\,2} < \alpha \, | \, b_{m-1}^* \, |^{\,2}$$

where $\alpha$ is a constant, $\frac{1}{4} < \alpha < 1$ (often chosen to be $\frac{3}{4}$), the vectors $b_m$ and $b_{m-1}$ are interchanged and $m$ is decreased by 1.

The final output of the algorithm is an LLL-reduced basis of the lattice, i. e. one for which

$$\mu_{ij} \leq \frac{1}{2} \quad (1 \leq j < i \leq n)$$

and

$$|b_i^* + \mu_{ii-1} b_{i-1}^* \, |^2 \; \geq \; \alpha |b_{i-1}^* \, |^2 \quad (2 \leq i \leq n)$$

holds. If $(b_1, \ldots, b_n)$ is LLL-reduced then it is known that for $\alpha = \frac{3}{4}$

$$|b_j \, |^2 \leq 2^{i-1} |b_i^*|^2 \quad \text{for} \quad 1 \leq j \leq i \leq n,$$

$$d(L) \le \Pi_{i=1}^{n}|b_i| \le 2^{\frac{n(n-1)}{4}} d(L),$$
$$|b_1| \le 2^{\frac{n-1}{4}} d(L)^{\frac{1}{n}},$$

where $d(L)$ is the determinant of the lattice. Often the norms of the vectors in an LLL-reduced basis are much smaller than those guaranteed by the above bounds.
In CAS the algorithm is slightly modified in order to allow it to be applied to a generating set of the lattice instead of a basis (see [Poh 87] and [Sei 88]). In any case the output is an LLL-reduced basis. A computational problem arises from the fact that one has to deal with the non-integral rational numbers $\mu_{ij}$. In the present version this is solved by computing the $\mu_{ij}$ from the exact values of the scalar products $(b_i, b_j)$, which are integers and the previous $\mu_{ij}$ using Fortran-double-precision. Of course this sets a limit to the number of vectors and also to the size of the norms, for which the programs might be successfully applied. One of the largest successful applications was a 150-dimensional lattice spanned by 180 reducible characters when the algorithm produced almost 100 irreducible characters in one run. The group in question was $O_{12}^{-}(2).2$ with 375 conjugacy classes and the computation of its character table was joint work with J.S. Frame.

Even if the LLL-algorithm does not produce irreducible characters but only characters of small norms this will in general be useful. Sometimes a new algorithm of W. Plesken [Ple 90] for finding orthogonal embeddings may be applied to obtain irreducible characters although the lattice spanned by the given characters does not contain any character of norm one. We demonstrate this by way of a simple example.

**Example 1.1.** Given the conjugacy classes, centralizer orders and the powermaps of the sporadic simple groups $G = J_3$ one may compute the complete character table of $G$ in the following way. First one may induce up the characters of the cyclic subgroups of $G$. This has been implemented in CAS and has proved to be useful for several applications. Of course one obtains reducible characters with norms (ranging from 139158 to $|J_3|$), which are much too large for backtrack algorithms as described in [Con 84] and [NPP 84]. But applying the LLL-algorithm one obtains (after first reducing with the trivial character) 3 irreducible characters and 17 characters of norm 2 and 3. Applying Plesken's algorithm one immediately finds, that there are just three orthogonal embeddings, two in dimension 17 and one in dimension 20, which all agree on a 14-dimensional sublattice (which is in fact an extension of a 4-dimensional lattice by the Weyl-lattice $D_{10}$). This embedding yields 14 irreducible characters (by producing twice an irreducible character as a $\mathbb{Z}$-linear combination of the spanning vectors of the lattice). The ambiguity of the embeddings comes from a sublattice orthogonal to the above spanned by three vectors $(Y_1, Y_2, Y_3)$ with matrix of mutual scalar products

$$\begin{pmatrix} 2 & -1 & 0 \\ -1 & 2 & 0 \\ 0 & 0 & 3 \end{pmatrix}.$$

The algorithm produces the two solutions corresponding to the three-dimensional embeddings, namely that $Y_1 - Y_2 + Y_3$ or $Y_1 - Y_2 - Y_3$ must be three times a norm 1 vector; the 6-dimensional embedding can, of course, be ruled out. Finally the above alternative can be settled by testing the two candidates; symmetrization gives in one case non-integral scalar products with the other characters. Thus all irreducible characters have been found, although the lattice of given reducible characters has index 12 in the lattice of all generalized characters.

Of course it is not always realistic to assume that the powermaps of a group $G$ are known when one starts to construct the character table of $G$. Very often one knows some characters of $G$ (and may-be even these only on a subset of the set of conjugacy classes without any ambiguity) and one wants to obtain information on the powermaps, perhaps in order to compute symmetrized characters and in this way to get additional information on values of irreducible characters. To deal algorithmically with problems of this sort, Th. Breuer [Bre 90] has introduced the data structure of a parametrized map into CAS (more precisely into the GAP version of CAS). This has also proved to be extremely useful for other problems as well, in particular for finding all possible fusions of one group into another one provided that the character table (or at least parts thereof) of the groups are known or for finding all possible transitive permutation characters.

A parametrized map $F$ from a set $D$ to a set $R$ is by definition just a map of $D$ into the power set $P(R)$ of $R$. It is interpreted as a set of maps from $D$ to $R$ and one writes for a map $f : D \to R$ that $f \in F$ if and only if $f(d) \in F(d)$ for all $d \in D$. The number

$$|F| := \Pi_{d \in D} |F(d)|$$

is called the indeterminateness of $F$. Parametrized maps of indeterminateness 1 may be identified with ordinary maps. Parametrized maps $F_1 : D \to P(R_1)$ , $F_2 : R_1 \to P(R_2)$ may be concatenated

$$F_2 \cdot F_1(d) := \cup_{r \in F_1(d)} F_2(r).$$

In practice parametrized powermaps, fusions, characters and representative orders occur in CAS. Here a parametrized powermap $Pow_{p,G}$ is defined to be any parametrized map $Cl(G) \to P(Cl(G))$ of the set $Cl(G)$ of representatives of the conjugacy classes of $G$ with $pow_{p,G}(g) \in Pow_{p,G}(g)$ for any $g \in Cl(G)$, where $pow_{p,G} : Cl(G) \to Cl(G)$ is the $p-th$ powermap of $G$, i. e. $pow_{p,G}(g) \in Cl(G)$ is conjugate to $g^p$ in $G$. Similarly parametrized fusions, characters and representative orders are defined.

Concatenation of a parametrized fusion $Fus_{H,G} : Cl(H) \to P(Cl(G))$, with a character $\chi$ of $G$ leads to a parametrized character, i. e. a class function

$$\chi \cdot Fus_{H,G} : Cl(H) \to P(C).$$

Likewise a parametrized $p$-powermap ($p$ a prime dividing $|G|$) $Pow_{p,G}$ of $G$ and a character $\chi$ of $G$ leads to a parametrized character $\chi_p^-$ of $G$ with

$$\chi_p^-(g) = \{ \frac{\chi(g)^p - \chi(h)}{p} \mid h \in Pow_{p,G}(g)\}.$$

Now the algorithms for computing powermaps and fusions aim at improving recursively parametrized powermaps and fusions, i. e. replacing them by ones with smaller indeterminateness. One starts with very basic conditions, e. g. for the $p$-powermap using that

1) if $p$ divides $|g|$ then $|g^p| = \frac{1}{p}|g|$ and $|C_G(g)|$ divides $|C_G(g^p)|$,

2) if $p$ does not divide $|g|$ then $|g^p| = |g|$ and $|C_G(g)| = |C_G(g^p)|$,

3) if $p$ divides $|g|$ and $\chi$ is a character of $G$, then $p$ divides $\chi(g^p) - \chi(g)^{\sim p}$, where $\sim p$ denotes Galois conjugation included by $\xi \to \xi^p$, $\xi$ a primitive $|G|$ -th root of unity,

4) if $p$ does not divide $|g|$, then $\chi(g^p) = \chi(g)^{\sim p}$, and

5) if $g \in Ker\chi$ then $g^p \in Ker\chi$ and if $g \notin Ker\chi$ and $p \not| [G : Ker\chi]$ then $g^p \notin Ker\chi$.

Similarly, for initializing a parametrized fusion map one just uses the obvious conditions, that if $i : H \to G$ is an embedding of groups one has
i) $|i(h)| = h$ and
ii) $|C_H(h)| = |C_G(i(h))|$.

To improve the parametrized powermaps and fusion maps the consistency of powermap and fusion map may be checked. Furthermore the concatenations of these parametrized maps with known characters as defined above are computed. The basic observation is that the indeterminateness of these parametrized characters may be much less than the indeterminateness of the parametrized powermap or fusion map. In fact the algorithm searches for a character $\chi$ for which this indeterminateness becomes minimal and tries to reduce this (hopefully to one) by examining the class functions determined by the parametrized character and eliminating all values which don't give rise to proper characters. For example assume that all the irreducible characters of the group in question are known. If the degree of $\chi$ is small then it might be feasible to find the possible character(s) $\psi$ in $\chi \cdot Pow_{p,G}$ or $\chi \cdot Res_{G,H}$ by solving the corresponding knapsack problem

$$\psi = \sum a_\Theta \Theta \quad a_\Theta \in \mathbb{Z}, \quad a_\Theta \geq 0$$

where the sum ranges over the irreducible characters of $G$ or $H$, respectively, of degree not bigger than $\chi(1)$. If this is not practicable one might first find those $\psi$, which are generalized characters. This is always possible for rational characters (provided again that the irreducible characters are known). To do this one computes a $\mathbb{Z}$-basis for the space of rational generalized characters, which consists of generalized characters in echelon form using a variation of the elementary divisor algorithm. The usual problem of the elementary divisor algorithm, entry explosion, can effectively be avoided by observing, that for any $g \in G$

$$\psi_g(h) = \begin{cases} |C_G(g)| & for \quad <h> conjugate \ to <g> \\ 0 & else \end{cases}$$

is a rational generalized character. Thus all entries can be bounded by the centralizer orders.

Once the indeterminateness of a parametrized character which is the concatenation of a character and a parametrized powermap or fusion has been substantially reduced, this in turn will in general yield information on the possible powermaps or fusion, respectively. Then the next character can be processed.

Often there are cases where there are good reasons why the indeterminateness of parametrized fusion maps cannot be reduced. This is due to the existence of automorphisms of the character table. To consider a typical case, if $G$ and a subgroup $H$ have a family of nonrational conjugacy classes conjugate under the operation of a Galois group of a cyclotomic field, then the fusion cannot be uniquely determined but one is usually allowed to make a choice which will possibly entail further consequences. To be more precise we make the following definition.

**Definition 1.2.** *If $G$ is a finite group, $t : Cl(G) \to Cl(G)$ a permutation of the set of conjugacy classes of $G$ with*

*a)* $\chi \circ t \in Irr(G)$ *for all* $\chi \in Irr(G)$,

*b)* $|t(g)| = |g|$ *for all* $g \in Cl(G)$,

*c)* $t(g^p) = t(g)^p$ *for all* $g \in Cl(G)$, $p \in \mathbb{Z}$,

*then $t$ is called a table-automorphism of $G$. The set of table-automorphisms is denoted by $T(G)$.*

Obviously, if $H$ is a subgroup of $G$, then $T(H) \times T(G)$ operates on the set of possible fusions of $H$ in $G$. Usually one is only interested in obtaining representatives of the orbits of $T(H) \times T(G)$ on the set of fusions of $H$ into $G$. In order to make use of this an algorithm for computing $T(G)$ has been developed and implemented in the GAP-version of CAS by Th. Breuer [Bre 90].

The use of parametrized characters and the algorithms for decreasing the indeterminateness have also proved to be extremely effective for computing candidates for transitive permutation characters of a group $G$ given its character table. Here a character $\pi$ of $G$ is called a candidate for a permutation character of $G$ if it fulfills the following conditions:

a) $\pi(1) \mid |G|$

b) $(\pi, \chi) \leq \chi(1)$ for all $\chi \in Irr(G)$

c) $(\pi, 1_G) = 1$

d) $\pi(g) \in \mathbb{N} \cup \{0\}$ for all $g \in G$

e) $\pi(g) < |C_G(g)|$ for $1 \neq g \in G$

f) $\pi(g) \leq \pi(g^m)$ for all $g \in G$, $m \in \mathbb{Z}$

g) $\pi(g) = 0$ if $|g| \nmid \frac{|G|}{\pi(1)}$

h) $\pi(1) \mid \frac{|G|\pi(g)}{|N_G(\langle g \rangle)|}$ for all $g \in G$.

To demonstrate the power of these algorithms we list some sporadic simple groups for which Th. Breuer [Bre 90] was able to compute the complete lists of candidates of transitive permutation characters: $J_1(44)$, $J_2(354)$, $J_3(529)$, $M_{11}(39)$, $M_{12}(285)$, $M_{22}(228)$, $M_{23}(209)$, $M_{24}(7737)$, $McL(2838)$,$ON(12523)$. Here the numbers in brackets give the total number of candidates for transitive permutation characters. We remark, that already the smallest of the above examples appeared to be almost out of reach for the previous algorithm as described in [NPP 84].

The algorithms discussed so far are applicable to arbitrary finite groups and are very often applied to simple groups. If a group $G$ has a proper normal subgroup $N$ such that the character tables of $G/N$ and certain subgroups of $G/N$ (the inertia factorgroups of the irreducible characters of $N$) are known the method of Clifford-matrices as developed by B. Fischer (see [Fis 82] and [Fis 85]) is of great help. Clifford matrices have frequently been used in particular for the construction of the character tables of maximal subgroups of sporadic simple groups. To utilize this method in CAS several programs have been included by O. Bonten [Bon 88], see [Pah 90a] for a brief description of the method and the corresponding programs in CAS and [PaP 88] for a generalization of the method to generic character tables.

## 2. Table of marks and the Möbius function

Finding the transitive permutation characters of a finite group $G$ is a useful step towards analyzing the subgroup structure of $G$. In fact, the matrix of transitive permutation characters is a submatrix of the table of marks $M(G)$ as introduced by Burnside (see [Bur 11], sometimes it is also called the Burnside matrix of $G$). The rows and columns of this matrix $M(G)$ are parametrized by the conjugacy classes $(K_i, i \in I)$ of subgroups of $G$ and the element at position $(i, j)$ is $|Fix_{G/G_i}(G_j)|$ for $G_i \in K_i$, where

$$Fix_{G/G_i}(G_j) = \{gG_i \mid xgG_i = gG_i \text{ for all } x \in G_j\}$$
$$= \{gG_i \mid G_j \leq gG_ig^{-1}\}$$

is the set of fixed points of $G_j$ on $G/G_i$. The matrix $M(G)$ can also be considered as the character table of the Burnside ring of $G$ (cf. [DrK 70]).

Of course, the table of marks can be computed from the complete subgroup lattice of $G$. Already in the late fifties J. Neubüser had designed and implemented an algorithm for computing all solvable subgroups of a moderately large group ([Neu 60], [Neu 71]). His method has been further developed and re-implemented several times since then and also extended by V. Felsch and includes also the determination of non-solvable subgroups ([Fel 84]). Felsch's programs are available e. g. through the group theory program system CAYLEY (see e. g. [Can 84]). In these programs subgroups are stored and identified by bit-strings, which are the characteristic functions on the set of cyclic subgroups of prime power order. This gives some indication for the scope of applicability of the method. Some of the largest simple groups, for which the complete subgroup lattices have been computed by V. Felsch, are the Mathieu group $M_{11}$ and the alternating group $A_8$, with 1915 and 4117 cyclic subgroups of prime power order, respectively; the total number of subgroups of these groups are 8651 and 48337, respectively. Although it is not always necessary to store all the subgroups but, at the expense of computing time, to keep only representatives of conjugacy classes of subgroups, storage problems as well as computer time present a severe obstacle to apply the programs directly for larger groups.

In [Bue 86] F. Buekenhout introduced the "subgroup pattern" of a finite group $G$, which is the poset of conjugacy classes of subgroups of $G$ endowed with additional numerical informations on the subgroups, i. e. for any pair of subgroups $H_1 > H_2$ of

$G$ with $H_2$ being maximal in $H_1$, the number of $G$-conjugates and $H_1$-conjugates of $H_2$ contained in $H_1$ and the number of $G$-conjugates of $H_1$ containing $H_2$ are given in addition to the number of conjugates of $H_1$ (and $H_2$). This information is almost equivalent to the table of marks of $G$.
In [Bue 86] the subgroup patterns of $M_{11}$ and of $J_1$ are presented; actually the latter contains a few errors: the Janko group $J_1$ contains, in fact two conjugacy classes of symmetric group $S_3$ and two conjugacy classes of dihedral groups of order 10, which can easily be seen from the character table of $J_1$, since the number of finite polyhedral subgroups of a finite group can always be computed from the character table (see e. g. [NPP 84]). The subgroup pattern of $M_{12}$, with 147 conjugacy classes of subgroups was determined by Buekenhout and Rees [BuR 88]. Furthermore the subgroup pattern of $J_2$ with 146 conjugacy classes of subgroups and the corresponding table of marks were computed in [Pah 87].

Recently G. Pfeiffer has developed a method to compute in an interactive way the table of marks of a finite group $G$ given the corresponding tables of the maximal subgroups of $G$. This has been implemented in the GAP-version of CAS (see [Pfe 90]). The main idea is to obtain the characters of the Burnside ring for $G$ by inducing up the characters of the Burnside rings of the maximal subgroups. But of course before being able to do this the main problem is to find the fusions and, first of all a list of representatives for the conjugacy classes of subgroups of $G$. So, to initialize the algorithm, at first the disjoint union of the $M_i$-conjugacy classes of subgroups of the maximal subgroups $M_i$ (of course taking just one $M_i$ for each set of $G$-conjugates) is constructed. This means that a list is constructed containing $[G : H_{ij}]$ and $[M_i : N_{M_i}(H_{ij})]$ (and a reference to the corresponding row in the table of marks of $M_i$) for each representative $H_{ij}$ of a $M_i$-conjugacy class of subgroups of $M_i$. Of course, several $H_{ij}$ might be conjugate in $G$. Therefore an equivalence relation is introduced and successively refined such that subgroups in different equivalence classes cannot be conjugate in $G$. One starts by just comparing the orders of the subgroups but for refining the classes one computes (using the table of marks of the $M_i$) for each $H_{ij}$ the number of subgroups of a particular equivalence class contained in $H_{ij}$. Also information on the fusion of cyclic subgroups which is usually available from the character tables of the groups can be invoked. Of course the aim of the process is to refine the equivalence classes until they coincide with the $G$-conjugacy classes.

If one finds sufficient conditions for two subgroups $H_{ij}$, $H_{kl}$ to be conjugate in $G$, these will be fused in the list mentioned above, $[M_i : N_{M_i}(H_{ij})]$ and $[M_k : N_{M_k}(H_{kl})]$ will be replaced by their least common multiple and the references to the rows of the table of marks of $M_i$ and $M_k$ are joined. Such sufficient conditions may be obtained e.g. by Sylow's Theorem, for cyclic subgroups and considering normalizers of subgroups known to be conjugate in $G$. Also other information available from the character table, such as the number of polyhedral subgroups of $G$ may be used.

Of course it is not guaranteed by this method, that the equivalence classes introduced above become singletons. But if they do, it is proved that a complete set of representatives of $G$-conjugacy classes of subgroups of $G$ has been found. Furthermore the fusion of the subgroups of the maximal subgroups into $G$ has been constructed along the way.

When a complete set of representatives of conjugacy classes of subgroups of $G$ has been determined this way, the marks of the maximal subgroups of $G$ may be induced using the following induction formula [Pfe 90]

$$|Fix_{G/U}(H)| = |N_G(H)| \sum_{H_i \sim H} \frac{1}{|N_M(H_i)|} |Fix_{M/U}(H_i)|,$$

where $U, H \leq M \leq G$, and the sum ranges over all representatives $H_i$ of $M$-conjugacy classes of subgroups of the (maximal) subgroup $M$, which are conjugate in $G$ to $H$.

Using the algorithm described above, G. Pfeiffer was able to compute the complete table of marks for the following sporadic simple groups

$$\begin{array}{ll} M_{12} & (147), \\ M_{22} & (156), \\ M_{23} & (204), \\ J_3 & (137), \end{array}$$

where the numbers in brackets give the total numbers of conjugacy classes of subgroups.

Given the table of marks of a finite group $G$ it is possible to compute values of the Möbius function $\mu(H,G) = \mu_G(H,G)$ for the lattice of subgroups of $G$ containing $H$ and $\lambda(H,G) = \lambda_G(H,G)$ for the poset of conjugacy classes of subgroups of $G$ containing a conjugate of a particular subgroup $H$ of $G$. In general the Möbius function (cf. [Rot 64]) associated with a finite poset $P$ is the map $\mu_P : P \times P \to \mathbb{Z}$ satisfying $\mu_P(x,y) = 0$ unless $x \leq y$, when it is defined recursively by the equations

$$\begin{array}{rcll} \mu_P(x,x) & = & 1 & \text{and} \\ \sum_{x \leq z \leq y} \mu_P(x,z) & = & 0 & \text{when } x < y; \end{array}$$

hence $(\mu_P(x_i, x_j))$ is the inverse of the incidence matrix of $P$ if one arranges the $x_i$ in such a way that $x_i \leq x_j$ implies $i \leq j$, where $P = \{x_1, \ldots, x_n\}$. To be more specific let $[H_1], \ldots, [H_n]$ be the conjugacy classes of subgroups of $G$ arranged in such a way, that if $H_i$ is contained in a conjugate of $H_j$ then $i \leq j$; in particular $H_1 = \{1\}$ and $H_n = G$. Let $E = (e_{ij})$ be the transpose of incidence matrix of the subgroup pattern of $G$, i.e.

$$e_{ij} = \begin{cases} 1 & \text{if } H_j \text{ is conjugate to a subgroup of } H_i \\ 0 & \text{else} \end{cases}$$

and let $B = (b_{ij})$ be the table of marks of $G$, hence

$$b_{ij} = [N_G(H_i) : H_i] \cdot a_{ij}$$

with

$$a_{ij} = \text{number of subgroups conjugate to } H_i \text{ containing } H_j.$$

Thus $E$ is obtained from the table of marks by replacing every non-zero entry by one and the value of the Möbius function $\lambda$ of the subgroup pattern on $([H_i],[H_j])$ is $\lambda(H_i, H_j)$, the $(j,i)$-element of $E^{-1}$. Furthermore, let $C = (c_{ij})$ with

$$c_{ij} = a_{i1}^{-1} a_{ij} a_{j1}.$$

Then

$$c_{ij} = \text{number of subgroups conjugate to } H_j \text{ contained in } H_i.$$

Let

$$M = (C^{-1})^{tr} = (m_{ij}).$$

Then the equation

$$\delta_{ik} = \sum_{j-1}^{n} c_{ij} m_{kj} = \sum_{U \leq H_i} m_{kU},$$

with $m_{kU} = m_{kj}$ if $U$ is conjugate to $H_j$, shows for $k = 1$, that

$$m_{1j} = \mu(1, H_j),$$

$\mu$ being the Möbius function of the subgroup lattice of $G$. Similarily, let $D = (d_{ij})$ be the transposed inverse of $A = (a_{ij})$; then

$$\delta_{ik} = \sum_{j=1}^{n} d_{ji} a_{jk} = \sum_{H_k \leq U} d_{Ui}$$

with $d_{Ui} = d_{ji}$ if $U$ is conjugate to $H_j$. Putting $i = n$ one obtains

$$\mu(H_j, G) = d_{jn}.$$

Thus one has

**Proposition 2.1.** *If $\mu$ is the Möbius function of the subgroup lattice, then for any subgroup $H$ of $G$ the values $\mu(1, H)$ and $\mu(H, G)$ can be computed from the table of marks of $G$.*

Of course, in general $\mu(H_i, H_j)$ depends on the choice of $H_i$ and $H_j$ in their conjugacy classes.

If $e_H$ is the idempotent in the rational Burnside algebra $\mathbb{Q} \otimes B(G)$ corresponding to $H$, then by a formula of Gluck and Yoshida (see e.g. [Glu 81])

$$e_H = \sum_{1 \leq S \leq H} \frac{|S|}{|N_G(H)|} \mu(S, H)[G/S].$$

On the other hand, if $H$ is conjugate to $H_j$ then

$$e_H = \sum_{i=1}^{j} x_i [G/H_i]$$

with

$$(x_1, \ldots, x_j) \cdot \begin{pmatrix} b_{11} & 0 & 0 \\ & & 0 \\ b_{j1} & & b_{jj} \end{pmatrix} = (0, \ldots, 0, 1)$$

which implies

$$x_i = \frac{1}{[N_G(H_i) : H_i]} d_{ij}.$$

Hence

$$d_{ij} = \frac{|N_G(H_i)|}{|N_G(H_j)|} \sum_{\substack{S \leq H_j \\ S \sim H_i}} \mu(S, H_j).$$

In [HIÖ 89] it is shown for solvable groups, that

$$\mu(1,G) = |G'|\lambda(1,G),$$

and the authors observe that the same relation holds also for many non-solvable groups, in fact for every example checked by them. Although this does not hold in general the above relation is just a special case of a much more general phenemenon, relating $\mu(H,G)$ to $\lambda(H,G)$ for arbitrary subgroups of a solvable group $G$ as follows, cf. [Pah 90b]:

**Theorem 2.2.** *If $G$ is a solvable group and $H$ a subgroup, then*
(*)
$$\mu(H,G) = [N_{G'}(H) : H \cap G']\lambda(H,G).$$

The assertion of the theorem seems to be true for many non-solvable groups as well, but not for all. For instance it holds for all groups $PSL(2,p)$,$p$ a prime (cf. [Pah 90b]) whereas the groups $PSU(3,3), M_{12}, J_2, M_{23}$ and $A_9$ present counterexamples. But even for these counterexamples there are only very few subgroups $H$ for which the equation $(*)$ is not satisfied.
In $M_{12}$ $(*)$ does also not hold for $H = \{1\}$, which was previously observed cf. [BMV 90]. It is easy to see, that for simple groups the assertion of the Theorem holds at least "on the average".

**Remark 2.3.** *If $G$ is not cyclic, then*

$$\sum_{i=1}^{n}(\lambda(H_i,G) - \frac{\mu(H_i,G)}{[N_G(H_i):H_i]}) = 0.$$

## 3. Modular representations and characters

Let $p$ be a prime and $(K,R,F)$ be a $p$-modular splitting system for the finite group $G$ and all its subgroups; this means that $R$ is a complete discrete valuation ring (of rank 1) with quotient field $K$ of characteristic 0. Furthermore if $\pi R$ denotes the maximal ideal of $R$, it follows that $F = R/\pi R$ has characteristic $p$ and both $K$ and $F$ are splitting fields for $G$ and its subgroups. Let $V$ be a (left) $KG$-module (finite dimensional over $K$) and choose an $RG$-lattice $M$ (an "$R$-form of V") such that $V = K \otimes_R M$. Then $\bar{V} = F \otimes_R M \cong M/\pi M$ has composition factors which are independent of the choice of $M$ according to a well-known theorem of R. Brauer, see for example [Dor 72].

Let $X_1,\ldots,X_n$ $(Y_1,\ldots,Y_r)$ be representatives of the isomorphism classes of irreducible $KG$-modules (resp. $FG$-modules) with characters $\chi_1,\ldots,\chi_n$ (resp. Brauer characters $\varphi_1, \ldots, \varphi_r$). Let $d_{ij}$ be the multiplicity of $Y_j$ as a composition factor in a composition series of $\bar{X}_i$. Thus

$$\chi_i' = \sum_{j=1}^{r} d_{ij}\varphi_j \quad \text{for } i = 1, \ldots, n,$$

where $\chi_i'$ denotes the restriction of $\chi_i$ to the set of $p$-regular elements of $G$. The $d_{ij}$ are the "$p$-decomposition numbers" of $G$. Given the ordinary character table of $G$ a knowledge of the $p$-decomposition numbers is equivalent to a knowledge of the (irreducible) $p$-Brauer characters, since the elementary divisors of the decomposition matrix $D = (d_{ij})$ are all one. As D. Benson writes ([Ben 84],p.59): " One of the most difficult problems in modular representation theory is to find the decomposition matrices for particular groups modulo particular primes."

For the computation of the decomposition numbers the following well-known facts are basic:

1.) The restriction map $\chi \mapsto \chi'$ introduced above induces a surjective homomorphism from the lattice $\langle \chi_1, \ldots, \chi_n \rangle$ of generalized characters onto the lattice

$$\mathcal{V} := \langle \varphi_1, \ldots, \varphi_r \rangle$$

of virtual Brauer characters.

2.) If $P_j = FGe_j$ is the projective cover of $Y_j$ generated by the primitive idempotent $e_j$, then $e_j$ can be lifted to an idempotent
$\hat{e}_j$ of $RG$. If $\hat{P}_j = KG\hat{e}_j$ has character $\Phi_j$ then

$$\Phi_j = \sum_{i=1}^{n} d_{ij}\chi_i \quad \text{for } j = 1, \ldots r.$$

Let $\mathcal{P}$ denote the lattice of virtual projective characters:

$$\mathcal{P} := \langle \Phi_1, \ldots, \Phi_r \rangle.$$

3.) The orthogonality relations give rise to a nonsingular pairing

$$(\ ,\ ) : \mathcal{V} \times \mathcal{P} \longrightarrow \mathbb{Z}$$

$$(\varphi, \Phi) := \frac{1}{|G|} \sum_{g \in G_{p'}} \varphi(g)\overline{\Phi(g)},$$

where $G_{p'}$ denotes the set of $p$-regular elements of $G$. $IBr(G) = \{\varphi_1, \ldots, \varphi_r\}$,the set of irreducible Brauer characters of $G$ and $IPr(G) = \{\Phi_1, \ldots, \Phi_r\}$, the set of projective indecomposable characters of $G$, are dual bases with respect to this pairing.

4.) If $B_1, \ldots, B_t$ are the $p$-blocks of $RG$ then the decomposition matrix splits up into submatrices, which are the decomposition matrices for the blocks $B_j$. Thus the whole computation can be done block-wise, starting with the block-decomposition of the ordinary characters. So in the sequel we will asume that $\mathcal{V}$ (resp. $\mathcal{P}$) is the $\mathbb{Z}$-lattice spanned by the irreducible Brauer characters (resp. projective indecomposable characters) belonging to a fixed block $B$.

In order to compute the decomposition numbers of finite groups a package of computer programs called MOC (modular characters) has been developed by R. Parker (Cambridge) and G. Hiß, K. Lux in Aachen. This package has been successfully applied to a large numbers of simple groups, their automorphism groups and their covering groups. At first the main ideas and concepts of the system will be described. For the general problems and techniques see for example [JaK 81].

First of all the program system is based on a new format of representing character tables. This is due to the fact that Brauer character tables tend to contain more irrationalities than ordinary character tables. One of the important consequences of the new MOC-format is that the basic algorithms involving Brauer characters can be reformulated as algorithms for $\mathbb{Z}$-lattices, avoiding explicit calculations in cyclotomic fields. The MOC-format also seems to be a more compact representation of a character table.

**Definition 3.1.** *For $x \in G$ define $\mathbb{Q}(x) = \mathbb{Q}(\chi(x)|\chi \in Irr(G))$ to be the column field of $x$.*

**Definition 3.2.** *Let $x_1, \ldots, x_r$ be elements in $G$ such that $\langle x_1\rangle, \ldots, \langle x_r\rangle$ are representatives for the $G$-conjugacy classes of cyclic subgroups of $G$. Note that two cyclic subgroups $\langle x\rangle, \langle y\rangle$ are conjugate if and only if $x$ is $G$-conjugate to a power $y^i$, where $i$ is prime to the order of $y$. Denote the dimension of $\mathbb{Q}(x_i)$ by $d_i$ and choose a basis $b_1(x_i), \ldots, b_{d_i}(x_i)$ of the ring of integers of $\mathbb{Q}(x_i)$. The MOC-character table format is now defined as follows:*

*The columns of the MOC-table are indexed by*

$$b_1(x_1), \ldots, b_{d_1}(x_1), b_1(x_2), \ldots, b_{d_2}(x_2), \ldots, b_1(x_r), \ldots, b_{d_r}(x_r)$$

*and the entries of $\chi \in Irr(G)$ at the columns $b_1(x_i), \ldots, b_{d_i}(x_i)$ are given by the coefficients of the decomposition of $\chi(x_i)$ in the basis $b_1(x_i), \ldots, b_{d_i}(x_i)$.*

**Lemma 3.3.** *The following statement holds:*

$$\sum_{i=1}^{r} d_i = \text{ Number of conjugacy classes of elements of } G$$

**Proof.** Observe that the number of $G$-conjugacy classes of elements of $G$ which contain a generator for a given cyclic subgroup $\langle y\rangle$ such that $\langle y\rangle = \langle x_i\rangle$ is just $d_i$. Q.E.D.

So by the above we get the following equation, if we have ordered the classes conveniently:

$$\text{Ordinary Table } = \text{ MOC Table } \cdot A,$$

where $A$ is a block matrix. The number of blocks is equal to the number of conjugacy classes of cyclic subgroups of $G$ and if $A(x)$ is the block corresponding to $\langle x\rangle$ it is defined to be

$$A(x) = (\sigma_i(b_j(x))), \quad \text{where} \quad Gal(\mathbb{Q}(x)/\mathbb{Q}) = \{\sigma_1, \sigma_2, \ldots, \sigma_s\}.$$

Note that

$$(det(A(x))^2 = disc(\mathbb{Q}(x)/\mathbb{Q}).$$

The MOC-system contains a database for subfields $L$ of cyclotomic fields (abelian extensions over $\mathbb{Q}$). In this database a field is characterized by the following information: The degree of a cyclotomic field $\mathbb{Q}_n$ which contains $L$ minimally, the degree of $L$ over $\mathbb{Q}$, generators for the subgroup of the Galoisgroup $Gal(\mathbb{Q}_n/\mathbb{Q})$ which fixes $L$ and an integral basis of $L$. This basis consists of orbitsums of nth-roots of unity under the fixgroup of $L$ and it is therefore only necessary to store one representative for each orbit. The choice of the basis is based upon the following observations:

**Lemma 3.4.** *Let $\epsilon$ be an nth-root of unity and let $Stab(\epsilon)$ be the stabilizer of $\epsilon$ in the fixgroup of $L$, then it follows:*

$$\bar{Tr}(\epsilon) := \frac{1}{|Stab(\epsilon)|} Tr_L^{\mathbb{Q}_n}(\epsilon)$$

*is an integral element of $L$, where $Tr_L^{\mathbb{Q}_n}$ denotes the usual trace map from $\mathbb{Q}_n$ to $L$.*

Since the character values are sums of roots of unity, it follows that we can write them in terms of $\bar{Tr}(\epsilon)$ even if the basis does not form an integral basis. The trace map however is quite often surjective as the following theorem shows, see for example [CuR 81].

**Theorem 3.5.** *If $8 \not| n$, and $L$ is minimally embedded into $\mathbb{Q}_n$, it follows that $Tr$ is a surjective linear map from the ring of integers of $\mathbb{Q}_n$ onto the ring of integers of $L$.*

In the above case we know that if we have $|L : \mathbb{Q}|$ linear independent sums of roots of unity in $L$, in which all the other sums of roots of unity decompose integrally, then they must form an integral basis.

**Remark 3.6.** *Actually we are in a better situation since we do not take $Tr$ but $\bar{Tr}$. For the fields in the database, which are not covered by the Theorem above, we still were able to construct an integral basis using $\bar{Tr}$.*

**Example 3.7.** Let $A_5$ be the alternating group on five letters. The ordinary character table is given by :

| | | | | | | | | | | | | |
|---|---|---|---|---|---|---|---|---|---|---|---|---|
| CAS | | | | | | | MOC | | | | | |
| | 2 | 2 | 2 | . | . | . | | | | | | |
| | 3 | 1 | . | 1 | . | . | | | | | | |
| | 5 | 1 | . | . | 1 | 1 | | | | | | |
| | | 1a | 2a | 3a | 5a | 5b | | | | | | |
| | 2P | 1a | 1a | 3a | 5b | 5a | | | | | | |
| | 3P | 1a | 2a | 1a | 5b | 5a | | | | | | |
| | 5P | 1a | 2a | 3a | 1a | 1a | | 1 | 2 | 3 | 5 | 5 |
| | 2 | | | | | | | | | | | |
| X.1 | + | 1 | 1 | 1 | 1 | 1 | | 1 | 1 | 1 | 1 | 0 |
| X.2 | + | 3 | -1 | . | A | *A | | 3 | -1 | 0 | 1 | 1 |
| X.3 | + | 3 | -1 | . | *A | A | | 3 | -1 | 0 | 0 | -1 |
| X.4 | + | 4 | . | 1 | -1 | -1 | | 4 | 0 | 1 | -1 | 0 |
| X.5 | + | 5 | 1 | -1 | . | . | | 5 | 1 | -1 | 0 | 0 |

The integral basis for the cyclic group of order 5 is given by 1 and $\epsilon + \epsilon^4$, where $\epsilon = e^{2\pi i/5}$ and $A = (1+\sqrt{5})/2, *A = (1-\sqrt{5})/2$.

It follows from the definition above that the MOC-table only contains integers. The sum of two characters is easily calculated in the MOC-format. For the tensor product of two characters we have to calculate the following information beforehand.

Given two elements $\eta_j, \eta_k$ of our basis for $\mathbb{Q}(x_i)$ say, determine:

$$\eta_j\eta_k = \sum_{m=1}^{d_i} a_m\eta_m, \quad a_m \in \mathbb{Z}.$$

This can be done once and for all and will be stored on the character table itself.

Given a Brauer character its decomposition in terms of a basic set of Brauer characters is part of a more general problem:

Given a basis for a $\mathbb{Z}$-lattice and a vector $w$ in the lattice, calculate the coefficients of the decomposition of $w$ in terms of the basis.

J. Dixon and R. Parker independently suggested an elegant way of solving this problem, see for example [Dix 82].

**Algorithm 3.8.** Given $m+1$ rows of integers

$$w, b_1, b_2, \ldots, b_m \in \mathbb{Z}_{1\times n}$$

where $b_1, b_2, \ldots, b_m$ are linearly independent over $\mathbb{Z}$. If possible write $w$ as a linear combination of these.

Let $T$ be the matrix

$$T = \begin{pmatrix} b_1 \\ \vdots \\ b_m \end{pmatrix}$$

Let $q$ be a (large) prime such that $\bar{T}$ (the reduction modulo $q$ of $T$) is of rank $m$.

We distinguish the following two cases.

(a) $\bar{w}$ is not in the $\mathbb{Z}_q$-span of $\bar{b}_1, \bar{b}_2, \ldots, \bar{b}_m$. Then it follows that $w$ is not in the $\mathbb{Z}$-span of $b_1, b_2, \ldots, b_m$.

(b) $\bar{w}$ is in the $\mathbb{Z}_q$-span of $\bar{b}_1, \bar{b}_2, \ldots, \bar{b}_m$. We define the integers $z_{i,j}$ recursively as follows:

Define the set $z_{i,0}$ by

$$\bar{w} = \sum_{i=1}^{m} \bar{z}_{i,0}\bar{b}_i, \quad \text{where} \quad -(q-1)/2 \leq z_{i,0} \leq (q-1)/2.$$

Suppose now that $z_{i,t}$ have been defined for $t \leq j-1$ such that if $w_j$ is defined as

$$w_j = \sum_{t=0}^{j-1}(z_{1,t}, \ldots, z_{m,t})q^t T$$

then every coefficient of $w - w_j$ is divisible by $q^j$. So let us write

$$\overline{((w - w_j)/q^j))} = \sum_{i=1}^{m} \bar{z}_{i,j}\bar{b}_j, \quad -(q-1)/2 \leq z_{i,j} \leq (q-1)/2.$$

This leads to an expression

$$w = (\sum_{j=0}^{\infty}(z_{1,j}, \ldots, z_{m,j})q^j)T$$

in the $q$-adic metric. Now if indeed $w$ is in the $\mathbb{Q}$-span of $b_1, b_2, \ldots, b_m$, i.e.

$$w = \sum_{i=1}^{m} r_i b_i, \quad r_i \in \mathbb{Q}$$

then

$$r_i = \sum_{j=0}^{\infty} z_{i,j}q^j.$$

This means that the integers $z_{i,j}$ approximate $r_i$ $q$-adically. If $r_i$ is an integer then the above series will be finite. So if in particular $w$ is in the $\mathbb{Z}$-span of our basis the algorithm will find the coefficients in a finite number of steps.

Let $\mathcal{V}$ and $\mathcal{P}$ be as above and let

$$\mathcal{V}^+ = \{\sum_{i=1}^{r} a_i\varphi_i | a_i \in \mathbb{Z}, a_i \geq 0\} \quad \text{and}$$

$$\mathcal{P}^+ = \{\sum_{i=1}^{r} a_i\Phi_i | a_i \in \mathbb{Z}, a_i \geq 0\}.$$

A "basic set" for $\mathcal{V}$ (resp. $\mathcal{P}$) is a $\mathbb{Z}$-basis of $\mathcal{V}$ (resp. $\mathcal{P}$) consisting of elements in $\mathcal{V}^+$ (resp. $\mathcal{P}^+$). Note that this definition differs from the one used by some other authors, where basic sets do not necessarily consist of proper characters.

The following Lemma on dual bases is obvious but fundamental for the procedures to follow.

**Lemma 3.9.** *If $B = (v_1, \ldots, v_r)$ is a basic set for $\mathcal{V}$ (or for $\mathcal{P}$) then the dual basis $B^* = (v_1^*, \ldots, v_r^*)$ of $\mathcal{P}$ (resp. $\mathcal{V}$) has the property that if*

$$\psi = \sum_{i=1}^{r} a_i v_i^* \in \mathcal{P}^+ \quad (\text{resp.} \mathcal{V}^+) \Longrightarrow a_i \geq 0 \quad (*).$$

*If $B^*$ is also a basic set then $B = IBr(G)$ (resp. $B = IPr(G)$).*

**Proof.** Observe that $a_i = (v_i, \psi)$ must be nonnegative for a proper projective character $\psi$. Furthermore, if $B^*$ is also a basic set (of $\mathcal{P}$) then any irreducible Brauer character must be a nonnegative linear combination of the Brauer characters in $B^{**} = B$. Q.E.D.

A $\mathbb{Z}$-basis of $\mathcal{V}$ or $\mathcal{P}$ having property $(*)$ is sometimes called an atom system.

The first step in the computation of the decomposition numbers is to find a basic set for $\mathcal{V}$. One starts with the restrictions of the ordinary irreducible characters (belonging to the block in question) to the set of $p$-regular elements and tries to find a $\mathbb{Z}$-basis of the span consisting of proper Brauer characters. There is an algorithm for this due to R. Parker, which is based upon the first algorithm described above:

**Algorithm 3.10.** Assume inductively that a subset $\tilde{B} = \{\beta_1, \ldots, \beta_{k-1}\}$ of proper Brauer characters has been found with span equal to $\langle \chi_1', \ldots, \chi_s' \rangle_{\mathbb{Z}}$, $s \leq n$ and such that $\beta_1, \ldots, \beta_{k-1}$ are linearly independent modulo some large prime $l$. If $s < n$ one takes the next Brauer character $\chi_{s+1}'$ in the generating set and tries to express it in terms of $\beta_1, \ldots, \beta_{k-1}$.

a) If $\chi_{s+1}' \in \langle \beta_1, \ldots \beta_{k-1} \rangle_{\mathbb{Z}}$ then the $s$ above may be replaced by $s+1$ and if this is still less than $n$, the next $\chi_i'$ in the list will be considered.

b) If on the other hand $\beta_1, \ldots, \beta_{k-1}, \chi_{s+1}'$ are linearly independent modulo $l$, then $\tilde{B}$ may be replaced by $\{\beta_1, \ldots, \beta_{k-1}, \chi_{s+1}'\}$ and $s$ by $s+1$.

c) It might happen however, that $\beta_1, \ldots, \beta_{k-1}, \chi_{s+1}'$ are linearly dependent modulo $l$ but linearly independent over $\mathbb{Z}$. In this case one replaces $l$ by another prime.

d) Finally, if $\beta := \chi_{s+1}' \in \langle \beta_1, \ldots, \beta_{k-1} \rangle_{\mathbb{Q}}$ but

$$\beta \notin \langle \beta_1, \ldots, \beta_{k-1} \rangle_{\mathbb{Z}}$$

then we use the $l$-adic expression for the coefficients of $\beta$ (see Algorithm 1) to find the smallest possible natural number $m_1$ with

$$m_1 \beta = \sum_{i=1}^{k-1} z_i \beta_i$$

with $z_i \in \mathbb{Z}$. Let $m = gcd(m_1, z_j) \geq 0$ be $\min\{gcd(m_1, z_i) | 1 \leq i \leq k-1\}$ and let $m = am_1 + bz_j$. For $i$ different from $j$ let $c_i$ be the smallest integer such that $c_i + bz_j m_1^{-1} \geq 0$ and define

$$\tilde{\beta}_j = b\beta + a\beta_j + \sum_{i \neq j} c_i \beta_i.$$

Then $m_1\tilde{\beta}_j$ is a proper Brauer character and hence $\tilde{\beta}_j$ is one, too. One finds

$$m\beta_j, m\beta \in \langle \beta_1, \ldots, \beta_{j-1}, \tilde{\beta}_j, \beta_{j+1}, \ldots, \beta_{k-1}\rangle_{\mathbb{Z}}.$$

So replacing $\beta_j$ by $\tilde{\beta}_j$ in $\tilde{B}$ one finds that the exponent of of the group

$$\langle \beta_1, \ldots, \beta_{k-1}, \tilde{\beta}_j\rangle / \langle \tilde{B}\rangle$$

has decreased. Continuing this way one finally obtains a new set of $k-1$ Brauer characters which span $\langle \beta_1, \ldots, \beta_{k-1}, \beta\rangle$ over $\mathbb{Z}$.

In the examples, which have been calculated by MOC so far, a basic set for $\mathcal{V}$ can be found among the restrictions of the ordinary irreducible characters, but it is an open question whether this is true in general.

Having found a basic set $B = (\beta_1, \ldots, \beta_r)$ of Brauer characters one starts producing projective characters by

tensoring defect 0 ordinary characters with Brauer characters,

inducing up projective characters of (maximal) subgroups

restricting known projective characters of supergroups, and projecting onto the block in question.

**Remark 3.11.** *In theory one is sure to obtain a basic set of projective characters if one induces up all projective characters of all elementary subgroups see* [Dor 72] *and hence of all maximal subgroups, but this is not of practical importance. In the case of the symmetric groups $S_n$ on $n$ letters for example inducing up from $S_{n-1}$ is sufficient, see* [JaK 81].

If $(\pi_1, \ldots, \pi_r)$ is any system of projective characters consider the matrix

$$A = ((\beta_i, \pi_j))$$

Since $(\beta_1, \ldots, \beta_r)$ is a basic set

$$\beta_i = \sum_{k=1}^{r} b_{ik}\varphi_k \quad \text{with } b_{ik} \geq 0.$$

and $(b_{ij})$ is unimodular; furthermore

$$\pi_j = \sum_{l=1}^{r} c_{jl}\Phi_l \quad \text{with } c_{jl} \geq 0$$

with $(c_{jl})$ unimodular if and only if $(\pi_1, \ldots, \pi_r)$ is a basic set. Hence

$$A = (b_{ij})(c_{ij})^{tr} \quad (**)$$

is unimodular if and only if $(\pi_1, \ldots, \pi_r)$ is a basic set. Furthermore if $A$ is the identity matrix or, more generally a permutation matrix, then $\{\beta_1, \ldots \beta_r\} = \{\varphi_1, \ldots, \varphi_r\}$ and $\{\pi_1, \ldots, \pi_r\} = \{\Phi_1, \ldots, \Phi_r\}$.

In general the problem is to find for a given unimodular matrix $A$ all possible decompositions (**) of $A$ as a product of two unimodular positive matrices. For any given proper Brauer character $\beta$ or proper projective character $\pi$ one obtains additional positivity conditions, namely

$$\beta = \sum_{i=1}^{r} a_i\beta_i = \sum_{k=1}^{r}(\sum_{i=1}^{r} a_i b_{ik})\varphi_k,$$

hence

$$\sum_{i=1}^{r} a_i b_{ik} \geq 0.$$

Of course this condition gives more than the positivity of $(b_{ik})$ only in case that some of the elements $a_i$ are negative. To make use of this for the subsequent computations one keeps together with the basic sets $B = (\beta_1, \ldots, \beta_r)$ and $B' = (\pi_1, \ldots, \pi_r)$ of Brauer characters and projective characters a (large) set $Br$ of proper Brauer characters and a (large) set $Pr$ of proper projective characters which are not positive linear combinations of the $\beta_i$ or the $\pi_i$ respectively. These sets can be utilized in the following irreducibility test.

**Lemma 3.12.** *Let $\beta \in B \cup Br$; then $\beta = \sum_{i=1}^{r} a_i\pi_i^*$ with $a_i \geq 0$. Assume that the $a_i$ are small, so that it is feasable to consider all combinations*

$$\beta' = \sum_{i=1}^{r} b_i\pi_i^* \quad \text{with} \quad 0 \leq b_i \leq a_i.$$

*If for any such nonzero $\beta' \neq \beta$ there is a $\pi \in B' \cup Pr$ such that*

$$(\beta', \pi) < 0 \quad \text{or} \quad (\beta - \beta', \pi) < 0$$

*then $\beta$ is an irreducible Brauer character.*

Of course there is a corresponding indecomposability test for projective characters. In addition to these tests for irreducibility and indecomposability MOC contains programs to improve the basic sets and the sets $Br$ and $Pr$ as well. These programs are based on the following:

**Proposition 3.13.** *Assume $\pi_1, \ldots, \pi_s$ are known to be projective indecomposable characters. If $\beta$ is a proper Brauer character $\beta$ let*

$$\beta = \sum_{i=1}^{r} a_i(\beta)\pi_i^*.$$

*Thus the $a_i(\beta)$ are all nonnegative integers. For any projective character $\pi$ let $m = m(\pi_1, \pi)$ be the minimal scalar product $(\alpha, \pi)$, where $\alpha$ runs through all (virtual) Brauer characters which fulfill the following conditions:*

*1.) $\alpha$ is of the form $\alpha = \pi_1^* + \sum_{i=s+1}^{r} a_i'\pi_i^*$, with $0 \leq a_i' \leq a_i(\beta)/a_1(\beta)$ for all $\beta \in B \cup Br$ with $a_1(\beta) \neq 0$.*

*2.) For all $\beta \in B \cup Br$ and all projectives $\pi' \in B' \cup Pr$ $(\alpha, \pi') \geq 0$ and $(\beta - a_1(\beta)\alpha, \pi') \geq 0$.*

*Then $\pi - m\pi_1$ is a proper projective character.*

**Proof.** For the proof one just has to observe that if $\varphi_1$ is the irreducible Brauer character with $(\varphi_1, \pi_1) = 1$, then $\varphi_1$ will be a constituent of $\beta$ with multiplicity $a_1(\beta)$, so its coefficients $a_i(\varphi_1)$ in the decomposition into the $\pi_i^*$s are less or equal to $a_i(\beta)/a_1(\beta)$ if $a_1(\beta) \neq 0$. Since $\varphi_1$ is a Brauer character $(\varphi_1, \pi') \geq 0$ for all projectives $\pi$ and since it is contained $a_1(\beta)$ times in $\beta$, $\beta - a_1(\beta)\varphi_1$ is a Brauer character and hence $(\beta - a_1(\beta)\varphi_1, \pi') \geq 0$ for all projective characters $\pi'$. Since $\varphi_1$ is amongst the $\alpha$, which fulfill 1.) and 2.) it follows that $(\varphi_1, \pi)$ is greater than or equal the minimum of the $(\alpha, \pi)$. Q.E.D.

After having found $B$ and $B'$ the MOC-system proceeds as follows.

1.) Try to prove irreducibility of elements in $B$ and the indecomposability of elements in $B'$ using the Lemma above.

2.) Use Proposition 3.13 to show that elements in $B$ or $B'$ are reducible and change the bases $B$ and $B'$ accordingly.

These two steps may be repeated until there is no further progress made. In this case the generation of new projective characters and Brauer characters by inducing, restricting and tensoring might help. If one is finally left with more than one possible Brauer character table, MOC can be used to generate these possible tables, if there are not too many. Since the tensor product of two Brauer characters is again a Brauer character, this can be checked for all possible tables using MOC and thereby one can rule out tables which do not fulfill this condition.

Using MOC G. Hiß and K. Lux could determine most of the decomposition matrices for blocks of cyclic defect in sporadic simple groups and their covering groups, see [HiL 89]. The following list gives some results for blocks of noncyclic defect, which have been obtained using MOC and possibly other methods:

| Group | Author | Methods |
|---|---|---|
| Tits | G. Hiß | MOC |
| $J_2$ | G. Hiß, K. Lux | MOC |
| He, p=7 | A. Ryba | MOC |
| McL, p=5 | Hiß,Parker, Lux | MOC, Meat-Axe |
| $Co_2$ ,p=5 | Lux | MOC |
| He, p=5 | Lux | MOC, Meat-Axe |
| $Sp_4(5)$,p=3 | Lux | MOC |
| $Sp_8(2)$,p=5 | Lux | MOC |
| $O_8^+(2) : S_3$,p=5 | Lux | MOC |
| $O_{10}^+(2)$,p=5 | Lux | MOC |
| $O_{10}^-(2)$,p=5 | Lux | MOC |
| $O_8^+(3)$,p=5 | Lux | MOC |

**Example 3.14.** We consider the non-principal block in characteristic 3 of the Held simple group with defect group $C_3 \times C_3$. Using MOC we can show that there are 9 ordinary irreducible characters in the block and 7 irreducible Brauer characters. MOC also gives us the following approximation to the decomposition matrix (The ordinary characters are given in ATLAS notation). The columns of the following matrix give the decomposition of 7 projective characters in the ordinary irreducible characters in the block. These projective characters form a basic set of projective characters in the block and if we take the restrictions of the first 7 ordinary irreducible characters of the block to the 3-regular classes we get a basic set of Brauer characters in the block.

| | | | | | | | |
|---|---|---|---|---|---|---|---|
| 51a | 1 | . | . | . | . | . | . |
| 51b | 1 | 1 | . | . | . | . | . |
| 1029a | . | 5 | 1 | . | . | . | . |
| 1029b | . | 5 | . | 1 | . | . | . |
| 1275a | . | 3 | . | . | 1 | . | . |
| 1275b | . | 4 | . | . | . | 1 | . |
| 1920a | . | 6 | . | . | . | . | 1 |
| 4080a | 2 | 17 | 1 | 1 | . | . | 1 |
| 6528a | . | 23 | 1 | 1 | 1 | 1 | 1 |

MOC also informs us how it determined the projective characters it used for the approximation:

1st column: induce projective $\chi_1+\chi_{12}+\chi_{14}$ from the maximal subgroup $Sp(4,4):2$.

2nd column: Tensor the first defect 0 character with $\chi_{16}$.

3rd column: Induce the defect 0 character $\chi_{27}$ from $Sp(4,4):2$.

4th column: Induce the defect 0 character $\chi_{28}$ from $Sp(4,4):2$.

5th column: Tensor the second defect 0 character with $\chi_3$.

6th column: Tensor the first defect 0 character with $\chi_3$.

5th column: Tensor the first defect 0 character with $\chi_2$.

Thus the results may be checked by hand. MOC provides full proofs for the results obtained by it. This holds true also for larger examples, where it is (almost) impossible to find the projective characters which give useful information without a computer; even so it is possible to check the results by hand due to this documenting management of MOC.

Examining the character table of He we see that there has to be exactly one real irreducible Brauer character in this block. It follows that there is exactly one real projective indecomposable character. This is the last projective in the above approximation, since it is real and indecomposable by the Lemma 3.12 above. Actually it is easy to see that all but the first two projectives are indecomposable using the Lemma above. The first column of the approximation is invariant under complex conjugation and cannot contain the only real projective indecomposable. So it has to be a sum of complex conjugate projective indecomposable characters. From this it follows easily that the only possible splitting of the first column is the sum of two projective indecomposable characters and we are done.

Observe that in this case all irreducible Brauer characters are liftable. The decomposition matrix is:

| | | | | | | | |
|---|---|---|---|---|---|---|---|
| 51a | 1 | . | . | . | . | . | . |
| 51b | . | 1 | . | . | . | . | . |
| 1029a | . | . | 1 | . | . | . | . |
| 1029b | . | . | . | 1 | . | . | . |
| 1275a | . | . | . | . | 1 | . | . |
| 1275b | . | . | . | . | . | 1 | . |
| 1920a | . | . | . | . | . | . | 1 |
| 4080a | 1 | 1 | 1 | 1 | . | . | 1 |
| 6528a | . | . | 1 | 1 | 1 | 1 | 1 |

There is no guarantee that MOC will finally give the decomposition matrix, but even in difficult cases it gives good bounds on the degrees for the modular irreducible representations. In the case that we do not succeed using MOC, a different approach, i.e. constructing the irreducible representations over $F$, might be useful. For this purpose R. Parker, see [Par 84], has developed an excellent tool called the Meat-Axe in order to be able of working with the representations of a group over a finite field. The Meat-Axe allows us to construct representations by tensoring, symmetrizing, inducing, restricting representations. On the other hand it can show the irreducibility of a representation or split it, if it is not irreducible. In contrast to MOC however it is only usable if the representations are not to large , i.e. not bigger than few thousand dimensional.

R. Parker and K. Lux decided to write a new Fortran version. This version consists of 22 independent Fortran programs, which call a common subroutine, which deals with the field operations and row operations on matrices. This subroutine is either available written in Fortran, C or assembler. Besides the basic algorithms described in [Par 84] the current version is also able of dealing with permutations (of quite large degree).

In the moment the new version runs on the following set of computers:

MASSCOMP 5300/5500, VAX under VMS, IBM 370, Atari 1040ST, SUN 3/60, SUN4/280, HP9000/825, Apollo DN10000, and IBM RT/6150.

There is also version of the Meat-Axe available, which has been written completely in C by a student , M. Ringe in Aachen.

As mentioned above the Meat-Axe is restricted to work with matrices with dimension of few thousand. Using a new method called condensation we are able to prove the irreducibilty of $FG$-modules using the Meat-Axe in the following way.

Let us suppose we are given a subgroup $H$ of $G$ which has order prime to $p$. Let $e := \frac{1}{|H|}\sum_{h\in H} h$ be the idempotent belonging to the trivial representation and let $M$ be an $FG$-module. Then $Me = M^H$, the fixed space of $M$ under $H$ inherits a $eFGe$-module structure in the obvious way.

If furthermore $M$ is a permutation module, a basis for $Me$ is given by the orbit sums, i.e. sums over the permutation basis vectors in $M$ where the indices lie in an orbit.

So in the case of a permutation module, it is straightforward to determine the matrix of $eae \in eFGe$ for $a \in G$ with respect to the above basis, if we know the permutation, which $a$ induces.

We now need the following Lemma see [Gre 80]:

**Lemma 3.15.** *If $S_1, \ldots, S_r$ are the irreducible $FG$-modules with $S_i e \neq 0$, then $S_1 e, \ldots, S_r e$ is a complete set of irreducible modules (up to isomorphism) for $eFGe$.*

The following example shows, how this theorem can be applied to determine the entries of the decomposition matrix.

Suppose we suspect that a $FG$-module $S$ is irreducible, but we cannot show this using MOC. To prove that $S$ is irreducible we want to apply the Lemma above. The idea is to find a subgroup $H$ such that $Se$ is not zero for the module $S$ and all the possible constituents of $S$. Depending on whether the module $S$ is irreducible or not the condensed module $Se$ will be irreducible or not. Using the Meat-Axe on $Me$ as $eFGe$ module, where $M$ is a permutation module in which $S$ is contained, we can try to prove that $Se$ is irreducible and thereby show that $S$ is irreducible.

**Example 3.16.** From the part of the decomposition matrix already calculated by G. Hiß and R. Parker we get the following irreducible Brauer characters for McL given in ATLAS notation: 1a, 21a, 210a, 230a, 560a, 896a, 896b, 3038a, 1200a , 1200b where 896a,896b and 1200a,1200b are complex conjugate pairs. The last pair of complex conjugate characters is contained in the ordinary characters of degree 8019 and the possible degrees are: 3245 or 2685. So in the following table the columns correspond to projective characters but it is not clear at the moment, whether one has to subtract the last two columns from the 5th column in order to get the correct decomposition matrix.

| | | | | | | | | | | | | |
|---|---|---|---|---|---|---|---|---|---|---|---|---|
| 1a | 1 | . | . | . | . | . | . | . | . | . | . | . |
| 22a | 1 | 1 | . | . | . | . | . | . | . | . | . | . |
| 231a | . | 1 | 1 | . | . | . | . | . | . | . | . | . |
| 252a | 1 | 1 | . | 1 | . | . | . | . | . | . | . | . |
| 770a | . | . | 1 | . | 1 | . | . | . | . | . | . | . |
| 770b | . | . | 1 | . | 1 | . | . | . | . | . | . | . |
| 896a | . | . | . | . | . | 1 | . | . | . | . | . | . |
| 896b | . | . | . | . | . | . | 1 | . | . | . | . | . |
| 3520a | . | 2 | 1 | 1 | . | . | . | 1 | . | . | . | . |
| 3520b | . | . | . | . | 2 | . | . | . | 1 | 1 | . | . |
| 4752a | . | . | . | . | 1 | 1 | 1 | . | 1 | 1 | . | . |
| 5103a | 1 | 2 | . | 1 | . | 1 | 1 | 1 | . | . | . | . |
| 5544a | 1 | 1 | 1 | . | 2 | 1 | 1 | . | 1 | 1 | . | . |
| 8019a | 1 | 1 | . | . | 2 | 1 | 1 | . | 1 | 1 | 1 | . |
| 8019b | 1 | 1 | . | . | 2 | 1 | 1 | . | 1 | 1 | . | 1 |
| 9856a | . | 1 | . | . | 2 | 1 | 1 | 1 | 1 | . | 1 | . |
| 9856b | . | 1 | . | . | 2 | 1 | 1 | 1 | . | 1 | . | 1 |
| 10395a | 1 | 2 | 1 | . | 2 | 2 | 2 | 1 | 1 | 1 | . | . |
| 10395b | 1 | 2 | 1 | . | 2 | 2 | 2 | 1 | 1 | 1 | . | . |

To apply condensation we first have to look for a suitable permutation module. Using the character table of the maximal subgroups, which are contained in the CAS-library, we find that the permutation module $M$ on $M_{11}$ with 113400 points actually contains the unkown pair of irreducible representations $\phi$ and $\bar{\phi}$. Since we will work over the prime field GF(5) we will get an irreducible constituent of twice the degree. Obviously a direct approach is neither suitable nor possible. We want to look at the condensed version of this permutation module for the subgroup $H = Sl(2,7)$, contained for example in 2A8, the centralizer of an involution. The dimensions for the condensed irreducible representations of McL are as follows:

| Degree | 1 | 21 | 210 | 230 | 560 | 896 | 896 | 3038 |
|---|---|---|---|---|---|---|---|---|
| Dimension | 1 | 0 | 1 | 2 | 2 | 2 | 2 | 9 |

| Degree | $1200 + \overline{1200}$ | $\phi + \bar{\phi}$ | 9625 | $8250 + \overline{8250}$ |
|---|---|---|---|---|
| Dimension | 8 | 18 or 14 | 28 | 28 |

| Degree | 1750 | 4500 |
|---|---|---|
| Dimension | 8 | 12 |

The dimensions of the fixed spaces can be calculated from the knowledge of the permutation character on $H$ and the matrix above.

Using the Meat-Axe to actually split the representation on $Me$ for $< eae, ebe >$, where a,b are certain elements in McL we get an irreducible $< eae, ebe >$-module of dimension 18.

Using permutation modules on fewer points we can actually prove beforehand that the fixed spaces for all irreducible representations involved in $M$ remain irreducible when restricted to $< eae, ebe >$. So we know that the fixed space is irreducible for the last two representations and it follows that 3245 is the correct degree.

## REFERENCES

[Ben 84] D. Benson, Modular representation theory: New trends and methods. Lecture Notes in Mathematics 1081. Springer, Berlin, Heidelberg, New York, Tokyo 1984.

[BMV 90] M. Bianchi, A. Gillio Berta Mauri, L.Verardi, On Hawkes-Isaacs-Özaydin's conjecture, preprint 1990.

[Bon 88] O. Bonten, Clifford-Matrizen, Diplomarbeit Aachen 1988

[Bre 90] Th. Breuer, Diplomarbeit, Aachen 1990.

[Bue 86] F. Buekenhout, The geometry of the finite simple groups; in "Buildings and the Geometry of Diagrams", ed. by L. A. Rosati, pp 1 - 78. Lecture Notes in Mathematics 1181, Springer, Berlin, Heidelberg, New York, Tokyo 1984.

[BuR 88] F. Buekenhout, S. Rees, The subgroup structure of the Mathieu group $M_{12}$, Mathematics of Computation, **50** (1988), 595-605.

[Bur 11] W. S. Burnside, Theory of groups of finite order, 2nd. ed. Cambridge 1911.

[Can 84] J. J. Cannon, An introduction to the group theory language Cayley, in: Computational Group Theory (ed. M. D. Atkinson) pp 145 - 183, London: Academic Press 1984.

[Con 84] J. H. Conway, Character Calisthenics; in: Computational Group Theory (ed. M. D. Atkinson) pp. 249-266, London: Academic Press 1984

[Con 85] J. Conway, R. Curtis, S. Norton, R. Parker, R. Wilson, Atlas of Finite Groups. Oxford: Clarendon Press, 1987.

[CuR 81] C.W. Curtis, I. Reiner. Methods of Respresentation Theory with Applications to Finite Groups and Orders Volume I. John Wiley & Sons, New York, 1981.

[Dix 82] J.D. Dixon. Exact Solution of Linear Equations using $P$-adic Expansions. Numer. Math. 40,p.137-140, 1982.

[Dor 72] L. Dornhoff. Group Representation Theory Part B. Marcel Dekker Inc., New York, 1972.

[DrK 70] A. Dress, M. Kückler, Zur Darstellungstheorie endlicher Gruppen I. Mimeographed notes, Bielefeld 1970.

[Fel 84] V. Felsch, G. Sandlöbes, An interactive program for computing subgroups, in:

Computational Group Theory (ed. M. D. Atkinson) pp. 137 - 143, London: Academic Press 1984.

[Fis 82] B. Fischer, Clifford Matrizen, manuscript 1982.

[Fis 85] B. Fischer, unpublished manuscript 1985.

[Glu 81] D. Gluck, Idempotent formula for the Burnside algebra with applications to the p-subgroup simplicial complex, Illinois J. Math. **25** (1981), 63-67.

[Gre 80] J.A. Green. Polynomial Representations of $Gl_n$. Springer Verlag,Berlin Heidelberg New York, 1980.

[HIÖ 89] T. Hawkes, I. M. Isaacs, M. Özaydin, On the Möbius function of a finite group, Rocky Mountain J. Math. **19** (1989), p.1003-1034.

[HiL 88] G. Hiss and K. Lux. The Brauer Characters of the Hall-Janko Group. Comm. Alg., 16(2), p.357-398, 1988.

[HiL 89] G. Hiss and K. Lux. Brauer Trees of Sporadic Groups. Oxford University Press, 1989.

[Isa 76] I. M. Isaacs, Character Theory of Finite Groups. New York: Academic Press 1976

[JaK 81] G.D. James and A. Kerber. The Representation Theory of the Symmetric Group. Addison Wesley, Reading, Massachusetts, 1981.

[Jan 66] G.J. Janusz, Primtive idempotents in group algebras. Proc. AMS **17** (1966), p. 520-523.

[LLL 82] A. K. Lenstra, H. W. Lenstra, L. Lovasz, Factoring polynomials with rational coefficients. Math. Ann. **261** (1982), 513-534.

[Mat 87] B. H. Matzat, Konstruktive Galoistheorie. Lecture Notes in Mathematics 1284 Springer, Berlin, Heidelberg, New York, Tokyo 1987.

[Mat 88] B. H. Matzat, Über das Umkehrproblem der Galoischen Theorie.
J. ber. d. Dt. Math. Verein. **90** (1988), 155-183

[Neu 60] J. Neubüser, Untersuchungen des Untergruppenverbandes endlicher Gruppen auf einer programmgesteuerten elektronischen Dualmaschine, Numerische Mathematik **2**, 280-292.

[Neu 71] J. Neubüser, Computing moderately large groups, SIAM-AMS Proc. **4** (1971), 183-190.

[NNS 88] W. Nickel, A. Niemeyer, M. Schönert, GAP, getting started and reference manual, Aachen 1988.

[NPP 84] J. Neubüser, H. Pahlings, W. Plesken, CAS; design and use of a system for the handling of characters of finite groups, in: Computational Group Theory (ed. M.

D. Atkinson) pp 195-247, London: Academic Press 1984.

[Pah 87] H. Pahlings, The subgroup structure of the Hall-Janko group $J_2$. Bayreuther Math. Schriften **23** (1987), 135-165.

[Pah 88] H. Pahlings, Some sporadic groups as Galois groups, Rend. Sem. Mat. Univ. Padova, **79** (1988), 97-107.

[Pah 89] H. Pahlings, Some sporadic groups as Galois groups II, Rend. Sem. Mat. Univ. Padova, **82** (1989), 163-171.

[Pah 90] H. Pahlings, Realizing finite groups as Galois groups, Bayreuther Math. Schriften **33** (1990), 137-152.

[Pah 90a] H. Pahlings, Computing with characters of finite groups, Proc. Conf. on Computational Algebra, Rome 1990, to appear.

[Pah 90b] H. Pahlings, On the Möbius function of a finite group, preprint, 1990.

[PaP 87] H. Pahlings, W. Plesken, Group actions on Cartesian powers with applications to representation theory. J. reine angew. Math., **380** (1987), 178-195.

[Par 84] R.A. Parker. The Computer Calculation of Modular Characters. The Meat-Axe. In M.D. Atkinson (ed.). Computational Group Theory. Academic Press New York, 1984.

[Pfe 90] G. Pfeiffer, Diplomarbeit Aachen 1990.

[Ple 90] W. Plesken, Additive decompositions of positive integral quadratic forms, preprint, 1990.

[Poh 87] M. Pohst, A modification of the LLL reduction algorithm, J. Symbolic Computation **4** (1987), 123-127

[Rot 64] G.-C. Rota, On the foundations of combinatorial theory I. Theory of Möbius Functions. Z. Wahrscheinlichkeitstheorie **2** (1964), 340-368.

[Sei 88] C. Seidler, der LLL-Algorithmus und seine Implementation in CAS. Diplomarbeit, Aachen 1988.

[Zum 89] M. Zumbroich, Grundlagen einer Arithmetik in Kreisteilungskörpern und deren Implementation in CAS. Diplomarbeit, Aachen 1989.

Progress in Mathematics, Vol. 95, © 1991 Birkhäuser Verlag Basel

# Der Kenntnisstand in der konstruktiven Galoisschen Theorie

B. HEINRICH MATZAT

## Einführung

Vorliegender Übersichtsartikel versteht sich als eine Fortsetzung der Zusammenfassung im Jahresbericht der Deutschen Mathematiker-Vereinigung (Matzat [1988]). Dort war der erstmals von Fried [1977] beschrittene mehrdimensionale Zugang auf Grund der ungelösten arithmetischen Probleme weitgehend ausgeklammert worden. In der Zwischenzeit konnten bei der Lösung dieser Probleme deutliche Fortschritte erzielt werden (sowohl bei der Ausgrenzung der äußeren Automorphismen der Galoisgruppe als auch bei der Suche nach rationalen Punkten), so daß sich nunmehr dieser Weg für einen Bericht über den derzeitigen Kenntnisstand in der konstruktiven Galoisschen Theorie anbietet. Darüber hinaus konnten inzwischen mit den neuen arithmetischen Rationalitätskriterien auch Gruppen als Galoisgruppen über $\mathbb{Q}$ nachgewiesen werden, für die der Nachweis mit den im DMV-Bericht dargestellten algebraischen Rationalitätskriterien nicht gelang.

## 1. Das Umkehrproblem der Galoisschen Theorie

(1.1) Unter dem Umkehrproblem der Galoisschen Theorie über einem Körper $K$ versteht man die Frage, ob es zu jeder endlichen Gruppe $G$ eine Galoiserweiterung $N/K$ gibt, deren Galoisgruppe $\mathrm{Gal}(N/K)$ isomorph zu $G$ ist.

Bekanntlich — dies folgt aus dem Riemannschen Existenzsatz — besitzt das Umkehrproblem eine positive Lösung über dem Körper der komplexwertigen rationalen Funktionen $\mathbb{C}(t)$ und damit auch über den rationalen Funktionenkörpern (einer und mehrerer Variablen) über jedem algebraisch abgeschlossenen Körper der Charakteristik 0. Hieraus kann das entsprechende Resultat auch für $\mathbb{R}(t)$ und allgemeiner rationale Funktionenkörper über reell abgeschlossenen Körpern abgeleitet werden (siehe Krull-Neukirch [1971] oder auch Serre [1991]). Unter Verwendung von algebraischen Scheinüberlagerungen anstelle der dem Riemannschen Existenzsatz zugrunde liegenden topologischen Überlagerungen Riemannscher Flächen ist es Harbater [1984,1987] gelungen, das Umkehrproblem der Galoisschen Theorie unter anderem auch über $\mathbb{Q}_p(t)$, dem rationalen Funktionenkörper über dem hinsichtlich der $p$-adischen Topologie vollständigen Körper der $p$-adischen Zahlen, zu lösen.

Für andere Körper hingegen ist die Lösung des Umkehrproblems negativ. So sind Galoisgruppen über dem Primkörper $\mathbb{F}_p$ in Charakteristik $p$ zyklisch und über $\mathbb{Q}_p$ stets auflösbar. In beiden Fällen konnte sogar die Struktur der absoluten Galoisgruppen vollständig bestimmt werden (siehe Jannsen-Wingberg [1982]). Weitere Körper mit zum Teil exotischen Eigenschaften, für die das Umkehrproblem der Galoisschen Theorie keine positive Lösung besitzt, sind von Jensen [1986] in Form einer Ankündigung mitgeteilt worden.

(1.2) Über dem Körper der rationalen Zahlen $\mathbb{Q}$ – dies ist das klassische Umkehrproblem der Galoisschen Theorie – ist die Frage bisher noch unbeantwortet. Das weitestgehende allgemeine Resultat geht hier auf Shafarevich [1954a,b,c] zurück, der zeigen konnte, daß zumindest jede auflösbare Gruppe als Galoisgruppe über $\mathbb{Q}$ vorkommt.

Fast alle weiteren Resultate über $\mathbb{Q}$ stützen sich auf den Hilbertschen Irreduzibilitätssatz. Bezeichnet man die Permutationsdarstellung von $\mathrm{Gal}(N/K)$ auf den Nullstellen eines $N/K$ erzeugenden (irreduziblen) Polynoms $f \in K[X]$ mit $\mathrm{Gal}(f)$, so lautet dieser in der hier passenden Form:

**Satz 1.1** (Hilbert [1982]): *Es sei $K$ ein über $\mathbb{Q}$ endlich erzeugter Körper und $f(t,X) \in K(t)[X]$ ein Polynom über $K(t)$. Dann gibt es unendlich viele $a \in K$ mit*

$$\mathrm{Gal}(f(a,X)) \cong \mathrm{Gal}(f(t,X)).$$

Allgemein nennt man heutzutage einen Körper $K$ einen Hilbertkörper, wenn der Hilbertsche Irreduzibilitätssatz für $K$ gültig ist. Zu diesen zählen nach Kuyk [1970] (siehe auch Fried-Jarden [1986], Th. 15.6) unter anderem auch die über dem vollen Kreisteilungskörper $\mathbb{Q}^{ab}$ endlich erzeugten Körper.

Hilbert [1892] selbst benutzte den Irreduzibilitätssatz, um die symmetrischen und alternierenden Gruppen $S_n$ und $A_n$ als Galoisgruppen über $\mathbb{Q}$ nachzuweisen. In der Theorie der Modulfunktionen besitzen die $n$-Teilungspolynome die Gruppe $GL_2(\mathbb{Z}/n\mathbb{Z})$ und die Transformationspolynome die Gruppe $PGL_2(\mathbb{Z}/n\mathbb{Z})$ als Galoisgruppen über $\mathbb{Q}(j)$ und führen so zu Galoiserweiterungen mit den entsprechenden Galoisgruppen über $\mathbb{Q}$ (siehe Weber [1908], Macbeath [1969]). Die in den Gruppen $GL_2(p)$ bzw. $PGL_2(p)$ enthaltenen einfachen Faktoren $PSL_2(p)$ konnten erst von Shih [1974] und bisher nur für Primzahlen $p$ mit $(\frac{a}{p}) = -1$ für ein $a \in \{2,3,7\}$ als Galoisgruppen über $\mathbb{Q}(t)$ und $\mathbb{Q}$ nachgewiesen werden. Ein weiteres aus der Theorie der Modulfunktionen stammendes Ergebnis geht auf Mestre [1988] zurück, der zeigen konnte, daß die Gruppen $PSL_2(p^2)$ für Primzahlen $p$ mit $(\frac{5}{p}) = -1$ Galoisgruppen über $\mathbb{Q}(t)$ und $\mathbb{Q}$ sind.

(1.3) Da das Umkehrproblem der Galoisschen Theorie über $\mathbb{C}(t)$ gelöst ist, drängt sich die Frage auf, über welchen Körpern $K(t)$ eine vorgegebene Galoiserweiterung $N/\mathbb{C}(t)$ mit der Gruppe $G$ (als Galoiserweiterung) definiert ist, das heißt, wann es eine über dem Konstantenkörper $K$ reguläre Körpererweiterung $N_K/K(t)$ gibt mit $\mathbb{C}N_K = N$ (und $\mathrm{Gal}(N_K/K(t)) \cong G$). Im Falle $K = \mathbb{Q}$ gibt es dann nämlich nach dem

Hilbertschen Irreduzibilitätssatz unendlich viele Galoiserweiterungen über $\mathbb{Q}$ mit einer zu $G$ isomorphen Galoisgruppe. Bei vorgegebenen algebraischen Verzweigungspunkten von $N/\mathbb{C}(t)$ wird diese Frage weitgehend durch die im DMV-Bericht zusammengefaßten (algebraischen) Rationalitätskriterien gelöst (siehe auch Matzat [1987]). Das einfachste von diesen lautet:

**Satz 1.2** (Belyi [1979], Matzat [1984], Thompson [1984a]): *Es seien $G$ eine endliche Gruppe mit trivialem Zentrum und $C_1, ..., C_s$ rationale Konjugiertenklassen von $G$. Weiter operiere $G$ durch Konjugation transitiv auf der Menge der Erzeugendensysteme $(\sigma_1, ..., \sigma_s)$ von $G$ mit $\sigma_j \in C_j$ und $\sigma_1 \cdot ... \cdot \sigma_s = \iota$. Dann gibt es eine reguläre Galoiserweiterung $N/\mathbb{Q}(t)$ mit der Galoisgruppe $G$.*

Bei Thompson [1984a] wurden solche Erzeugendensysteme starr genannt. Durch Nachweis der Existenz solcher starren Erzeugendensysteme konnte eine Reihe einfacher Gruppen, darunter auch die meisten aller sporadischen Gruppen, als Galoisgruppen über $\mathbb{Q}(t)$ und $\mathbb{Q}$ dargestellt werden (siehe z.B. Matzat [1988]).

Bei transzendenten Verzweigungspunkten ist $\mathbb{Q}(t)$ häufig algebraisch abgeschlossen in $K(t)$. Im Fall eines über $\mathbb{Q}$ rationalen Funktionenkörpers $K/\mathbb{Q}$ kann dann der Hilbertsche Irreduzibilitätssatz auch auf $K$ bzw. $K(t)$ angewandt werden, um Galoiserweiterungen über $\mathbb{Q}(t)$ und $\mathbb{Q}$ zu erhalten. Die Frage nach der Rationalität von $K/\mathbb{Q}$ oder schwächer nach der Existenz rationaler Stellen von $K/\mathbb{Q}$ führt auf arithmetische Rationalitätskriterien. Solche werden hier in den nächsten Abschnitten beschrieben.

## 2. Die Hurwitzsche Zopfgruppe als Fundamentalgruppe

(2.1) Um bei der Untersuchung der Struktur der Definitionskörper von Galoiserweiterungen $N/\mathbb{C}(t)$ mit transzendenten Verzweigungspunkten topologische Hilfsmittel einsetzen zu können, wird hier wie bei Fried-Biggers [1982] anstatt von einer an endlich vielen Punkten gelochten Riemannschen Zahlenkugel (bzw. einer projektiven Geraden über $\mathbb{C}$) vom *$r$-fachen unvollständigen Produkt von* $\mathbf{P}^1(\mathbb{C})$ ausgegangen, das durch

$$\mathbf{P}^1(\mathbb{C})_r^\bullet := \{(x_1, ..., x_r) \in \mathbf{P}^1(\mathbb{C})^r \mid x_i \neq x_j \text{ für } i \neq j\}$$

definiert ist. Der Bahnenraum unter der Operation der symmetrischen Gruppe $S_r$ auf den Komponenten, das heißt

$$\tilde{\mathbf{P}}^1(\mathbb{C})_r^\bullet := \mathbf{P}^1(\mathbb{C})_r^\bullet / S_r,$$

wird dann als das *$r$-fache unvollständige symmetrische Produkt von* $\mathbf{P}^1(\mathbb{C})$ bezeichnet.

Für beide Räume ist die Fundamentalgruppe wohlbekannt und kann als Faktorgruppe der (reinen) Artinschen Zopfgruppe (durch Hinzunahme einer weiteren Relation) beschrieben werden. Nennt man nämlich

$$\begin{aligned} \tilde{\mathcal{H}}_r \ := \ & \langle \beta_1, ..., \beta_{r-1} \mid \beta_i\beta_j = \beta_j\beta_i \text{ für } |i-j| \neq 1, \ \beta_i\beta_{i+1}\beta_i = \beta_{i+1}\beta_i\beta_{i+1}, \\ & \beta_1 \cdot ... \cdot \beta_{r-1}\beta_{r-1} \cdot ... \cdot \beta_1 = \iota \rangle \end{aligned}$$

*(volle) Hurwitzsche Zopfgruppe* und den Kern des kanonischen Epimorphismus $q$ von $\tilde{\mathcal{H}}_r$ auf $S_r$ *reine Hurwitzsche Zopfgruppe (auf r Strängen)*:

$$\mathcal{H}_r := \mathrm{Kern}(q) \text{ mit } q : \tilde{\mathcal{H}}_r \to S_r,\ \beta_i \mapsto (i, i+1),$$

— letzere wird von den Elementen

$$\beta_{ij} = (\beta_{j-1}^2)^{\beta_{j-2}\cdot\ldots\cdot\beta_i} = (\beta_i^2)^{\beta_{i+1}^{-1}\cdot\ldots\cdot\beta_{j-1}^{-1}} \text{ für } 1 \le i < j \le r$$

erzeugt — , so gilt der

**Satz 2.1** (Fadell - Van Buskirk [1962]): *Die Fundamentalgruppe des unvollständigen symmetrischen Produkts von* $\mathbf{P}^1(\mathbb{C})$ *ist die volle Hurwitzsche Zopfgruppe:*

$$\pi_1(\tilde{\mathbf{P}}^1(\mathbb{C})_r^\bullet) = \tilde{\mathcal{H}}_r.$$

*In der hierdurch gegebenen Galoiskorrespondenz für Überlagerungen entspricht der* $S_r$*-Überlagerung* $\mathbf{P}^1(\mathbb{C})_r^\bullet \to \tilde{\mathbf{P}}^1(\mathbb{C})_r^\bullet$ *die reine Hurwitzsche Zopfgruppe:*

$$\pi_1(\mathbf{P}^1(\mathbb{C})_r^\bullet) = \mathcal{H}_r.$$

Die erzeugenden Elemente $\beta_i$ beziehungsweise $\beta_{ij}$ können vermöge der Koordinatengraphen von Schleifen in $\tilde{\mathbf{P}}^1(\mathbb{C})_r^\bullet$ bzw. $\mathbf{P}^1(\mathbb{C})_r^\bullet$ wie folgt als Zöpfe veranschaulicht werden:

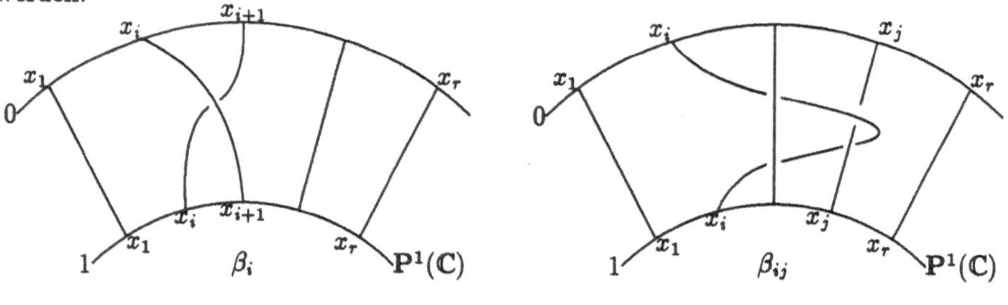


(2.2) Für $r \le 3$ sind die so definierten Hurwitzschen Zopfgruppen endlich, genauer gelten

$$\tilde{\mathcal{H}}_2 \cong Z_2,\ \tilde{\mathcal{H}}_3 \cong Z_3 \rtimes Z_4 \text{ (nicht abelsch)}$$

und demzufolge $\mathcal{H}_2 = I$, $\mathcal{H}_3 \cong Z_2$. Für $r \ge 4$ hingegen sind sie unendliche Gruppen, deren Struktur induktiv durch den folgenden Satz beschrieben wird:

**Satz 2.2** (Fadell - Van Buskirk [1962], Gilette - Van Buskirk [1968]): *Es seien* $r \ge 4$,

$$\tilde{\mathcal{H}}_r^\bullet := \{\beta \in \tilde{\mathcal{H}}_r \mid q(\beta) \in S_{r-1}\} \text{ und}$$

$$\mathcal{G}_{r-1} := \langle \gamma_1, \ldots, \gamma_{r-1} \mid \gamma_1 \cdot \ldots \cdot \gamma_{r-1} = \iota \rangle \text{ mit } \gamma_i := \beta_{ir}.$$

*Dann ist* $\mathcal{G}_{r-1}$ *ein freier Normalteiler von* $\tilde{\mathcal{H}}_r^\bullet$ *(vom Rang* $r-2$*), und* $\tilde{\mathcal{H}}_r^\bullet$ *ist eine Erweiterung von* $\mathcal{G}_{r-1}$ *mit einer zu* $\tilde{\mathcal{H}}_{r-1}$ *isomorphen Gruppe. Die Operation der*

*Erzeugenden $\mathcal{G}_{r-1}\beta_j$ der Faktorgruppe auf $\mathcal{G}_{r-1}$ ist durch die folgende Formel (bis auf simultane Konjugation) festgelegt:*

$$[\gamma_1, ..., \gamma_{r-1}]^{\beta_j} = [\gamma_1, ..., \gamma_{j-1}, \gamma_j\gamma_{j+1}\gamma_j^{-1}, \gamma_j, \gamma_{j+2}, ..., \gamma_{r-1}] \text{ für } 1 \leq j \leq r-2.$$

*Darüber hinaus besitzt $\mathcal{G}_{r-1}$ in der reinen Hurwitzschen Zopfgruppe $\mathcal{H}_r$ ein zu $\mathcal{H}_{r-1}$ isomorphes Komplement.*

(2.3) Nennt man nun eine Gruppe $G$ eine *koendliche Gruppe*, wenn der Durchschnitt der Normalteiler von endlichem Index in $G$ die Einsgruppe ist, so erhält man aus dem obigen Struktursatz noch die

**Folgerung 2.3** (Matzat [1990,1991a]): *Die Hurwitzschen Zopfgruppen $\tilde{\mathcal{H}}_r$ und $\mathcal{H}_r$ sind endlich erzeugte koendliche Gruppen.*

Im Falle $r = 4$ ist noch bekannt, daß $\tilde{\mathcal{H}}_4$ einen zur Quaternionengruppe $Q_8$ isomorphen Normalteiler besitzt, erzeugt von $(\beta_1\beta_2\beta_3)^2$ und $\beta_1\beta_3^{-1}$, so daß die zugehörige Faktorgruppe isomorph zu $PSL_2(\mathbb{Z})$ ist (Fried-Thompson [1991]). Weitere Resultate über die Hurwitzschen Zopfgruppen sind bei Birman [1974] zusammengetragen.

## 3. Die proendliche Hurwitzsche Zopfgruppe als Galoisgruppe

(3.1) Das $r$-fache unvollständige (symmetrische) Produkt von $\mathbf{P}^1(\mathbb{C})$ ist eine quasiprojektive normale Mannigfaltigkeit über $\mathbb{C}$. Bedeutet $\mathcal{X}$ eine solche und $\mathbb{C}(\mathcal{X})$ deren Funktionenkörper, so ist die Vereinigung aller endlichen, über $\mathcal{X}$ unverzweigten Erweiterungskörper von $\mathbb{C}(\mathcal{X})$ ein über $\mathbb{C}(\mathcal{X})$ galoisscher Körper $M$, dessen Galoisgruppe zur proendlichen Komplettierung der Fundamentalgruppe $\pi_1(\mathcal{X})$ isomorph ist. Hierbei heißt ein Körper unverzweigt über $\mathcal{X}$, wenn in ihm die Normalisierung von $\mathcal{X}$ unverzweigt ist (siehe etwa Grothendieck [1971], Exp. XII, oder auch Popp [1970], erste Vorlesung).

Der Funktionenkörper von $\mathbf{P}^1(\mathbb{C})^{\bullet}_r$ (bzw. von $\tilde{\mathbf{P}}^1(\mathbb{C})^{\bullet}_r$) wird durch die Koordinatenfunktionen $t_1, ..., t_r$ (bzw. deren elementarsymmetrische Funktionen $\tilde{t}_1, ..., \tilde{t}_r$) erzeugt und ist damit ein rationaler Funktionenkörper über $\mathbb{C}$. Nennt man die proendliche Komplettierung von $\tilde{\mathcal{H}}_r$ (bzw. von $\mathcal{H}_r$ *reine*) *proendliche Hurwitzsche Zopfgruppe* und bezeichnet diese mit $\tilde{H}_r$ (bzw. mit $H_r$), so erhält man gewissermaßen als Korollar zum Satz 2.1 den

**Satz 3.1** (Matzat [1989]): *Die Galoisgruppe der maximalen über $\tilde{\mathbf{P}}_1(\mathbb{C})^{\bullet}_r$ unverzweigten Körpererweiterung $M_r/\mathbb{C}(\tilde{t}_1, ..., \tilde{t}_r)$ ist die volle proendliche Hurwitzsche Zopfgruppe:*

$$\mathrm{Gal}(M_r/\mathbb{C}(\tilde{\underline{t}})) = \tilde{H}_r.$$

*Die reine proendliche Hurwitzsche Zopfgruppe erhält man als Galoisgruppe von $M_r/\mathbb{C}(t_1, ..., t_r)$:*

$$\mathrm{Gal}(M_r/\mathbb{C}(\underline{t})) = H_r.$$

(3.2) Mit dem Resultat in (2.3) überträgt sich die Struktur der diskreten Hurwitzschen Zopfgruppe unverändert auf deren proendliche Komplettierung, dies besagt der

**Satz 3.2** (Matzat [1990,1991a]): *Ersetzt man im Satz 2.2 die Gruppen $\tilde{\mathcal{H}}_r$, $\tilde{\mathcal{H}}_r^\bullet$, $\mathcal{H}_r$, $\mathcal{G}_{r-1}$ durch ihre proendliche Komplettierungen $\tilde{H}_r$, $\tilde{H}_r^\bullet$, $H_r, G_{r-1}$, so bleibt die Aussage des Struktursatzes gültig. Überdies ist die natürliche Abbildung von $\tilde{\mathcal{H}}_r$ in $\tilde{H}_r$ injektiv.*

(3.3) Die gemäß dem Satz 3.2 in $\tilde{H}_r$ eigebetteten Erzeugenden $\beta_{ij}$ (und auch $\beta_i$) bekommen in der Galoiserweiterung $M_r/\mathbb{C}(\underline{t})$ (bzw. $M_r/\mathbb{C}(\underline{\tilde{t}})$) die Bedeutung von Trägheitsgruppenerzeugenden. Für $M_r/\mathbb{C}(\underline{t})$ gilt zum Beispiel der

**Zusatz 3.3** (Matzat [1990,1991a]): *Das durch $(t_i - t_j)$ in $\mathbb{C}(\underline{t})$ erzeugte Bewertungsideal $\mathfrak{D}_{ij}$ besitzt eine Fortsetzung $\widehat{\mathfrak{D}}_{ij}$ in $M_r$, derart daß die Trägheitsgruppe von $\widehat{\mathfrak{D}}_{ij}/\mathfrak{D}_{ij}$ als prozyklische Gruppe durch $\beta_{ij}$ erzeugt wird.*

(3.4) Unter Verwendung dieses Zusatzes erhält man aus dem Struktursatz für die proendliche Hurwitzsche Zopfgruppe für die Galoiserweiterung $M_r/\mathbb{C}(\underline{t})$ den

**Satz 3.4** (Matzat [1989,1991a]): *Für $r \geq 4$ ist der Fixkörper von $M_r$ unter dem abgeschlossenen Normalteiler $G_{r-1}$ von $H_r$ gerade $M_{r-1}(t_r)$:*

$$M_r^{G_{r-1}} = M_{r-1}(t_r).$$

*Ferner ist $M_{r-1}$ algebraisch abgeschlossen in $M_r$. Die Körpererweiterung $M_r/M_{r-1}(t_r)$ ist eine maximale, außerhalb der Menge $\{\mathfrak{P}_1, \dots, \mathfrak{P}_{r-1}\}$ der von $(t_i - t_r)$ erzeugten Bewertungsideale $\mathfrak{P}_i$ von $M_{r-1}(t_r)/M_{r-1}$ unverzweigte Galoiserweiterung.*

Durch Induktion ergibt sich hieraus noch die

**Folgerung 3.5** (Matzat [1990,1991a]): *Für $3 \leq s < r$ ist*

$$G_s^{(r)} := \langle \beta_{ij} \mid 1 \leq i < j,\ s < j \leq r \rangle^\wedge \ (proendlicheKomplettierung)$$

*ein abgeschlossener Normalteiler von $H_r$. Mit diesem gilt*

$$M_r^{G_s^{(r)}} = M_s(t_{s+1}, \dots, t_r).$$

(3.5) Nach der Grothendieck'schen Deformationstheorie für Überlagerungen (siehe Grothendieck [1971], Exp. X) bleiben alle Resultate dieses Abschnitts gültig, wenn der Körper $\mathbb{C}$ durch irgendeinen algebraisch abgeschlossenen Körper der Charakteristik 0 ersetzt wird, also auch zum Beispiel durch den Körper aller algebraischen Zahlen $\bar{\mathbb{Q}}$ (in $\mathbb{C}$). Die entsprechenden Körper werden hier mit $\bar{M}_r, \dots$ anstelle von $M_r, \dots$ bezeichnet. Da der Verzweigungsort von $\bar{M}_r/\bar{\mathbb{Q}}(\underline{t})$ bereits über $\mathbb{Q}(\underline{t})$ definiert ist, ist der Körper $\bar{M}_r$ auch über $\mathbb{Q}(\underline{t})$ galoissch, und für die zugehörige Galoisgruppe gilt der

**Satz 3.6** (Matzat [1990,1991a]): *Die Galoisgruppe von* $\bar{M}_r/\mathbb{Q}(\underline{t})$ *ist ein semidirektes Produkt von* $\mathrm{Gal}(\bar{M}_r/\bar{\mathbb{Q}}(\underline{t})) \cong H_r$ *mit einem zur Gruppe* $\Lambda := \mathrm{Gal}(\bar{\mathbb{Q}}(\underline{t})/\mathbb{Q}(\underline{t}))$ *isomorphen Komplement:*

$$\mathrm{Gal}(\bar{M}_r/\mathbb{Q}(\underline{t})) \cong H_r \rtimes \Lambda.$$

*Hierin operiert* $\Lambda$ *auf den Konjugiertenklassen* $[\beta_{ij}]$ *der Erzeugenden* $\beta_{ij}$ *von* $H_r$ *über den Kreisteilungscharakter* $c : \Lambda \mapsto \hat{\mathbb{Z}}^\times$*:*

$$[\beta_{ij}]^\lambda = [\beta_{ij}^{c(\lambda)}].$$

Durch diesen Satz wird insbesondere eine Operation von $\Lambda$ auf der Kommutatorfaktorgruppe $H_r/H_r'$ erklärt (vergl. Matzat [1987], I. §5.3). Die explizite Operation auf $H_r$ ist bisher nur für die komplexe Konjugation von $\bar{\mathbb{Q}}(\underline{t})/\mathbb{Q}(\underline{t})$ bekannt (siehe Hurwitz [1891], Dèbes - Fried [1990]).

## 4. Klassenzahlen von Erzeugendensystemen

(4.1) Nun seien $s := r-1$ und $t := t_r$. Gemäß dem Satz 3.4 ist dann $\bar{M}_s(t)$ der Fixkörper von $G_s$ in $\bar{M}_{s+1}$. Wegen $G_s \trianglelefteq \mathrm{Gal}(\bar{M}_{s+1}/\mathbb{Q}(\underline{t}))$ operiert $\Delta := \mathrm{Gal}(\bar{M}_s(t)/\mathbb{Q}(\underline{t}))$ durch Fortsetzung seiner Elemente auf den galoisschen Zwischenkörpern von $\bar{M}_{s+1}/\bar{M}_s(t)$. Dabei entsprechen den Fixpunkten unter dieser Operation Körper, die nach dem Satz 3.6 zwar über $\mathbb{Q}(\underline{t})$ definiert, aber nicht notwendig als Galoiserweiterungen definiert sind.

Eine geeignete Beschreibung der Operation von $\Delta$ auf den Zwischenkörpern von $\bar{M}_{s+1}/\bar{M}_s(t)$ mit einer vorgegebenen Galoisgruppe $G$ erhält man durch die Hurwitz-Klassifikation. Offenbar ist jeder Normalteiler von $G_s$ mit einer zu $G$ isomorphen Faktorgruppe der Kern eines Epimorphismus von $G_s$ auf $G$. Ein solcher bildet das Erzeugendensystem $\gamma_1, ..., \gamma_s$ (gemäß Satz 3.2 aus Satz 2.2) von $G_s$ ab auf ein Erzeugendensystem $\sigma_1, ..., \sigma_s$ von $G$ mit $\sigma_1 \cdot ... \cdot \sigma_s = \iota$. Die Menge dieser Erzeugendensysteme sei

$$\Sigma_s(G) := \{\underline{\sigma} \in G^s \mid \sigma_1 \cdot ... \cdot \sigma_s = \iota, \langle\underline{\sigma}\rangle = G\}.$$

Umgekehrt bestimmt jedes $\underline{\sigma} \in \Sigma_s(G)$ vermöge $\gamma_j \mapsto \sigma_j$ einen Epimorphismus von $G_s$ auf $G$, dessen Kern mit $\mathrm{Kern}(\underline{\sigma})$ bezeichnet werde. Da sich zwei Epimorphismen von $G_s$ auf $G$ mit demselben Kern um einen Automorphismus von $G$ unterscheiden, erhält man den

**Satz 4.1** (Matzat [1989,1991a]): *Es seien* $G$ *eine endliche Gruppe,* $\Sigma_s^a(G) := \Sigma_s(G)/\mathrm{Aut}(G)$ *der Bahnenraum von* $\mathrm{Aut}(G)$ *auf* $\Sigma_s(G)$ *und* $\mathbf{N}_{\bar{\mathbb{S}}}(G)$ *die Menge der außerhalb von* $\bar{\mathbb{S}} = \{\bar{\mathfrak{P}}_1, ..., \bar{\mathfrak{P}}_s\}$ *unverzweigten Galoiserweiterungen von* $\bar{M}_s(t)$ *mit einer zu* $G$ *isomorphen Galoisgruppe. Dann ist*

$$N_{\bar{\mathbb{S}}} : \Sigma_s^a(G) \to \mathbf{N}_{\bar{\mathbb{S}}}(G),\ [[\underline{\sigma}]] \mapsto N_{\bar{\mathbb{S}}}(\underline{\sigma}) := \bar{M}_{s+1}^{\mathrm{Kern}(\underline{\sigma})}$$

*eine bijektive Abbildung.*

(4.2) Eine solche Zuordnung von Erzeugendensystemklassen von $G$ und Galoiserweiterungen mit der Galoisgruppe $G$ heißt eine *Hurwitz-Klassifikation*. Da sich die Fortsetzungen von $\delta \in \Delta$ auf $\bar{N} \in \mathbf{N}_{\mathfrak{s}}(G)$ jeweils nur um ein Element aus $\mathrm{Gal}(\bar{N}/\bar{M}_s(t))$ unterscheiden, läßt sich die hier vorliegende Hurwitz-Klassifikation verfeinern zu einer surjektiven Abbildung des Bahnenraums $\Sigma_s^i(G) := \Sigma_s(G)/\mathrm{Inn}(G)$ auf $\mathbf{N}_{\mathfrak{s}}(G)$. Für diese ergibt sich aus dem Satz 3.6 noch der

**Zusatz 4.2** (Matzat [1990,1991a]): *Die Elemente $\delta \in \Delta$ operieren auf $\Sigma_s^i(G)$ über den Kreisteilungscharakter:*

$$[\tilde{\sigma}_j] = [\sigma_j^{c(\delta)}] \quad \textit{für} \quad [\underline{\tilde{\sigma}}] = [\underline{\sigma}]^\delta.$$

Durch diesen Zusatz sind die Konjugiertenklassen der Komponenten von $[\underline{\sigma}]^\delta$ bekannt. Daher ist es sinnvoll, die Erzeugendensystemklassen aus $\Sigma_s^i(G)$ nach der Klassenzugehörigkeit der Komponenten einzuteilen: Für einen Vektor $\mathbf{C} = (C_1, ..., C_s)$ von $s$ vorgegebenen Konjugiertenklassen $C_j$ von $G$ sei dementsprechend

$$\Sigma^i(\mathbf{C}) := \{[\underline{\sigma}] \in \Sigma_s^i(G) \mid \sigma_j \in C_j\}.$$

Die Elementenzahl von $\Sigma^i(\mathbf{C})$ ist eine *Klassenzahl von Erzeugendensystemen von $G$* und wird mit $l^i(\mathbf{C})$ bezeichnet. Ferner heißt die Länge der Bahn von $\mathbf{C}$ unter der Operation von $\Delta$ (über den Kreisteilungscharakter) der *Irrationalitätsgrad von* $\mathbf{C}$ und wird durch $d^i(\mathbf{C})$ symbolisiert. Offenbar ist $\mathbf{C}$ genau dann ein *rationaler Klassenvektor* (Vektor aus rationalen Konjugiertenklassen), wenn $d^i(\mathbf{C}) = 1$ gilt.

(4.3) Im folgenden seien

$$\Delta_{\underline{\sigma}}^i := \{\delta \in \Delta \mid [\underline{\sigma}]^\delta = [\underline{\sigma}]\} \quad \text{und} \quad K_{\underline{\sigma}}^i(t) := \bar{M}_s(t)^{\Delta_{\underline{\sigma}}^i}$$

die Fixgruppe von $[\underline{\sigma}]$ in $\Delta$ beziehungsweise der Fixkörper von $\Delta_{\underline{\sigma}}^i$ in $\bar{M}_s(t)/\mathbb{Q}(\underline{t})$. Dann ist $\bar{N}_{\underline{\sigma}} := N_{\mathfrak{s}}(\underline{\sigma})$ über $K_{\underline{\sigma}}^i(t)$ galoissch mit etwa der Galoisgruppe $\Gamma$, und jedes Element von $\Gamma$ operiert als innerer Automorphismus auf $G = \mathrm{Gal}(\bar{N}_{\underline{\sigma}}/\bar{M}_s(t))$. Besitzt also $G$ ein triviales Zentrum, so ist der Fixkörper des Zentralisators von $G$ in $\Gamma$ über $K_{\underline{\sigma}}^i(t)$ galoissch mit einer zu $G$ isomorphen Galoisgruppe. Allgemeiner gilt der

**Satz 4.3** (Matzat [1989,1991a]): *Es seien $G$ eine endliche Gruppe, in der das Zentrum ein Komplement besitzt, und $[\underline{\sigma}] \in \Sigma_s^i(G)$. Dann existiert eine reguläre Galoiserweiterung $N_{\underline{\sigma}}^i/K_{\underline{\sigma}}^i(t)$ mit*

$$\mathrm{Gal}(N_{\underline{\sigma}}^i/K_{\underline{\sigma}}^i(t)) \cong G \quad \textit{und} \quad \bar{M}_s N_{\underline{\sigma}}^i = \bar{N}_{\underline{\sigma}} \ (= N_{\mathfrak{s}}(\underline{\sigma})).$$

(4.4) Unter den obigen Voraussetzungen ist also die Köpererweiterung $\bar{N}_{\underline{\sigma}}/\bar{M}_s(t)$ als Galoiserweiterung über $K_{\underline{\sigma}}^i(t)$ definiert. Nach dem Zusatz 4.2 enthält der Definitionskörper $K_{\underline{\sigma}}^i(t)$ den über $\mathbb{Q}$ durch die Charakterwerte auf $C_1 := [\sigma_1], ..., C_s := [\sigma_s]$ erzeugten Körper $k_{\mathbf{C}}$. Mit diesen Bezeichnungen erhält man damit den

**Zusatz 4.4** (Matzat [1990,1991a]): *Der Körper $K_{\underline{\sigma}}^i(t)$ umfaßt den Körper $k_{\mathbf{C}}(t)$ bezüglich $\mathbf{C} = ([\sigma_1], ..., [\sigma_s])$, und es gelten*

$$[K_{\underline{\sigma}}^i(t) : k_{\mathbf{C}}(\underline{t})] \leq l^i(\mathbf{C}) \quad \textit{sowie} \quad [k_{\mathbf{C}}(\underline{t}) : \mathbb{Q}(\underline{t})] = d^i(\mathbf{C}).$$

Der Satz 4.3 zusammen mit dem Zusatz 4.4 ergibt eine mehrdimensionale Variante des ersten Rationalitätskriteriums (vergleiche Matzat [1987], II. §4 oder auch Matzat [1988], §3). Im Spezialfall eines rationalen Klassenvektors bekommt man daraus die zum Satz 1.2 äquivalente

**Folgerung 4.5** (Matzat [1990,1991a]): *Besitzt eine Gruppe mit trivialem Zentrum einen rationalen Klassenvektor* $\mathbf{C}$ *mit* $l^i(\mathbf{C}) = 1$, *so gibt es eine reguläre Galoiserweiterung* $N/\mathbb{Q}(\underline{t})$ *mit*

$$\mathrm{Gal}(N/\mathbb{Q}(\underline{t})) \cong G.$$

(4.5) Demnach ist es nützlich, Methoden zur Berechnung von $l^i(\mathbf{C})$ beziehungsweise zum Nachweis der Starrheit von $\mathbf{C}$, das heißt von $l^i(\mathbf{C}) = 1$, zu kennen. Eine erste geht auf Belyi [1979] zurück. Diese benutzt geeignete Matrizendarstellungen von $G$ und ist besonders für klassische Gruppen vom Lie-Typ geeignet (siehe Belyi [1979,1983] sowie Walter [1984]). Eine zweite führt die Berechnung von $l^i(\mathbf{C})$ auf die Berechnung von Strukturkonstanten zurück und wurde zum Beispiel von Thompson [1984a] und anderen Autoren bei den sporadischen Gruppen und von Malle [1988b,1991a] bei den exzeptionellen Gruppen vom Lie-Typ verwandt. Beide Methoden sind bei Matzat [1987], II. §5 und §6, dargestellt und auch schon in Matzat [1988] als Zusätze 1 und 2 referiert worden.

Das in diesem Rahmen vielleicht aufsehenerregendste Resultat ist Thompson [1984a] zu verdanken, der nachwies, daß der Klassenvektor $(2A, 3B, 29A)$ der größten sporadischen Gruppe $F_1$ (in der Notation des Gruppenatlas, siehe Conway et al. [1985]) rational und starr ist.

## 5. Zopfbahnen und deren Geschlechter

(5.1) In den Satz 4.4 und in die Folgerung 4.5 geht nur der Grad der Körpererweiterung $K^i_{\underline{\sigma}}(t)/\mathbb{Q}(\underline{t})$ ein. Genauere und damit bessere Resultate sind zu erwarten, wenn man Kenntnisse über die Struktur von $K^i_{\underline{\sigma}}(t)/\mathbb{Q}(\underline{t})$ einsetzt. Solche kann man sich aus der Operation der Zopfgruppe auf den Erzeugendensystemklassen verschaffen.

**Bemerkung 5.1** (Hurwitz [1891]): *Die Erzeugenden* $\beta_1, ..., \beta_{s-1}$ *von* $\mathrm{Gal}(\bar{M}_s(t)/\bar{\mathbb{Q}}(\underline{t})) \cong \tilde{H}_s$ *operieren auf* $\Sigma^i_s(G)$ *folgendermaßen:*

$$[\sigma_1, ..., \sigma_s]^{\beta_j} = [\sigma_1, ..., \sigma_{j-1}, \sigma_j\sigma_{j+1}\sigma_j^{-1}, \sigma_j, \sigma_{j+2}, ..., \sigma_s].$$

Die Formel erhält man durch Übertragung aus dem Satz 2.2 unter Einsatz der Sätze 3.2 und 4.1. Da die reine Hurwitzsche Zopfgruppe $H_s \cong \mathrm{Gal}(\bar{M}_s(t)/\bar{\mathbb{Q}}(\underline{t}))$ die Konjugiertenklassen der Komponenten von $[\underline{\sigma}]$ nicht ändert, läßt sich die Operation von $H_s$ für jeden Klassenvektor $\mathbf{C} = (C_1, ..., C_s)$ von $G$ auf $\Sigma^i(\mathbf{C})$ einschränken und zerlegt damit $\Sigma^i(\mathbf{C})$ in Bahnen. Die Anzahl $h^i(\mathbf{C})$ dieser Bahnen wird als *Zopfklassenzahl von* $\Sigma^i(\mathbf{C})$ bezeichnet.

(5.2) Unter den Zopfbahnen auf $\Sigma^i(\mathbf{C})$ sind nun diejenigen von besonderem Interesse, die unter der vollen Gruppe $\Delta$ invariant bleiben. Diese Eigenschaft wird durch den Begriff der Starrheit sichergestellt: Eine Bahn unter der Operation einer Gruppe auf einer Menge heißt *starr*, wenn sie charakterisiert ist durch die Stabilisatoren modulo $\mathrm{Aut}(G)$ ihrer Elemente. Insbesondere ist $\Sigma^i(\mathbf{C})$ selbst eine starre Zopfbahn, wenn $h^i(\mathbf{C}) = 1$ ist. Damit gilt der

**Satz 5.2** (Matzat [1990,1991a]): *Es seien $G$ eine endliche Gruppe, $\mathbf{C} = (C_1, ..., C_s)$ ein Klassenvektor von $G$ und $[\underline{\sigma}] \in \Sigma^i(\mathbf{C})$. Ist dann die Zopfbahn $Z(\underline{\sigma}) := [\underline{\sigma}]^{H_s}$ starr in $\Sigma^i(\mathbf{C})$, so ist der Körper der Charakterwerte $k_{\mathbf{C}}$ algebraisch abgeschlossen in $K^i_{\underline{\sigma}}(t)$, und es gilt*

$$[K^i_{\underline{\sigma}}(t) : k_{\mathbf{C}}(\underline{t})] = |Z(\underline{\sigma})|.$$

Dieses Regularitätskriterium verallgemeinert den Hauptsatz bei Fried [1977]. Es folgt hier einfach aus der Tatsache, daß $\Delta_{\mathbf{C}} := \mathrm{Gal}(\bar{M}_s(t)/k_{\mathbf{C}}(\underline{t}))$ wegen $H_s \trianglelefteq \Delta_{\mathbf{C}}$ eine in $\Sigma^i(\mathbf{C})$ starre $H_s$-Bahn nicht vergrößert und somit der Index der Fixgruppe von $[\underline{\sigma}]$ in $\Delta_{\mathbf{C}}$ mit dem in $H_s$ übereinstimmt. Im Falle eines rationalen Klassenvektors $\mathbf{C}$ ist also dann der Körper $K^i_{\underline{\sigma}}(t)$ über $\mathbb{Q}$ regulär.

(5.3) Im folgenden seien

$$Z_s(\underline{\sigma}) := [\underline{\sigma}]^{G_{s-1}} \quad \text{und} \quad Z_j(\underline{\sigma}) := [Z_{j+1}(\underline{\sigma})]^{G_{j-1}} \quad \text{für} \quad 0 < j < s$$

die Bahnen von $Z_{j+1}(\underline{\sigma})$ unter $G^{(s)}_{j-1}/G^{(s)}_j \cong G_{j-1}$ (mit der in der Folgerung 3.5 festgelegten Bedeutung). Bezeichnet man die Anzahl der Zyklen von $\beta_{ij}G^{(s)}_j$ auf $Z_j(\underline{\sigma})$ mit $z_{ij}$, so heißt

$$g_j(\underline{\sigma}) := 1 - |Z_j(\underline{\sigma})| + \frac{1}{2}\sum_{i=1}^{j-1}(|Z_j(\underline{\sigma})| - z_{ij})$$

das *$j$-te Zopfbahnengeschlecht von* $[\underline{\sigma}]$. Dabei ist $g_j(\underline{\sigma}) = 0$ für $j \leq 3$ wegen $|Z_j(\underline{\sigma})| = 1$. Die auf diese Weise kombinatorisch definierten Zopfbahnengeschlechter lassen sich folgendermaßen als Geschlechter von Körpern kennzeichnen:

**Satz 5.3** (Matzat [1990,1991a]): *Bedeutet $K_j$ die algebraisch abgeschlossene Hülle von $\mathbb{Q}(t_1, ..., t_j)$ in $K^i_{\underline{\sigma}}(t)$, so gelten*

$$[K_j : K_{j-1}(t_j)] = |Z_j(\underline{\sigma})| \quad \textit{und} \quad g(K_j/K_{j-1}) = g_j(\underline{\sigma}) \quad \textit{für} \quad 1 \leq j \leq s.$$

Insbesondere ist $K^i_{\underline{\sigma}} = K_s$ ein rationaler Funktionenkörper über $K_0$, wenn $g_j(\underline{\sigma}) = 0$ ist für $4 \leq j \leq s$ und zudem $K_j/K_{j-1}$ jeweils einen Primdivisor ungeraden Grades besitzt. Dies führt zu der

**Folgerung 5.4** (Matzat [1990,1991a]): *Für $j = 4, ..., s$ gelte:*

(1) *Das $j$-te Zopfbahnengeschlecht ist Null: $g_j(\underline{\sigma}) = 0$.*

(2) *Bei der Operation von $\beta_{ij}G^{(s)}_j$ auf $Z_j(\underline{\sigma})$ tritt für ein $i < j$ eine der Zyklenlängen in ungerader Anzahl auf.*

*Dann ist $K^i_{\underline{\sigma}} = K_s$ ein über $K_0$ rationaler Funktionenkörper.*

(5.4) Durch Zusammenfassung der Sätze 4.3, 5.2 und der Folgerung 5.4 erhält man als ein erstes arithmetisches Rationalitätskriterium den

**Satz 5.5** (Matzat [1990,1991a]): *Es sei $G$ eine endliche Gruppe, in der das Zentrum ein Komplement besitzt. Ferner seien $\mathbf{C} = (C_1, ..., C_s)$ ein Klassenvektor von $G$ und $[\underline{\sigma}]$ ein Element einer starren $H_s$-Bahn in $\Sigma^i(\mathbf{C})$ mit den Eigenschaften* (1) *und* (2) *in der Folgerung 5.4. Dann ist $K^i_{\underline{\sigma}}(t)/k_{\mathbf{C}}$ ein rationaler Funktionenkörper, etwa $K^i_{\underline{\sigma}}(t) = k_{\mathbf{C}}(\underline{u}, t)$. Des weiteren existiert eine Galoiserweiterung $N^i_{\underline{\sigma}}/k_{\mathbf{C}}(\underline{u}, t)$ mit*

$$\mathrm{Gal}(N^i_{\underline{\sigma}}/k_{\mathbf{C}}(\underline{u}, t)) \cong G \quad \textit{und} \quad \bar{M}_s N^i_{\underline{\sigma}} = \bar{N}_{\underline{\sigma}}.$$

*Dabei gilt im Falle eines rationalen Klassenvektors noch $k_{\mathbf{C}} = \mathbb{Q}$.*

Wenn im Satz 5.5 die $H_s$-Bahn $Z(\underline{\sigma})$ sogar eine starre $G_{s-1}$-Bahn ist, so stimmt $Z(\underline{\sigma})$ mit $Z_s(\underline{\sigma})$ überein, und die Rationalitätsbedingungen (1) und (2) sind für $j < s$ automatisch erfüllt. Die so entstehende speziellere Version des obigen Rationalitätskriteriums wurde schon in Matzat [1989] vorgestellt (vergleiche auch Fried [1984], Th. 4.2).

(5.5) Als Animierungsbeispiel wird hier die Gruppe $G = P\Gamma L_3(4)$ gewählt. Hierin ist $\mathbf{C} = (2A, 2C, 6E, 3B)$ (in der Notation des Gruppenatlas) ein rationaler Klassenvektor mit $l^i(\mathbf{C}) = 54$ und $h^i(\mathbf{C}) = 1$. Damit bildet $\Sigma^i(\mathbf{C})$ eine starre Zopfbahn der Länge 54. Für die Permutationstypen von $\beta_{i4}$ auf $\Sigma^i(\mathbf{C})$ gelten

$$\mathrm{Typ}(\beta_{14}) = (5^4, 3^8, 2^5), \ \mathrm{Typ}(\beta_{24}) = (4^{10}, 2^7), \ \mathrm{Typ}(\beta_{34}) = (3^{12}, 2^8, 1^2).$$

Demnach ist $g_4(\underline{\sigma}) = 0$ für $[\underline{\sigma}] \in \Sigma^i(\mathbf{C})$. Da überdies die Rationalitätsbedingung (2) der Folgerung 5.4 für $s = 4$ erfüllt ist, existiert nach dem Satz 5.5 zunächst eine Galoiserweiterung $N/K_4(t)$ über dem rationalen Funktionenkörper $K_4 = \mathbb{Q}(u_1, ..., u_4)$ mit $\mathrm{Gal}(N/K_4(t)) \cong G$. In dieser Körpererweiterung ist der Fixkörper von $G' = PSL_3(4)$ ein rationaler Funktionenkörper, etwa $N^{G'} = K_4(z)$. Also ist $N/K_4(z)$ eine Galoiserweiterung mit $\mathrm{Gal}(N/K_4(z)) \cong PSL_3(4)$, für die der zugehörige Klassenvektor die Gestalt $\mathbf{C}' = (2A, 2A, 2A, 2A, 2A, 2A, 3A, 3A, 3A)$ hat (siehe Przywara [1991] oder auch Matzat [1991a], §9.3).

## 6. Abstieg des Konstantenkörpers

(6.1) Nach der Bemerkung 5.1 operiert nicht nur die reine Hurwitzsche Zopfgruppe $H_s$ sondern sogar die volle Hurwitzsche Zopfgruppe $\tilde{H}_s$ auf $\Sigma^i_s(G)$. Die Elemente der Faktorgruppe $\tilde{H}_s/H_s \cong S_s$ können nun sowohl dazu benutzt werden, den Konstantenkörper von $K^i_{\underline{\sigma}}(t)/\mathbb{Q}$ zu verkleinern, als auch dazu, die Zopfbahnengeschlechter zu erniedrigen. Da $\beta H_s \in \tilde{H}_s/H_s$ die Komponenten eines Klassenvektors $\mathbf{C}$ über die Abbildung $q$ in (2.1) permutiert, sind hierzu aber nur diejenigen Elemente von $\tilde{H}_s/H_s$ von Nutzen, die $\mathbf{C}$ in der Bahn des Kreisteilungscharakters belassen.

Dies führt zu den folgenden Definitionen: Eine Untergruppe $V$ von $S_s$ heißt eine *Gruppe von Symmetrien von* $\mathbf{C}$, wenn es zu jedem $\omega \in V$ ein zu $|G|$ teilerfremdes $\nu \in \mathbb{N}$ gibt mit $\mathbf{C}^\omega = \mathbf{C}^\nu$, das heißt mit $(C_{1^\omega}, ..., C_{s^\omega}) = (C_1^\nu, ..., C_s^\nu)$. Für eine solche seien künftig

$$\tilde{H}_V := \{\beta \in \tilde{H}_s | \; q(\beta) \in V\} \quad \text{sowie} \quad \tilde{\Delta}_V := \langle \Delta, \tilde{H}_V \rangle$$

(innerhalb von $\mathrm{Gal}(\bar{M}_s(t)/\mathbb{Q}(\underline{t}))$). Offenbar operiert $\tilde{H}_V$ auf

$$\Sigma_V^i(\mathbf{C}) := \{[\underline{\sigma}] \in \Sigma_s^i(G) \mid [\underline{\sigma}] \in \Sigma^i(\mathbf{C}^\omega), \omega \in V\}.$$

Die Anzahl der Bahnen heißt *V-symmetrisierte Zopfklassenzahl von* $\mathbf{C}$ und wird mit $\tilde{h}_V^i(\mathbf{C})$ bezeichnet. Die Länge der Bahn von $\mathbf{C}^V := \{\mathbf{C}^\omega \mid \omega \in V\}$ unter der Operation von $\Delta$ (über den Kreisteilungscharakter) wird der *V-symmetrisierte Irrationalitätsgrad von* $\mathbf{C}$ genannt und durch $\tilde{d}_V^i(\mathbf{C})$ abgekürzt. Dieser stimmt offenbar mit dem Grad des Fixkörpers $\tilde{k}_\mathbf{C}$ aller Automorphismen von $k_\mathbf{C}/\mathbb{Q}$ überein, die $\mathbf{C}^V$ (über den Kreisteilungscharakter) auf sich abbilden. Schließlich heißt $\mathbf{C}$ ein *V-symmetrischer Klassenvektor*, wenn $\tilde{d}_V^i(\mathbf{C}) = 1$ ist beziehungsweise wenn $\tilde{k}_\mathbf{C} = \mathbb{Q}$ gilt.

(6.2) Gemäß dem Vorgehen im Abschnitt (4.3) seien

$$\tilde{\Delta}_{\underline{\sigma}}^i := \{\delta \in \tilde{\Delta}_V \mid [\underline{\sigma}]^\delta = [\underline{\sigma}]\} \quad \text{und} \quad \tilde{K}_{\underline{\sigma}}^i(t) := \bar{M}_s(t)^{\tilde{\Delta}_{\underline{\sigma}}^i}.$$

Da $\bar{N}_{\underline{\sigma}}$ auch über $\tilde{K}_{\underline{\sigma}}^i(t)$ galoissch ist, erhält man als Analogon zum Satz 4.3 den

**Satz 6.1** (Matzat [1990,1991a]): *Es seien $G$ eine endliche Gruppe, in der das Zentrum ein Komplement besitzt, $\mathbf{C} = (C_1, ..., C_s)$ ein Klassenvektor von $G$, $[\underline{\sigma}] \in \Sigma^i(\mathbf{C})$ und $V$ eine Symmetriegruppe von $\mathbf{C}$. Dann gibt es eine reguläre Galoiserweiterung $\tilde{N}_{\underline{\sigma}}^i/\tilde{K}_{\underline{\sigma}}^i(t)$ mit*

$$\mathrm{Gal}(\tilde{N}_{\underline{\sigma}}^i/\tilde{K}_{\underline{\sigma}}^i(t)) \cong G \quad \textit{und} \quad \bar{M}_s\tilde{N}_{\underline{\sigma}}^i = \bar{N}_{\underline{\sigma}}.$$

Hierzu erhält man als Analogon zum Satz 5.2 weiter den

**Zusatz 6.2** (Matzat [1990,1991a]): *Ist im Satz 6.1 die Bahn $\tilde{Z}(\underline{\sigma}) := [\underline{\sigma}]^{\tilde{H}_V}$ eine in $\Sigma_V^i(\mathbf{C})$ starre $\tilde{H}_V$-Bahn, so ist $\tilde{K}_{\underline{\sigma}}^i(t)$ über dem Körper der V-symmetrisierten Charakterwerte $\tilde{k}_\mathbf{C}$ regulär.*

(6.3) So konzentriert sich jetzt die Frage darauf, unter welchen Voraussetzungen $\tilde{K}_{\underline{\sigma}}^i(t)$ oder auch ein Zwischenkörper von $\bar{M}_s(t)/\tilde{K}_{\underline{\sigma}}^i(t)$ ein rationaler Funktionenkörper über $\tilde{k}_\mathbf{C}$ ist. Diese wird in diesem Abschnitt für den Fall beantwortet, daß $K_{\underline{\sigma}}^i$ über dem algebraischen Abschluß $k_{\underline{\sigma}}^i$ von $\mathbb{Q}$ in $K_{\underline{\sigma}}^i$ rational ist. Hierzu heißt $[\underline{\sigma}] \in \Sigma_s^i(G)$ eine *$H_j$-invariante Erzeugendensystemklasse*, wenn die $G_{j-1}$-Bahn $Z_j(\underline{\sigma})$ unter $H_j$ abgeschlossen ist (vergleiche Abschnitt (5.3)). Nach dem Satz 5.3 ist dann der Körper $k_{\underline{\sigma}}^i(t_1, ..., t_j)$ in $K_{\underline{\sigma}}^i$ algebraisch abgeschlossen.

**Satz 6.3** (Matzat [1990,1991a]): *Es seien $G$ eine endliche Gruppe, in der das Zentrum ein Komplement besitzt, $\mathbf{C} = (C_1, \ldots, C_s)$ ein Klassenvektor von $G$ und $[\underline{\sigma}] \in \Sigma^i(\mathbf{C})$ eine $H_j$-invariante Erzeugendensystemklasse. Weiter seien $V \leq S_j$ eine Symmetriegruppe von $\mathbf{C}$, $\Delta_V^i := \mathrm{Gal}(\bar{M}_s(t)/\tilde{k}_{\mathbf{C}}(\underline{t}))$, $\tilde{Z}(\underline{\sigma}) := [\underline{\sigma}]^{\tilde{H}_V}$ eine in $\Sigma_V^i(\mathbf{C})$ starre $\tilde{H}_V$-Bahn sowie $\tilde{\mathbf{Z}}$ die Menge der $H_s$-Bahnen innerhalb von $\tilde{Z}(\underline{\sigma})$. Ferner gelte*

(1) $K_{\underline{\sigma}}^i(t)/k_{\underline{\sigma}}^i$ *ist ein rationaler Funktionenkörper,*

(2) $\tilde{H}_V/H_s$ *und* $\Delta_V^i/H_s$ *operieren simultan wie* $V$ *auf* $\tilde{\mathbf{Z}}$.

*Dann existieren ein über $\tilde{k}_{\mathbf{C}}$ rationaler Zwischenkörper $\tilde{K}_V^i(t)$ von $\bar{M}_s(t)/\tilde{K}_{\underline{\sigma}}^i(t)$ und eine reguläre Galoiserweiterung $\tilde{N}_{\underline{\sigma}}^i/\tilde{K}_V^i(t)$ mit*

$$\mathrm{Gal}(\tilde{N}_{\underline{\sigma}}^i/\tilde{K}_V^i(t)) \cong G \quad \textit{und} \quad \bar{M}_s\tilde{N}_{\underline{\sigma}}^i = \bar{N}_{\underline{\sigma}}.$$

*Hierin ist $\tilde{k}_{\mathbf{C}} = \mathbb{Q}$, falls $\mathbf{C}$ ein $V$-symmetrischer Klassenvektor ist.*

(6.4) Für $s \leq 3$ ist $K_{\underline{\sigma}}^i = k_{\underline{\sigma}}^i(t_1, \ldots, t_s)$ nach (5.3) stets ein rationaler Funktionenkörper, für $s \geq 4$ kann die Rationalität von $K_{\underline{\sigma}}^i(t)/k_{\underline{\sigma}}^i$ zum Beispiel mit Hilfe der Folgerung 5.4 nachgewiesen werden. Leider ist die Bedingung (2) des Satzes in der Regel nicht direkt nachzuprüfen. Häufig kann sie aber für eine hinreichend große Untergruppe von $V$ sichergestellt werden.

**Zusatz 6.4** (Matzat [1990,1991a]): *Im Satz 6.3 werde statt* (2)*vorausgesetzt, daß für $\tilde{V} := \{\omega \in V \mid \mathbf{C}^\omega = \mathbf{C}\}$ entweder*

(2a) $\tilde{V} = I$ *gilt und* $\tilde{H}_V/H_s$ *(über den Kreisteilungscharakter) treu auf* $\tilde{\mathbf{Z}}$ *operiert oder*

(2b) $\tilde{V} = V$ *ist und* $\tilde{H}_V/H_s$ *als volle symmetrische Gruppe auf* $\tilde{\mathbf{Z}}$ *operiert.*

*Dann ist die Aussage des Satzes 6.3 für eine Untergruppe $U$ von $V$ richtig, für die $\mathbf{C}^U = \mathbf{C}^V$ gilt und die damit zu demselben Konstantenkörper $\tilde{k}_{\mathbf{C}}$ führt.*

Im Spezialfall einer $H_s$-invarianten Erzeugendensystemklasse $[\underline{\sigma}]$ und damit jedenfalls für $s \leq 3$ kann der Satz 6.3 daher als eine mehrdimensionale Version des zweiten Rationalitätskriteriums angesehen werden (siehe Matzat [1987], III. §3, bzw. Matzat [1988], §4). Übrigens läßt er sich noch dadurch verallgemeinern, daß man den Körper $K_{\underline{\sigma}}^i(t)$ durch den Fixkörper des Stabilisators von $[\underline{\sigma}]$ in $\tilde{\Delta}_W$ für eine die Ziffern $1, \ldots, j$ festlassende Symmetriegruppe $W$ von $\mathbf{C}$ mit $[\underline{\sigma}]^{\tilde{H}_W} = [\underline{\sigma}]^{H_s}$ ersetzt (siehe Matzat [1991a], Zusatz 6.3).

(6.5) Zum Abschluß dieses Abschnitts wird noch gezeigt, wie man mit der einfachsten Form des Satzes 6.3 beweisen kann, daß jede endliche abelsche Gruppe $G$ als Galoisgruppe einer regulären Köpererweiterung über $\mathbb{Q}(t)$ vorkommt. Hierzu wird $G$ in ein direktes Produkt von etwa $r$ zyklischen Gruppen $G_j = <\sigma_j>$ der Ordnung $n_j > 1$

zerlegt. Für jeden Faktor bildet $\underline{\sigma}_j = (\sigma_j, \sigma_j^{-1})$ im Falle von $n_j = 2$ beziehungsweise $\underline{\sigma}_j = (\sigma_j^\nu | \nu \in Z_{n_j}^\times)$ im Falle von $n_j > 2$ ein Erzeugendensystem mit dem Produkt $\iota$. Durch Zusammenfügen erhält man ein Erzeugendensystem $\underline{\sigma} = (\underline{\sigma}_j | j = 1, ..., r)$ von $G$ mit ebenfalls dem Produkt $\iota$, das aus $s$ Komponenten bestehen möge. Da in einer abelschen Gruppe jedes Erzeugendensystem der Länge $s$ mit dem Produkt $\iota$ eine $H_s$-invariante Klasse bildet, ist $K_{\underline{\sigma}}^i(t)/k_{\underline{\sigma}}^i$ ein rationaler Funktionenkörper. Wählt man nun die durch die Operation über den Kreisteilungscharakter erzeugte Gruppe $V$ als Symmetriegruppe des Klassenvektors $\mathbf{C}$ von $\underline{\sigma}$, so wird damit die Bedingung (2a) erfüllt, und $\mathbf{C}$ wird überdies zu einem $V$-symmetrischen Klassenvektor. Da hiermit alle Voraussetzungen im Satz 6.3 erfüllt sind, gibt es einen rationalen Funktionenkörper $\tilde{K}_V^i = \mathbb{Q}(u_1, ..., u_s)$ und eine reguläre Galoiserweiterung $\tilde{N}_{\underline{\sigma}}^i/\tilde{K}_V^i(t)$ mit einer zu $G$ isomorphen Galoisgruppe. Durch Anwendung des Hilbertschen Irreduzibilitätssatzes bekommt man die eingangs gewählte Formulierung dieses Resultats. (Andere Beweise sind bei Saltman [1982], Th. 3.12(c), Coombes-Harbater [1985], Ex. 3.1 mit Ex. 3.2, Thompson [1986] und auch in Matzat [1987], IV. §2, zu finden.)

## 7. Defektfreie rationale Stellen

(7.1) Um reguläre Erweiterungen $\tilde{N}_{\underline{\sigma}}^i/\mathbb{Q}$ mit $\mathrm{Gal}(\tilde{N}_{\underline{\sigma}}^i/\mathbb{Q}(t)) \cong G$ zu bekommen, braucht $\tilde{K}_{\underline{\sigma}}^i/\mathbb{Q}$ nicht unbedingt rational zu sein. Es genügt, daß $\tilde{K}_{\underline{\sigma}}^i/\mathbb{Q}$ geeignete rationale Stellen besitzt. Dabei heißt eine Stelle $P$ eine *rationale Stelle von* $\tilde{K}_{\underline{\sigma}}^i$, wenn das Bild des Stellenrings von $P$ der Konstantenkörper $\tilde{k}_{\underline{\sigma}}^i$ von $\tilde{K}_{\underline{\sigma}}^i$ ist. Besitzt $P$ eine Fortsetzung $\tilde{P}$ auf $\tilde{N}_{\underline{\sigma}}^i$ mit $\tilde{P}(t) = t$, derart daß

$$\mathrm{Gal}(\tilde{P}(\tilde{N}_{\underline{\sigma}}^i)/\tilde{P}(\tilde{K}_{\underline{\sigma}}^i(t))) \cong \mathrm{Gal}(\tilde{N}_{\underline{\sigma}}^i/\tilde{K}_{\underline{\sigma}}^i(t))$$

gilt, so heißt $P$ eine *Stelle ohne Galoisdefekt* oder kurz eine defektfreie Stelle. Rationale Stellen ohne Galoisdefekt erzeugen also definitionsgemäß reguläre Galoiserweiterungen über $\tilde{k}_{\underline{\sigma}}^i(t)$ mit der gewünschten Galoisgruppe.

Im folgenden wird $P$ eine *Stelle außerhalb des Verzweigungsortes von* $\tilde{K}_{\underline{\sigma}}^i/\tilde{k}_{\underline{\sigma}}^i$ genannt, wenn $P(t_i) \neq P(t_j)$ für $1 \leq i < j \leq s$ gilt (im Kompositum mit $\bar{\mathbb{Q}}(t_1, ..., t_s)$). Dann ist $\bar{\mathbb{Q}}\tilde{P}(\tilde{N}_{\underline{\sigma}}^i)/\bar{\mathbb{Q}}(t)$ eine außerhalb der Menge $\bar{\mathbf{S}}'$ der Zählerdivision $\mathfrak{P}_j'$ von $(t - P(t_j))$ unverzweigte Körpererweiterung. Bei defektfreien Stellen stimmt der Körper $\bar{\mathbb{Q}}\tilde{P}(\tilde{N}_{\underline{\sigma}}^i)$ hinsichtlich der Hurwitz-Klassifikation

$$N_{\bar{\mathbf{S}}'}' : \Sigma_s^a(G) \to \mathbf{N}_{\bar{\mathbf{S}}'}(G), \ [[\underline{\sigma}]] \mapsto N_{\bar{\mathbf{S}}'}'(\underline{\sigma}) := \bar{M}^{\mathrm{Kern}(\underline{\sigma})}$$

— hierbei ist $\bar{M}/\bar{\mathbb{Q}}(t)$ die maximale außerhalb $\bar{\mathbf{S}}'$ unverzweigte algebraische Körpererweiterung — mit dem Körper $\bar{N}_{\underline{\sigma}}' := N_{\bar{\mathbf{S}}'}'(\underline{\sigma})$ überein (vergleiche Matzat [1987], II. §1.1).

(7.2) Mit diesen Bezeichnungen gewinnt man den

**Satz 7.1** (Fried-Völklein [1991a], Matzat [1991a]): *Mit den Bezeichnungen des Satzes 6.1 gelten für eine Gruppe mit trivialem Zentrum:*

(a) *Besitzt* $\tilde{K}_{\underline{\sigma}}^{i}/\tilde{k}_{\underline{\sigma}}^{i}$ *eine rationale Stelle* $P$ *außerhalb des Verzweigungsortes, so gilt für die (einzige) Fortsetzung* $\tilde{P}$ *von* $P$ *auf* $\tilde{N}_{\underline{\sigma}}^{i}/\tilde{k}_{\underline{\sigma}}^{i}(t)$ *mit* $\tilde{P}(t) = t$

$$\mathrm{Gal}(\tilde{P}(\tilde{N}_{\underline{\sigma}}^{i})/\tilde{k}_{\underline{\sigma}}^{i}(t)) \cong G \quad \textit{und} \quad \bar{\mathbb{Q}}\tilde{N}_{\underline{\sigma}}^{i} = \bar{N}_{\underline{\sigma}}'.$$

(b) *Existiert eine Galoiserweiterung* $\tilde{N}/k(t)$ *mit* $\bar{\mathbb{Q}}\tilde{N} = \bar{N}_{\underline{\sigma}}'$*, so gilt* $k \geq \tilde{k}_{\underline{\sigma}}^{i}$*, und* $\tilde{K}_{\underline{\sigma}}^{i}/\tilde{k}_{\underline{\sigma}}^{i}$ *besitzt eine rationale Stelle* $P$ *außerhalb des Verzweigungsortes mit* $k\tilde{P}(\tilde{N}_{\underline{\sigma}}^{i}) = \tilde{N}$.

Insbesondere besitzt $\tilde{K}_{\underline{\sigma}}^{i}/\tilde{k}_{\underline{\sigma}}^{i}$ genau dann eine rationale Stelle außerhalb des Verzweigungsortes, wenn $\bar{N}_{\underline{\sigma}}'/\bar{\mathbb{Q}}(t)$ als Galoiserweiterung über $\tilde{k}_{\underline{\sigma}}^{i}(t)$ definiert ist.

**Zusatz 7.2** (Matzat [1991a]): *Liegen im Satz 7.1 die Bilder* $P(t_i)$ *in* $\tilde{k}_{\underline{\sigma}}^{i}$*, so kann dort* $\tilde{K}_{\underline{\sigma}}^{i}/\tilde{k}_{\underline{\sigma}}^{i}$ *durch die Körpererweiterung* $K_{\underline{\sigma}}^{i}/k_{\underline{\sigma}}^{i}$ *ersetzt werden.*

Nach obigem Satz besitzen rationale Stellen außerhalb des Verzweigungsortes von $\tilde{K}_{\underline{\sigma}}^{i}/\tilde{k}_{\underline{\sigma}}^{i}$ keinen Galoisdefekt. Damit braucht man nur hinreichend viele rationale Stellen von $\tilde{K}_{\underline{\sigma}}^{i}/\tilde{k}_{\underline{\sigma}}^{i}$ zu finden. Solche existieren auch außerhalb des Verzweigungsortes stets dann, wenn $\tilde{K}_{\underline{\sigma}}^{i}/\tilde{k}_{\underline{\sigma}}^{i}$ ein rationaler — oder allgemeiner — ein unirationaler Funktionenkörper ist (vergleiche Satz 6.3). Im Verzweigungsort von $\tilde{K}_{\underline{\sigma}}^{i}/\tilde{k}_{\underline{\sigma}}^{i}$ gibt es dagegen immer rationale Stellen, diese sind aber nur unter zusätzlichen Voraussetzungen defektfrei (siehe Matzat [1991a], §8.3).

(7.3) Eine andere Methode, rationale Stellen von $\tilde{K}_{\underline{\sigma}}^{i}/\tilde{k}_{\underline{\sigma}}^{i}$ nachzuweisen, ist es, zunächst nur $t_1, ..., t_{s-1}$ zu geeigneten $a_j \in \bar{\mathbb{Q}}$ zu spezialisieren und dann rationale Stellen im Bildkörper zu suchen. Hierzu seien eine $H_{s-1}$-invariante Erzeugendensystemklasse $[\underline{\sigma}] \in \Sigma_s^{i}(G)$ und eine Symmetriegruppe $V \leq S_{s-1}$ des Klassenvektors $\mathbf{C}$ von $[\underline{\sigma}]$ vorgegeben. Weiter bilde $\underline{a} = (a_1, ..., a_{s-1}) \in \mathbf{P}^1(\bar{\mathbb{Q}})_{s-1}^{\bullet}$ eine *V-Konfiguration*, das heißt, es gebe eine zu $V$ isomorphe Gruppe $\tilde{V}$ von Automorphismen von $\bar{\mathbb{Q}}(u)/\bar{\mathbb{Q}}$, die zusätzlich die Zählerdivisoren $\bar{\mathfrak{Q}}_j$ von $(u - a_j)$ via $V$ permutiert. Dann ist für die (eindeutig bestimmte) Stelle $Q$ von $\tilde{K}_{\underline{\sigma}}^{i}/\tilde{k}_{\underline{\sigma}}^{i}$ mit $Q(t_j) = a_j$ für $j = 1, ..., s-1$ und $Q(t_s) = u$ (in einem Kompositum mit $\bar{\mathbb{Q}}(t_1, ..., t_s)$) der Körper $Q(\tilde{K}_{\underline{\sigma}}^{i})$ ein Funktionenkörper einer Variablen über $\tilde{k}_{\underline{\sigma}}^{i}$, dessen Geschlecht mit $\tilde{g}_V(\underline{\sigma})$ bezeichnet wird und das *V-symmetrisierte Zopfbahnengeschlecht von* $[\underline{\sigma}]$ heißt. Dieses läßt sich berechnen, wenn man die Erzeugenden der Trägheitsgruppen von $\tilde{K}_{\underline{\sigma}}^{i}$ über $\tilde{k}_{\underline{\sigma}}^{i}(\tilde{u}) := \tilde{k}_{\underline{\sigma}}^{i}(u)^{\tilde{V}}$ kennt. Hierzu seien etwa $\tilde{\beta}_1, ..., \tilde{\beta}_{\tilde{s}}$ gewisse Erzeugende von Trägheitsgruppen der in $\bar{M}/\bar{\mathbb{Q}}(\tilde{u})$ verzweigten Primdivisoren $\bar{\tilde{\mathfrak{Q}}}_1, ..., \bar{\tilde{\mathfrak{Q}}}_{\tilde{s}}$ von $\bar{\mathbb{Q}}(\tilde{u})/\bar{\mathbb{Q}}$. Diese operieren auf den Nebenklassen von $\mathrm{Gal}(\bar{M}/\bar{\mathbb{Q}}Q(\tilde{K}_{\underline{\sigma}}^{i}))$ in $\mathrm{Gal}(\bar{M}/\bar{\mathbb{Q}}(\tilde{u}))$ und damit auf der $V$-symmetrisierten Zopfbahn $\tilde{Z}(\underline{\sigma}) = [\underline{\sigma}]^{\tilde{H}_V}$. Dies führt zu dem

**Satz 7.3** (Matzat [1989,1991a]): *Es seien* $G$ *eine endliche Gruppe mit trivialem Zentrum,* $\mathbf{C}$ *ein Klassenvektor von* $G$ *und* $[\underline{\sigma}] \in \Sigma^{i}(\mathbf{C})$ *eine* $H_{s-1}$*-invariante Erzeugendensystemklasse. Ferner seien* $V \leq S_{s-1}$ *eine Symmetriegruppe von* $\mathbf{C}$*,* $\tilde{Z}(\underline{\sigma}) = [\underline{\sigma}]^{\tilde{H}_V}$ *und*

$\tilde{\beta}_i$ *für* $i = 1, ..., \tilde{s}$ *Erzeugende der in* $\bar{M}/\bar{\mathbb{Q}}(\tilde{u})$ *verzweigten Primdivisoren von* $\bar{\mathbb{Q}}(\tilde{u})/\bar{\mathbb{Q}}$. *Dann gilt bezüglich einer vorgegebenen V-Konfiguration* $\underline{a}$

$$\tilde{g}_V(\underline{\sigma}) = 1 - |\tilde{Z}(\underline{\sigma})| + \frac{1}{2}\sum_{i=1}^{\tilde{s}}(|\tilde{Z}(\underline{\sigma})| - \tilde{z}_i),$$

*wobei* $\tilde{z}_i$ *die Anzahl der Zyklen von* $\tilde{\beta}_i$ *auf* $\tilde{Z}(\underline{\sigma})$ *bedeutet.*

**Zusatz 7.4** (Matzat [1991a]): *Im Falle* $s = 4$ *ist* $\tilde{g}_V(\underline{\sigma})$ *unabhängig von der gegebenen V-Konfiguration* $\underline{a}$. *Für die Permutationstypen von* $\tilde{\beta}_i$ *auf* $\tilde{Z}(\sigma)$ *gelten:*

(a) $\mathrm{Typ}(\tilde{\beta}_1) = \mathrm{Typ}(\beta_{14}), \mathrm{Typ}(\tilde{\beta}_2) = \mathrm{Typ}(\beta_1), \mathrm{Typ}(\tilde{\beta}_3) = \mathrm{Typ}(\beta_1\beta_{14})$ *für* $V = \langle(12)\rangle$,

(b) $\mathrm{Typ}(\tilde{\beta}_1) = \mathrm{Typ}(\beta_{14}), \mathrm{Typ}(\tilde{\beta}_2) = \mathrm{Typ}(\beta_1\beta_2) = \mathrm{Typ}(\tilde{\beta}_3)$ *für* $V = \langle(123)\rangle$,

(c) $\mathrm{Typ}(\tilde{\beta}_1) = \mathrm{Typ}(\beta_1), \mathrm{Typ}(\tilde{\beta}_2) = \mathrm{Typ}(\beta_1\beta_{14}), \mathrm{Typ}(\tilde{\beta}_3) = \mathrm{Typ}(\beta_1\beta_2)$ *für* $V = S_3$.

Entsprechende Formeln für $s = 5$ sind bei Przywara [1991] zu finden.

(7.4) Mit diesen Sätzen läßt sich nachprüfen, ob $Q(\tilde{K}^i_{\underline{\sigma}})/\tilde{k}^i_{\underline{\sigma}}$ ein rationaler Funktionenkörper ist. Als Resultat ergibt sich der

**Satz 7.5** (Matzat [1989,1991a]): *Es seien G,* $\mathbf{C}$, $[\underline{\sigma}]$ *wie im Satz 7.3 und* $\tilde{Z}(\underline{\sigma})$ *eine starre* $\tilde{H}_V$*-Bahn. Ferner gelte:*

(1) *Es gibt eine Untergruppe V von* $S_{s-1}$ *und eine V-Konfiguration* $\underline{a}$ *mit* $\tilde{g}_V(\underline{\sigma}) = 0$.

(2) *Bei der Operation von* $\tilde{\beta}_1, ..., \tilde{\beta}_{\tilde{s}}$ *auf* $\tilde{Z}(\underline{\sigma})$ *tritt eine der Zyklenlängen der* $\tilde{\beta}_i$, *aufsummiert über alle* $\tilde{\beta}_i$ *vom selben Permutationstyp, in ungerader Anzahl auf.*

*Dann ist* $Q(\tilde{K}^i_{\underline{\sigma}})/\tilde{k}_{\mathbf{C}}$ *ein rationaler Funktionenkörper, etwa* $Q(\tilde{K}^i_{\underline{\sigma}}) = \tilde{k}_{\mathbf{C}}(\tilde{x})$.

Insbesondere besitzt dann $\tilde{K}^i_{\underline{\sigma}}/\tilde{k}_{\mathbf{C}}$ sogar unendlich viele rationale Stellen außerhalb des Verzweigungsortes. Unter Verwendung der (eindeutig bestimmten) Fortsetzungen $\tilde{Q}$ von $Q$ auf $\tilde{N}^i_{\underline{\sigma}}$ mit $\tilde{Q}(t) = \tilde{t}$ gilt schärfer die

**Folgerung 7.6** (Matzat [1989,1991a]): *Unter den Voraussetzungen zum Satz 7.5 ist* $\tilde{Q}(\tilde{N}^i_{\underline{\sigma}})/\tilde{k}_{\mathbf{C}}(\tilde{x}, \tilde{t})$ *eine reguläre Galoiserweiterung über dem rationalen Funktionenkörper* $\tilde{k}_{\mathbf{C}}(\tilde{x}, \tilde{t})$ *mit*

$$\mathrm{Gal}(\tilde{Q}(\tilde{N}^i_{\underline{\sigma}})/\tilde{k}_{\mathbf{C}}(\tilde{x}, \tilde{t})) \cong G.$$

*Hierin ist* $\tilde{k}_{\mathbf{C}} = \mathbb{Q}$, *falls* $\mathbf{C}$ *ein V-symmetrischer Klassenvektor ist.*

(7.5) Das Paradebeispiel für die Anwendung der Folgerung 7.6 ist noch immer die Mathieugruppe $M_{24}$ : Für den Klassenvektor $\mathbf{C} = (2A,2A,2A,12B)$ gelten $l^i(\mathbf{C}) = 144$ und $h^i(\mathbf{C}) = 1$. Aus dem Zusatz 7.4 (a) ergibt sich für $V = \langle(12)\rangle$

$$\mathrm{Typ}(\tilde{\beta}_1) = (5^3, 3^{39}, 2^6),\ \mathrm{Typ}(\tilde{\beta}_2) = (6^{19}, 5^3, 4^3, 3),\ \mathrm{Typ}(\tilde{\beta}_3) = (2^{72}).$$

Damit sind beide Bedingungen (1) und (2) des Satzes 7.5 erfüllt. Es gibt also zumindest eine zweiparametrige Schar von $M_{24}$-Körpern über $\mathbb{Q}$ (siehe Matzat [1989,1991a]).

## 8. Einfache Gruppen als Galoisgruppen über $\mathbb{Q}(t)$ und $\mathbb{Q}^{ab}(t)$

(8.1) Mit den Rationalitätskriterien in den Abschnitten 4 bis 7 konnten neben den zyklischen Gruppen (siehe Abschnitt (6.5)) bisher folgende einfache endliche Gruppen als Galoisgruppen über $k_{\mathbb{C}}$ und damit über dem vollen Kreisteilungskörper $\mathbb{Q}^{ab}$ nachgewiesen werden:

**Satz 8.1** *Die folgenden einfachen Gruppen lassen sich als Galoisgruppen regulärer Körpererweiterungen $N/\mathbb{Q}^{ab}(t)$ darstellen:*

(a) *Die einfachen alternierenden Gruppen $A_n$* (Hilbert [1892], Shih [1974]).

(b) *Die einfachen klassischen Gruppen vom Lie-Typ:*

$A_n(q), {}^2A_n(q), B_n(q), C_n(q), D_n(q), {}^2D_n(q)$ (Belyi [1979,1983], Walter [1984]).

(c) *Die folgenden exzeptionellen Gruppen vom Lie-Typ:*

$G_2(q), {}^2G_2(q), F_4(q), {}^3D_4(q)$,
$E_6(q)$ *für* $p \geq 3$, $E_7(q)$ *für* $p \geq 5$, $E_8(q)$ *für* $p \geq 7$ *sowie*
${}^2E_6(q)$ *für* $q \not\equiv -1 \pmod 3$ (Malle [1988b,1991a]).

(d) *Die 26 sporadischen einfachen Gruppen* (Hoyden-Siedersleben, Hunt, Matzat, Pahlings, Thompson [1984-1988]).

*Hierbei bedeutet wie üblich $q = p^f$ die Potenz einer Primzahl $p \in \mathbb{P}$.*

Damit kommen zumindest alle einfachen Gruppen bis auf ${}^2B_2(2^{2n+1})$, ${}^2F_4(2^{2n+1})$, $E_6(2^n)$, ${}^2E_6(q)$ für $q \equiv -1 \pmod 3$ sowie $E_7(q)$ und $E_8(q)$ jeweils in schlechter Charakteristik als Galoisgruppen über $\mathbb{Q}^{ab}(t)$ und $\mathbb{Q}^{ab}$ vor. (Die kleinste einfache Gruppe der Suzukikette $Sz(8) = {}^2B_2(8)$ und die Titsgruppe $T = {}^2F_4(2)'$ sind noch bei Malle [1988b] als Galoisgruppen über $\mathbb{Q}^{ab}(t)$ nachgewiesen worden.)

(8.2) Über dem Körper der rationalen Zahlen $\mathbb{Q}$ ist das Bild bei weitem nicht so vollständig. Hier hat sich bisher die nachstehende Liste von Resultaten ergeben:

**Satz 8.2** *Die folgenden einfachen Gruppen lassen sich als Galoisgruppen regulärer Körpererweiterungen $N/\mathbb{Q}(t)$ darstellen:*

(a) *Die einfachen alternierenden Gruppen* (Hilbert [1892], Shih [1974]).

(b) *Die folgenden klassischen Gruppen vom Lie-Typ:*

$PSL_2(p)$ *für* $p \not\equiv \pm 1 \pmod{24}$ (Shih [1974], Matzat [1984]),
*und für* $(\frac{7}{p}) = -1$ (Shih [1974]),

| | | |
|---|---|---|
| $PSL_2(p^2)$ | *für* $p \equiv \pm 2 \pmod 5$ | (Feit [1984], Mestre [1988]), |
| $PSL_3(p)$ | *für* $p \equiv 1 \pmod 4$ | (Thompson [1984b]), |
| $PSU_3(p)$ | *für* $p \equiv -1 \pmod 4, p \geq 5$ | (Malle [1990]), |
| *und* | *für* $p \equiv 3, 5 \pmod 7, p \geq 7$ | (Malle [1990]), |
| $PSp_4(p)$ | *für* $p \equiv \pm 2 \pmod 5, p \geq 3$ | (Dentzer [1989]), |
| $PSp_{l-1}(2)$ | *für* $l \in \mathbb{P}$ *mit P.w.* 2, $l \geq 7$ | (Häfner [1991]), |
| $PSO_7(p)$ | *für* $p \equiv 3, 5 \pmod{14}, p \in \mathbb{P}_1(g)$ | (Häfner [1991]), |
| $PSO^+_{l+1}(2)$ | *für* $l \in \mathbb{P}$ *mit P.w.* 2, $l \equiv -1 \pmod 4, l \geq 11$ | (Häfner [1991]), |
| $PSO^-_{l-1}(2)$ | *für* $l \in \mathbb{P}$ *mit P.w.* 2, $l \geq 11$ | (Thompson [1984d]). |

(c) *Die folgenden exzeptionellen Gruppen vom Lie-Typ:*

| | | |
|---|---|---|
| $G_2(p)$ | *für* $p \geq 5$ | (Feit-Fong [1984], Thompson [1984c]), |
| $F_4(p)$ | *für* $p \equiv \pm 2, \pm 6 \pmod{13}, p \geq 19$ | (Malle [1988b]), |
| $E_6(p)$ | *für* $p \equiv 4, 5, 6, 9, 16, 17 \pmod{19}$ | (Malle [1991a]), |
| $E_8(p)$ | *für* $p \equiv \pm 3, \pm 7, \pm 9, \pm 10, \pm 11, \pm 12, \pm 13, \pm 14 \pmod{31}, p \geq 131$ | (Malle [1988b]). |

(d) *Die sporadischen einfachen Gruppen mit höchstens der Ausnahme* $M_{23}$ (Hoyden-Siedersleben, Hunt, Matzat, Pahlings, Thompson [1984-1989]).

Eine Beweislücke bei Feit [1984] wurde inzwischen von Serre geschlossen (siehe Matzat [1990]). Das Symbol *P.w.* wird als Abkürzung für Primitivwurzel verwendet. Ferner bedeutet $\mathbb{P}_1(g)$ die Menge der Primzahlen, für die das in $\mathbb{Q}[X]$ irreduzible Polynom

$$g(X) = X^8 + 14X^7 + 70X^6 + 124X^5 - 280X^4 - 1736X^3 \\ -12424X^2 - 74704X + 348880$$

modulo $p$ einen Linearfaktor abspaltet.

Daneben sind noch reguläre Galoiserweiterungen über $\mathbb{Q}(t)$ bekannt für die einfachen Gruppen $PSL_2(8)$ und $PSL_3(3)$ (Matzat [1984]), $PSL_2(25)$ und $PSL_3(4)$ (Przywara [1991], Matzat [1991a]) sowie $PSp_6(2)$, $PSp_8(2)$, $PSO_7(3)$, $PSO_6^-(2)$ und $PSO_8^-(2)$ (Häfner [1991]).

(8.3) Mit den aus den Existenzsätzen gewonnenen Daten lassen sich wenigstens grundsätzlich auch Polynome berechnen, deren Nullstellen den in Frage stehenden Körper erzeugen. Hierbei besteht die Hauptschwierigkeit darin, algebraische Gleichungssysteme exakt zu lösen. Dieses Problem wird durch den Buchbergerschen Algorithmus auf die Lösung jeweils einer Polynomgleichung zurückgeführt (siehe Buchberger [1970], Trinks [1978]). Bei umfangreicheren Gleichungssystemen hat sich die von Trinks [1984] vorgeschlagene modulare Version des Buchbergerschen Algorithmus bewährt (siehe auch Winkler [1988]). Hierfür ist ein Resultat von Beckmann [1989] hilfreich, nach dem schlechte Reduktion höchstens bei Primteilern der Verzweigungspunktdifferenzen und der Gruppenordnung auftritt. Bisher konnten damit die folgenden Resultate gewonnen werden:

**Satz 8.3** *Für die folgenden Permutationsgruppen $G$ vom Grad $n$ sind Polynome $f(t,X) \in \mathbb{Q}(t)[X]$ mit einer zu $G$ isomorphen Galoisgruppe bekannt:*

(a) *Die symmetrischen und alternierenden Gruppen $S_n$ und $A_n$* (Hilbert [1892], Matzat [1984]).

(b) *Alle primitiven, nicht auflösbaren Gruppen für $n \leq 15$* (Malle, Matzat, Shih, Zeh-Marschke [1978-1987]).

(c) *Die folgenden primitiven, nicht auflösbaren Gruppen für $16 \leq n \leq 30$:*

| | |
|---|---|
| $PSL_2(p)$ *und* $PGL_2(p)$ *für* $p = 17, 19, 23, 29$ *und* $n = p + 1$ | (Malle [1991b]), |
| $P\Sigma L_3(4)$ *für* $n = 21$ | (Malle [1988a]), |
| $M_{22}$ *und* $\mathrm{Aut}(M_{22})$ *für* $n = 22$ | (Malle [1988a]), |
| $PSU_4(2)$ *und* $\mathrm{Aut}(PSU_4(2))$ *für* $n = 27$ | (Häfner [1991]), |
| $PSp_6(2)$ *für* $n = 28$ | (Häfner [1991]). |

(8.4) Als Kostprobe und damit stellvertretend für die anderen Resultate wird hier nur das erst kürzlich konstruierte Polynom von Grad 28 mit der Galoisgruppe $PSp_6(2)$ über $\mathbb{Q}(t)$ angegeben:

$$f(t,X) = (X^4 - 10X^2 - 8X + 1)^7 - X^3(X^2 + 3X + 1)^5 t.$$

Entsprechende Polynome für einige auflösbare Gruppen kleinen Permutationsgrades sind noch bei Seidelmann [1918] für $n \leq 4$, Matzat [1987] für $n = 5$, Malle [1987] für $n = 6$ sowie bei Jensen-Yui [1982], Bruen-Jensen-Yui [1986] für eine Reihe von Frobeniusgruppen und bei Jensen-Yui [1987], Schneps [1989] für gewisse Gruppen der Ordnung 8 und 16 zu finden.

## 9. Zerfallende Einbettungsprobleme mit nilpotentem Kern

(9.1) Um zusammengesetzte Gruppen als Galoisgruppen über einem Körper $K$ darzustellen, ist es häufig zweckmäßig, die Galoiserweiterung induktiv über eine Normalreihe aufzubauen. Dies führt zu einer Kette von Einbettungsproblemen. Hierunter versteht man die Frage, ob sich eine vorhandene Galoiserweiterung $N/K$ mit der Gruppe $G$ für eine gegebene Gruppenerweiterung $\tilde{G} = H \cdot G$ in eine Galoiserweiterung $\tilde{N}/K$ mit $\mathrm{Gal}(\tilde{N}/K) \cong \tilde{G}$ einbetten läßt. Dies läßt sich wie folgt präzisieren: Bedeuten $\gamma : \Gamma_K \to G$ einen durch $N$ gegebenen Epimorphismus der absoluten Galoisgruppe von $K$ auf $G$ und

$$I \longrightarrow H \xrightarrow[\varepsilon]{} \tilde{G} \xrightarrow[\kappa]{} G \longrightarrow I$$

eine durch die Gruppenerweiterung definierte exakte Sequenz, so wird ein Epimorphismus $\tilde{\gamma} : \Gamma_K \to \tilde{G}$ gesucht mit $\kappa \circ \tilde{\gamma} = \gamma$. Dieser heißt dann eine *Lösung des Einbettungsproblems*, und der Fixkörper $\tilde{N}$ des Kerns von $\tilde{\gamma}$ heißt ein *Lösungskörper*.

(9.2) Man nennt ein solches Einbettungsproblem ein *zerfallendes Einbettungsproblem*, falls $\tilde{G} = H \cdot G$ eine zerfallende Gruppenerweiterung, das heißt ein semidirektes Produkt, ist beziehungsweise ein *Frattini-Einbettungsproblem*, wenn $H$ in der Frattinigruppe $\Phi(\tilde{G})$ liegt. Eine grobe Aufteilung bewirkt die

**Bemerkung 9.1** (Nobusawa [1961], Sonn [1972]): *Jedes endliche Einbettungsproblem über einem Körper $K$ läßt sich zerlegen in ein Frattini-Einbettungsproblem, gefolgt von einem zerfallenden Einbettungsproblem, jeweils über $K$.*

Dies ergibt sich aus der folgenden einfachen Überlegung: Für eine minimale Untergruppe $U$ von $\tilde{G}$ mit $\langle H, U\rangle = \tilde{G}$ gilt $H \cap U \leq \Phi(U)$, und man erhält ein Frattini-Einbettungsproblem für $U$. Weiter wird durch die Operation von $U$ auf $H$ innerhalb von $\tilde{G}$ ein semidirektes Produkt $E = H \rtimes U$ definiert, das $\tilde{G}$ als Faktorgruppe besitzt. Eine Lösung der Einbettungsprobleme für $U$ und $E$ zieht also eine Lösung des Einbettungsproblems für $\tilde{G}$ nach sich.

(9.3) Der Abkürzung halber wird für den Rest des Artikels die folgende Sprachregelung getroffen: Eine endliche Gruppe besitzt eine G(alois)-*Realisierung über* $K(\underline{t}) = K(t_1, ..., t_r)$, falls es eine reguläre Körpererweiterung $N/K$ mit $\mathrm{Gal}(N/K(\underline{t})) \cong G$ gibt. Insbesondere besitzen damit die in den Sätzen 8.1 und 8.2 aufgezählten Gruppen G-Realisierungen über $\mathbb{Q}^{ab}(t)$ beziehungsweise über $\mathbb{Q}(t)$. Weiter heißt im Falle eines rationalen Funktionenkörpers $K = k(t_0)$ ein Einbettungsproblem über $K$ *reguläres Einbettungsproblem*, wenn $N/k$ regulär ist; eine Lösung eines solchen Einbettungsproblems wird regulär genannt, wenn auch der Lösungskörper $\tilde{N}$ regulär über $k$ ist. Damit gilt der

**Satz 9.2** (Kuyk [1970], Matzat [1987]): *Es seien $K(= k(t_0))$ ein Hilbertkörper, $N/K$ eine endliche (reguläre) Galoiserweiterung mit der Galoisgruppe $G \leq S_r$, $H$ eine endliche Gruppe mit einer G-Realisierung über $K(t)$ sowie $\tilde{G} = H \wr G$ das durch die gegebene Permutationsdarstellung definierte Kranzprodukt. Dann läßt sich $N$ in einen (über $k$ regulären) Körper $\tilde{N}$ mit* $\mathrm{Gal}(\tilde{N}/K) \cong \tilde{G}$ *einbetten.*

Insbesondere besitzen damit reguläre Einbettungsprobleme für Kranzprodukte auch reguläre Lösungen. Eine andere Variante dieses Satzes ist bei Saltman [1982] zu finden.

(9.4) Da sich semidirekte Produkte $H \rtimes G$ mit abelschem Normalteiler als Faktorgruppen regulärer Kranzprodukte $H \wr G$ darstellen lassen, folgt aus dem Satz 9.2 unter Verwendung des Resultats in (6.5) unmittelbar der

**Satz 9.3** (Scholz [1929], Saltman [1982]): *Jedes (reguläre) endliche zerfallende Einbettungsproblem mit abelschem Kern über einem Hilbertkörper $K$ der Charakteristik* 0 *besitzt eine (reguläre) Lösung.*

Ausgehend von abelschen Gruppen lassen sich mit diesem Satz alle Gruppen als Galoisgruppen regulärer Galoiserweiterungen über zum Beispiel $\mathbb{Q}(t)$ darstellen, die Faktorgruppen eines semidirekten Produkts mit abelschem Kern und einem bereits realisierten Kokern sind. Diese Gruppen werden hier kurz *semiabelsche Gruppen* genannt, sie lassen sich charakterisieren durch den

**Zusatz 9.4** (Dentzer [1991]): *Eine endliche Gruppe ist genau dann semiabelsch, wenn sie einen abelschen Normalteiler mit einem echten semiabelschen Supplement besitzt.*

Zu diesen semiabelschen Gruppen zählen nach Dentzer [1991] unter anderem alle $p$-Gruppen, deren Ordnung ein Teiler von $2p^4$ ist. Weiter folgt hieraus auch die

**Folgerung 9.5** (Thompson [1986]): *Jede endliche nilpotente Gruppe der Klasse* 2 *ist als Galoisgruppe einer regulären Körpererweiterung über* $\mathbb{Q}(t)$ *darstellbar.*

Die Frage, inwieweit andere auflösbare Gruppen als Galoisgruppen regulärer Körpererweiterungen über $\mathbb{Q}(t)$ vorkommen, ist bisher ungelöst.

(9.5) Bei Einbettungsproblemen über Zahlkörpern ist hier der Kenntnisstand sehr viel besser: Nach einer Ankündigung von Shafarevich [1958] (siehe auch Išhanov [1976]) ist nämlich über einem Körper von endlichem Grad über $\mathbb{Q}$ jedes endliche zerfallende Einbettungsproblem mit nilpotentem Kern lösbar. Bekanntlich ist in einer auflösbaren Gruppe $G$ die Frattinigruppe $\Phi(G)$ eine echte Untergruppe der Fittinggruppe $F$, die somit ein Supplement $U$ in $G$ besitzt (siehe zum Beispiel Huppert [1967], Kap. III, Satz 4.2). Demgemäß ist $G$ darstellbar als Faktorgruppe von $F \rtimes U$ mit $|U| \leq |G|$. Durch Induktion nach $|G|$ erhält man damit den eingangs erwähnten Satz von Shafarevich [1954c] in der folgenden Formulierung: Über einem gegebenen Zahlkörper $k$ kommt jede endliche auflösbare Gruppe als Galoisgruppe vor.

## 10. Frattini-Einbettungsprobleme

(10.1) Über reguläre Frattini-Einbettungsprobleme — soweit man diese nicht wie bei den auflösbaren Gruppen umgehen kann — weiß man bisher noch sehr wenig. Ein speziell hierfür geeigneter Lösungsansatz besteht darin, die Kenntnis der durch $N = \bar{N}^i_{\underline{\sigma}}$ gegebenen invarianten Erzeugendensystemklasse $[\underline{\sigma}]$ von $G$ zu nutzen und nach invarianten Urbildern $[\tilde{\underline{\sigma}}]$ in $\tilde{G}$ zu suchen. Um dies zu präzisieren, sei $\mathbf{C} = (C_1, ..., C_s)$ der Klassenvektor von $G$ mit $[\underline{\sigma}] \in \Sigma^i(\mathbf{C})$, wobei noch einige der Komponenten von $\underline{\sigma}$ trivial sein dürfen. Eine solche Erzeugendensystemklasse $[\underline{\sigma}] \in \Sigma^i(\mathbf{C})$ heißt *eindeutig liftbar* zu einer Erzeugendensystemklasse $[\tilde{\underline{\sigma}}]$ von $\tilde{G} = H \cdot G$, wenn es einen Klassenvektor $\tilde{\mathbf{C}} = (\tilde{C}_1, ..., \tilde{C}_s)$ von $\tilde{G}$ mit $\tilde{\mathbf{C}}^\kappa = \mathbf{C}$ gibt, so daß $[\underline{\sigma}]$ unter $\kappa$ genau ein Urbild $[\tilde{\underline{\sigma}}] \in \Sigma^i(\tilde{\mathbf{C}})$ besitzt.

Bei Frattini-Erweiterungen läßt sich die eindeutige Liftbarkeit allein aus der Klassenzerlegung ablesen, es gilt nämlich die

**Bemerkung 10.1** (Matzat [1991b]): *Es seien* $\tilde{G} = H{\cdot}G$ *eine Frattini-Erweiterung,* $\mathbf{C} = (C_1, ..., C_s)$ *ein Klassenvektor von* $G$*, und es gebe für* $j = 1, ..., s-1$ *Konjugierten-*

*klassen $\tilde{C}_j$ von $\tilde{G}$ mit $\tilde{C}_j^\kappa = C_j$ und*

$$\prod_{j=1}^{s-1} \frac{|\tilde{C}_j|}{|C_j|} = \frac{|\mathrm{Inn}(\tilde{G})|}{|\mathrm{Inn}(G)|}.$$

*Dann gibt es zu jeder Erzeugendensystemklasse $[\underline{\sigma}] \in \Sigma^i(\mathbf{C})$ von $G$ genau eine Konjugiertenklasse $\tilde{C}_s$ von $\tilde{G}$ mit $\tilde{C}_s^\kappa = C_s$, derart daß $\Sigma^i(\tilde{\mathbf{C}})$ ein und damit genau ein Urbild $[\underline{\tilde{\sigma}}]$ von $[\underline{\sigma}]$ enthält.*

Dadurch ist die Frage nach der eindeutigen Liftbarkeit auf die Bestimmung der Quotienten $f_j := |\tilde{C}_j|/|C_j|$ zurückgeführt. Dabei gilt zum Beispiel $f_j = 1$, wenn $H$ eine Untergruppe des Zentralisators $C_{\tilde{G}}(\tilde{\sigma}_j)$ ist und diese eine zu $o(\tilde{\sigma}_j)$ teilerfremde Ordnung hat, beziehungsweise $f_j = |H|$, falls $H$ einen trivialen Durchschnitt mit $C_{\tilde{G}}(\tilde{\sigma}_j)$ besitzt.

(10.2) Im folgenden heißt $\mathbf{C}$ ein *K-symmetrischer Klassenvektor* (bezüglich eines Zwischenkörpers $K$ von $\bar{M}_s/\mathbb{Q}(\underline{t})$), wenn $\mathbf{C}$ ein Fixpunkt unter der im Zusatz 4.2 gegebenen Operation von $\tilde{\Delta}_K := \mathrm{Gal}(\bar{M}_s(t)/K(t))$ ist, beziehungsweise ein *K-rationaler Klassenvektor*, falls $\mathbf{C}$ invariant unter $\tilde{\Delta}_K \cap \Delta$ ist. (Bei dieser Definition sind genau die $\mathbb{Q}(\underline{t})$-rationalen Klassenvektoren rational.)

**Satz 10.2** (Matzat [1991b]): *Es sei $N/K(t)$ eine reguläre Galoiserweiterung mit der Galoisgruppe $G$ und $\bar{M}_sN = \bar{N}_{\underline{\sigma}}$. Weiter gelte für eine endliche Gruppenerweiterung:*

(1) *Die Erzeugendensystemklasse $[\underline{\sigma}]$ von $G$ ist eindeutig liftbar zu einer Erzeugendensystemklasse $[\underline{\tilde{\sigma}}]$ von $\tilde{G}$ in einem K-symmetrischen Klassenvektor $\tilde{\mathbf{C}}$.*

(2) *Es existieren ein Primdivisor $\mathfrak{P}$ von $K(t)/K$ vom Grad 1 und eine Fortsetzung $\mathfrak{P}'$ von $\mathfrak{P}$ auf $N' := N^{\mathcal{Z}(G)}$ derart, daß das Zentrum $\mathcal{Z}(\tilde{G})$ ein Komplement im Urbild der Zerlegungsgruppe von $\mathfrak{P}'/\mathfrak{P}$ in $\tilde{G}$ besitzt.*

*Dann läßt sich $N'/K(t)$ einbetten in eine reguläre Galoiserweiterung $\tilde{N}/K(t)$ mit $\mathrm{Gal}(\tilde{N}/K(t)) \cong \tilde{G}$.*

Ist in diesem Satz nur die Bedingung (1) erfüllt, so läßt sich $N'/K(t)$ immerhin noch in eine reguläre Galoiserweiterung $\tilde{N}'/K(t)$ mit $\mathrm{Gal}(\tilde{N}'/K(t)) \cong \tilde{G}/\mathcal{Z}(\tilde{G})$ einbetten. Insbesondere ist im Falle $\mathcal{Z}(\tilde{G}) = I$ die Voraussetzung (2) überflüssig.

(10.3) Im Spezialfall einer Frattini-Erweiterung mit trivialem Zentrum erhält man aus dem Satz 10.2 die folgende Verallgemeinerung eines Resultats von Feit:

**Zusatz 10.3** (Feit [1989]): *Im Satz 10.2 sei $\tilde{G} = H \cdot G$ eine Frattini-Erweiterung mit $\mathcal{Z}(\tilde{G}) \cong \mathcal{Z}(G) \cong I$. Ferner mögen die Konjugiertenklassen $C_j = [\sigma_j]$ von $G$ jeweils K-rationale Urbildklassen $\tilde{C}_1, \ldots, \tilde{C}_s$ besitzen mit $\mathrm{Sym}(\tilde{\mathbf{C}}) = \mathrm{Sym}(\mathbf{C})$ und $f_j = 1$ für $j = 1, \ldots, s-2$ sowie $f_{s-1} = |H|$. Dann sind die Bedingungen (1) und (2) des Satzes erfüllt.*

Hierin bedeutet Sym($\tilde{\mathbf{C}}$) bzw. Sym($\mathbf{C}$) die größtmögliche Gruppe von Symmetrien von $\tilde{\mathbf{C}}$ bzw. von $\mathbf{C}$. Mit diesem Zusatz konnte Feit [1989] unter Verwendung der Resultate im Satz 8.2, (a) und (d), nachweisen, daß für $S \in \{A_6, A_7, M_{22}, Mc, Sz, ON, Fi_{22}, Fi'_{24}\}$ die Überlagerungsgruppe $\tilde{S} = Z_3 \cdot S$ als Galoisgruppe einer regulären Galoiserweiterung $N/\mathbb{Q}(t)$ vorkommt. Dasselbe stimmt auch für die Gruppen $S = PSL_3(p)$ mit $p \equiv 1 \pmod{12}$ nach Thompson [1984b] und $S = PSU_3(p)$ mit $p \equiv -1 \pmod{12}$ nach Malle [1990]. Ausgehend von der $PSL_3(4)$-Erweiterung im Abschnitt (5.5) erhält man aus dem Zusatz 10.3 weiter, daß auch die zentrale, nicht zerfallende Gruppenerweiterung $E_4{\cdot}PSL_3(4)$ Galoisgruppe einer regulären Körpererweiterung $N/\mathbb{Q}(t)$ ist (siehe Matzat [1991b]).

(10.4) Für zentrale Frattini-Erweiterungen bekommt man unter anderem die folgende Version eines Satzes von Völklein:

**Zusatz 10.4** (Völklein [1991]): *Im Satz 10.2 sei $\tilde{G} = H \cdot G$ eine zentrale Frattini-Erweiterung mit $\mathcal{Z}(G) = I$. Ferner gelte:*

(1') *Für $j = 1, ..., s$ ist der größte gemeinsame Teiler von $o(\sigma_j)$ und $|H|$ jeweils* 1.

(2') *Der Körper $N/K$ besitzt einen Primdivisor vom Grad* 1.

*Dann sind die Bedingungen* (1) *und* (2) *des Satzes erfüllt.*

Notwendige Voraussetzungen für die Erfüllbarkeit von (2') sind von Beckmann [1991] angegeben worden.

Dieser Zusatz läßt sich zum Beispiel dazu verwenden, zu zeigen, daß sich die von Mestre [1990] konstruierten regulären $A_n$-Erweiterungen $N/\mathbb{Q}(t)$ in reguläre $\tilde{A}_n$-Erweiterungen $\tilde{N}/\mathbb{Q}(t)$ einbetten lassen (siehe Völklein [1991]). ($\tilde{A}_n$ bedeutet die zentrale, nicht zerfallende Erweiterung $Z_2{\cdot}A_n$.) Im ursprünglichen Beweis von Mestre wurde hierzu der Einbettungssatz von Serre [1984] für Einbettungsprobleme mit dem Kern $Z_2$ benutzt.

(10.5) Bei diesem Satz geht man von einem Stammkörper $L$ der Galoiserweiterung $N/K$ aus mit etwa dem Grad $n = [L : K]$ und der Diskriminante $D_{L/K}$. Durch $L$ wird eine nicht ausgeartete quadratische Form definiert, nämlich

$$S_{L/K} : L \longrightarrow K, \quad x \mapsto \mathrm{Spur}_{L/K}(x^2),$$

deren *Hasse-Witt-Invariante* hier mit $s_{L/K}$ bezeichnet wird (vergleiche Serre [1984,1988], Fröhlich [1985]).

Nun sei $f \in K[X]$ das Minimalpolynom eines primitiven Elements von $L/K$. Dann ist $G = \mathrm{Gal}(f)$ eine Untergruppe von $S_n$. Bekanntlich hat der Schurmultiplikator von $S_n$ für $n \geq 4$ die Ordnung 2, und $S_n$ besitzt 2 Überlagerungsgruppen $\tilde{S}_n^+$ und $\tilde{S}_n^-$ (siehe Schur [1911]):

$$I \longrightarrow Z_2 \longrightarrow \tilde{S}_n^{\pm} \longrightarrow S_n \longrightarrow I.$$

Im folgenden bedeute $\tilde{G}^{\pm}$ das Urbild von $G \leq S_n$ in $\tilde{S}_n^{\pm}$. Dann gilt für das hierdurch gegebene Einbettungsproblem der

**Satz 10.5** (Serre [1984], Fröhlich [1985]): *Es sei $K$ ein Hilbertkörper mit einer von 2 verschiedenen Charakteristik, $L = K(x)$ eine separable Körpererweiterung vom Grad $n \geq 4$ und $G = \mathrm{Gal}(f)$ die Galoisgruppe des Minimalpolynoms von $x$ über $K$. Ferner bedeute $N$ die galoissche Hülle von $L/K$. Dann sind für nicht zerfallende Gruppenerweiterungen $\tilde{G}^{\pm} = Z_2 \cdot G$ die folgenden beiden Aussagen äquivalent:*

(1) *$N/K$ läßt sich einbetten in eine Galoiserweiterung $\tilde{N}^{\pm}/K$ mit* $\mathrm{Gal}(\tilde{N}^{\pm}/K) \cong \tilde{G}^{\pm}$.

(2) *In der Gruppe $Br_2(K)$ gilt $s_{L/K} = (\pm 2, D_{L/K})$.*

Hierin bedeutet $Br_2(K)$ die Gruppe der Involutionen in der Brauergruppe $Br(K)$. Für Untergruppen der $A_n$ vereinfacht sich das obige Kriterium zum

**Zusatz 10.6** (Serre [1984], Fröhlich [1985]): *Im Falle $G \leq A_n$ gilt $\tilde{G}^+ = \tilde{G}^-$, und (2) vereinfacht sich zu $s_{L/K} = 1$.*

(10.6) Mit dem Satz 10.5 beziehungsweise dem Zusatz 10.6 konnte — wie bereits erwähnt — Mestre [1990] die Gruppen $\tilde{A}_n$ für alle $n \geq 4$ sowie Vila [1988] und Sonn [1989a] die Gruppen $\tilde{S}_n^+$ für $n \not\equiv 6, 7 \pmod 8$ und $\tilde{S}_n^-$ für $n \not\equiv 4, 5 \pmod 8$ und jeweils $n \geq 5$ als Galoisgruppen regulärer Körpererweiterungen über $\mathbb{Q}(t)$ nachweisen. Explizite Erzeugende für diese Galoiserweiterungen sind von Crespo [1990] angegeben worden. Ferner wurden damit auch die Gruppen $SL_2(7)$ (ausgehend vom Polynom von LaMacchia [1980]) und $\tilde{M}_{12}$ (mit dem Polynom bei Matzat [1987]) von Mestre (persönliche Mitteilung) sowie die Überlagerungsgruppe $Z_2{\cdot}Sp_6(2)$ von Häfner [1991] als Gruppe einer regulären Galoiserweiterung über $\mathbb{Q}(t)$ erkannt.

## 11. Einbettungsprobleme mit zentrumsfreiem Kern

(11.1) Nachdem in den letzten beiden Abschnitten Einbettungsprobleme mit nilpotentem Kern behandelt wurden, stehen hier solche mit nicht auflösbarem Kern im Vordergrund. Aktuell benötigt wird allerdings nur, daß das Zentrum des Kerns trivial ist.

Eine G-Realisierung $N/K(\underline{t})$ von $G$ (vergl. Abschnitt (9.3)) wird eine GA-*Realisierung* genannt, wenn die folgende Automorphismenbedingung erfüllt ist: $\mathrm{Aut}(N/K)$ besitzt eine zu $\mathrm{Aut}(G)$ isomorphe Untergruppe $A$, wobei $K(\underline{t})$ der Fixkörper von der zu $\mathrm{Inn}(G)$ isomorphen Untergruppe von $A$ ist. Hierdurch werden offenbar Gruppen mit nicht trivialem Zentrum von vornherein ausgeschlossen.

Eine GA-Realisierung heißt schließlich eine GAR-*Realisierung*, wenn zusätzlich die folgende Rationalitätsbedingung erfüllt ist: Jede über $K$ reguläre Körpererweiterung $R/N^A$ mit $\bar{K}R = \bar{K}(\underline{t})$ ist ein rationaler Funktionenkörper über $K$. (Dabei bedeutet $\bar{K}$ die algebraisch abgeschlossene Hülle von $K$ in einer algebraisch abgeschlossenen Hülle von $K(\underline{t})$). Im Falle $K(\underline{t}) = K(t)$ ist diese Bedingung beispielsweise erfüllt, wenn $\mathrm{Out}(G) = I$ ist oder wenn $s \equiv 1 \pmod 2$ ist (in der bisherigen Bedeutung) oder auch wenn die kohomologische Dimension von $K$ höchstens 1 ist. Letzeres ist zum

Beispiel für alle Zwischenkörper von $\bar{\mathbb{Q}}/\mathbb{Q}^{ab}$ erfüllt (siehe Serre [1964], Ch. II, Prop. 9).

(11.2) Das Hauptresultat für Gruppen mit GAR-Realisierungen über $K(\underline{t})$ ist der folgende Einbettungssatz:

**Satz 11.1** (Matzat [1985b,1987]): *Es seien $K(= k(t_0))$ ein Hilbertkörper der Charakteristik 0 und $H$ eine endliche Gruppe, deren Kompositionsfaktoren* GAR*-Realisierungen über $K(\underline{t})$ besitzen. Dann besitzt jedes endliche (reguläre) Einbettungsproblem mit dem Kern $H$ eine (reguläre) Lösung.*

Im Spezialfall $G = I$ wird hieraus die

**Folgerung 11.2** (Matzat [1985b,1987]): *Es seien $K$ ein Hilbertkörper der Charakteristik 0 und $\tilde{G}$ eine Gruppe, deren Kompositionsfaktoren* GAR*-Realisierungen über $K(\underline{t})$ besitzen. Dann existiert eine reguläre Galoiserweiterung $\tilde{N}/K(\underline{t})$ mit* $\mathrm{Gal}(\tilde{N}/K(\underline{t})) \cong \tilde{G}$.

(11.3) Nach diesen Resultaten stellt sich die Frage, für welche einfachen Gruppen GAR-Realisierungen existieren. Im Falle des vollen Kreisteilungskörpers $K = \mathbb{Q}^{ab}$ sind die bisherigen Ergebnisse zusammengestellt im

**Satz 11.3** *Die folgenden einfachen Gruppen besitzen* GAR*-Realisierungen über $\mathbb{Q}^{ab}(t)$:*

(a) *Die einfachen alternierenden Gruppen* (Matzat[1985b]).

(b) *Die folgenden klassischen Gruppen vom Lie-Typ:*

| | |
|---|---|
| $PSL_2(p)$ | (Belyi [1987], Matzat [1987]), |
| $PSL_2(p^2)$ *für* $p \equiv \pm 2 \pmod 5$ | (Feit [1984]*), |
| $PSL_3(p)$ *für* $p \equiv 5 \pmod{12}$ | (Thompson [1984b]*), |
| $PSU_3(p)$ *für* $p \equiv 7 \pmod{12}$ | (Malle [1990]), |
| *und für* $p \equiv 10, 19 \pmod{21}$ | (Malle [1990]), |
| $PSp_{2n}(p)$ | (Belyi [1987], Matzat [1987]), |
| $PSO_{2n+1}(p)$ | (Belyi [1987], Matzat [1987]), |
| $PSO^+_{l+1}(2)$ *für* $l \in \mathbb{P}$ *mit P.w.* 2, $l \equiv -1 \pmod 4$, $l \geq 11$ | (Häfner [1991]), |
| $PSO^-_{l-1}(2)$ *für* $l \in \mathbb{P}$ *mit P.w.* 2, $l \geq 11$ | (Thompson [1984d]*). |

(c) *Die folgenden exzeptionellen Gruppen vom Lie-Typ:*

| | |
|---|---|
| $G_2(p)$ *für* $p \geq 5$ | (Belyi [1987], Matzat [1987]), |
| $F_4(p)$ *für* $p \geq 3$ | (Malle [1988b]), |
| $E_6(p)$ *für* $p \not\equiv 1 \pmod 3$, $p \geq 3$ | (Malle [1991a]*), |
| $E_8(p)$ *für* $p \geq 7$ | (Malle [1988b]). |

(d) *Die sporadischen Gruppen* (Matzat [1985b], Pahlings [1988]).

Hierbei bedeutet der Stern hinter der Referenz, daß das Resultat in der Arbeit zwar bewiesen, aber nicht explizit formuliert wurde, und *P.w.* wird wieder als Abkürzung für Primitivwurzel benutzt.

Einige weitere Ergebnisse hinsichtlich der Gruppen $PSO_{2n}^{\pm}(p)$ sind noch von Belyi [1987] ohne Beweis angekündigt worden.

(11.4) Über dem Körper $\mathbb{Q}$ hat sich bisher das folgende Bild ergeben:

**Satz 11.4** *Die folgenden einfachen Gruppen besitzen* GAR-*Realisierungen über* $\mathbb{Q}(t)$:

(a) *Die einfachen alternierenden Gruppen* $A_n$ *mit* $n \neq 6$ (Matzat [1985b]).

(b) *Die folgenden klassischen Gruppen vom Lie-Typ:*

| | | | |
|---|---|---|---|
| $PSL_2(p)$ | *für* | $p \not\equiv \pm 1 \pmod{24}$ | (Malle-Matzat [1985]), |
| $PSL_3(p)$ | *für* | $p \equiv 5 \pmod{12}$ | (Thompson [1984b]*), |
| $PSU_3(p)$ | *für* | $p \equiv 7 \pmod{12}$ | (Malle [1990]), |
| *und* | *für* | $p \equiv 10, 19 \pmod{21}$ | (Malle [1990]), |
| $PSp_4(p)$ | *für* | $p \equiv 13, 17 \pmod{20}$ | (Häfner [1991]), |
| *und* | *für* | *12 Kongruenzklassen* (mod 280) | (Häfner [1991]), |
| $PSp_{l-1}(2)$ | *für* | $l \in \mathbb{P}$ *mit P.w.* $2$, $l \geq 7$ | (Häfner [1991]), |
| $PSO_7(p)$ | *für* | $p \equiv 3, 5 \pmod{14}$, $p \in \mathbb{P}_1(g)$ | (Häfner [1991]), |
| $PSO_{l+1}^{+}(2)$ | *für* | $l \in \mathbb{P}$ *mit P.w.* $2$, $l \equiv -1 \pmod 4$, $l \geq 11$ | (Häfner [1991]), |
| $PSO_{l-1}^{-}(2)$ | *für* | $l \in \mathbb{P}$ *mit P.w.* $2$, $l \geq 11$ | (Thompson [1984d]*). |

(c) *Die folgenden exzeptionellen Gruppen vom Lie-Typ:*

| | | | |
|---|---|---|---|
| $G_2(p)$ | *für* | $p \geq 5$ | (Matzat [1987]), |
| $F_4(p)$ | *für* | $p \equiv \pm 2, \pm 6 \pmod{13}, p \geq 19$ | (Malle [1988b]), |
| $E_6(p)$ | *für* | $p \equiv 5, 17, 23, 35, 44, 47 \pmod{57}$ | (Malle [1991a]*), |
| $E_8(p)$ | *für* | $p \equiv \pm 3, \pm 7, \pm 9, \pm 10, \pm 11, \pm 12, \pm 13, \pm 14 \pmod{31}, p \geq 131$ | (Malle [1988b]). |

(d) *Die sporadischen Gruppen mit höchstens der Ausnahme* $M_{23}$ (Matzat, Pahlings [1985-1989]).

Die Primzahlenmenge $\mathbb{P}_1(g)$ wurde in (8.2) definiert, und ein Stern hinter der Referenz hat wieder dieselbe Bedeutung wie im Satz 11.3.

Als Einzelresultate sind noch GAR-Realisierungen über $\mathbb{Q}(t)$ für $PSp_6(2)$, $PSp_8(2)$, $PSO_7(3)$ und $PSO_8^-(2)$ bekannt (siehe Häfner [1991]).

(11.5) Mit den bisherigen Sätzen kann man nun versuchen, alle Gruppen $G$ bis zu einer vorgegebenen Schranke für die Ordnung als Galoisgruppen über $\mathbb{Q}$ darzustellen. Dies wurde von Sonn [1989b] bis $|G| < 672$ durchgeführt. Dabei können die meisten der nicht auflösbaren unter diesen Gruppen sogar als Galoisgruppen regulärer Körpererweiterungen über $\mathbb{Q}(t)$ realisiert werden.

## 12. Hilbertkörper der kohomologischen Dimension 1

(12.1) Sehr viel einfacher gestalten sich die Einbettungsprobleme, wenn der zugrunde liegende Körper $K$ die kohomologische Dimension 1 besitzt. Dann nämlich ist die absolute Galoisgruppe $\Gamma$ von $K$ eine projektive Gruppe, das heißt, zu jedem durch $\gamma$ und $\kappa$ gegebenen Einbettungsproblem gibt es zumindest einen (nicht notwendig surjektiven) Homomorphismus $\tilde{\gamma} : \Gamma \to \tilde{G}$ mit $\kappa \circ \tilde{\gamma} = \gamma$; ein solcher heißt dann auch eine *schwache Lösung des Einbettungsproblems.* Wenn der Kern des Einbettungsproblems abelsch ist, folgt aus der Existenz einer schwachen Lösung die Existenz einer Lösung, die in diesem Zusammenhang auch als *starke Lösung des Einbettungsproblems* bezeichnet wird. Dies ergibt sich aus dem Satz 9.3 zusammen mit der

**Bemerkung 12.1** (Ikeda [1960], Nobusawa [1961]): *Besitzt ein endliches (reguläres) Einbettungsproblem mit dem Kern $H$ über einem Körper $K(= k(t_0))$ eine schwache (reguläre) Lösung und ist über $K$ jedes zerfallende (reguläre) Einbettungsproblem mit dem Kern $H$ (regulär) lösbar, so besitzt das gegebene Einbettungsproblem auch eine starke (reguläre) Lösung über $K$.*

(12.2) Damit ist über einem Hilbertkörper der kohomologischen Dimension 1 jedes endliche Einbettungsproblem mit abelschem Kern lösbar. Durch Induktion nach einer Hauptreihe des Kerns erhält man hieraus den folgenden Einbettungssatz von Iwasawa [1953] in der Fassung bei Matzat [1987] (siehe auch Fried-Jarden [1986], Th. 24.50):

**Satz 12.2** (Iwasawa [1953]): *Über einem Hilbertkörper der Charakteristik* 0 *und der kohomologischen Dimension* 1 *besitzt jedes (reguläre) endliche Einbettungsproblem mit auflösbarem Kern eine (reguläre) Lösung.*

Da abelsche Gruppen auch als Galoisgrupen über Hilbertkörpern der Charakteristik $p$ realisiert werden können (siehe Serre [1959], Ch.VI.29, Th.4, oder auch Fried-Jarden [1986], Th.24.48), kann die Charakteristikvoraussetzung im Satz 12.2. entfallen. Insbesondere besitzt also auch jedes endliche Einbettungsproblem über $\bar{\mathbb{F}}_p(t)$ mit auflösbarem Kern eine Lösung.

(12.3) Wenn über einem Körper $K$ mit abzählbar vielen Elementen nicht nur Einbettungsprobleme mit auflösbarem Kern sondern sogar mit beliebigem Kern lösbar sind, so ist die absolute Galoisgruppe von $K$ frei. Dieses Kriterium wurde von Iwasawa [1953] gefunden (siehe auch Fried-Jarden [1986], Cor. 24.2):

**Satz 12.3** (Iwasawa [1953]): *Es sei $K$ ein abzählbarer Körper, über dem jedes endliche Einbettungsproblem lösbar ist. Dann ist die absolute Galoisgruppe von $K$ eine freie proendliche Gruppe von abzählbar unendlichem Rang.*

Letzteres wurde von Shafarevich auch für $\mathrm{Gal}(\bar{\mathbb{Q}}/\mathbb{Q}^{ab})$ vermutet (siehe Belyi [1983]) und wäre nach den Sätzen 11.1 und 12.2 richtig, falls jede einfache, nicht abelsche

Gruppe eine GAR-Realisierung über $\mathbb{Q}^{ab}(\underline{t})$ besäße.

(12.4) Beweisen läßt sich diese Aussage bisher nur für eine sehr spezielle Klasse von Körpern: Ein Körper $k$ heißt *pseudo-algebraisch abgeschlossen* oder kurz ein PAC-*Körper*, wenn jede über $k$ definierte absolut irreduzible Mannigfaltigkeit $\mathcal{V}$ einen $k$-rationalen Punkt beziehungsweise der zugehörige Funktionenkörper $k(\mathcal{V})$ eine rationale Stelle besitzt. Dann ist die Menge $\mathcal{V}(k)$ der $k$-rationalen Punkte von $\mathcal{V}$ sogar dicht in der Zariskitopologie von $\mathcal{V}$ (siehe Fried-Jarden [1986], Prop. 10.1), und der Körper $k(\mathcal{V})$ besitzt rationale Stellen außerhalb einer jeden echten Teilmannigfaltigkeit. Insbesondere besitzt dann auch die aus der Körpererweiterung $\tilde{K}^i_{\underline{\sigma}}/\tilde{k}^i_{\underline{\sigma}}$ durch Konstantenerweiterung mit einem PAC-Körper $k$ gewonnene Körpererweiterung $k\tilde{K}^i_{\underline{\sigma}}/k\tilde{k}^i_{\underline{\sigma}}$ rationale Stellen außerhalb des Verzweigungsortes, so daß für diese die Hauptvoraussetzung zum Satz 7.1 erfüllt ist.

Dies wirft die Frage auf, ob es zu einem vorgegebenem PAC-Körper $k$ und einer vorgegebenen Gruppe $G$ eine Erzeugendensystemklasse $[\underline{\sigma}] \in \Sigma^i_s(G)$ gibt, so daß $\tilde{k}^i_{\underline{\sigma}}$ in $k$ enthalten ist. Diese ist gleichbedeutend mit der Frage nach der Existenz einer starren Zopfbahn in der Menge $\Sigma^i(\mathbf{C})$ von Erzeugendensystemklassen in einem $k(\underline{t})$-symmetrischen Klassenvektor $\mathbf{C}$ von $G$ (vergl. die Definition in (10.2)). Letztere wird durch einen Satz von Conway und Parker für die Gruppen bejaht, deren Schurmultiplikator durch Kommutatoren erzeugt wird, und alle Klassenstrukturen, in denen jede Konjugiertenklasse von $G$ hinreichend und gleich oft vorkommt (siehe Fried-Völklein [1991a], Appendix). Da nach Fried-Völklein [1991a], Lemma 3, jede endliche Gruppe $G$ Faktorgruppe einer solchen Gruppe ist, besitzt sogar jede endliche Gruppe $G$ einen solchen Klassenvektor $\mathbf{C}$, für den dann $\Sigma^i_V(\mathbf{C})$ bezüglich $V = \mathrm{Sym}(\mathbf{C})$ eine starre $V$-symmetrisierte Zopfbahn enthält. Hieraus folgt mit dem Satz 7.1 unter Beachtung von Satz 12.2. — die kohomologische Dimension eines PAC-Körpers ist durch 1 beschränkt — der

**Satz 12.4** (Fried-Völklein [1991a]): *Es sei* $K = k(t_0)$ *ein rationaler Funktionenkörper über einem* PAC-*Körper* $k$ *der Charakteristik* 0. *Dann ist jede endliche Gruppe als Galoisgruppe einer regulären Galoiserweiterung* $N/K$ *darstellbar.*

Insbesondere hat das Umkehrproblem der Galoisschen Theorie über jedem hilbertschen PAC-Körper $k$ der Charakteristik 0 eine positive Lösung. Beispiele für solche Körper sind bei Fried-Jarden [1986], Theorem 16.46, zu finden.

(12.5) Nach dem Satz 12.2 besitzt jedes endliche Einbettungsproblem mit auflösbarem Kern über einem hilbertschen PAC-Körper $k$ eine Lösung.

Nun seien $n/k$ eine endliche Galoiserweiterung mit der Gruppe $G$ und $\tilde{G} = H{\cdot}G$ eine Gruppenerweiterung von einer zentrumsfreien endlichen Gruppe $H$ mit $G$. Nach (12.4) existieren ein Klassenvektor $\mathbf{C}$ von $H$, in dem jede Konjugiertenklasse hinreichend und gleich oft vorkommt, und eine Erzeugendensystemklasse $[\underline{\sigma}] \in \Sigma^i(\mathbf{C})$, deren $V$-symmetrisierte Zopfbahn $\tilde{Z}(\underline{\sigma}) = [\underline{\sigma}]^{\tilde{H}_V}$ in $\Sigma^i_V(\mathbf{C})$ starr ist. Weiter seien $V^a := \mathrm{Sym}^A(\mathbf{C})$ die durch die Permutationen von $A := \mathrm{Aut}(H)$ auf $\mathbf{C}$ erweiterte Gruppe von Symmetrien und $\tilde{K}^a_{\underline{\sigma}}(t)$ der Fixkörper des Stabilisators der zugehörigen groben Erzeugendensystemklasse $[[\underline{\sigma}]] \in \Sigma_s^a(H)$ in $\tilde{\Delta}_{V^a}$ (vergl. Matzat [1991a], §6.1). Dann gilt $[k\tilde{K}_{\underline{\sigma}}^i(t) : k\tilde{K}_{\underline{\sigma}}^a(t)] = |\mathrm{Out}(H)|$, und $k\tilde{N}_{\underline{\sigma}}^i/k\tilde{K}_{\underline{\sigma}}^a(t)$ ist galoissch mit

$$\mathrm{Gal}(k\tilde{N}_{\underline{\sigma}}^i/k\tilde{K}_{\underline{\sigma}}^a(t)) \cong \mathrm{Aut}(H).$$

Demgemäß ist auch $n\tilde{N}_{\underline{\sigma}}^i/k\tilde{K}_{\underline{\sigma}}^a(t)$ eine Galoiserweiterung mit

$$\mathrm{Gal}(n\tilde{N}_{\underline{\sigma}}^i/k\tilde{K}_{\underline{\sigma}}^a(t)) \cong \mathrm{Aut}(H) \times G.$$

Diese Gruppe besitzt nun eine zu $\tilde{G}$ isomorphe Untergruppe $U$, deren Durchschnitt mit $\mathrm{Gal}(n\tilde{N}_{\underline{\sigma}}^i/n\tilde{K}_{\underline{\sigma}}^a(t)) \cong \mathrm{Aut}(H)$ gerade $\mathrm{Gal}(n\tilde{N}_{\underline{\sigma}}^i/n\tilde{K}_{\underline{\sigma}}^i(t)) \cong H$ ist (vergl. Matzat [1987], IV. §4.1). Der Fixkörper $R(t)$ von $U$ ist also über $k$ regulär, und es gilt zudem $nR(t) = n\tilde{K}_{\underline{\sigma}}^i(t)$. Da $k$ ein PAC-Körper ist, besitzt $R/k$ eine rationale Stelle $P$, deren Fortsetzung auf $nR = n\tilde{K}_{\underline{\sigma}}^i$ ohne Galoisdefekt für $n\tilde{N}_{\underline{\sigma}}^i/n\tilde{K}_{\underline{\sigma}}^i(t)$ ist. (Nach dem Satz 7.1 liegen alle Stellen mit Galoisdefekt im Verzweigungsort von $n\tilde{K}_{\underline{\sigma}}^i(t)/n(\underline{t})$.) Das Bild von $n\tilde{N}_{\underline{\sigma}}^i/R(t)$ unter dieser Stelle $P$ ist daher eine Galoiserweiterung $\tilde{N}/k(t)$ mit $\mathrm{Gal}(\tilde{N}/k(t)) \cong \tilde{G}$ und $\tilde{N} \geq n(t)$. Auf Grund des Hilbertschen Irreduzibilitätssatzes ist somit auch jedes endliche Einbettungsproblem über $k$ mit zentrumsfreiem Kern lösbar.

Insgesamt ergibt sich aus diesen Überlegungen und dem Satz 12.3 der

**Satz 12.5** (Fried-Völklein [1991b]): *Die absolute Galoisgruppe eines abzählbaren hilbertschen* PAC-*Körpers der Charakteristik* 0 *ist eine freie proendliche Gruppe von abzählbar unendlichem Rang.*

Damit ist das Analogon der Vermutung von Shafarevich für hilbertsche PAC-Körper bewiesen, für den vollen Kreisteilungskörper $\mathbb{Q}^{ab}$ bleibt sie jedoch nach wie vor offen (siehe Fried-Jarden [1986], Cor. 10.15).

## LITERATURVERZEICHNIS


ASCHBACHER, M. et al. eds. [1984]: *Proceedings of the Rutgers group theory year, 1983-1984.* Cambridge University Press 1984

BECKMANN, S. [1989]: Ramified primes in the field of moduli of branched coverings of curves. J. Algebra **125**, 236-255 (1989)

BECKMANN, S. [1991]: On extensions of number fields obtained by specializing branched coverings. (Erscheint demnächst)

BELYI, G.V. [1979]: On Galois extensions of a maximal cyclotomic field. Izv. Akad. Nauk SSSR Ser. Mat. **43**, 267-276 (1979); Math. USSR Izv. **14**, 247-256 (1980)

BELYI, G.V. [1983]: On extensions of the maximal cyclotomic field having a given classical Galois group. J. reine angew. Math. **341**, 147-156 (1983)

BELYI, G.V. [1987]: On the commutator of the absolute Galois group. Pp. 346-449 in *Proc. Int. Congr. Math. Berkeley 1986.* Amer. Math. Soc. 1987.

BIRMAN, J.S. [1974]: *Braids, links and mapping class groups.* Princeton University

Press 1974

BRUEN, A.A.; JENSEN, C.U.; YUI, N. [1986]: Polynomials with Frobenius groups of prime degree as Galois groups II. J. Number Theory **24**, 305-359 (1986)

BUCHBERGER, B. [1970]: Ein algorithmisches Kriterium für die Lösbarkeit eines algebraischen Gleichungssystems. Aequationes Math. **4**, 374-383 (1970).

CONWAY, J.H. et al. [1985]: *Atlas of finite groups.* Clarendon Press, Oxford 1985

COOMBES, K.; HARBATER, D. [1985]: Hurwitz families and arithmetic Galois groups. Duke Math. J. **52**, 821-839 (1985)

CRESPO, T. [1990]: Explicit construction of $2S_n$ Galois extensions. J. Algebra **129**, 312-319 (1990)

DÈBES, P.; FRIED, M.D. [1990]: Rigidity and real residue class fields. Acta Arithmetica **56**, 1-32 (1990)

DENTZER, R. [1989]: Projektive symplektische Gruppen $PSp_4(p)$ als Galoisgruppen über $\mathbb{Q}(t)$. Arch. Math. **53**, 337-346 (1989)

DENTZER, R. [1991]: On split embedding problems with abelian kernel. (In Vorbereitung)

FADELL, E.; VAN BUSKIRK, J. [1962]: The braid groups of $E^2$ and $S^2$. Duke Math. J. **29**, 243-257 (1962)

FEIT, W. [1984]: Rigidity of $\mathrm{Aut}(PSL_2(p^2)), p \equiv \pm 2 \pmod 5, p \neq 2$. Pp. 351-356 in ASCHBACHER et al. eds. [1984]

FEIT, W. [1989]: Some finite groups with nontrivial centers which are Galois groups. Pp. 87-109 in: *Group Theory, Proceedings of the 1987 Singapore Conference.* W. de Gruyter, Berlin-New York 1989

FEIT, W.; FONG, P. [1984]: Rational rigidity of $G_2(p)$ for any prime $p > 5$. Pp. 323-326 in ASCHBACHER et al. eds. [1984]

FRIED, M.D. [1977]: Fields of definition of function fields and Hurwitz families — Groups as Galois groups. Commun. Alg. **5**, 17-82 (1977)

FRIED, M.D. [1984]: On reduction of the inverse Galois group problem to simple groups. Pp. 289-301 in ASCHBACHER et al. eds. [1984]

FRIED, M.D.; BIGGERS, R. [1982]: Moduli spaces of covers and the Hurwitz monodromy group. J. reine angew. Math. **335**, 87-121 (1982)

FRIED, M.D.; JARDEN, M. [1986]: *Field arithmetic.* Springer-Verlag, Berlin etc. 1986

FRIED, M.D.; THOMPSON, J.G. [1991]: The Hurwitz monodromy group $H(4)$ and modular curves. (Erscheint demnächst)

FRIED, M.D.; VÖLKLEIN, H. [1991a]: The inverse Galois problem and rational points on moduli spaces. (Erscheint demnächst)

FRIED, M.D.; VÖLKLEIN, H. [1991b]: The embedding problem over a Hilbertian PAC-field. (Erscheint demnächst)

FRÖHLICH, A. [1985]: Orthogonal representations of Galois groups, Stiefel-Witney classes and Hasse-Witt invariants. J. reine angew. Math. **360**, 84-123 (1985)

GILETTE, R., VAN BUSKIRK, J. [1968]: The word problem and consequences for the braid groups and mapping class groups of the 2-sphere. Trans. AMS **131**, 277-296

(1968)

GROTHENDIECK, A. [1971]: *Revêtements étales et groupe fondamental.* Springer-Verlag, Berlin etc. 1971

HÄFNER, F. [1991]: Einige orthogonale und symplektische Gruppen als Galoisgruppen über $\mathbb{Q}$. (Erscheint demnächst)

HARBATER, D. [1984]: Mock covers and Galois extensions. J. Algebra **91**, 281-293 (1984)

HARBATER, D. [1987]: Galois coverings of the arithmetic line. In CHUDNOVSKY, D.V. et al. eds.: *Number Theory, New York 1984-1985.* Springer-Verlag, Berlin etc. 1987

HILBERT, D. [1892]: Über die Irreduzibilität ganzer rationaler Funktionen mit ganzzahligen Koeffizienten. J. reine angew. Math. **110**, 104-129 (1892)

HOYDEN-SIEDERSLEBEN, G. [1985]: Realisierung der Jankogruppen $J_1$ und $J_2$ als Galoisgrupen über $\mathbb{Q}$. J. Algebra **97**, 14-22 (1985)

HOYDEN-SIEDERSLEBEN, G.; MATZAT, B.H. [1986]: Realisierung sporadischer einfacher Gruppen als Galoisgruppen über Kreisteilungskörpern. J. Algebra **101**, 273-285 (1986)

HUNT, D.C. [1986]: Rational rigidity and the sporadic groups. J. Algebra **99**, 577-592 (1986)

HUPPERT, B. [1967]: *Endliche Gruppen I.* Springer-Verlag, Berlin etc. 1967

HURWITZ, A. [1891]: Über Riemann'sche Flächen mit gegebenen Verzweigungspunkten. Math. Ann. **39**, 1-61 (1891)

IKEDA, M. [1960]: Zur Existenz eigentlicher galoisscher Körper beim Einbettungsproblem für galoissche Algebren. Abh. Math. Sem. Univ. Hamburg **24**, 126-131 (1960)

IŠHANOV, V.V. [1976]: On the semidirect imbedding problem with nilpotent kernel. Izv. Akad. Nauk SSSR Ser. Mat **40**, 3-25 (1976); Math. USSR Izv. **10**, 1-23 (1976)

IWASAWA, K. [1953]: On solvable extensions of algebraic number fields. Ann. Math. **58**, 548-572 (1953)

JANNSEN, U.; WINGBERG, K. [1982]: Die Struktur der absoluten Galoisgruppe $p$-adischer Zahlkörper. Invent. math. **70**, 71-98 (1982)

JENSEN, C.U. [1986]: On the general inverse problem of Galois theory. C.R. Math. Rep. Acad. Sci. Canada **8**, 145-149 (1986)

JENSEN, C.U.; YUI, N. [1982]: Polynomials with $D_p$ as Galois group. J. Number Theory **15**, 347-375 (1982)

JENSEN, C.U.; YUI, N. [1987]: Quaternion extensions. Pp. 155-182 in *Algebraic Geometry and Commutative Algebra in Honor of Masayoshi Nagata.*

KRULL, W.; NEUKIRCH, J. [1971]: Die Struktur der absoluten Galoisgruppe über dem Körper $\mathbb{R}(t)$. Math. Ann. **193**, 197-209 (1971)

KUYK, W. [1970]: Extensions de corps Hilbertiens. J. Algebra **14**, 112-124 (1970)

LAMACCHIA, M.E. [1980]: Polynomials with Galois group PSL(2,7). Commun. Algebra **8**, 983-992 (1980)

MACBEATH, A.M. [1969]: Extensions of the rationals with Galois group $PGL(2, Z_n)$. Bull. London Math. Soc. **1**, 332-338 (1969)

MALLE, G. [1987]: Polynomials for primitive nonsolvable permutation groups of degree $d \leq 15$. J. Symb. Comput. **4**, 83-92 (1987)

MALLE, G. [1988a]: Polynomials with Galois groups $\mathrm{Aut}(M_{22}), M_{22}$, and $PSL_3(\mathbb{F}_4)\cdot 2_2$ over $\mathbb{Q}$. Math. Comput. **51**, 761-768 (1988)

MALLE, G. [1988b]: Exceptional groups of Lie type as Galois groups. J. reine angew. Math. **392**, 70-109 (1988)

MALLE, G. [1990]: Some unitary groups as Galois groups over $\mathbb{Q}$. J. Algebra **131**, 476-482 (1990)

MALLE, G. [1991a]: Darstellungstheoretische Methoden bei der Realisierung einfacher Gruppen vom Lie Typ als Galoisgruppen. Dieser Band.

MALLE, G. [1991b]: Polynome mit Galoisgruppen $PGL_2(p)$ und $PSL_2(p)$ für Primzahlen $p \leq 29$ über $\mathbb{Q}(t)$. (Erscheint demnächst)

MALLE, G.; MATZAT, B.H. [1985]: Realisierung von Gruppen $PSL_2(\mathbb{F}_p)$ als Galoisgruppen über $\mathbb{Q}$. Math. Ann. **272**, 549-565 (1985)

MATZAT, B.H. [1984]: Konstruktion von Zahl- und Funktionenkörpern mit vorgegebener Galoisgruppe. J. reine angew. Math. **349**, 179-220 (1984)

MATZAT, B.H. [1985a]: Zwei Aspekte konstruktiver Galoistheorie. J. Algebra **96**, 499-531 (1985)

MATZAT, B.H. [1985b]: Zum Einbettungsproblem der algebraischen Zahlentheorie mit nicht abelschem Kern. Invent. math. **80**, 365-374 (1985)

MATZAT, B.H. [1986]: Topologische Automorphismen in der konstruktiven Galoistheorie. J. reine angew. Math. **371**, 16-45 (1986)

MATZAT, B.H. [1987]: *Konstruktive Galoistheorie*. Springer-Verlag, Berlin etc. 1987

MATZAT, B.H. [1988]: Über das Umkehrproblem der Galoisschen Theorie. Jber. d. Dt. Math.-Ver. **90**, 155-183 (1988)

MATZAT, B.H. [1989]: Rationality criteria for Galois extensions. Pp. 361-383 in Y. IHARA et al. eds.: *Galois groups over* $\mathbb{Q}$. Springer-Verlag, New York etc. 1989

MATZAT, B.H. [1990]: Braids and Galois groups. Doğa — Tr. J. of Mathematics **14**, 57-69 (1990)

MATZAT, B.H. [1991a]: Zöpfe und Galoissche Gruppen. J. reine angew. Math. (erscheint demnächst)

MATZAT, B.H. [1991b]: Frattini-Einbettungsprobleme über Hilbertkörpern. (Erscheint demnächst)

MATZAT, B.H.; ZEH-MARSCHKE, A. [1986]: Realisierung der Mathieugruppen $M_{11}$ und $M_{12}$ als Galoisgruppen über $\mathbb{Q}$. J. Number Theory **23**, 195-202 (1986)

MESTRE, J.-F. [1988]: Courbes hyperelliptiques à multiplications réelles. C.R. Acad. Sci. Paris **307**, 721-724 (1988)

MESTRE, J.-F. [1990]: Extensions régulières de $\mathbb{Q}$(t) de groupe de Galois $\tilde{A}_n$. J. Algebra **131**, 483-495 (1990)

NOBUSAWA, N. [1961]: On the imbedding problem of fields and Galois algebras. Abh. Math. Sem. Univ. Hamburg **25**, 89-92 (1961)

PAHLINGS, H. [1988]: Some sporadic groups as Galois groups. Rend. Sem. Math. Univ. Padova **79**, 97-107 (1988)

PAHLINGS, H. [1989]: Some sporadic groups as Galois groups II. Rend. Sem. Math. Univ. Padova **82**, 163-171 (1989)

POPP, H. [1970]: *Fundamentalgruppen algebraischer Mannigfaltigkeiten.* Springer-Verlag, Berlin etc. 1970

PRZYWARA, B. [1991]: Zopfbahnen und Galoisgruppen. (In Vorbereitung)

SALTMAN, D.J. [1982]: Generic Galois extensions and problems in field theory. Adv. Math. **43**, 250-283 (1982)

SCHNEPS, L. [1989]: $\tilde{D}_4$ et $\hat{D}_4$ comme groupe de Galois. C.R. Acad. Sci. Paris **308**. I, 33-36 (1989)

SCHOLZ, A. [1929]: Über die Bildung algebraischer Zahlkörper mit auflösbarer galoisscher Gruppe. Math. Z. **30**, 332-356 (1929)

SCHUR, I. [1911]: Über die Darstellung der symmetrischen und alternierenden Gruppen durch gebrochene lineare Substitutionen. J. reine angew. Math. **139**, 155-250 (1911)

SEIDELMANN, F. [1918]: Die Gesamtheit der kubischen und biquadratischen Gleichungen mit Affekt bei beliebigem Rationalitätsbereich. Math.Ann **78**, 230-233 (1918)

SERRE, J.-P. [1959]: *Groupes algébriques et corps de classes.* Hermann, Paris 1959

SERRE, J.-P. [1964]: *Cohomologie galoisienne.* Springer-Verlag, Berlin etc. 1964

SERRE, J.-P. [1984]: L'invariant de Witt de la forme $Tr(x^2)$. Comment. Math. Helvetici **59**, 651-676 (1984)

SERRE, J.-P. [1988]: Groupes de Galois sur $\mathbb{Q}$. Astérisque **161** –**162**, 73-85 (1988)

SERRE, J.-P. [1991]: Topics in Galois Theory. (Erscheint demnächst)

SHAFAREVICH, I.R. [1954a]: On the construction of fields with a given Galois group of order $l^a$. Izv. Akad. Nauk SSSR. Ser. Mat. **18**, 216-296 (1954); Amer. Math. Soc. Transl. **4**, 107-142 (1956)

SHAFAREVICH, I.R. [1954b]: On the problem of imbedding fields. Izv. Akad. Nauk SSSR. Ser. Mat. **18**, 389-418 (1954); Amer. Math. Soc. Transl. **4**, 151-183 (1956)

SHAFAREVICH, I.R. [1954c]: Construction of fields of algebraic numbers with given solvable Galois group. Izv. Akad. Nauk SSSR. Ser. Mat. **18**, 525-578 (1954); Amer. Math. Soc. Transl. **4**, 185-237 (1956)

SHAFAREVICH, I.R. [1958]: The imbedding problem for split extensions (russ.). Dokl. Akad. Nauk SSSR. **120**, 1217-1219 (1958)

SHIH, K.-Y. [1974]: On the construction of Galois extensions of function fields and number fields. Math. Ann. **207**, 99-120 (1974)

SHIH, K.-Y. [1978]: P-division points on certain elliptic curves. Compositio Math. **36**, 113-129 (1978)

SONN, J. [1972]: On the embedding problem for nonsolvable Galois groups of algebraic number fields: reduction theorems. J. Number Theory **4**, 411-436 (1972)

SONN, J. [1989a]: Central extensions of $S_n$ as Galois groups via trinomials. J. Algebra **125**, 320-330 (1989)

SONN, J. [1989b]: Groups of small order as Galois groups over $\mathbb{Q}$. Rocky Mountain J. Math. **19**, 947-956 (1989)

THOMPSON, J.G. [1984a]: Some finite groups which appear as Gal$(L/K)$ where

$K \leq \mathbb{Q}(\mu_n)$. J. Algebra **89**, 437-499 (1984)

THOMPSON, J.G. [1984b]: $PSL_3$ and Galois groups over $\mathbb{Q}$. Pp. 309-319 in ASCHBACHER et al. eds. [1984]

THOMPSON, J.G. [1984c]: Rational rigidity of $G_2(5)$. Pp. 321-322 in ASCHBACHER et al. eds. [1984]

THOMPSON, J.G. [1984d]: Primitive roots and rigidity. Pp. 327-350 in ASCHBACHER et. al eds. [1984]

THOMPSON, J.G. [1986]: Regular Galois extensions of $\mathbb{Q}(x)$. Pp. 210-220 in TUAN HSIO-FU ed.: *Group theory, Beijing 1984*. Springer-Verlag, Berlin etc. 1986

TRINKS, W. [1978]: Über B. Buchbergers Verfahren, Systeme algebraischer Gleichungen zu lösen. J. Number Theory **10**, 475-488 (1978)

TRINKS, W. [1984]: On improving approximate results of Buchberger's algorithm by Newton's method. ACM SIGSAM Bull. **18**, 3, Pp. 7-11 (1984)

VILA, N. [1988]: On stem extensions of $S_n$ as Galois group over number fields. J. Algebra **116**, 251-260 (1988)

VÖLKLEIN, H. [1991]: Central extensions as Galois groups. (Erscheint demnächst)

WALTER, J.H. [1984]: Classical groups as Galois groups. Pp. 357-383 in ASCHBACHER et al. eds. [1984]

WEBER, H. [1908]: Lehrbuch der Algebra III. Vieweg, Braunschweig 1908

WINKLER, F. [1988]: A p-adic approach to the computation of Gröbner bases. J. Symb. Comput. **6**, 287-304 (1988)

Progress in Mathematics, Vol. 95, © 1991 Birkhäuser Verlag Basel

# Contributions to Modular Representation Theory of Finite Groups

GERHARD O. MICHLER

## Introduction

Since all finite simple groups have been classified [43], it is a natural question whether the major conjectures in modular representation theory are consequences of this important and deep classification theorem. In this article, a survey is given about the progress which has been achieved on the famous open conjectures of J. Alperin [2] and R. Brauer [9] during the last decade. It turns out that all these problems have rather complete affirmative answers for almost all infinite series of finite groups like the symmetric, classical or exceptional groups of Lie type $G$ for which there is a good parametrization of the irreducible characters $\chi$ of $G$ into $p$-blocks $B$, where $p$ is a prime divisor of the order of $G$. Using techniques from Clifford theory as described in Berger [4] and chapter 10 of Feit [32] it is often possible to reduce the proof of a general conjecture to the case of the automorphism groups of all the covering groups of a finite simple group. In particular, such reduction theorems exist for one direction of Brauer's height zero conjecture and for Alperin's weight conjecture as has been shown by Berger and Knörr [5] and Dade [22], respectively.

Another successful application of this method is the proof of Alperin's T.I. conjecture made by Blau and the author [8] which is described in section 4. There a rather complete report on the recent results of the modular representation theory of finite groups with a T.I. Sylow $p$-subgroup is given.

Because of these reduction theorems it is important to verify the major conjectures of modular representation theory for the infinite series of the symmetric groups, alternating groups and their covering groups. For these groups there are complete results for almost all the conjectures. They are presented in sections 5 and 6. In section 7 the corresponding results on the $p$-blocks of the general linear, unitary, orthogonal, symplectic and exceptional groups of Lie type are explained and stated. Of course, the main tool here is the Deligne-Lusztig theory of the irreducible complex characters of finite groups of Lie type, see Carter [16] and Lusztig [65]. All the important new results of this section are due to Fong and Srinivasan or they build on their work on the distribution of the irreducible characters into blocks of classical groups. The final short section is concerned with the modular representation theory of the finite sporadic simple groups.

In order to show the importance of the major conjectures for the further development of the general theory of modular representations of finite groups the basic definitions and theorems of Brauer's block theory and Green's theory of indecomposable group modules are presented in sections 1 and 2. In the third section the central open conjectures of modular representation theory are stated. It is shown that several of them are closely related, expecially in the case of blocks with abelian defect groups. Furthermore, the theorems of Gluck and Wolf, Külshammer, Okuyama and Okuyama-Wajima are mentioned which show that all but Brauer's conjecture on the bound of the number $k(B)$ of ordinary irreducible characters in a $p$-block $B$ hold for $p$-solvable groups.

The main obstacle for completing this program at present lies in the difficulty of finding the complete character table of the exceptional groups $F_4(q)$, $E_6(q)$, $E_7(q)$ and $E_8(q)$ and their twisted types. At least the distribution of the irreducible characters of these groups into blocks has to be found. For this task more theoretical and computational work has to be done, expecially for the primes dividing the Weyl groups of these groups.

All the conjectures are true for blocks with cyclic defect groups, see Feit [32] and Hiß and Lux [47]. They also hold for 2-blocks with dihedral, semidihedral or quaternion defect groups by Erdmann's work on the structure of the blocks of tame representation type. It is presented in her recent book [31]. Therefore, this report concentrates on the cases where the blocks of the finite groups have wild representation type, which means that their defect groups are neither cyclic $p$-groups nor dihedral, semidihedral or quaternion 2-groups.

Concerning notation and terminology we refer to the books of Carter [16], Feit [32], Gorenstein [43], James and Kerber [52], Lusztig [65], the author's lecture notes [72], and the Atlas [18].

## 1. Blocks of Finite Groups and Their Invariants.

Let $p > 0$ be a fixed prime number dividing the order $|G|$ of the finite group $G$ of order $|G| = p^a q$, $(p,q) = 1$. Let $R$ be a complete discrete rank one valuation ring with maximal ideal $J(R) = \pi R$ such that its residue class field $F = R/\pi R$ has characteristic $p$ and its quotient field $S = \mathrm{quot}(R)$ has characteristic zero. Then the triple $(F, R, S)$ is called a *p-modular system for G*.

Let $A \in \{F, R, S\}$. Then $AG = \{\sum_{g \in G} a_g g | a_g \in A\}$ denotes the *group ring of G over A*. Its *Jacobson radical* is denoted by $J(AG)$.

Two elements of $AG$ are added componentwise. The multiplication of $AG$ is the one of $G$ extended distributively to $AG$.

*Definition.* Let $A \in \{F, S\}$. If $A \cong \mathrm{End}_{AG}(M)$ for every simple $AG$-module $M$ then $A$ is called a *splitting field* for $G$.

*Definition.* The *exponent* $m$ of the finite group $G$ is the smallest positive integer satisfying $g^m = 1$ for all $g \in G$.

**Theorem 1.1.** *Let $G$ be a finite group with exponent $m$ and let $(F, R, S)$ be any p-modular system for $G$. Then:*

a) *The cyclotomic field $\mathbf{Q}(\omega_m)$ generated by a primitive $m$-th root of unity $\omega_m$ over the field $\mathbf{Q}$ of rational numbers is a splitting field for $G$ and its subgroups.*
b) *$S(\omega_m)$ is a splitting field for $G$ and its subgroups.*
c) *There exists a finite $R$-algebra $R_1$ with quotient field $S_1 = \mathrm{quot}(R_1)$ and residue class field $F_1 = R_1/J(R_1)$ which is a finite extension of $F$ such that $S_1$ and $F_1$ are both splitting fields for $G$.*

Assertions a) and *b*) of Theorem 1.1 are proved in Curtis-Reiner I [21], p. 385 and p. 386, respectively. c) follows from b) and Theorem 17.1 of Curtis-Reiner I [21], p. 418.

*Definition.* A $p$-modular system $(F, R, S)$ is a *splitting $p$-modular system* for the finite group $G$ if $F$ and $S$ are both splitting fields for $G$.

Let $(F, R, S)$ be a $p$-modular system for the finite group $G$. The group algebra $AG$ is a semi-perfect ring for each ring $A \in \{F, R, S\}$. Let $\rho : RG \to FG \to 0$ be the canonical ring epimorphism from the group algebra $RG$ onto $FG$ with kernel $\ker \rho = \pi RG$. Then $\rho$ induces a bijection between the blocks $\hat{B} = \hat{e}RG$ of $RG$ and the blocks $B = eFG$ given by $e = \rho(\hat{e})$ for every block idempotent $\hat{e} = (\hat{e})^2$ of the center $Z(RG)$ of $RG$. Furthermore, $RG$ has only finitely many centrally primitive idempotents $\hat{e} \neq 0$. They are uniquely determined by $RG$.

*Definition.* A *block of the finite group $G$* is a triple $B = (B = eFG,\ \hat{B} = \hat{e}RG,\ B_S = \hat{B} \otimes_R S = \hat{e}SG)$ of algebras with identity elements $e = \rho(\hat{e})$ or $\hat{e}$ where $\hat{e} = (\hat{e})^2 \neq 0$ is a centrally primitive idempotent of $RG$.

For each $A \in \{F, R, S\}$ an *indecomposable $AG$-module $M$ belongs to the block $B$* if $M = M\hat{e}$.

In case $A = F$ observe that $M\hat{e} = Me$, where $e = \rho(\hat{e})$ because $\pi RG$ annihilates $M$.

*Notation.* Let $B = (B, \hat{B}, B_S)$ be a block of $G$. Then:

$k(B)$ denotes the number of non-isomorphic simple $SG$-modules $L_i$ of $B_S$, $1 \leq i \leq k(B)$.

$l(B)$ denotes the number of non-isomorphic simple $FG$-modules $M_j$ of $B$, $1 \leq j \leq l(B)$.

*Remark 1.2.* Let $B = (B, \hat{B}, B_S)$ be a block of $G$. Then:

a) $l(B)$ is the number of non-isomorphic indecomposable projective $FG$-modules $P_j$ of $B$ which can be indexed such that $P_j$ is the projective cover of the simple FG-module $M_j$.

b) $l(B)$ is the number of non-isomorphic indecomposable projective $RG$-lattices $\hat{P}_j$ of $\hat{B}$ which can be indexed such that $\hat{P}_j$ is the projective cover of the simple $RG$-module $M_j$.

c)

$$P_j \cong \hat{P}_j \otimes_R F \cong \hat{P}_j / \hat{P}_j \pi RG\ ,$$

$$\hat{P}_j / \hat{P}_j J(RG) \cong M_j \cong P_j / P_j J(FG) \text{ for } 1 \leq j \leq l(B)$$

*Definition.* Let $(F, R, S)$ be a splitting $p$-modular system for the finite group $G$. Let $B = (B, \hat{B}, B_S)$ be a block of $G$. With the notations of 1.1 and 1.2 let $c_{ij} = \dim_F(\mathrm{Hom}_{FG}(P_i, P_j))$ for $1 \leq i, j \leq l(B)$. The $l(B) \times l(B)$ matrix $C = C(B) = (c_{ij})$ is the *Cartan matrix* of the block $B$ and its entries $c_{ij}$ are called *Cartan invariants* of $B$.

Since $SG$ is a semi-simple $S$-algebra, for each indecomposable projective $RG$-module $\hat{P}_j$ there are uniquely determined natural numbers $d_{ij}$ such that

$$\hat{P}_j \otimes_R S \cong \sum_{i=1}^{k(B)} d_{ij} L_i \text{ for } 1 \leq j \leq l(B)\ .$$

The $k(B) \times l(B)$ matrix $D = D(B) = (d_{ij})$ is the *decomposition matrix* of the block $B$ and its entries $d_{ij}$ are called *decomposition numbers* of $B$.

**Lemma 1.3.** *Let $K_i (i = 1, 2, \dots, k)$ denote the different conjugacy classes of $G$ and put $c_i = \sum_{g \in K_i} g \in FG$. Let $A \in \{F, R, S\}$.*

Then for each $A \in \{F, R, S\}$ the class sums $c_i$ of the $k$ conjugacy classes $K_i$ of $G$ form a free $A$-module basis of the center $ZAG$ of the group algebra $AG$.

**Lemma 1.4.** *Let $(F, R, S)$ be a splitting $p$-modular system for the finite group $G$. Let $B = (B, \hat{B}, B_S = \hat{B} \otimes_R S)$ be a block of $G$, and let $k(B)$ be the number of non-isomorphic simple $SG$-modules of $B_S$. If $Z(B)$ denotes the center of the block ideal $B = eFG = \hat{B} \otimes_R F$, then $k(B) = \dim_F Z(B)$.*

*Definition.* Let $g$ be a fixed element of the conjugacy class $K$ of $G$. Then a *defect group* of $K$ is a Sylow $p$-subgroup $\delta(K)$ of $C_G(g)$. If $p^{d(K)} = |\delta(K)|$ then $d(K)$ is *the defect of* $K$.

*Definition.* Since by Lemma 1.3 every $z \in ZFG$ has the unique representation $z = \sum_{i-1}^{k} r_i c_i$, where $r_i \in F$, the set $\operatorname{Sup} z = \{c_i | r_i \neq 0\}$ is called the *central support* of $z$.

Using the notation of Lemma 1.3 we state the following results and definitions.

**Lemma 1.5 (Osima).** *For every $p$-subgroup $D$ of $G$ the sets $J_D = \{z \in ZFG | c_i \in \operatorname{Sup} z$ implies $\delta(K_i) \leq_G D\}$, and $\hat{J}_d = \{z \in ZFG | c_i \in \operatorname{Sup} z\}$ implies $\delta(K_i) <_G D\}$ are ideals of the center $ZFG$ of the group algebra $FG$.*

*Definition.* A $p$-subgroup $D$ of $G$ is a *defect group* $\delta(B)$ of the block $B \leftrightarrow e \leftrightarrow \lambda$ if $e \in J_D$ but $e \notin \hat{J}_D$. Then $\delta(B)$ $(= \delta(e))$ is uniquely determined by the block $B \leftrightarrow e \leftrightarrow \lambda$ up to $G$-conjugacy. The exponent $d(e) = d(B)$ of the order $|\delta(B)|$ of $\delta(B)$ is called the *defect* of the block $B \leftrightarrow e \leftrightarrow \lambda$.

*Definition.* Let $D$ be a $p$-subgroup of the finite group $G$. Then the map $\sigma : ZFG \to ZFC_G(D)$ defined by

$$\sigma(c_i) = \begin{cases} \sum_{g \in C_i} g & \text{if } C_i = K_i \cap C_G(D) \neq \emptyset \\ 0 & \text{if } C_i = K_i \cap C_G(D) = \emptyset \end{cases}$$

is the *Brauer homomorphism* from the center $ZFG$ of the group algebra $FG$ into the center $ZFC_G(D)$ of the group algebra $FC_G(D)$.

It is clear that $\sigma$ is an $F$-vector space homomorphism. But in fact it is an $F$-algebra homomorphism by the following

**Theorem 1.6.** *Let $D$ be a $p$-subgroup of $G$ and let $H$ be any subgroup of $G$ satisfying $C_G(D) \leq H \leq N_G(D)$. Then the Brauer homomorphism $\sigma$ is an $F$-algebra homomorphism from the center $ZFG$ of $FG$ into the center $ZFH$ of the group algebra $FH$ such that $\ker \sigma = \{z \in ZFG | c_i \in \operatorname{Sup} z \Rightarrow K_i \cap C_G(D) = \emptyset\}$.*

**Theorem 1.7 (First Main Theorem on Blocks, [32]).** *Let $D$ be a $p$-subgroup of the finite group $G$ with order $|D| = p^d$, let $H = N_G(D)$ and $F$ be a field of characteristic $p > 0$.*

*Then the Brauer homomorphism $\sigma$ induces a one-to-one correspondence between blocks $B \leftrightarrow e \leftrightarrow \lambda$ of the group algebra $FG$ with defect groups $\delta(B) =_G D$ and the blocks $b \leftrightarrow f \leftrightarrow \mu$ of the group algebra $FH$ with defect $d$.*

The following result yields a slightly stronger form of the first main theorem.

**Corollary 1.8.** *Let $D$ be a $p$-subgroup of $G$ and $H$ a subgroup of $G$ satisfying $H \geq N_G(D)$.*

*Then there is a bijection between the blocks $B \leftrightarrow e \leftrightarrow \lambda$ of $G$ with defect groups $\delta(B) =_G D$ and the blocks $B_1 \leftrightarrow e_1 \leftrightarrow \lambda_1$ of $H$ with defect groups $\delta(B_1) =_H D$.*

Although the Brauer homomorphism $\sigma$ is only an algebra homomorphism in characteristic $p > 0$ it induces a block correspondence $\hat{\sigma}$ in characteristic zero.

**Corollary 1.9.** *Let $(F, R, S)$ be a $p$-modular system for $G$. Let $D$ be a $p$-subgroup of $G$ and $\sigma$ the Brauer homomorphism from $ZFG$ into $ZFC_G(D)$. Denote the epimorphism from $RG$ onto $FG$ with kernel $\pi RG$ by $\rho$. Let $C_G(D) \leq H \leq N_G(D)$.*

*Suppose that $\hat{e}$ is a central idempotent of $ZRG$ such that $\sigma\rho(\hat{e}) \neq 0$. Then there is precisely one central idempotent $\hat{f} \in ZRC_G(D)$ such that $\sigma\rho(\hat{e}) = \hat{f}$.*

*Therefore, if $H = N_G(D)$ then $\sigma$ induces a bijection $\hat{\sigma}$ between the blocks $\overline{B}$ of $RG$ with defect groups $\delta(\hat{B}) =_G D$ and the blocks $\hat{b}$ of $RH$ with defect group $\delta(\hat{b}) = D$.*

*In particular, the following diagram is commutative*

$$\begin{array}{ccc} \hat{B} & \xrightarrow{\hat{\sigma}} & \hat{b} \\ \rho\downarrow & & \downarrow\rho \\ B & \xrightarrow{\sigma} & b \end{array}$$

R. Brauer has also proved an extended first main theorem. Its purpose is to count the number of blocks $B$ of a finite group $G$ with a given $p$-subgroup $\delta(B) =_G D$ as defect group by means of the $\overline{H}$-conjugacy classes of blocks $\overline{b}$ of $\overline{K}$ with defect $d(\overline{b}) = 0$, where $H = N_G(D)$, $K = DC_G(D)$, $\overline{H} = H/D$ and $\overline{K} = K/D$.

For blocks of defect zero the ordinary and the modular representations coincide. This follows from R. Brauer's

**Theorem 1.10.** *Let $B = (B, \hat{B}, B_S)$ be a block of $G$ with defect group $\delta(B) =_G D$. Then the following assertions are equivalent:*

*1.* $B \cong (F)_n$ *2.* $\hat{B} \cong (R)_n$ *3.* $B_s \cong (S)_n$
*4.* $d(B) = 0$ *5.* $D = 1$

Projective $FG$-modules have complexity zero in the sense of the following definition of J.L. Alperin.

*Definition.* Let $M$ be a finitely generated $FG$-module with minimal projective resolution $0 \leftarrow M \leftarrow P_0 \leftarrow P_1 \leftarrow \ldots \leftarrow P_{n-1} \leftarrow P_n \leftarrow \ldots$ Then its *complexity* $c_G(M)$ is the least integer $s \geq 0$ such that $\lim_{n\to\infty}(n^{-s} \dim_F P_n) = 0$.

The indecomposable projective $FG$-modules $M$ with complexity $c_G(M) = 1$ are precisely the *periodic* $FG$-modules, where $M$ is called *periodic* if $M$ has a periodic minimal projective resolution.

Using this concept of complexity Bessenrodt introduces in [7] the following new invariants of a $p$-block $B$ of a finite group $G$.

*Definition.* For every integer $i \geq 0$ let $M_i(B) = \{V|V$ an indecomposable $B$-module with $c_G(V) \leq i\}$. Let

$$n_1 = n_1(B) = \text{Min}\,\{n \in \mathbf{N}|p^{a-n}|\dim_F V \text{ for all } V \in M_1(B)\} \ ,$$
$$n_i = n_i(B) = \text{Min}\left\{n \in M|p^{(a-\sum_{j=1}^{i} n_j)-n}|\dim_F V \text{ for all } V \in M_i(B)\right\} \quad \text{for } i \geq 2 \ .$$

In [7] Bessenrodt shows that the sequence $\{n_1, n_2, \ldots\}$ of invariants of a $p$-block $B$ with defect group $\delta(B) =_G D$ of $p$-rank $r$ has the following two properties:
a) $n_i = 0$ for all $i > r$ and $n_r > 0$, and
b) $\sum_{i=1}^{r} n_i = d = d(B)$ the defect of $B$.

In [7] she shows then with this notation

**Theorem 1.11.** *Let $B$ be a $p$-block of $G$ with an abelian defect group $D$ of rank $r$. Then the abelian group $D$ has elementary divisors $n_1(B) \geq n_2(B) \geq \ldots \geq n_r(B) > 0$.*

As the main application of this result Bessenrodt proves in [6].

**Theorem 1.12.** *Let $G$ and $H$ are two finite groups with abelian Sylow $p$-subgroups $P$ and $Q$, respectively. If $\mathbf{Z}G \cong \mathbf{Z}H$ then $P \cong Q$.*

## 2. Green Correspondence.

Throughout this section $G$ denotes a finite group. Let $(F, R, S)$ be a $p$-modular system for the finite group $G$ and let $A \in \{F, R\}$. If $A = R$ then all finitely generated right $AG$-modules are assumed to be $AG$-lattices.

*Definition.* Let $I$ be an index set and $\mathcal{S} = \{S_i|i \in I\}$ be a set of subgroups $S_i \leq G$ of $G$. A right $AG$-module $M$ is called *$\mathcal{S}$-projective* if for each $i \in I$ there is a right $AS_i$-module $M_i$ such that $M$ is isomorphic to a direct summand of $\bigoplus_{i\in I}(M_i)^g$.

A right $AG$-module $M$ is *$\mathcal{S}$-projective-free* if $M$ does not contain any $\mathcal{S}$-projective direct summand.

If $\mathcal{S}$ consists just of one subgroup $S$ of $G$ then one writes *$S$-projective* and *$S$-projective-free* instead of $\mathcal{S}$-projective and $\mathcal{S}$-projective-free, respectively.

*Notation.* Let $U, V$ be right $AG$-modules and let $H$ be a subgroup of $G$.

**Lemma 2.1.** *Let $M$ be an indecomposable right $AG$-module. Let $V(M) = \{H \leq G|M$ is $H$-projective$\}$. Then:*
a) *Each pair $H_1$, $H_2$ of minimal elements of $V(M)$ is $G$-conjugate.*
b) *Each minimal element $V \in V(M)$ is a $p$-subgroup of $G$. It is called vertex $\text{vx}(M) =_G V$ of $M$.*

*Definition.* Let $M$ be an indecomposable right $AG$-module with vertex $\text{vx}(M) =_G V$. An indecomposable $AV$-module $W$ with $M|W^G$ is called a *source* of $M$.

The following result is due to J. A. Green, see [32].

**Theorem 2.2.** *Let $M$ be an indecomposable right $AG$-module with vertex $\mathrm{vx}(M) =_G V$. Then the following statement holds.*

*The sources $W$ of $M$ are uniquely determined by $M$ up to $N_G(V)$-conjugacy where $V =_G \mathrm{vx}(M)$ is a vertex of $M$. In particular, if $W_1$ is another source of $M$ then $W_1 \cong W \otimes_{AV} x$ as $AV$-modules for some $x \in N_G(V)$.*

**Theorem 2.3.** *Let $M$ be an indecomposable right $FG$-module with $vx(M) =_G V$. Suppose that $M$ belongs to the block $B$ of $G$ with defect group $\delta(B) =_G D$. Then $V \leq_G D$.*

**Corollary 2.4.** *The vertex of the trivial $FG$-module $1_G$ is a Sylow $p$-subgroup and hence the defect group $\delta(B_0)$ of the principal block $B_0$ of a finite group $G$ is a Sylow $p$-subgroup of $G$.*

**Corollary 2.5.** *Let $B$ be a block of $G$ with a normal defect group $\delta(B) = D$. Then each simple $FG$-module $M$ of $B$ has vertex $vx(M) = D$.*

If $\mathcal{S}$ is a set of subgroups $L$ of the finite group $G$ then $L \in_G \mathcal{S}$ means that $L^g \in \mathcal{S}$ for some $g \in G$. Throughout this section the following notation is kept.

*Notation 2.6.* Let $(F, R, S)$ be a $p$-modular system for the finite group $G$ and let $A \in \{F, R\}$. All right $AG$-modules are assumed to be finitely generated and free as $A$-modules.

Let $P$ be a fixed $p$-subgroup of $G$ and let $H$ denote a subgroup of $G$ containing $N_G(P)$. Consider the following sets of subgroups of $G$:

$$\mathcal{X} = \mathcal{X}(P,H) = \{L \leq G | L \leq P \cap P^x \text{ for some } x \in G \setminus H\}$$
$$\mathcal{Y} = \mathcal{Y}(P,H) = \{L \leq G | L \leq H \cap P^x \text{ for some } x \in G \setminus H\}$$
$$\mathcal{A} = \mathcal{A}(P,H) = \{L \leq G | L \leq P \text{ and } L \notin_G \mathcal{X}\}$$

Let $M$ be a right $AG$-module and $W$ a right $AH$-module. Then

$$M_H \equiv W \bmod \mathcal{Y}$$

means that $M_H \cong W \oplus W'$ where $W'$ is a $\mathcal{Y}$-projective $AH$-module.

$$W^G \equiv M \bmod \mathcal{X}$$

means that $W^G \cong M \oplus M'$ where $M'$ is an $\mathcal{X}$-projective $AG$-module.

"$M$ mod $\mathcal{X}$" and "$W$ mod $\mathcal{Y}$" are also abbreviated by $M$ $(\mathcal{X})$ and $W$ $(\mathcal{Y})$, respectively.

Since $N_G(P) \leq H$ all elements of $\mathcal{X}$ are proper subgroups of $P$. Furthermore, $P \in \mathcal{A}$ but $P \notin \mathcal{Y}$. Clearly, $\mathcal{X}$ is a subset of $\mathcal{Y}$.

If $U$, $V$ are right $AG$-modules, then $(U,V)_{\mathcal{X},G}$ denotes the set of all $AG$-module homomorphisms which factor through an $\mathcal{X}$-projective $AG$-module. Furthermore, $(U,V)_G^{\mathcal{X}} \cong (U,V)_G/(U,V)_{\mathcal{X},G}$, where $(U,V)_G = \mathrm{Hom}_{AG}(U,V)$. J. A. Green's correspondence theorem [32], p. 117, is restated as

**Theorem 2.7.** *Let $P$ be a $p$-subgroup of $G$ and $H$ a subgroup of $G$ containing $N_G(P)$. Then there is a one-to-one correspondence $f$ between the isomorphism classes of indecomposable right $AG$-modules $V$ with vertices $\mathrm{vx}(V) \in \mathcal{A}$ and the isomorphism classes of indecomposable right $AH$-modules $W$ with vertices $\mathrm{vx}(W) \in \mathcal{A}$ which is characterized as follows:*

a) *If $V$ is an indecomposable right $AG$-module with vertex $\mathrm{vx}(V) \in \mathcal{A}$ then $V_H$ has a unique indecomposable direct summand $f(V)$ with vertex $\mathrm{vx}(f(V)) =_H \mathrm{vx}(V)$ and*

$$V_H \equiv f(V)\ (\mathcal{Y}) .$$

*Furthermore, $V$ and $f(V)$ have a common source which is uniquely determined by $V$ up to $N_G(\mathrm{vx}(V))$-conjugation.*

b) *If $W$ is an indecomposable right $AH$-module with vertex $\mathrm{vx}(W) \in \mathcal{A}$ then $W^G$ has a unique indecomposable direct summand $g(W)$ with vertex $\mathrm{vx}(g(W)) = \mathrm{vx}(W)$ and*

$$W^G \equiv g(W)\ (\mathcal{X}) .$$

c) *In particular, $gf(V) \cong V$ and $fg(W) \cong W$.*

**Theorem 2.8.** *Let $P$ be a $p$-subgroup of $G$ and $N_G(P) \leq H \leq G$. Let $f$ be the Green correspondence defined in Theorem 2.7. Suppose that $U, V$ are indecomposable right $AG$-modules such that $\mathrm{vx}(V) \in \mathcal{A}$. Then the following assertions hold:*

a) *$(U,V)^{\mathcal{X}}_G \cong (U_H, f(V))^{\mathcal{X}}_H$ as $A$-modules.*

b) *If $\mathrm{vx}(U) \in \mathcal{A}$ also then $(U,V)^{\mathcal{X}}_G \cong (f(U), f(V))^{\mathcal{X}}_H$ as $A$-modules.*

**Theorem 2.9.** *Let $P$ be a $p$-subgroup of $G$ and $H = N_G(P)$. Let $A \in \{F, R\}$ and $\hat{\sigma}$ be the Brauer correspondence between the blocks $B \leftrightarrow e$ of $G$ with defect groups $\delta(B) =_G P$ and the blocks $b \leftrightarrow \hat{\sigma}(e)$ of $H$ with defect group $\delta(b) = P$. Let $f$ be the Green correspondence between the non-projective indecomposable $AG$-modules $U$ with vertex $\mathrm{vx}(U) \in \mathcal{A}$ and the non-projective indecomposable $AH$-modules $W$ with vertex $vx(W) \in \mathcal{A}$.*

*Then the Brauer correspondence $\hat{\sigma}$ and the Green correspondence $f$ are related in the following sense: $U$ belongs to $B$ if and only if $f(U)$ belongs to $b = \hat{\sigma}(B)$.*

There is another fruitful relation between the Brauer and the Green correspondences.

By means of the multiplication $v(g_1, g_2) = g_1^{-1} v g_2$, $v \in AG$, $(g_1, g_2) \in G \times G$, the group ring $AG$ becomes a right $A[G \times G]$-module. For each subset $X$ of $G$ let $\Delta X = \{(x,x) \in G \times G | x \in X\}$. With this notation J. A. Green's theorem, see Feit [32, p. 134], can be restated as

**Theorem 2.10.** *Let $B$ be a $p$-block of the finite group $G$ with defect group $\delta(B) =_G D$ and Brauer correspondent $b$ in $H = N_G(D)$. Then the following assertion holds:*

*The indecomposable right $F[G \times G]$-module $B$ has vertex $\mathrm{vx}(B) =_{G \times G} \Delta D$, and the indecomposable right $F[H \times H]$-module $b$ is isomorphic to the Green correspondent $f(B)$.*

Using this result the author gave in [73] a short proof of the following theorem of R. Brauer [10].

**Theorem 2.11.** *Let $G$ be a finite group of order $|G| = p^a q$, $(p,q) = 1$. Let $B = eFG$ be a block ideal of the group algebra $FG$ with defect $d(B) = d$. If $\nu$ denotes the p-adic exponential valuation of the field $\mathbf{Q}$ of rational numbers, then $\nu(\dim_F B) = 2a - d$.*

## 3. The Conjectures of J. Alperin and R. Brauer.

Most of the important long standing conjectures in modular representation theory of finite groups are due to R. Brauer [9] and J. L. Alperin [1], [2]. According to J. L. Alperin [1], all these questions are part of the main problem of block theory: If $G$ is a finite group, then give simple rules which determine, in terms of $p$-local subgroups, the values of the characters on the $p$-singular elements.

In order to state Alperin's and Brauer's conjectures the following definition and notation are needed.

*Definition.* Let $(F, R, S)$ be a splitting $p$-modular system for the finite group $G$. Let $B$ be a block of $G$ with defect $d(B) = d$, and $\chi \in \mathrm{Irr}_S(B)$. If $p^{a-d+h}$ is the precise power of $p$ dividing the degree $\chi(1)$ of $\chi$, then $h = \mathrm{ht}(\chi)$ is called the *height* of $\chi$.

$$\mathrm{Irr}_S^0(B) = \{\chi \in \mathrm{Irr}_S(B) |\, \mathrm{ht}(\chi) = 0\} ,$$
$$k_0(B) = |\, \mathrm{Irr}_S^0(B)| \; .$$

*Remark.* $\mathrm{Irr}_S^0(B) \neq \emptyset$ for every $p$-block $B$ of a finite group $G$.

We now state some important open conjectures. Almost all of them hold for $p$-solvable groups $G$, where $G$ is called *$p$-solvable*, if all the simple composition factors of $G$ are either $p$-groups or have order prime to $p$.

In [9] R. Brauer conjectured that a $p$-block $B$ of a finite group $G$ has an abelian defect group $\delta(B) =_G D$ if and only if $k(B) = k_0(B)$ *(Height zero conjecture)*.
Gluck and Wolf [41] have verified this for $p$-solvable groups.

This conjecture is related to the *Alperin-McKay conjecture* $k_0(B) = k_0(b)$, where $b$ denotes the Brauer correspondent of $B$ in $H \geq N_G(D)$. See Alperin [1].
It holds for $p$-solvable groups by Okuyama and Wajima [78].

R. Brauer's *$k(B)$-conjecture* [9] asserts that $k(B) \leq |D|$.
It is open even for $p$-solvable groups, see Knörr [56].

In [80] J.B. Olsson conjectured that for each $p$-block $B$ of a finite group $G$ with defect group $\delta(B) =_G D$, $k_0(B) \leq |D : D'|$.

These three conjectures are closely related. In fact, in [59] B. Külshammer has shown that Olsson's conjecture is a consequence of Brauer's $k(B)$-conjecture and the Alperin-McKay conjecture. In particular, Olsson's conjecture is open for $p$-solvable groups.

**Theorem 3.1.** *Let $B$ be a p-block of the finite group $G$ with defect group $\delta(B) =_G D$ and Brauer correspondent $b$ in $N_G(D)$. If $k_0(B) = k_0(b)$ for all groups $G$, then $k_0(B) \leq |D : D'|$, provided Brauer's $k(B)$-conjecture holds for all p-blocks $B$ of all finite p-solvable groups.*

The proof uses W. Reynold's result [86] confirming the Height zero conjecture for blocks with normal defect groups.

In his famous list of conjectures [9] R. Brauer poses the following two questions.

*Brauer's Problem 21.* For every positive integer $k$ are there only finitely many isomorphism classes of $p$-groups $D$ which can occur as defect groups $\delta(B)$ of $p$-blocks $B$ of finite groups $G$ with exactly $k$ irreducible complex characters?

*Brauer's Problem 19.* Can the number $z(G)$ of blocks of defect zero of a finite group $G$ be described in terms of group theoretic invariants of $G$?

In [87] Robinson gave a solution for this problem. However, it seems to be not very practical when $z(G)$ has to be computed for a specific group which might even be $p$-solvable. Recently, Külshammer has verified Problem 21 for $p$-solvable groups in his papers [60], [61] and [62].

In order to state *Alperin's weight conjecture* the following definitions are necessary.

*Definition.* Let $U$ be a *radical $p$-subgroup* of $G$, i.e. $U = O_p(N_G(U))$. Then an irreducible character $\varphi$ of $N_G(U)$ is called a *weight character* if $\varphi$ is trivial on $U$ and belongs to a block of defect zero of $N_G(U)/U$. The $G$-conjugacy class of the pair $(U, \varphi)$ is called a *weight of $G$*.

*Definition.* Let $B$ be a block of $G$. A weight $(U, \varphi)$ of $G$ is a *$B$-weight* if $\varphi$ belongs to a block $b$ of $N_G(U)$ such that $B$ and $b$ correspond under the Brauer homomorphism with respect to $U$, i.e. $B = b^G$. Let $l^*(B)$ be the number of $B$-weights of $G$.

*Alperin's Conjecture* [2]. $l(B) = l^*(B)$.

A proof of Alperin's conjecture would immediately imply the following results, which at present are also open problems.

*Consequence* (Alperin). Let $B$ be a block of $G$ with an abelian defect group $D$. Let $b$ be the Brauer correspondent of $B$ in $N_G(D)$. Then $l(B) = l(b)$ and $k(B) = k(b)$.

Okuyama proved Alperin's weight conjecture for $p$-solvable groups in [77].

In [1, Conjecture $M$] J. L. Alperin restates another open problem in modular representation theory as

*Donovan's Conjecture.* Up to Morita equivalence there are only a finite number of $p$-blocks $B$ of finite groups $G$ with a defect group $\delta(B)$ isomorphic to a given $p$-subgroup $D$.

L. Puig has verified this conjecture for $p$-solvable groups. The result is unpublished.

All these conjectures are not completely independent from each other. As an example, the following recent result of Knörr and Robinson [58] is mentioned.

**Theorem 3.2.** *Any two of the following assertions imply the third:*

a) *Alperin's conjecture is valid for every $p$-block which has an abelian defect group.*
b) *The Alperin-McKay conjecture is valid for every $p$-block which has an abelian defect group.*

c) *All irreducible characters* $\chi$ *in p-blocks with abelian defect groups have height* $\mathrm{ht}(\chi) = 0$.

The following beautiful result of Knörr [57] is perhaps the most general contribution to Brauer's Height zero conjecture.

**Theorem 3.3.** *The p-block B of a finite group G has an abelian defect group D if and only if all virtually irreducible RG-lattices of B have D as a vertex.*

## 4. Modular representation theory of finite groups with a T.I. Sylow $p$-subgroup

Let $p > 0$ be a fixed prime, and let $G$ be a finite group with a T.I. Sylow $p$-subgroup $P$. That is, $P \cap P^x = 1$ for all $x \in G \setminus N_G(P)$. For this class of finite groups Alperin's conjecture, the Alperin-McKay conjecture and Brauer's Problem 19 have been verified by H. Blau and the author [8] using the classification of the finite simple groups.

**Theorem 4.1.** *Let G be a finite group with a T.I. Sylow p-subgroup P. Let B be a p-block of G with defect group* $\delta(B) =_G P$ *and Brauer correspondent b in* $N = N_G(P)$. *Then*

a) $k(B) = k(b)$
b) $k_0(B) = k_0(b)$
c) $l(B) = l(b)$
d) $k(G) = k(N) + z(G)$, *where* $z(G)$ *denotes the number of p-blocks of defect zero in G.*

This result has several other consequences. It implies that Brauer's height zero conjecture holds for groups with T.I. Sylow $p$-subgroups.

**Corollary 4.2.** *Let G be a finite group with a T.I. Sylow p-subgroup P. Let B be a p-block of G with defect group* $\delta(B) =_G P$. *Then* $k(B) = k_0(B)$ *if and only if P is abelian.*

**Corollary 4.3.** *Let G be a finite group with a T.I. Sylow p-subgroup P such that P is not normal in G. Let B be a p-block of G with defect group* $\delta(B) =_G P$. *Then*

a) $k(B) \leq |P|$
b) $k_0(B) \leq |P/P'|$

Brauer's $k(B)$-conjecture appears to be very difficult even for groups with a normal Sylow $p$-subgroup $P$. For this case Gluck and Knörr [56] have found solutions when $P$ has a complement $U$ in $G$ which is of odd order or supersolvable.

Theorem 4.1 also supplies some further support for a long-standing conjecture in the theory of finite-dimensional algebras.

Two finite-dimensional $F$-algebras $B$ and $b$ are called *stably equivalent* if their module categories Mod $B$ and Mod $b$ are equivalent modulo their projective objects and projective homomorphisms. It has been conjectured for a long time that two such algebras have the same number $l(B) = l(b)$ of non-isomorphic non-projective simple modules. Generalizing work by P. Gabriel and C. Riedtmann [40], R. Martinez-Villa showed in [68] that two stably equivalent algebras of finite representation type have the same number of nonprojective simple modules.

Block ideals $B$ of group algebras with a T.I. Sylow $p$-subgroup $P$ of a finite group $G$ as defect group and their Brauer correspondent $b$ in $N = N_G(P)$ yield important examples of stably equivalent indecomposable algebras (see Feit [32, p. 118]), which are not necessarily of finite representation type. Of course, the verification of the above conjecture in the theory of algebras would imply assertions (a), (c), and (d) of Theorem 4.1 immediately. Unfortunately, we are very far from a general solution of this conjecture.

The proof of Theorem 4.1 depends heavily on the following group theoretical results due to Suzuki, Puig, and Gorenstein and Lyons, see [44] for precise references. Together with Clifford's theory they yield efficient reduction theorems.

**Lemma 4.4.** *Let $X$ be a finite group with a nonnormal T.I. Sylow $p$-subgroup $P$. Then the following assertions hold:*

a) *If $X$ is $p$-solvable, then $P$ is either cyclic or a generalized quaternion 2-group.*
b) *If $P$ is a generalized quaternion 2-group, then either $X = O(X)P$ or $X/O(X) \simeq \mathrm{SL}_2(3)$.*

**Proposition 4.5.** *Let $X$ be a finite group with a non-normal T.I. Sylow $p$-subgroup $P$. Suppose that $P$ is neither cyclic nor a generalized quaternion 2-group. Let $H = O^{p'}(X)$ and $U = O^p(H)$. Then the following assertions hold:*

a) $O_{p'}(X) \leq N_X(P)$.
b) $O_{p'}(H) < U \leq UP = H$.
c) $X = UN_X(P)$.
d) $G = U/O_{p'}(H)$ *is a nonabelian simple group with a T.I. Sylow $p$-subgroup.*
e) $X/O_{p'}(X)$ *acts faithfully on $G$ via conjugation.*
f) $O_{p'}(X) = C_X(H)$.
g) $O_{p'}(H) = Z(H)$.
h) $H = U$ *unless* $p = 3$ *and* $G \simeq \mathrm{PSL}(2,8)$ *or* $p = 5$ *and* $G \simeq {}^2B_2(2^5)$.
i) $Z(U) = Z(H)$ *is a factor group of the Schur multiplier of $G$.*

**Proposition 4.6.** *Let $G$ be a non-abelian simple group with a non-cyclic T.I. Sylow $p$-subgroup $P$. Then $G$ is isomorphic to one of the following groups:*

| | | |
|---|---|---|
| *a)* $\mathrm{PSL}_2(q)$, | *where* $q = p^n$ | *and* $n \geq 2$. |
| *b)* $\mathrm{PSU}_3(q^2)$, | *where* $q = p^n$. | |
| *c)* $p = 2$ | *and* $G \simeq {}^2B_2(2^{2m+1})$. | |
| *d)* $p = 3$ | *and* $G \simeq {}^2G_2(3^{2m+1})$ | *and* $m \geq 1$. |
| *e)* $p = 3$ | *and* $G \simeq \mathrm{PSL}_3(4)$ | *or* $M_{11}$. |
| *f)* $p = 5$ | *and* $G \simeq {}^2F_4(2)'$ | *or* McL. |
| *g)* $p = 11$ | *and* $G \simeq J_4$. | |

It is perhaps worthwile mentioning that the classification of the finite simple groups with a noncyclic T.I. Sylow $p$-subgroup given in Proposition 4.6 requires the classification of all finite groups at present; see Gorenstein-Lyons [44].

All assertions of Theorem 4.1 and its Corollaries 4.2 and 4.3 can be considered as generalizations of the theory of blocks $B$ with cyclic defect groups. However, the proof of the cyclic case does not generalize at all. There a crucial role is played by the Green correspondents $f(M)$ of the simple $B$-modules $M$. Each $f(M)$ is a uniserial $FN$-module, where $N = N_G(P)$. In particular, the head $\operatorname{hd} f(M) = f(M)/f(M)J(FN)$ and the socle $\operatorname{soc} f(M)$ of $f(M)$ are simple $FN$-modules.

At the Santa Cruz Conference on Finite Groups in 1979 already, J. L. Alperin had pointed out that there are 3-modular simple modules $M$ of $\mathrm{PSL}_3(4)$ whose Green correspondents $f(M)$ have non-simple heads. For details see Schneider [88].

Let $G$ be a finite group with a T.I. Sylow $p$-subgroup $P \neq 1$ and normalizer $N = N_G(P)$. Let $M$ be a simple $FG$-module with vertex $\mathrm{vx}(M) =_G P$ and Green correspondent $f(M)$ in $N$. Since $P$ is a T.I. set, $f(M)$ is the unique non-projective indecomposable direct summand of the restriction

$$M{\downarrow}_N = f(M) \oplus \text{projective} .$$

Since the projective part of $M{\downarrow}_N$ has a longer Loewy series than the non-projective direct summand $f(M)$, and since the structure of the indecomposable projective $FN$-modules is known, it is fairly easy to determine the structure of the socle and the head of the Green correspondent $f(M)$ whenever the matrices of the generators of $G$ corresponding to the representation of the $FG$-module $M$ are known. In order to give one example at least, the socle series of the Green correspondents of the non-projective 5-modular simple modules of the simple Tits group ${}^2F_4(2)'$ are now restated from H. Gollan's recent paper [42].

The *socle series* of an $FG$-module $M$ is a series

$$0 = \operatorname{soc}_0(M) < \operatorname{soc}_1(M) < \operatorname{soc}_2(M) < \ldots < \operatorname{soc}_s(M) = M$$

of submodules $\operatorname{soc}_i(M)$ of $M$ such that $\operatorname{soc}_i(M)/\operatorname{soc}_{i-1}(M)$ is the largest semi-simple submodule of $M/\operatorname{soc}_{i-1}(M)$ for $1 \leq i \leq s$. The integer $s$ is called the *socle length* of $M$.

**Theorem 4.7.** *Let $G = {}^2F_4(2)'$ be the simple Tits group. Then $F = \mathrm{GF}(25)$ is a splitting field for $G$, and a Sylow 5-subgroup $P$ of $G$ is a T.I. set with its normalizer $N = N_G(P) \cong (\mathbf{Z}_5 \times \mathbf{Z}_5) : 4A_4$.*

*Let $B_0$ be the principal 5-block of $G$ and let $b_0$ be its Brauer correspondent in $N$. Then the following assertions hold:*

a) *$B_0$ has 14 simple modules denoted by their dimensions: $1_G$, 26, 26*, 27, 27*, 78, 109, 109′, 351, 351*, 460, 460′, 593, 593′, where $n^*$ denotes the dual of the $FG$- module $n$, and where $n'$ is an algebraic conjugate of $n$.*
b) *Each simple $FG$-module $n$ of $B_0$ has vertex $P$.*
c) *$b_0$ has 14 simple modules denoted by their dimensions: $I = 1_N$, $1$, $1_a$, $1_a^*$, $1_b$, $1_b^*$, $2_a$, $2_a^*$, $2_b$, $2_b^*$, $2_c$, $2_c^*$, $3_a$, $3_b$.*
d) *The Green correspondents $f(n)$ of the simple $FG$-modules $n$ have the following socle series:*

$$f(1_G) = I = 1_N \ , \quad f(26) = 1_a^* \ , \quad f(26^*) = 1_a \ ,$$

$$f(27) = \begin{array}{ccccc} & & 1 & & \\ & & 2_a & & \\ & 3_b & & 3_b & \\ 2_a^* & & 2_b & & 2_c^* \\ 1_a & & 1_a^* & & 3_a \\ & 2_b^* & & 2_c & \\ & & 3_b & & \end{array} , \qquad f(27^*) = \begin{array}{ccccc} & & 3_b & & \\ & 2_b & & 2_c^* & \\ 1_a & & 1_a^* & & 3_a \\ & 2_b^* & & 2_c & \\ & & 3_b & & \\ & 2_a & & 2_a^* & \\ & 1 & & 3_b & \end{array} ,$$

$$f(78) = \begin{array}{ccccc} & & 2_a^* & & \\ & & 3_a & & \\ 2_a & & 2_b^* & & 2_c \\ 1_b & & 1_b^* & & 3_b \\ & 2_b & & 2_c^* & \\ & I & & 3_a & \\ & 2_a & & 2_a^* & \end{array},$$

$$f(109) = \begin{array}{c} 1_b \\ 2_b \\ 3_a \\ 2_b^* \\ 1_b^* \end{array}, \qquad f(109') = \begin{array}{c} 1_b^* \\ 2_c^* \\ 3_a \\ 2_c \\ 1_b \end{array},$$

$$f(351) = \begin{array}{ccccccccccccccc} & & & & & & & 3_a & & & & & & & \\ & & & & & 2_a & & 2_b & & 2_c & & & & & \\ & & & I & & 1_b & & 1_b^* & & 3_b & & 3_b & & & \\ 2_a^* & & 2_a^* & & 2_b & & 2_b & & 2_b & & 2_c^* & & 2_c^* & & 2_c^* \\ & 1 & & 1_a & & 1_a^* & & 3_a & & 3_a & & 3_a & & 3_a & \\ & & 2_a & & 2_a & & 2_b^* & & 2_b^* & & 2_c & & 2_c & & \\ & & & I & & 1_b & & 1_b^* & & 3_b & & 3_b & & & \\ & & & & & 2_a^* & & 2_b & & 2_c^* & & & & & \end{array},$$

$$f(351^*) = \begin{array}{ccccccccccccc} & & & & 2_a & & 2_b^* & & 2_c & & & & \\ & & I & & 1_b & & 1_b^* & & 3_b & & 3_b & & \\ & 2_a^* & & 2_a^* & & 2_b & & 2_b & & 2_c^* & & 2_c^* & \\ 1 & & 1_a & & 1_a^* & & 3_a & & 3_a & & 3_a & & 3_a \\ & 2_a & & 2_a & & 2_b^* & & 2_b^* & & 2_c & & 2_c & \\ & & I & & 1_b & & 1_b^* & & 3_b & & 3_b & & \\ & & & & 2_a^* & & 2_b & & 2_c^* & & & & \\ & & & & 2_b^* & & 2_c & & 3_a & & & & \end{array},$$

$$f(460) = \begin{array}{cccccc} & & 2_a^* & 2_b & & \\ & 1 & 1_a & 3_a & 3_a & \\ 2_a & 2_a & 2_b^* & 2_b^* & 2_c & 2_c \\ I & 1_b & 1_b^* & 3_b & 3_b & 3_b \\ 2_a^* & 2_a^* & 2_b & 2_b & 2_c^* & 2_c^* \\ & 1 & 1_a^* & 3_a & 3_a & \\ & & 2_a & 2_b^* & & \end{array}, \quad f(460') = \begin{array}{cccccc} & & 2_a^* & 2_c^* & & \\ & 1 & 1_a^* & 1_a & 3_a & \\ 2_a & 2_a & 2_b^* & 2_b^* & 2_c & 2_c \\ I & 1_b & 1_b^* & 3_b & 3_b & 3_b \\ 2_a^* & 2_a^* & 2_b & 2_b & 2_c^* & 2_c^* \\ & 1 & 1_a & 3_a & 3_a & \\ & & 2_a & 2_c & & \end{array},$$

$$f(593) = \begin{array}{ccc} & 3_a & \\ 2_a & & 2_b^* \\ I & & 3_b \\ 2_a^* & & 2_b \\ & 3_a & \end{array}, \qquad f(593') = \begin{array}{ccc} & 3_a & \\ 2_a & & 2_c \\ I & & 3_b \\ 2_a^* & & 2_c^* \\ & 3_a & \end{array}.$$

**Proof.** Since $P$ is elementary abelian of order 25, assertion b) follows immediately from Knörr's theorem, see [55]. Assertion c) is an easy consequence of the group structure of $N$ explained in the Atlas [18], p. 74. Assertion a) is due to G. Hiss [45], and assertion d) is a restatement of the main result of H. Gollan [42]. Both papers depend on intensive computer work.

**Corollary 4.8.** *Let $G$ be a finite group with a T.I. Sylow $p$-subgroup $P$ such that neither the simple Janko group $J_4$ nor the simple McLaughlin group* McL *is a composition factor. Let $B$ be a $p$-block of $G$ with defect group $\delta(B) =_G P$. Then for every simple $FG$-module $M$ of $B$ with Green correspondent $f(M)$ the following assertions hold:*

a) *$M$ has vertex $\mathrm{vx}(M) =_G P$.*
b) *The head $\mathrm{hd}(f(M))$ and the socle $\mathrm{soc}(f(M))$ of the Green correspondent $f(M)$ of $M$ are multiplicity-free semisimple $FN$-modules, where $N = N_G(P)$.*

**Proof.** Using the methods of the paper [8] of Blau and the author, the proof can be reduced to the case, where $G$ is a perfect central extension of a simple group $S$ with a T.I. Sylow $p$-subgroup $P$.

If $S$ is a finite simple group of Lie type, then $p$ is the describing prime of $S$ by Proposition 4.6. Therefore, both assertions a) and b) are true by the main results of Dipper [24] and [25]. In fact, Dipper has shown that $\mathrm{hd}(f(M))$ is a simple $FN$-module in this case.

In all the remaining cases Proposition 4.6 asserts that $P$ is abelian. Thus a) holds by Knörr's theorem [55].

Now Theorem 4.7 and Proposition 4.6 imply that assertion b) has to be checked for $p = 3$ and $S \in \{\mathrm{PSL}_3(4), M_{11}\}$. This follows from the results of Schneider [88] and [89].

*Remark 4.9.* Assertion b) of Corollary 4.8 does not hold for the simple McLaughlin group McL in characteristic 5. This has been observed by W. Lempken and R. Staszewski [64] recently.

By the Atlas [18], p. 100, the simple group $G =$ McL contains a subgroup $U \cong U_4(3)$ with permutation character

$$(1_U)^G = 1_G \oplus 22 \oplus 252 \ .$$

From this permutation representation Lempken and Staszewski constructed a simple $FG$-module $M = 21$ of dimension 21 over the field $F = \mathrm{GF}(5)$. In its third wedge power $\bigwedge_3 21$ there is a simple composition factor of dimension 560 for which GF(5) is also a splitting field.

Let $P$ be a Sylow 5-subgroup of $G$, and let $N = N_G(P)$ be its normalizer. Then by the Atlas [18], p. 100, $N = PK$, where $P \cap K = 1$ and $K = \mathbf{Z}_3 : \mathbf{Z}_8$ with $\mathbf{Z}_8$ acting invertingly on $\mathbf{Z}_3 \triangleleft K$. The ordinary characters of $K$ are precisely the irreducible modular characters of $N$. It follows from the character table of $K$ that $N$ has eight linear characters denoted by $\lambda_1 = 1_N, \lambda_2, \ldots, \lambda_8$, and it has four 2-dimensional characters $\mu_1, \mu_2, \mu_3$ and $\mu_4$ such that $\mu_1$ and $\mu_2$ are self-dual, and $\mu_3 = \mu_4^*$.

Let $P(\mu_i)$ denote the projective cover of the simple $FN$-module $\mu_i$.

With this notation Lempken and Staszewski [64] have shown:

**Theorem 4.10.** *Let $P$ be a Sylow 5-subgroup of the simple McLaughlin group* McL, *and let $N = N_G(P)$. Let $F =$ GF(5). Then the 560-dimensional simple $FG$-module 560 has the following properties:*

a) *Its vertex* vx(560) *is $G$-conjugate to $P$.*
b) *Its restriction to $N$ decomposes as $560\downarrow_N = f(560) \oplus P(\mu_2)$, where the 310-dimensional $FN$-module $f(560)$ is the Green correspondent of 560.*
c) $\mathrm{hd}(f(560)) \cong 2 \cdot \mu_3 \oplus 2 \cdot \mu_4 \cong \mathrm{soc}(f(560))$. *In particular, the head and the socle of the Green correspondent $f(560)$ are not multiplicity-free semisimple $FN$-modules.*

Their work involves the explicit construction of all 5-modular irreducible representations of the group McL, and so obviously requires a thorough use of computational work. Schneider's computational methods [89] for finding Green correspondents, vertices and sources of indecomposable modules were helpful.

In view of Theorem 4.8.a) we state the following open

*Question.* Let $G$ be a finite group with a T.I. Sylow $p$-subgroup $P \neq 1$, and let $F$ be a field of characteristic $p > 0$. Is $P$ then a vertex of every non-projective simple $FG$-module?

## 5. Modular representation theory of the symmetric groups.

Throughout this section $S(n)$ denotes the symmetric group on $n$ letters, and $p > 0$ a fixed prime number. To each irreducible character $\chi$ of $S(n)$ there corresponds a unique partition $\lambda = (\lambda_1, \lambda_2, \ldots, \lambda_r)$ of $n$, where $\lambda_1 \geq \lambda_2 \geq \ldots \geq \lambda_r$, and $n = \sum_{i=1}^{n} \lambda_i$. Therefore, $\chi$ is also denoted by $\chi_\lambda$.

To each partition $\lambda = (\lambda_1, \lambda_2, \ldots, \lambda_r)$ of $n$ there belongs also a Young diagram. It consists of $n$ nodes placed in $r$ rows such that the $i$-th row consists of $\lambda_i$ nodes, and all the rows start in the same column. The node of the Young diagram $\lambda$ which lies in the $i$-th row and $j$-th column is denoted by $\lambda_{ij}$. It is the corner of the $(i,j)$-hook $H_{ij}^\lambda$ of $\lambda$, which consists of $\lambda_{ij}$, and all the nodes to the right of it in the same row together with all the nodes lower down and in the same column as the corner. The number $h_{ij}^\lambda$ of the nodes of the hook $H_{ij}^\lambda$ is called the hook length of $H_{ij}^\lambda$.

Let $q$ be any positive integer. A partition $\lambda$ without hooks of length $q$ is called a $q$-core. For any $q$ and every partition $\lambda$ of $n$ there is a uniquely determined $q$- core $\lambda_{(q)}$ which is obtained from $\lambda$ by successive removals of hooks of length $q$ as often as possible, see James-Kerber [52], p. 79.

With these notations and definitions we can now state the fundamental theorem in the modular representation theory of the symmetric groups. It is due to R. Brauer and G. de B. Robinson, and it confirms a conjecture of T. Nakayama, see [52], p. 245.

**Theorem 5.1.** *Two ordinary irreducible characters $\chi_\lambda$ and $\chi_\mu$ of the symmetric group $S(n)$ belong to the same $p$-block $B$ of $S(n)$ if and only if their $p$-cores $\lambda_{(p)}$ and $\mu_{(p)}$ are equal.*

The common $p$-core partition $\kappa_{(p)}$ of all irreducible characters $\chi_\lambda$ of a $p$-block $B$ of $S(n)$ is called the *p-core* of $B$. In particular, $B$ has defect zero if and only if its $p$-core is a partition of $n$.

Each $p$-block $B$ of $S(n)$ has a *weight* $w(B) = w$ defined by $w = p^{-1}(n - |\kappa_{(p)}|)$, where $|\kappa_{(p)}|$ denotes the sum of the parts $\kappa_i$ of the $p$-core partition $\kappa_{(p)} = (\kappa_1, \kappa_2, \ldots, \kappa_t)$ of $B$.

In 1976 J. B. Olsson verified the Alperin-McKay conjecture and the height zero conjecture for the $p$-blocks of $S(n)$ in his fundamental paper [79]. Without stating it explicitely he proved there the following reduction theorem which has become a powerful tool in the modular representation theory of $S(n)$, see also Theorem 1.10 of [74].

**Theorem 5.2.** *Let $B$ be a $p$-block of the symmetric group $S(n)$ with weight $w(B) = w$. Then there is a canonical height preserving bijection between the characters of $B$ and the characters of the principal $p$-block $B_0$ of the symmetric group $S(pw)$. Furthermore:*

a) *Each defect group $\delta(B)$ of $B$ is isomorphic to a Sylow $p$-subgroup $\delta(B_0)$ of $S(pw)$.*
b) *$B$ and $B_0$ have the same number $l(B) = l(B_0)$ of irreducible modular characters.*

All the conjectures mentioned in section 3 have been proved for the $p$-blocks $B$ of $S(n)$ by determining the precise values of the block invariants $k(B)$, $k_0(B)$ and $l(B)$. The same methods were applied later for the alternating groups and the covering groups.

For every integer $n \geq 0$ let $\pi(n)$ be the number of partitions $\lambda \vdash n$ of $n$, and $\pi(0) = 1$. Let

$$P(X) = \sum_{n \geq 0} \pi(n) X^n$$

be the generating function in the indeterminate $X$. For a pair of integers $s, t \geq 0$ the integer $k(s,t)$ is defined by

$$P(X)^s = \sum_{t \geq 0} k(s,t) X^t \ .$$

In particular, $k(1,t) = \pi(t)$ and $k(s,1) = s$ for all $s, t \geq 0$.

An *s-split* of $t$ is an $s$-tuple $(t_1, t_2, \ldots, t_s)$ of non-negative integers $t_i$, $1 \leq i \leq s$, such that $t = \sum_{i=1}^{s} t_i$. If $\Phi(s,t)$ denotes the set of $s$-splits of $t$, then its cardinality $|\Phi(s,t)| = \binom{s+t-1}{t}$, a binomial coefficient. In particular,

$$k(s,t) = \sum_{(t_1, t_2, \ldots, t_s) \in \Phi(2,t)} \pi(t_1)\pi(t_2)\ldots\pi(t_s) \ .$$

With these notations we can now state the local version of the reduction theorem. It is due to Olsson and the author, see [74], p. 224.

**Theorem 5.3.** *Let $B$ be a $p$-block of the symmetric group $G = S(n)$ with weight $w(B) = w$ and defect group $\delta(B) =_G D$. Let $w = \sum_{\beta \geq 0} t_\beta p^\beta$ be the $p$-adic expansion of $w$. Let $b$ be the Brauer correspondent of $B$ in $N = N_G(D)$. Let $\widetilde{G} = S(pw)$, $B_0$ be the principal $p$-block of $\widetilde{G}$ and $\widetilde{b}_0$ its Brauer correspondent in $\widetilde{N} = N_{\widetilde{G}}(\widetilde{D})$.*

*Then the block ideals $b$ and $b_0$ are Morita equivalent. In particular, they have the same Cartan matrix, decomposition matrix and the same block invariants:*

a) $l(b) = l(\widetilde{b}_0) = \prod_{\beta \geq 0} k\left((p-1)^{\beta+1}, t_\beta\right)$,

b) $k(b) = k(\widetilde{b}_0)$,

c) $k_0(b) = k_0(\widetilde{b}_0) = \prod_{\beta \geq 0} (p^{\beta+1}, t_\beta)$.

*Remark 5.4.* It is *not* true that the $p$-block $B$ of $G = S(n)$ and $B_0$ of $\widetilde{G} = S(pw)$ are Morita equivalent, in general. By Theorem 5.2 their block invariants

$$k(B) = k(B_0) = k(p, w) ,$$

$$l(B) = l(B_0) = k(p-1, w)$$

are the same. Therefore, the shape of the decomposition matrices $D(B)$ and $D(B_0)$ of $B$ and $B_0$, respectively, is the same. But in general, $D(B) \neq D(B_0)$.

For a specific counterexample consider the principal 3-block $B$ of the symmetric group $G = S(11)$ on 11 letters. It has weight $w(B) = w = 3$. Thus $B_0$ is the principal 3-block of $\widetilde{G} = S(9)$. By James [51], p. 145 all decomposition numbers $d_{ij}$ of $D(B_0)$ are zero or one, whereas $D(B)$ has some decomposition numbers $d_{ij} = 2$, see [51], p. 147.

In [79] and [80] Olsson proved the height zero conjecture, the Alperin-McKay conjecture, and Brauer's $k(B)$-conjecture for the $p$-blocks of $S(n)$. His results are collected in

**Theorem 5.5.** *Let $B$ be a $p$-block of $G = S(n)$ with defect group $\delta(B) =_G D$ and Brauer correspondent $b$ in $N = N_G(D)$. If $w = \sum_{\beta \geq 0} t_\beta p^\beta$ is a $p$-adic expansion of the weight $w(B) = w$ of $B$, then the following assertions hold:*

a) $k(B) = k(p, w) \leq |D|$

b) $k_0(B) = k_0(b) = \prod_{\beta \geq 0} k\left(p^{\beta+1}, t_\beta\right)$

c) $k(B) = k_0(B)$ *if and only if $D$ is abelian*

d) $k_0(B) \leq \prod_{\beta \geq 0} p^{\beta+1} t_\beta = |D : D'|$

In [80] Olsson proves assertion a) of Theorem 5.5 by showing first that for any pair $s \geq 2$, $t \geq 1$ of integers $k(s,t) \leq s^t$, unless $s = 2$ and $2 \leq t \leq 6$. For $p = 2$ and $w \in \{2, 3, 4, 5, 6\}$ assertion a) can be checked easily. In all the other cases the result follows then from the inequalities

$$k(B) = k(p, w) \leq p^{w + \nu(w!)} = p^{d(B)} = |D| ,$$

where $\nu$ denotes the $p$-adic valuation of the integers.

In [3] J. Alperin and P. Fong have verified Alperin's weight conjecture for the $p$-blocks of the symmetric groups $S(n)$. The bulk of their proof consists of the determination of the structure of the radical $p$-subgroups of $S(n)$ and their normalizers.

Let $S(n) = S(V)$ be the symmetric group of degree $n$ acting on a set $V$ with $n = |V|$ elements. For each positive integer $c$, let $A_c$ be an elementary abelian $p$-subgroup of $S(n)$ with order $|A_c| = p^c$, embedded regularly as a subgroup of $S(p^c)$. Then

$$A_c = C_{S(p^c)}(A_c), \text{ and } N_{S(p^c)}(A_c)/A_c \cong \mathrm{GL}(c,p) .$$

For each sequence $r = (c_1, c_2, \ldots, c_{s(r)})$ of positive integers, let $A_r$ be the $s(r)$-fold wreathed product

$$A_r = A_{c_1} \wr A_{c_2} \wr \ldots \wr A_{c_{s(r)}}, \text{ and let } d(r) = \sum_{i=1}^{s(r)} c_i .$$

Then $A_r$ is embedded as a transitive subgroup of the symmetric group $S\left(p^{d(r)}\right)$, and

$$N_{S(p^{d(r)})}(A_r)/A_r \cong \mathrm{GL}(c_1,p) \times \mathrm{GL}(c_2,p) \times \ldots \times \mathrm{GL}(c_{s(r)},p)$$

by Alperin-Fong [3]. If $d = d(r)$, then $A_r$ is called a *basic $p$-subgroup* of the symmetric group $S(p^d)$ with *degree* $\deg(A_r) = p^d$ and length $l(A_r) = s(r)$.

Using these basic $p$-subgroups $A_r$ Alperin and Fong [3] describe the structure of the radical subgroups and their normalizers in Lemmas (2A) and (2B) which are restated together in

**Proposition 5.6.** *Let $C$ be the set of sequences $r = (c_1, c_2, \ldots, c_{s(r)})$ of positive integers. Let $R$ be a radical $p$-subgroup of $G = S(n) = S(V)$. Then the following assertions hold:*

a) *There exist decompositions $V = V_0 \cup V_1 \cup V_2 \cup \ldots \cup V_u$ $R = R_0 \times R_1 \times R_2 \times \ldots \times R_u$ such that $R_0$ is the identity subgroup of $S(V_0)$, and for each $i \in \{1, 2, \ldots, u\}$ $R_i \neq 1$ is a basic $p$-subgroup $A_r$ of $S(V_i)$ for some sequence $r \in C$.*

b) *For each $r \in C$ let $V(r) = \bigcup_i V_i$, $R(r) = \prod_i R_i$, where $i$ runs over all the indices $i$ such that $R_i = A_r$. Let $\zeta(r)$ be the multiplicity of $A_r$ in $R(r)$. Then $\zeta$ is a function $C \to \mathbb{N} \cup \{0\}$ satisfying $\sum_r \zeta(r) p^{d(r)} \leq n$ and the following assertions hold:*

$$R = R_0 \times \prod_r R(r) ,$$

$$N_G(R) = S(V_0) \times \prod_r N_{S(V(r))}(R(r)) ,$$

$$N_G(R)/R = S(V_0) \times \prod_r N_{S(V(r))}(R(r))/R(r) .$$

*$\zeta$ is called the multiplicity function of $R$.*

c) *If $V_r$ denotes the underlying set of $A_r$ in $V$ then*

$$N_{S(V(r))}(R(r)) \cong [N_{S(V_r)}(A_r)] \wr S(\zeta(r)) ,$$

$$N_{S(V(r))}(R(r))/R(r) \cong [N_{S(V_r)}(A_r)/A_r] \wr S(\zeta(r)) .$$

d) *For each $r \in C$ $A_r$ is a basic $p$-subgroup of $S\left(p^{d(r)}\right)$ with length $l(A_r) = s(r)$ and degree $\deg(A_r) = p^{d(r)}$, and*

$$R \cong \prod_{d \geq 1} \prod_{\{r | d(r) = d\}} (A_r)^{\zeta(r)} .$$

e) *The G-conjugacy class of the radical p-subgroup R is uniquely determined by the multiplicity function* $\zeta : C \to \mathbf{N} \cup \{0\}$, *i.e.*

$$R =_G R_\zeta = \prod_{d \geq 1} \prod_{\{r|d(r)=d\}} (A_r)^{\zeta(r)} .$$

*Definition.* For every radical $p$-subgroup $R$ with multiplicity function $\zeta$ the number

$$w(R) = \sum_{d \geq 1} \sum_{\{r|d(r)=d\}} \zeta(r) p^{d-1}$$

is called the *width* of $R$.

By the reduction theorem 5.2 it suffices to verify Alperin's weight conjecture for the principal $p$-block $B$ of $S(pw)$, where $w$ denotes the weight of $B$. In this situation the following result holds by Alperin-Fong [3].

**Proposition 5.7.** *Let B be the principal p-block of S(pw). Let R be a radical p-subgroup of G with width*

$$w(R) = \sum_{d \geq 1} \sum_{\{r|d(r)=d\}} \zeta(r) p^{d-1} .$$

*For each* $r \in C$ *let* $N_r = [N_{S(p^{d(r)})}(A_r)/A_r] \wr S(\zeta(r))$, $e(r) = (p-1)^{s(r)}$, *and let* $d_0(N_r)$ *be the number of* $e(r)$*-tuples* $(\kappa_1, \kappa_2, \ldots, \kappa_{e(r)})$ *of p-core partitions* $\kappa_i$ *such that*

$$\sum_{i=1}^{e(r)} |\kappa_i| = \zeta(r) .$$

*Then* $\prod_{r \in C} d_0(N_r)$ *equals the number of B-weights* $(R, \varphi)$ *with radical p-subgroup R.*

Using these subsidiary results Alperin and Fong [3] give a fairly short proof of the following result confirming Alperin's weight conjecture for the symmetric groups.

**Theorem 5.9.** *Let B be a p-block of the symmetric group S(n) with weight* $w(B) = w$. *Let* $l^*(B)$ *be the number of B-weights* $(R, \varphi)$ *of G. Then*

$$l^*(B) = k(p-1, w) = l(B) .$$

Recently, J. Scopes, a former student of K. Erdmann at Oxford University, has verified Donovan's conjecture for the symmetric groups. In [91] she proves the following beautiful result.

**Theorem 5.10.** *The p-blocks B of the symmetric groups S(n),* $n = 1, 2, \ldots$, *with a given weight* $w(B) = w > 0$ *can be collected into families according to the shapes of their p-cores* $\kappa_{(p)}$. *Each family consists of Morita equivalent blocks. Furthermore,*

a) *the number of such families is* $\prod_{i=1}^{p} [(i-1)(w-1)+1]$ *at most;*
b) *every p-block B of a symmetric group S(n) with weight* $w(B) = w$ *is a Morita equivalent to a p-block B' of some symmetric group S(m) with weight* $w(B') = w$, *where* $m \leq wp + \frac{1}{4}p^2(p-1)^2(w-1)^2$.

**Corollary 5.11.** *a) Donovan's conjecture holds for p-blocks of the symmetric groups.*

b) *The number of Cartan matrices of p-blocks of the symmetric groups with a given defect group is finite.*

## 6. Modular representation theory of the covering groups of the alternating and symmetric groups

In his fundamental paper [90] I. Schur constructed the irreducible characters of the covering groups $S^+(n)$ and $A^+(n)$ of the finite symmetric and alternating groups $S(n)$ and $A(n)$ of degree $n$, respectively. In fact, $S^+(n)$ denotes one of the following two non-isomorphic groups

$$\hat{S}(n) = \left\{ a_1, a_2, \ldots, a_{n-1}, z \,\middle|\, \begin{array}{c} z^2 = 1,\ a_i^2 = z,\ (a_i a_{i+1})^3 = z \\ \cdot\ [a_i, a_j] = z \text{ if } |i-j| \geq 2 \end{array} \right\},$$

$$\widetilde{S}(n) = \left\{ a'_1, a'_2, \ldots, a'_{n-1}, z \,\middle|\, \begin{array}{c} z^2 = 1,\ a_i'^2 = 1,\ (a'_i a'_{i+1})^3 = 1 \\ [a'_i, z] = 1,\ [a'_i, a'_j] = z \text{ if } |i-j| \geq 2 \end{array} \right\}.$$

In particular, $\hat{S}(1) = \widetilde{S}(1) = \langle z \rangle$, $\hat{S}(2) \cong \mathbf{Z}_4$, $\hat{S}(2) = \mathbf{Z}_2 \times \mathbf{Z}_2$. If $n \geq 3$ there is a nonsplit exact sequence

$$1 \longrightarrow \langle z \rangle \longrightarrow S^+(n) \xrightarrow{\pi} S(N) \longrightarrow 1,$$

where $|z| = 2$, and $\pi(a_i) = (i, i+1)$.

$A^+(n)$ is the commutator subgroup of $S^+(n)$. All the results mentioned here hold for $\hat{S}(n)$ and $\widetilde{S}(n)$. Therefore, we do not distinguish these covering groups. Using the following notations the modular representation theory of the covering groups $S^+(n)$ and $A^+(n)$ can be developed simultaneously. When $H$ is a subgroup of $S(n)$ we define

$$H^+ = \pi^{-1}(H), \quad H^- = \pi^{-1}(H \cap A(n)).$$

Moreover, $S^-(n) = A^+(n) = A^-(n)$ is the covering group of $A(n)$. Thus $H^\delta$ is defined for each sign $\delta \in \{+1, -1\}$. The exceptional 6-fold covers of $A(6)$ and $A(7)$ are denoted by $C_6$ and $C_7$, respectively.

If $H \subseteq S(n)$ and $P$ is a normal $p$-subgroup of $H$ then $P$ may also be considered as normal $p$-subgroup of $H^+$. Therefore, for notational convenience $[H/P]^+$ is written as $H^+/P$.

An irreducible character $\chi$ of $S^\delta(n)$, which does not have $\langle z \rangle$ in its kernel, is called a *spin* character. A partition

$$\lambda = (\lambda_1, \lambda_2, \ldots, \lambda_m),\ \lambda_1 > \lambda_2 > \ldots > \lambda_m > 0$$

of $n$ such that all parts $\lambda_i$ of $\lambda$ are different is called a *bar partition*. Each such partition $\lambda$ has a sign defined by $\epsilon(\lambda) = (-1)^{n-m}\delta$.

By Schur [90] there is a correspondence between the bar partitions $\lambda$ of $n$ and the irreducible spin characters $\chi_\lambda$ of $S^\delta(n)$, where $\chi_\lambda$ denotes one spin character if $\epsilon(\lambda) = \delta$ and a pair of two characters if $\epsilon(\lambda) = -\delta$.

Proving a conjecture of A. O. Morris the distribution of the irreducible spin characters of $S^\delta(n)$ into $p$-blocks was given by J. F. Humphreys [50] and M. Cabanes [14] for all primes $p \neq 2$. So far, the case $p = 2$ yields a difficult open problem in the representation theory of the covering groups. In order to state Humphreys' theorem the following notations and subsidiary results are needed.

Let $\lambda = (\lambda_1, \lambda_2, \ldots, \lambda_m)$ be a bar partition of $n$. The *shifted Young diagram* $S(\lambda)$ of $\lambda$ is obtained from the usual Young diagram by shifting the $i$-th row $(i-1)$-positions to the right. The $j$-th node in the $i$-th row is called $(i,j)$-node. To each node of $S(\lambda)$ one associates a *bar length*, such that the bar lengths of the $(i,j)$-nodes in the $i$-th row are the integers in the set

$$(\{1, 2, \ldots, \lambda_i\} \cup \{\lambda_i + \lambda_j | j > i\}) \setminus \{\lambda_i - \lambda_j | j > i\}$$

written in decreasing order.

*Example.* $\lambda = (6, 4, 2)$

$S(\lambda) =$ 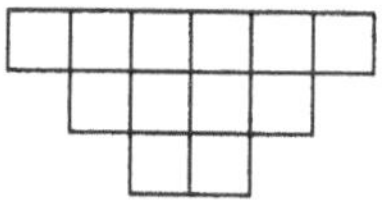

Bar lengths

| | | | | | |
|---|---|---|---|---|---|
| 10 | 8 | 6 | 5 | 3 | 1 |
| | 6 | 4 | 3 | 1 | |
| | | 2 | 1 | | |

Denote the $(i,j)$-bar length by $l_{ij}^\lambda$. In [82] Olsson introduces the following definitions.

*Definition.* Let $\lambda = (\lambda_1, \lambda_2, \ldots, \lambda_m)$ be a bar partition of $n$. To each node $(i,j)$ of the shifted Young diagram $S(\lambda)$ belongs a subdiagram called *bar*. It consists of $l_{ij}^\lambda$ nodes such that:

a) If $i + j > m$, then the $(i,j)$-bar consists of the last $l_{ij}^\lambda$ nodes in the $i$-th row of $S(\lambda)$. Such a bar is called *unmixed*.

b) If $i + j \leq m$, the $(i,j)$-bar consists of all the nodes in the $i$-th and all the nodes in the $(i+j)$-th row of $S(\lambda)$. Such a bar is called *mixed*.

A bar of length $l$ is called an *$l$-bar*.

*Definition.* Let $\lambda = (\lambda_1, \lambda_2, \ldots, \lambda_m)$ be a bar partition of $n$. The bar partition $\mu$ obtained from $\lambda$ by removing the $(i,j)$-bar of $S(\lambda)$ of length $l_{ij}^\lambda$ is the partition $\mu \vdash (n - l_{ij}^\lambda)$ the parts of which are

$$\begin{array}{ll} \lambda_1, \lambda_2, \ldots, \lambda_{i-1}, \lambda_i - l_{ij}^\lambda, \lambda_{i+1}, \ldots, \lambda_m & \text{if } i + j > m, \text{ and} \\ \lambda_1, \lambda_2, \ldots, \lambda_{i-1}, \lambda_{i+1}, \ldots, \lambda_{i+j-1}, \lambda_{i+j+1}, \ldots, \lambda_m & \text{if } i + j \leq m. \end{array}$$

In [82] Olsson gives a short proof of the following result due to Morris.

**Proposition 6.1.** *Let $p \neq 2$ be a prime number, and let $\lambda$ be a bar partition. Let $\lambda_{(\overline{p})}$ be the bar partition obtained from $\lambda$ by removing all $p$-bars from $\lambda$. Then $\lambda_{(\overline{p})}$ is uniquely determined by $\lambda$, and it is called the $p$-bar core of $\lambda$.*

After all these preparations Humphreys' theorem [50] called Morris conjecture can be stated.

**Theorem 6.2.** *Let $p$ be an odd prime, and let $\lambda$ and $\mu$ be bar partitions of $n$, which are not $p$-bar cores. Then the irreducible spin characters $\chi_\lambda$ and $\chi_\mu$ of $S^\delta(n)$ are in the same $p$-block $B$ of $S^\delta(n)$ if and only if their $p$-bar cores $\lambda_{(\overline{p})}$ and $\mu_{(\overline{p})}$ are equal.*

*If $\lambda$ is a $p$-bar core, then each irreducible spin character labelled by $\lambda$ forms a $p$- block of defect zero. Furthermore, a $p$-block $B$ of $S^+(n)$ contains either only spin characters or no spin characters at all.*

A $p$-block $B$ of $S^\delta(n)$ consisting only of irreducible spin characters is called *spin block* of $S^\delta(n)$. The common $p$-bar core partition $\kappa_{(\overline{p})}$ of all irreducible spin characters $\chi_\lambda$ of $B$ is called the *$p$-bar core* of $B$. If $\kappa_{(\overline{p})} = (n)_{(\overline{p})}$, then $B$ is called the *principal spin* block of $S^\delta(n)$.

Each spin $p$-block $B$ of $S^\delta(n)$ has a *$p$-bar weight* $w_{(\overline{p})}(B) = w$ defined by $w = p^{-1}(n - |\kappa_{(\overline{p})}|)$, where $|\kappa_{(\overline{p})}|$ denotes the sum of the parts $\kappa_i$ of the $p$-bar core partition $\kappa_{(\overline{p})} = (\kappa_1, \dots, \kappa_t)$. The *sign* $\sigma(B)$ of the spin $p$-block $B$ of $S^\delta(n)$ is defined by $\sigma(B) = \delta(-1)^{|\kappa_{(\overline{p})}|-t} = \epsilon(\kappa_{(\overline{p})})\delta$.

Let $H \subseteq S(n)$ and $\delta$ be a sign. Two irreducible characters $\chi, \psi$ of $H^\delta$ are called *associated* if $\chi^{H^+} = \psi^{H^+}$ for $\delta = -1$ or $\chi_{|H^-} = \psi_{|H^-}$ for $\delta = 1$. If $\chi$ has only itself as associated character then $\chi$ is *selfassociate* (s.a.). Otherwise, $\chi$ is called non selfassociate (n.s.a.), and $\chi^a$ denotes the unique associated character of $H^\delta$ different from $\chi$. Each irreducible spin character $\chi$ of $H^\delta$ has a *sign* $\sigma(\chi)$ defined by

$$\sigma(\chi) = \begin{cases} 1 & \text{if } \chi = \chi^a \\ -1 & \text{if } \chi \neq \chi^a \end{cases}.$$

Suppose that $H^\delta \neq H^{-\delta}$. Let $B$ be a $p$-block of $H^\delta$. Then there is a unique $p$-block $B^*$ of $H^\delta$ such that $B^*$ covers $B$ if $\delta = -1$, or $B^*$ is covered by $B$ if $\delta = 1$. The blocks $B$ and $B^*$ are called corresponding blocks.

For each $p$-block $B$ of $H^\delta$ let $k(B)_+$ and $k(B)_-$ be the number of s.a. and the number of *pairs* of n.s.a. irreducible characters of $B$, respectively. In particular, $k(B) = k(B)_+ + 2k(B)_-$.

Similarly, $k_0(B) = k_0(B)_+ + 2k_0(B)_-$, and $l(B) = l(B)_+ + 2l(B)_-$, where $k_0(B)$ and $l(B)$ denote the irreducible height zero and modular characters of $B$, respectively. By Lemma 1.4 of Michler-Olsson [76] $l(B)_- = l(B^*)_-$ and $l(B^*)_+ = l(B)_-$.

The last assertion of the following reduction theorem is due to Cabanes [14], all other parts are due to Olsson and the author [75].

**Theorem 6.3.** *Let $B$ be a spin $p$-block of $G = S^\delta(n)$ with weight $w = w(B) > 0$ and sign $\sigma(B) = \sigma$. Let $b$ be its Brauer correspondent in $N = N_G(D)$, where $D$ is a defect group of $B$. Let $B_0$ be the principal spin block of $S^{\sigma\delta}(pw)$ and $b_0$ its Brauer correspondent. Then the following assertions hold:*

a) *$k(B)_\epsilon = k(B_0)_\epsilon$ and $k(b)_\epsilon = k(b_0)_\epsilon$ for $\epsilon \in \{+1, -1\}$*
b) *$k_0(B)_\epsilon = k_0(B_0)_\epsilon$ and $k_0(b)_\epsilon = k_0(b_0)_\epsilon$ for $\epsilon \in \{+1, -1\}$*
c) *$l(B)_\epsilon = l(B_0)_\epsilon$ and $l(b)_\epsilon = l(b_0)_\epsilon$ for $\epsilon \in \{+1, -1\}$*
d) *The defect group $D$ of $B$ is a Sylow $p$-subgroup of $S^{\sigma\delta}(pw)$*

If $\lambda$ is a partition, then $\lambda^0$ denotes its dual partition, and $\lambda$ is called *symmetric* if $\lambda = \lambda^0$. For every pair of integers $r, w$ let

$$K(r,w) = \left\{ (\lambda_1, \lambda_2, \ldots, \lambda_r) \middle| \lambda_i \text{ partition, and } \sum_{i=1}^{r} |\lambda_i| = w \right\} .$$

The cardinality of $K(r,w)$ is denoted by $k(r,w)$. For every $\underline{\lambda} = (\lambda_1, \lambda_2, \ldots, \lambda_r) \in K(r,w)$ define the *dual* by $\underline{\lambda}^0 = (\lambda_r^0, \lambda_{r-1}^0, \ldots, \lambda_2^0, \lambda_1^0)$.

An $r$-tuple $\underline{\lambda}$ of partitions is called *self-dual* if $\underline{\lambda} = \underline{\lambda}^0$. The set of all such selfdual $\underline{\lambda}$ is denoted by $K^s(r,w) = \{\underline{\lambda} \in K(r,w) | \underline{\lambda} = \underline{\lambda}^0\}$, and $k^s(r,w) = |K^s(r,w)|$. In particular, $k^s(p,w)$ denotes the number of self-dual (or symmetric) $p$-quotients of weight $w$, see Olsson [83], p. 196.

Let $w$ be a positive integer and $t = \frac{1}{2}(p-1)$. A $\overline{p}$-quotient of weight $w$ is a $(t+1)$-tuple $(\lambda_0, \lambda_1, \ldots, \lambda_t)$ of partitions such that $\lambda_0$ is a bar partition and

$$|\lambda_0| + |\lambda_1| + \ldots + |\lambda_t| = w .$$

The sign of this $\overline{p}$-quotient is defined by

$$\sigma(\lambda_0, \lambda_1, \ldots, \lambda_t) = (-1)^{w-|\lambda_0|}\epsilon(\lambda_0) ,$$

where $\epsilon(\lambda_0)$ denotes the sign of the bar partition $\lambda_0$.

The number of $\overline{p}$-quotients of weight $w$ is denoted by $q(\overline{p}, w)$, and $q^\sigma(\overline{p}, w)$ denotes the number of $\overline{p}$-quotients of weight $w$ with sign $\sigma$.

The block invariants $l(B)$, $k(B)$ and $k_0(B)$ of a $p$-block $B$ of $A(n)$, $A^+(n)$ and $S^+(n)$ were determined by Olsson in his papers [83], [84]. Using the above notations we now can state his results. They also contain proofs for Brauer's $k(B)$-conjecture, height zero conjecture and Olsson's conjecture for the $p$-blocks $B$ of $A(n)$, $A^+(n)$ and $S^+(n)$.

**Theorem 6.4.** *Let $B^*$ be a $p$-block of $A(n)$ covered by a $p$-block $B$ of $S(n)$ with weight $w(B) = w > 0$ and $p$-core $\kappa_{(p)}$. Let $w = \sum_{\beta \geq 0} t_\beta p^\beta$ be a $p$-adic expansion of $w$. Then $B$ and $B^*$ have a common defect group $\delta(B^*) = \delta(B) = D$, and the following assertions hold:*

a) *If $\kappa_{(p)}$ is non-symmetric, then*

$$l(B^*) = l(B) = k(p-1, w) ,$$
$$k(B^*) = k(B) = k(p, w) \leq |D| ,$$
$$k_0(B^*) = k_0(B) = \prod_{\beta \geq 0} k(p^{\beta+1}, t_\beta) \leq |D : D'| .$$

b) *If $\kappa_{(p)}$ is symmetric, then*

$$l(B^*) = \frac{1}{2}k(p-1, w) = \frac{3}{2}k^s(p-1, w) ,$$
$$k(B^*) = \frac{1}{2}k(p, w) + \frac{3}{2}k^s(p, w) ,$$
$$k_0(B^*) = \frac{1}{2}\prod_{\beta \geq 0} k(p^{\beta+1}, t_\beta) + \frac{3}{2}\prod_{\beta \geq 0} k^s(p^{\beta+1}, t_\beta) \leq |D : D'| .$$

c) *$k(B^*) = k_0(B^*)$ if and only if $D$ is abelian.*

Concerning assertion c) of the previous result we remark that $w \geq p$ if and only if $D$ is non-abelian, see Olsson [83], Proposition 4.8.

**Theorem 6.5.** *Let $B$ be a spin p-block of $S^\delta(n)$ with weight $w(B) = w > 0$ and sign $\sigma(B) = \sigma$. If $w = \sum_{\beta=1}^{k} t_\beta p^{\beta-1}$ is the p-adic expansion of $w$, then the following assertions hold:*

a) $l(B) = l(B)_+ + 2l(B)_-$, *where for each* $\epsilon \in \{+1, -1\}$

$$l(B)_\epsilon = \begin{cases} k(\frac{1}{2}(p-1), w) & \text{if } \epsilon\delta\sigma = (-1)^w \\ 0 & \text{otherwise} \end{cases}$$

b) $k(B) = q^{\delta\sigma}(\overline{p}, w) + 2q^{\delta\sigma}(\overline{p}, w) \leq |D|$

c) $k_0(B) = k_0(B)_+ + 2k_0(B)_- \leq |D : D'|$, *where for each* $\epsilon \in \{+1, -1\}$

$$k_0(B)_\epsilon = \sum_{\{(\sigma_1, \sigma_2, \ldots, \sigma_k)\}} \prod_{\beta=1}^{k} q^{\sigma_\beta}(\overline{p}^\beta, t_\beta) ,$$

*where $(\sigma_1, \sigma_2, \ldots, \sigma_k)$ runs through all k-tuples of signs $\sigma_i$ satisfying $\sigma_1, \sigma_2, \ldots, \sigma_k = \epsilon$, and where $q^{\sigma_\beta}(\overline{p}^\beta, t_\beta)$ denotes the number of all $\overline{p}^\beta$-quotients with sign $\sigma_\beta$ and weight $t_\beta$.*

d) $k(B) = k_0(B)$ *if and only if $D$ is abelian.*

In [75] Olsson and the author verified the Alperin-McKay conjecture for the spin $p$- blocks $B$ of $S^\delta(n)$ with weight $w > 0$ and sign $\sigma = \sigma(B)$. By the reduction theorem 6.3 $k_0(B)$ equals the number $k_0(b)$ of height zero characters of the principal spin block $b$ of the Sylow $p$-normalizer $N^{\delta\sigma}$ of $S^{\delta\sigma}(pw)$. The duality lemma 2.1 of Michler-Olsson [75] asserts that one may restrict oneself to the case $N^+$ in $S^+(pw)$. Let $X$ be a Sylow $p$-subgroup of $S^+(pw)$. Let $w = \sum_{\beta=1}^{k} t_\beta p^{\beta-1}$, $0 \leq t_\beta \leq p-1$ be the $p$-adic decomposition of $w$. Then $X$ may be considered as a Sylow $p$-subgroup $X$ of $S = S(pw)$. Hence

$$X = X_1^{t_1} \times \ldots \times X_k^{t_k} \text{ , and}$$

$$N = N_S(X) \cong (N_1 \wr S(t_1)) \times \ldots \times (N_k \wr S(t_k))$$

is a direct product of wreath products. The subgroups $N_i \wr S(t_i)$ of $S$ operate on disjoint sets. If $\hat{\times}$ denotes the twisted central product defined by Humphreys [49], then

$$N^+ = (N_1 \wr S(t_1))^+ \hat{\times} \ldots \hat{\times} (N_k \wr S(t_k))^+ .$$

For any subgroup $H$ of $S(n)$ let $s_0(H^\delta)$ be the number of height zero irreducible characters $\chi$ of $H^\delta$ such that the central involution $z \neq 1$ of $H^\delta \leq S^\delta(n)$ is not in the kernel of $\chi$. Let $s_0(H^\delta)_+$ and $s_0(H^\delta)_-$ denote the numbers of s.a. spin height zero characters and pairs of n.s.a. height zero spin characters of $H^\delta$, respectively. Then Proposition 1.5 of Michler-Olsson [75] asserts that

$$k_0(b)_\sigma = s_0(N^+)_\sigma = \sum_{\substack{\{(\sigma_1, \ldots, \sigma_k)\} \\ \sigma_1, \ldots, \sigma_k = \sigma}} \prod_{\beta=1}^{k} s_0\left((N_\beta \wr S(t_\beta))^+\right)_{\sigma_\beta}$$

for every sign $\sigma \in \{+1, -1\}$.

Thus it remains to consider the case where $w = tp^\beta$ for some $t$ with $0 \le t \le p-1$. In order to determine the numbers of s.a. and n.s.a. spin characters of $p'$-degree in a group on the form $(N_\beta \wr S(t))^+$, where $1 \le t \le p-1$, one has to study the structure of $(N_\beta \wr S(t))^+$ more closely.

If $N_\beta$ denotes the normalizer of a $p$-Sylow subgroup $X_\beta$ of $S(p^\beta)$ then $N = N_\beta \wr S(t)$ is a $p$-Sylow normalizer in $S(p^\beta t)$, $1 \le tp-1$, $\beta \ge 1$. In order to determine the irreducible height zero spin characters of $N^+$ Proposition 3.9 of Michler-Olsson [75] is restated as

**Lemma 6.6.** *For all $\beta \ge 1$, $N_\beta^+$ has one s.a. spin representation $D$ of degree $(p-1)^\beta$ and $(p^\beta - 1)/2$ pairs of n.s.a. spin characters of degree 1. Thus*

$$s_0(N_\beta^+)_+ = 1 \ , \ s_0(N_\beta^+)_- = (p^\beta - 1)/2 \ .$$

Now the structure of $N^+ = (N_\beta \wr S(t))^+$ is described. The wreath product $N_\beta \wr S(t)$ is a semidirect product

$$N_\beta \wr S(t) = M : S_t$$

where
$$M = N_\beta^{(1)} \times \ldots \times N_\beta^{(t)}$$

is the base subgroup and $S_t$ is a subgroup which is conjugate to the $p^\beta$-fold diagonalization $\Delta_{p^\beta} S(t) \subseteq S = S(p^\beta t)$. Here $N_\beta^{(i)}$ operates like $N_\beta$ on the set $\{(i-1)p^\beta + 1, (i-1)p^\beta + 2, \ldots, ip^\beta\}$. Therefore, the subgroups $N_\beta^{(i)}$ operate on disjoint sets. If $\hat{\times}$ denotes the twisted central product defined by Humphreys [49], then the covering group $M^+$ of $M$ in $N^+$ is given by

$$M^+ = N_\beta^{(1)} + \hat{\times} N_\beta^{(2)} + \hat{\times} \ldots \hat{\times} N_\beta^{(t)} \triangleleft N^+ \ ,$$

$$N^+ = M^+ S^+(t) \ , \text{ where } M^+ \cap S^+(t) = \langle z \rangle \ .$$

Let $D$ be the unique s.a. spin representation of $N_\beta^+ \cong N_\beta^{(i)+}$, $1 \le i \le t$, mentioned in Lemma 6.6. Then using a method of I. Schur [90, p. 209], it is possible to construct an irreducible s.a. spin representation $\hat{D}$ of $M^+$ by means of $D$, which is called the $t$-fold twisted power of $D$. It follows that $\hat{D}$ is stable in $N^+ = M^+ S^+(t)$. The irreducible constituents of the induced representation $(\hat{D})^{N^+}$ are determined by Proposition 4.8 of Michler-Olsson [75] which is restated as

**Proposition 6.7.** *Let $t \ge 4$ and let $D$ be the s.a. spin representation of $N_\beta^+$ with degree $(p-1)^\beta$. Then the $t$-fold twisted power $D^t = \hat{D} \in \mathrm{Irr}_{\mathbb{C}}(M^+)$ of $D$ can neither be extended to an irreducible representation of $N^- = M^+ A_t^+$ nor to one of $N^+ = M^+ S_t^+$. Furthermore*

a) *Every irreducible constituent $V$ of $(\hat{D}^{N^+})$ is of the form $V \cong \hat{D} \otimes T$, where $t$ is an irreducible spin representation of $S_t^+$.*
b) *Every irreducible constituent $V$ of $(\hat{D})^{N^-}$ is of the form $V \cong \hat{D} \otimes T$, where $t$ is an irreducible spin representation of $A_t^+$.*

*In each case, $V$ is s.a. if and only if $T$ is s.a.*

Let $\lambda$ be one of the $\frac{1}{2}(p^\beta - 1)$ representatives of the pairs of n.s.a. linear spin representations of $N_\beta^+ \cong N_\beta^{(i)+}$, $1 \leq i \leq t$. Let $\hat{\lambda}$ be its $t$-fold twisted power. Then $\hat{\lambda}$ is an irreducible spin representation of $M^+$, which is stable in $N^+ = M^+S^+(t)$. Using ingenious methods of Schur's paper [90] Olsson and the author ([75], Proposition 4.4) were able to show the following result.

**Proposition 6.8.** *Let $\lambda$ be a n.s.a. linear spin representation of $N_\beta^+$. Then the $t-fold$ twisted power $\hat{\lambda} \in \mathrm{Irr}_{\mathbf{C}}(M^+)$ of $\lambda$ can be extended to an irreducible spin representation $D_\lambda$ of $N^+ = M^+S^+(t)$, and every irreducible constituent $V$ of $(\hat{\lambda})^{N^+}$ is of the form $V = D_\lambda \otimes T$, where $T$ is an irreducible ordinary representation of $N^+/M^+ \cong S(t)$. Furthermore:*

a) *If $t$ is odd then every irreducible constituent $V$ of $(\hat{\lambda})^{N^+}$ is n.s.a.*
b) *If $t$ is even then every irreducible constituent $V$ of $(\hat{\lambda})^{N^+}$ is s.a.*

Using Theorem 6.5c), Proposition 6.7 and Proposition 6.8, Olsson and the author proved the Alperin-McKay conjecture for spin $p$-blocks of $S^\delta(n)$ for all odd prime numbers $p > 2$ in [75].

**Theorem 6.9** *Let $p$ be an odd prime, and let $B$ be a $p$-block of $G \in \{S^+(n), A^+(n)\}$ with defect group $X$ and Brauer correspondent $b$ in $N_G(X)$. Then $k_0(B) = k_0(b)$.*

The block invariants of the 2-blocks $B$ of the alternating groups $A(n)$ were determined by Olsson [83]. Combining his result with Theorem 6.4 and Theorem 6.9, Olsson and the author were able to verify the Alperin-McKay conjecture for all primes in [75].

**Theorem 6.10.** *The Alperin-McKay conjecture holds for all $p$-blocks of the alternating groups $A(n)$.*

Alperin's weight conjecture was proved by Olsson and the author in [75] for $p$-blocks $B$ of $A(n)$, $A^+(n)$ and $S^+(n)$ for all *odd* primes $p$. Certainly the group structure of a radical $p$-subgroup $R$ of $S^+(n)$ and of $S(n)$ is the same. The structure of its normalizer $N^\delta = N_{s^\delta(n)}(R)$ in $S^\delta(n)$, $\delta \in \{+1, -1\}$, is described in Proposition 2.6 of Michler-Olsson [75]. This result is analoguous to Proposition 5.6 of the previous section except for the following non-trivial isomorphism holding for each positive integer $c$.

$$\mathrm{GL}(c,p)^+ \cong \mathrm{SL}(c,p) : C^+ \text{ , where}$$

$$C^+ = \begin{cases} C \times \mathbf{Z}/2\mathbf{Z} & \text{if } p \equiv 1(8), p \equiv 3(8) \text{ and } c \text{ is even, if } p \equiv 7(8) \text{ and } c \text{ is odd,} \\ \mathbf{Z}/2(p-1)\mathbf{Z} & \text{otherwise.} \end{cases}$$

By the reduction theorem 6.3 it suffices to verify Alperin's weight conjecture for the principal spin $p$-block $B$ of $S^+(pw)$, where $w$ denotes the weight of $B$. As $p$ is odd, the width $w(R)$ of a radical $p$-subgroup is defined as in section 5. For each sequence $r = (c_1, c_2, \ldots, c_{s(r)})$ of positive integers $c_i$ let $A_r$ be he basic $p$-subgroup of $S(p^d)$ with degree $\deg(A_r) = p^d$ and length $l(A_r) = s(r)$, where $d = \sum_{i=1}^{s} c_i$.

Let $R$ be a radical $p$-subgroup of $G^+ = S^+(pw)$ with width

$$w(R) = \sum_{d \geq 1} \sum_{\{r \in C | d(r) = d\}} \zeta(r) p^{d-1} .$$

Then by Proposition 2.6 of Michler-Olsson [76]

$$N_{G^+}(R)/R = \hat{\prod_{r\in C}} \left[(N_{S(p^{d(r)})}(A_r)/A_r) \wr S(\zeta(r))\right]^+ ,$$

$$\left[N_{S(p^{d(r)})}(A_r)/A_r\right]^+ \cong \mathrm{GL}(c_1,p)^+ \curlyvee \mathrm{GL}(c_2,p)^+ \curlyvee \ldots \curlyvee \mathrm{GL}(c_{s(r)},p)^+ ,$$

for each sequence $r = (c_1, c_2, \ldots, c_{s(r)}) \in C$, where $\curlyvee$ denotes the untwisted central product. Now each

$$\mathrm{GL}(c_i,p)^+ \cong \mathrm{SL}(c_i,p) : C_i^+ \text{ for } 1 \le i \le s(r) ,$$

where $C_i$ is a cyclic group of order $p-1$. By Lemma 4.6 of [76] each irreducible defect zero spin representation $\Theta$ of $[N_{s(p^{d(r)})}(A_r)/A_r]^+$ is of the form

$$\Theta = \bigotimes_{i=1}^{s(r)} (\mathrm{St}_i \otimes \lambda_i) = \left(\bigotimes_{i=1}^{s(r)} \mathrm{St}_i\right) \otimes \lambda ,$$

where $\mathrm{St}_i$ denotes the Steinberg representation of $\mathrm{SL}(c_i,p)$, $\lambda_i$ is a n.s.a. linear spin representation of $C_i^+$ and $\lambda = \bigotimes_{i=1}^{s} \lambda_i$. Furthermore, each irreducible defect zero spin representation $\mu$ of $[N_{S(p^{d(r)})}(A_r)/A_r]^+$ is equivalent to one of these irreducible characters $\Theta$.

Each $\Theta$ plays now the role of the n.s.a. linear character $\lambda$ of $N_\beta^+$ in Proposition 6.8. Using similar methods of proof, Olsson and the author prove in Proposition 4.8. of [76] the corresponding result for the extensions of the stable $\zeta(r)$-fold twisted power $\hat{\Theta}$ of $M_r^+$, where $M_r$ denotes the base subgroup of the wreath product

$$(N_{S(p^{d(r)})}(A_r)/A_r) \wr S(\zeta(r)) , \; r \in C .$$

By means of these technical results the number of $B$-weights $(R,\varphi)$ of $G^+ = S^+(pw)$ having the same radical $p$-subgroup $R$ can be computed as follows.

**Proposition 6.11.** *Let $B$ be the principal spin $p$-block of $G^+ = S^+(pw)$. Let $R$ be a radical $p$-subgroup of $G^+$ with width*

$$w(R) = \sum_{d\ge 1} \sum_{\{r\in C \mid d(r)=d\}} \zeta(r)p^{d-1} .$$

*For each $r \in C$ let*

$$N_r^+ = \left[(N_{S(p^{d(r)})}(A_r)/A_r) \wr S(\zeta(r))\right]^+ ,$$

*and $e(r) = \frac{1}{2}(p-1)^{s(r)}$. For every sign $\epsilon \in \{+1,-1\}$ let*

$$d_0(N_{G^+}(R)/R)_\epsilon = \begin{cases} \prod_{r\in C} d_0(N_r^+)_{\sigma(r)} & \text{if } \epsilon = (-1)^w \\ 0 & \text{otherwise} \end{cases} ,$$

*where $\sigma(r) = (-1)^{\zeta(r)}$ for every $r \in C$.*

*Then the following assertions hold:*

a) *The number of $B$-weights $(R,\varphi)$ with radical $p$-subgroup $R$ is*

$$d_0(N_{G^+}(R)/R) = d_0(N_{G^+}(R)/R)_+ + 2d_0(N_{G^+}(R)/R)_- .$$

b) *For every $r \in C$ the number $d_0(N_r^+)_{\sigma(r)}$ equals the number of $e(r)$-tuples $(\kappa_1, \kappa_2, \ldots, \kappa_{e(r)})$ of $p$-core partitions $\kappa_i$ such that $\sum_{i=1}^{e(r)} |\kappa_i| = \zeta(r)$.*

Using the counting techniques of Alperin-Fong [3], Olsson and the author determine the number of $B$-weights of a spin $p$-block of $S^\delta(n)$ in Theorem 5.2 of [76]. From Theorem 6.5 then follows Alperin's weight conjecture for odd primes $p$.

**Theorem 6.12.** *Let $B$ be a spin $p$-block of $S^\delta(n)$ with weight $w > 0$ and sign $\sigma(B) = \sigma$. Then the number $l^*(B)$ of $B$-weight $(R, \varphi)$ is given by*

$$l^*(B) = l^*(B)_+ + 2l^*(B)_- \ , \text{ where for each } \epsilon \in \{+1, -1\}$$

$$l^*(B)_\epsilon = \begin{cases} k(\frac{1}{2}(p-1), w) & \text{if } \epsilon\delta\sigma = (-1)^w \\ 0 & \text{otherwise} \end{cases} .$$

*In particular, $l(B)_\epsilon = l^*(B)_\epsilon$ for each sign $\epsilon$. Hence $l(B) = l^*(B)$.*

Olsson and the author [76] derive from this result the following

**Corollary 6.13.** *Let $p \neq 2$.*

a) *Let $B$ be a $p$-block of $A(n)$ with positive weight $w > 0$. Then $l(B)_\sigma = l^*(B)_\sigma$ for each sign $\sigma$.*

b) *Alperin's weight conjecture holds for all $p$-blocks $B$ of the covering groups $A^+(n)$ of the alternating groups $A(n)$ and of the exceptional 6-fold covers $C_6$ and $C_7$ of $A(6)$ and $A(7)$, respectively.*

So far, Donovan's conjecture has not been proved for the spin $p$-blocks of $S^+(n)$ and $A^+(n)$, but it seems possible to generalize J. Scopes' methods of proof [91] for Corollary 5.11. Absolutely open is Brauer's problem 19 asking for an efficient group theoretical description of the number $z(G)$ of $p$-blocks of defect zero in an alternating group $G = A(n)$ or symmetric group $G = S(n)$. One does not even know an answer to the following

*Question.* Let $p \geq 5$ be a prime number. Does every alternating group $A(n)$ with $n \geq p$ have a $p$-block of defect zero?

It has been mentioned already that there is no analogue to Theorem 6.2 for the prime $p = 2$. However, there is a conjecture by Knörr and Olsson, how the spin characters should distribute into 2-blocks. For details see Olsson [82], p. 246.

## 7. Modular representation theory of finite groups of Lie type.

For a short introduction into the representation theory of finite groups of Lie type the reader is referred to the lecture notes of R. W. Carter [15] and C. W. Curtis [20].

Let $K$ be an algebraically closed field of characteristic $p > 0$. Let $q = p^s$ for some integer $s > 0$. Throughout this section $G$ denotes a connected reductive affine algebraic group defined over $K$. Let $\sigma : G \to G$ be a Frobenius map and $G_\sigma = \{g \in G | \sigma(g) = g\}$. Then $G_\sigma$ is called a finite group of Lie type.

Let $r$ be a prime number, and let $(F, R, S)$ be a splitting $r$-modular system for the finite group $G_\sigma$. At first, we consider the modular representations of the finite group $G_\sigma$ of Lie type. We distinguish two cases, $r = p$ and $r \neq p$.

### 1) The natural characteristic $r = p$.

In this case, Cabanes [13] has proved Alperin's conjecture in general. According to Alperin [2], the notion of a weight is closely related to the notion of a Lie weight. Because of the importance of Alperin's weight conjecture this relation is explained in detail.

Let $B$ be a Borel subgroup of the finite group $G_\sigma$ of Lie type with Lie rank $l$, $U = O_p(B)$ and $T$ a complement of $U$ in $B$. Let $W$ be the Weyl group of $G_\sigma$. Let $I = \{s_1, s_2, \ldots, s_l\}$ be the set of distinguished generators of $W = W_i$. The involutions $s_i \in I$ are the fundamental reflections with respect to a set of fundamental roots $\Pi = \{\alpha_1, \alpha_2, \ldots, \alpha_l\}$ of $G$. For each subset $J$ of $I$ let $P_J = BW_JB$ be the corresponding standard parabolic subgroup.

The simple $FG_\sigma$-modules are classified by Curtis' theorem [19], p. 94 which is stated as

**Theorem 7.1.** *Let $G_\sigma$ be a finite group of Lie type of rank $l$ and $I = \{s_1, s_2, \ldots, s_l\}$ be the set of distinguished involutive generators of the Weyl group $W = W_I$ of $G_\sigma$. If $M$ is a simple $FG_\sigma$-module, then the following assertions hold:*

a) *$M$ contains a unique 1-dimensional $F$-subspace $\mathrm{inv}_U(M) = \{m \in M | mu = m$ for all $u \in U\} = Fm_0$ stabilized by the Borel subgroup $B$, where $U = O_p(B)$.*
b) *If $J = \{s \in I | m_0 s = m_0\}$, then the parabolic subgroup $P_J = BW_JB$ is the full stabilizer of the line $Fm_0$.*
c) *$M$ determines the subset $J$ of $I$ and a linear character $\chi$ from the parabolic subgroup $P_j$ into $F^*$ uniquely. The pair $(\chi, J)$ is called the Lie weight of the simple $FG_\sigma$-module.*

In particular, two simple $FG_\sigma$-modules are isomorphic if and only if they have the same Lie weight.

Using the notation of Theorem 7.1 we now can state Cabanes' theorem [13].

**Theorem 7.2.** *Let $G_\sigma$ be a finite group of Lie type. Then:*

a) *If $M$ is a simple $FG_\sigma$-module with weight $(\chi, J)$, then $U_J = O_p(P_J)$ is a radical $p$-subgroup of $G_\sigma$, and up to conjugation there is one Steinberg character $\mathrm{St}_J$ of the Levi complement $L_J$ of the parabolic subgroup $P_J = U_J L_J$ which is a weight character of $M$. The $G_\sigma$-conjugacy class $(U_J, \mathrm{St}_J)$ is uniquely determined by $M$, it is the weight of $M$.*
b) *There is a bijection between the isomorphism classes of simple $FG_\sigma$-modules $M$ and the weights $(U_J, \mathrm{St}_J)$ of $G_\sigma$, $J \leq I$.*

*In particular, $l(FG_\sigma) = l^*(G_\sigma)$.*

The most important problem of the modular representation theory of finite groups of Lie type $G_\sigma$ in the natural characteristic $p > 0$ is the determination of the dimensions $\dim_F M$ of the simple $FG_\sigma$-modules $M$. If $G$ is a simply-connected, reductive algebraic group with Lie rank $l$ such that $G_\sigma$ is defined over a field with $q$ elements, then $G_\sigma$ has $q^l$ non-isomorphic simple $FG_\sigma$-modules $M$, and $\dim_F M \leq |P|$, where $P$ is a Sylow $p$-subgroup of $G_\sigma$. Furthermore, there is precisely one simple $FG_\sigma$-module St of $G_\sigma$ which is projective and has $\dim_F \mathrm{St} = |P|$. All this follows from Steinberg's tensor product theorem, see [93], p. 35. St is called the *Steinberg module* of $G_\sigma$. J. E. Humphreys [48] showed that all non-projective simple $FG_\sigma$-modules are contained in the principal $p$-block $B_0$. In [24] and [25] R. Dipper proved the following

**Theorem 7.3.** *Let $G_\sigma$ be a finite group of Lie type. Then every non-projective $FG_\sigma$-module $M$ has a Sylow $p$-subgroup $P$ of $G_\sigma$ as a vertex.*

Using J. Carlson's theory of module varieties [15] Fleischmann, Jantzen and Janiszczak determined all the periodic, non-projective simple $FG_\sigma$- modules in their articles [33], [34], [35], [53] and [54].

**Theorem 7.4.** *Let $G_\sigma$ be a finite group of Lie type and $F$ an algebraically closed field of characteristic $p > 0$. Then the group algebra $FG_\sigma$ has periodic, non-projective simple $FG_\sigma$-modules $M$ if and only if $G_\sigma$ is isomorphic to* $\mathrm{SL}_2(q)$, $\mathrm{SU}_3(q^2)$ *or $p = 2$ and $G_\sigma \cong {}^2B_2(2^{2m+1})$.*

In each case the periodic non-projective simple $FG_\sigma$-modules are classified by Jeyakumur, Fleischmann and Jantzen.

Since the Sylow $p$-subgroups of the finite groups of Lie type $G_\sigma$ are only abelian for $\mathrm{SL}_2(q)$ and the $p$-blocks $B$ of $G_\sigma$ have either full defect or defect zero by Steinberg's tensor product theorem, Brauer's height zero conjecture is easily checked for these $p$-blocks.

Also the Alperin-McKay conjecture holds for finite groups of Lie type in the natural characteristic. This was observed by W. Feit in [32], p. 171. He mentions that the conjecture follows from the theorem of Green, Lehrer and Lusztig stated as Proposition 8.3.4 in Carter [16], p. 279.

**Theorem 7.5.** *Let $G_\sigma$ be a finite group of Lie type and natural characteristic $p > 5$. Let $P$ be a Sylow $p$-subgroup of $G_\sigma$ and $H = N_{G_\sigma}(P)$. Let $b$ be a Brauer correspondent of the principal $p$-block $B$ of $G$ in $H$. Then $k_0(B) = k_0(b)$.*

The author does not know whether the Green-Lehrer-Lusztig theorem quoted in the proof of Theorem 7.3 has an analogue for the bad primes $p \in \{2, 3, 5\}$.

Since one does not know the precise number of conjugacy classes for almost all finite groups of Lie type $G_\sigma$, in particular not for $E_7(q)$ and $E_8(q)$, Brauer's $k(B)$-conjecture is completely open.

Although there is a large literature on the determination of the semisimple and unipotent conjugacy classes of a finite group of Lie type $G_\sigma$, it is a rather difficult and highly computational problem to find the precise polynomial in $q$ with rational coefficients which describes the number $k(G_\sigma)$ of all conjugacy classes.

Each element $x \in G_\sigma$ has a Jordan decomposition $x = x_s \cdot x_u = x_u \cdot x_s$ into its semisimple part $x_s$ and its unipotent part $x_u$. If the algebraic group $G$ is simply connected, a semisimple element $x = x_s \in G_\sigma$ is called *regular* if its centralizer is a maximal torus $T_w$, $w \in W$. In [36] Fleischmann and Janiszczak have given algorithms for the computation of the *polynomial* in $q$ which describes the number of $G_\sigma$-conjugacy classes of regular elements having a fixed representative in a given maximal torus of a classical group $G_\sigma$ of type $A_l(q)$, ${}^2A_l(q^2)$, $B_l$, $D_l(q)$, and ${}^2D_l(q)$. As an example their Corollary 7.11 of [36] is restated as

**Theorem 7.6.** *Let $G_\sigma \in \{\mathrm{SL}_l(q)$ or $\mathrm{SU}_l(q)\}$. Let $T_1$ be the maximal split torus and $T_{w_0}$ be the Coxeter torus of $G_\sigma$. Then the number $f(T_w, q)$ of regular $G_\sigma$- classes $x^{G_\sigma}$ having a representative in the torus $T_w$ is given by*

$$a)\quad f(T_1, q) = \begin{cases} (q-1)^{-1} \sum_{d|(q-1,l)} (-1)^{l-ld^{-1}} \varphi(d) \binom{d^{-1}(q-1)}{d^{-1}l} & \text{if } G_\sigma = \mathrm{SL}_l(q) \\ (q+1)^{-1} \sum_{d|(q+1,l)} (-1)^{l+ld^{-1}} \varphi(d) \binom{d^{-1}(q+1)}{d^{-1}l} & \text{if } G_\sigma = \mathrm{SU}_l(q) \end{cases}$$

$$b)\quad f(T_{w_0}, q) = \begin{cases} (q-1)^{-1} l^{-1} \sum_{d|l} \mu(d)(q-1, l, d)(q^{d^{-1}l} - 1) & \text{if } G_\sigma = \mathrm{SL}_l(q) \\ (q+1)^{-1} l^{-1} \sum_{d|l} \mu(d)(q+1, l, d)(q^{d^{-1}l} + 1) & \text{if } G_\sigma = \mathrm{SU}_l(q) \end{cases}$$

*where $(q-\epsilon, l, d)$ denotes the greatest common divisor of the three numbers in the bracket, and where $\mu$ and $\varphi$ denote the Möbius and Euler functions, respectively.*

**2) The unnatural characteristic case $r \neq p$.**

Since the important paper [38] of Fong and Srinivasan has appeared the study of modular representations of finite groups of Lie type $G_\sigma$ over splitting fields $F$ with characteristic $r$ unequal to the defining characteristic $p$ has become a very active area of research. All the relevant papers are based on Lusztig's and Deligne-Lusztig's fundamental work on the character theory of finite groups of Lie type as described in the books by Carter [16] and Lusztig [65].

In the following we assume that $G$ is a connected reductive affine algebraic group over the algebraic closure $K$ of the field $F(p)$ with $p$ elements such that its center $Z(G)$ is also connected. Let $\sigma$ be a Frobenius endomorphism of $G$ such that $G_\sigma$ is finite, and let $G^*$ be the dual group with Frobenius endomorphism $\sigma^*$. Let $R_{t,\Theta}$ denote the generalized Deligne-Lusztig character of $G_\sigma$ with respect to the linear character $\Theta$ of the $\sigma$-stable maximal torus $T$ of $G$, see Carter [17], p. 43. Let $\langle\, , \rangle$ denote the inner product of generalized characters. Two irreducible characters $\chi$, $\chi'$ of $G_\sigma$ are in the same *geometric conjugacy class* of $\mathrm{Irr}_\mathbb{C}(G_\sigma)$ if there exists a sequence $\chi = \chi_1, \chi_2, \ldots, \chi_n = \chi'$ such that for each $i$ there is some generalized Deligne-Lusztig character $R_{t,\Theta}$ satisfying $\langle \chi_i, R_{t,\Theta}\rangle \neq 0 \neq \langle \chi_{i+1}, R_{t,\Theta}\rangle$.

A fundamental theorem of G. Lusztig asserts that the irreducible characters of $G_\sigma$ have a "Jordan form", see [17], p. 51. It is restated as

**Theorem 7.7.** *Let $\chi_s$ be the semi-simple character of $G_\sigma$ corresponding to the conjugacy class of the semi-simple element $s^*$ in $G^*_{\sigma^*}$. Then there is a bijection $\chi \to \chi_u$ between the characters $\chi$ in the geometric conjugacy class of $\chi_s$ and the unipotent characters $\chi_u$ of $C_{G^*_{\sigma^*}}(s^*)$ such that $\chi(1) = \chi_s(1)\chi_u(1)$.*

The correspondence $\chi \to \chi_u$ in general can be chosen in more than one way. Nevertheless, we write $\chi = \chi_{s,u}$. If $s$ is an $r'$-element of $G^*_{\sigma^*}$ in $G^*_{s^*}$, then Broué and Michel have shown in [12] that this correspondence preserves unions of $r$-blocks. This leads to the notion of geometric conjugacy classes of $r$-blocks.

The centers of general linear and unitary groups are connected, and both algebraic groups are selfdual, i.e. $G \cong G^*$. Hence we write $G = G^*$ and $\sigma = \sigma^*$, in this case.

Let $\nu$ be the $r$-adic valuation of $\mathbf{Q}$. Let $e$ be the order of $q^2$ modulo $r$, and let $t = \nu(q^{2e}-1)$. Then $r$ is a *linear prime* if $r^t$ divides $q^e - 1$, a *unitary* prime if $r^t$ divides $q^e + 1$. In the linear case $e$ is necessarily odd. Let $\epsilon = 1$ or -1 accordingly as $r$ is linear or unitary.

Let $\mathcal{F}_0$ be the set of all monic irreducible polynomials in $\mathrm{GF}(q)[X]$ which are different from $X$. If $q = q_0^2$, then $\mathrm{GF}(q)$ has a unique automorphism $\alpha$ of order 2. For every polynomial $\Delta(X) = X^m + f_{m-1}X^{m-1} + \ldots + f_1X + f_0 \in \mathcal{F}_0$ let $\widetilde{\Delta}(X) = \alpha(f_0^{-1})X^m\alpha(\Delta(X^{-1}))$. Then $\widetilde{\ }$ is a permutation of $\mathcal{F}_0$. Let $\mathcal{F}_1 = \{\Delta \in \mathcal{F}_0 | \Delta = \widetilde{\Delta}\}$ and $\mathcal{F}_2 = \{\Delta\widetilde{\Delta} | \Delta \in \mathcal{F}_0 \text{ with } \Delta \neq \widetilde{\Delta}\}$. Each $\Delta \in \mathcal{F}_2$ is called *unitary*. In order to have a notation which applies to both the linear and unitary cases let $\mathcal{F} = \mathcal{F}_0$ or $\mathcal{F}_1 \cup \mathcal{F}_2$ according as $G$ is a linear or unitary group.

Each semisimple element $s$ of a finite general linear group $G_\sigma = \mathrm{GL}(n,q)$ or unitary group $G_\sigma = U(n,q^2)$ has a primary decomposition $s = \prod_{\Delta\in\mathcal{F}} s_\Delta$, where $s_\Delta$ is the factor of $s$ corresponding to the elementary divisors $\Delta$ of $s$. Let

$$C_{G_\sigma}(s) = \prod_{\Delta\in\mathcal{F}} C_{G_\sigma}(s)_\Delta = \prod_{\Delta\in\mathcal{F}} G_\Delta\left(m_\Delta(s), q^{\delta_\Delta}\right)$$

be the corresponding direct decomposition of $C_{G_\sigma}(s)$, where

$$G_\Delta = \begin{cases} U & \text{if } G \text{ is unitary and } \Delta \text{ is unitary} \\ G & \text{otherwise} \end{cases},$$

$m_\Delta(s)$ is the multiplicity of $\Delta$ and $\delta_\Delta$ is the reduced degree of the elementary divisor $\Delta$. See also Fong-Srinivasan [38], p. 112. Each subgroup $G_\Delta(m_\Delta(s), q^{\delta_\Delta})$ can be embedded as a subgroup of $G(m_\Delta(s)d_\Delta, q)$, where $d_\Delta$ denotes the degree of $\Delta$. Therefore $\sum_\Delta m_\Delta(s)d_\Delta = n$.

Each unipotent character $\chi_u$ of $C_{G_\sigma}(s)$ is a product

$$\chi_u = \prod_{\Delta\in\mathcal{F}} \chi_\Delta \text{ of unipotent characters } \chi_\Delta \text{ of } G_\Delta(m_\Delta(s), q^{\delta_\Delta}).$$

Thus $\chi_\Delta$ corresponds uniquely to a partition $\nu_\Delta \vdash m_\Delta(s)$ by Carter [17], p. 53. Let $\nu = \prod_{\Delta\in\mathcal{F}} \nu_\Delta$. Then each irreducible character $\chi = \chi_{s,u}$ can also be written as $\chi_{s,u} = \chi_{s,\nu}$. For each $\Gamma \in \mathcal{F}$ let $e_\Gamma = e_\Gamma(G) = \min\left\{m \in \mathbb{N} \middle| r \middle| |G_\Gamma(m, q^{\delta_\Gamma})|\right\}$. Let $\mathcal{F}'$ be the set of polynomials whose roots have multiplicative order prime to $r$.

Given a semisimple $r'$-element $s$ of $G_\sigma \in \{\mathrm{GL}(n,q), U(n,q^2)\}$ and $\Gamma \in \mathcal{F}'$, let $\mathcal{C}_\Gamma(s)$ be the set of all $e_\Gamma$-core partitions $\mu_\Gamma$ of $m_\Gamma(s)$. Let $\mathcal{C}(s) = \prod_\Gamma \mathcal{C}_\Gamma(s)$.

With this notation the fundamental theorem of Fong and Srinivasan [38] can be stated.

**Theorem 7.8.** *For each $r$-block $B$ with defect group $D$ of $G_\sigma \in \{\mathrm{GL}(n,q), U(n,q^2)\}$ there is a semisimple $r'$-element $s \in G$ and a product $\nu = \prod_{\Delta\in\mathcal{F}'} \nu_\Delta \in \mathcal{C}(s)$ such that an irreducible character $\chi_{t,\lambda}$ of $G_\sigma$ belongs to $B$ if and only if*
a) *$t$ is $G_\sigma$-conjugate to $sy$ for some $y \in D$, and*
b) *for all $\Delta \in \mathcal{F}$, $\lambda_\Delta$ has $e_\Delta$-core $\nu_\Delta$.*

Therefore, the $r$-block $B = B_{s,\nu}$ is uniquely determined by the $G_\sigma$-conjugacy class $(s,\nu)$.

In fact, Fong and Srinivasan proved this theorem in [38] with the extra hypothesis $r \neq 2$, which finally was removed by M. Broué [11]. As an immediate consequence they obtained the following result.

**Theorem 7.9.** *Brauer's height zero conjecture holds for the $r$-blocks of* $\mathrm{GL}(n,q)$ *and $U(n,q^2)$.*

The Alperin-McKay conjecture was proved by Olsson and the author [74] by means of a reduction theorem similar to Theorem 5.2. In order to state it the following definitions and notations are necessary.

Let $B = B_{s,\nu}$ be an $r$-block of $G_\sigma \in \{\mathrm{GL}(n,q), U(u,q^2)\}$, let $\nu = \prod_{\Gamma\in\mathcal{F}'} \nu_\Gamma$. Let $\mu_\Gamma$ be the $e_\Gamma$-core of some partition of $m_\Gamma(s)$. For each $\Gamma \in \mathcal{F}'$ let

$$w_\Gamma(B) = e_\Gamma^{-1}\left(m_\Gamma(s) - |\mu_\Gamma|\right) \text{ be the } \Gamma\text{-weight of } B.$$

**Theorem 7.10.** *Let $B = B_{s,\nu}$ be an r-block of $G_\sigma \in \{GL(n,q), U(u,q^2)\}$ with defect groups $\delta(B) =_{G_\sigma} D$ and Brauer correspondent b in $N = N_{G_\sigma}(D)$. Then*

a) *B has the same block invariants $k(B)$, $k_0(B)$ and $l(B)$ as the principal r- block $\widetilde{B}_0 = \Pi \widetilde{B}_\Gamma^0$ of the subgroup*

$$\widetilde{G} = \prod_{\Gamma \in \mathcal{F}'} G_\Gamma \left(m_\Gamma(s) - |\mu_\Gamma|, q^{\delta_\Gamma}\right) \text{ of } G .$$

*In particular, there is a canonical height preserving bijection between the characters of the blocks B and $\widetilde{B}_0$ which respects geometric conjugacy classes.*

b) *If $\widetilde{D}$ is a Sylow r-subgroup of $\widetilde{G}$, then $D \cong \widetilde{D}$.*

c) *If $\widetilde{b}_0$ is the Brauer correspondent of $\widetilde{B}_0$ in $\widetilde{N} = N_{\widetilde{G}}(\widetilde{D})$, then the block ideals b and $\widetilde{b}_0$ are Morita equivalent. In particular, $k(b) = k(\widetilde{b}_0)$, $k_0(b) = k_0(\widetilde{b}_0)$ and $l(b) = l(\widetilde{b}_0)$.*

Using this reduction theorem Olsson and the author show in [74] that there are bijective character correspondences $\psi$, $\sigma$, $\widetilde{\psi}$ and $\widetilde{\sigma}$ such that the following diagram commutes:

$$\begin{array}{ccc} \mathrm{Irr}^0(B) & \stackrel{\psi}{\longleftrightarrow} & \mathrm{Irr}^0(b) \\ \downarrow\sigma & & \downarrow\widetilde{\sigma} \\ \mathrm{Irr}^0(\widetilde{B}_0) & \stackrel{\widetilde{\psi}}{\longleftrightarrow} & \mathrm{Irr}^0(\widetilde{b}_0) \end{array}$$

In particular the Alperin-McKay conjecture holds.

**Theorem 7.11.** *Let $B = B_{s,\nu}$ be an r-block of $G_\sigma \in \{\mathrm{GL}(n,q), U(u,q^2)\}$ with defect group D and Brauer correspondent b in $N = N_{G_\sigma}(D)$. Then*

$$k_0(B) = k_0(b) = \prod_{\Gamma \in \mathcal{F}'} \prod_{\beta \geq 0} k\left((e_\Gamma + \frac{r^{\alpha_\Gamma + t} - 1}{e_\Gamma})r^a, t^{\Gamma,\beta}\right) ,$$

*where $t = \nu(q^{2e} - 1)$, and where for each $\Gamma \in \mathcal{F}'$ the $\Gamma$-weight w of B has the r-adic expansion $w_\Gamma(B) = \prod_{\beta \geq 0} t^{\Gamma,\beta} r^\beta$, and $\alpha_\Gamma = \nu(d_\Gamma)$.*

Using these results and methods Olsson showed in [80]

**Theorem 7.12.** *Let $B = B_{s,\nu}$ be an r-block of $G_\sigma \in \{\mathrm{GL}(n,q), U(u,q^2)\}$ with defect group $\delta(B) =_{G_\sigma} D$.*

a) $k(B) \leq |D|$.

b) $k_0(B) \leq |D : D'|$.

Recently, Alperin and Fong [3] have determined the radical r-subgroups of $\mathrm{GL}(n,q)$. Using then Theorem 7.8 of Fong and Srinivasan, they verified the Alperin weight conjecture for the r-blocks $B = B_{s,\nu}$ of $G_\sigma = \mathrm{GL}(n,q)$ in [3].

**Theorem 7.13.** *Let B be an r-block of $G_\sigma = \mathrm{GL}(n,q)$. Then $l(B) = l^*(B)$.*

For all other classical groups $G_\sigma$ with connected center $Z(G)$ Alperin's weight conjecture has not been proved yet for the r-blocks of $G_\sigma$. Donovan's conjecture is open for all infinite series of classical groups even for $\mathrm{GL}(n,q)$.

The irreducible characters $\chi$ of the conformal symplectic groups $\mathrm{CSp}(2n,q)$, conformal orthogonal groups $\mathrm{CSO}(2n,q)$ and special orthogonal groups $\mathrm{SO}(2n+1,q)$ have been classified by Lusztig, see [65]. Since the corresponding algebraic groups $G$ have a connected center $Z(G)$, each such character $\chi$ has a Jordan normal form $\chi = \chi_{s,u}$, where $\chi_u$ is a unipotent irreducible character of $C_{G^*_{\sigma^*}}(s^*)$. By Lusztig [65] the unipotent irreducible characters $\chi_{1,u}$ of these classical groups are parametrized by symbols, which are unordered pairs $[X, Y]$ of finite subsets $X$, $Y$ of non-negative integers. Lusztig defines an equivalence relation on these symbols. The equivalence classes can then be represented by triples $(\lambda, \mu; z)$, where $\lambda$ and $\mu$ are partitions of positive integers, and where $z \in \mathbf{Z}$. In [81] Olsson generalizes the theory of $e$-hooks, $e$-cores and quotients of partitions to symbols $(\lambda, \mu; z)$. Using these combinatorial notions each irreducible character $\chi_{s,u}$ of one of the classical groups $\mathrm{CSp}(2n,q)$, $\mathrm{CSO}(2n,q)$ and $\mathrm{SO}(2n+1,q)$ can be parametrized by $\chi_{s,u} = \chi_{s,\nu}$, where $\nu$ is a product of symbols $\nu_\Delta$, where $\Delta$ is again an elementary divisor of the semisimple element $s$. Given a product $\nu = \prod_{\Delta\in\mathcal{F}} \nu_\Delta$, where $\nu_\Delta$ is a symbol or a partition according to $\Delta \in \mathcal{F}_0$ or $\Delta \in \mathcal{F}_1 \cup \mathcal{F}_2$. Then the *core* of $\nu$ is defined by

$$\mathcal{K} = \{\kappa = \prod_{\Delta\in\mathcal{F}} \kappa_\Delta | \kappa_\Delta \text{ is } e_\Delta\text{-core of } \nu_\Delta \text{ for all } \Delta \in \mathcal{F}\} .$$

By Fong and Srinivasan [39], p. 160 $\mathcal{K}$ contains one, two or four elements. If $|\mathcal{K}| = 1$, then one writes $\mathcal{K} = \kappa$. This is the case when $G_\sigma = \mathrm{SO}(2n+1,q)$. In particular, to each $r$-block $B$ of $G_\sigma$ corresponds an $r'$-element $s$ and a core $\kappa$ which are uniquely determined up to $G_\sigma$-conjugacy. Therefore, one writes $B = B_{s,\kappa}$. The pair $(s,\kappa)$ is called the *label* of $B$.

In analogy to their Theorem 7.8 Fong and Srinivasan were able to determine the distribution of the irreducible characters $\chi_{s,\nu}$ of $G_\sigma \in \{\mathrm{CSp}(2n,q), \mathrm{CSO}(2n,q), \mathrm{SO}(2n+1,q)\}$ into $r$-blocks $B = B_{s,\kappa}$ in their deep and rather technical paper [39]. In fact, they treated only the cases where $p$ and $r$ are odd primes. As an example we restate their Theorem (12A) of [39] as

**Theorem 7.14.** *Let $B = B_{s,\kappa}$ be an $r$-block of $G_\sigma = \mathrm{SO}(2n+1,q)$ with label $(s,\kappa)$. Then an irreducible character $\chi_{t,\nu}$ belongs to $B$ if and only if the following two conditions hold:*

a) *$t$ is $G_\sigma$-conjugate to $sy$, where $y$ belongs to a dual defect group $D^*$ of $D$,*

b) *$\kappa$ is the core of $\nu = \prod_{\Delta\in\mathcal{F}} \nu_\Delta$.*

The other cases $G_\sigma \in \{\mathrm{CSp}(2n,q), \mathrm{CSO}(2n,q)\}$ are more technical. As an application of their deep theorems Fong and Srinivasan prove Brauer's height zero conjecture in the following cases in [39], p. 182 and 183.

**Theorem 7.15.** *Let $r \neq 2$ and $p \neq 2$. Let $B$ be an $r$-block of $G_\sigma \in \{\mathrm{CSp}(2n,q), \mathrm{CSO}(2n,q), \mathrm{SO}(2n+1,q)\}$ with defect group $D$. Then $k(B) = k_0(B)$ if and only if $D$ is abelian.*

*Remark 7.16.* Using Theorem 7.14 and the methods of [74] Olsson and the author have verified the Alperin-McKay conjecture for the $r$-blocks $B$ with defect group $D$ of $G_\sigma = \mathrm{SO}(2n+1,q)$, where $r \neq 2$ and $2 \not| q$. This result has not been published so far. Olsson's methods of [80] and [81] can be used to prove also that $k(B) \leq |D|$ and $k_0(B) \leq |D : D'|$.

Not every simple alternating or sporadic group has 2-blocks or 3-blocks of defect zero, see [71], p. 228. The finite simple groups of Lie type $G_\sigma$ behave much better. Using the Deligne-Lusztig generalized characters, the author [69] and W. Willems [95] proved the following result for odd primes $p$ and for $p = 2$, respectively.

**Theorem 7.17.** *Let $G$ be a finite simple group of Lie type. Then $G$ has a $p$-block of defect zero for every prime $p$.*

Although the alternating groups $A_n$, $n \geq 5$, do not necessarily contain $p$-blocks of defect zero, they always have $p$-blocks $B$ whose defect groups $\delta(B)$ are not Sylow $p$-subgroups of $A_n$. Therefore, as shown by the author [71], the classification theorem of finite simple groups and Theorem 7.17 imply

**Theorem 7.18.** *Let $p \neq 2$. Then every finite simple group $G$ has at least one $p$-block $B$ whose defect group $\delta(B)$ is not a Sylow $p$-subgroup of $G$.*

In [70] the author gives a proof for the following result which answers a question of N. Ito.

**Theorem 7.19.** *The finite group $G$ has an abelian normal Sylow $p$-subgroup $P$ if and only if $p \not| \chi(1)$ for every irreducible ordinary character $\chi$ of $G$.*

Another consequence of Theorem 7.17 proved by the author [70] is

**Theorem 7.20.** *The finite group $G$ has a normal Sylow $p$-subgroup $P$ if and only if $p \not| \varphi(1)$ for every irreducible modular character $\varphi$ of $G$.*

In his survey articles [70], [71] the author gives further applications of Theorem 7.17.

The modular representation theory of the exceptional groups of Lie type is rather complete for the groups ${}^2B_2(2^{2m+1})$, ${}^2G_2(3^{2m+1})$, $G_2(q)$ and ${}^3D_4(q)$. In these groups the Sylow $r$-subgroups are cyclic or abelian except for the primes 2 and 3.

The distribution of the irreducible characters $\chi$ of the Suzuki groups ${}^2B_2(2^{m+1})$ into $r$-blocks follows easily from the work of Martineau [67]. In particular, all the conjectures of Brauer, Alperin, Alperin-McKay are verified. The same assertions are true for the $r$-blocks of the Ree groups ${}^2G_2(3^{2m+1})$ by the work of Ward [94], Fong [37], and Landrock and the author [63].

In [92] Shamash has given the distribution of the irreducible characters $\chi$ of $G_2(q)$ into $r$-blocks. It follows that Brauer's $k(B)$-conjecure and Olsson's $k_0(B)$-conjecture hold. In [46] G. Hiß has verified Brauer's height zero conjecture and the Alperin-McKay conjecture. Therefore, Alperin's weight conjecture holds by Theorem 3.2 of Knörr and Robinson for all primes $r \neq 2$.

In [23] Deriziotis and the author have computed the character table of the simple triality groups ${}^3D_4(q)$. It enabled us to verify Brauer's height zero conjecture, Brauer's $k(B)$-conjecture, Olsson's $k_0(B)$-conjecture, and the Alperin-McKay conjecture. Therefore, Alperin's weight conjecture has only to be checked for the primes $r \in \{2,3\}$ by Theorem 3.2.

According to G. Hiß [46], p. 37, Malle's work [66] on the unipotent irreducible characters of ${}^2F_4(2^{2m+1})$ enables one to find the distribution of irreducible characters of this group into $r$-blocks. Therefore, it is possible to check all the major conjectures also for this series of groups of Lie type.

For all other exceptional groups of Lie type $F_4(q)$, $E_6(q)$, ${}^2E_6(q)$, $E_7(q)$, ${}^2E_7(q)$ and $E_8(q)$ there are no complete answers on the conjectures under review in this report.

In the case of the general linear groups $\mathrm{GL}(n,q)$ Dipper and James have given another proof of Theorem 7.8 which is independent of the Deligne-Lusztig theory. It follows from their work on the Hecke algebras of these groups over fields $F$ with positive characteristic $r \neq p$, see [28], [29], [30]. Its main aim is to relate the difficult problem of finding important parts of the decomposition matrices $D(B)$ of $r$-blocks $B$ of $\mathrm{GL}(n,q)$ by means of the decomposition matrices of the $r$-blocks of the Weyl group $S(n)$. This work generalizes Dipper's work [26], [27] in which he showed that $D(B)$ is a triangular matrix with all diagonal entries equal to 1.

## 8. Modular representation theory of the sporadic simple groups.

Most of the conjectures can easily be checked for the 26 sporadic simple groups $G$ by their character tables which can be found in the Atlas [18]. In particular, Brauer's height zero conjecture, his $k(B)$-conjecture and Olsson's $k_0(B)$-conjecture are true for all $p$-blocks $B$ of such a group $G$. In fact they also hold for all automorphism groups of all covering groups of $G$.

In order to check the Alperin-McKay conjecture Ostermann [85] computed the character tables of most of the normalizers $N_G(D)$, where $D$ is a defect group of a $p$-block $B$ of a sporadic simple group $G$. Whenever he was unable to get the complete character table of $N_G(D)$, e.g. for the Sylow 2-subgroups of the Monster and some other large sporadic groups, then he at least computed the degrees of all irreducible characters $\chi$ of $N_G(D)$ with height $\mathrm{ht}(\chi) = 0$. All together he showed

**Theorem 8.1.** *Let $p > 0$ be a prime. The Alperin-McKay conjecture holds for all $p$- blocks $B$ of all sporadic simple groups $G$.*

This result was independently obtained by R. A. Wilson.

Alperin's weight conjecture has not been checked yet for all sporadic groups. It is rather difficult to find all the $G$-conjugacy classes of radical $p$-subgroups of a large sporadic simple group $G$ for the prime 2.

Whenever one has a complete character table it is easy to count $z(G)$ the number of $p$-blocks of defect zero. Therefore, Brauer's problem 19 has a complete answer for the sporadic simple groups. Instead of Donovan's conjecture one should compute all the Cartan matrices of the sporadic groups $G$. Again we are very far from a complete solution of this problem.

## REFERENCES


[1] J. L. Alperin, "Local representation theory", Proc. Symposia Pure Math. 37 (1980), 364–384.

[2] J. L. Alperin, "Weights for finite groups", The Arcata Conference on Represen-

tations of Finite Groups, Proc. Symposia in Pure Math. 47 (1987), 369–379.

[3] J. L. Alperin, P. Fong, "Weights for symmetric and general linear groups", J. Alg. 131 (1990), 2–22.

[4] T. Berger, "On the structure of a representation of a solvable group", Vorlesungen aus dem Fachbereich Mathematik der Universität GHS Essen, Heft 12 (1985).

[5] T. Berger, R. Knörr, "On Brauer's height 0 conjecture", Nagoya Math. J. 109 (1988), 109–116.

[6] C. Bessenrodt, "The isomorphism type of an abelian defect group of a block is determined by its modules", J. London Math. Soc. (2) 39 (1989), 61–66.

[7] C. Bessenrodt, "Some new block invariants coming from cohomology", Astérisque 181/182 (1990), 11–29.

[8] H. Blau, G. Michler, "Modular representation theory of finite groups with T. I. Sylow $p$-subgroups", Trans. Amer. Math. Soc. 319 (1990), 417–468.

[9] R. Brauer, "Representations of finite groups", Lect. on Modern Math., Vol. 1, Wiley, New York (1963), 133–175.

[10] R. Brauer, "Notes on representations of finite groups", J. London Math. Soc. 13 (1976), 162–166.

[11] M. Broué, "Les $l$-blocs des groupes $\mathrm{GL}(n,q)$ et $U(n,q^2)$ et leurs structures locales", Astérisque 133/134 (1986), 159–188.

[12] M. Broué, J. Michel, "Blocs et séries de Lusztig dans un groupe fini", J. reine angew. Math. 395 (1990), 56–67.

[13] M. Cabanes, "Brauer morphisms between Hecke algebras", J. Alg. 115 (1988), 1–31.

[14] M. Cabanes, "Local structure of the $p$-blocks of $\widetilde{S}_n$", Math. Z. 198 (1988), 519–543.

[15] J. Carlson, "Module varieties and cohomology rings of finite groups", Vorlesungen aus dem Fachbereich Mathematik der Universität GHS Essen, Heft 13 (1985).

[16] R. Carter, "Finite groups of Lie type: Conjugacy classes and complex characters", J. Wiley, New York (1985).

[17] R. Carter, "On the structure and complex representation theory of finite groups of Lie type", Vorlesungen Fachb. Mathematik der Universität Essen, Heft 16 (1987).

[18] J. Conway, R. Curtis, S. Norton, R. Parker, R. Wilson, "Atlas of finite groups", Clarendon Press, Oxford (1985).

[19] C. Curtis, "Modular representations of finite groups with split $(B,N)$-pairs" Lect. Notes Math. 131 (1970), 57–95.

[20] C. Curtis, "Representation theory of Hecke algebras and complex representations of finite groups with split BN-pairs", Vorlesungen Fachb. Mathematik der Universität Essen, Heft 15 (1987).

[21] C. Curtis, I. Reiner, "Methods of representation theory I and II", J. Wiley, New York (1981 and 1987).

[22] E. C. Dade, "A possible approach to Alperin's weight conjecture", Letter (1988).

[23] D. Deriziotis, G. Michler, "Character table and blocks of finite simple triality groups ${}^3D_4(q)$", Trans. Amer. Math. Soc. 303 (1987), 39–70.

[24] R. Dipper, "Vertices of irreducible representations of finite Chevalley groups in the describing characteristic", Math. Z. 175 (1980), 143–159.

[25] R. Dipper, "On irreducible modules of twisted groups of Lie type", J. Algebra 81 (1983), 370–389.

[26] R. Dipper, "On the decomposition numbers of the finite general linear groups", Trans. Amer. Math. Soc. 290 (1985), 315–343.

[27] R. Dipper, "On the decomposition numbers of the finite general linear groups II", Trans. Amer. Math. Soc. 292 (1985), 123–133.

[28] R. Dipper, G. James, "Representations of Hecke algebras of general linear groups", Proc. London Math. Soc. 52 (1986), 20–52.

[29] R. Dipper, G. James, "Blocks and idempotents of Hecke algebras of general linear groups", Proc. London Math. Soc. 54 (1987), 57–82.

[30] R. Dipper, G. James, "Identification of the irreducible modules of $GL_n(q)$", J. Alg. 104 (1986), 266–288.

[31] K. Erdmann, "Blocks of tame representation type and related algebras", Lect. Notes Math. 1428, Springer, Heidelberg (1990).

[32] W. Feit, "The representation theory of finite groups", North Holland, Amsterdam (1982).

[33] P. Fleischmann, "Periodic simple modules for $SU_3(q^2)$ in the describing characteristic $p \neq 2$", Math. Z. 198 (1988), 555–568.

[34] P. Fleischmann, "The complexities and rank varieties of the simple modules of $({}^2A_2)(q^2)$ in the natural characteristic", J. Algebra 121 (1989), 399–408.

[35] P. Fleischmann, J. C. Jantzen, "Simple periodic modules of twisted Chevalley groups", Pacific J. Math. 143 (1990), 229–242.

[36] P. Fleischmann, I. Janiszczak, "Semisimple conjugacy classes of finite groups of Lie type I", Preprint, Essen (1990).

[37] P. Fong, "On decomposition numbers of $J_1$ and $R(q)$". In Symposia Mathematica 23, Academic Press, London (1974), 415–422.

[38] P. Fong, B. Srinivasan, "Blocks of finite general linear and unitary groups", Invent. Math. 69 (1982), 109–153.

[39] P. Fong, B. Srinivasan, "The blocks of finite classical groups.", J. reine angew. Math. 396 (1989), 122–191.

[40] P. Gabriel, C. Riedtmann, "Group representations without groups", Comment. Math. Helv. 54 (1979), 240–287.

[41] D. Gluck, T. Wolf, "Brauer's height zero conjecture for $p$-solvable groups", Transact. Amer. Math. Soc. 282 (1984), 137–152.

[42] H. Gollan, "The 5-modular representations of ${}^2F_4(2)'$, their Green correspondents and sources", Math. Comp. (to appear).

[43] D. Gorenstein, "Finite simple groups, an introduction to their classification", Plenum Press, New York (1982).

[44] D. Gorenstein, R. Lyons, " On finite groups of characteristic 2-type", Mem. Amer. Math. Soc. 276 (1982).

[45] G. Hiß, "The modular characters of the Tits simple group and its automorphism group", Comm. Alg. 14 (1986), 125–154.

[46] G. Hiß, "Zerlegungszahlen endlicher Gruppen vom Lie-Typ in nicht-definierender Charakteristik", Habilitationsschrift TH Aachen (1990).

[47] G. Hiß, K. Lux, "Brauer trees of sporadic simple groups", Oxford Science Publ., Clarendon Press (1989).

[48] J. E. Humphreys, "Defect groups for finite groups", Math. Z. 119 (1971), 149–152.

[49] J. E. Humphreys, "On certain projective modular representations of direct products", J. London Math. Soc. 32 (1985), 449–459.

[50] J. F. Humphreys, "Blocks of projective representations of the symmetric groups", J. London Math. Soc. 33 (1986), 441–452.

[51] G. James, "The representation theory of the symmetric groups", Lect. Notes Math. 682, Springer, Heidelberg (1978).

[52] G. James, A. Kerber, "The representation theory of the symmetric group", London, Addison-Wesley (1981).

[53] I. Janiszczak, "Irreducible periodic modules over $\mathrm{SL}(m, q)$ in the describing characteristic", Comm. Alg. 15 (1987), 1375–1391.

[54] I. Janiszczak, J.C. Jantzen, "Simple periodic modules over Chevalley groups", J. London Math. Soc. 41 (1990), 217–230.

[55] R. Knörr, "On the vertices of irreducible modules", Annals of Math. 110 (1979), 487–499.

[56] R. Knörr, "On the number of characters in a $p$-block of $p$-solvable group", Illinois J. Maths. 28 (1984), 181–210.

[57] R. Knörr, "Virtually irreducible lattices", Proc. London Math. Soc. (3) 59 (1989), 99–132.

[58] R. Knörr, G. Robinson, "Some remarks on a conjecture of Alperin", J. London Math. Soc. (2) 39 (1989), 48–60.

[59] B. Külshammer, "A remark on conjectures in modular representation theory", Arch. Math. 49 (1987), 396–399.

[60] B. Külshammer, "Blocks, solvable permutation groups, and Landau's theorem", J. reine angew. Math. 398 (1989), 180–186.

[61] B. Külshammer, "Landau's theorem for $p$-blocks of $p$-solvable groups", J. reine angew. Math. 404 (1990), 189–191.

[62] B. Külshammer, "Solvable subgroups of $p$-solvable semilinear groups", J. reine angew. Math. 404 (1990), 171–188.

[63] P. Landrock, G. Michler, "Principal 2-blocks of the simple groups of Ree type", Transact. Amer. Math. Soc. 260 (1980), 83–111.

[64] W. Lempken, R. Staszewski, "Construction of the 5-modular irreducible representations of the simple McLaughlin group", In preparation.

[65] G. Lusztig, "Characters of reductive groups over a finite field", Annals Math. Studies, Princeton Univ. Press (1984).

[66] G. Malle, "Die unipotenten Charaktere von ${}^2F_4(q^2)$", Comm. Algebra (to appear).

[67] R. P. Martineau, "On the representations of the Suzuki groups over fields of odd characteristics", J. London Math. Soc. 6 (1972), 153–160.

[68] R. Martinez-Villa, "The stable equivalence for algebras of finite representation type", Comm. Alg. 13 (1985), 991–1018.

[69] G. Michler, "A finite simple group of Lie type has $p$-blocks with different defects, $p \neq 2$", J. Alg. 104 (1986), 220–230.

[70] G. Michler, "Brauer's conjectures and the classification of finite simple groups", Proc. International Conference on Representations of Algebras, Ottawa 1984, Lect. Notes Math., Springer, Berlin, 1178 (1986), 129–142.

[71] G. Michler, "Modular representation theory and the classification of finite simple groups", Summer Research Inst. on Representations of Finite groups and related topics, Arcata, July 1986, Proc. Symposia Pure Math. (1987), 223–232.

[72] G. Michler, "Ring theoretical and computational methods in group representation theory", Vorlesungen Fachb. Mathematik der Universität Essen, Heft 18 (1989).

[73] G. Michler, "Trace and defect of a block idempotent", J. Alg. 131 (1990), 496–501.

[74] G. Michler, J. Olsson, "Character correspondences in finite general linear, unitary and symmetric groups", Math. Z. 184 (1983), 203–233.

[75] G. Michler, J. Olsson, "The Alperin-McKay conjecture holds in the covering groups of symmetric and alternating groups, $p \neq 2$", J. reine angew. Math. 405 (1990), 78–111.

[76] G. Michler, J. Olsson, "Weights for covering groups of symmetric and alternating groups, $p \neq 2$", Canad. J. Math. (to appear).

[77] T. Okuyama, "Module correspondence in finite groups", Hokkaido Math. J. 10 (1981), 299–318.

[78] T. Okuyama, M. Wajima, "Character correspondence and $p$-blocks of $p$-solvable groups", Proc. Japan Acad. Sciences, Ser. A 55 (1979), 309–312.

[79] J. Olsson, "McKay numbers and heights of characters", Math. Scand. 38 (1976), 25–42.

[80] J. Olsson, "On the number of characters in blocks of finite general linear, unitary and symmetric groups", Math. Z. 186 (1984), 41–47.

[81] J. Olsson, "Remarks on symbols, hooks and degrees of unipotent characters", J. Comb. Theory, Series A, 42 (1986), 223–238.

[82] J. Olsson, "Frobenius symbols for partitions and degrees of spin characters", Math. Scand. 61 (1987), 223–247.

[83] J. Olsson, "On the $p$-blocks of symmetric and alternating groups and their covering groups", J. Alg. 128 (1990), 188–213.

[84] J. Olsson, "The number of modular characters of certain $p$-blocks", Preprint.

[85] T. Ostermann, "Charaktertafeln von Sylownormalisatoren sporadischer einfacher Gruppen", Vorlesungen Fachb. Mathematik der Universität Essen, Heft 14 (1986).

[86] W. F. Reynolds, "Blocks and normal subgroups of finite groups", Nagoya Math. J. 22 (1963), 15–32.

[87] G. Robinson, "The number of blocks with a given defect group", J. Algebra 84 (1983), 493–502.

[88] G. Schneider, "PSL(3,4) in characteristic 3", Comm. Alg. 15 (1987), 1543–1547.

[89] G. Schneider, "Computing with modular representations", J. Symb. Comp. 9 (1990), 607–636.

[90] I. Schur, "Über die Darstellung der symmetrischen und der alternierenden Gruppe durch gebrochene lineare Substitutionen", J. reine angew. Math. 139 (1911), 115–250.

[91] J. Scopes, "Cartan matrices and Morita equivalence for blocks of the symmetric groups", J. Algebra (to appear).

[92] J. Shamash, "Blocks and Brauer trees for groups of type $G_2(q)$". Proc. of the Symposia in Pure Maths. 47 (1987), 283–295.

[93] R. Steinberg, "Representations of algebraic groups", Nagoya J. Math. 22 (1963), 33–56.

[94] H. N. Ward, "On Ree's series of simple groups", Trans. Amer. Math. Soc. 121 (1966), 62–89.
[95] W. Willems, "Blocks of defect zero in finite simple groups of Lie-type", J. Alg. 113 (1988), 511–522.

Progress in Mathematics, Vol. 95, © 1991 Birkhäuser Verlag Basel

# Recent Advances in the Representation Theory of Finite Dimensional Algebras

CLAUS MICHAEL RINGEL

This is a report on advances in the representation theory of finite dimensional algebras in the years 1984 – 1990. During these years, the German research council (DFG) has sponsered a Forschungsschwerpunkt devoted to the representation theory of finite groups and finite dimensional algebras; it started in 1984 and will be finished by 1991.

The topics we have chosen for this report are those related to investigations carried out in the Forschungsschwerpunkt. However, we will not restrict our attention to these investigations, but try to cover the topics in full generality. The reader will observe that two special classes of algebras reappear throughout the report: the hereditary algebras, and the canonical algebras. The module categories of these algebras are quite well understood, as we will outline below. These algebras serve as an important source of inspiration; dealing with them, one may hope to get an answer even to questions which in general may be impossible to attack. Despite of being rather special, one should keep in mind that these classes comprise some of the most important algebras. Also, quite surprising contacts to other parts of mathematics have been found in the last years involving such algebras.

There are many subjects which we have to omit at all. We will refrain from dealing with degenerations of modules. Also, infinite dimensional modules will be discussed only when they shed light on questions dealing with finite dimensional ones. As the title indicates, we will deal with representations of algebras which are *finite dimensional* over some field. Of course, we know that algebras over higher dimensional commutative rings have attracted a lot of interest in the last years, and they have been discussed in the Forschungsschwerpunkt. But we will be able to mention them only in case there is a direct relationship to finite dimensional algebras. We will try to restrict our attention to those results where a full proof is available, at least as a preprint; all other claims may be considered as mere conjectures. Similarly, we regret that we can cover the Russian literature only so far as translations do exist. Anyway, we had to be selective, and the results presented here are those which are not too technical, and which should be of interest to a wider audience.

The reader is advised that several reports are available dealing with the development of the representation theory before 1984: in particular, there were Riedtmann's Bourbaki talk in 1985, and the lectures by Auslander, and Gabriel at the ICM 1986 in Berkeley, see also the Proceedings of the Durham conference 1985.

In spite of its length, the list of references does not try to be complete; besides the papers quoted in the text, we have included only a few additional ones dealing with related questions.

The author would like to thank D. Happel for his constant help when preparing this survey.

## General conventions

We denote by $k$ a commutative field, it will be the base field for all our algebras. Quite often we will restrict to the case of an algebraically closed base field: many problems of representation theory tend to get more lucid under this assumption. All the rings we will consider are supposed to have sufficiently many idempotents, nearly always they even will have a unit element (for example, in case we deal with a finite dimensional algebra), however a subring $B$ of a ring $A$ is not supposed to have the same unit element.

Modules usually will be left modules; module homomorphisms will be written on the opposite side of the scalars, thus usually on the right, so that in this case, the composition of $f : M_1 \to M_2$, $g : M_2 \to M_3$ has to be denoted by $fg$. For any ring $R$, we denote by $R$–mod the category of finitely generated $R$–modules, by $R$–Mod the category of all $R$–modules. If not otherwise stated, any algebra $A$ and any $A$–module considered will be finite dimensional (over our base field $k$), and we denote by rad $A$ the radical of $A$. We write $s(A)$ for the number of isomorphism classes of simple $A$–modules, let $E(1), \dots, E(s(A))$, be the simple $A$–modules; they may be indexed by the vertices of the (Gabriel) quiver $Q(A)$ of $A$, note that there is an arrow $x \to y$ in $Q(A)$ if and only if $\mathrm{Ext}^1(E(x), E(y)) \neq 0$ (actually, we may consider $Q(A)$ as a valued quiver by attaching to any arrow the corresponding dimensions of $\mathrm{Ext}^1$ over the endomorphism rings of the given simple modules). The projective cover of $E(x)$ will be denoted by $P(x)$. We denote by $\Gamma(A)$ the Auslander–Reiten quiver of $A$, and $\tau$ or $\tau_A$ is the Auslander–Reiten translation.

## 1. Tame and wild

**1.1.** The wild behaviour of what now are called wild algebras was first exhibited by Corner, and Brenner. Donovan and Freislich conjectured that there should be a clear distinction between the tame and the wild algebras. In 1979, Drozd presented his tame-and-wild theorem: any finite dimensional algebra over an algebraically closed field is either tame or wild. We refer to **[C1]** for a complete proof.

Recall that a finite dimensional algebra $A$ over an algebraically closed field $k$ is said to be *tame* provided for any $d \in \mathbb{N}$, there is a finite number of $A$–$k[T]$–bimodules $M_1, \dots, M_n$ which are free of rank $d$ as right $k[T]$–modules, such that almost all indecomposable $A$–modules of dimension $d$ are of the form $M_i \otimes_{k[T]} k[T]/(T-\lambda)$ for some $1 \leq i \leq n$, and $\lambda \in k$. If $A$ is tame, and $d \in \mathbb{N}$, the smallest possible number $n$ of such bimodules is denoted by $\mu_A(d)$.

**1.2.** In general, we may consider $A$–$k[T]$–bimodules $M$ which are free of finite rank as right $k[T]$–modules, and the corresponding functors $F_M = M \otimes_{k[T]} -$ from $k[T]$–mod to $A$–mod. In case almost all the modules $F_M(k[T]/(T-\lambda))$ are indecomposable, and pairwise non–isomorphic, we may call these modules an *affine one-parameter family* of indecomposable modules. Different affine one–parameter families may intersect non–trivially. In case there are infinitely many (isomorphism classes of) indecomposable modules which belong to two affine one–parameter families, these families will be said to be equivalent. The union of all modules belonging to the affine one–parameter families in one equivalence class may be called a *complete one–parameter family.*

In order to avoid the difficulties of dealing with families of modules (see, for example, the clumsy definition of a complete one–parameter family of indecomposable modules), Crawley–Boevey **[C8]** has proposed to consider instead generic modules. A *generic* $R$–module $M$ over an arbitrary ring $R$ is by definition an indecomposable $R$–module of infinite length, such that $M$ considered as an $\mathrm{End}(M)$–module, is of finite length (its *endolength*). Of course, the generic modules with endomorphism ring a division ring just, form the vertices of the (Cohn) spectrum of $R$. Note that given a functor of the form $F_M$ as considered above, we obtain a generic module $F_M(k(T))$, where $k(T)$ is the rational function field in one variable over $k$. The endomorphism ring of a generic module always is a local ring **[C9]**, the proof of this result uses concepts from model theory: a generic module satisfies the descending chain condition on the socalled *pp*–definable subgroups.

**Theorem (Crawley–Boevey).** *Let $A$ be a finite dimensional algebra over an algebraically closed field. The algebra $A$ is representation finite if and only if there are no generic modules, and $A$ is tame if and only if for any $d \in \mathbb{N}$, there are only finitely many generic modules of endolength $d$, if and only if for any generic module $M$, the algebra $\mathrm{End}(M)/\operatorname{rad}\mathrm{End}(M)$ is isomorphic to $k(T)$. Also, in case $A$ is tame, $\mathrm{End}(M)$ is split over its radical, and any two splittings are conjugate.*

The concept of a generic module seems to be so natural that one wonders why it was not considered earlier. Obviously, generic modules should be of interest for arbitrary rings, not just finite dimensional algebras or artinian rings.

Of particular interest will be the generic modules $M$ without selfextensions. For example, any tame hereditary algebra has precisely one generic module, and this module does not have selfextensions. For the generalized Kronecker algebras $K(r) = \begin{bmatrix} k & k^r \\ 0 & k \end{bmatrix}$ with $r \geq 3$, Happel and Unger **[HU2]** have constructed infinite dimensional generic modules without selfextensions such that the endomorphism ring is a universal division ring of fractions for a free associative $k$–algebra in finitely many variables.

**1.3** Some remarks concerning methods of proof may be appropriate. The use of bocses as introduced by Klejner and Rojter has turned out to be very essential. Drozd's tame–and–wild theorem and the modifications due to Crawley–Boevey deal with bocses, and no other proof seems to be in sight. Of course, the theory of bocses is now more accessible, see **[C1]** and **[C9]**. Some of the usual techniques of the representation theory of finite dimensional algebras have been copied for bocses: in particular, the existence of almost split sequences for bocses has been established by Bautista and Klejner **[BK]**, see also **[BB]**. Methods similar to the usual bocs reduction may be applied also to the case of algebras over fields which are not necessarily algebraically closed, or even to artinian rings. In particular, Crawley-Boevey has shown in this way that given an indecomposable $R$–module $M$ over some representation finite artinian ring $R$, there is a simple $R$–module $S$ such that the division rings $\mathrm{End}(M)/\operatorname{rad}\mathrm{End}(M)$ and $\mathrm{End}(S)$ are isomorphic **[C7]**.

Let $A$ be a finite dimensional algebra over some algebraically closed field, and assume that $A$ is not representation finite. The existence of a generic module for $A$ is derived from the positive solution of the second Brauer–Thrall conjecture. We recall that a solution had been announced by Nazarova and Rojter in 1973. Our optimistic report [7] does not seem to be appropriate. The usual strategy for attacking the problem is to choose a minimal non–zero ideal $I$, so that by induction, one may assume that $A/I$ is representation finite. The main difficulties arise in the case when the module ${}_AI$ has selfextensions, and it is this case which has been treated insufficiently by Nazarova and Rojter. In the proceedings of the Ottawa conference 1984, Nazarova and Rojter [**NR1**] have presented another approach to the second Brauer–Thrall conjecture: they claim to provide a new reduction of the representations of an algebra to the representations of a completed poset, and to attach to each completed poset a non–completed one of the same representation type. However, *both* reductions do not work! It is not diificult, to exhibit counter examples to the proposed methods of proof, as well as to the actual stated assertions. In the meanwhile, a corrected version of part of the second reduction has been published [**NR2**].

For fields of characteristic different from 2, the first complete proof for the second Brauer–Thrall conjecture has been given by Bautista [**Bau**]. The assumption on the characteristic of the base field has later been removed by Bongartz [**Bo**], and modifications of the proof have been published by Bretscher–Todorov [**BT**] and Fischbacher [**Fi**]. All these proofs rely on the existence theorem for a multiplicative basis. The problem of finding a rather direct proof of the second Brauer–Thrall conjecture still exists. Also, it would be of interest to have a proof for arbitrary base fields. Of course, the case of a perfect base field $k$ follows from that of an algebraically closed field, so the non–perfect base fields remain to be considered.

**1.4** There still is the problem of finding a convenient definition of tameness. The definition used by Drozd, as well as Crawley–Boevey's characterization in terms of generic modules involve infinite dimensional modules. One may use instead concepts from algebraic geometry, namely one may consider the sheets of indecomposable modules. Is there a definition of tameness which only involves finite dimensional modules, and avoids any reference to algebraic geometry?

Our survey [8] tried to present such a definition, but without success: we have asked that for any dimension $d$, there is a finite number of embedding functors $F_i$ from $k[T]$–mod to $A$–mod such that all but a finite number of indecomposable $A$–modules of dimension $d$ are of the form $F_i(L)$ for some $i$ and some indecomposable $k[T]$–module $L$. However, any wild hereditary algebra $A$ over an algebraically closed field $k$ satisfies this condition: Consider the set $\mathcal{M}(x)$ of isomorphism classes of indecomposable $A$–modules with dimension vector $x$. We may assume that $x$ is an imaginary root, thus the cardinality of $\mathcal{M}(x)$ is equal to that of $k$, let $\phi : k \to \mathcal{M}(x)$ be any bijection. Now, define a functor $F$ from $k[T]$–mod to $A$–mod by sending the $k[T]$–module $L_\lambda[n] = k[T]/(T-\lambda)^n$ to the direct sum $n\phi(\lambda)$ of $n$ copies of $\phi(\lambda)$. The inclusion and projection maps between the various modules $L_\lambda[n]$ with fixed $\lambda$ shall be sent under $F$ to the inclusion and projection maps between the corresponding direct sums (identify $(n-1)\phi(\lambda)$ with $(n-1)\phi(\lambda) \oplus 0 \subseteq n\phi(\lambda)$.) Clearly, $F$ is an embedding functor, and, by construction, any indecomposable $A$–module with dimension vector $x$ is of the form $F(X)$ for some simple $k[T]$–module. Of course, the functor $F$ not at all is well-behaved, since $\phi$ is just a bijection of sets. Also note that under $F$ all exact sequences go to split exact ones.

Two recent results may lead to intrinsic definitions of tameness. First of all, Crawley–Boevey **[C1]** has shown that for a tame algebra $A$ over an algebraically closed field, almost all indecomposable $A$–modules $M$ of fixed dimension satisfy $\tau M \cong M$, thus almost all indecomposable $A$–modules $M$ of fixed dimension belong to tubes of rank one. Bautista has conjectured that this property may characterize the tame algebras. Even for group algebras, this conjecture was solved only recently by Erdmann **[E2]**. On the other hand, in the category of finitely generated Cohen–Macaulay modules over an isolated hypersurface singularity, all objects are $\tau$–periodic, even in the wild case, as Eisenbud has shown (see **[4]**).

Second, one may consider the possible endomorphism rings of indecomposable modules. There is a common feeling that the wild algebras may be characterized by the property that any finite dimensional algebra can be realized as a factor ring of the endomorphism ring of some module modulo some ideal. However, first of all no proof that the wild algebras have this property, has been published yet. Second, there are modules over certain tame algebras, for example string modules, which have rather large endomorphism rings. Fortunately, for many tame algebras, there are only finitely many isomorphism classes of algebras which occur as endomorphism rings of indecomposable modules of fixed dimension. (However, the example of the biserial algebra $A = k\langle X,Y\rangle/(X^2,Y^2)$ shows that the indecomposable modules in a one–parameter family may yield a one-parameter family of endomorphism rings: consider the factor rings $A_\lambda = k\langle X,Y\rangle/(X^2,Y^2,XY-\lambda YX)$ as $A$–modules: this is a one–parameter family of indecomposable $A$–modules, and $\operatorname{End}_A(A_\lambda) = A_\lambda$.) But even if there are only countably many isomorphism classes of algebras which occur as endomorphism rings of indecomposable modules, it may be conceivable that there are indecomposable modules $M_n$ such that the algebra $k\langle X,Y\rangle/(X,Y)^n$ is isomorphic to a factor ring of $\operatorname{End}(M_n)$, and then at least all finite dimensional local algebras generated by two elements could be realized as factor rings of endomorphism rings. One typical class of indecomposable modules over tame algebras has been studied in detail by Krause **[Kr2]**, the string modules. As Crawley–Boevey **[C5]** has shown, the maps between string modules may be described combinatorially. Krause shows that the class of factor rings of endomorphism rings of string modules is very restricted: if $\operatorname{rad} A$ is generated by 2 elements, then $A$ can be realized as a factor ring of $\operatorname{End}(M)$ for some string module $M$ only in case

$$\dim_k A/(\operatorname{rad} A)^n \le 2n^2 - 2n + 1,$$

and even the algebra $k\langle X,Y\rangle/(X,Y)^4$ cannot be realized as factor ring of the endomorphism ring of a string module. One may ask whether there is a polynomial $p$ such that for any indecomposable module $M$ over a tame algebra $A$, any factor ring $A$ of $\operatorname{End}(M)$ with $\operatorname{rad} A$ generated by two elements satisfies

$$\dim_k A/(\operatorname{rad} A)^n \le p(n).$$

**1.5** In dealing with affine one–parameter families, say given by a functor $F_M = M \otimes_{k[T]} - : k[T]\text{–mod} \to A\text{–mod}$, it sometimes seems to be convenient to look for factorizations of $F_M$ of the form

$$k[T]\text{–mod} \longrightarrow K\text{–mod} \xrightarrow{G} A\text{–mod}$$

where $K = K(2)$ is the usual Kronecker algebra and $G$ again is exact. First of all, we may ask whether it always will be possible to find such a factorization (for $A$ finite dimensional!). Second, we should remark that $G$, if it exists, may not be uniquely determined, a typical example are the two embeddings $G_1, G_2$ of $K$–mod into $A$–mod, where $A$ is the hereditary algebra of type $\tilde{A}_{1,2}$

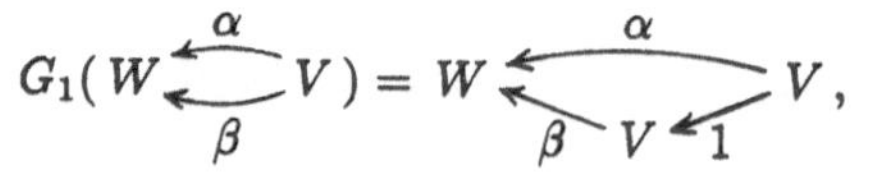


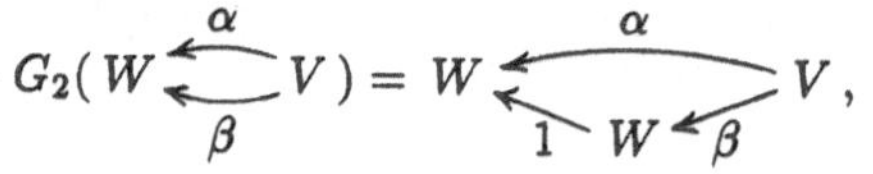


where $V, W$ are $k$–vectorspaces, and $\alpha, \beta : V \to W$ are linear maps. The functors $G_1, G_2$ coincide on the subcategory of $K$–mod given by all objects $(V, W, \alpha, 1)$, and this subcategory is equivalent to $k[T]$–mod. We deal here with the situation of a one–parameter family indexed by $\mathbf{P}_1$, where the point $(1 : 0) \in \mathbf{P}_1$ occurs with multiplicity two. In general, it often happens that we deal with a one–parameter family indexed by the projective line $\mathbf{P}_1$, where finitely many points of $\mathbf{P}_1$ occur with a multiplicity greater than 1.

In order to study this phenomenon, consider pairwise different points $\lambda_1, \dots, \lambda_r$ of $\mathbf{P}_1$, and attach to each $\lambda_i$ some multiplicity $p_i = p(\lambda_i) \geq 1$ (or, equivalently, take a function $p : \mathbf{P}_1 \to \mathbb{N}_1$ such that $p - 1$ has finite support, thus $p - 1$ is an effective divisor). We can assume that $r \geq 2$, and that $\lambda_1 = \infty, \lambda_2 = 0$, and therefore $\lambda_i \in k \setminus \{0\}$, for all $i \geq 3$. The corresponding canonical algebra $C(p)$ is given by the quiver

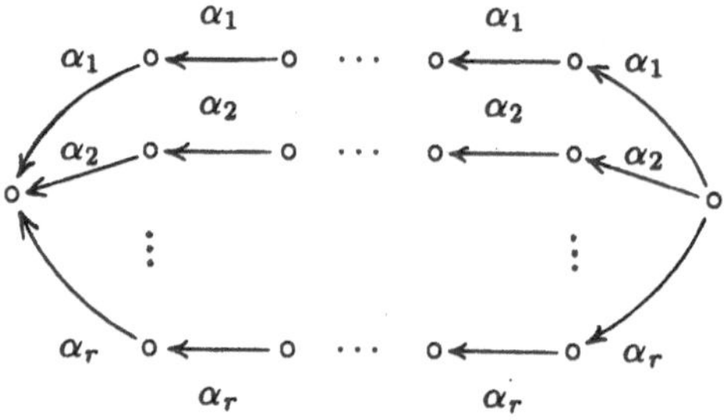


with $p_i$ arrows labelled $\alpha_i$, and the relations

$$\alpha_i^{p_i} = \lambda_i \alpha_1^{p_1} + \alpha_2^{p_2} \quad \text{for} \quad i \geq 3.$$

Let 0 be the sink of the quiver, and $\omega$ the source, and let $\Delta(p)$ be the quiver obtained by deleting $\omega$, it is a star with several arms.

The defect of a representation $M$ is by definition $\delta(M) = \dim_k M_\omega - \dim_k M_0$. Denote by $\mathcal{T}$ the representations which are direct sums of indecomposable representations of defect zero. Then we have shown **[R1]** that $\mathcal{T}$ is a standard tubular family of tubular type $(p_1, \dots, p_n)$, it separates the full subcategory $C(p)$–mod$^-$ of all indecomposable representations of negative defect from the full subcategory $C(p)$–mod$^+$ of all indecomposable representations of positive defect. A more concise proof has been given in **[R3]**.

Geigle and Lenzing **[GL1]**, see also **[DGL]**, have related the category $C(p)$–mod to the category of coherent sheaves coh $\mathbb{X}(p)$ over what they call the 'weighted projective line' $\mathbb{X}(p)$ of type $p : \mathbb{P}_1 \to \mathbb{N}_1$. In fact, they have shown that the derived categories $D^b(C(p)\text{–mod})$ and $D^b(\operatorname{coh}\mathbb{X}(p))$ are equivalent. The category coh $\mathbb{X}(p)$ is an abelian category of global dimension 1, thus the structure of $D^b(\operatorname{coh}\mathbb{X}(p))$ is known as soon as we know coh $\mathbb{X}(p)$. Their direct description of coh $\mathbb{X}(p)$ therefore yields a completely different, and very illuminating proof for the structure of the category $C(p)$–mod.

Lenzing has stressed the importance of the rank–one modules over a canonical algebra. Here, a module $M$ is said to be a *rank–one module*, provided $M$ is indecomposable, and $\delta(M) = -1$. For example, the radical $Q$ of the injective hull of $E(0)$ is a rank–one module, and will examine the $\tau$–orbit of this module in detail.

First, let $M$ be an arbitrary rank–one module. We claim that $\tau M$ is a rank–one module, too, and the Auslander–Reiten sequence ending in $M$ is conservative (this means that proj. dim. $M = 1 =$ inj. dim. $\tau M$). (Let us outline the proof: In case $Z$ is an indecomposable non–projective module of negative defect, and the Auslander–Reiten sequence ending in $Z$ is conservative, then the dimension vector of $\tau Z$ can be calculated by applying the Coxeter transformation of $A$ to the dimension vector of $Z$, and therefore $Z$ and $\tau Z$ have the same defect. Now, all indecomposable modules of negative defect have projective dimension at most one. Let us show that for any non–projective rank–one module $M$, we have $\operatorname{Hom}(M, {}_AA) = 0$. Assume we have a non–zero map $\phi : M \to P(a)$, for some indecomposable projective module $P(a)$. Any non-zero submodule $U$ of $P(a)$ has defect $\delta(U) \leq -1$, thus the kernel of $\phi$ has non–negative defect. But this is possible only in case the kernel is zero, thus $M$ is a submodule of $P(a)$. It is easy to see that the rank–one submodules of any $P(a)$ are projective. This yields a contradiction. As a consequence, the Auslander–Reiten sequence ending in $M$ is conservative, and $\tau M$ again is a rank–one module.)

As a consequence, we see: In case $\Delta(p)$ is Dynkin, the modules $\tau^{-n}Q$ are rank-one modules, for all $n \in \mathbb{N}$. In case $\Delta(p)$ is wild, the modules $\tau^{n}Q$ are rank–one modules, for all $n \in \mathbb{N}$. (Consider first the Dynkin case. If $r \leq 2$, then the algebra $A$ is hereditary, thus $\tau^-$ respects the defect. Let $r \geq 3$. In this case, $\operatorname{rad} P(\omega)$ is an indecomposable module and a predecessor of $Q$. It follows that $\operatorname{Hom}(\tau^{-n}Q, P(\omega)) = 0$ for all $n \in \mathbb{N}$. As a consequence, the Auslander–Reiten sequences starting with $\tau^{-n}Q$, for $n \geq 0$, are conservative. For $r \geq 1$, the module $\tau^-P(\omega)$ is of Loewy length 2, its socle is the direct sum of $r-1$ copies of $E(0)$, and its top is multiplicity free (of length $r$). In the wild case, we have $r \geq 3$, therefore $\tau^-P(\omega)$ is not a rank–one module. Since the set of rank–one modules together with the zero–module is closed under $\tau$, it follows that $\tau^nQ$ cannot be isomorphic to $P(\omega)$, for any $n \in \mathbb{N}$, thus $\tau^nQ$ is a rank–one module, for any $n \in \mathbb{N}$.)

Given an endofunctor $F : \mathcal{U} \to \mathcal{U}$ of some full subcategory $\mathcal{U}$ of $A$–mod, and a module $M$ in $\mathcal{U}$, Lenzing has introduced the ring

$$\mathcal{A}(F; U) = \bigoplus_{n=0}^{\infty} \operatorname{Hom}(U, F^nU),$$

the product of $f : U \to F^nU$, and $g : U \to F^mU$ is given by the composition $f \cdot F^n(g)$.

**Theorem (Geigle–Lenzing)** *If $\Delta(p)$ is a Dynkin diagram, then the ring $\mathcal{A}(\tau^-; Q)$ is the simple surface singularity of type $\Delta(p)$.*

Recall that the simple surface singularities are of the form $k[X,Y,Z]/(f)$, where $f$ is given es follows:

| | |
|---|---|
| (p,q) | $X^{p+q}+YZ$ |
| (2,2,2s) | $X(Y^2+YX^s)+Z^2$ |
| (2,2,2s+1) | $X(Y^2+ZX^s)+Z^2$ |
| (2,3,3) | $Y^3+X^2Z+Z^2$ |
| (2,3,4) | $Y^3+X^3Y+Z^2$ |
| (2,3,5) | $Y^3+X^5+Z^2.$ |

**Theorem (Lenzing)** *If $\Delta(p)$ is wild, then the ring $\mathcal{A}(\tau;Q)$ is a ring of automorphic forms.*

For example, for the 14 cases $p=(2,3,7),(2,3,8),\dots,(4,4,4)$, and the base field $k=\mathbb{C}$, one just obtains the 14 rings of automorphic forms with three generators, thus Arnold's 14 exceptional unimodal singularities. Note that we obtain corresponding rings for every base field $k$, even independent of the characteristic; for the 14 cases, the rings are of the form $k[X,Y,Z]/(f)$, where $f$ is a polynomial which has been calculated explicitly by Hübner **[Hü]**.

Let us add that canonical algebras may be defined more generally for an arbitrary base field. We start with an arbitrary tame hereditary algebra $A$ with $s(A)=2$, thus the $A$-modules are just the representations of a tame bimodule ${}_FM_G$. Here $F,G$ are division $k$-algebras, ${}_FM_G$ is an $F$–$G$–bimodule on which $k$ operates centrally, and $(\dim_F M)(\dim M_G)=4$; the algebra A being given by $\begin{bmatrix} F & M \\ 0 & G\end{bmatrix}$. Let $\Omega$ be the set of isomorphism classes of simple regular $A$-modules, and take a function $p:\Omega\to\mathbb{N}_1$, such that $p-1$ has finite support. In **[R3]**, we have defined the corresponding canonical algebra $C({}_FM_G,p)$, it has a standard tubular family indexed by $\Omega$, such that the tube with index $\lambda$ is stable of rank $p(\lambda)$, as we want to have it, and again, this tubular familly is separating. Note that the index set $\Omega$ for an arbitrary tame bimodule ${}_FM_G$ has been studied by Crawley–Boevey **[C6]**.

**1.6.** A tame algebra may be said to be *domestic* with at most $n$ one–parameter families of tubes, provided $\mu_A(d)\le n$ for all $d\in\mathbb{N}$. Typical examples are the tame concealed algebras, these are the endomorphism rings of preprojective (or preinjective) tilting modules over a tame hereditary algebra. They have been classified by Happel and Vossieck. Note that any tilting equivalence class contains precisely one canonical algebra (this is one of the reason for the term 'canonical'). The algebras which are tilting equivalent to tame hereditary algebras have been studied in detail by Assem and Skowroński. The algebras which are tilting equivalent to a hereditary algebra of type $\tilde{A}_n$ can be characterized in terms of quivers and relations: we obtain in this way certain special biserial algebras **[AS1]**. Also, they have shown that any representation infinite algebra which is tilting equivalent to a tame hereditary algebra contains a unique convex subalgebra which is tame concealed, thus it can be obtained from this tame concealed algebra by forming extensions and coextensions (see **[AS4]**).

We recall that a *cycle* in $A$–mod is a sequence $X=X_0\to X_1\to\cdots\to X_{n-1}\to X_n=X$ of non–zero and non–invertible maps between indecomposable modules $X_i$. An indecomposable module $M$ is said to be *directing* provided it does not belong to a cycle, and $M$ is said to be *sincere*, provided every simple $A$–module appears as a composition factor of $M$. Note that an algebra with a sincere directing module $M$ always is a tilted algebra. A tame algebra $A$ with a sincere directing module is domestic with at most two one–parameter families of tubes, as de la Peña **[P3]** has shown, and either $A$ is representation directed, or a finite enlargement of a tame concealed algebra, or a glueing of two tame concealed algebras.

Let rad($A$–mod) be the radical of the category $A$–mod, and let $\mathrm{rad}^\infty(A\text{–mod})$ denote the intersection of the powers $\mathrm{rad}^n(A\text{–mod})$, with $n \in \mathbb{N}$. Consider any non-zero map $f : X \to Y$, between indecomposable modules $X, Y$. If $X, Y$ belong to different components of the Auslander–Reiten quiver $\Gamma(A)$, then $f \in \mathrm{rad}^\infty(A\text{–mod})$, whereas in case $X, Y$ belong to one component, and this component is standard (or finite), then $f \notin \mathrm{rad}^\infty(A\text{–mod})$. It follows easily (see **[KSk]**) that $\mathrm{rad}^\infty(A\text{–mod}) = 0$ if and only if $A$ is of finite representation type. Kerner and Skowroński conjecture that $\mathrm{rad}^\infty(A\text{–mod})$ can be nilpotent (or even T-nilpotent) only in case $A$ is domestic, and they prove this for 'standard' selfinjective algebras.

One may call an algebra $A$ *cycle finite,* provided no map in a cycle of $A$–mod belongs to $\mathrm{rad}^\infty(A\text{–mod})$. Cycle finite algebras have been considered by Assem and Skowroński, they are tame, and maybe they are always of finite growth. But there are even domestic algebras which are not cycle finite, for example

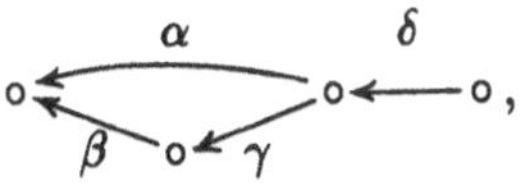


, with $\beta\gamma\delta = 0$.

Assem and Skowroński have generalized the notion of a tube to that of a *coil,* these are certain finite enlargements of tubes, where additional projective–injective vertices and nodes are allowed. They call an algebra $A$ a *coil algebra* provided every cycle in $A$–mod is inside a standard coil, and they show that the minimal representation infinite coil algebras are just the tame concealed ones **[AS5]**.

**1.7.** The difference between finite growth and infinite growth representation type was first observed by Nazarova and Zavadskij when dealing with representations of posets. Note that an algebra $A$ is said to be of *finite growth* provided $A$ is tame and there exists an $n$ such that $\mu_A(d) \leq d^n$ for all $d \in \mathbb{N}$.

Typical examples of algebras of finite growth which are not domestic, are the tubular algebras, in particular the canonical algebras of Euclidean type. The main goal of the lecture notes **[R1]** was to present the structure theory for the module category of a tubular algebra. Some mathematicians have complaint that the book does not contain a definition of tameness: it really only deals with examples of algebras, and the tame ones occurring there are obviously 'tame', so no definition was necessary. One should see that only slowly a general theory of tame algebras is emerging, a theory which seems to show that the examples studied in detail before are the typical building blocks for general tame algebras.

The structure of $A$–mod, for $A$ a tubular algebra, may be visualized by the following picture:

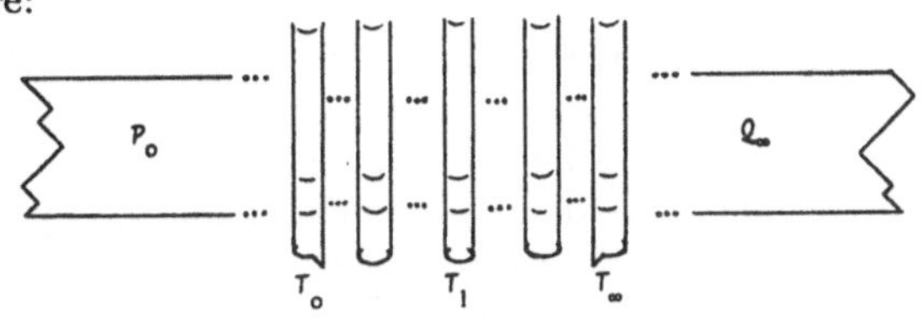

Here, $\mathcal{P}_0$ is a preprojective component, $\mathcal{Q}_\infty$ is a preinjective component, and all the families $\mathcal{T}_i$ are one–parameter families of tubes, the index set for $i$ being the set of non–negative rational numbers including the symbol $\infty$. The tubular family $\mathcal{T}_0$ contains projective modules, and $\mathcal{T}_\infty$ contains injective modules, the remaining families $\mathcal{T}_i$ are regular. All the components of a tubular algebra are standard, and the picture indicates the possible maps between components: there are only maps going from left to right. It should be mentioned that this description relies on an understanding of complete one–parameter families, and so it was stimulated by Gabriel's interpretation of the regular modules for the four subspace quiver in terms of tubes, and it was guided by suggestions of Brenner to use tilting modules for getting a complete classification of the indecomposable modules for a tubular algebra. It is sufficient to consider the tubular algebras which are canonical, and the Geigle–Lenzing approach via coh $\mathbb{X}(p)$ again may be used.

A personal comment: Skowroński has started to rename the tubular algebras. Of course, it is always nice to see the own name in print, especially when it appears parallel to names like Dynkin and Euclid (but they may not even know what their algebras are about). However, it seems that the name 'tubular' is very suggestive, whereas I hope not to look like being made up from tubes. Also, one should keep in mind the guiding intuition of Gabriel, and of Brenner, and the parallel investigations of Zavadskij, so one may speak of Gabriel–Brenner–Ringel–Zavadskij–algebras, but may–be this sounds a little odd? And, as Lenzing has noted, this all relates to Atiyah's classification of vector bundles over elliptic curves, so Atiyah–Gabriel–B–R–Z–algebras!

One may construct further examples of algebras of finite growth by making inductively suitable one–point extensions, see **[HR]**, **[PTo]**, in particular, one may consider convex subalgebras of the socalled repetitive algebra $\hat{A}$ of a tubular algebra $A$, or also algebras which have $\hat{A}$ as a Galois covering. In section 5, we will deal with $\hat{A}$ in more detail. Here we only note that Skowronski (**[Sk2]**, see also **[NS]**) has classified the standard selfinjective algebras $B$ of finite growth: the representation finite ones have been described by Riedtmann, so we can assume that $B$ is representation infinite, and then $B$ has a Galois covering $\hat{A}$ with $A$ being a tame tubular extension of a tame concealed algebra.

There are corresponding tubular vectorspace categories, see **[R1]**, the representations of those which are posets have also been described by Zavadskij **[Za]**. There are other categories occurring in representation theory which may be described in a similar way by reducing to tubular algebras or tubular vectorspace categories. The most prominent one seems to be the category of lattices for a cyclic group of order 3 over a complete discrete valuation ring where 3 is 4-fold ramified. This problem was solved by Dieterich **[Di]**, he showed that the lattice type is tame, and that we deal with a tubular problem of tubular type $\tilde{D}_4$. Note that in all other cases, the lattice type for finite groups and complete discrete valuation rings had been known before, according to the work of Gudivok and others. The remaining cases are either domestic, or of infinite growth, or wild.

**1.8.** Let us consider now the tame algebras of infinite growth. The Gelfand–Ponomarev paper on the representations of the Lorentz group has put forward a method which has turned out to be very fruitful, since it can be used for all special biserial algebras. We should remark that Dowbor and Skowroński **[DS]** have clarified the procedure by putting it into the context of covering theory. Of course, what Gelfand and Ponomarev have called the modules of the first kind, the strings, are just those indecomposable modules which are obtained by using a related covering functor. Dowbor and Skowroński show that the remaining modules, the modules of the second kind, or bands, may be considered as corresponding to modules over the group algebra $k[T, T^{-1}]$ of the infinite cyclic group $\mathbf{Z}$, and that this is really a categorical description: let $A$ be special biserial, then the category obtained from $A$–mod by factoring out all maps which factor through strings is the categorical sum of copies of $k[T, T^{-1}]$–mod, one copy for each equivalence class of primitive cyclic words.

The main advance concerning tame algebras of infinite growth has been the study of Crawley–Boevey **[C3]** of what he calls 'clans'. In this way, he gave a solution to the Gelfand problem of classifying the indecomposable representations of the quiver

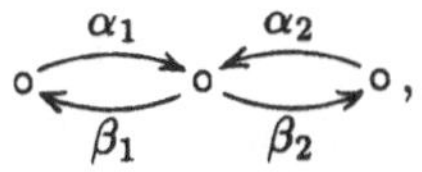


with $\alpha_1\beta_1 = \alpha_2\beta_2$ nilpotent,

posed at the ICM in Nice, 1970. We recall that in 1973, Nazarova and Rojter have shown that this is a tame problem, and they gave a partial solution to the classification problem; however the normal forms which they proposed didn't work. The only assumption needed by Crawley–Boevey is that we deal with a base field $k$ with at least three elements. Starting point of these investigations was his description **[C2]** of the category of finite dimensional $A$–modules for the algebra

$$A = k\langle X, Y\rangle/(X^2 - X, Y^2),$$

of course the $A$–modules are just given by pairs of square matrices, one being idempotent, the other having square zero.

We have mentioned above that a general theory of tame algebras seems to be emerging. But some crucial questions, even on the level of dealing with examples, are still open. Let us mention at least two: what can be said about the biserial algebras which are not special, for example, only few of the local biserial algebras are special (but all have been conjectured to be tame in **[6]**), and what about the representations of the algebras

$$k\langle X, Y\rangle/(X^2 - (YX)^n Y, Y^2 - (XY)^n X),$$

with $n \geq 1$. In characteristic 2, the group algebras of the quaternion 2-groups are, when we disregard socles, of this form, and they are known to be tame by the theorem of Drozd and Bondarenko: the proof uses the realization of a quaternion group as a subgroup of index 2 in a semidihedral group. But for characteristic different from 2, no such trick seems to work.

**1.9.** Investigations of Kerner on wild hereditary algebras give a completely new interpretation of what may be called the 'wild' behaviour of wild algebras. This seems to be the most spectacular development in representation theory in the last years. In order to pin-point the result, let us recall the main features of wild algebras known before. The intuitive notion of 'wild' algebras was built on investigations of Corner and Brenner who showed that there are finite dimensional $k$–algebras $A$, such that for *any* finite dimensional $k$–algebra $B$, there is a full exact embedding of the module category $B$–mod into $A$–mod; such an algebra $A$ is said to be *strictly wild*. Clearly, in order to show that $A$ is strictly wild, we only have to find for any $n \in \mathbb{N}$ a full exact embedding of $F_n$–mod into $A$–mod, where $F_n = k\langle X_1, \ldots, X_n\rangle$, the free algebra in $n$ (non–commuting) generators $X_i$, since we may write any finite dimensional $k$–algebra as a factor algebra of some $F_n$. For example, the algebra $F_2$ is strictly wild, a full exact embedding $\iota_n$ of $F_n$–mod into $F_2$–mod is constructed as follows: the $F_n$–modules are of the form $(V, \varphi_1, \ldots, \varphi_n)$, where $V$ is a $k$–space, and $\varphi_i$ are $k$–linear endomorphisms of $V$, and $\iota_n(V, \varphi_1, \ldots, \varphi_n) = (V^{n+2}, \alpha, \beta)$, where

$$\alpha = \begin{bmatrix} 0 & 1 & & & \\ & 0 & 1 & & \\ & & \ddots & \ddots & \\ & & & 0 & 1 \\ & & & & 0 \end{bmatrix} \quad \text{and} \quad \beta = \begin{bmatrix} 0 & & & & \\ 1 & 0 & & & \\ \varphi_1 & 1 & 0 & & \\ & \ddots & \ddots & \ddots & \\ & & \varphi_n & 1 & 0 \end{bmatrix}.$$

Of course, with $F_2$ all algebras $F_n$, where $n \geq 2$, are strictly wild. The mutual embeddings of the module categories of the various strictly wild algebras indicate that these module categories are similarly complicated: a complete classification of the indecomposable $A$–modules for one strictly wild algebra $A$ would yield a complete classification for any algebra (provided we can control the embeddings effectively, but this often seems to be the case, see the example above): so it is not surprising that no such classification is known (there are some papers who claim to provide one, but they just outline an inductive procedure of what should be done for any fixed dimension). Some features of a strictly wild $k$–algebra $A$ should be stressed: given any finite dimensional $k$–algebra $B$, there is an $A$–module with endomorphism ring $B$, in particular, there are indecomposable $A$–modules which are arbitrarily complicated, and a classification of the $A$–modules would also yield sort of a classification of all finite dimensional $k$–algebras.

Assume that $A$ is a representation infinite hereditary finite dimensional $k$–algebra, where $k$ is an algebraically closed field. We can assume that $A$ is, in addition, basic and connected, thus $A$ is the path algebra of some finite connected quiver $\Delta$ without oriented cycles, and $\Delta$ is not a Dynkin diagram. Now, $A$–mod has a preprojective, and a preinjective component. The preprojective modules, as well as the preinjective modules, are easily classified, so it remains to consider the remaining indecomposable modules: they are called the regular modules. One knows that $A$ is tame, if and only if $\Delta$ is a Euclidean diagram, and, in this case, the components containing regular modules are stable tubes, and the set of these components is indexed by the projective line $\mathbb{P}_1(k)$ over $k$. In case $\Delta$ is neither a Dynkin nor a Euclidean diagram, it is well–known that $A$ is strictly wild, and the components containing regular modules are of the form $\mathbb{Z}A_\infty$. In both cases, the modules on the boundary of these components are called quasi–simple. Any indecomposable regular module $M$ has a filtration (unique up to isomorphism) $M = M_0 \supset M_1 \supset \cdots \supset M_{t-1} \supset M_t = 0$ with all factors $M_{i-1}/M_i$ quasi–simple, and $M_{i-1}/M_i \cong \tau^i(M/M_1)$, the number t is called the quasi–length, the module $M/M_1$ the quasi–top of $M$. Any indecomposable regular module is uniquely determined by its quasi–length and its quasi–top, and, conversely, given a quasi–simple module and a positive integer, there is an indecomposable module with this quasi–top and this quasi–length.

Let us denote the set of regular components of $A$–mod by $\Omega(A)$. We may use $\Omega(A)$ also as index set for the $\tau$–orbits of the (isomorphism classes of) quasi–simple

$A$–modules, keeping in mind that the modules belonging to a fixed regular component are determined by the quasi–simple modules in this component.

The main, but rather hopeless problem is to obtain a reasonable description of $\Omega(A)$. Of course, for $A$ tame, $\Omega(A)$ may be identified with $\mathbf{P}_1(k)$ (and for $A$ representation finite, $\Omega(A) = \emptyset$). Assume now that $A$ is wild. It is easy to see ([**Z1**]) that in this case any Auslander–Reiten component contains at most one isomorphism class of indecomposable $A$–modules with a fixed dimension vector. Thus if we denote by $\mathcal{R}(d)$ the set of isomorphism classes of indecomposable $A$–modules with dimension vector $d$, where $d \in \mathbb{N}_0^n$ is an imaginary root, then we may embed $\mathcal{R}(d)$ into $\Omega(A)$, sending the isomorphism class of a module to the component containing it.

The investigations of Kerner center around the problem of describing $\Omega(A)$ in the wild case. What he shows, and this is very amazing, is that the set $\Omega(A)$ may be considered as being independent of $A$: given two wild hereditary algebras $A, A'$ over the same (algebraically closed) field, there are intrinsic bijections between $\Omega(A)$ and $\Omega(A')$. We stress here the word 'intrinsic': it is rather easy to see (and trivial in case $k$ is an uncountable field) that the set $\Omega(A)$ has the same cardinality as $k$, thus there are many bijections between $\Omega(A)$ and $\Omega(A')$. The essential steps for constructing Kerner's bijections will be reviewed below. For the algebras $A, A'$, there are at most a countable number of such bijections, indexed by finite sequences of tilting modules. At the moment it is not clear at all, whether bijections with different indices are actually different or not: thus either there is even a unique such bijection, or else, fixing one of these sets $\Omega(A)$, there is a countable group $G(A)$ of automorphisms operating on it, so that at least the orbit space $\Omega(A)/G(A)$ is an intrinsic set, independent of $A$ (note that in case $k$ is uncountable, the set $\Omega(A)/G(A)$ still is uncountable). The reader should keep in mind that the classification problem for wild hereditary algebras deals with many important mathematical problems, such as the $n$–subspace problems, for $n \geq 5$, or the problem of classifying $n$-tuples of matrices of the same size with respect to simultaneous multiplication from the left and simultaneous multiplication from the right, for $n \geq 3$, and the assertion is that for all these problems (after deleting the preprojective and the preinjective modules) there are natural bijections!

Let us outline the construction in some detail. Let $A$ be a representation-infinite connected hereditary $k$–algebra, and ${}_AT$ a tilting module which has no indecomposable preinjective direct summand. As Strauß [**St**] has shown, there is a unique preprojective component in End$({}_AT)$–mod, and if $I$ denotes its annihilator in End$({}_AT)$, then $C = \text{End}({}_AT)/I$ is a concealed algebra of the same representation type as $A$, and we say that $A$ *dominates* $C$ via $T$. For tame hereditary algebras $A, C$ the domination relation can be read from the Euclidean diagrams, and it coincides with the degeneration relation for the corresponding singularities. Consider the generalized Kronecker algebras $K(r)$, they are wild for $r \geq 3$, and the algebra $H$ defined by the quiver

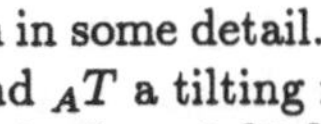

According to Unger [**U3**], any connected wild hereditary algebra dominates one of the form $K(r)$, with $r \geq 3$, and the algebra $H$ dominates any $K(r)$, with $r \geq 3$; this we will outline below. It follows that the equivalence relation generated by the dominance relation consists of a unique equivalence class. Thus, in order to construct the Kerner bijections, we may restrict to the case of a pair of algebras, one dominating the other.

Let $A, C$ be connected wild hereditary algebras, and assume $C$ is dominated by $A$ via some tilting module ${}_AT$. Let $B = \mathrm{End}({}_AT)$. We are going to construct Kerner's map $\eta_T : \Omega(A) \longrightarrow \Omega(C)$.

Let $M$ be a regular $A$-module. Then Kerner (**[K2]**, **[K4]**) shows that there are integers $s(M), t(M)$ such that $\tau_A^t M$ is generated by ${}_AT$, for all $t \geq t(M)$, and such that $\tau_B^{-s} \mathrm{Hom}_A({}_AT, M)$ is a $C$-module, for all $s \geq s(M)$.

**Theorem (Kerner)** *Let $M$ be an indecomposable regular $A$-module. Let*

$$\eta_T(M) = \tau_B^{-s(\tau_A^{t(M)}M)} \mathrm{Hom}_A({}_AT, \tau_A^{t(M)}M).$$

*Then for indecomposable regular modules $M, M'$ in the same component of $A$-mod, the $C$-modules $\eta_T(M)$, and $\eta_T(M')$ belong to the same component of $C$-mod, and the induced map $\eta_T : \Omega(A) \longrightarrow \Omega(C)$ is bijective.*

The situation is as follows: altogether we deal with countably many (isomorphism classes of) finite dimensional algebras, the path algebras of finite connected wild quivers without cycles over some fixed algebraically closed field $k$. For any such algebra $A$, the set $\Omega(A)$ is defined, and there are at most countably many tilting $A$-modules, thus there are at most countably many bijections of the form $\eta_T$.

The further aim will be to look for properties of the sets $\Omega(A)$ which are invariant under the maps $\eta_T$. If we fix some connected wild hereditary algebra $A$ as the basic example (say $K(3)$, or the 5-subspace quiver) and denote $\Omega = \Omega(A)$, we may consider this set, together with its yet unknown additional structure of properties which are invariant under the Kerner bijections, as a universal index set for handling the representations of wild hereditary algebras.

One may be curious to know what additional wild algebras have $\Omega$ as index set for their regular components. Lenzing and de la Peña **[LP]** have shown that given a wild canonical algebra $C(p)$, there are again intrinsic bijections between $\Omega$ and the set of all components of $C(p)$-mod$^-$, as well as between $\Omega$ and the set of components of $C(p)$-mod$^+$.

It seems worthwhile to indicate part of the proofs: The existence of the number $t(M)$ is an easy consequence of the following lemma **[K2]**: If $X, Y$ are regular modules, then there exists an integer $t(X, Y)$ such that $\mathrm{Hom}_A(\tau^t X, Y) = 0$ for all $t \geq t(X, Y)$. Note that this is the opposite assertion a lemma due to Baer **[B2]**: If $X, Y$ are regular modules, then there exists an integer $s(X, Y)$ such that $\mathrm{Hom}_A(X, \tau^s Y) \neq 0$ for all $s \geq s(X, Y)$. Taking both assertions together, we see that in a regular component, globally, the maps are going in one direction, and this is the opposite direction of the arrows in the Auslander–Reiten quiver. (The two statements are actually interrelated: Kerner uses in his proof Baer's lemma, and one derives from his lemma the existence of the bound $t(M)$ which may be considered as a partial strengthening of Baer's lemma: Let $T = T_p \oplus T_r$, be a tilting module, with $T_p$ preprojective, and $T_r$ regular, let $M$ be a regular $A$-module, and define $t(M) = t(M, \tau T_r)$. Then, for $t \geq t(M)$, $0 = \mathrm{Hom}_A(\tau^t M, \tau T_r) \cong \mathrm{Ext}_A^1(T_r, \tau^t M)$. Since $T_p$ is preprojective, and $\tau^t M$ is regular, we also have $\mathrm{Ext}_A^1(T_p, \tau^t M) = 0$, thus $\tau^t M$ is generated by ${}_AT$.)

Let us exhibit in detail bijections $\eta_T : \Omega(H) \longrightarrow \Omega(K(n))$, for $n \geq 3$. The indecomposable $H$-module with dimension vector $(n+1, n, 0)$ will be denoted by $M(n)$. Since $\dim_k \mathrm{Ext}_A^1(E(3), M(n)) = n$, there is a universal extension of the form $0 \to M(n)^n \to N(n) \to E(3) \to 0$, and clearly $N(n)$ is indecomposable and not preinjective. For $n \geq 1$, we can embed $P(3)$ into $M(n)$, and the cokernel is the direct sum of $n-1$ copies of $M(n+1)$. It follows that $T(n) = M(n) \oplus N(n) \oplus M(n+1)$ is a tilting module, and its endomorphism ring $B(n)$ is given by the quiver

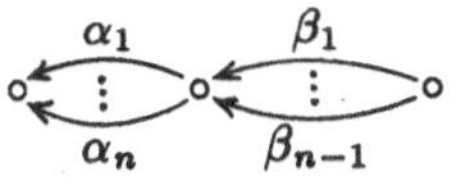


and the relations

$$\alpha_i\beta_i = \alpha_1\beta_1, \quad \alpha_{i+1}\beta_i = \alpha_2\beta_1, \quad \alpha_j\beta_i = 0 \quad \text{for } j \notin \{i, i+1\}.$$

Note that the algebra $B(n)$ is a one–point extension of $K(n)$ by an indecomposable $K(n)$–module $X(n)$ with dimension vector $(2, n-1)$. We have $B(2) = H$. So consider now the cases $n \geq 3$. Under this assumption, all indecomposable $K(n)$–modules with dimension vector $(2, n-1)$ are regular, thus, the preprojective component of $K(n)$–mod is a component of $B(n)$–mod, and therefore $H$ dominates $K(n)$ via $T(n)$. What can we say about components of $B(n)$? By the Kerner lemma, any regular component of $K(n)$–mod has many indecomposable modules $M$ such that both $\mathrm{Hom}_{K(n)}(X(r), M) = 0$ and $\mathrm{Hom}_{K(n)}(X(r), \tau_{K(n)}M) = 0$, and for these modules $M$ we have $\tau_{B(n)}M = \tau_{K(n)}M$; in particular, the $B(n)$–component containing $M$ again will be of the form $\mathbb{Z}A_\infty$, provided it does not contain the module $X(n)$. It is surprising that we obtain in this way all regular components of $B(n)$–mod.

## 2. Combinatorial Methods I: The Structure of Auslander–Reiten Components

**2.1.** Recall that the stable Auslander–Reiten quiver $\Gamma_s(A)$ of an artin algebra $A$ is obtained from the Auslander–Reiten quiver $\Gamma(A)$ of $A$ by deleting all translates of projective or injective vertices (and the corresponding arrows). It is a stable valued translation quiver and the length function yields a subadditive function with values in $\mathbb{N}_1$. By definition, the *regular* components of the Auslander–Reiten quiver of an artin algebra are those components which do not contain projective or injective vertices. Thus, a regular component $\Gamma$ is always a component of the stable Auslander–Reiten quiver, and the length function is an unbounded additive function on $\Gamma$ with values in $\mathbb{N}_1$.

A stable valued translation quiver will be called *smooth* provided the valuation is trivial (i.e. $d(\alpha) = d'(\alpha) = 1$, for all arrows $\alpha$), and any vertex is end point of precisely two arrows. Note that the last condition just means that the corresponding topological realization is a manifold without boundary.

Given any valued quiver $Q = (Q_0, Q_1, s, e, d, d')$, we may follow Riedtmann in order to define a stable valued translation quiver $\mathbb{Z}Q$: the vertex set of $\mathbb{Z}Q$ is given by $\mathbb{Z} \times Q_0$, for any arrow $\alpha : x \to y$ in $Q$, there are arrows $(z, \alpha) : (z, x) \to (z, y)$ and $\sigma(z, \alpha) : (z-1, y) \to (z, x)$ for any $z \in \mathbb{Z}$, the translation $\tau$ is defined by $\tau(z, x) = (z-1, x)$, and $d(z, \alpha) = d'\sigma(z, \alpha) = d(\alpha), d'(z, \alpha) = d\sigma(z, \alpha) = d'(\alpha)$. The structure of connected periodic stable translation quivers with subadditive functions with values in $\mathbb{N}_1$ is known: their universal covering is of the form $\mathbb{Z}Q$, with $Q$ a valued quiver whose underlying graph is either a Dynkin diagram, a Euclidean diagram, or of the form $A_\infty, B_\infty, C_\infty, D_\infty$, or $A_\infty^\infty$. There is the following general structure theorem for non–periodic stable translation quivers with subadditive functions:

**Theorem (Zhang)** *Let $\Gamma$ be a connected non–periodic valued stable translation quiver with a non–zero subadditive function $f$ with values in $\mathbb{N}_0$. Then either $\Gamma$ is smooth and $f$ is additive and bounded, or else $\Gamma = \mathbb{Z}Q$ for some valued quiver $Q$.*

The proof [**Z2**] is rather complicated. One has to consider the first homology group of the orbit graph of $\Gamma$ and some additive function on it measuring the difference between the numbers of forward and of backward arrows in any walk. In order to write $\Gamma$ in the form $\mathbb{Z}Q$, one has to find a suitable orientation on the orbit graph of $\Gamma$. Even in case the orbit graph is countable (and this is the case encountered in representation theory), one needs transfinite induction in order to construct such an orientation.

**Corollary** *Let $\Gamma$ be a component of the stable Auslander–Reiten quiver $\Gamma_s(A)$ of an artin algebra $A$. Then either $\Gamma$ is periodic, or else $\Gamma = \mathbb{Z}Q$ for some valued quiver $Q$ without oriented cycles.*

On $\Gamma$, there is the length function $f$. Note that it is impossible that $f$ is both additive and unbounded, since $f$ is only additive in case $\Gamma$ is a regular component, und then $f$ is unbounded, according to Auslander. Thus, in case $\Gamma$ is non–periodic, we see that $\Gamma = \mathbb{Z}Q$ for some valued quiver $Q$, and $Q$ cannot have an oriented cycle, since this would yield sectional cyclic paths in $\Gamma(A)$, which is impossible according to Bautista–Smalø.

**2.2.** It seems to be of interest to know what kind of valued quivers $Q$ actually can occur in $\Gamma = \mathbb{Z}Q$, where $\Gamma$ is a component of $\Gamma_s(A)$ for some artin algebra $A$. Of course, $Q$ cannot have oriented cycles. Also, $Q$ has to be symmetrizable.

Let $Q$ be a connected valued quiver without oriented cycles. If $Q$ is a Dynkin or a Euclidean quiver, then any additive function on $\mathbb{Z}Q$ with values in $\mathbb{N}_1$ is bounded, thus $\mathbb{Z}Q$ cannot arise as a regular component of the Auslander–Reiten quiver of an artin algebra.

Consider now the case where $Q$ is neither a Dynkin, nor a Euclidean diagram. In [**R2**], we have shown that a connected wild hereditary algebra $H$ has a regular tilting module if and only if there are at least three simple $H$ modules. Let $H$ be a connected wild hereditary algebra with at least three simple modules, and let $Q(H)$ be its valued quiver. Let ${}_HT$ be a regular tilting module. Then the connecting component of $B = \mathrm{End}({}_HT)$ is regular and of the form $\mathbb{Z}Q(H)$.

Also, let $Q$ be a connected symmetrizable valued quiver without oriented cycles and assume that after deletion of finitely many vertices and arrows we obtain a disjoint union of quivers of type $A_\infty$ (with trivial valuation). Then there are algebras $R$ with regular components of the form $\mathbb{Z}Q$, see [**CR**].

**2.3.** We have asked in [**10**] whether an Auslander–Reiten component $\Gamma$ may contain infinitely many isomorphism classes of indecomposable modules of the same dimension. Zhang [**Z1**] has observed that this is impossible in case we deal with a hereditary algebra. Consider now an arbitrary algebra, and let $\Gamma$ be a regular component. In case $\Gamma$ is periodic, it is a regular tube, and therefore for any non–zero additive function $f$ on $\Gamma$, the fibres are finite. Thus assume that $\Gamma$ is non–periodic. By Zhang's theorem, $\Gamma = \mathbb{Z}\Delta$ for some valued quiver $\Delta$. In case $\Delta$ is finite, or of the form $A_\infty^\infty, B_\infty, C_\infty$ or $D_\infty$, we have shown in a joint paper with Marmolejo [**MR**] that

$\Gamma$ can contain only finitely many isomorphism classes of indecomposable modules of the same dimension. In all other cases, the question seems to be open. Let us remark that it should be difficult to use only combinatorial properties of translation quivers to answer the question. For, let $\Delta$ be a finite connected symmetrizable valued quiver with no oriented cycle, but with at least one (non-oriented) cycle. Assume in addition that there are at least three vertices and that $\Delta$ is not of type $\tilde{A}_n$. Let $\Delta'$ be some infinite covering of $\Delta$. Then, we claim that $\mathbb{Z}\Delta'$ has an additive function $f'$ with values in $\mathbb{N}_1$, such that all fibres of $f'$ are infinite. Namely, as we have mentioned above, there exists a tilted algebra with a regular component of the form $\mathbb{Z}\Delta$. Let $f$ be the length function of this component. Denote by $\pi : \mathbb{Z}\Delta' \to \mathbb{Z}\Delta$ a covering map, and let $f' = f \circ \pi$.

**2.4.** Let us consider the special case of the group algebra $kG$ of some finite group $G$ over an algebraically closed field $k$ of characteristic $p$. Let $B$ be a block of $kG$, and $\delta(B)$ its defect group. According to Webb, any non-periodic component $\Gamma$ of the stable Auslander–Reiten quiver $\Gamma_s(kG)$ is of the form $\mathbb{Z}\Delta$, where $\Delta$ is a Euclidean diagram, or else of the form $A_\infty, D_\infty$, or $A_\infty^\infty$.

First, let us consider the case when $\Delta$ is Euclidean (in particular, $\Gamma$ is not a regular component). Then the defect group $\delta(B)$ has to be elementary abelian of order 4, and $\Gamma$ is of the form $\mathbb{Z}\tilde{A}_{1,1}$, or $\mathbb{Z}\tilde{A}_{3,3}$ (see **[Ok]**, **[Bs]**, **[ES]**).

Of course, the defect group $\delta(B)$ is cyclic if and only if the block $B$ is representation finite. Bondarenko and Drozd have shown that the block $B$ is tame, and not representation finite, if and only if $p = 2$, and $\delta(B)$ is dihedral, semidihedral or quaternion. In this case, there is no component of the form $\mathbb{Z}A_\infty$ (if $\delta(B)$ is dihedral, there are also no components $\mathbb{Z}D_\infty$, and for $\delta(B)$ elementary abelian of order 4, or quaternion, all regular components are periodic), as Erdmann **[E1]** has shown.

One may use the knowledge on Auslander–Reiten components in order to obtain information about the block and its representations. This is the main philosophy of Erdmann's treatment of tame blocks **[E1]**. She determines the possible structure of all symmetric algebras with non-singular Cartan matrix which have prescribed components similar to those which are known to exist for tame blocks. In this way, she copies the approach of Riedtmann of classifying the representation finite selfinjective algebras. However, one should observe that there is an intrinsic difference: if $A$ is a representation finite algebra, we know that $\operatorname{rad}^\infty(A\text{–mod}) = 0$, thus we can recover all maps in $A$–mod from the Auslander–Reiten quiver; in particular, this holds true for maps between indecomposable projective modules. Since the algebra is given by maps between the indecomposable projective modules, it does not seem to be surprising that we are able to recover the algebra. Actually, we know that non–standard representation finite algebras do exist only in characteristic 2. In the representation–infinite case, the situation is different: the maps between indecomposable projective modules usually will belong to $\operatorname{rad}^\infty(A\text{–mod})$, even if we deal with modules in one component, thus there is no way to recover these maps directly from the mesh category. Let us explain one detail of her advance: In order to recover $A$ from $\Gamma(A)$, one needs to know in particular the quiver $Q(A)$. As mentioned above, we may consider the arrows in $Q(A)$ as being special maps $f : P(y) \to P(x)$ between indecomposable projective modules. We also may ask whether they define special meshes in the Auslander–Reiten quiver, and indeed, they do. Consider the cokernel $M$ of $f$. The middle term of the Auslander–Reiten sequence ending in $M$ is indecomposable **[BR]**, so certain of the meshes with a unique middle term will correspond to the arrows of $Q(A)$. For example, for $A$ any string algebra, all meshes which have a unique middle term, and which do not belong to homogeneous tubes, arise in this way.

For $B$ a wild block, there are infinitely many components of the form $\mathbb{Z}A_\infty$, **[E2]**. But also there are always many tubes. In fact, an indecomposable non-projective $B$-module $M$ is $\tau$-periodic if and only if its complexity is equal to one, thus if and only if the subvariety $X_G(M)$ of the maximal spectrum of the even cohomology ring $H^{ev}(G,k)$ is one-dimensional, and Carleson has shown how to construct indecomposable modules with prescribed irreducible subvarieties, see the book by Benson **[Be]**.

**2.5.** Directing modules are quite rare. Of course, all the indecomposable modules which belong to a preprojective, or a preinjective component are directing, and also the indecomposable modules in the connecting component of a tilted algebra are directing. Skowroński and Smalø**[SS]** have shown that an Auslander-Reiten component $\Gamma$ of a finite dimensional algebra $A$ which consists entirely of directing modules, can have only finitely many $\tau$-orbits. In case $\Gamma$ is in addition regular, then $\Gamma$ is the connecting component of some convex subalgebra $B$ of $A$ which is a tilted algebra. It follows that an algebra can have only finitely many components which contain directing modules only.

Recall that a finite dimensional algebra $A$ is said to be *representation directed,* provided it is representation finite, and we can order the indecomposable representations $M_1, \ldots, M_n$ in such a way that we have $\mathrm{Hom}(M_i, M_j) = 0$ for $i > j$, or, equivalently, provided the Auslander-Reiten quiver is finite and does not have oriented cycles. The position of sincere modules in the Auslander-Reiten quiver of a representation directed algebra has been studied further by de la Peña **[P1]**.

## 3. Combinatorial Methods II: Quadratic Forms and Roots

Recall that a polynomial $\chi = \chi(X_1, \ldots, X_n)$ of the form

$$\chi(X_1, \ldots, X_n) = \sum_i X_i^2 + \sum_{i<j} \chi_{ij} X_i X_j$$

with integer coefficients $\chi_{ij}$ is called an *integral quadratic form* in $n$ variables. The quadratic forms which are of interest in the representation theory of finite dimensional algebras are integral, at least if we deal with an algebraically closed base field. Let $\chi$ be an integral quadratic form in $n$ variables. An $n$-tuple $x = (x_1, \ldots, x_n)$ of integers is called a *root* provided $\chi(x_1, \ldots, x_n) = 1$. Note that the canonical base vectors $e(i)$ with $e(i)_i = 1$, and $e(i)_j = 0$ for $j \neq i$, are roots. The quadratic form $\chi$ defines a symmetric bilinear form $\langle -, - \rangle$ on $\mathbb{Z}^n$, thus any root $x$ defines a reflection $\sigma_x$ by $\sigma_x(y) = y - \langle y, x \rangle \cdot x$. The group generated by the reflections $\sigma_{e(i)}$ with $1 \leq i \leq n$ is called the Weyl group for $\chi$, and the images of the base vectors $e(i)$ under the elements of the Weyl group are called *Weyl roots*. Of course, Weyl roots are roots. In case the coefficients $\chi_{ij}$ all are non-positive, so that the quadratic form $\chi$ is given by a generalized Cartan matrix, then a corresponding Kac-Moody Lie-algebra is defined, and therefore also the socalled *imaginary roots,* but we stress that for an imaginary root $x$, we have $\chi(x) \leq 0$, thus imaginary roots are not roots in the sense defined above.

**3.1.** Let $A$ be a finite dimensional hereditary algebra, say over an algebraically closed field $k$, thus $A$ is the path algebra of a quiver $Q(A)$ without oriented cycles. The theorem of Kac asserts that the dimension vectors of the indecomposable $A$–modules are just the positive roots $x$ of the corresponding quadratic form $\chi_{Q(A)}$. Note that we may distinguish between the real roots (with $\chi_{Q(A)}(x) = 1$) and the imaginary roots (with $\chi_{Q(A)}(x) \leq 0$). If $x$ is a positive real root, there is precisely one indecomposable $A$–module with dimension vector $x$. If $x$ is an imaginary root, there is an $n$–parameter family of indecomposable $A$–modules with dimension vector $x$, where $n = 1 - \chi_{Q(A)}(x)$. A positive root $x$ is called a *Schur root* provided there exists a module $M$ with endomorphism ring $k$ and dimension vector $x$. Observe that $x$ is a real Schur root if and only if there exists an indecomposable module $M$ with dimension vector $x$ such that $\mathrm{Ext}^1(M,M) = 0$.

There is the following inductive procedure for constructing indecomposable modules without selfextensions: Let $M_1, M_2$ be indecomposable $A$–modules without selfextensions, assume that

$$\mathrm{Hom}(M_1,M_2) = 0, \quad \mathrm{Hom}(M_2,M_1) = 0, \quad \text{and} \quad \mathrm{Ext}^1(M_2,M_1) = 0,$$

and let $\dim_k \mathrm{Ext}^1(M_1,M_2) = n$. Let $(c,d)$ be a real root for the generalized Cartan matrix $\begin{bmatrix} 2 & -n \\ -n & 2 \end{bmatrix}$. Then there is a unique indecomposable module $M$ with an exact sequence $0 \to dM_2 \to M \to cM_1 \to 0$, and $M$ has no self extensions. The set of modules obtained in this way will be called the *line* determined by $M_1, M_2$.

**Theorem (Schofield)** *Let $M$ be an indecomposable $A$–module $M$ without selfextensions, and let $s$ be the number of isomorphism classes of composition factors of $M$. Then $M$ belongs to precisely $s-1$ lines.*

In particular, we see that we obtain all indecomposable modules without selfextensions starting from the simple modules and forming inductively lines.

The main tool for this investigation is the perpendicular category $M^{\perp}$. Given a set $\mathcal{S}$ of $A$–modules of projective dimension at most 1, the subcategory

$$\mathcal{S}^{\perp} = \{M \mid \quad \mathrm{Hom}(S,M) = 0, \quad \mathrm{Ext}^1(S,M) = 0, \quad \text{for all} \quad S \in \mathcal{S}\}$$

is called the *right perpendicular category* to $\mathcal{S}$, it is an abelian subcategory and the inclusion $\mathcal{S}^{\perp} \subset A$–mod is exact. This construction is very useful, it has been considered in detail by Geigle–Lenzing **[GL2]**. Note that in case we have in addition $\mathrm{Ext}^1(\mathcal{S},\mathcal{S}) = 0$ (so that $\mathcal{S} \subseteq \mathrm{add}\, S$ for a basic partial tilting module $S$), the category $\mathcal{S}^{\perp}$ is equivalent to a module category $B$–mod, for some finite dimensional algebra $B$. Happel has pointed out that always $s(A) = s(B) + s(\mathrm{End}\, S)$. Indeed, take indecomposable $A$–modules $S_1, \ldots, S_s$ such that $S = \bigoplus S_i$, and let $E_1, \ldots, E_t$ be the Bongartz complement. Let $E_i'$ be the trace of $S$ in $E_i$. According to **[RS]**, this is a proper submodule of $E_i$, and we let $Q_i = E_i/E_i'$. On the one hand, $\bigoplus Q_i$ is a progenerator for $\mathcal{S}^{\perp}$ (see **[H7]**), on the other hand, one can show easily that the modules $Q_i$ are indecomposable and pairwise non–isomorphic.

There are several conjectures due to Kac **[Kc]** dealing with representations of quivers. As we have mentioned above, given a real root $x$, and $k$ an algebraically closed field, there exists a unique indecomposable representation $M(x)$ with dimension vector $x$. In case the characteristic of $k$ is non–zero, Kac had shown that $M(x)$ is defined already over the prime field, and Schofield could show that the same is true in characteristic zero, thus solving Conjecture 4 of Kac. On the other hand, Conjecture 9 asserted that given a representation $M$ with endomorphism ring $k$, there is no non–trivial way of writing the dimension vector of $M$ in the form $y+z$, with $y, z \in \mathbb{N}_0^n$, and $\langle y,z\rangle \geq 0$, $\langle z,y\rangle \geq 0$, however, as Le Brujn [Le] has observed, there are obvious counter examples: consider the quiver $\circ \rightleftarrows \circ \rightleftarrows \circ$, there is a unique indecomposable representation $M$ with dimension vector $(2,1,2)$, and its endomorphism ring is $k$. Let $y=(1,1,1)$, and $z=(1,0,1)$.

**3.2** Let $A$ be a finite dimensional algebra over an algebraically closed field, and assume its quiver has no oriented cycles. Then we may attach to $A$ two integral quadratic forms, namely the Tits form and the Euler form. These forms often will determine the representation type of $A$, and we even may obtain complete information about the dimension vectors of the indecomposable $A$–modules in terms of these forms. We refer to the report **[P2]** of de la Peña which surveys the known results: he calls an algebra *good* provided its quiver has no oriented cycles, it satisfies the separation condition, it is Schurian (i.e. $\dim_k \operatorname{Hom}(P,P') \leq 1$, for $P, P'$ indecomposable projective), and there is no full subalgebra which is hereditary of type $\tilde{A}_n$. He conjectures that the good algebras are controlled by the Tits form. Note that the Bongartz criterion asserts that a good algebra is representation finite if and only if it does not contain as a convex subalgebra an algebra from the Happel–Vossieck list.

The integral quadratic form $\chi$ in $n$ variables is said to be *weakly positive*, provided we have $\chi(x) > 0$ for all positive $x \in \mathbb{Z}^n$ (here, $x \in \mathbb{Z}^n$ is said to be positive, written $x > 0$, provided all coordinates of $x$ are non–negative, and $x \neq 0$). An element $x \in \mathbb{Z}^n$ with all coordinates $x_i \neq 0$ will be said to be *sincere*.

If $A$ is a representation directed algebra, then, as Bongartz has shown, the Tits form is weakly positive, and the dimension vectors of the indecomposable $A$–modules are just the positive roots. In order to study an indecomposable module, we may assume that its dimension vector is sincere. We recall that $A$ is said to be *sincere* provided there exists a sincere indecomposable $A$–module.

Bongartz has exhibited a list of 24 series of sincere directed algebras which include all sincere directed algebras $A$ with $s(A) \geq 14$. The list of the remaining sincere directed algebras has been published by Dräxler **[D2]**, it should be of interest to obtain a better understanding of these exceptional algebras and their indecomposable modules, or, combinatorially, of the corresponding quadratic forms and their positive roots. For example, Dräxler has observed that any sincere directed algebra has a unique minimal sincere indecomposable module: the coefficients of the corresponding root $x_A$ should be a measure for the zero relations needed to define the algebra (for example, $A$ is given by a fully commutative quiver if and only if all the coefficients of $x_A$ are equal to 1). On the other hand, there usually will be several maximal sincere positive roots, and it is an interesting question to determine the number of such roots in advance. For example, the form

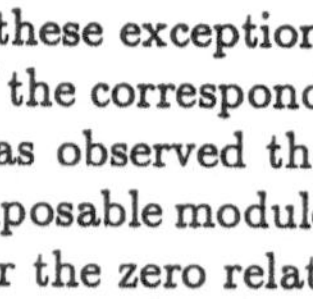

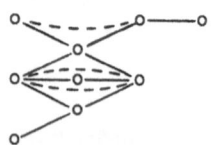

has precisely 10 maximal roots, this is the largest number which can occur. Note that Unger **[U5]** also has shown that there is only one sincere directed algebra $A$ with indefinite quadratic form $\chi_A$ which has more than one maximal sincere positive root, namely

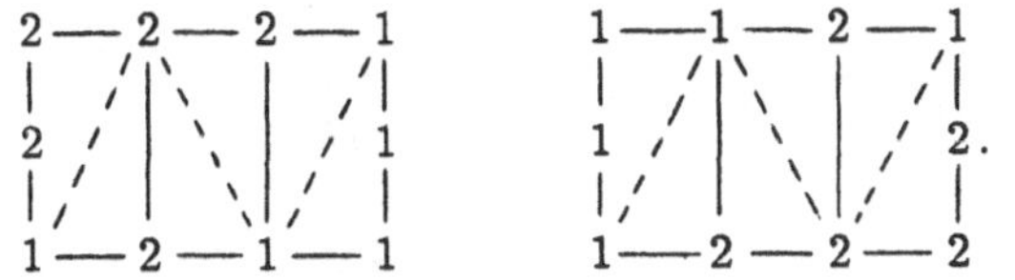

The list of all tame concealed algebras as presented by Happel and Vossieck is a list of integral quadratic forms. This list (as the one given by Dräxler) was produced with the help of a computer. A purely combinatorial approach to this list which abandons the use of a computer, has been given by von Höhne **[Hö]**. He classifies certain integral quadratic forms which are $\mathbb{Z}$–equivalent to to a quadratic form of type $\tilde{A}_n, \tilde{D}_n, \tilde{E}_6, \tilde{E}_7, \tilde{E}_8$, and singles those out which appear as quadratic forms for tame concealed algebras.

An integral quadratic form $\chi$ in $n$ variables is said to be *critical* provided $\chi(x) > 0$ for every non–sincere vector $x \in \mathbb{N}_0^n$, but there is a vector $x \in \mathbb{N}_1^n$ with $\chi(x) \leq 0$. Similarly, $\chi$ is said to be *hypercritical* provided $\chi(x) \geq 0$ for every non–sincere vector $x \in \mathbb{N}_0^n$, and there is a vector $x \in \mathbb{N}_1^n$ with $\chi(x) < 0$. The forms which occur in the Happel–Vossieck list are typical critical forms. Unger **[U2]** has calculated the list of all minimal wild concealed algebras, clearly the corresponding quadratic forms are hypercritical; this list should be of interest for a further study of tame algebras (see **[P2]**). Note that for every wild concealed algebra $A$, say of type $\Delta$, with $s(A) \geq 3$, there exists a representation infinite concealed factor algebra $B$ obtained from $A$ by factoring out the twosided ideal generated by some primitive idempotent, such that the type of $B$ is a connected full subquiver of $\Delta$, see **[HU1]**.

**3.3.** A famous result of Ovsienko asserts that the coordinates of a positive root of a weakly positive integral quadratic form $f$ are bounded by 6. Also, he has shown that if $\chi$ is weakly positive, and there exists a positive root for $\chi$ with at least one of the coordinates equal to 6, then $\chi$ is a form in at least 8 variables. Note that the quadratic form $\chi_8$ for the Lie algebra $E_8$ has a positive root with coordinates

$$2-4-\overset{\overset{3}{|}}{6}-5-4-3-2,$$

and is even positive definite.

Ostermann and Pott **[OP]** have shown that a weakly positive quadratic form with a sincere positive root with at least one of the coordinates equal to 6 is a form in at most 24 variables. Also, there is a unique such form $\chi_{24}$ with 24 variables:

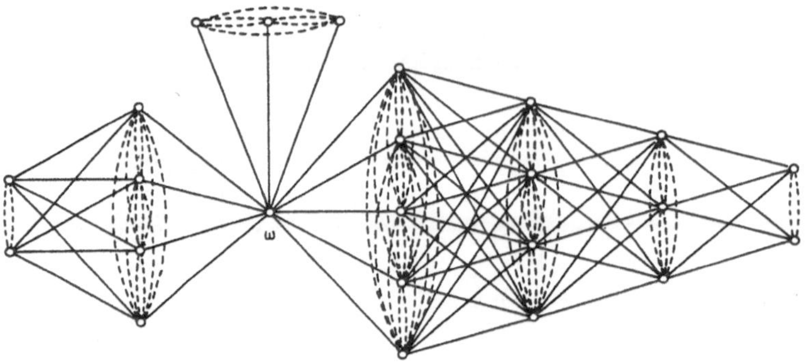

Any weakly positive quadratic form with a sincere positive root with at least one of the coordinates equal to 6 is a radical extension of the form $\chi_8$ and has $\chi_{24}$ as a radical extension. This shows that there is a full control over all such forms.

This result can be applied to representation theory. Assume that $A$ is a representation directed algebra with a sincere indecomposable module $X$ with dimension vector $x$. Then $x$ is a positive root for the quadratic form $\chi_A$, and $\chi_A$ is weakly positive. Ovsienko's theorem asserts that all the coefficients of $x$, thus all Jordan Hölder multiplicities of $X$, are bounded by 6. In case one of these multiplicities is equal to 6, we see that $\chi_A$ is a radical extension of the quadratic form $\chi_8$, in particular, $\chi$ is positive semidefinite. But this implies that $\chi$ is either of type $E_8$ or of type $\tilde{E}_8$, and therefore the number of variables is 8 or 9. Thus $A$ is an algebra with 8 or 9 simple modules. (Actually, the case of 9 simple modules cannot occur).

Let us denote by $\mathcal{F}_n$ the set of weakly positive integral quadratic forms $\chi$ which have a sincere positive root with at least one component equal to $n$, and such that all components of positive roots of $\chi$ are bounded by $n$. Ovsienko's result means that $\mathcal{F}_n$ is empty for $n > 6$, and, as we have seen above, the forms in $\mathcal{F}_6$ are forms in at most 24 variables, and are positive semidefinite. For $n < 6$, It is easy to construct a form in $\mathcal{F}_n$ in $30 - n$ variables (by splitting the central vertex of the form $\chi_{24}$ into $7 - n$ vertices). For $n \leq 3$, there are forms in $\mathcal{F}_n$ with arbitrarily many variables: take the forms $A_n, D_n$, and the following ones (exhibited together with the maximal root)

$$
\begin{array}{c}
1-\cdots-1 \\
1\!\!\diagup\!\!-\,-\,-\,-\,-\,-\,-\,-\!\!\diagdown\!\!1 \\
\diagdown 2-3-2\diagup \\
\overset{|}{2}
\end{array}\quad,
$$

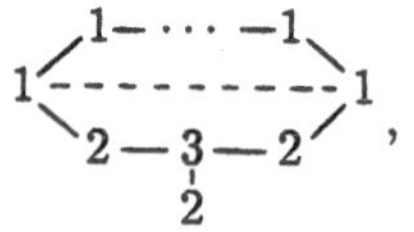

all these forms are realized by finite dimensional algebras. For $n = 4$ and $n = 5$, we do not know whether $\mathcal{F}_n$ contains finitely or infinitely many forms; but according to Bongartz, only finitely many can be realized by algebras.

**3.4.**  One may use the Hochschild cohomology in order to single out some combinatorial properties of algebras. We assume that the base field is algebraically closed. For a representation directed algebra $A$, Happel has shown that the Hochschild cohomology groups $H^i(A)$ are zero, for all $i \geq 2$, and, in addition, $H^1(A) = 0$, if and only if the quiver of $A$ is a tree [H5].

For further assertions concerning the interplay of the combinatorics of the quiver (and the relation) defining an algebra on the one hand, and the Hochschild cohomology on the other, we refer to [H5]; in [H10], there are several results dealing with the

Auslander algebra of a representation directed algebra, thus with the combinatorial behaviour of the Auslander–Reiten quiver.

**3.5.** Let $A$ be a connected finite dimensional hereditary algebra, and let $C_A$ be its Cartan matrix. The Coxeter matrix $\Phi_A = -C_A^{-t}C_A$ is an important tool in representation theory, since for any indecomposable non–projective $A$–module $M$, with dimension vector $\mathbf{dim}\, M$, we have $\mathbf{dim}\, \tau M = \Phi_A(\mathbf{dim}\, M)$. Let $\rho = \rho_A$ be the spectral radius of $\Phi_A$, it is called the *growth number* for $A$.

We note that in case $A$ is representation infinite, then $\rho$ is an eigenvalue of $\Phi_A$. Of course, for $A$ tame, we have $\rho = 1$, and we will assume that $A$ is wild.

Let us assume in addition that $\mathrm{rad}^2 A = 0$, and let $(u_{xy}, v_{xy})$ be the valuation on the arrow $x \to y$ in $Q(A)$. Then we obtain an $n \times m$ matrix $U = (u_{xy})$, and an $n \times m$ matrix $V = (v_{xy})$, where $n$ is the number of sources, and $m$ the number of sinks of $Q(A)$. In this case, it is very easy to calculate all the eigenvalues of $\Phi_A$, since they are naturally related to the eigenvalues of non–negative matrix $UV^t$; in particular A'Campo and Subbotin–Stekolshchik have shown that any eigenvector of $\Phi_A$ is real or of absolute value 1. This has been used by de la Peña and Takane **[PTa]** and by Zhang **[Z3]** in order to show that there are vectors $p, q$ with positive coefficients such that for any preprojective or regular module $X$ and for any preinjective or regular module $Y$, we have real numbers $\alpha_p(X) > 0, \alpha_q(X) > 0$, with

$$\lim_{n\to\infty} \frac{1}{\rho^n}\mathbf{dim}\, \tau^{-n}X = \alpha_p(X)p, \quad \text{and} \quad \lim_{n\to\infty} \frac{1}{\rho^n}\mathbf{dim}\, \tau^{n}Y = \alpha_q(X)q.$$

Also, one can use the vectors $p, q$ in order to obtain a numerical criterion for an indecomposable module $M$ for being preprojective, regular or preinjective: $M$ is preprojective if and only if $\langle \mathbf{dim}\, M, p\rangle > 0$, and preinjective if and only if $\langle q, \mathbf{dim}\, M\rangle > 0$; also, if $M$ is regular, then $\langle p, \mathbf{dim}\, M\rangle > 0$, and $\langle \mathbf{dim}\, M, q\rangle > 0$, see **[PT]**. Here, $\langle -, -\rangle$ is the usual (non–symmetric) bilinear form on $\mathbf{Z}^{s(A)}$ which encodes the homological behaviour of $A$–modules. It seems to us that all these assertions should be true also in the general case of $\mathrm{rad}^2 A$ being arbitrary.

Xi **[X4]** has shown that the smallest possible growth number $\rho_A$ for $A$ wild, occurs for the path algebra of the quiver

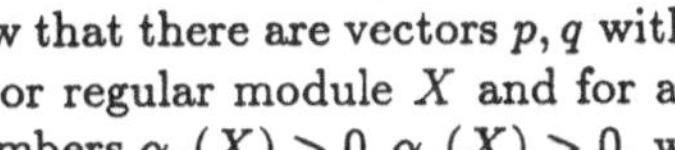

it is the largest root of the polynomial $x^{10} + x^9 - x^7 - x^6 - x^5 - x^4 - x^3 + x + 1$, thus approximately 1.176. Note that growth numbers can be used to deal with the structure of Auslander–Reiten components, see **[Z3]**.

**3.6.** These investigations have also other applications. Chains of finite dimensional semisimple algebras (say defined over $\mathbf{Z}$ or $\mathbf{C}$)

$$A_0 \subseteq A_1 \subseteq A_2 \cdots,$$

the inductive limits, and their completions, have been considered in detail in the theory of $C^*$–algebras, for a survey see **[GHJ]**.

In particular, starting with a semisimple algebra $A_1$ and a semisimple subalgebra $A_0$, containing the unit element, there is the socalled fundamental construction of Jones of taking as $A_2$ the endomorphism ring $\mathrm{End}_{A_0}(A_1)$, note that $A_1$ embeds into

$A_2$ as the set of right multiplications. We use induction in order to obtain the tower $(A_i)$ corresponding to the pair $A_0, A_1$, namely let $A_{i+1} = \mathrm{End}_{A_{i-1}}(A_i)$.

Let $B = A_0, C = A_1$, and form the matrix ring $A = \begin{bmatrix} B & C \\ 0 & C \end{bmatrix}$. This is a hereditary algebra with $\mathrm{rad}^2 A = 0$. Note that every hereditary algebra $A'$ with $\mathrm{rad}^2 A' = 0$ is Morita equivalent to one of the form $\begin{bmatrix} B & C \\ 0 & C \end{bmatrix}$. By definition, the Jones index of the tower is just the spectral radius of $U_A V_A^t$, and the Bratelli graph is the underlying quiver of the preprojective component of $A$. In this way, we see that one may relate questions concerning such towers of algebras to those arising in the representation theory of hereditary algebras. In particular, the values $> 4$ of the Jones index are related to the growth numbers of wild hereditary algebras: for example, the algebra exhibited by Xi yields a tower with Jones index approximately 4.026, this is the smallest possibe value $> 4$. See the joint paper **[DR8]** with Dlab.

## 4. Combinatorial Methods III: Representations of Posets

It was one of the first devices of the new representation theory of finite dimensional algebras, to reduce problems to the consideration of vectorspaces with prescribed subspaces. This is the method of the earlier papers of the Kiev school of Nazarova–Rojter, as well as of Gabriel's work on quivers of finite representation type; but at the same time, we also should mention Corner's and Brenner's investigations on wild behaviour. The school of Nazarova–Rojter showed the importance of the representation theory of posets, in particular, all their attempts to solve the second Brauer–Thrall conjecture use reductions from representations of algebras to representations of posets. The representation theory of posets disposes of rather striking results, there is the Klejner criterion for representation finiteness, Nazarova's criterion for tameness, the Nazarova–Zavadskij criterion for being of finite growth: always, there is given a small number of 'bad' posets which have to be excluded. There are also corresponding lists of the 'good' posets, say the Klejner list of sincere representation finite posets, or Zavadskij's list of the sincere posets of finite growth; they are longer, but still it is possible to overlook them. In contrast, similar results for finite–dimensional algebras cannot be expected, as already the algebras with two simple modules show: there are so many algebras that any list will tend to be useless.

Representations of a poset $S$ were defined by Nazarova–Rojter in terms of partitioned matrices, and Gabriel has introduced the concept of an $S$–space; the corresponding categories of representations of $S$ and of $S$–spaces are very similar, the precise relationship was determined by Drozd and Simson.

**4.1.** Let us assume that the base field $k$ is algebraically closed, and assume we are dealing with a representation finite $k$–algebra $A$. The use of $S$–spaces can best be demonstrated when we add the assumption that $A$ is representation directed. In addition, we may assume that $A$ is basic. Let $A$–ind be a complete set of indecomposable $A$–modules, one from each isomorphism class. Let $Q$ be the quiver of $A$. For any vertex $x \in Q$, let $E(x) \in A$–ind be the corresponding simple $A$–module, and let $P(x) \in A$-ind be its projective cover. Given an $A$–module $M$, and $x \in Q$, we denote by $M_x$ the vectorspace at the vertex $x$, note that we can identify $M_x$ and

$\operatorname{Hom}_A(P(x), M)$; also, we may identify $M$ and $\bigoplus_{x\in Q_0} M_x$.

For any vertex $x$ of $Q$, we will define a poset $S(x) = S_A(x)$, and for any $A$-module $M$, we will endow the $k$-vectorspace $M_x$ with subspaces so that it becomes an $S(x)$-space $\tilde{M}_x$. Let $A$-ind$_x$ be the set of $A$-modules $M \in A$-ind with $M_x \neq 0$. Then, the $S(x)$-spaces $\tilde{M}_x$, with $M \in A$-ind$_x$ are indecomposable, pairwise non-isomorphic, and, up to isomorphism, any indecomposable $S(x)$-space occurs in this way. We may reformulate this as follows: Consider the direct sum

$$\bigoplus_{M\in A\text{-ind}} M = \bigoplus_{M\in A\text{-ind}} \bigoplus_{x\in Q_0} M_x$$

and identify for any $x \in Q_0$ the direct sum $\bigoplus_{M\in A\text{-ind}} M_x$ with the direct sum of the total spaces of all indecomposable $S(x)$-spaces (one from each isomorphism class).

Let us define $S(x)$, for $x \in Q_0$. Vertices of $S(x)$ are the modules $U \in A$-ind$_x$, which are different from $P(x)$, such that $(\tau U)_x = 0$. Given two modules $U, U'$ in $S(x)$, let $U \leq U'$ it and only if there is a map $f : U' \to U$ such that $\operatorname{Hom}(P(x), f) \neq 0$. (According to v.Höhne **[Hö]**, it follows from $(\tau U)_x = 0$ that $\dim_k \operatorname{Hom}(P(x), U) \leq 1$, and therefore the relation $\leq$ defined on $S(x)$ is transitive.) Also, any $A$-module $M$, the vectorspace $M_x$ may be endowed in a canonical way with subspaces indexed by the elements of $S(x)$ : for $U \in S(x)$, take as subspace of $M_x = \operatorname{Hom}(P(x), M)$ the set of maps $P(x) \to M$, which factor through $U$.

For a proof, we refer to the joint paper **[RV]** with Vossieck, further information, as well as examples, may be found in **[R10]**. The notion behind the result mentioned above is that of a hammock, namely the Auslander–Reiten quiver of the category of $S(x)$-spaces, or, equivalently, the hammock of all modules in $A$-ind$_x$, thus of all indecomposable $A$-modules $M$ with $M_x \neq 0$. Hammocks of this kind have been considered by Brenner **[Br]** in order to give a numerical characterization of finite Auslander–Reiten quivers (the intuition is the following: the modules in $A$-ind$_x$ strech from $P(x)$ to $Q(x)$, the injective envelope of $E(x)$, these are the pegs to which the hammock is fastened. In between, there are the (Auslander–Reiten) meshes, in this way, the hammock is knitted.) In **[RV]**, we have given a combinatorial definition of *hammocks* as the finite translation quivers $\Gamma$ with a unique source $\omega$, and no oriented cycle, such that there is an additive function $h$ on $\Gamma$ (called the *hammock function*) with $h(\omega) = 1$, $h(p) = \sum_{y\to p} h(y)$, for all projective vertices different from $\omega$, and with $h(q) \geq \sum_{q\to y} h(y)$, for all injective vertices. The hammocks turn out to be just the Auslander–Reiten quivers of the categories of $S$-spaces, where $S$ is a representation finite poset. Given the hammock $\Gamma$, we can recover the corresponding poset as follows: its vertices are the projective vertices of $\Gamma$ different from the unique source, and the partial ordering ist derived from the existence of maps in the mesh category of $\Gamma$.

It is not difficult to see that the full translation subquiver of $\Gamma_A$ given by the vertices $[M]$ with $M \in A$-ind$_x$ form a hammock, with hammock function $\dim_k \operatorname{Hom}(P(x), -)$, the hammock function just counts the Jordan–Hölder multiplicity of the simple module $E(x)$.

In case $x$ is a source of $Q$, we deal with a one–point extension, say $A = B[M]$, for some algebra $B$ and $M = \operatorname{rad} P_A(x)$. Then $S_A(x)$ is given by $\operatorname{Hom}_B(M, -)$. A similar description exists in case $x$ is a sink for $Q$. The general case where $x$ is an arbitrary vertex of $Q$ can be reduced to the consideration of these two special cases: namely, as Scheuer **[Sr2]** has shown, $S(x)$ has a canonical decomposition into an ideal $S(x)^-$ and a coideal $S(x)^+$. Let $S(x)^+$ be the subset of $S(x)$ given by all $U$ with $\text{Hom}(U, E(x)) \neq 0$. Then $S(x)^+$ is a coideal, and we denote by $S(x)^-$ its complement. Let $A_x^+$ be the restriction of $A$ to the subquiver of $Q$ obtained by deleting all proper predecessors of $x$, similarly, let $A_x^-$ be the restriction of $A$ to the subquiver of $Q$ obtained by deleting all proper successors of $x$ (thus, $x$ is a source for $A_x^+$, and a sink for $A_x^-$.) Then $S(x)^+$ can be identified with $S_{A_x^+}(x)$, and $S(x)^-$ with $S_{A_x^-}(x)$.

**4.2.** It is rather difficult to decide whether a given algebra is representation finite or not. The usual procedure is to use covering theory, and then to invoke the Bongartz criterion. In this way, we have to deal with the Happel–Vossieck list of critical algebras. Of course, it would be easier if we only would have to invoke the Klejner list of critical posets. As we have seen above, the hammock approach allows to attach to each vertex $x$ of a representation directed algebra the poset $S(x)$, but we note that the elements of $S(x)$ already may be rather complicated modules. Dräxler **[D3]** has proposed to consider a smaller poset, namely the full subposet $S'(x)$ of all thin modules, a module $M$ being called *thin* provided $\dim_k M_y \leq 1$, for all vertices $y$. Actually, for any good algebra $A$, we may define $S'(x)$ as the set of isomorphism classes of thin indecomposable modules $M$ which are not isomorphic to $P(x)$, such that $M_x \neq 0$, and $(\tau M)_x = 0$. This set becomes a poset by defining $[M] \leq [M']$ provided there is a map $f : M \to M'$ such such that $f_x : M_x \to M'_x$ is non-zero. Dräxler shows that a good algebra is representation finite if and only if all the posets $S'(x)$ are representation finite. Let us stress that this is a very effective criterion: clearly, the thin modules always are easy to write down, thus the sets $S'(x)$ can be computed without difficulty, and then we can use the Klejner list.

**4.3.** We have considered above representation directed, or, more generally, good algebras. For an arbitrary representation finite algebra $A$, we can define hammocks $S(x)$ for every vertex $x$ of the quiver of $A$, by going to the universal covering $\tilde{A}$ of $A$, or also directly by considering the socalled radical layers in $A$-mod. If we work with $A$ itself, and not with $\tilde{A}$, and consider the full translation subquiver $\Gamma(x)$ of the Auslander–Reiten quiver $\Gamma(A)$, given by all indecomposable modules $M$ with $M_x \neq 0$, then this will be a hammock if and only if all indecomposable $A$-modules have $k$ as endomorphism ring **[D4]**.

There are other hammocks which appear rather naturally in representation theory, see Scheuer **[Sr1]**; also, we will see in section 6 that for quasi-hereditary algebras, we similarly obtain hammocks such that the hammock function counts the multiplicity of the standard module $\Delta(x)$ in a $\Delta$-good filtration.

For general algebras, one has to replace posets by vectorspace categories. A detailed study of the vectorspace categories $\mathcal{S}(x)$ attached to the vertices of appropriate algebras has been carried out by Xi. His general observations may be found in **[X1]**. For a tame concealed algebra **[X2]**, the vectorspace categories $\mathcal{S}(x)$ all contain precisely one critical subset, this subset is convex in $\mathcal{S}(x)$, and leads to a filtration of $\mathcal{S}(x)$ by ideals, which may be described in terms of the defect. Also we should note that one obtains in this way partial tilting modules. Xi also has studied the corresponding problem for tubular algebras **[X3]**, and obtains (what one would expect) convex subsets of $\mathcal{S}(x)$ which are tubular, and which may be embedded as convex subsets into the pattern exhibited in **[R1]**. However, what seems to be astonishing, is that these subsets may be rather large, they usually contain several critical subsets.

**4.4.** Let us add some remarks concerning the situation when the base field $k$ is not algebraically closed. In this case, we have to take field extensions into account; in particular, we cannot expect that we may deal with posets. A generalization of the notion of a hammock to this setting has not yet been considered, but will be needed. The calculations of Hall polynomials mentioned in section 7 rely on one-point extensions of representation directed algebras over finite fields, so a substitute of the Kleiner list is needed. Fortunately, Klemp and Simson **[KSi]** have exhibited the full list of all sincere directed vectorspace categories of finite type which may be used. For a general report on the interplay between the representation theory of algebras and of vectorspace categories, we refer to Simson **[Si]**.

## 5. Modules with Finite Projective Dimension

**5.1.** Let $A$ be a finite dimensional $k$–algebra over some field $k$. It is not known whether a simple $A$–module $E$ with finite projective dimension, has to satisfy $\operatorname{Ext}^1(E,E)=0$. However, if $k$ is algebraically closed and all simple $A$–modules have finite projective dimension (thus, if $A$ has finite global dimension), then no simple $A$–module has self–extensions. In terms of quivers, it may be reformulated as follows: the quiver of a finite dimensional algebra of finite global dimension has no loops. Essentially, this is an old result of Lenzing **[L1]**. A recent paper by Igusa **[Ig]** reproves it in the context of algebraic K-theory. Let us remark that Lenzing actually has shown the following stronger theorem: Assume that $\operatorname{Ext}^1(E,E)\neq 0$ for some simple $A$–module $E$. Then there exists a module $M$ with infinite projective dimension such that $M/\operatorname{rad}M = E$. The assertion stated in Lenzing's paper is the following: Assume that $a$ is a nilpotent element in some noetherian ring $R$, such that all left ideals $Ra^n$ have finite projective dimension. Then $a$ belongs to the commutator subgroup $[R,R]$. In order to apply this result, let us assumme that $A$ is a basic finite dimensional $k$–algebra over some algebraically closed field $k$. Consider an indecomposable length 2 module $X$ with top and socle isomorphic to $E = A/\operatorname{rad}Ae$, for some primitive idempotent $e$. Then $[A,A]$ annihilates $X$. (Note that $1-e$ annihilates $X$, since $A$ is assumed to be basic. Now, take elements $b,c\in A, x\in X$. Then $[b,c]x = [ebe,ece]x$, since $1-e$ annihilates $X$. But $[ebe,ece]\in[eAe,eAe]\subseteq \operatorname{rad}^2 eAe\subseteq \operatorname{rad}^2 A$, since for any local algebra $B$ over an algebraically closed field, we have $[B,B]\subseteq\operatorname{rad}^2 B$. Of course, $\operatorname{rad}^2 A$ annihilates $X$.) On the other hand, there exists an element $a = eae\in\operatorname{rad}A$ with $aX\neq 0$. Since we have seen that $a$ cannot belong to $[A,A]$, it follows that one of the modules $Aa^n$ has infinite projective dimension, therefore also $Ae/Aa^n$.

Schofield **[Sf1]** has shown that that there is a function $f$ such that the global dimension of the $k$–algebras of finite global dimension with $k$–dimension $d$ is bounded by $f(d)$. In general, nothing is known about this bound $f(d)$, however, there are several results dealing with special classes of algebras. There are some classes of algebras $A$ for which the global dimension may be bounded by the number $s(A)$ of simple modules. For example, the quasi–hereditary algebras considered in the next section all have global dimension at most $2s(A)-2$. On the other hand, there are examples of algebras with $s(A)=2$ and arbitrarily large global dimension. The first such examples have been exhibited by E. Green, they have been studied in detail by Happel [**H6**]. Kirkman and Kuzmanovich [**KK**] have shown that there are even algebras $A$ with $s(A) = 2$ and $\operatorname{rad}^4 A = 0$ of arbitrarily large finite global dimension. On the other hand, in case $\operatorname{rad}^3 A = 0$, the finitistic dimension is bounded by $s(A)^2$, see [**Zi1**].

**5.2.** Given a finite dimensional algebra $A$, we denote by $\operatorname{fd} A$ the *finitistic dimension* of $A$, by definition, this is the supremum of proj. dim. $M$, for all (finite dimensional) $A$–modules $M$ of finite projective dimension. It has been conjectured a long time ago that the finitistic dimension $\operatorname{fd} A$ is always finite; this conjecture is usually contributed to Bass, to M. Auslander, or to Rosenberg and it has attracted a lot of interest in the last years. The conjecture has been verified for special classes of algebras: by Green, Kirkman and Kuzmanovich [**GKK**] and by Igusa and Zacharia [**IZ**] for monomial algebras, by Green and Zimmermann–Huisgen [**GZ**] for algebras $A$ with $\operatorname{rad}^3 A = 0$.

Note that we also may consider the supremum of proj. dim. $M$, for arbitrarily, not necessarily finite dimensional $A$–modules $M$ of finite projective dimension, it is called the *big finististic dimension* and denoted by $\operatorname{Fd} A$. Of course, $\operatorname{fd} A \leq \operatorname{Fd} A$, and a second conjecture asserts that we should have equality. For monomial algebras, Zimmermann–Huisgen [**Zi2**] has shown that $\operatorname{fd} A$ and $\operatorname{Fd} A$ may differ by at most one. Her proof rests on a structure theorem for modules which occur as second syzygy of some module, thus for kernels of maps between projective modules: they are isomorphic to direct sums of left ideals of $A$, each of which being generated by a path (note that a monomial algebra $A$ is defined as the factor algebra of the path algebra of a quiver by an ideal $I$ generated by some paths, thus the paths which do not belong to $I$ yield a multiplicative basis of $A$).

**5.3.** Following Miyashita [**My**], an $A$–module $M$ will be called a tilting module provided proj. dim. $M$ is finite, $\operatorname{Ext}^i(M, M) = 0$ for all $i \geq 1$, and there is an exact sequence $0 \to {}_AA \to M_0 \to \cdots \to M_n \to 0$, where all modules $M_i$, with $1 \leq i \leq n$ belong to $\operatorname{add} M$. (Actually, Miyashita has considered arbitrary rings and modules, thus he had to add suitable finiteness conditions.) The case of tilting modules of projective dimension at most 1 had been discussed in detail before; these modules have turned out to be very useful.

We recall that Bongartz has shown that any module $M$ with projective dimension at most 1, and $\operatorname{Ext}^1(M, M) = 0$, can be written as a direct summand of a tilting module of projective dimension at most 1. The corresponding assertion for tilting modules of arbitrary projective dimension is no longer true: Rickard and Schofield [**RS**] have exhibited a module $M$ with finite projective dimension, and $\operatorname{Ext}^i(M, M) = 0$, for all $i \geq 1$, which cannot be written as direct summand of a tilting module: Let $K$ be the Kronecker algebra, let $X$ be an indecomposable $A$–module of length 2, form the one–point extension $C = K[X]$, let P be the indecomposable projective module corresponding to the extension vertex; now form the one–point coextension $B = [P]C$, and identify the sink and the source of $B$ in order to obtain an algebra $A$ with a node. The algebra $A$ together with the simple module $E$ corresponding to the node is the desired example: $E$ has projective dimension 2, and $\operatorname{Ext}^i(E, E) = 0$, for all $i \geq 1$, and if $\operatorname{Ext}^2(X, E) = 0 = \operatorname{Ext}^2(E, X)$ for some $A$–module $X$, then $X$ is in fact a regular $K$–module, and therefore $\operatorname{Ext}^1(X, X) \neq 0$ in case $X \neq 0$. This shows that $E$ cannot be a direct summand of a tilting module.

Given a module $M$, let $s(M)$ denote the number of isomorphism classes of indecomposable direct summands of $M$. There still is the problem whether a module

$M$ with finite projective dimension, with $\operatorname{Ext}^i(M,M) = 0$, for all $i \geq 1$, and with $s(M) = s(A)$ has to be a tilting module.

For example, assume that any indecomposable injective $A$-module has finite projective dimension, and consider the injective cogenerator $D({}_AA)$, where $D = \operatorname{Hom}_k(-,k)$ denotes the duality with respect to the base field $k$. Then $D({}_AA)$ has finite projective dimension, $\operatorname{Ext}^i(D({}_AA), D({}_AA)) = 0$, for all $i \geq 1$, and $s(D({}_AA)) = s(A)$. Note that in this case $D({}_AA)$ will be a tilting module if and only ${}_AA$ has also finite injective dimension, but already here, it is not known whether this always has to be so. Algebras $A$, for which $D({}_AA)$ has finite projective dimension, and ${}_AA$ has finite injective dimension, have been called *Gorenstein algebras*. They form a convenient generalization of algebras of finite global dimension, as well as of self–injective algebras. For an outline of their properties, we refer to the papers by Auslander–Reiten **[AR2]** and Happel **[H9]** in this volume.

Given an $A$–module $M$, we denote by $\operatorname{ann}(M)$ its annihilator in $A$. The modules $M$ which are tilting $A/\operatorname{ann}(M)$–modules, have been considered by D'Este and Happel **[DH]**. They show that these modules yield the representable equivalences which are represented by faithful modules in the sense of Menini and Orsatti **[MO]**.

Let $\mathcal{E}$ be the set of isomorphism classes of basic modules $M$ with projective dimension 1 and without self–extensions, thus the modules $M$ in $\mathcal{E}$ are just the direct summands of tilting modules with projective dimension at most 1. In case we deal with a wild finite quiver without oriented cycles, the elements $\mathcal{E}$ just correspond to the real Schur roots.

Note that $\mathcal{E}$ may be considered as a simplicial complex; here, $[M_1]$ will be a face of $[M_2]$ if and only if $M_1$ is a direct summand of $M_2$. Thus, a module $M$ in $\mathcal{E}$ will be a simplex of dimension $s(M) - 1$. We can rephrase the result of Bongartz by saying that all maximal simplices are of dimension $s - 1$, where $s = s(A)$, (this means that $\mathcal{E}$ is a pure simplicial complex).

Consider a module $M$ in $\mathcal{E}$ which is an $(s-2)$–simplex, thus $s(M) = s(A) - 1$. Then there are at most two isomorphism classes of indecomposable modules $M'$, such that $M \oplus M'$ is a tilting module, and there are two such classes if and only if $M$ is faithful, see **[RS]** and **[H7]**. Also, in case $M'$ and $M''$ are non–isomorphic indecomposable modules such that $M \oplus M'$ and $M \oplus M''$ are tilting modules, then we can assume that $\operatorname{Ext}^1(M'', M') \neq 0$, and then there exists a non–split exact sequence of the form $0 \to M' \to E \to M'' \to 0$, with $E \in \operatorname{add} M$. It follws that $\mathcal{E}$ is unramified, and its boundary is given by the non–faithful modules in $\mathcal{E}$. We see that $\mathcal{E}$ is nearly a pseudo–manifold with boundary. In case $\mathcal{E}$ is finite, Riedtmann and Schofield **[RS]** have shown that its geometric realization is just a ball.

In case $\mathcal{E}$ is infinite, Unger **[U6]** shows that $\mathcal{E}$ usually is not locally finite, it may be non–connected, and also inside a connected component, it may be impossible to connect two $(s-1)$–simplices by an alternating sequence of simplices of dimension $s-1$ and $s-2$, such that consecutive simplices are incident. She has considered in detail certain links of simplices. Recall that given a simplex $\sigma$ of some simplicial complex $\Sigma$, its link $\operatorname{lk}(\sigma)$ is the subcomplex of $\Sigma$ consisting of those simplices $\tau$ for which $\sigma \cup \tau$ is a simplex, and $\sigma \cap \tau$ is empty. Thus, given a module $M$ in $\mathcal{E}$, the link $\operatorname{lk}(M)$ is the simplicial complex of all isomorphism classes of modules $M'$ such that $M \oplus M'$ is a basic tilting module of projective dimension at most 1. Assume that $M$ is an $(s-3)$–simplex in $\mathcal{E}$, thus $\operatorname{lk}(M)$ is a graph and any vertex has at most two neighbors. If $M$ is faithful, then $\operatorname{lk}(M)$ is connected, thus it is of the form $A_\infty^\infty$ or

$\tilde{A}_r$. If $M$ is not faithful, then $\text{lk}(M)$ is either of the form $A_r$ or the disjoint union of two graphs $A_\infty$.

**5.4.** Let $\mathcal{X}$ be a subcategory of $A$–mod. Recall that $\mathcal{X}$ is said to be *resolving* provided it is closed under extensions, under kernels of surjective maps, and contains all projective modules. For example, the classes $\mathcal{P}^i$ of all modules $M$ with $\text{proj. dim.} M \leq i$, and the class $\mathcal{P}^\infty$ of all modules with finite projective dimension are resolving. Given an $A$–module $M$, a *right $\mathcal{X}$–approximation* of $M$ is a map $g : X \to M$ with $X \in \mathcal{X}$ such that for any map $h : X' \to M$ with $X' \in \mathcal{X}$, there is a map $h' : X' \to X$ such that $h = h'g$. In case every $A$–module has a right $\mathcal{X}$–approximation, $\mathcal{X}$ is said to be *contravariantly finite in* $A$–mod.

Subcategories of $A$–mod which are both resolving and contravariantly finite have been studied by Auslander and Reiten **[AR1]**, on the basis of previous investigations of Auslander and Buchweitz **[AB]**. On the one hand, there are many important classes of modules which have these properties, on the other hand, the results of Auslander and Reiten give a clear picture of the behaviour of such subcategories.

Assume that $\mathcal{X}$ is a resolving and contravariantly finite subcategory. Let $X_i \to E_i$ be right $\mathcal{X}$–approximations of the simple $A$–modules $E_1, \dots, E_n$. Then $\mathcal{X}$ can be recovered from knowing the modules $X_i$, namely $\mathcal{X} = \text{add}\,\mathcal{F}(X_1, \dots, X_n)$, where $\mathcal{F}(X_1, \dots, X_n)$ denotes the class of modules which have a filtration with factors of the form $X_j$. This result has the following consequence: Assume $\mathcal{P}^\infty$ is contravariantly finite, for some algebra $A$, and let $X_i \to E_i$ be right $\mathcal{P}^\infty$–approximations. Then $\text{fd}\,A$ is just the maximum of $\text{proj. dim.} X_i$, in particular, $\text{fd}\,A$ is finite. We should remark that there do exist examples of algebras such that $\mathcal{P}^\infty$ is not contravariantly finite **[IST]**: Take a one–point coextension $B = [R]K$, where $K$ is the Kronecker algebra, and [R] is indecomposable of length 2, and identify the sink and the source in order to obtain an algebra $A$ with a node. Then $\mathcal{P}^\infty = \mathcal{P}^1$, and these modules are obtained from preprojective and regular $B$–modules, therefore the simple $A$–module corresponding to the node cannot have a right $\mathcal{P}^\infty$–approximation.

If $\mathcal{Z}$ is a subcategory of $A$–mod, let ${}^\circ\mathcal{Z}$ be the full subcategory of all $X$ satisfying $\text{Ext}^i(X, \mathcal{Z}) = 0$, for all $i \geq 1$, and $\mathcal{Z}^\circ$ that of all $Y$ satisfying $\text{Ext}^i(\mathcal{Z}, Y) = 0$, for all $i \geq 1$. If $\mathcal{X}$ is a contravariantly finite and resolving subcategory of $A$–mod, then $\mathcal{X}^\circ$ has the dual properties: it is a covariantly finite and coresolving subcategory. In this way, we obtain a bijection between the contravariantly finite and resolving subcategories and the covariantly finite and coresolving subcategories.

Auslander and Reiten have related these concepts to tilting theory. If $T$ is a tilting module, then $T^\circ$ is always a covariantly finite and coresolving subcategory. For simplicity, let us now assume that $A$ is of finite global dimension. We obtain a bijection between the isomorphism classes of basic tilting modules, and the covariantly finite and coresolving subcategories, since we can recover the tilting module $T$ from $\mathcal{X} = T^\circ$, due to the fact that $\mathcal{X} \cap \mathcal{X}^\circ = \text{add}\,T$. Also, we obtain $\mathcal{Y} = \mathcal{X}^\circ$ from $T$ as ${}^\circ T$. Alternatively, we can describe the modules in $\mathcal{X}$ as those modules which have a finite $T$–coresolution, and the modules in $\mathcal{Y}$ as those ones which have a finite $T$–resolution. Duality shows that the given tilting module $T$ is also a cotilting module, thus for algebras of finite global dimension, the tilting modules and the cotilting modules coincide. An interesting example of the correspondence between tilting modules, contravariantly finite and resolving subcategories and the covariantly finite and coresolving subcategories will be exhibited when we deal with quasi–hereditary algebras.

**5.5.** Given a finite dimensional $k$–algebra $A$, we denote by $\widehat{A}$ the corresponding repetitive algebra, it is constructed as follows: Take copies $A_i$ of $A$, indexed by the integers, and consider $Q_i = \mathrm{Hom}_k(A, k)$ as an $A_{i-1}$–$A_i$–bimodule. Then $\widehat{A}$ is the trivial extension of the ring $\bigoplus_{i\in\mathbf{Z}} A_i$ (with componentwise multiplication) by the bimodule $\bigoplus_{i\in\mathbf{Z}} Q_i$. Note that $\widehat{A}$ is an infinite dimensional algebra (without unit!), the indecomposable projective $\widehat{A}$–modules are just the indecomposable injective $\widehat{A}$–modules, and our main interest lies in the stable category $\widehat{A}$–$\underline{\mathrm{mod}}$ (its objects are the $\widehat{A}$–modules, and the morphisms are the residue classes of module maps modulo maps which factor through projective modules). Note that $\widehat{A}$–$\underline{\mathrm{mod}}$ is a triangulated category. This is a result which essentially is due to Heller, see **[H1]**, **[H4]**.

**Theorem (Happel)** *For any finite dimensional algebra $A$, there is an exact embedding (of triangulated categories)*

$$D^b(A\text{-}mod) \longrightarrow \widehat{A}\text{-}\underline{\mathrm{mod}}.$$

*The categories $D^b(A\text{-}mod)$ and $\widehat{A}$-$\underline{\mathrm{mod}}$ are equivalent if and only if $A$ has finite global dimension.*

The main difficulty had been the definition of an appropriate embedding functor. Happel's original proof uses induction on the width of a complex. Keller and Vossieck **[KV]**, **[Kl]** have proposed a different construction: they define such an embedding as the composition of four rather natural functors. This may be used for a better understanding, but it is still quite complicated. A very natural and straight forward definition has recently been given by Happel **[H8]**: The canonical embedding functor from $A$–mod to $\widehat{A}$–mod is exact, thus it extends to an exact functor from $D^b(A\text{–mod})$ to $D^b(\widehat{A}\text{–mod})$. Compose this functor with Rickard's functor considered below:

$$D^b(A\text{–mod}) \longrightarrow D^b(\widehat{A}\text{–mod}) \longrightarrow \widehat{A}\text{–}\underline{\mathrm{mod}},$$

in order to obtain the desired embedding. For $A$ of finite global dimension, this embedding functor actually is an equivalence of triangulated categories. In order to decide whether the triangulated categories $D^b(\widehat{A}\text{–mod})$ and $\widehat{A}$–$\underline{\mathrm{mod}}$ may be equivalent also for other algebras, Tachikawa and Wakamatsu **[TW]** have considered Grothendieck groups. Always, the Grothendieck groups of $A$–mod (modulo exact sequences) and the Grothendieck group of $D^b(\widehat{A}\text{–mod})$ (modulo triangles) are isomorphic, and they show that the Grothendieck groups of $A$–mod and $\widehat{A}$–$\underline{\mathrm{mod}}$ (modulo triangles) are isomorphic if and only if the determinant of the Cartan matrix of $A$ is equal to $\pm 1$. Since there do exist algebras with infinite global dimension such that the determinant of $C_A$ is $\pm 1$, one needs other methods to decide the question: Happel **[H8]** has shown that for $A$ of infinite global dimension, not all indecomposable objects in the category $D^b(A\text{–mod})$ do have sink or source maps, whereas in $\widehat{A}$–$\underline{\mathrm{mod}}$, all have, thus the categories $D^b(A\text{–mod})$ and $\widehat{A}$–$\underline{\mathrm{mod}}$ cannot be equivalent (even if we forget the triangular structure).

**5.6.** Algebras $A, B$ will be said to be *tilting equivalent*, provided there is a finite sequence of algebras $A = A_0, A_1, \ldots, A_n = B$, such that for $1 \le i \le n$, one of the algebras $A_{i-1}$ and $A_i$ is the endomorphism ring of a tilting module over the other ring. And $A$ and $B$ are said to be *derived equivalent*, provided $D^b(A\text{–Mod})$ and $D^b(B\text{–Mod})$ are equivalent as triangulated categories. One knows that a tilting module ${}_AT$ with endomorphism ring $B$ yields an equivalence of the derived categories $D^b(A\text{–mod})$ and $D^b(B\text{–mod})$, and also between $D^b(A\text{–Mod})$ and $D^b(B\text{–Mod})$, thus tilting equivalent algebras are derived equivalent. The converse is not true: let $A, B$ be tilting equivalent, such that the trivial extension algebras $T(A), T(B)$ are not Morita equivalent, for example, take the two algebras given as follows:

$$\circ \leftarrow \circ \leftarrow \circ \qquad \text{and} \qquad \circ \leftarrow \circ \leftarrow \circ,$$

it is easy to see that the second algebra $B$ is the endomorphism ring of a tilting $A$–module, and $\dim_k T(A) = 12$, $\dim_k T(B) = 10$, thus $T(A), T(B)$ cannot be isomorphic; they are basic, so they are not Morita equivalent. The only tilting modules for selfinjective algebras are the progenerators, thus they cannot be tilting equivalent. But Rickart [**Rc2**] has shown that with $A, B$ also $T(A), T(B)$ are derived equivalent.

An algebra has been called *piecewise hereditary* and *piecewise tubular* provided it is derived equivalent to a hereditary of a tubular algebra, respectively. These algebras have the advantage that one has full control over the corresponding derived categories. Clearly, if $A$ is hereditary, then the derived category $D^b(A\text{–mod})$ is obtained by taking the disjoint union of countable many copies of $A$–mod, and adding maps in one direction (using $\mathrm{Ext}^1$), the derived category of any canonical algebra can be calculated by using Happel's theorem, see [**HR**], [**10**], or else using the Geigle–Lenzing equivalence $D^b(C(p)\text{–mod}) \cong D^b(\mathrm{coh}\,\mathbb{X}(p))$, and the fact that $\mathrm{coh}\,\mathbb{X}(p)$ has global dimension 1, so that again we just take countably many copies of $\mathrm{coh}\,\mathbb{X}(p)$ and add maps in one direction. Piecewise hereditary and piecewise tubular algebras are derived equivalent if and and if they are tilting equivalent, see [**HRS**], and [**AS3**]. The algebras which are derived equivalent to a tame hereditary or a tame canonical algebra can be characterized as follows [**AS3**]:

**Theorem (Assem–Skowroński)** *A finite dimensional algebra $A$ is derived equivalent to a tame hereditary or a tubular algebra if and only if $\widehat{A}$ is locally support finite and cycle finite.*

Here, an infinite dimensional algebra $B$ whose indecomposable projective modules are finite dimensional (like $\widehat{A}$) is said to be *locally support finite* provided for every indecomposable projective $B$–module $P$, there are only finitely many simple $B$–modules which can occur as composition factors of indecomposable $B$–modules $M$ with $\mathrm{Hom}(P, M) \neq 0$. One may conjecture that $\widehat{A}$ can only be cycle finite, if it is also locally support finite. Assem and Skowroński have shown that in case $\widehat{A}$ is cycle finite, then $A$ itself is either 'simply connected' or else tilting equivalent to a hereditary algebra of type $\tilde{A}_n$.

**5.7.** Rickard [**Rc1**] has determined precise conditions for derived categories to be equivalent: Let $R$ be an arbitrary rings. Let $R$-proj be the category of finitely generated projective $R$–modules. If $\mathcal{R}$ is a full subcategory of $R$–Mod, let $K^b(\mathcal{R})$ denote the category of bounded complexes over $\mathcal{R}$ modulo homotopy.

**Theorem (Rickard).** *The rings $R, R'$ are derived equivalent if and only of there is a bounded complex $T^\bullet$ of finitely generated projective left $R$–modules such that*

*(i) $R'$ is the endomorphism ring of $T^\bullet$ in $K^b(R\text{-proj})$,*
*(ii)* $\mathrm{Hom}_{K^b(R\text{-proj})}(T, T[i]) = 0$ *for $i \neq 0$,*
*(iii)* $\mathrm{add}(T^\bullet)$ *generates $K^b(R\text{-proj})$ as a triangulated category.*

A bounded complex $T^\bullet$ with the properties listed in the Theorem may be called a *tilting complex*. König [**Kö**] has given a corresponding characterization for recollements of derived categories of module categories in terms of the existence of suitable partial tilting complexes.

**5.8.** Rickard [**Rc2**] has shown that two selfinjective algebras $A, B$ which are derived equivalent, also are stable equivalent. Namely, we can obtain the stable category $A\text{-}\underline{\text{mod}}$ as a quotient category of the derived category $D^b(A\text{-mod})$ as follows: An object of $D^b(A\text{-mod})$ can be represented by a complex $P^\bullet = (P^i, \delta^i)$ of finitely generated projective modules bounded to the right, with bounded cohomology. Assume that $H^i(P^\bullet) = 0$ for all $i \leq n$, where $n \leq 0$. Let

$$\pi(P^\bullet) = \Sigma^n(\mathrm{Cok}(P^{n-1} \to P^n)),$$

then this defines a functor $\pi : D^b(A\text{-mod}) \to A\text{-}\underline{\text{mod}}$, and this is the quotient functor for the embedding of $K^b(A\text{-proj})$ into $D^b(A\text{-mod})$.

Broué [**Br**] has exhibited examples of selfinjective algebras which are stable equivalent, but not derived equivalent. Alperin and Auslander–Reiten have conjectured that algebras which are stably equivalent, should have the same number of non-projective simple modules. This conjecture has been settled for representation finite algebras by Martinez [**Ma1**], but is open otherwise, even in the case where one of the algebras is local. Martinez [**Ma2**] has shown that it is sufficient to verify the conjecture for selfinjective algebras.

Let us add that algebras which are derived equivalent, always have the same Hochschild cohomology ring [**Ha5**], [**Rc3**].

**5.9.** In 1979, papers by Bernstein–Gelfand–Gelfand and Beilinson have exhibited algebraic descriptions of the derived category $\mathrm{coh}\, D^b(\mathbf{P}_n)$ of the category of coherent sheaves on the projective complex $n$-space $\mathbf{P}_n$, namely they construct a finite dimensional algebra $\Lambda_n$ of finite global dimension and equivalences of categories

$$D^b(\mathrm{coh}\,\mathbf{P}_n) \cong D^b(\Lambda_n\text{-mod}) \quad \text{and} \quad D^b(\mathrm{coh}\,\mathbf{P}_n) \cong \widehat{\Lambda}_n\text{-}\underline{\text{mod}}.$$

The precise relationship between the these equivalences and the one given by Happel has been determined by Dowbor and Meltzer [**DM**]. As an application, one may pass rather freely between the categories $D^b(\mathrm{coh}\,\mathbf{P}_n)$, $D^b(\Lambda_n\text{-mod})$ and $\widehat{\Lambda}_n\text{-}\underline{\text{mod}}$. This may be of interest for explicit calculations of vector bundles over $\mathbf{P}_n$.

Similar descriptions of the categories of coherent sheaves over other algebraic varieties are by now available.

Let $X$ be a nonsingular projective variety. Baer [**B3**] has introduced the concept of a *tilting sheaf* on $X$, this is a coherent sheaf without selfextensions that generates $C^b(\mathrm{coh}\,X)$ as a triangulated category, and such that its endomorphism ring has finite global dimension. (She even has considered nonsingular 'weighted' projective varieties, in order to cover also the weighted projective lines $\mathbf{X}(p)$, as considered by Geigle and Lenzing.) A tilting sheaf $\mathcal{T}$ induces an equivalence $D^b(\mathrm{coh}\,X) \to D^b(B\text{-mod})$, where $B$ is the endomorphism ring of $\mathcal{T}$. Of course, $B$ is a finite dimensional algebra, thus we see that the existence of a tilting sheaf means that questions concerning coherent sheaves over $X$ may be reduced to questions dealing with modules over some finite dimensional algebra. Meltzer [**Me**] has exhibited tilting sheaves for $X = \mathbf{P}_1 \times \mathbf{P}_1$, and for $X$ the flag variety $\mathsf{F}(1,2)$, and Kapranov [**Ka**] has considered the case of an arbitrary flag variety. Tilting sheaves and related objects have been studied extensively by Russian mathematicians, we refer to a collection of papers and surveys presented in the Seminaire Rudakov **[Ru]**. Let us note that in all known cases, the quiver $Q(B)$ (and even $Q(\widehat{B})$) may be endowed with an integral value function $s$ such that arrows $x \to y$ only exist in case $s(x)+1 = s(y)$ (in particular, $Q(B)$ is directed), often these algebras $B$ are 'quadratic', so that we can form $B^!$ and $B$ and $B^!$ tend to be Ext–dual to each other.

## 6. Quasi–hereditary algebras

Quasi–hereditary algebras have been defined by Cline, Parshall, and Scott (**[S]**, **[CPS2]**, **[PS]**) in order to deal with highest weight categories as they arise in the representation theory of semisimple complex Lie–algebras and algebraic groups. Many algebras which arise rather natural have been shown to be quasi–hereditary: the Schur algebras, the Auslander algebras, and it seems surprising that this class of algebras (which is defined purely in ring theoretical terms) has not been studied before by mathematicians devoted to ring theory. Even when Scott started to propagate quasi–hereditary algebras, it took him some while to find some ring theory resonance.

**6.1.** The definition of a quasi–hereditary algebra which we will give, follows a suggestion of Soergel **[Soe3]**. We have to start with an algebra $A$ and an ordering of the simple $A$–modules, thus let $E(i)$, with $i \in \Lambda$, be the set of simple $A$–modules, where $\Lambda$ is a (totally) ordered set. For $i \in \Lambda$, let $P(i)$ be the projective cover, and $Q(i)$ the injective envelope of $E(i)$. We denote by $\Delta(i)$ the maximal quotient of $P(i)$ with composition factors of the form $E(j)$, where $j \leq i$, and similarly, let $\nabla(i)$ be the maximal submodule of $Q(i)$ with composition factors of the form $E(j)$, where $j \leq i$. Our notation $\Delta(i), \nabla(i)$ should indicate the shape of the modules: by definition, $\Delta(i)$ has simple top, and $\nabla(i)$ has simple socle.

The algebra $A$ is said to be *quasi–hereditary* with respect to the ordering $\Lambda$ provided: (a) $\mathrm{End}(\Delta(i))$ is a division ring, for all $i \in \Lambda$, and (b) $\mathrm{Ext}^2(\Delta(i), \nabla(j)) = 0$ for all $i, j$.

For a quasi–hereditary algebra $A$, the modules $\Delta(i)$ are called the *standard* modules (in some special cases, they are also called Verma modules, or Weyl modules,) the modules $\nabla(i)$ the *costandard* modules. It turns out that for a quasi–hereditary algebra $A$, also the dual conditions for the costandard modules are satisfied, since clearly $\mathrm{End}(\Delta(i)) \cong \mathrm{End}(\nabla(i))$. In particular, we see that the opposite of a quasi–hereditary algebra is quasi–hereditary, again.

If $\mathcal{X}$ is a set of modules, we denote by $\mathcal{F}(\mathcal{X})$ the set of modules which have a filtration with factors in $\mathcal{X}$, these modules may be said to be $\mathcal{X}$*-good*. For any $M \in \mathcal{F}(\mathcal{X})$ and $X \in \mathcal{X}$, let $[M : X]$ denote the number of factors isomorphic to $X$ in some $\mathcal{X}$–filtration of $M$, provided this is well–defined. We will be interested in the $\Delta$–good and the $\nabla$–good modules. Note that in case the condition (a) is satisfied, condition (b) is equivalent to the requirement that ${}_AA$ belongs to $\mathcal{F}(\Delta)$. In this way, we see that for a quasi–hereditary algebra $A$, the category $A$–mod together with the set $\Delta$ of standard modules becomes a 'highest weight category', with weight set $\Lambda$, as defined by Cline, Parshall, Scott, and conversely, any highest weight category with finite weight set is the module category of a quasi–hereditary algebra.

**6.2.** The usual (and equivalent) definition of quasi–hereditary algebras uses heredity ideals. A ***heredity ideal*** of an algebra $A$ is an idempotent ideal $I$, with $I(\operatorname{rad} A)I = 0$, and such that ${}_AI$ is a projective left module (or, equivalently, $I_A$ is a projective right module). A chain of ideals

$$A = I_0 \supseteq I_1 \supseteq \cdots \supseteq I_n = 0$$

is called a ***heredity chain*** provided $I_{t-1}/I_t$ is a heredity ideal of $R/I_t$, for $1 \leq t \leq n$. A finite dimensional algebra is quasi–hereditary if and only if it has a heredity chain. (Note that a heredity chain may always be refined so that for every $t$, the indecomposable summands of a fixed module ${}_A(I_{t-1}/I_t)$ are all isomorphic, say isomorphic to some $\Delta(t)$, and then these modules are just the standard modules. Conversely, given an algebra $A$ which satisfies the conditions (a) and (b), then for every $t$ there exists a maximal left ideal $I_t$ of $A$ which belongs to $\mathcal{F}(\{\Delta(t+1), \ldots, \Delta(s(A))\}$, it clearly will be a twosided ideal, and, in this way, we obtain a heredity chain $(I_t)$ for $A$. Note that the $A/I_t$–modules are just those $A$–modules which only have composition factors of the form $E(i)$, with $i \leq t$.

Given a heredity chain $(I_t)_t$ for $A$, the factor algebras $A/I_t$ again will be quasi–hereditary. An algebra $A$ such that all factor algebras are quasi–hereditary, is necessarily hereditary **[DR1]**. A quasi–hereditary algebra usually will have many ideals, and even heredity ideals, such that the corresponding factor algebras are not quasi–hereditary. Examples of the latter have been given by Agoston **[Ag]** and Wiedemann **[Wi]**. On the other hand, Xi **[X6]** has shown that an algebra $A$ with $A/\operatorname{rad}^n A$ quasi–hereditary for some $n \geq 2$, is quasi–hereditary itself.

An ideal $I_t$ which belongs to a heredity chain is always idempotent, thus we may choose an idempotent $e$ which generates $I$ as an ideal. Then the subalgebra $eAe$ is quasi–hereditary **[DR1]**. On the other hand, for an arbitrary idempotent $f$ in a quasi–hereditary algebra $A$, the subalgebra $fAf$ may be far away from being quasi–hereditary. In fact, using a construction due to Auslander, a joint paper with Dlab **[DR3]** shows that given an arbitrary finite dimensional algebra $B$, there exists an idempotent $f$ in some quasi–hereditary algebra $A$ with $B = fAf$. We will see below that there are important classes of quasi–hereditary algebras which are endomorphism rings of faithful modules over selfinjective algebras.

Any ideal $I$ which belongs to a heredity chain of an algebra $A$ gives rise to a recollement: choose an idempotent $e$ which generates $I$ as a twosided ideal:

$$D^b(A/I\text{–mod}) \overset{\longleftarrow}{\underset{\longleftarrow}{\longrightarrow}} D^b(A\text{–mod}) \overset{\longleftarrow}{\underset{\longleftarrow}{\longrightarrow}} D^b(eAe\text{–mod}),$$

thus $D^b(A\text{–mod})$ is built up (via recollments) from the rather trivial derived categories $D^b(D_i\text{–mod})$, where $D_i$ are division rings. This was one of the reasons for Cline, Parshall, Scott to introduce heredity chains, see **[PS]**, **[CPS1]**.

**6.3.** Any algebra with directed quiver $Q(A)$ is quasi–hereditary in two ways which are essentially different (provided $A$ is not semisimple): we can take all simple modules as standard modules, then $\mathcal{F}(\Delta) = A$–mod, or else, we may take the indecomposable projective modules as standard modules, then $\mathcal{F}(\Delta)$ contains only the projective modules.

All algebras $A$ of global dimension at most 2 are quasi–hereditary **[DR1]**: any idempotent ideal $I$ of $A$ with minimal possible Loewy length is a heredity ideal, and the global dimension of $A/I$ again is at most 2. Recall that any representation finite algebra $B$ gives rise to an algebra of global dimension at most 2, take the endomorphism ring of an additive generator of $B$–mod. For these Auslander–algebras, there are several choices of standard modules which reflect properties of the $B$–modules, and are related to the preprojective or preinjective partitions introduced by Auslander and Smalø, or to the Rojter measure on $B$–mod, see **[DR2]**. On the other hand, Uematsu and Yamagata **[UY]** have exhibited examples of algebras of global dimension 3, which are not quasi–hereditary. It is easy to see that a serial algebra $A$ is quasi–hereditary provided there exists just one heredity ideal, and this happens if and only if there is a simple module with projective dimension 0 or 2 **[UY]**.

There are some classes of quasi–hereditary algebras which may be considered more natural, and which Cline, Parshall, Scott had in mind when they introduced this concept: see the famous 'moose'–notes by Parshall and Scott **[PS]**. Let us direct the attention at least to some of these algebras.

We start with a semisimple finite dimensional complex Lie algebra $\mathfrak{g}$, with a Cartan subalgebra $\mathfrak{h}$ and a Borel subalgebra $\mathfrak{b} \supset \mathfrak{h}$, and consider the corresponding category $\mathcal{O}$ as defined by Bernstein, Gelfand, Gelfand, it is the category of all finitely generated $\mathfrak{g}$–modules, which are locally $\mathfrak{b}$–finite, and semisimple as $\mathfrak{h}$–modules. The simple objects in $\mathcal{O}$ are indexed by the dual space $\mathfrak{h}^*$, and $\mathcal{O}$ is a highest weight categories with standard modules the Verma modules. Also note that $\mathcal{O}$ is the categorical sum of blocks which are abelian length categories with only finitely many simple modules, and all of them are equivalent to module categories for (finite dimensional) quasi–hereditary algebras. The most interesting case is the principal block $\mathcal{O}_0$ containing the trivial representation of $\mathfrak{g}$. The block $\mathcal{O}_0$ has a unique indecomposable module $P$ which is projective and injective. Its endomorphism ring has been calculated by Soergel **[Soe1]**: Consider the Weyl group action on $\mathfrak{h}^*$, let $I$ be the ideal in the ring $R$ of regular functions on $\mathfrak{h}^*$ generated by the homogeneous invariants of degree at least one; it is known that the factor algebra $R/I$ is a finite dimensional, local, selfinjective $\mathbf{C}$–algebra. Soergel shows that the center of the universal enveloping algebra $U(\mathfrak{g})$ maps surjectively on $\operatorname{End}(P)$, and in this way, there is a canonical identification of $R/I$ with $\operatorname{End}(P)$. Also, he constructs certain $R/I$–modules $M(w)$, indexed by the elements $w$ of the Weyl group, forms $A = \operatorname{End}(\bigoplus_w M(w))$, and obtains a categorical equivalence between the category $\mathcal{O}_0$ and $A$–mod. Since $A$ is the endomorphism ring of a faithful module over a selfinjective algebra, the dominant dimension of $A$ is at least 2. Soergel shows that $A$ is 'quadratic', so that we can form $A^!$, and $A$ and $A^!$ are Ext–dual to each other (this was conjectured before by Beilinson and Ginsburg **[BG]**). These considerations are extended in **[Soe2]** to other blocks.

Another important class are the Schur algebras. Let us consider the classical case of the Schur algebras of $G = \mathbf{gl}_n(k)$, they are defined as follows: let $V$ be the canonical $n$–dimensional $\mathbf{gl}_n$–module, and form the $r$–fold tensor power $V^{\otimes r}$. There is a diagonal action of $G$ on $V^{\otimes r}$, thus an algebra homomorphism

$$T_r : kG \to \operatorname{End}(V^{\otimes r}),$$

where $kG$ is the group algebra of $G$, and the Schur algebra $S(n,r)$ is just the image of this map. In **[Pa]**, Parshall has shown that the Schur algebras are quasi–hereditary, with standard modules the Weyl modules. Note that for any subgroup $H$ of $G$, the image of $kH$ under $T_r$ will be a subalgebra of $S(n,r)$, denoted by $S(H)$. Let $B^-$ and $B^+$ be the Borel subgroups of $G$ of all lower, or upper triangular matrices, respectively. Then $S(B^-)S(B^+) = S(n,r)$, the algebras $S(B^-)$ and $(B^+)$ are quasi-hereditary and isomorphic (since $B^-$ and $B^+$ are conjugate), see J.A.Green **[Gr]**.

Note that there is the obvious permutation action of the symmetric group $\Sigma_r$ on $V^{\otimes r}$ on the right; in this way, $V^{\otimes r}$ becomes an $S(n,r)$–$\Sigma_r$–bimodule, and it is known that $S(n,r)$ is the full endomorphism ring of $V^{\otimes r}$, considered as a right $\Sigma_r$–module. For $n \geq r$, the right $\Sigma_r$–module $V^{\otimes r}$ clearly is faithful, and since the group algebra $k\Sigma_r$ is selfinjective, we see that the dominant dimension of the Schur algebra $S(n,r)$ is at least 2.

We remark that Xi **[X6]** has exhibited the structure of the Schur algebras $S(p,p)$, for $p$ a prime, using only the fact that $s(S(p,p)) = s(\Sigma_p) + 1$. In fact, he shows that there are only few quasi–hereditary algebras $A$ with an idempotent $e$ such that $eAe$ is selfinjective and $s(A) = s(eAe) + 1$. Of course, in this way, he also gets a new proof for the structure of $k\Sigma_p$.

**6.4.** The quasi–hereditary algebras satisfy some rather restrictive conditions. Let $A$ be quasi–hereditary.

First of all, as we have noted above, the global dimension of $A$ is always finite **[PS]**, in fact it is bounded by $2s(A) - 2$, see **[DR1]**. For certain examples, like the Schur algebras, it had been considered a mystery that they are of finite global dimension (for Schur algebras, this had been established by Akin–Buchsbaum and Donkin), before it was realized that this is really a trivial consequence of being quasi–hereditary.

Second, the Loewy length of $A$ is bounded by $2^{s(A)} - 1$. This is easy to see: Let $I$ be a heredity ideal, then $\operatorname{rad} I$ is an $A/I$–module, thus the Loewy length $LL(A)$ of $A$ is bounded by $1 + 2LL(A/I)$.

Also, we note that given a quasi–hereditary algebra $A$, we can bound its dimension if we know $A/\operatorname{rad}^2 A$, and the ordering $\Lambda$. A quasi–hereditary algebra $A$ is said to be *shallow,* provided for any standard module $\Delta(i)$, its radical is semisimple. And, $A$ is called *deep* provided $\operatorname{rad}\Delta(i)$ is projective when considered as an $A/J_{i-1}$–module. For any algebra $B$ with $\operatorname{rad}^2 B = 0$, and without loops (this means, $\operatorname{Ext}^1(E,E) = 0$, for any simple module $E$), and any total ordering $\Lambda$ of the simple modules, there exists a corresponding shallow algebra $S(B,\Lambda)$ and a deep algebra $D(B,\Lambda)$, such that for any quasi–hereditary algebra $A$ with $B = A/\operatorname{rad}^2 A$, and this ordering $\Lambda$, we have **[DR5]**

$$\dim_k S(B,\Lambda) \leq \dim_k A \leq \dim_k D(B,\Lambda).$$

**6.5.** Important problems of representation theory center around the Jordan–Hölder multiplicities of the standard modules (for example, the Kazhdan–Lusztig conjectures are questions of this kind). For simplicity, let us assume that the base field is algebraically closed. There is the following reciprocity formula

$$[P(j):\Delta(i)] = [\nabla(i):E(j)]$$

(it is usually referred to as the Bernstein–Gelfand–Gelfand–reciprocity law), see **[CPS2]**. Recall that the Cartan matrix $C$ of $A$ is by definition the transpose of the matrix $\mathbf{dim}\, P$, thus we can reformulate the reciprocity formula as follows:

$$C^t = (\mathbf{dim}\,\Delta)^t \cdot (\mathbf{dim}\,\nabla).$$

In particular, we see that the determinant of the Cartan matrix is equal to 1. Of particular interest will be the case when there exists a duality * on $A$–mod with

$E(i)^* = E(i)$, for all $i$. In this case, we have $\Delta^* \cong \nabla$, thus the reciprocity formula becoms $[P(j) : \Delta(i)] = [\Delta(i) : E(j)]$.

**6.6.** There is an inductive procedure for obtaining all quasi–hereditary algebras, the 'not–so–trivial–extension' method by Parshall and Scott **[PS]**. Let $B$ be a ring, $D$ a division ring. Recall that the one–point extension of $B$ by a bimodule ${}_BM_D$ is the algebra $\begin{bmatrix} B & M \\ 0 & D \end{bmatrix}$. Similarly, the one–point coextension of $B$ by the bimodule ${}_DN_B$ is the algebra $\begin{bmatrix} B & 0 \\ N & D \end{bmatrix}$. Now assume both bimodules ${}_BM_D, {}_DN_B$ are given, thus we can consider the tensor product ${}_BM \otimes_D N_B$. Let $\tilde{B}$ be a Hochschild extension of $B$ by the bimodule ${}_BM \otimes_D N_B$, thus $\tilde{B}$ is a ring with an ideal $J$ with $J^2 = 0$, and $\tilde{B}/J = B$, so that we may consider $J$ as a $B$–$B$–bimodule, and we require in addition that this $J$ is isomorphic to ${}_BM \otimes_D N_B$. The algebra we are looking for is

$$A = \begin{bmatrix} \tilde{B} & M \\ N & D \end{bmatrix}.$$

There is the ideal $I = \begin{bmatrix} J & M \\ N & D \end{bmatrix} \cong \begin{bmatrix} M \otimes N & M \\ N & D \end{bmatrix} \cong \begin{bmatrix} M \\ D \end{bmatrix} \otimes_D [N \;\; D]$, clearly, this is a heredity ideal of $A$, and $A/I = B$. On the other hand, given a minimal non–zero heredity ideal of a basic finite dimensional algebra $A$, with a heredity ideal $I$, choose an idempotent $e$ which generates $I$ as a twosided ideal; note that $D = eAe$ is a division ring, and let $B = A/I, M = \operatorname{rad} Ae, N = \operatorname{rad}(eA)_A$, then clearly we are in the situation depicted above.

Another inductive procedure has been introduced by Mirollo and Vilonen **[MV]** in the context of categories of perverse sheaves; it avoids the use of Hochschild extensions, and deals instead with tensor products and bimodule maps. Whereas the not–so–trivial–extension method relates $A$ and $A/I$, where $I$ is a heredity ideal, here we deal with an idempotent $e$ such that the multiplication map $Ae \otimes_{eAe} eA \to AeA$ is bijective, and such that there exists a division subring $D$ which complements $(1-e)AeA(1-e)$ in $(1-e)A(1-e)$. We should remark that for any ideal $AeA$ in a heredity chain, the corresponding multiplication map is bijective, and the second condition will be satisfied in case $A/AeA$ is a division ring, and the base field is perfect. Thus, for algebras over a perfect base field, we obtain in this way an alternative way for constructing all quasi–hereditary algebras, see **[DR4]**. For some time, we had thought that tensor products and bimodule maps are easier to handle than Hochschild extensions, however actual calculations show that sometimes this is not the case. The Hochschild cohomology group $H^2(B, M \otimes N)$ which plays the decisive role in the not–so–trivial–extension approach is just $\operatorname{Ext}^2_B(\operatorname{Hom}_k(N,k), M)$, and its elements are often very easy to handle, see **[DR7]**.

**6.7.** Let us consider the module category of a quasi–hereditary algebra in more detail, following **[R4]** and **[R6]**. The category $\mathcal{F}(\Delta)$ of $\Delta$–good modules has (relative) almost split sequences, thus we can deal with the corresponding (relative) Auslander–Reiten quiver $\Gamma_{\mathcal{F}(\Delta)}$. The relative projective modules in $\mathcal{F}(\Delta)$ are just the projective modules, and for any index $i \in \Lambda$, there is precisely one indecomposable, relative injective module $T(i)$, with $\Delta(i)$ embedded into $T(i)$, and such that $T(i)/\Delta(i)$ has only composition factors of the form $E(j)$, with $j < i$.

There are no loops or sectional cycles in $\Gamma_{\mathcal{F}(\Delta)}$ (however, in contrast to a full module category, the composition of maps along a sectional path may be zero). It follows that in case the indecomposable $\Delta$–good modules are of bounded length, then there are only finitely many isomorphism classes of indecomposable $\Delta$–good modules: so the analogue of the first Brauer–Thrall conjecture holds. Also, for the stable components for $\Gamma_{\mathcal{F}(\Delta)}$, there are the same restrictions as in the case of a full module category: periodic components will be of the form $\mathbb{Z}\Delta/G$, where $\Delta$ is either a Dynkin diagram or else of the form $A_\infty$, and $G$ is a non–trivial group of automorphisms of $\mathbb{Z}\Delta$, and non–periodic components are of the form $\mathbb{Z}\Delta$, where $\Delta$ is a connected valued quiver without cyclic paths.

Let us assume that $\mathcal{F}(\Delta)$ is finite, and that the base field is algebraically closed. In this case, we can define for any $i \in \Lambda$ a corresponding hammock (inside the universal covering of $\Gamma_{\mathcal{F}(\Delta)}$), such that the hammock function counts the multiplicities $[M : \Delta(i)]$, for $\Delta$–good modules $M$. Note that in $\Gamma_{\mathcal{F}(\Delta)}$, we have to take into account multiple arrows: they actually may occur even under our assumption that $\mathcal{F}(\Delta)$ is finite, but fortunately only from an injective vertex to a projective vertex, thus never inside a mesh.

There are corresponding assertions for the category $\mathcal{F}(\nabla)$ of $\nabla$–good modules. Note that we have

$$\mathcal{F}(\Delta)^\circ = \mathcal{F}(\nabla) \quad \text{and} \quad {}^\circ\mathcal{F}(\nabla) = \mathcal{F}(\Delta),$$

and $\mathcal{F}(\Delta)$ is a contravariantly finite and resolving subcategory, whereas $\mathcal{F}(\nabla)$ is a covariantly finite and coresolving subcategory, so we are in the situation studied by Auslander and Reiten, see section 5. The intersection $\mathcal{F}(\Delta) \cap \mathcal{F}(\nabla)$ is just $\operatorname{add} T$, where $T = \bigoplus T(i)$, thus this is a tilting module, and we can recover from $T$ the structure of $A$ as a quasi–hereditary algebra. This $A$–module $T$, and its direct summands seem to be of special interest. In the case of the category $\mathcal{O}$, these modules have been described by Collingwood and Irving **[CI]**: clearly, the $T(i)$ are just the indecomposable self–dual modules which have a Verma–filtration.

The endomorphism ring $A'$ of $T$ is in a canonical way again a quasi–hereditary algebra, the $A'$–modules $\Delta'(i) = \operatorname{Hom}(T, \nabla(i))$ being the standard modules, but we have to reverse the ordering of $\Lambda$. Then the category of $\nabla$–good $A$–modules is equivalent (under the functor $\operatorname{Hom}(T, -)$) to the category of $\Delta'$–good $A'$–modules. Thus, we see that there is no difference between categories which arise as categories of $\Delta$–good or $\nabla$–good modules. Also, the category of $\Delta$–good $A$–modules is equivalent to the category of $\nabla'$–good $A'$–modules, thus we see that the two subcategories are really exchanged in $A$–mod and in $A'$–mod. If we repeat these considerations, we will return to $A$ in the next step: the endomorphism ring of an additive generator of $\mathcal{F}(\Delta') \cap \mathcal{F}(\nabla')$ is always Morita equivalent to $A$.

## 7. Combinatorial Methods IV:

## Hall Algebras

The free abelian group with basis indexed by the isomorphism classes of finite $p$–groups may be endowed with a product by counting filtrations of finite $p$–groups: we obtain what is called the Hall algebra $\mathcal{H}(\mathbb{Z}_p)$ of the ring $\mathbb{Z}_p$ of $p$–adic integers. It is a commutative and associative ring with identity element and plays an important role in algebra and combinatorics. It first was considered by Steinitz in 1900, and later, in 1959, Ph. Hall started a detailled investigation. The basic concept may be generalized to fairly arbitrary rings. Under a mild finiteness condition on the ring $R$, one may define a similar product on the free abelian group with basis indexed by the set $\mathcal{B}$ of isomorphism classes of finite $R$–modules (where finite means to have finitely many elements, not just finite length), and one obtains an associative ring $\mathcal{H}(R)$ with identity element, the integral Hall algebra of $R$. In contrast to the case $R = \mathbb{Z}_p$, the Hall algebras in general do not need to be commutative, in fact the main concern are the corresponding Lie–algebras.

**7.1.** Let $R$ be any ring, and $N_1, \cdots, N_t$, and $M$ finite $R$–modules. Let $F^M_{N_1 \cdots N_t}$ be the number of filtrations

$$M = U_0 \supseteq U_1 \supseteq \cdots \supseteq U_t = 0$$

of $M$ such that $U_{i-1}/U_i \cong N_i$, for $1 \leq i \leq t$. (Note that in case $N_1, \ldots, N_t$ are in addition simple, we just count the number of composition series with prescribed composition factors).

We call a ring $R$ *finitary,* provided for all finite $R$–modules $M, M'$ also the extension group $\mathrm{Ext}^1(M, M')$ is finite. Note that all noetherian rings as well as all rings which are finitely generated (over $\mathbb{Z}$) are finitary. Assume that $R$ is a finitary ring. Let $\mathcal{H}(R)$ be the free abelian group with basis $(u_{[M]})_{[M]}$, indexed by the set of isomorphism classes of finite $R$–modules. We define on $\mathcal{H}(R)$ a multiplication by the following rule

$$u_{[N_1]} u_{[N_2]} = \sum_{[M]} F^M_{N_1 N_2} u_{[M]};$$

note that on the right, we deal with a finite sum, since $R$ is assumed to be finitary. Clearly, $\mathcal{H}(R)$ is an associative ring with 1, the identity element is $u_{[0]}$, and the associativity of this multiplication follows from the fact that the coefficient of $u_{[M]}$ in either $u_{[N_1]}(u_{[N_2]} u_{[N_3]})$ or $(u_{[N_1]} u_{[N_2]}) u_{[N_3]}$ is just $F^M_{N_1 N_2 N_3}$. We call $\mathcal{H}(R)$ the *integral Hall algebra* of $R$. The special case of $R = \mathbb{Z}_p$ is the one considered by Steinitz and Ph. Hall. In contrast to $R = \mathbb{Z}_p$, the Hall algebras in general are not commutative. For example, let $R = \begin{bmatrix} k & k \\ 0 & k \end{bmatrix}$, the ring of upper triangular $2 \times 2$ matrices over the finite field $k$. Then there are two non–isomorphic simple $R$–modules $E_1, E_2$, and a non–split exact sequence $0 \to E_2 \to P \to E_1 \to 0$, whereas $\mathrm{Ext}^1_R(E_2, E_1) = 0$. It follows that in $\mathcal{H}(R)$, we have

$$u_{[E_1]} u_{[E_2]} = u_{[E_1 \oplus E_2]} + u_{[P]},$$

but

$$u_{[E_2]} u_{[E_1]} = u_{[E_1 \oplus E_2]}.$$

In particular, we can write $u_{[P]}$ as a commutator:

$$u_{[P]} = [u_{[E_1]}, u_{[E_2]}].$$

**7.2.** Let us assume that $A$ is a representation directed algebra, thus $A$ is representation finite, and we may index the indecomposable $A$–modules in such a way that $\mathrm{Hom}(X_i, X_j) = 0$ for $i > j$. Also, we assume that $A$ is connected. Then Guo **[G2]** has shown that the center of $\mathcal{H}(A)$ is trivial (i.e. equal to $\mathbb{Z}$) except in case all indecomposable $A$-modules are of length at most 2, thus the center of $\mathcal{H}(A)$ is non-trivial only in case $A$ is serial and $\operatorname{rad}^2 A = 0$. Also, Guo **[G1]** has shown that the Hall algebra $\mathcal{H}(A)$ determines $A$ to a large extend. Namely, if also $A'$ is representation directed, then the rings $\mathcal{H}(A)$, and $\mathcal{H}(A')$ are isomorphic if and only if there is a bijection $E_i \mapsto E_i'$ between the simple $A$-modules $E_i$ and the simple $A'$-modules $E_i'$ such that $|\operatorname{Ext}^t_A(E_i, E_j)| = |Ext^t_{A'}(E_i', E_j')|$ for $t = 0, 1, 2$, and all $i, j$.

**7.3.** Recall that in the classical case $\mathcal{H}(\mathbb{Z}_p)$, the multiplication coefficients are evaluations of polynomials at $p$, and these Hall polynomials play a decisive role. The same is true for the Hall algebras $\mathcal{H}(A)$, where $A$ is a representation directed algebra. In the classical case, the Hall polynomials are indexed by a tripel of partitions, since the isomorphism classes of abelian $p$-groups correspond bijectivly to partitions. Consider now the case of an algebra $A$ with a connected Auslander–Reiten quiver $\Gamma(A)$ (we have in mind the finite dimensional representation finite algebras, and suitable localizations of the path algebras of the cyclic quivers). Let $\mathcal{B}$ be the set of functions $b : \Gamma(A)_0 \to \mathbb{N}_0$ with finite support. The finite dimensional $A$-modules correspond bijectively to the elements of $\mathcal{B}$: for any vertex $i \in \Gamma(A)_0$, let $M(A, i)$ be a representative of the isomorphism class $i$, then $b \in \mathcal{B}$ will be attached to the isomorphism class of the module $M(A, b) = \bigoplus b(i)M(A, i)$. Note that the set $\mathcal{B}$ may have different interpretations for certain rings. For example, for $A$ representation directed, we may identify $\Gamma(A)_0$ with the set of positive roots for a corresponding quadratic form, whereas in case we deal with the cyclic quiver with $N$ vertices, we may identify $\mathcal{B}$ with the set of all $N$-tuples of partitions. In case $A$ is representation directed **[R7]** as well as in case we deal with a cyclic quiver **[R11]**, given $a, b, c \in \mathcal{B}$, there exists a monic polynomial $\phi^b_{ac} \in \mathbb{Z}[T]$ such that

$$F^{M(A,b)}_{M(A,a)M(A,c)} = \phi^b_{ac}(q_A)$$

for some fixed number $q_A$. These polynomials may be called *Hall polynomials,* they depend on the Auslander–Reiten quiver $\Gamma(A)$, but not on $A$ itself.

Some Hall polynomials have been calculated explicitly in **[R8]**. Note that in the classical case, for $M$ indecomposable, we always have $F^M_{N_1 N_2} = 0$ or 1. In the case of a representation directed algebra, the situation is much more complicated: even if we assume that all three modules $M(A, a), M(A, b), M(A, c)$ are indecomposable, the polynomial $\phi^b_{ac}$ may have degree 5. Here is the list of all polynomials $\phi^b_{ac}$ different from 0 and 1 which occur for a representation directed algebra $A$ which is given by a quiver with relations:

$$\begin{gathered} T-2, \\ (T-2)^2, \\ (T-2)^3, \\ T^3 - 5T^2 + 10T - 7, \\ (T-2)(T^3 - 4T^2 + 8T - 6), \\ T^5 - 6T^4 + 15T^3 - 23T^2 + 25T - 13. \end{gathered}$$

There are similar polynomials in case $A$ is given by a species with relations. The polynomials have been calculated using suitable one–point extension algebras, and counting rational points of certain algebraic varieties. It may be of interest to obtain an interpretation of the various coefficients. Let us remark that for the polynomials exhibited above we have $|\phi^b_{ac}(1)| = 1$, in the general case, we have $|\phi^b_{ac}(1)| \leq 3$.

**7.4.** In those cases where Hall polynomials do exist, we may introduce the *generic Hall algebra* $\mathcal{H}(A, \mathbb{Z}[T])$ and the *degenerate Hall algebra* $\mathcal{H}(A)_1$ as follows: In order to obtain $\mathcal{H}(A, \mathbb{Z}[T])$, take the free $\mathbb{Z}[T]$-module with basis $(u_b)_{b\in\mathcal{B}}$, and define the multiplication by

$$u_a u_c = \sum_{b\in\mathcal{B}} \phi^b_{ac} u_b.$$

In order to obtain $\mathcal{H}(A)_1$, take the free abelian group with basis $(u_b)_{b\in\mathcal{B}}$, and define the multiplication by

$$u_a u_c = \sum_{b\in\mathcal{B}} \phi^b_{ac}(1) u_b.$$

Of course, both are associative rings with 1.

There is the following interesting property of the numbers $F^M_{N_1N_2}$, for $A$–modules $N_1, N_2, M$, where $A$ is a $k$–algebra over a finite field $k$. If $N_1, N_2$ are indecomposable, and $M$ is decomposable, then $|k| - 1$ divides $F^M_{N_1N_2} - F^M_{N_2N_1}$. This has the following consequence: We may consider $\Gamma(A)_0$ as a subset of $\mathcal{B}$, by identifying an element of $\Gamma(A)_0$ with its characteristic function. Then $T - 1$ divides $\phi^b_{ac}(1) - \phi^b_{ca}(1)$ in case $a, c$ belong to $\Gamma(A)_0$, and $b$ not. As a consequence, the subgroup of $\mathcal{H}(A)_1$ with basis $(u_b)_{b\in\Gamma(A)_0}$ is a Lie subring of $\mathcal{H}(A)_1$. By definition, this subgroup is the free abelian group on the set of isomorphism classes of indecomposable $A$–modules, thus we may consider it as the Grothendieck group $K(A\text{–mod})$ of all finite $A$–modules modulo split exact sequences. In particular, we see that $K(A\text{–mod})$ is, in a natural way, a Lie ring.

**Theorem** *Let $A$ be a hereditary algebra of Dynkin type $\Delta$. Let $\mathfrak{g}$ be the semisimple complex Lie algebra of type $\Delta$, with triangular decomposition $\mathfrak{g} = \mathfrak{n}_- \oplus \mathfrak{h} \oplus \mathfrak{n}_+$. Then $K(A\text{–mod})$ is a Chevalley $\mathbb{Z}$–form of $\mathfrak{n}_+$, and $\mathcal{H}(A)_1$ may be identified with the corresponding Kostant $\mathbb{Z}$–form of the universal enveloping algebra $U(\mathfrak{n}_+)$.*

Indeed, the elements $u_b$ with $b \in \Gamma(A)_0$ themselves form a Chevalley $\mathbb{Z}$–basis of $\mathfrak{n}_+$. This is a consequence of the explicit determination of certain Hall polynomials. In particular, it is necessary to know the corresponding values $|\phi^b_{ac}(1)|$, as mentioned above.

Let us add that the rather technical form of the basis elements of the Kostant $\mathbb{Z}$–form of $U(\mathfrak{n}_+)$ gets a very natural interpretation in terms of the Hall algebra. Since $A$ is representation directed, we may assume that we have indexed the indecomposable $A$–modules $M(1), M(2), \ldots, M(m)$, such that $\mathrm{Hom}(M(i), M(j)) = 0$ for $i > j$. Let $b : \{1, 2, \ldots, m\} \to \mathbb{N}_0$ be an element of $\mathcal{B}$, and consider $u^{b(1)}_{M(1)} u^{b(2)}_{M(2)} \cdots u^{b(m)}_{M(m)}$ in $\mathcal{H}(A)$. We want to write it as a linear combination in our basis $(u_{[M]})_{[M]}$, and we see that the only non–zero coefficient can occur for $M = \bigoplus b(i)M(i) = M(b)$, since any filtration of the type considered has to split. Also, the only remaining coefficient can be calculated without difficulty; in $\mathcal{H}(A)_1$, it is the evaluation of a polynomial $\phi$, with $\phi(1) = \prod b(i)!$. It follows that

$$u_1^{b(1)} u_2^{b(2)} \cdots u_m^{b(m)} = \prod b(i)! \cdot u_b,$$

and therefore

$$\frac{u_1^{b(1)}}{b(1)!}\frac{u_2^{b(2)}}{b(2)!}\cdots\frac{u_m^{b(m)}}{b(m)!} = u_b.$$

The theorem brings to an end investigations started by Gabriel: In 1972, he showed that for a hereditary algebra $A$ of type $A_n, D_n, E_6, E_7, E_8$, the indecomposable modules correspond bijectively to the positive roots of the corresponding simple complex Lie algebra. This result was extended to all hereditary algebras of Dynkin type in joint work with Dlab. Thus, it was known for a long time that we may identify $K(A\text{-mod}) \otimes_{\mathbb{Z}} \mathbb{C}$ with $\mathfrak{n}_+$ as $\mathbb{C}$-spaces, and there was the problem whether it is possible to recover the Lie multiplication of $\mathfrak{n}_+$ in terms of the representation theory of $A$. We see that the Hall algebras provide a possibility to do so.

**7.5.** One may ask whether it is possible to use representations of algebras to recover the whole Lie algebra $\mathfrak{g}$ and not only $\mathfrak{n}_+$. It is easy to define an extended Hall algebra $\mathcal{H}'(A)$ in order to obtain $U(\mathfrak{b})$ as the corresponding degenerate extended Hall algebra: We form the twisted polynomial ring over $\mathcal{H}(A)$ by adjoining variables $X_1, \dots, X_{s(A)}$, such that $[X_i, u_{[M]}] = (\mathbf{dim}\ M)_i u_{[M]}$. Schofield **[Sc4]** has proposed another way by dealing with the varieties of all composition series of modules with fixed dimension, and he is able to construct in this way the complete Lie algebra $\mathfrak{g}$.

**7.6.** The Hall algebra approach presented above yields, in fact, a $q$-analogue of the enveloping algebra of $U(\mathfrak{n}_+)$, and it turns out that this is really the quantization of the universal enveloping algebra as defined by Jimbo and Drinfeld **[Dr]**. For details, we refer to our survey **[R10]**. One may consider in the same way any hereditary algebra $A$ say of type $\Delta$, and one obtains a canonical ring homomorphism from the quantization of the universal enveloping algebra of the Kac–Moody Lie algebra of type $\Delta$, or better from a $\mathbb{Z}$-form, into the Hall algebra of $A$. The image will be the subalgebra generated by the simple modules, or the semisimple modules, we call this the composition algebra, or the Loewy algebra of $A$. At least in the case of a Euclidean diagram, we can show that we obtain in this way a realization of the corresponding quantum groups. Let us add that Lusztig **[Lu]** has used this approach in order to define canonical bases for the universal enveloping algebras of all finite dimensional semisimple Lie algebras over $\mathbb{C}$.

If we want to handle the Euclidean quivers, the main difficulties arise already in the special case of an oriented cycle. Of course, the corresponding path algebra is not a finite dimensional algebra, but even if we only are interested in the remaining cases (which give finite dimensional algebras), we have to consider the modules which belong to non-homogeneous tubes, and this is just the case of dealing with the (locally nilpotent) representations of an oriented cycle. The corresponding composition algebra has been exhibited in detail in **[R11]**, and we think that the combinatorial methods introduced should be of interest elsewhere. Let us remark that Guo **[G3]** has considered the structure of the complete Hall algebra of an oriented cycle.

It seems surprising that the parameter $q$ used for the quantization of the universal enveloping algebra $U(\mathfrak{b})$ has an interpretation as a variable which stands for the cardinality of a finite field, but this is what we encounter when identifying the quantization with a Hall or Loewy algebra.

## REFERENCES


[Ag] Agoston, I.: Quotients of quasi-hereditary algebras. C. R. Math. Rep. Acad. Sci. Canada 11 (1989), 99-103.

[AS1] Assem, I.; Skowronski, A.: Iterated tilted algebras of type $\tilde{A}_n$, Math. Z. 195 (1987), 269-290

[AS2] Assem, I.; Skowronski, A.: On some classes of simply connected algebras. Proc. London Math. Soc. 56 (1988), 417-450.

[AS3] Assem, I.; Skowronski, A.: Algebras with cycle finite derived categories. Math. Ann. 280 (1988), 441-463.

[AS4] Assem, I.; Skowronski, A.: Algèbres pré-inclinées et catégories dérivées. In: Sém. d'Algèbre Dubreil–Malliavin. Springer LNM. 1404 (1989), 1-34.

[AS5] Assem, I.; Skowronski, A.: Minimal representation–infinite coil algebras. Manuscr. Math. 67 (1990), 305-331.

[AB] Auslander, M.; Buchweitz, R.–O.: The homology theory of maximal Cohen–Macaulay approximations. Mém. Soc. Math. France N.S. (1989).

[APT] Auslander, M.; Platzeck, M. I.; Todorov, G.: Homological theory of idempotent ideals. (To appear).

[AR1] Auslander, M.; Reiten, I.: Applications of contravariantly finite subcategories. Preprint 8/1989 Univ. Trondheim.

[AR2] Auslander, M.; Reiten, I.: Cohen–Macaulay and Gorenstein Artin algebras. This volume.

[BS] Bakke, Ø.; Smalø, S.O.: Modules with the same socles and tops as a directing module are isomorphic. Comm. Algebra 15 (1987), 1-10.

[Bau] Bautista, R.: On algebras of strongly unbounded representation type. Comment. Math. Helv. 60 (1985), 392-399.

[BK] Bautista, R.; Kleiner, M.: Almost split sequences for relatively projective modules. J.Algebra 135 (1990), 19-56

[B1] Baer, D.: Homological properties of wild hereditary Artin algebras. In: Representation Theory I. Springer LNM 1177 (1986), 1-12

[B2] Baer, D.: Wild hereditary Artin algebras and linear methods. Manuscr. Math. 55 (1986), 69-82.

[B3] Baer, D.: Tilting sheaves in representation theory of algebras. Manuscr. Math. 60 (1988), 323-347.

[B4] Baer, D.: A note on wild quiver algebras and tilting modules. Comm. Algebra 17 (1989), 751-757

[BGL] Baer, D.; Geigle, W.; Lenzing,H.: The preprojective algebra of a tame hereditary Artin algebra. Comm. Algebra 15 (1987), 425-457.

[BG] Beilinson, A.; Ginsburg, V.: Mixed categories, Ext-duality abd representations (Results and conjectures). (To appear).

[Be] Benson, D.: *Modular Representation Theory: New Trends and Methods.* Springer LNM 1081 (1984).

[Bs] Bessenrodt–Timmerscheidt, Ch.: The Auslander–Reiten quiver of a modular group algebra revisited. Preprint SM-DU-159 Duisburg (1989).

[Bo] Bongartz, K.: Indecomposables are standard. Comment. Math. Helv. 60 (1985), 400-410.

[BS] Bongartz, K.; Smalø, S.O.: Modules determined by their tops and socles. Proc. Amer. Math. Soc. 96 (1986), 34-38.

[Br] Brenner. Sh.: A combinatorial characterization of finite Auslander–Reiten quivers. In: Representation Theory I. Springer LNM 1177 (1986), 13-49.

[BT] Bretscher, O; Todorov, G.: On a theorem of Nazarova and Roiter. In: Representation Theory I. Springer LNM 1177 (1986), 50-54.

[Bé] Broué, M.: News about perfect isometries. (Unpublished).

[BB] Burt, W.L.; Butler, M.C.R.: Almost split sequences for bocses. (To appear).

[BR] Butler, M.C.R.; Ringel, C.M.: Auslander–Reiten sequences with few middle terms,with applications to string algebras. Comm.Alg.15 (1987),145-179.

[CPS1] Cline, E.; Parshall, B.; Scott, L.: Algebraic statification in representation categories. J. Algebra. 117 (1988), 504-521

[CPS2] Cline, E.; Parshall, B.; Scott, L.: Finite dimensional algebras and highest weight categories. J. reine angew. Math. 391 (1988), 85-99

[CPS3] Cline, E.; Parshall, B.; Scott, L.: Duality in highest weight categories. In: Classical Groups and Related Topics. Contemp. Math. 82 (1989), 7-22.

[CI] Collingwood, D.H.; Irving, R.S.: A decomposition theorem for certain self-dual modules in the categorie $\mathcal{O}$. Duke Math. J. 58 (1989), 89–102.

[C1] Crawley–Boevey, W.: On tame algebras and bocses. Proc. London Math. Soc.(3), 56 (1988),451-483.

[C2] Crawley–Boevey, W.: Functorial filtrations and the problem of an idempotent and a square–zero matrix. J. London Math. Soc. (2) 38 (1988), 385-402.

[C3] Crawley–Boevey, W.: Functorial filtrations II: Clans and the Gelfand problem. J. London Math. Soc. (2) 40 (1989), 9-30.

[C4] Crawley–Boevey, W.: Functorial filtrations III: Semi–dihedral algebras. J. London Math. Soc. (2) 40 (1989), 31-39.

[C5] Crawley–Boevey, W.: Maps between representations of zero–relation algebras. J. Algebra 126 (1989), 259-263

[C6] Crawley–Boevey, W.: Tame hereditary algebras, hereditary orders and their curves. (To appear).

[C7] Crawley–Boevey, W.: Matrix reduction for artinian rings, and an application to rings of finite representation type. Preprint 90-020 SFB 343 Bielefeld.

[C8] Crawley–Boevey, W.: Tame algebras and generic modules. Preprint 89-028 SFB 343 Bielefeld.

[C9] Crawley–Boevey, W.: Lectures at the Tsukuba Workshop 1990. (To appear).

[CR] Crawley–Boevey, W.; Ringel, C.M.: Algebras whose Auslander–Reiten components have large regular components. Preprint 90-052 SFB 343 Bielefeld.

[CU] Crawley–Boevey, W.; Unger, L.: Dimensions of Auslander–Reiten translates for representation–finite algebras. Comm. Algebra 17 (1989), 837-842.

[DH] D'Este, G.; Happel, D.: Representable equivalences are represented by tilting modules. Rend. Sem. Mat. Univ. Padova 82 (1990), 77-80.

[Di] Dieterich, E.: *Solution of a non–domestic tame classification problem from intergal representation theory of finite groups* ($\Lambda = RC_3, v(3) = 4.$) Memoirs Amer.Math.Soc. (To appear).

[DR1] Dlab, V.; Ringel, C.M.: Quasi–hereditary algebras. Illinois J. Math. 33 (1989), 280-291

[DR2] Dlab, V.; Ringel, C.M.: Auslander algebras as quasi-hereditary algebras. J. London Math. Soc. (2) 39 (1989), 457-466.

[DR3] Dlab, V.; Ringel, C.M.: Every semiprimary ring is the endomorphism ring of a projective module over a quasi-hereditary ring. Proc. Amer.Math.Soc. 107 (1989), 1-5.

[DR4] Dlab, V.; Ringel, C.M.: A construction for quasi-hereditary algebras. Compositio Math. 70 (1989), 155-175.

[DR5] Dlab, V.; Ringel, C.M.: The dimension of a quasi-hereditary algebra. In: Topics in Algebra. Banach Center Publ. 26. (To appear).

[DR6] Dlab, V.; Ringel, C.M.Filtrations of right ideals related to projectivity of left ideals. In: Sém. d'Algèbre Dubreil-Malliavin. Springer LNM. 1404 (1989), 95–107.

[DR7] Dlab, V.; Ringel, C.M.: The Hochschild cocycle corresponding to a long exact sequence. Tsukuba J. Math (To appear).

[DR8] Dlab, V.; Ringel, C.M.: Towers of semi-simple algebras. J. Funct. Anal. (To appear)

[DGL] Dowbor, P.; Geigle, W.; Lenzing, H.: *Graded Sheaf Theory and Group Quotients, with Applications to Representations of Finite Dimensional Algebras.* (To appear).

[DM] Dowbor, P.; Meltzer, H.: On equivalences of Bernstein–Gelfand–Gelfand, Beilinson and Happel. Preprint 257 Humboldt–Universität Berlin (1990).

[DS] Dowbor, P.; Skowroński, A.: Galois coverings of representation–infinite algebras. Comment. Math. Helv. 62 (1987), 311-337.

[D1] Dräxler, P.: U-Fasersummen in darstellungsendlichen Algebren. J. Algebra 113 (1988), 430-437.

[D2] Dräxler, P.: *Aufrichtige gerichtete Ausnahmealgebren.* Bayreuther Math. Schriften 29 (1989).

[D3] Dräxler, P.: Fasersummen über dünnen s–Startmoduln. Arch.Math. 54 (1990), 252-257.

[D4] Dräxler, P.: On indecomposable modules over directed algebras. (To appear).

[Df] Drinfeld, V.G.: Quantum groups. In: Proc. Intern. Congr. Math. 1986. Amer. Math. Soc. (1987), Vol.1, 798–820.

[E1] Erdmann, K.: *Tame Blocks and Related Algebras.* Springer LNM 1428 (1990).

[E2] Erdmann, K.: On Auslander–Reiten components for wild blocks. This volume.

[ES] Erdmann, K.; Skowroński, A.: On Auslander–Reiten components of blocks and self–injective biserial algebras. (To appear).

[Fi] Fischbacher, U.: Une nouvelle preuve d'un théorème de Nazarova et Roiter. C.R.Acad.Sc. Paris I 300 (1985), 259-262.

[FP] Fischbacher, U.; de la Peña, J.: Algorithms in representation theory of algebras. In: Representation Theory I. Springer LNM 1177 (1986), 94-114.

[GP] Gabriel, P.; de la Peña, J.: Quotients of representation–finite algebras. Comm. Algebra 15 (1987), 279-308.

[GL1] Geigle, W.; Lenzing, H.: A class of weighted projective curves arising in the representation theory of finite dimensional algebras. In: Singularities, Representations of Algebras, and Vector Bundles. Springer LNM 1273 (1987), 265-297.

[GL2] Geigle, W.; Lenzing, H.: Perpendicular categories with applications to representations and sheaves. J. Algebra (To appear).

[Gr] Green, J.A.: On certain subalgebras of the Schur algebra. (To appear).

[GKK] Green, E.L.; Kirkman, E.; Kuzmanovich, J.: Finitistic dimension of finite dimensional monomial algebras. J. Algebra 136 (1991), 37-50.

[GZ] Green, E.L.; Zimmermann–Huisgen, B.: Finitistic dimension of artinian rings with vanishing radical cube zero. Math. Z. (To appear).

[GHJ] Goodman, F.M., de la Harpe, P.; Jones, V.F.R.: *Coxeter Graphs of Algebras.* Math. Sci. Res. Inst. Publ. 14 (1989).

[G1] Guo, J.Y.: The isomorphism of Hall algebras. Preprint 90-060 SFB 343 Bielefeld.

[G2] Guo, J.Y.: The center of a Hall algebra. Preprint 90-061 SFB 343 Bielefeld.

[G3] Guo, J.Y.: The Hall algebra of a cyclic serial algebra. (In preparation)

[H1] Happel, D.: On the derived category of a finite–dimensional algebra. Comment. Math. Helv. 62 (1987), 339-389.

[H2] Happel, D.: Iterated tilted algebras of affine type. Comm. Alg. 15 (1987), 29-46

[H3] Happel, D.: Repetitive categories. In: Singularities, Representations of Algebras, and Vector Bundles. Springer LNM 1273 (1987), 298-317

[H4] Happel, D.: *Triangulated Categories in the Representation Theory of Finite-dimensional Algebras.* London Math. Soc. LNS 119 (1988).

[H5] Happel, D.: Hochschild cohomology of finite–dimensional algebras. In: Sém. d'Algb́re. Dubreil–Malliavin. Springer LNM 1404 (1989), 108-126.

[H6] Happel, D.: A family of algebras with two simple modules and Fibonacci numbers. Archiv Math. (To appear)

[H7] Happel, D.: Partial tilting modules and recollement. Preprint 89-016 SFB 343 Bielefeld.

[H8] Happel, D.: Auslander–Reiten triangles in derived categories of finite–dimensional algebras. Proc. Amer.Math.Soc. (To appear).

[H9] Happel, D.: On Gorenstein algebras. This volume.

[H10] Happel, D.: Hochschild cohomology of Auslander algebras. In: Topics in Algebra. Banach Center Publ. 26. (To appear).

[HRS] Happel, D.; Rickard, J.; Schofield, A.: Piecewise hereditary algebras. Bull. London Math.Soc. 20 (1988), 23-28.

[HR] Happel, D.; Ringel, C.M.: The derived category of a tubular algebra. In: Representation Theory I. Springer LNM 1177 (1986). 156-180.

[HU1] Happel, D.; Unger, L.: Factors of concealed algebras. Math. Z. 201 (1989), 477-483.

[HU2] Happel, D.; Unger, L.: A family of infinite–dimensional non–selfextending bricks for wild hereditary algebras. Proc. Tsukuba Conf. 1990 (To appear).

[Hi] Hille, L.: Assoziative gestufte Algebren und Kippfolgen mit dim($X$)+1 Stufen auf projektiven glatten algebraischen Mannigfaltigkeiten. Diplomarbeit Humboldt Univ. Berlin (1990).

[HM] Hoshino, M.; Miyachi, J.: Tame triangular matrix algebras over self–injective algebras. Tsukuba J. Math. 11 (1987), 383-391.

[Hö] v. Höhne; H. J.: On weakly positive unit forms. Comm. Math. Helv. 63 (1988), 312-336.

[Hü] Hübner, T.: (In preparation).

[Ig] Igusa, K.: Notes on the loop conjecture. J. Pure and Appl. Algebra. (To appear).

[IST] Igusa, K.; Smalø, S.; Todorov, G.: Finite projectivity and contravariant finiteness. (To appear).

[IT] Igusa, K.; Todorov, G.: A numerical characterization of finite Auslander–Reiten quivers. In: Representation Theory I. Springer LNM 1177 (1986), 181-198.

[IZ] Igusa, K.; Zacharia, D.: Syzygy pairs in a monomial algebra. Proc. Amer. Math. Soc. 108 (1990) 601-604.

[Ir] Irving, R.: BGG algebras and the BGG reciprocity principle. J. Algebra 135 (1990), 363-380.

[Kc] Kac, V. G.: Root systems, representations of quivers and invariant theory. In: Invariant Theory. Springer LNM 996 (1983), 74-108.

[Kp] Kapranov, M.M.: On the derived categories of coherent sheaves on some homogeneous spaces. Inv. Math. 92 (1988), 479-508.

[Kl] Keller, B.: Chain complexes and stable categories. Manuscr. Math. 67 (1990), 379-417.

[KV] Keller, B.; Vossieck, D: Sous les catégories dérivées. C.R.Acad. Sci. Paris I 305 (1987) 225-228.

[K1] Kerner, O.: Preprojective components of wild hereditary algebras. Manuscr. Math. 61 (1988), 429-445.

[K2] Kerner, O.: Tilting wild algebras. J.London Math.Soc.(2) 39 (1989), 29-47.

[K3] Kerner, O.: Universal exact sequences for torsion theories. In: Topics in Algebra. Banach Center Publ. 26. (To appear).

[K4] Kerner, O.: Stable components of wild tilted algebras. J.Algebra (To appear).

[KSk] Kerner, O.; Skowroński, A.: On module categories with nilpotent infinite radical. (To appear).

[KK] Kirkman, E.; Kuzmanovich, J.: Algebras with large homological dimension. (To appear)

[KSi] Klemp,B.; Simson, D.: Schurian sp–representation–finite right peak PI–rings and their indecomposable socle projective modules. J. Algebra 134 (1990), 390-468.

[Kö] König, S.: Tilting complexes, perpendicular categories, and recollement of derived module categories of rings. (To appear)

[Kr1] Krause, H.: Maps between tree and band modules. J.Algebra (To appear).

[Kr2] Krause, H.: Endomorphismem von Worten in einem Köcher. Dissertation. Bielefeld 1991.

[Le] Le Bruyn, L.: Counterexamples to the Kac–conjecture on Schur roots. Bull. Sci. Math. 110 (1986), 437-448.

[L1] Lenzing, H.: Nilpotente Elemente in Ringen von endlicher globaler Dimension. Math. Z. 108 (1969), 313-324.

[L2] Lenzing, H.: Curve singularities arising from the representation theory of tame hereditary Artin algebras. In: Representation Theory I. Springer LNM 1177 (1986), 199-231.

[L3] Lenzing, H.: Canonical algebras and rings of automorphic forms. (In preparation).

[LP] Lenzing, H.; de la Peña, J. A.: The Auslander–Reiten components of a canonical algebra. (In preparation).

[Lu] Lusztig, G.: Canonical bases arising from quantized enveloping algebras. J. Amer. Math. Soc. 3 (1990), 447-498

[MP] Marmaridis, N.; de la Peña, J.A.: Quadratic forms and preinjective modules. J. Algebra 134 (1990), 326-343.

[MR] Marmolejo, E.; Ringel, C.M.: Modules of bounded length in Auslander–Reiten components. Arch. Math. 50 (1988), 128-133.

[Ma1] Martínez–Villa, R.: The stable equivalence for algebras of finite representation type. Comm. Algebra 13 (1985), 991-1018.

[Ma2] Martínez–Villa, R.: Some remarks on stably equivalent algebras. Comm. Algebra (To appear).

[Me] Meltzer, H.: Tilting bundles, repetitive algebras, and derived categories of coherent sheaves. Preprint 193 Humboldt–Universität Berlin (1988)

[MO] Menini, C.; Orsatti, A.: Representable equivalences between categories of modules and applications. Rend. Sem. Mat. Univ. Padova 82 (1989), 203-231.

[MV] Mirollo,R.; Vilonen,K.: Bernstein–Gelfand–Gelfand reciprocity on perverse sheaves. Ann. Sci. E.N.S. 20 (1987), 311-324.

[My] Miyashita, Y.: Tilting modules of finite projective dimension. Math. Z. 193 (1986), 113-146.

[NR1] Nazarova, L.A.; Rojter, A.V.: Representations of completed partially ordered sets. In: Proc. Fourth Intern. Conf. Representations of Algebras. Vol.1. Carleton–Ottawa LNS 1 (1985).

[NR2] Nazarova, L.A.; Rojter, A.V.: Representations of bipartite completed posets. Comment. Math. Helv. 63 (1988), 498-526.

[NS] Nehring, J.; Skowroński, A.: Polynomial growth trivial extensions of simply connected algebras. Fund. Math. 132 (1989), 117-134.

[Ok] Okuyama, T.: On the Auslander–Reiten quiver of a finite group. J. Algebra 110 (1987), 425-430.

[OP] Ostermann, A.; Pott, A.: Schwach positive ganze quadratische Formen, die eine aufrichtige, positive Wurzel mit einem Koeffizienten 6 besitzen. J. Algebra 126 (1989), 80-118.

[Pa] Parshall, B.: Finite dimensional algebras and algebraic groups. In: Classical Groups and Related Topics. Contemp. Math. 82 (1989).

[PS] Parshall, B; Scott, L.L.: *Derived Categories, Quasi-hereditary Algebras, and Algebraic Groups.* Proc. Ottawa–Moosonee Workshop Algebra. Carleton–Ottawa Math. LNS 3 (1988), 1-105

[Pt] Partharasarathy, R.: t-structures dans la catégorie dérivée associée aux représentations d'un carquois. C.R. Acad. Sci. Paris I 304 (1987), 355-357.

[P1] de la Peña, J.: On omnipresent modules in simply connected algebras. J. London Math. Soc. (2) 36 (1987), 385-392.

[P2] de la Peña, J.: *Quadratic forms and the representation type of an algebra.* Ergänzungsreihe 90-003 SFB 343 Bielefeld.

[P3] de la Peña, J.: Tame algebras with sincere directing modules. (To appear).

[PS] de la Peña, J., Simson, D.: Prinjective modules, reflection functors, quadratic forms and Auslander–Reiten sequences. (To appear).

[PTa] de la Peña, J.; Takane, M.: Spectral properties of Coxeter transformations and applications. Arch. Math. 55 (1990), 120-134.

[PTo] de la Peña, J.; Tomé, B.: Iterated tubular algebras. J. Pure Appl. Algbera (To appear).

[Rc1] Rickard, J.: Morita theory for derived categories. J. London Math.Soc. 39 (1989), 436-456

[Rc2] Rickard, J.: Derived categories and stable equivalence. J. Pure Appl. Algebra 61 (1989), 436-456.

[Rc3] Rickard, J.: Derived equivalences as derived functors. (To appear).

[RS] Rickard, J.; Schofield, A.: Cocovers and tilting modules. Proc. Cambridge Phil.Soc. 106 (1989) 1-5.

[RS] Riedtmann, Ch.; Schofield, A.: On a simplicial complex associated with tilting modules. Prépubl. l'Inst. Fourier. 137 Grenoble (1989)

[R1] Ringel, C.M.: *Tame Algebras and Integral Quadratic Forms.* Springer LNM 1099 (1984).

[R2] Ringel, C.M.: The regular components of the Auslander–Reiten quiver of a tilted algebra. Chinese Ann. Math. B 9 (1988), 1-18.

[R3] Ringel, C.M.: The canonical algebras. (With an appendix by W. Crawley-Boevey). In: Topics in Algebra. Banach Center Publ. 26. (To appear).

[R4] Ringel, C.M.: The category of modules with good filtrations over a quasi-hereditary algebra has almost split sequences. Math.Z. (To appear).

[R5] Ringel, C.M.: On contravariantly finite subcategories. In: Proc. Tsukuba Conf. 1990 (To appear)

[R6] Ringel, C.M.: The category of good modules over a quasi-hereditary algebra. In: Proc. Tsukuba Conf. 1990 (To appear)

[R7] Ringel, C.M.: Hall algebras. In: Topics in Algebra. Banach Centre Publ. 26. Warszawa (To appear).

[R8] Ringel, C.M.: Hall polynomials for the representation–finite hereditary algebras. Adv. Math. 84 (1990), 137-178.

[R9] Ringel, C.M.: From representations of quivers via Hall and Loewy algebras to quantum groups. In: Proc. Novosibirsk Conf. Algebra 1989 (To appear)

[R10] Ringel, C.M.: Hall algebras and quantum groups. Invent. Math. 101 (1990), 583-592

[R11] Ringel, C.M.: The composition algebra of a cyclic quiver. (To appear)

[RV] Ringel, C.M.; Vossieck, D.: Hammocks. Proc. London Math. Soc. (3) 54 (1987), 216-246.

[Ru] Rudakov, A. N. et. al.: *Helices and Vector Bundles. Seminaire Rudakov.* London Math. Soc. LNS 148 (1990).

[Sr1] Scheuer, T.: More hammocks. In: Topics in Algebra. Banach Centre Publ. 26. Warszawa (To appear).

[Sr2] Scheuer, T.: The canonical decomposition of the poset of a hammock. Proc. London Math. Soc. (To appear).

[Sw1] Schewe, W.: The set of $\mathbb{Z}$–coverings for a finite translation quiver. Dissertation Bielefeld 1989.

[Sw2] Schewe, W.: Fundamental domains for representation–finite algebras. Rapport 74 Univ. Sherbrooke (1990).

[Sf1] Schofield, A.: Bounding the global dimension in terms of the dimension. Bull. London Math. Soc. 17 (1985), 393-394.

[Sf2] Schofield, A.: The field of definition of a real representation of a quiver Q. (To appear)

[Sf3] Schofield, A.: The internal structure of real Schur representations. (To appear)

[Sf4] Schofield, A.: (To appear).

[Sc] Scott, L.L.: Simulating algebraic geometry with algebra I: The algebraic theory of derived categories. Proc. Symp. Pure Math. 47 (1987), 271-281.

[Si] Simson, D.: Module categories and adjusted modules over traced rings. Diss. Math. 269 (1990).

[Sk1] Skowroński, A.: Group algebras of polynomial growth. Manuscr. Math. 59 (1987), 499-516.

[Sk2] Skowroński, A.: Selfinjective algebras of polynomial growth. Math. Ann. 285 (1989), 177-199.

[Sk3] Skowroński, A.: Algebras of polynomial growth. Topics in Algebra. Banach centre publ. 26. (To appear).

[SS] Skowroński, A.; Smalø, S.O.: Directing modules. Preprint 8/1990 Univ. Trondheim.

[Sm] Smalø, S.O.: Functorial finite subcategories over triangular matrix rings. Preprint 6/1989 Trondheim.

[Soe1] Soergel, W.: Kategorie $\mathcal{O}$, perverse Garben und Moduln über den Koinvarianten zur Weylgruppe. J. Amer. Math. Soc. 3 (1990), 421-445.

[Soe2] Soergel, W.: Parabolisch–singuläre Dualität für Kategorie $\mathcal{O}$. Preprint MPI Bonn 89-68.

[Soe3] Soergel, W.: Construction of projectives and reciprocity in an abstract setting. (To appear).

[St] Strauß, H.: Tilting modules over wild hereditary algebras. Thesis. Carleton University 1986. Abstract: C.R.Math., Acad.Sci.Canada 9 (1987), 161-166.

[TW] Tachikawa, H., Wakamatsu, T.: Cartan matrices and Grothendieck groups of stable categories. (To appear).

[UY] Uematsu, M.; Yamagata, K.: On serial quasi–hereditary rings. Hokkaido Math. J. 19 (1990), 165-174.

[U1] Unger, L.: Lower bounds for indecomposable, faithful, preinjective modules. Manuscr. Math. 57 (1986), 1-31.

[U2] Unger, L.: The concealed algebras of the minimal wild hereditary algebras. Bayreuther Math. Schriften 31 (1990), 145-154.

[U3] Unger, L.: On wild tilted algebras which are squids. Archiv Mathematik (To appear).

[U4] Unger, L.: Schur modules over wild, finite dimensional path algebras with three non isomorphic simple modules. J. Pure Appl. Algebra. (To appear).

[U5] Unger, L.: On the number of maximal sincere modules over sincere directed algebras. J.Algebra 133 (1990), 211-231.

[U6] Unger, L.: One–dimensional links of the simplicial complex of partial tilting modules. (In preparation).

[W1] Wakamatsu, T.: On modules with trivial selfextensions. J.Algebra 114 (1988), 106-114.

[W2] Wakamatsu, T.: Stable equivalences for selfinjective algebras and a generalization of tilting modules. J. Algebra 134 (1990), 298-325.

[Wi] Wiedemann, A.: Quotients of quasi–hereditary algebras. (To appear).

[X1] Xi, Ch.: Die Vektorraumkategorie zu einem unzerlegbaren projektiven Modul einer Algebra. J. Algebra (To appear).

[X2] Xi, Ch.: Die Vektorraumkategorie zu einem Punkt einer zahmen verkleideten Algebra. J. Algebra (To appear).

[X3] Xi, Ch.: Die Vektorraumkategorie zu einem unzerlegbaren projektiven Modul einer tubularen Algebra. Manuscr. Math. 69 (1990), 223-235.

[X4] Xi, Ch.: On wild hereditary algebras with small growth numbers. Comm. Algebra 18 (1990), 3413-3422.

[X5] Xi, Ch.: Minimal elements of the poset of a hammock. Preprint 90-016 SFB 343 Bielefeld.

[X6] Xi, Ch.: Quasi-heredity of algebras and their factor algebras. Preprint 90-59 SFB 343 Bielefeld.

[X7] Xi, Ch.: Symmetric algebras as endomorphism rings of large projective modules over quasi-hereditary algebras. Preprint 90-59 SFB 343 Bielefeld.

[Za] Zavadskij, A. G.: Classification of representations of posets of finite growth. in: Proc. Fourth Intern. Conf. Representations of Algebras. Vol.2. Carleton-Ottawa LNS 2 (1984), 36.01-36.15

[Z1] Zhang, Y.: The modules in any component of the AR-quivaer of a wild hereditary artin algebra are uniquely determined by their composition factors. Archiv Math. 53 (1989), 250-251.

[Z2] Zhang, Y.: Eigenvalues of Coxeter transformations and the structure of the regular components of the Auslander-Reiten quiver. Comm. Algebra. 17 (1989), 2347-2362.

[Z3] Zhang, Y.: The structure of stable components. Can. J. Math. (To appear).

[Zi1] Zimmermann-Huisgen, B.: Bounds on finitistic and global dimension for artinian rings with vanishing radical cube. (To appear).

[Zi2] Zimmermann-Huisgen, B.: Predicting syzygies over monomial algebras. (To appear).

## Older surveys

[1] Auslander, M.: The what, where and why of almost split sequences. Proc. Intern. Congr. Math. 1986. Amer. Math. Soc.(1987), Vol.1, 338-345.

[2] Gabriel, P.: Darstellungen endlichdimensionaler Algebren Proc. Intern. Congr. Math. 1986. Amer. Math. Soc.(1987), Vol.1, 378-388.

[3] Reiten, I.: An introduction to the representation theory of Artin algebras. Bull. London Math. Soc. 17 (1985), 209-233.

[4] Reiten, I.: Finite dimensional algebras and singularities. In: Singularities, Representations of Algebras, and Vector Bundles. Springer LNM 1273 (1987), 35-57.

[5] Riedtmann, Ch.: Algèbre de type de représentation fini. Séminaire Bourbaki. 37e année 1984-1985, n. 650. Astérisque 133-134 (1986), 335-350.

[6] Ringel, C.M.: The representation type of local algebras. In: Representations of Algebras. Springer LNM 488 (1975), 282-305.

[7] Ringel, C.M.: Report on the Brauer-Thrall conjectures. In: Representation Theory I. Springer LNM 831 (1980), 104-136

[8] Ringel, C.M.: Tame algebras. In: Representation Theory I. Springer LNM 831 (1980), 137-287

[9] Ringel, C.M.: Indecomposable representations of finite dimensional algebras. Proceedings Intern. Conf. Math. Warszawa 1983, (1984), 425-436.

[10] Ringel, C.M.: Representation theory of finite dimensional algebras. Durham Lectures 1985. London Math. Soc. Lecture Note Series 116 (1986), 7-79.

Progress in Mathematics, Vol. 95, © 1991 Birkhäuser Verlag Basel

# The isomorphism problem for integral group rings of finite groups

K. W. ROGGENKAMP

## § 1 The History of the isomorphism problem

The modern theory of groups originated with the treatments of Galois (1811-1832), Cauchy (1789-1857) and Serret (1819-1885) on finite discontinuous substitution groups.

However, it was in the spring of 1870, when Sophus Lie and Felix Klein were studying C. Jordan's book on "*substitutions and algebraic equations*", in which Jordan was explaining the importance of the theory of groups for the solutions of equations by radicals, as it was developed by E. Galois.

Both Lie and Klein noted the importance of group theory not only for solving equations, but for the whole of mathematics, and in particular for the theory of invariants. They developed group theory further, and later on demonstrated its importance in continuous and discontinuous geometry.

At that time the theory of abstract groups was only little developed. The mathematicians at the end of last century were used to work with permutation groups and linear groups. So it was natural to study abstract groups by letting them act on sets or linearly on vectorspaces. Time has shown that studying linear actions of groups on vectorspaces is a much richer theory than letting groups act on sets, since one can invoke the arithmetic of the general linear groups.

In this spirit W. Burnside, F.G. Frobenius and J. Schur developed the ordinary (complex) *representation theory of finite groups* in the years 1896 – 1910, after Frobenius had defined ordinary characters of finite groups in 1896.

Representation theory of the finite group $G$ is the study of homomorphisms

$$\varphi : G \to Gl(n, R), \tag{1.1}$$

where $R$ is a commutative ring with identity; originally $R$ was the ring of complex numbers. The idea of Frobenius, Burnside and Schur was to study the matrix groups $Im(\varphi)$ for various homomorphisms $\varphi$ ($R = \mathbf{C}$), and use the informations on these matrix groups to obtain informations on the abstract finite group $G$.

An important demonstration of the power of complex representation theory, where one could use the arithmetical properties of $Im(\varphi) \subset Gl(n, \mathbf{C})$, was the proof of the following group theoretical statement

**Burnside's Theorem (1911)**: *Let $G$ be a finite group of order $|G| = p^a \cdot q^b$, where $p$ and $q$ are different prime numbers. Then the group $G$ is soluble, i.e. there is a chain of normal subgroups*

$$1 = N_0 \leq N_1 \leq N_2 \leq \ldots \leq N_k = G$$

*with* $N_i/N_{i-1}$, *an abelian group*, $1 \leq i \leq k$.

Since abelian groups were well understood at that time, solvable groups seemed to be "easy". We point out that a purely group theoretical proof was found only in 1967 by H. Bender [Be].

In the years 1902 – 1907 L.E. Dickson considered representations in fields; i.e. homomorphisms

$$\varphi : G \to Gl(n, K) ,$$

$K$ a field, and noted that the theory was similar to the complex representation theory, provided $|G|$ and Char($K$) were relatively prime. If, however Char($K$) was a divisor of $|G|$ – this is called *modular representation theory* –, then he showed, that the theory was quite different.

Modular representation theory lay dormant until R. Brauer in 1935 – at the suggestion of I. Schur – developed the theory further, using *ring theory*, which enters as follows:

Given a representation

$$\begin{aligned} \varphi : G &\to Gl(n, R) \subset Mat(n, R) , \\ g &\mapsto \varphi(g) . \end{aligned}$$

Since $Mat(n, R)$, the ring of $n \times n$ matrices over $R$, is an $R$-algebra, we may associate with $\varphi$ the $R$-algebra

$$\Lambda_\varphi = \left\{ \sum_{g \in G} r_g \cdot \varphi(g) \mid r_g \in R \right\} , \tag{1.2}$$

generated by $Im(\varphi)$ over $R$. Studying the representation $\varphi$ is tantamount to studying the $R$-algebra $\Lambda_\varphi$. There is a universal $R$-algebra, the *group ring*, where one takes *formal linear combinations*,

$$RG = \left\{ \sum_{g \in G} r_g \cdot g | r_g \in R \right\} ; \tag{1.3}$$

addition is componentwise, and the $R$-linear multiplication is induced from the multiplication in $G$. This group ring maps onto each of the various $\Lambda_\varphi$.

*The original question of Burnside, Frobenius and Schur can thus be rephrased in the following way:*

*Which properties of the finite group $G$ are reflected in $RG$?*

At the beginning of the century one knew, that $\mathbb{C}G$ does not reflect all of the properties of $G$, since *two finite abelian groups $G$ and $H$ have isomorphic complex group algebras if and only if* $|G| = |H|$, in fact by Maschke's theorem $\mathbb{C}G$ is semisimple.

However, if one considers the group rings $FG$ for all fields, then the various isomorphism types of abelian groups can be distinguished by their group algebras. Thus we come to the

**(1.4) Problem 1**: *Let $G$ and $H$ be finite groups such that $KG \simeq KH$ for all fields $K$ (i.e. ordinary and modular representation theory coincide for both groups).*

*Is then $G \simeq H$?*

Though this problem arose early, when modular representation theory was developed, it was solved only in 1971 by E. Dade: Recall that a finite group $G$ is called *metabelian*, if $G$ has an abelian normal subgroup $A$ such that $G/A$ is abelian – these are so to speak "almost" abelian groups.

**(1.5) Theorem (Dade):** *There are two non-isomorphic finite metabelian groups $G$ and $H$ of order $p^6 \cdot q^6$, $p$ and $q$ different rational prime numbers, such that $KG \simeq KH$ for every field $K$.*

The importance here is the isomorphism of the group rings for the fields $K$ with $\mathrm{Char}(K) = p, q$. The construction of the isomorphism of Dade is a different one in case of characteristic $p$ than in characteristic $q$.

Since the representation theory of the finite group $G$ over a field with characteristic $p$ is closely related to the structure of the Sylow $p$-subgroup, a *still open problem* is:

**(1.6) Problem 2:** *Let* $\mathrm{Char}(K) = p > 0$, *and let $G$ and $H$ be $p$-groups, such that $KG \simeq KH$.*

*Is then $G \simeq H$?*

Very little is known here, except, that by computer-analysis in Stuttgart it was checked, that for 2-group of order $\leq 2^6$, the problem has a positive answer.

Now the ring of integers $\mathbb{Z}$ is the universal commutative ring, and if for two finite groups the integral group rings $\mathbb{Z}G$ and $\mathbb{Z}H$ are isomorphic, then $RG \simeq RH$ for all commutative rings; i.e., the representation theory for all commutative rings cannot distinguish the two groups. Thus we come to

**(1.7) Problem 3 (Isomorphismproblem):** *Assume that for two finite groups, the integral group rings $\mathbb{Z}G$ and $\mathbb{Z}H$ are isomorphic.*

*Is then $G \simeq H$?*

A positive answer would imply that the integral representation theory of a finite group would determine the group up to isomorphism.

*Example*: Let $S_3$ be the symmetric group on 3 letters, then the rational group algebra $\mathbb{Q}S_3$ has the form

$$\begin{bmatrix} \mathbb{Q} & \mathbb{Q} \\ \mathbb{Q} & \mathbb{Q} \end{bmatrix} \quad \Pi\, \mathbb{Q}^+ \,\Pi\, \mathbb{Q}^-,$$

and the integral group ring $\mathbb{Z}S_3$ is embedded into $\mathbb{C}S_3$ in the following way

$$\begin{bmatrix} \mathbb{Z} & \mathbb{Z} \\ 3\cdot\mathbb{Z} & \mathbb{Z} \end{bmatrix} \quad \Pi\, \mathbb{Z}^+ \,\Pi\, \mathbb{Z}^-,$$

in terms of elements:

$$\begin{bmatrix} e+3\cdot a & b \\ 3\cdot c & e+2\cdot f+3\cdot d \end{bmatrix}, \qquad e,\, e+2\cdot f\,.$$

In general, the rational group algebra, say for a $p$-group $G$ of odd order has the form:

$$\mathbf{Q}G = \mathbf{Q}^{+} \,\Pi\, (\Pi_{i=1}^{v}\, \mathrm{Mat}(n_i, K_i))\,,$$

where $K_i$ are algebraic number fields, $1 \leq i \leq v$, and $\mathbf{Q}^{+}$ comes from the trivial representation. *The integral group ring* $\mathbf{Z}G$ *is then a subring of finite index in*

$$\mathbf{Z}G \subset \mathbf{Z}^{+} \,\Pi\, (\Pi_{i=1}^{v}\, \mathrm{Mat}(n_i, K_i))\,,$$

*where* $R_i$ *are the algebraic integers in the number fields* $K_i$, $1 \leq i \leq v$.

I shall report here on some *recent progress on the isomorphism problem* (1.7), which was obtained in **joint work with Leonard L. Scott** in the last 10 years, 1980-1990.

In attacking the isomorphism problem one should expect, that the following subjects are invoked:

1. *GROUP THEORY*
2. *REPRESENTATION THEORY*
3. *NUMBER THEORY AND ARITHMETIC*

It turned out during our long struggle, that we also had to use

4. *COHOMOLOGY OF GROUPS AND OF RINGS*
5. *P-ADIC NON COMMUTATIVE ANALYSIS*
6. *ALGEBRAIC K-THEORY, GROTHENDIECK GROUPS AND WHITEHEAD GROUPS.*

## § 2 Major results until 1980

As mentioned above, the isomorphism problem was not considered in the beginning of representation theory – too many non isomorphic groups have isomorphic complex group algebras. It was first considered by *Graham Higman* in his thesis in 1939. Higman was a student of G. Whitehead, who investigated at that time what are now called "Whitehead" groups, and he told his student Higman to calculate some examples. This means calculations in the group of units in rings. The integral group rings turn up in topology in connection with cohomology, and so Higman studied units in integral group rings of finite groups. His most spectacular result was:

**(2.1) Theorem** (G. Higman [Hi]): *Let* $u$ *be a unit of finite order in* $\mathbf{Z}A$ *for a finite abelian – i.e. commutative – group* $A$. *Then* $u = \pm a$ *for some group element* $a \in A$.

We shall *sketch the proof*: This argument uses the fact, that the integral group ring of a finite group, $\mathbf{Z}G$, is not only a $\mathbf{Z}$-algebra, but also a Hopf-algebra, where the antipode is induced from the *anti-involution*

$$*_G : \mathbb{Z}G \to \mathbb{Z}G, \tag{2.2}$$
$$\sum_{g\in G} z_g \cdot g \mapsto \sum_{g \in G} z_g \cdot g^{-1}.$$

For the proof of Higman's result, let – more generally – $G$ be a finite group and $u \in U(\mathbb{Z}G)$, the group of units in $\mathbb{Z}G$, be a central unit of finite order $n$. (We shall here only deal with the case where $n$ *is odd.*) Then $u^*$, the image of $u$ under the anti-involution $*_G$ is also a central unit of order $n$, and thus

$$v = u \cdot u^* \quad \text{is a central unit in} \quad \mathbb{Z}G$$

with $v^n = 1$, using the fact that $u$ is central; moreover, $v$ is fixed under $*_G$. It is easily seen, that on the centre $Z$ of $\mathbb{Q}G$, $Z = \Pi_{i=1}^{s} K_i$, where $K_i$ are algebraic number fields, $*_G$ induces the complex conjugation. Thus $v$ lies in a product of real fields. Since the order of $v$ is odd, we conclude $v = 1$. Thus

$$1 = \Big(\sum_{g\in G} z_g \cdot g\Big) \cdot \Big(\sum_{g\in G} z_g \cdot g^{-1}\Big),$$

provided $u = \sum_{g\in G} z_g \cdot g$. We now consider the coefficient of 1 in the above product and conclude

$$1 = \sum z_g^2 .$$

Since $z_g \in \mathbb{Z}$, we have $u = \pm g_0$ for some $g_0 \in G$. (The proof for $n$ even is done similarly.)

In this connection we point out, that this argument has heavily used the fact, that in $\mathbb{Z}$ a sum of squares is one if and only if there is only one non zero summand. So the ring of rational integers plays an exceptional role.

In order to appreciate this result, let us look at an *example*:

Let $A = C_5 \times C_5$ be the product to two cyclic groups of order 5. Then the group ring $\mathbb{Z}A$ is of finite index in

$$\Gamma = \mathbb{Z} \,\Pi\, \mathbb{Z}[\zeta_5] \,\Pi\, \mathbb{Z}[\zeta_5] \,\Pi\, \mathbb{Z}[\zeta_5] \,\Pi\, \mathbb{Z}[\zeta_5] \,\Pi\, \mathbb{Z}[\zeta_5] \,\Pi\, \mathbb{Z}[\zeta_5],$$

and the group of the units of finite order in $\Gamma$ is isomorphic to

$$C_2 \times C_5 \times C_5 \times C_5 \times C_5 \times C_5 \times C_5,$$

whileas the group of units of finite order in $\mathbb{Z}A$ is isomorphic to

$$C_2 \times C_5 \times C_5 .$$

**(2.3) Remark**: The group ring $RG$ of the finite group $G$ over the commutative ring $R$ is not only an $R$-algebra, but even an augmented $R$-algebra with augmentation map

$$\epsilon_G : RG \to R,$$
$$\sum_{g\in G} r_g \cdot g \to \sum_{g\in G} r_g .$$

In order to reformulate Higman's result, we have to introduce some more notation:

**(2.4) Definition**: 1. $U(RG)$ stands for the units in $RG$ – i.e. the elements, which have a multiplicative inverse.

2. $V(RG) = \{u \in U(RG) \mid \epsilon(u) = 1\}$ are the *units of augmentation one*. Then

$$U(RG) = V(RG) \times U(R).$$

*Since the units in R are assumed to be known, it suffices to study V(RG).*

Let us turn to our aim, the isomorphism problem (1.7):

If $\mathbb{Z}G = \mathbb{Z}H$, then $H \subset U(\mathbb{Z}G)$, in general however, it will not be contained in $V(\mathbb{Z}G)$. But this can easily be remedied:

**(2.5)** Replace $H$ by $H' = \{h \cdot \epsilon_G(h)^{-1} \mid h \in H\}$. Then $H \simeq H'$ and $\mathbb{Z}H = \mathbb{Z}H'$; moreover,

$$H' \subset V(\mathbb{Z}G).$$

*Thus from now on we will always assume that an equality of group rings $\mathbb{Z}G = \mathbb{Z}H$ is always augmented; i.e. $H \subset V(\mathbb{Z}G)$.*

In view of this, the most important consequence of Higman's theorem is:

**(2.6) Corollary**: *Let $A$ be a finite abelian group. If $\mathbb{Z}A = \mathbb{Z}G$ (augmented), then also $G$ is abelian and moreover, $A = G$ in $\mathbb{Z}A = \mathbb{Z}G$.*

This means that the group $A$ is *unique* in $\mathbb{Z}G$, as augmented group of units. *This result is a very elegant positive answer to the isomorphism problem.*

Actually, Higman, in his thesis "*Units in group rings*", was the first to speculate about the isomorphism problem: "*Whether it is possible for two non isomorphic groups to have isomorphic integral group rings I do not know, but the results... here... suggested that it is unlikely*". So, Higman was a bit in favour of a positive answer to the isomorphism problem.

In this connection I should also mention Richard Brauer, who in his "*Harvard-Lecture on Modern Mathematics, Representations of finite groups*" in 1963 listed the isomorphism problem as one of the important open questions in representation theory; however, he did not commit himself in either direction. Contrary to Hans Zassenhaus, who firmly believes in a positive answer to the isomorphism problem and even made a much stronger conjecture in 1974 — cf. below.

The isomorphism problem, $\mathbb{Z}G = \mathbb{Z}H$ as augmented algebras, appears in a new light, if we assume that $*_G = *_H$ (cf. 2.2); i.e., if

$$g \in \sum_{h \in H} z_h h \text{ , then } g^{-1} = \sum_{h \in H} z_h h^{-1}.$$

This means that $\mathbb{Z}G = \mathbb{Z}H$ as $\mathbb{Z}$-algebras and as Hopf-algebras. In this case we have a very strong result:

**(2.7) Proposition** (Banachevski [Ba]): Let $\mathbb{Z}G = \mathbb{Z}H$ as augmented algebras and as Hopf-algebras; i.e. $*_G = *_H$. Then $G = H$ in $\mathbb{Z}G = \mathbb{Z}H$.

**Proof**: For $g \in G$, let $g = \sum z_h \cdot h$; then

$$1 = g \cdot g^{-1} = \Big(\sum_{h \in H} z_h \cdot h\Big) \cdot \Big(\sum_{h \in H} z_h \cdot h^{-1}\Big),$$

and the same argument as in the proof of (2.1) above shows $g = h$ for some $h \in H$.

As was colported to us, this result apparently led J. Dieudonné, when he said in his talk "*Lie groups, classical, algebraic and formal*" at the British Mathematical Colloquium in Birmingham, 1969, that the isomorphism problem is an *ill posed problem*.

A consequence of Higman's result is thus the following:

**(2.8) Corollary**: Let $\mathbb{Z}A = \mathbb{Z}B$ as augmented algebras with $A$ and $B$ abelian. Then $\mathbb{Z}G = \mathbb{Z}H$ as Hopf-algebras.

After these rather special observations we turn to the general situation. Given an equality of augmented algebras $\mathbb{Z}G = \mathbb{Z}H$. Then the elements $\{g \mid g \in G\}$ form a finite subgroup in $V(\mathbb{Z}H)$, the units of augmentation one in $\mathbb{Z}H$, consisting of *$\mathbb{Z}$-linearly independent elements*. Conversely, given any finite subgroup $U$ in $V(\mathbb{Z}H)$ of order $|H|$, the order of $H$, the elements of which are linearly independent over $\mathbb{Z}$, then we have an augmented identification $\mathbb{Z}U = \mathbb{Z}H$. S.D. Berman has observed, that the linear independence is often automatic:

The above definitions for the integral group ring $\mathbb{Z}G$ can obviously be extended to the group ring $RG$, with $R$ an integral domain.

**(2.9) Proposition** (S.D. Berman [Ber]): Let $R$ be an integral domain of characteristic zero and $G$ a finite group, such that no rational prime divisor of $|G|$ is a unit in $R$. Let $U \leq V(RG)$ be a finite subgroup. Then the elements in $U$ are $R$-linearly independent in $RG$.

This result allows to *phrase the isomorphism problem differently*:

**(2.10)** *Let $U$ be a finite subgroup in $V(\mathbb{Z}G)$, with $|G| = |U|$. Is then $U \simeq G$?*

One might go even *one step further than the isomorphism problem*, and ask: Let $U \leq V(\mathbb{Z}G)$ and assume $|V| = |G|$. *How is $U$ embedded in $V(\mathbb{Z}G)$?* G. Higman gave the answer for abelian $G$: In this case $U = G$.

However, in general, one cannot expect that $U = G$, since $V(\mathbb{Z}G)$ need not be abelian. Thus conjugation with the units in $V(\mathbb{Z}G)$ do not necessarily stabilize $G$, since $G$ will not be normal in $V(\mathbb{Z}G)$. Moreover, even for the dihedral group $D$ of order 8, there exists a unit $u \in \mathbb{Q}D \setminus V(\mathbb{Z}D)$, such that

$$D \neq u \cdot D \cdot u^{-1} \subset \mathbb{Z}D,$$

and this conjugation is not inner in $\mathbb{Z}D$. So for $G$ not abelian, the obstruction to $U = G$ are not just the inner automorphisms of $V(\mathbb{Z}G)$.

In this connection, H. Zassenhaus [Za] made a far reaching conjecture:

**(2.11) Zassenhaus conjecture I**: Let $U \leq V(\mathbb{Z}G)$ be a finite subgroup with $|V| = |G|$. Then there exists a unit $u \in \mathbb{Q}G$ with $u \cdot V \cdot u^{-1} = G$.

Because of the Skolem-Noether theorem this is equivalent to

**(2.12) Zassenhaus conjecture II**: Let $\mathbb{Z}G = \mathbb{Z}H$ as augmented algebras. Then there exists a group isomorphism $\rho : G \to H$ such that

$$\alpha \cdot \rho^{-1} : \mathbb{Z}G \to \mathbb{Z}G$$

is a central automorphism; i.e. an automorphism leaving the centre of $\mathbb{Z}G$ elementwise fixed.

In view of the structure of the centre $Z$ of $\mathbb{Q}G$, $Z = \prod_{i=1}^{n} K_i$, $K_i$ algebraic number fields, $1 \leq i \leq n$, this conjecture would imply that every automorphism $\sigma$ of $Z$, which can be extended to all of $\mathbb{Z}G$ is induced there from a group automorphism; this is a very strong statement.

Prior to (1985) there were some special classes of groups, for which the Zassenhaus conjecture was verified:

1) Higman's result (2.2) shows that it is true to abelian groups,
2) Sehgal [Se] proved it for nilpotent groups of class 2,
3) Ritter-Sehgal [RiSe] proved it for certain metacyclic groups.

Let us return to the original question of (1.7): A direct extension of Higman's result was given by Glaubermann and Berman [Ber2].

**(2.13) Theorem**: *Let $\mathbb{Z}G = \mathbb{Z}H$ as augmented algebras and let $K_g = \sum_{x \in G/C_G(g)} x_g$ be a class sum in $G$. $C_G(g)$ denotes the centralizer of $g$ in $G$. Then $K_g = K_h$ is a class sum in $\mathbb{Z}H$.*

Again the proof uses the properties of sums of squares of rational integers.

In a similar spirit, the lattices of normal subgroups of $G$ and $H$ are isomorphic:

**(2.14) Theorem** (Berman, Glauberman, Sehgal [Se]): *Let $N$ be a normal subgroup of $G$, and let $\mathbb{Z}G = \mathbb{Z}H$ as augmented algebras. Then $\sum_{n \in N} n = \sum_{m \in M} m$ for a normal subgroup $M$ in $H$.*

These results also hold for $RG$ if $R$ is a Dedekind domain of characteristic zero, in which no prime divisor of $|G|$ is invertible. *It would be important* for our results on the conjugacy problem for defect groups, to know whether the hypotheses in the above theorems (2.13) and (2.14) could be weakened as follows: if the rational prime $p$ is not a unit in $R$, is there still a correspondence between the normal $p$-subgroups in $G$ and those in $H$?

A stronger statement than "$\mathbb{Z}G = \mathbb{Z}H$ implies the existence of an isomorphism $\varphi : G \to H$" is the following natural extension, motivated by (2.13):

**(2.15) Isomorphism problem over the class sums**: *$\mathbb{Z}G = \mathbb{Z}H$ implies that there exists an isomorphism $\chi : G \to H$ inducing the class sum correspondence; i.e. for $g \in G$, $K_g = K_{g\chi}$. We shall call this an isomorphism over the class sums.*

**(2.16) Observation:** *The existence of an isomorphism over the class sums is equivalent to the Zassenhaus conjecture.*

In fact, assume that (2.15) holds, and let $\alpha : \mathbb{Z}G \to \mathbb{Z}G$ be an automorphism. We put $H = \Im(\alpha|_G)$. By (2.15) there exists an isomorphism $\varphi : G \to H = \alpha(G)$

such that if $K_g = K_{g'\alpha}$, then $K_{g\varphi} = K_{g'\alpha}$. Thus $K_g = K_{g\varphi}$, and hence the automorphism on $\mathbb{Z}G$, induced by $\varphi$ is a central ring automorphism. Since $\alpha \cdot \varphi^{-1}$ is an automorphism of $G$, the Zassenhaus conjecture follows.

Conversely, assume that the Zassenhaus conjecture is true, and let $\mathbb{Z}G = \mathbb{Z}H$, then an isomorphism $\alpha : G \to H$, which exists, since the Zassenhaus conjecture implies a positive answer to the isomorphism problem, has the structure $\alpha = \varrho \cdot \gamma$, where $\varrho$ is an automorphism of $G$ and $\gamma$ is a central automorphism of $\mathbb{Z}G$. Then $\varrho^{-1} \cdot \alpha : G \to H$ has the property required in (2.15), as one sees by reversing the above arguments.

Then next result of Whitcomb pushes the isomorphism problem further to metabelian groups, and it seems to give rise to a possible induction for the solvable case:

**(2.17) Theorem** (Whitcomb [WH]): *Let $\mathbb{Z}G = \mathbb{Z}H$ as augmented algebras, and let $A$ be an abelian normal subgroup of $G$ corresponding to $B$ in $H$, which is then automatically abelian normal in $H$ (2.14). Then we have a commutative diagramm with exact rows:*

$$\begin{array}{ccccccccc} 0 & \longrightarrow & I(A)G & \longrightarrow & \mathbb{Z}G & \longrightarrow & \mathbb{Z}G/A & \longrightarrow & 0 \\ & & \| & & \| & & \| & & \\ 0 & \longrightarrow & I(B)G & \longrightarrow & \mathbb{Z}G & \longrightarrow & \mathbb{Z}G/B & \longrightarrow & 0 \end{array}$$

*Assume that $G/A = H/B$ in $\mathbb{Z}G/A = \mathbb{Z}G/B$. Then $G \simeq H$.*

**(2.18) Corollary**: *In (2.17) assume that $G$ is metabelian with $G/A$ abelian. Then $G \simeq H$.*

In fact, by Higman's result (2.6), $G/A = H/B$ in $\mathbb{Z}G/A = \mathbb{Z}H/B$.

**Proof of (2.17)**: $I(A)G$ is the kernel of the natural map $\mathbb{Z}G \to \mathbb{Z}G/A$. How Heinz Hopf has observed that

$$\begin{gathered} \gamma : I(A)G/I(A)I(G) \to A, \\ (a-1)\cdot g + 1(A)I(G) \mapsto a \end{gathered}$$

is an isomorphism of *left* $\mathbb{Z}G$-modules, where $G$ acts on $A$ via conjugation. Hence the diagram in (2.14) gives rise to the diagram

$$\begin{array}{ccccccccc} 0 & \longrightarrow & A & \longrightarrow & \overline{\mathbb{Z}G} & \longrightarrow & \mathbb{Z}G/A & \longrightarrow & 0 \\ & & \| & & \| & & \| & & \\ 0 & \longrightarrow & B & \longrightarrow & \overline{\mathbb{Z}G} & \longrightarrow & \mathbb{Z}G/B & \longrightarrow & 0 \end{array} .$$

We also have the commutative diagram — arising from the natural embeddings —

$$\begin{array}{ccc} G/A & \xrightarrow{\kappa_1} & \mathbb{Z}G/A \\ \| & & \| \\ H/B & \xrightarrow{\kappa_2} & \mathbb{Z}H/B . \end{array}$$

An easy cohomology argument now shows that the group extensions

$$\begin{array}{ccccccccc} 1 & \longrightarrow & A & \longrightarrow & G & \longrightarrow & G/A & \longrightarrow & 1 \\ & & \| & & \| & & \| & & \\ 1 & \longrightarrow & B & \longrightarrow & H & \longrightarrow & H/B & \longrightarrow & 1 \end{array}$$

are the same, via the pullbacks along $\kappa_1$ and $\kappa_2$.

Note, however, that the proof does not give any information of how $G$ is embedded into $V(\mathbb{Z}H)$; also, it does not give any clue for the Zassenhaus conjecture. As a matter of fact, the Zassenhaus conjecture (2.11) is not at all suited for an induction as (2.17) might suggest.

The following result was recently obtained by Kimmerle, Lyons and Sandling, who showed — using heavily the classification of the finite simple groups:

**(2.19) Theorem** [KLS]: *Let $\mathbb{Z}G = \mathbb{Z}H$ as augmented algebras, and let*

$$1 = N_0 < \ldots < N_t = G$$

*be a chief series of $G$. Then there exists a chief series*

$$1 = M_0 < \ldots < M_\tau = H$$

*of $H$, such that $t = \tau$ and the chief factors $N_i/N_{i-1} \simeq M_i/M_{i-1}$, $1 \leq i \leq t$, are isomorphic even with the same indices.*

This shows in particular that finite simple groups are determined by their integral group rings.

## § 3 Recent progress by L. L. Scott and myself

When we started to consider the isomorphism problem first in 1980, the Zassenhaus conjecture was at first a tempting target for a counterexample. At that time we were not successful, and eventually started to believe, that the Zassenhaus conjecture might be true for certain classes of groups; in fact, it was a guide for our work on $p$-groups.

Roughly speaking, the Zassenhaus conjecture states, that there are only few automorphisms of the integral group ring $\mathbb{Z}G$ of the finite group $G$. After having done plenty of calculations, we were looking for some *heuristic evidence* for the Zassenhaus conjecture. We noted that

(3.1) *every derivation $\delta : \mathbb{Z}G \longrightarrow \mathbb{Z}G$ is inner,*

i.e. every $\mathbb{Z}$-linear map $\delta : \mathbb{Z}G \longrightarrow \mathbb{Z}G$ with $\delta(x \cdot y) = x \cdot \delta(y) + \delta(x) \cdot y$, $x, y \in \mathbb{Z}G$ is of the form $\delta(x) = ad_a(x) =: a \cdot x - x \cdot a$ for some fixed $a \in \mathbb{Z}G$.

Borrowing ideas from Lie theory, we found a *strategy*, to show that an automorphism $\alpha : \mathbb{Z}G \longrightarrow \mathbb{Z}G$ is an inner automorphism — using the fact that $H^1(\mathbb{Z}G, \mathbb{Z}G) = 0$:

1. Form $\log(\alpha) = -\sum_{i=0}^{\infty} \frac{(1-\alpha)^i}{i} =: D$, if it exists.
2. By formal reasoning, $\log(\alpha) = D$ is a derivation from $\mathbb{Z}G$ to $\mathbb{Z}G$.
3. Since every derivation from $\mathbb{Z}G$ to $\mathbb{Z}G$ is an inner derivation, we conclude that $D(x) = ad_a(x)$ for some fixed $a \in \mathbb{Z}G$.
4. Form $\exp(D) = \sum_{i=0}^{\infty} \frac{D^i}{i!}$, if possible.

5. Since $D = ad_a = \lambda_a - \rho_a$, where $\lambda_a$ is left multiplication by $a$ and $\rho_a$ is right multiplication with $a$, we conclude $\exp(ad_a)(x) = \exp(a)\cdot x\cdot \exp(a)^{-1}$ is conjugation by $\exp(a)$, if all the terms do exist.
6. Show that $\alpha(x) = \exp(ad_a)$, thus proving that $\alpha$ is an inner automorphism.

Though $\log(\alpha)$ and $\exp(D)$ do not make sense for $\mathbb{Z}G$, infinite series make sense over the complete topological ring $\mathbb{Z}_p^\wedge G$, where $\mathbb{Z}_p^\wedge = pro.\lim.(\mathbb{Z}/p^i\mathbb{Z})$ is the ring of $p$-adic integers, $p$ a rational prime number.

*Following through the steps 1.) — 6.) above was a vital part in our proof of the conjugacy of group bases in $\mathbb{Z}_p^\wedge G$, $P$ a p-group.*

One might even go one step further than Zassenhaus and ask — this question was briefly raised by Berman and Rossa [BR] for complete rings:

**(3.2) Conjugacy problem**: *If $V(\mathbb{Z}_p^\wedge G) = V(\mathbb{Z}_p^\wedge G)$, are $G$ and $H$ conjugate in $V(\mathbb{Z}_p^\wedge G)$?*

The results, which we have obtained so far let it appear reasonable to ask the following *question*:

**(3.3)** Conjugacy of defect groups: Let $R$ be a complete Dedekind domain of finite rank over $\mathbb{Z}_p^\wedge$, with residue field of characteristic $p > 0$. Let $B_0$ be the principal block of $RG$. Is the "defect group" of $B_0$ uniquely determined up to conjugacy in $B_0$?

## A. Progress on the isomorphism problem

The next results were obtained by Leonard Scott and myself in 1985 [RS1] and [RS3]:

**(3.4) Theorem**: *Let $G$ be a finite group such that there exists an exact sequence*

$$1 \longrightarrow A \longrightarrow G \longrightarrow N \longrightarrow 1,$$

*where $A$ is abelian and $N$ is nilpotent. If $\mathbb{Z}G \simeq \mathbb{Z}H$, then $G$ is isomorphic to $H$.*

Again, this is a statement about an abstract isomorphism, and it does not give any information about the embedding of $H$ into $V(\mathbb{Z}G)$.

**(3.5) Theorem**: *Let $G$ be a finite group, $p \in \mathbb{Z}$ a prime, such that $G$ has a normal p-subgroup $P$ with $C_G(P)$ a p-group, and assume that $\mathbb{Z}G \simeq \mathbb{Z}H$. Then $G$ is isomorphic to $H$.*

**(3.6) Remark**: For a solvable group $G$ the above implies

a) For every prime $p$, the group $G/O_{p'}(G)$ is determined by $\mathbb{Z}G$, where $O_{p'}(G)$ is the largest normal subgroup of $G$ of order prime to $p$.
b) The Sylow subgroups of $G$ are determined — up to isomorphism — by $\mathbb{Z}G$.

We shall see below, that in this case one can say more about the embedding of $H$ into $V(\mathbb{Z}G)$, since the Theorem is a consequence of a much deeper result.

### B. Drawbacks and progress on the Zassenhaus conjecture

Leonard Scott and I have tried hard, to prove the Zassenhaus conjecture (2.12) for abelian by nilpotent groups, even for metabelian groups, however unsuccessfully.

In connection with the above problems we have proved the following positive results:

**(3.7) Theorem**: *The Zassenhaus conjecture is true for the following classes of groups:*

(i) *$G$ is nilpotent,*
(ii) *there exists prime $p \in \mathbb{Z}$, such that $G$ has a normal $p$-subgroup $P$ with $C_G(P)$ a $p$-group.*

Let us turn now to a *solvable group* $G$. In that case $G/O_{p'}(G)$ satisfies the condition (ii) above, and so the Zassenhaus conjecture is true. Thus there is an injective homomorphism

$$\varphi : G \longrightarrow \prod_{i=1}^{n} G/O_{p'_i}(G);$$

i.e., $G$ is a subdirect product of groups, for which the Zassenhaus conjecture holds (2.12). We would like to describe an *obstruction* for the Zassenhaus conjecture to hold for $\mathbb{Z}G$. We also note that the groups $O_{p'_i}(G)$ are characteristic; but what is more, the induced augmentation ideal $I(O_{p'_i}(G)) \cdot \mathbb{Z}G$ is determined by $\mathbb{Z}G$.

As a general *setup* we assume

**(3.8)** Let $G$ be a finite group with normal subgroups $\{K_\alpha\}_{\alpha \in I}$, where $I$ is a finite index set, such that
(i) $\cap_{\alpha \in I} K_\alpha = 1$,
(ii) the augmentation ideals $I(K_\alpha) \cdot \mathbb{Z}G$ are determined by $\mathbb{Z}G$.

We then let

$$\varphi_\alpha : G \longrightarrow G/K_\alpha =: G_\alpha$$

be the natural map, and put for a subset $S \subset I$

$$\varphi_S : G \longrightarrow G/(\prod_{\alpha \in S} K_\alpha) =: G_S$$

to be the natural projection. Then

$$G = \varprojlim_{S}(G_s).$$

However, the conjugacy classes do not behave well for projective limits, and so for the sets of conjugacy classes

$$Cl(G) \neq \varprojlim_{S}(Cl(G_s)) =: X_g.$$

Similarly, group rings do not behave well with respect to projective limits, and so

$$\mathbb{Z}G \neq \varprojlim_{S}(\mathbb{Z}G_s) =: \Gamma.$$

If we *assume that the Zassenhaus conjecture holds for the groups* $G_S$, then a weaker form of the Zassenhaus conjecture, which is adapted to the projective limit is the following:

**(3.9) Weak form of the Zassenhaus conjecture** (with respect to the family of subgroups $\{G_\alpha\}_{\alpha \in i}$):
*If* $\mathbb{Z}G = \mathbb{Z}H$, *then there exists an automorphism* $\rho$ *from* $G$ *to* $H$, *which is an automorphism over* $X$. (Note that because of (2.13), $X_g = X_h$.) *We even could replace* $\mathbb{Z}G$ *by* $\Gamma$.

Let us look at the special case, where above $I = \{1, 2\}$. (This is also the case of our counterexample!) $G$ is the pullback

$$\begin{array}{ccc} G & \longrightarrow & G_1 \\ \downarrow & & \downarrow \\ G_2 & \longrightarrow & G' \end{array} .$$

We assume now, that $\sigma : \mathbb{Z}G \longrightarrow \mathbb{Z}G$ is an augmented isomorphism. Because of (3.8,ii) $\sigma$ induces augmented isomorphisms $\sigma_i : \mathbb{Z}G_i \longrightarrow \mathbb{Z}G_i$, $i = 1, 2$ and $\sigma' : \mathbb{Z}G' \longrightarrow \mathbb{Z}G'$. According to the hypothesis, the Zassenhaus conjecture is true for $G_i$, $i = 1, 2$ and $G'$; i.e.,

$$\sigma_i|_{\mathbb{Z}G_i} = conj(a_i) \cdot \rho_i \ , \ i = 1, 2 \ ,$$

where $a_i$ is a unit in $\mathbb{Q}G_i$ stabilizing $\mathbb{Z}G_i$ and $\rho_i \in Aut(G_i)$, $i = 1, 2$. *Note* however, that neither $\rho_i$ nor $conj(a_i)$ are uniquely determined. They are only unique up to central automorphisms: a central automorphism $\gamma_i$ of $G_i$ (i.e. $\gamma_i(x)$ and $x$ are conjugate in $G_i$) can also be written as $conj(c_i)$, $c_i \in \mathbb{Q}G_i$.

On $\mathbb{Z}G'$ we have $\sigma_1 = \sigma_2$; thus $conj(a_1) \cdot \rho_1 = conj(a_2) \cdot \rho_2$ on $\mathbb{Z}G'$, and so $\rho_1 \cdot \rho_2^{-1} = conj(a_1^{-1}) \cdot conj(a_2) =: \gamma_{12}$ on $\mathbb{Z}G'$.

Thus $\rho_1 \cdot \rho_2^{-1} = \gamma_{12}$ is a *central group automorphism of* $\mathbb{Z}G'$; note that though the above construction is not unique, the automorphism $\gamma_{12}$ is independent of the chosen "Zassenhaus decomposition".

**(3.10) Lemma:** *If* $\gamma_{12} = \gamma_1 \cdot \gamma_2^{-1}$ *for central group automorphisms* $\gamma_i$ *of* $G_i$, $i = 1, 2$ *then the Zassenhaus conjecture is true for the pullback*

$$\begin{array}{ccc} \Gamma & \longrightarrow & \mathbb{Z}G_1 \\ \downarrow & & \downarrow \\ \mathbb{Z}G_2 & \longrightarrow & \mathbb{Z}G' \end{array} ;$$

*i.e.* $\sigma = conj(a) \cdot \rho$ *on* $\Gamma$ *for a group automorphism* $\rho$ *of* $G$ *and a unit* $a$ *in* $\mathbb{Q}G$, *centralizing* $\Gamma$, *and conversely.*

*Proof:* If the Zassenhaus conjecture holds for $\Gamma$, then on $\Gamma$ we have $\sigma = conj(a) \cdot \rho$. This induces $\sigma_i = conj(a_i) \cdot \rho_i$ on $\mathbb{Z}G_i$, where $conj(a_i) = conj(a)$ on $\mathbb{Z}G_i$. Thus $\gamma_{12} = 1$. Conversely, assume that $\varphi_{12} = \gamma_1 \cdot \gamma_2^{-1}$ for central group automorphisms $\gamma_i$ of $G_i$. Then we have $\rho_1 \cdot \rho_2^{-1} = \gamma_1 \cdot \gamma_2^{-1}$. Hence $\gamma_1^{-1} \cdot \rho_1 = \gamma_2^{-1} \cdot \rho_2$ is a group automorphism. If now $\rho$ is the pullback of the group automorphisms $(\gamma_1^{1} \cdot \rho_1, \gamma_2^{-1} \cdot \rho_2)$, then $\rho$ is an automorphism of $G$. Moreover,

$$\begin{aligned} \sigma \cdot \rho^{-1} &= (conj(a_1) \cdot \rho_1 \cdot \rho_1^{-1} \cdot \gamma_1, conj(a_2) \cdot \rho_2 \cdot \rho_2^{-1} \cdot \gamma_2) = \\ &= (conj(a_1) \cdot \gamma_1, conj(a_2) \cdot \gamma_2) \end{aligned}$$

is a central automorphism of $\Gamma$.

**(3.11) Remark**: 1) From the above result it follows, that the Zassenhaus conjecture holds for $\Gamma$ if every central automorphism of $G'$ is an inner automorphism.

2) If every central automorphism $\gamma'$ of $G'$ can be written as $\gamma' = \gamma_1 \cdot \gamma_2^{-1}$ for central automorphisms $\gamma_i$ of $G_i$, then the Zassenhaus conjecture holds for $\Gamma$.

3) We note that the above condition $\gamma_{12} = \gamma_1 \cdot \gamma_2^{-1}$ is the condition that a 1-cocycle is a 1-coboundary for each Cech cohomology with respect to a covering of a topological space by two open sets.

4) For the construction of a counterexample to the Zassenhaus conjecture it is thus necessary to construct a group $G$ as a pullback, which does not satisfy 3).

5) For the projective limits as above, the validity of the weak form of the Zassenhaus conjecture is precisely the condition, that a Cech 1-cocycle is a coboundary.

We constructed a counterexample to the Zassenhaus conjecture (2.12):

**(3.12) Theorem [RS2]**: *There exists a finite metabelian group $G$, and an automorphism $\alpha$ of $\mathbb{Z}G$, commuting with the augmentation $\epsilon_G$, such that $\alpha \cdot \rho$ is not a central automorphism for any group automorphism $\rho$ of $G$.*

This shows that the Zassenhaus conjecture is false, even for metabelian groups.

## C: Progress on the conjugacy problem

As we have remarked above, $\mathbb{Z}D_8$, the group ring of the dihedral group of order 8, has another subgroup $U \simeq D_8$ in $V(\mathbb{Z}D_8)$, which is not conjugate in $V(\mathbb{Z}D_8)$ to $D_8$; however $U$ and $D_8$ are conjugate in $\mathbb{Q}D_8$ by a matrix, the determinant of which is 3. Now 3 becomes a unit in $\mathbb{Z}_2^\wedge$, the 2-adic integers. Moreover, we have the philosophy, that for a $p$-group $P$, the natural group ring is not $\mathbb{Z}P$, but rather $\mathbb{Z}_p^\wedge P$, the $p$-adic group ring. This was one of the reasons, why we have concentrated on $\mathbb{Z}_p^\wedge P$. Another reason is the following: In order to prove that the isomorphism problem has a positive answer for $p$-groups $P$, one wants to use induction on the order of $P$. Neither the isomorphism problem, nor the Zassenhaus conjecture are suited for induction. In fact, if the automorphism $\alpha$ of $\mathbb{Z}G$ is conjugation with a unit $a \in \mathbb{Q}G/N$ on the quotient $\mathbb{Z}G/N$, there is no reason, to believe that a can be lifted to a unit in $\mathbb{Q}G$ normalizing $\mathbb{Z}G$. However, if one considers $p$-groups over $\mathbb{Z}_p^\wedge$, and if $a$ is in $V(\mathbb{Z}_p^\wedge P/N)$, then it can be lifted to a unit in $V(\mathbb{Z}_p^\wedge P)$, and we can use induction. Following this philosophy, we were able to extend our result on $p$-groups [RS], and have obtained in 1986 the following result [RS3]:

**(3.13) Theorem**: *Let $G$ be a finite group with normal Sylow $p$-subgroup, such that $O_{p'}(G) = 1$. Let $\alpha$ be an augmented automorphism of $\mathbb{Z}_p^\wedge G$. Then there exists a unit $u \in V(\mathbb{Z}_p^\wedge G)$, such that $uGu^{-1} = \alpha(G)$.*

The theorem we have proved is actually more general: Let $G$ be a finite $p$-constrained group with $O_{p'}(G) = 1$ and let $N = O_p(G)$. If $\alpha$ is an augmented automorphism of $\mathbb{Z}_p^\wedge G$, stabilizing $I_{\mathbb{Z}_p^\wedge}(N)G$, then $G$ and $\alpha(G)$ are conjugate in $V(\mathbb{Z}_p^\wedge G)$.

This result applies in particular to $p$-groups, as one might have expected according to our philosophy. However, what is really surprising is, that the above groups are by no means all $p$-groups (the symmetric group on 3 elements satisfies that hypotheses of (3.13) for $p = 3$), and nevertheless, even $p'$-parts of these groups can be detected $p$-adically. Note that for a $q$-group $Q$, $q \neq p$ rational primes,

$$V(\mathbb{Z}_p^\wedge Q) = \prod_{i=1}^{n} GL(n_i, R_i),$$

where $R_i$ are unramified extensions of $\mathbb{Z}_p^\wedge$.

Also, for groups which satisfy the hypotheses of (3.13), the Zassenhaus conjecture is true. Again, comparing (3.13) with our counterexample to the Zassenhaus conjecture, (3.12) — recall that our counterexample is metabelian — one sees how delicate these problems are.

We have stated in (3.7), that the Zassenhaus conjecture is true for $\mathbb{Z}N$, provided $N$ is a nilpotent group. Our result (3.13) is so strong, that it allows to compute the Piccard group of $\mathbb{Z}N$ semilocally [RS], and from that we have:

**(3.14) Theorem**: *For a nilpotent group $N$, the conjugacy problem has a negative answer both for $V(\mathbb{Z}N)$ and for $V(\mathbb{Z}_p^\wedge N)$.*

Let us pause for a moment to contemplate about further possibilities: Let $G$ be a finite group with a normal Sylow $p$-subgroup and with $O_{p'}(G) = 1$. Then a consequence of (3.13) is the following:

Let $U \leq V(\mathbb{Z}_p^\wedge G)$, with $U \simeq G$, and such that the elements of $U$ are linearly independent over $\mathbb{Z}_p^\wedge$. Then $U$ and $G$ are conjugate in $(\mathbb{Z}_p^\wedge G)$. I do not know, whether the hypothesis, that the elements of $U$ are linearly independent, can be dropped. Our results, and the evidence we have gathered up to now, make it reasonable to ask, whether $(\mathbb{Z}_p^\wedge G)$ has the

**(3.15) Sylow property**: *Let $G$ be a finite group with a normal Sylow $p$-subgroup $P$ such that $O_{p'}(G) = 1$. Let $U$ be a finite $p$-subgroup of $V(\mathbb{Z}_p^\wedge G)$. Is $U$ conjugate in $V(\mathbb{Z}_p^\wedge G)$ to a subgroup of $P$?*

One consequence of this would be that for groups as above, the vertices of indecomposable $\mathbb{Z}_p^\wedge G$-lattices would be unique up to conjugacy in $V(\mathbb{Z}_p^\wedge G)$.

The result (3.13) can be interpreted of a first step to prove Sylow's theorems for finite subgroups of $V(RG)$, provided $G$ is a $p$-group. In an attempt to give an answer to this question we were guided by the fact that for a finite group $H$, the connectedness of the spectrum of the cohomology ring $H^*(H, \mathbb{F}_p)$ is equivalent to Sylow's theorems for the finite $p$-subgroups of $H$. We tried unsuccessfully to mimic John Carlson's proof that the variety of an indecomposable module is connected [Ca].

More precisely, let $V$ be a profinite $p$-group, and denote by $H^*(V, \mathbb{F}_p)$ the continuous (even-dimensional for $p$ odd) cohomology ring of $V$ with coefficients in $\mathbb{F}_p$, the field with $p$ elements. If the spectrum of $H^*(V, \mathbb{F}_p)$ is connected, we say that the *variety of $V$ is connected.* We shall write $VC(V, \mathbb{F}_p)$ for the variety of $H^*(V, \mathbb{F}_p)$.

Though we could not reach our original goal, we were able to prove for $R$ a complete Dedekind domain of characteristic zero with residue field of characteristic $p$:

**(3.16) Theorem:** *The following conditions are equivalent for a p-group G.*

(i) *Every finite p-subgroup U of $V(RG)$ is conjugate in $V(RG)$ to a subgroup of G.*

(ii) *For every p-subgroup P of G, the natural inclusion*

$$N_G(P)/P \longrightarrow N_{V(RG)}(P)/P$$

*induces a continuous map $VC(N_G(P)/P, \mathsf{F}_p) \to VC(N_{V(RG)}(P)/P, \mathsf{F}_p)$, which is a bijection.*

(iii) *The variety of $N_{V(RG)}(P)/P$ is connected for every p-subgroup P of G.*

(In the statement of the result we have used $N_A(B)$ to denote the normalizer of $B$ in $A$.)

We want to point out, that the statements (i) and (iii) are not true in general for profinite $p$-groups; in fact we have examples of unit groups of orders, where (i) is false. It would be interesting to have a group theoretical criterion for when (i) is true for profinite $p$-groups.

Let us return to a discussion of (3.15) in case $G$ is $p$-group. Some years ago we found a proof of (3.15) for $G$ a 2-group [R1]. Later on we were able to handle groups of order $p^3$, and we developed a sketch of the proof in the general $p$-group case in Nov. 1985 [S1]. Since in this proof we had not worked out *all* the details, we only made at the Arcatameeting in 1986 the conjecture, that (3.15) is true for $p$-groups. (In the meantime, Gary Thompson, a student of Leonard Scott, has worked out the details in [S1].) In October 1986 we learnt that Al Weiss from the University of Alberta [W] had a different proof of (3.15) for $p$-groups:

**(3.17) Subgroup rigidity theorem:** *Let G be a finite p-group, and U a finite subgroup of $V(\mathbb{Z}_p^{\wedge}G)$. Then U is conjugate in $V(\mathbb{Z}_p^{\wedge}G)$ to a subgroup of G.*

Recently, (3.16), which was proved by A. Weiss [W] also in case of unramified extensions, was proved for a complete Dedekind domain of finite rank over $\mathbb{Z}_p^{\wedge}$ in [GT] and in [R6].

## D: Units of finite order in $\mathbb{Z}G$

We next turn to questions about global units and conjugacy classes of finite subgroups in $V(\mathbb{Z}G)$.

The theorem (3.16) leads to another question about the cohomology rings: Since for a $p$-group $G$, the integral cohomology is determined $p$-adically, where its variety coincides with the variety of the $p$-adic cohomology of $V(\mathbb{Z}_p^{\wedge}G)$, is it possible, that for $p$-groups the varieties of $H^*(G, \mathsf{F}_p)$ and of $H^*(V(\mathbb{Z}G), \mathsf{F}_p)$ are also related? It should be noted though, that there is no obvious connection between $V(\mathbb{Z}G)$ and $V(\mathbb{Z}_p^{\wedge}G)$. By a result of Quillen [Qu 1,2], the irreducible components of the variety of $H^*(V(\mathbb{Z}G), \mathsf{F}_p)$ are in bijection to the conjugacy classes of maximal elementary abelian $p$-subgroups of $V(\mathbb{Z}G)$. The above question relates to:

**(3.18)** *For G a p-group, how are the conjugacy classes of maximal elementary abelian p-groups of $V(\mathbb{Z}G)$ related to those of G?*

The next example shows, that the cohomology varieties of the $p$-group $G$ and that of the unit group of its integral group ring $\mathbb{Z}G$ are not as intimately related as they are $p$-adically:

**(3.19) Lemma:** *The maximal elementary abelian 2-subgroups of the dihedral group $D$ of order 8 are all Klein's four groups. In $V(\mathbb{Z}D)$ there are three conjugacy classes of Klein's four groups. However, in $V(\mathbb{Z}_2^\wedge D)$ there are only two conjugacy classes of Klein's four groups.*

Now we describe shortly a theory, closely related to the theory of invertible bimodules by Fröhlich [Fr], in order to describe conjugacy classes of finite subgroups in $V(\mathbb{Z}G)$. Let $G$ be a finite group and let $U_0$ be a fixed subgroup of $G$. If

$$\alpha : U_0 \longrightarrow U(\mathbb{Z}G)$$

is a homomorphism, then we may view $\mathbb{Z}G$ as a $(U_0 \times G^{op})$-bimodule via $\alpha$, denoted by ${}_\alpha\mathbb{Z}G_G$:

$$(u \times g) \cdot x := \alpha(u) \cdot x \cdot g \ , \ u \in U_0 \ , \ g \in G \ , \ x \in \mathbb{Z}G \ .$$

If $U_0 = G$, then ${}_\alpha\mathbb{Z}G_G$ is an invertible bimodule, and we shall next translate some of the results on invertible bimodules to the present situation:

**(3.20) Lemma:** *Let $M$ be a $(U_0 \times G^{op})$-bimodule. $M$ is of the form ${}_\alpha\mathbb{Z}G_G$ for some $\alpha$ as above, iff $M|_{\mathbb{Z}G}$ is isomorphic to $\mathbb{Z}G$ as right $\mathbb{Z}G$-module.*

Among the augmented automorphisms of $\mathbb{Z}G$, the central automorphisms play an important role: For such an automorphism, the image of $G$ is of the form $a \cdot G \cdot a^{-1}$ for some unit $a$ in $\mathbb{Q}G$, which normalizes $\mathbb{Z}G$. In view of the Zassenhaus conjecture (2.11), the $(U_0 \times G^{op})$-bimodules $M$ are of particular importance, where the action of $U_0$ on $M$ is given via left multiplication with $a \cdot U_0 \cdot a^{-1} \leq U(\mathbb{Z}G)$; with other words the subgroups $U$ of $U(\mathbb{Z}G)$, which are of the form $a \cdot U_0 \cdot a^{-1}$ for some unit $a$ in $\mathbb{Q}G$; here the element $a$ needs not normalize $\mathbb{Z}G$; if it does normalize $\mathbb{Z}G$, then $a \cdot U_0 \cdot a^{-1}$ is part of the group basis $a \cdot G \cdot a^{-1}$. We characterize these bimodules in the following way:

**(3.21) Lemma:** *Let $M$ be a $(U_0 \times G^{op})$-bimodule. $M$ is isomorphic to a bimodule of the form ${}_{a \cdot U_0 \cdot a^{-1}}\mathbb{Z}G_g$, for a unit $a \in \mathbb{Q}G$ and $a \cdot U_0 \cdot a^{-1} = U \leq V(\mathbb{Z}G)$ iff*

1. *$M|_{\mathbb{Z}G}$ is right $\mathbb{Z}G$-free of rank one,*
2. *$\mathbb{Q}M$ is isomorphic as $(U_0 \times G^{op})$-bimodule to $\mathbb{Q}G$.*

*Remark*: The unit $a$ in $\mathbb{Q}G$ is by 1. and 2. only uniquely determined up to multiplication with a unit in $V(\mathbb{Z}G)$. More precisely, the bimodule class of ${}_{a \cdot U_0 \cdot a^{-1}}\mathbb{Z}G_g$ determines $a \cdot U_0 \cdot a^{-1}$ only up to conjugation with a unit in $V(\mathbb{Z}G)$. Thus the isomorphism classes of $(U_0 \times G^{op})$-bimodules satisfying 1. and 2. classify the conjugacy classes of the subgroups $U \leq V(\mathbb{Z}G)$, such that there exists a unit $a \in \mathbb{Q}G$ with $a \cdot U \cdot a^{-1} = U_0$.

In order to develop a theory similar to that one of central invertible bimodules for the above $U_0 \times G^{op}$-bimodules, we make the following definitions (cf. [Fr]).

**(3.22) Definition:** *Let $U_0$ be a fixed subgroup of $G$. Then*

$\mathrm{Pic}(U_0, \mathbb{Z}G)$ *are the isomorphism classes of $(U_0 \times G^{op})$-bimodules $M$ with $M|_{\mathbb{Z}G}$ projective of rank one.*

| | |
|---|---|
| $\mathrm{Piccent}(U_0, \mathbb{Z}G)$ | *are the isomorphism classes of* $(U_0 \times G^{op})$*-bimodules* $M$ *with* $M\|_{\mathbb{Z}G}$ *projective of rank one and with* $\mathbb{Q}M \simeq \mathbb{Q}G$ *as* $(U_0 \times G^{op})$*bimodule.* |
| $Out(U_0, \mathbb{Z}G)$ | *are the isomorphism classes of* $(U_0 \times G^{op})$*-bimodules* $M$ *with* $M\|_{\mathbb{Z}G}$ *free of rank one.* |
| $Out_c(U_0, \mathbb{Z}G)$ | *are the isomorphism classes of* $(U_0 \times G^{op})$*-bimodules* $M$ *with* $M\|_{\mathbb{Z}G}$ *free of rank one and with* $\mathbb{Q}M \simeq \mathbb{Q}G$ *as* $(U_0 \times G^{op})$*-bimodule.* |
| $C_{\mathbb{Z}G}(U_0)$ | *is the centralizer in* $\mathbb{Z}G$ *of* $U_0$. |
| $LFR(C_{\mathbb{Z}G}(U_0))$ | *are the isomorphism classes of locally free right* $C_{\mathbb{Z}G}(U_0)$*-ideals in* $C_{\mathbb{Z}G}(U_0)$. |
| $LFR_{\mathbb{Z}G}(C_{\mathbb{Z}G}(U_0))$ | *are the isomorphism classes of locally free right* $C_{\mathbb{Z}G}(U_0)$*-ideals* $A$ *in* $C_{\mathbb{Z}G}(U_0)$ *with* $A \cdot \mathbb{Z}G \simeq \mathbb{Z}G$ *as right* $\mathbb{Z}G$*-module.* |

In case $C_{\mathbb{Z}G}(U_0)$ is commutative, these are just the class groups with the appropriate conditions (cf. [Fr]). The above comments give the appropriate interpretations for subgroups of $U(\mathbb{Z}G)$. Similar definitions will be used for other coefficient domains.

Extending Fröhlich's localization sequence, we have

**(3.23) Lemma**: 1) *Let* $M$ *be a* $(U_0 \times G^{op})$*-bimodule with isomorphism class lying in* $\mathrm{Piccent}(U_0, \mathbb{Z}G)$ . *Then the elements* $(N) \in Piccent(U_0, \mathbb{Z}G)$*, which are locally isomorphic to* $M$ *as* $(U_0 \times G^{op})$*-bimodules are precisely the bimodules of the form* $N \simeq A \cdot M$*, where* $(A) \in LFR(C_{\mathbb{Z}G}(U_0))$.
2) *If* $(M) \in Out_c(U_0, \mathbb{Z}G)$*, then the bimodule classes* $(N) \in Out_c(U_0, \mathbb{Z}G)$*, which are locally isomorphic to* $M$ *are precisely the bimodules of the form* $N \simeq A \cdot M$ *for* $A \in LFR_{\mathbb{Z}G}(C_{\mathbb{Z}G}(U_0))$.

Since projective $\mathbb{Z}G$-modules are locally free, we have a natural isomorphism $\mathrm{Piccent}(U_0, \mathbb{Z}_p^{\wedge}G) \simeq Out_c(U_0, \mathbb{Z}_p^{\wedge}G)$. If $U_0 = G$, then the above result specializes to Fröhlich's localization sequence for invertible central bimodules.

The *proof of Lemma* (3.23) consists of an application of Jacobinski's theory of genera [Ja].

In order to apply (3.23) to special groups, we have to recall *Milnor's MayerVietoris sequence*, which is the algebraic analogue of Eilenberg-Steenrod's topological Mayer-Vietoris sequence:

Let

$$\begin{array}{ccc} \Lambda & \xrightarrow{\pi_1} & \Lambda_1 \\ \downarrow{\scriptstyle \pi_2} & & \downarrow{\scriptstyle \varphi_1} \\ \Lambda_2 & \xrightarrow[\varphi_2]{} & \Lambda' \end{array}$$

be a pullback diagram of rings and ring homomorphisms, and assume that at least one of the maps $\varphi_i$, $i = 1, 2$ is surjective.

**(3.24) Theorem (Milnor [MI])**: *Under the above assumptions the following sequence is exact:*

$$K_1(\Lambda) \to K_1(\Lambda_1) \oplus K_1(\Lambda_2) \to K_1(\Lambda') \\ \to K_0(\Lambda) \to K_0(\Lambda_1) \oplus K_0(\Lambda_2) \to K_0(\Lambda')\ ,$$

*where $K_1(\Lambda)$ is the Whitehead group of $\Lambda$ and $K_0(\Lambda)$ is the Grothendieck group of projective modules.*

In case of orders the result has been simplified by Reiner and Ullom [Re-Ul]:

**(3.25) Corollary:** *Let $R$ be a Dedekind domain with field of fractions $K$, and let $\Lambda$ be an $R$-order in a semi-simple $K$-algebra $A$, satisfying the Eichler condition. Assume further, that in the above pullback diagram (3.1) $\Lambda_1$ and $\Lambda_2$ are $R$-orders in semisimple $K$-algebras. $\Lambda'$ is a finite ring and either $\varphi_1$ or $\varphi_2$ is surjective. We put $U^*(\Lambda_i) = \varphi_i(U(\Lambda_i))$, where $U(\Lambda_i)$ denotes the unit group of $\Lambda_i$. Then there is an exact sequence*

$$\begin{aligned} &1 \longrightarrow U^*(\Lambda_1)\cdot U^*(\Lambda_2) \longrightarrow U(\Lambda') \\ &\xrightarrow{\delta} CL(\Lambda) \longrightarrow CL(\Lambda_1)\oplus CL(\Lambda_2) \longrightarrow 1 \quad . \end{aligned} \tag{3.26}$$

The map $\delta$ needs some explanations: If $\delta(u) = (\Lambda_u)$, $u \in U(\Lambda')$, then

$$\Lambda_u = \{(\lambda_1,\lambda_2) : \lambda_i \in \Lambda_i,\ \varphi_1(\lambda_1) = u\cdot\varphi_2(\lambda_2)\}\ .$$

Here $Cl(\Lambda)$ is the class group of projective one sided $\Lambda$-modules.

We point out, that a similar Mayer-Vietoris sequence for invertible bimodules has been used in [RS], and it has been formalized in [GR].

Before we come to explicit examples let us point out some consequences.

**(3.27) Remark:** Let $P$ be a finite $p$-group, and let $U$ be a finite subgroup of $V(\mathbf{Z}P)$; we want to classify the conjugacy classes in $V(\mathbf{Z}P)$ of such subgroups. Then $U$ is also a finite subgroup of $V(\mathbf{Z}_p^\wedge P)$, and so by Theorem I, $U$ is conjugate in $V(\mathbf{Z}_p^\wedge P)$ to a subgroup $U_0$ of $G$. The bimodules ${}_{U_0}\mathbf{Z}_p^\wedge G_G$ and ${}_U\mathbf{Z}_p^\wedge G_G$ thus differ by an automorphism of $U_0$. Now two bimodules $M_p$ and $N_p$ for $U_0 \times G^{op}$ over the local ring $\mathbf{Z}_p$ are isomorphic iff the corresponding $p$-adic completions are isomorphic. Consequently — up to automorphisms of $U_0$ — the conjugacy classes of finite subgroups $U$ of $V(\mathbf{Z}P)$ are parameterized by $Out_c(U_0, \mathbf{Z}P)$. Summarizing this discussion, we have

**(3.28) Lemma:** *Let $P$ be a $p$-group, then the conjugacy classes of finite subgroups $U$ of $V(\mathbf{Z}P)$ isomorphic to a finite subgroup $U_0$ of $G$ are parameterized — modulo automorphisms of $G$ and of $U$ — by $LFR_{\mathbf{Z}P}(C_{\mathbf{Z}G}(U_0))$.*

Using these observations, one can show:

**(3.29)** 1. Let $\Sigma_3$ be the symmetric group on three letters. In $V(\mathbf{Z}\Sigma_3)$ there is an involution $\iota$, which is not conjugate in $V(\mathbf{Z}\Sigma_3)$ to a group element in $\Sigma_3$; not even 2-adically $\iota$ is conjugate to a group element.

2. Let $D = \langle a, b : a^4 = 1, bab = a^{-1}\rangle$ be the dihedral group of order 8. In $V(\mathbf{Z}D)$ there are two conjugacy classes of group bases. Moreover, every finite subgroup of $V(\mathbf{Z}D)$ is part of a group basis.
3. Let $D = \langle a, b | a^8 = b^2 = 1, b\cdot a\cdot b = a^{-1}\rangle$ be the dihedral group of order 16. There exists an involution $\iota \in V(\mathbf{Z}D)$ which is not conjugate in $V(\mathbf{Z}D)$ to a group element in $D$, and which is not part of a group basis.

## § 4 Some remarks to the proofs

Questions about isomorphisms of group rings reduce to questions of automorphisms, by Kimmerle's trick [Ki]. The main result (3.13) is a consequence of the following statement:

Let $G$ be a finite group and $R$ an unramified extension of $\mathbb{Z}_p^\wedge$. Put $N = O_{p'}(G)$.

**(4.1) Theorem:** *Assume that $C_G(N) \subset N$ (note that this automatically implies $N \neq 1$). Let $\alpha$ be an augmented automorphism of $RG$ with $\alpha(N) \subset I_R(N) \cdot G$ — the induced augmentation ideal of $N$. Then $\alpha(G)$ is conjugate to $G$ in the units of $RG$.*

**(4.2) Remarks:** 1.) It turns out, that in the proof of (4.1) we do not really require that $\alpha$ stabilizes $I_R(N) \cdot G$, but only that it stabilizes $I_{\mathsf{F}}(N) \cdot G$, where $\mathsf{F} = R/rad\ R$.

2.) We do not know, whether the stabilization of either ideal in 1.) is automatic for an augmented automorphism $\alpha$ of $RG$. *This is indeed the case when* $N$ *is a Sylow p-subgroup*, since then $RG/N$ is the largest $R$-algebra homomorphic image of $RG$, which is torsionfree and whose reduction over the residue field $\mathsf{F}$ of $R$ is semi-simple. The obstruction lies in the fact, that we do not know, whether a class sum correspondence holds for classes of $p$-power elements.

3.) The theorem (4.1) is likely to be true for any complete discrete valuation domain of characteristic zero with residue field of characteristic $p$, assuming appropriate generalizations of the results of Weiss [W], which have been obtained by Gary Thompson [Th] and [R].

4.) The following criterion of when a group $G$ satisfies the hypotheses of (4.1) can be found in [HB, (13.5)].

**(4.3) Lemma:** *Let $G$ be a finite group. If there exists a normal p-subgroup $N$ with $C_G(N) \subset N$, then $O_{p'}(G) = 1$. Conversely, if $O_{p'}(G) = 1$, then for $N = O_p(G)$ we have $C_G(N) \subset N$, provided $G$ is p-constrained.*

**(4.4) Corollary 1:** *Let $G$ be a p-constrained group such that $O_{p'}(G) = 1$ for some prime p. Then the isomorphism problem and the Zassenhaus conjecture have a positive answer for $RG$* . (Any two normalized group bases are even $p$-adically conjugate.)

**(4.5) Corollary 2:** *Let $G$ be a finite group with a normal Sylow p-subgroup for some prime p. Then the defect group of the principal R-block of $G$ – $R$ is an unramified extension of $\mathbb{Z}_p^\wedge$ is uniquely determined, up to conjugacy in the block, by the principal block* (i.e., if $\alpha : RG \longrightarrow RG$ is an augmented automorphism, then $\alpha(P)$ and $P$ are conjugate in the principal block).

Again the assumption that $G$ has a normal Sylow $p$-subgroup is only to ensure that the augmentation ideal is stabilized by $\alpha$.

We shall now state the main ingredients of the proof of the theorem (4.1). There are three main ingredients in the proof.

The *first* one, which we already proved in April 1986 is:

**(4.6) Theorem:** *$G$ is a p-constrained group with $P$ a Sylow p-subgroup. $R$ is a complete Dedekind domain of characteristic zero with $pR \neq R$. Let $\alpha$ be a normalized automorphism of the principal block $B_0 = R(G/O_{p'}(G))$ of $RG$ — note that $B_0$ is an augmented algebra, the augmentation being induced from that of $RG$. Assume* *that $\alpha$ stabilizes the image of $P$ in $B_0$; i.e. $\alpha(P) = P$ in $B_0$ . Then $\alpha|_P$ — the restriction of $\alpha$ to $P$ — is induced from an automorphism $\rho$ of $G/O_{p'}(G)$, such that the automorphism induced from $\rho$ on $B_0$ agrees with $\alpha$, up to inner automorphisms centralizing $P$.*

The *proof is based* on the following two results:

**(4.7) Lemma** (Generalized Coleman result): *Let $G$ be a finite group, $P$ a $p$-subgroup of $G$, and $S$ an integral domain, in which $p$ is not invertible. Let $V = V(SG)$. Then we have for the normalizers*

$$N_V(P) = N_G(P) \cdot C_V(P) .$$

We shall be using the following *notation:* Given an automorphism $\alpha$ of $RG$, and an RG-bimodule $M$, we denote by ${}_\alpha M_1$ the bimodule, which has the original action on the right, but the left action is *twisted by* $\alpha$, i.e. $x \cdot m \cdot y = \alpha(x)my$, $x, y \in RG$, $m \in M$ [RMO, §37].

**(4.8) Lemma:** *$R$ is a complete Dedekind domain of characteristic zero with $pR \neq R$, $G$ is a finite group and $B$ a block of $RG$ with defect group $D$. Let $\alpha$ be an automorphism of $B$, which stabilizes the image of $D$ in $B$, $b$ is the Brauer correspondent in $N = N_G(D)$ to $B$. Then the Green correspondent to the twisted module ${}_\alpha B_1$ on $G \times G$, is ${}_\beta b_1$ on $N \times N$ for some automorphism $\beta$ of $RN$. Moreover, $\beta|_D = \alpha|_D$, where we regard $D \subset b \subset B$.*

The *second main ingredient* is the following: Let $R$ be an unramified extension of $\mathbb{Z}_p^\wedge$ with residue field $\mathsf{F}$. Let $N$ be a normal $p$-subgroup of a finite group $G$, and assume $C_G(N) \subset N$.

**(4.9) Theorem:** *Let $G$ be an augmented automorphism of $RG$, with $N$ and $G$ as above and assume that $\alpha$ stabilizes the ideal $I_{\mathsf{F}}(N) \cdot G$ in $\mathsf{F}G$. Then $\alpha(N)$ is conjugate to $N$ by a unit of $RG$.*

The *third main ingredient* is an extension of results of Weiss [W] combined with a criterion on the indecomposability of permutation modules, which is the main tool to show that group bases are conjugate. We let $R$ be an unramified extension of $\mathbb{Z}_p^\wedge$ the $p$-adic integers. $\mathsf{F} = R/pR$ is the residue field of $R$, and we denote by $x \longrightarrow \bar{x}$ the natural projection from $R$ to $\mathsf{F}$, and also for any $R$-module $X$ we put $\bar{X} = X/pX$.

**(4.10) Theorem:** *Assume that $P$ and $Q$ are $p$-groups. Let $M$ be an $R(P \times Q)$-module* (with the action $(x, y) \cdot m = x \cdot m \cdot y^{-1}$, $x \in P$, $y \in Q$, $m \in M$), *which is free of finite rank as $RQ$-module. If $\overline{M/(M \cdot I_R(Q))}$ is a permutation module for $\mathsf{F}P$, then $\bar{M}$ is a permutation module for $\mathsf{F}(P \oplus Q)$.*

The next result is the place where we use that $C_G(N) \subset N$.

**(4.11) Lemma:** *Suppose $C_G(N) \subset N$. Then the $\mathsf{F}(N \times G)$-module $\mathsf{F}G$ — with the action $(n, g) \cdot x = n \cdot x \cdot g^{-1}$, $n \in N$, $g \in G$, $x \in \mathsf{F}G$ — is absolutely indecomposable. Moreover, the same is true if the action of $N$ on the left is twisted by an augmented $\mathsf{F}$-algebra automorphism $\alpha$ of $G$, or by a group automorphism $\rho$ of $N$ — i.e., $(n, g) \cdot x = \rho(n) \cdot x \cdot g^{-1}$ for $n \in N$, $g \in G$ and $x \in \mathsf{F}G$.*

**(4.12) Corollary:** *If the finite group $G$ has a normal $p$-subgroup $N$ with $C_G(N) \subset N$, then $RG$ consists of a single $p$-block.*

We shall next come to the proof of the "Subgroup rigidity theorem" (3.16) as it was generalized in [R6]. We shall use the following notation:

$R$ is a complete Dedekind domain of finite rank over $\mathbf{Z}_p^\wedge$, the $p$-adic integers,
$K$ is the field of fractions of $R$,
$\pi$ is a parameter of $R$; i.e. $\mathrm{rad}(R) = \pi \cdot R$,
$R^\times$ is the multiplicative group of units of $R$,
$R/\pi^t$ is the reduction of $R$ modulo $\pi^t$ for some fixed $t \in \mathbb{N}$,
$M/\pi^t$ is the reduction of an $R$-module $M$ modulo $\pi^t$,
$G$ is a finite $p$-group,
$\chi_i^U$ are the various linear characters from the subgroup $U$ of $G$ to $R^\times$; i.e. homomorphisms $U \to R^\times$, $1 \leq i \leq n(U)$,
$\chi_i^U/\pi^t$ are the various characters from $U$ to $(R/\pi^t)^\times$, *which are induced from the* $\chi_i^U$ (note that different $\chi_i^U$'s may reduce to the same $\chi_i^U/\pi^t$),
$I_i^U(R)$ is the kernel of the map from $RU$ to $R$ induced by $\chi_i^U$. This kernel is generated over $R$ by $\{u - \chi_i^U(u)\}_{u \in U\setminus\{1\}}$,
$R\chi_i^U \uparrow_U^G$ is the corresponding transitive generalized permutation module for $G$.

The notation *transitive generalized permutation* module for $R/\pi^t G$ implies by the above, *that the $U$-action is induced from an action on $R$. A generalized permutation module* is a direct sum of transitive generalized permutation modules, for various subgroups. A similar notation is used for other rings of coefficients.

**(4.13) Theorem** (Subgroup rigidity): *Let $V(RG)$ be the units in $RG$ of augmentation one, and $G$ a $p$-group. If $V$ is a finite subgroup of $V(RG)$, then*

$$u \cdot V \cdot u^{-1} \subset G \text{ for some unit } \quad u \text{ in } RG.$$

The above result will be an immediate consequence of

**(4.14) Theorem:** *Let $M$ be an $RG$-lattice, and let $N$ be a normal subgroup of $G$. Assume that*
a) *$M \downarrow_n$ is a free $RN$-module*
b) *$M/N \cdot M$ is a permutation module for $G/N$, then $M$ is an $RG$-permutation module.*

We shall derive now (4.13) from (4.14), since the argument is very short [W]):

Let $V$ be a finite subgroup in $V(RG)$. We let $G^\vee = V \times G$, and consider $M = RG$ as an $RG^\vee$-module via the action

$$(v, g) \cdot m = v \cdot m \cdot g^{-1}.$$

Then (4.14) can be applied ($N = G$); note the point, where we use that $V$ is a group of normalized units: as $V$-module $M/(I)G \cdot M = R$ is trivial only if $V$ is normalized. Thus so $M$ must be a transitive permutation module, $M \simeq R\uparrow_H^{G^\vee}$, where $H \cap G = 1$, and $|H| = |V|$; thus $h = (v_h, g_h)$, $v_h \in H$, $h_h \in G$, and so, if $m_0$ corresponds in $M$ to $1 \in R$, then $m_0$ is a unit in $M = RG$, and $v_h \cdot m_0 \cdot g_h^{-1} = m_0$; i.e. $m_0^{-1} \cdot v_h \cdot m_0 \in G$ for every $v_h \in V$; but $\{v_h\}_{h \in h} = V$. Whence the statement.

The next result plays a major role in the proof of (4.14), though it is also of interest for its own sake.

**(4.15) Theorem:** 1.) *Assume that $K$ is a splitting field for $G$ and all of its subgroups. Let $\zeta$ be a primitive p-th root of unity, then we require $\pi^t \cdot R \subset (1-\zeta)\cdot R$. Then an RG-lattice $M$ is a generalized permutation module if and only if $M/\pi^t$ is a generalized permutation module; note that we are always assuming, that the various U-actions on $R/\pi^t$ are induced from the U-actions on $R$.*

2.) *If $\pi^t \cdot R \not\subset (1-\zeta)\cdot R$, then there is an RG-lattice $M$, which is not a generalized permutation module, but $M/\pi^t$ is a generalized permutation module.*

Let us point out, where the restrictive condition on $R/\pi^t$ becomes apparent:

**(4.16) Lemma:** *Let $R$ be a complete Dedekind domain of finite rank over $\mathbb{Z}_p^\wedge$. The induced map*

$$\kappa_g : Ext^1_{RG}(M,N) \longrightarrow Ext^1_{R/\pi^t G}(M/\pi^t, N/\pi^t)$$

*is injective for all generalized permutation lattices $M$ and $N$, if and only if either $R$ does not contain a primitive p-th root of unity or $\pi^t \cdot R \subset (1-\zeta)\cdot R$, where $\zeta$ is a primitive p-th root of unity.*

## REFERENCES


[Ba] Banachevski, B. , Integral group rings of finite groups. Can. Math. Bull. 10 (1967), pp.635-642.

[Bas] Bass, H. , Algebraic $K$-theory. W.A. Benjamin, New York, 1968.

[Be] Bender, H. , A group theoretic proof of Burnside's $p^\alpha \cdot q^\beta$ theorem. Math. Z. 126 (1972), pp.327-338.

[BR] Berman, S.D.; Rossa, A.R. , Integral group rings of finite and periodic groups. Algebra and Math. Logic: Studies in Algebra, pp. 44-53, Izdat. Kiew (1966).

[Ber1] Berman, S.D. , On a necessary condition for isomorphisms of integral group rings. Dopovidi Akad. Nauk Ukrain. RSR (1953), pp. 313-316.

[Ber2] Berman, S.D. , On certain properties of integral group rings. (Russian) Dokl. Akad. Nauk SSSR (N.S.) 91 (1953), pp.7-9.

[Ber3] Berman, S.D. , On a necessary condition for isomorphism of integral group rings. (Ukrainian) Dopovidi Akad. Nauk Ukrain. RSR (1953), pp.313-316.

[Ber4] Berman, S.D. , On the equation $x^m = 1$ in an integral group ring. (Russian) Ukrain. Math. Z. 7 (1955), pp.253-261.

[Ber5] Berman, S.D. , On automorphisms of the centre of an integral group ring. (Russian) Dikl. Soobshch. Uzhgorod. Gos. Univ. Ser. Fiz.-Mat. Nauk (1960) no.3, 55.

[Bo1] Bovdi, A.A. , Remarks on the automorphisms of a group ring. (Russian) Latv. Mat. Ezhegodnik 18 (1976), pp.19-26.

[Bo2] Bovdi, A.A. , The structure of group bases of a group ring. (Russian) Mat. Zametki 32 (1982), $pp.459 - 468 = Math.$ Notes 32 (1982), pp.709-713.

[BP] Bovdi, A.A.; Patai, Z.F. , The structure of the centre of the multiplicative group of the group ring of a $p$-group over a ring of characteristic $p$. (Russian) Vestsi Akad. Nauk BSSR Ser. Fiz.-Mat. Nauk (1987), N=no.1, pp.5-11.

[BS1] Bovdi, A.A.; Semirot, M.C. , Involution and isomorphism of group rings. (Russian) Twelfth All-Union Algebra Colloquium, Sverdlovsk (1973).

[BS2] Bovdi, A.A.; Semirot, M.C. , On the question of the isomorphism of modular group algebras. (Ukrainian) Abstracts. Second Conf. of Young Scientists, Akad. Nauk USSR, Uzhgorod (1975), dep VINITI, No.1734-76, pp.48-54.

[Br] Brauer, R. , Representations of finite groups. Lectures on Modern Mathematics. Vol.1, Wiley, New York (1963), pp.133-175. Reproduced in Vol.1 and 2 of: Richard Brauer, Collected Papers, MIT Press, Cambridge, MA (1980).

[Bro] Brown, C.F. , Automorphisms of integral group rings. Ph.D. thesis, Michigan State University (1971), DAI32 (1972), p.5302B.

[Ca] Carlson, J. , The variety of a module. In: Lecture Notes in Maths. Springer Verlag 1142 (1985), pp.88-95.

[CSW] Cliff, G.H.; Sehgal, S.K.; Weiss, A.R. , Units of integral group rings of metabelian groups. J. Algebra 73 (1981), 167-185.

[Co1] Coleman, D.B. , Finite groups with isomorphic group algebras. Ph.D. thesis, Purdue Univ. (1961), DA22 (1962), p.2405.

[Co2] Coleman, D.B. , Finite groups with isomorphic group algebras. Trans. Amer. Math. Soc. 105 (1962), pp.1-8.

[Co3] Coleman, D.B. , On the modular group ring of a $p$-group. Proc. Amer. Math. Soc. 15 (1964), pp.511-514.

[CR] Curtis, C.W.; Reiner, I. , Methods in representation theory I. John Wiley (1981).

[Da] Dade, E. , Deux groupes finis distintes ayant la même algèbre de groupes sur tout corps. Math. Z. 119 (1971), pp.345-348.

[De] Dennis, R.K. , The structure of the unit group of group rings. Ring theory II, Dekker, New York (1977), pp.103-130.

[Di] Dieudonné, J. , Lie groups; classical, algebraic and formal. Lecture, 21st British Mathematical Colloquium, Birmingham, 28 March 1969.

[EMS] Endo, S.; Miyata, T.; Sekiguchi, K. , Picard groups and automorphism groups of integral group rings of metacyclic groups. J. Algebra 77 (1982), 286-310.

[F] Feit, W. , The representation theory of finite groups. North Holland, New York (1982).

[Fr] Froehlich, A. , The Picard group of non-commutative rings, in particular of orders. Trans. Amer. Math. Soc. 180 (1973), 1-45.

[Hi] Higman, G. , The units of group-rings. Proc. London Math. Soc. (2) 46 (1940), pp.231-248.

[Hi1] Higman, G. , Units in group rings. D. Phil. theses Oxford University (1940).

[Hp] Hughes, I.; Pearson, K.E. , The group of units of the integral group ring $\mathbb{Z}S_3$. Canad. Math. Bull 15 (1972), 529-543.

[HB] Huppert, B.; Blackburn, N. , Finite groups II. Springer Verlag (1982).

[J] Jackson, D.A. , The groups of units of the integral group rings of finite metabelian and finite nilpotent groups. Quat. J. Math. Oxford, Ser.2, 20 (1969), pp.319-331.

[Ki] Kimmerle, W. , Personal communication 1985.

[KLS] Kimmerle, W.; Lyons, R.; Sandling, R. , Composition factors for group rings and Artin's theorem on orders of simple groups. Preprint (1987), to appear in Proc. London Math. Soc.

[P] Puig, L. , Pointed groups and construction of characters. Math. Z. 176 (1981), pp.265-292.

[PS] Passman, D.S.; Smith, P.F. , Units in integral group rings. J. Algebra 69 (1981), pp.213-239.

[Pe1] Peterson, G.L. , Automorphisms of the integral group ring of $S_n$. Proc. Amer. Math. Soc. 59 (1976), pp.14-18.

[Pe2] Peterson, G.L. , On the automorphisms group of an integral group ring I. Arch. Math. (Basel) 28 (1977), 577-583.

[Pe3] Peterson, G.L. , On the automorphisms group of an integral group ring II. Illinois J. Math. 21 (1977), 836-844.

[Qu1] Quillen, D.G. , On the associated graded ring of a group ring. J. Algebra 10 (1968), pp.411-418.

[Qu2] Quillen, D.G. , The spectrum of an equivariant cohomology ring I. Ann. of Math. (2) 94 (1971), pp.549-572.

[Qu3] Quillen, D.G. , The spectrum of an equivariant cohomology ring II. Ann. of Math. (2) 94 (1971), pp.573-602,MR457743.

[Qu4] Quillen, D.G. , A cohomological criterion for $p$-nilpotence. J. Pure Appl. Algebra 1 (1971), 361-372.

[QV] Quillen, D.G.; Venkov, B.B. , Cohomology of finite groups and elementary abelian subgroups. Topology 11 (1972), 317-318.

[Rei] Reiner, I. , Maximal orders, Academic Press (1975).

[Rey] Reynolds, W.F. , Blocks and normal subgroups of finite groups, Nagoya Math. J. 22 (1963), pp.15-32.

[Ri] Ritter, J. , On a Zassenhaus conjecture about units in group rings. Lecture, Orders and Their Applications, Oberwolfach, 7 June 1984.

[RiS] Ritter, J.; Sehgal, S. , Isomorphism of group rings. Arch. Math. (Basel) 40 (1983), pp.32-39.

[RiSe] Ritter, J.; Sehgal, S. , On a conjecture of Zassenhaus on torsion units in integral group rings. Preprint Univ. Augsburg (1983).

[RMO] Reiner, I. , Maximal orders. Academic Press (1975).

[R1] Roggenkamp, K.W. , Picard groups of integral group rings of nilpotent groups. MS (Nov.1986), to appear in the Proceedings of the Arcata meeting of the AMS, July 1986.

[R2] Roggenkamp, K.W. , Units in integral metabelian group rings I, Jackson's unit theorem revisited. Quat. J. Math. 23 (1981), pp.209-224.

[R3] Roggenkamp, K.W. , The isomorphism problem and units in group rings of finite groups. Groups – St. Andrews 1981, pp.313-327, Cambridge (1982).

[R4] Roggenkamp, K.W. , Automorphisms and isomorphisms of integral group rings of finite groups. Groups – Korea 1983, pp.118-135, Lecture Notes in Math. 1098, Springer, Berlin (1984).

[R5] Roggenkamp, K.W. , Isomorphisms of $p$-adic group rings I. Lecture, Orders and Their Applications, Oberwolfach, 4 June 1984.

[R6] Roggenkamp, K.W. , $p$-adic rigidity of group rings. Weiss' arguments revisited. MS (1990).

[RS] Roggenkamp, K.W.; L.L. Scott , Isomorphisms of $p$-adic group rings. MS, Sept. 1985, Annals of Mathematics 126 (1987), pp.593-647.

[RS1] Roggenkamp, K.W.; L.L. Scott , The isomorphism theorem for integral group rings of nilpotent by abelian groups. MS, May 1986, pp.1-14.

[RS2] Roggenkamp, K.W.; L.L. Scott , On a conjecture on group rings by H. Zassenhaus. MS (March 1987), pp.1-39.

[RS3] Roggenkamp, K.W.; L.L. Scott , A strong answer to the isomorphism problem for finite $p$-solvable groups with a normal $p$-subgroup containing its centralizer. MS (1987).

[RS4] Roggenkamp, K.W.; L.L. Scott , Non-splitting examples for normalized units in integral group rings of metacyclic Frobenius groups. C.R. Math. Rep. Acad. Sci. Canada 3 (1981), 29-32.

[RS5] Roggenkamp, K.W.; L.L. Scott , Units in metabelian group rings; non-splitting examples for normalized units. J. Pure Appl. Algebra 27 (1983), pp.299-314.

[RS6] Roggenkamp, K.W.; L.L. Scott , Units in metabelian group rings; non-splitting examples for normalized units. Addendum. J. Pure Appl. Algebra 28 (1983), pp.109.

[RS7] Roggenkamp, K.W.; L.L. Scott , Units in group rings; Splittings and the isomorphism problem. To appear (1985).

[Sa1] Sandling, R. , The isomorphism problem for group rings: A survey. Springer Lect. Notes Math. 1142, (1985), pp.239-255.

[Sa2] Sandling, R. , Note on the integral group ring problem. Math. Z. 124 (1972), pp.255-258.

[Sa3] Sandling, R. , Group rings of circle and unit groups. Math. Z. 140 (1974), pp.195-202

[Sa4] Sandling, R. , Graham Higman's thesis "Units in group rings". Integral Representations and Applications. Lect. Notes in Math. 882, Springer, Berlin (1981), pp.93-116.

[S1] Scott, L.L. , Report on the isomorphism problem. to appear in the Proceedings of the Arcata Meeting of the AMS, July 1986.

[S1] Scott, L.L. , The modular theory of permutation representations. Representation theory of finite groups and related topics. AMS Meeting at Madison, April 1970, pp.137-144.

[S1] Scott, L.L. , Isomorphisms of $p$-adic group rings II. Lecture, Orders and Their Applications, Oberwolfach, 5 June 1984.

[Se1] Sehgal, S.K. , Topics in group rings. Marcel Dekker, New York (1978).

[Se2] Sehgal, S.K. , On the isomorphism of group algebras. Math. Z. 95 (1967), pp.71-75.

[Se3] Sehgal, S.K. , On the isomorphism of integral group rings I. Cand. J. Math. 21 (1969), pp.410-413.

[Se4] Sehgal, S.K. , On the isomorphism of integral group rings II. Cand. J. Math. 21 (1969), pp.1182-1188.

[Se5] Sehgal, S.K. , Isomorphism of $p$-adic group rings. J. Number Theory 2 (1970), pp.500-508.

[Se6] Sehgal, S.K. , Topics in group rings. Dekker, New York (1978).

[Se1] Sehgal, Sudarshan K.; Sehgal, Surinder K.; Zassenhaus, H.J. , Isomorphism of integral group rings of abelian by nilpotent class two groups. Comm. Algebra 12 (1984), pp.2401-2407.

[Sek] Sekiguchi, K. , On the units of integral group rings. Tokyo J. Math. 3 (1980), pp.149-162.

[Ta] Takahashi, S. , A characterization of group rings as a special class of Hopf algebras. Com. Math. Bull. 8 (1965), pp.465-475.

[Wa] Ward, H.N. , Some results on the group algebra of a group over a prime field. Seminar on Finite Groups and Related Topics, Mimeographed Notes, Harvard Univ. (1960-1961), pp.13-19.

[W] Weiss, A. , $p$-adic rigidity of $p$-torsion. To appear Ann. Math.

[We] Weller, W.R. , The units of the integral group ring $ZD_4$ . D.Ed. thesis, Pennsylvania State Univ., 1972. DAI33 (1973), pp.4921B.

[Wh] Whitcomb, A. , The group ring problem. Ph.D. thesis, Chicago (1968).

[Za] Zassenhaus, H. , On the torsion units of finite group rings. Studies in Math., Instituto de Alta Cultura, Lisboa (1974), pp.119-126.

Progress in Mathematics, Vol. 95, © 1991 Birkhäuser Verlag Basel

# Cohen-Macaulay and Gorenstein Artin algebras

MAURICE AUSLANDER AND IDUN REITEN

## Introduction

The notions of Cohen-Macaulay and Gorenstein rings, as well as Cohen-Macaulay modules, are well established in commutative noetherian ring theory. In [6] we introduced similar notions for artin algebras. This paper is devoted to studying further the module theory of Cohen-Macaulay and Gorenstein artin algebras. As might be expected, much of the motivation for our work comes from the better developed theory for commutative rings. In fact, it is somewhat surprising how much of the commutative theory carries over to artin algebras with only minor changes.

We begin by giving a definition of Cohen-Macaulay artin algebras which is somewhat different, but equivalent, to the earlier one given in [6]. This new definition is motivated by the fact that a commutative noetherian complete local ring $R$ is Cohen-Macaulay if and only if $R$ has a dualizing module $\omega$. If $R$ is a Cohen-Macaulay ring, then the dualizing module $\omega$ has the following properties which uniquely determine it up to isomorphism. The functors $\mathrm{Hom}(\omega, \ ) : \mathrm{mod}R \to \mathrm{mod}R$ and $\omega\otimes : \mathrm{mod}R \to \mathrm{mod}R$, where $\mathrm{mod}R$ is the category of finitely generated $R$-modules, induce inverse equivalences between the full subcategories of $\mathrm{mod}R$ consisting of the modules of finite injective dimension and the modules of finite projective dimension. This suggests defining an artin algebra $\Lambda$ to be Cohen-Macaulay if there is a pair of adjoint functors $(F, G)$ from $\mathrm{mod}\Lambda$ to $\mathrm{mod}\Lambda$ which induce inverse equivalences between the full subcategories of $\mathrm{mod}\Lambda$ consisting of the $\Lambda$-modules of finite injective dimension and the $\Lambda$-modules of finite projective dimension. It follows that an artin algebra $\Lambda$ is Cohen-Macaulay if and only if there is a bimodule ${}_\Lambda\omega_\Lambda$ such that the pair of adjoint functors $(\omega_\Lambda\otimes\ ,\ \mathrm{Hom}_\Lambda({}_\Lambda\omega,\ ))$ has our desired properties. Such a bimodule ${}_\Lambda\omega_\Lambda$, if it exists, is called a dualizing module for $\Lambda$.

For artin algebras there is an extensive theory of tilting and cotilting modules which plays an important role in their representation theory. Dualizing modules can be characterized as bimodules which are special types of cotilting modules viewed both as left and as right modules. This connection between dualizing and cotilting modules gives an interesting interplay between cotilting theory for artin algebras and the module theory for commutative Cohen-Macaulay rings. This interplay provides much of the inspiration for this paper.

For a commutative Cohen-Macaulay ring $R$, the Cohen-Macaulay modules play a crucial role. These modules $C$ are characterized by $\mathrm{Ext}^i_R(C,\omega) = 0$ for $i > 0$, where $\omega$ is the dualizing module for $R$. This suggests the following definition. Suppose $\Lambda$ is a Cohen-Macaulay artin algebra with dualizing module ${}_\Lambda\omega_\Lambda$. A $\Lambda$-module $C$ is said to be a Cohen-Macaulay module if $\mathrm{Ext}^i_\Lambda(C,\omega) = 0$ for $i > 0$. As in the commutative theory, Cohen-Macaulay modules play a special role in the study of Cohen-Macaulay artin algebras.

While Cohen-Macaulay rings play an important role in commutative ring theory, it is the Gorenstein rings, those rings $R$ for which $R$ is a dualizing module, for which the module theory is most highly developed. It is therefore not too surprising that most of this paper is devoted to studying Gorenstein artin algebras, namely those artin algebras $\Lambda$ for which $\Lambda$, viewed as a twosided $\Lambda$-module, is a dualizing module.

We now turn our attention to giving a more detailed description of the various sections.

Section 1 is devoted to basic definitions and results which are used throughout the paper. Cohen-Macaulay and Gorenstein artin algebras are defined as well as dualizing and Cohen-Macaulay modules. How these concepts are connected with commutative ring theory and cotilting modules is explained.

In section 2 we give examples of Cohen-Macaulay and Gorenstein artin algebras, including constructions which preserve these properties.

Section 3 is devoted to studying Cohen-Macaulay modules over Gorenstein artin algebras. It is pointed out that the full subcategory of Cohen-Macaulay modules has almost split sequences since it is functorially finite and closed under extensions, and we describe the ends of these types of sequences in terms of each other. Motivated by the commutative theory of Gorenstein rings, we investigate when the Cohen-Macaulay approximation of a module as introduced by Auslander - Buchweitz [3], has no nonzero projective summands. We also point out that for Gorenstein artin algebras, contrary to the commutative situation, the full subcategory of mod$\Lambda$ consisting of modules of finite projective dimension is functorially finite.

In section 4 we recall the notion of finite $G$-dimension from commutative ring theory and define this notion for artin algebras. Using tilting theory, it is shown that an artin algebra is Gorenstein if and only if the finitely generated modules have finite $G$-dimension, a result already shown for commutative rings in [2], and with a different proof in [12].

We finally remark that other aspects of Gorenstein artin algebras are studied in [11], and noncommutative Gorenstein rings are studied in [8][13].

## 1. Definition of Cohen-Macaulay artin algebras and modules

In [6] the notions of Cohen-Macaulay and Gorenstein artin algebras were introduced. This section is primarily devoted to giving alternative definitions of these notions which seem more natural. Since the previous as well as the present approach to this subject were motivated by the theory of commutative noetherian complete local Cohen-Macaulay and Gorenstein rings, we begin by recalling some of the basic features of this theory. After giving the new definitions we establish connections with tilting theory.

Let R be a commutative noetherian complete local ring, and denote by mod$R$ the category of finitely generated R-modules. For ease of exposition we assume that R contains a formal power series ring S = $k[[X_1, \dots, X_n]]$ with k a field, such that R is a finitely generated S-module, even though this assumption is not essential to the theory. The ring R is said to be a Cohen-Macaulay ring if it is a free S-module. Assume from now on that R is Cohen-Macaulay. A finitely generated R-module is said to be a Cohen-Macaulay module if it is free when viewed as an S-module. We denote the full subcategory of mod R consisting of the Cohen-Macaulay R-modules by CM(R). In particular, the R-module $\omega = \mathrm{Hom}_S(R, S)$ is a Cohen-Macaulay R-module, called the dualizing module, and it plays a critical role in the study of Cohen-Macaulay rings. For example, R is said to be Gorenstein if $\omega$ is isomorphic to R.

The following properties of Cohen-Macaulay rings and modules serve as the principal motivation for our definitions of Cohen-Macaulay artin algebras and Cohen-Macaulay modules over such algebras.

**Proposition 1.1** Let R be a Cohen-Macaulay ring as above, and $\omega$ the dualizing module.

(a) $CM(R) = \{C; \mathrm{Ext}^i_R(C, \omega) = 0 \text{ for } i > 0\}$, and $\mathrm{Hom}_R(\,, \omega) : CM(R) \to CM(R)$ is a duality.

(b) $\mathrm{Hom}_R(\omega, \,) : \mathrm{mod} R \to \mathrm{mod} R$ induces an equivalence between the category $\mathcal{I}^\infty\,(R)$ of finitely generated R-modules of finite injective dimension and the category $\mathcal{P}^\infty\,(R)$ of finitely generated R-modules of finite projective dimension.

(c) R is Gorenstein if and only if $id_R R < \infty$.

Property (b) of Proposition 1.1 suggests the following definition. We say that an artin algebra $\Lambda$ is *Cohen-Macaulay* if there is a pair of adjoint functors (G,F) between mod$\Lambda$ and mod$\Lambda$, inducing inverse equivalences

$$\mathcal{I}^\infty\,(\Lambda) \underset{G}{\overset{F}{\rightleftarrows}} \mathcal{P}^\infty\,(\Lambda).$$

If $\Lambda$ is Cohen-Macaulay and (G,F) is an associated pair of adjoint functors, then F is left exact and given by $F = \mathrm{Hom}_\Lambda({}_\Lambda\omega_\Lambda, \,)$. We say that the bimodule ${}_\Lambda\omega_\Lambda$ is a *dualizing module.* Then G is given by ${}_\Lambda\omega_\Lambda \otimes_\Lambda$ . Further we say that $\Lambda$ is *Gorenstein* if ${}_\Lambda\Lambda_\Lambda$ is a dualizing module. This is equivalent to $\mathcal{I}^\infty\,(\Lambda) = \mathcal{P}^\infty\,(\Lambda)$, which again is clearly equivalent to $id_\Lambda \Lambda < \infty$ and $id \Lambda_\Lambda < \infty$, which was taken as a definition of Gorenstein algebra in [6].

Finally, for a Cohen-Macaulay algebra $\Lambda$ we define the category $CM(\Lambda)$ of *Cohen-Macaulay modules* to be ${}^\perp\,\mathcal{I}^\infty\,(\Lambda)$. For a subcategory $\mathcal{X}$ of mod$\Lambda$ we denote by ${}^\perp\,\mathcal{X}$ the category whose objects are the C with $\mathrm{Ext}^i_\Lambda(C, X) = 0$ for $i > 0$ and $X$ in $\mathcal{X}$. We shall see later that $CM(\Lambda) = {}^\perp\omega$ when ${}_\Lambda\omega_\Lambda$ is a dualizing module.

Our aim now is to show that dualizing modules for Cohen-Macaulay artin algebras are special types of cotilting modules. We begin by recalling that a $\Lambda$-module T is a cotilting module if (a) $\mathrm{Ext}^i_\Lambda(T, T) = 0$ for $i > 0$, (b) $id_\Lambda T < \infty$ and (c) $D\Lambda$ is in $\widehat{\mathrm{add}}\, T$. Here $id_\Lambda T$ denotes the injective dimension of T, D denotes the ordinary duality for artin algebras and add$T$ denotes the full subcategory of mod$\Lambda$ whose objects are the direct summands of finite direct sums of copies of T. For a subcategory $\mathcal{X}$ of mod$\Lambda$, $\hat{\mathcal{X}}$ denotes the category whose objects are the C in mod$\Lambda$ such that there is an exact sequence $0 \to X_n \to \ldots \to X_0 \to C \to 0$ with $X_i$ in $\mathcal{X}$.

As noted in [6], it is easily seen that condition (c) is equivalent to condition (c′): the subcategory $\widehat{\operatorname{add}} T$ is a cogenerator for modΛ. This means that for every C in $\widehat{\operatorname{add}} T$ there is a exact sequence $0 \to C \to T_0 \to B \to 0$ with $T_0$ in add$T$ and B in $\widehat{\operatorname{add}} T$.

Using conditions (a), (b) and (c′) as a description of cotilting modules, we observed in [6] that the dualizing module $\omega$ for a Cohen-Macaulay commutative local ring R is a cotilting module. This suggests that dualizing modules for Cohen-Macaulay artin algebras are also cotilting modules. To show that this is indeed the case, it is helpful to recall from [6] that a cotilting module T over an artin algebra Λ is said to be a strong cotilting module if $\widehat{\operatorname{add}} T = \mathcal{I}^{\infty}(\Lambda)$. Dually, a Λ-module T is a tilting module if DT is a cotilting module, and T is a strong tilting module if DT is a strong cotilting $\Lambda^{op}$-module. Before we characterize dualizing modules over artin algebras in terms of strong cotilting modules we prove the following general result. We recall that a subcategory of modΛ is resolving if it contains the projectives and is closed under extensions and kernels of epimorphisms, and is coresolving if it contains the injectives and is closed under extensions and cokernels of monomorphisms.

**Proposition 1.2.** Let $\mathcal{X}$ be a resolving subcategory of mod Λ closed under cokernels of monomorphisms and $\mathcal{Y}$ a coresolving subcategory of modΛ closed under kernels of epimorphisms.

Let (G,F) be an adjoint pair of functors between modΛ and modΛ, and assume that there are induced inverse equivalences of categories $F : \mathcal{Y} \to \mathcal{X}$ and $G : \mathcal{X} \to \mathcal{Y}$. Then we have the following.

(a) F is exact on exact sequences whose terms are in $\mathcal{Y}$, and G is exact on exact sequences whose terms are in $\mathcal{X}$.

(b) $\operatorname{Ext}^i_\Lambda(\mathcal{X}, F(D\Lambda)) = 0$ for $i > 0$.

(c) $\operatorname{Ext}^i_\Lambda(G(\Lambda), \mathcal{X}) = 0$ for $i > 0$.

(d) We have natural isomorphisms $\operatorname{Ext}^i_\Lambda(B, C) \xrightarrow{\sim} \operatorname{Ext}^i_\Lambda(FB, FC)$ for all $i > 0$ and B,C in $\mathcal{Y}$.

(e) We have natural isomorphisms $\operatorname{Ext}^i_\Lambda(B, C) \xrightarrow{\sim} \operatorname{Ext}^i_\Lambda(GB, GC)$ for all $i > 0$ and $B, C$ in $\mathcal{X}$.

**Proof** (a) Let $0 \to A \to B \to C \to 0$ be an exact sequence in modΛ, with A, B and C in $\mathcal{Y}$. Since F is left exact, we have the exact sequences $0 \to FA \to FB \to X \to 0$ and $0 \to X \to FC \to Z \to 0$. Here X and Z are in $\mathcal{X}$ since $\mathcal{X}$ is closed under cokernels of monomorphisms. Since G is right exact and F and G are inverse equivalences between $\mathcal{X}$ and $\mathcal{Y}$, we have the exact commutative diagram

$$\begin{array}{ccccccccc} & & GFA & \longrightarrow & GFB & \longrightarrow & GX & \longrightarrow & 0 \\ & & \downarrow\wr & & \downarrow\wr & & \downarrow\wr & & \\ 0 & \longrightarrow & A & \longrightarrow & B & \longrightarrow & C & \longrightarrow & 0. \end{array}$$

Since $GX \to GFC \to GZ \to 0$ is exact and $GX \to GFC \to C$ and $GFC \to C$ are isomorphisms, we get $GZ = 0$. Then we have $Z \simeq FGZ = 0$, so that $0 \to FA \to$

$FB \to FC \to 0$ is exact.

If $0 \to A \to B \to C \to 0$ is exact in mod$\Lambda$ with A,B and C in $\mathcal{X}$, it follows similarly that $0 \to GA \to GB \to GC \to 0$ is exact.

(b) Let $0 \to F(D\Lambda) \to E \to X \to 0$ be an exact sequence with X in $\mathcal{X}$. Since $\mathcal{Y}$ is coresolving, D$\Lambda$ is in $\mathcal{Y}$, so that F(D$\Lambda$) is in $\mathcal{X}$. Using that $\mathcal{X}$ is closed under extensions, we have that E is in $\mathcal{X}$. By (a) we then get an exact sequence $0 \to GF(D\Lambda) \to GE \to GX \to 0$, which splits since $GF(D\Lambda) \simeq D\Lambda$ is injective. From the exact commutative diagram

$$\begin{array}{ccccccccc} 0 & \longrightarrow & FGF(D\Lambda) & \longrightarrow & FGE & \longrightarrow & FGX & \longrightarrow & 0 \\ & & \downarrow\wr & & \downarrow\wr & & \downarrow\wr & & \\ 0 & \longrightarrow & F(D\Lambda) & \longrightarrow & E & \longrightarrow & X & \longrightarrow & 0 \end{array}$$

we see that $0 \longrightarrow F(D\Lambda) \longrightarrow E \longrightarrow X \longrightarrow 0$ splits. Hence we have $\mathrm{Ext}^1_\Lambda(\mathcal{X}, F(D\Lambda)) = 0$, and consequently $\mathrm{Ext}^i_\Lambda(\mathcal{X}, F(D\Lambda)) = 0$ for $i > 0$ since $\mathcal{X}$ is resolving.

(c) This is analogous to the proof of (b).

(d) Let C be in $\mathcal{Y}$ and $0 \to C \to I_0 \to I_1 \to \cdots$ an injective resolution of C. Denoting by $I^\bullet$ the complex $0 \to I_0 \to I_1 \to \cdots$ we have $H^i(\mathrm{Hom}_\Lambda(B, I^\bullet) = \mathrm{Ext}^i_\Lambda(B, C)$ for all $i > 0$ and B in $\mathcal{Y}$. Since $\mathcal{Y}$ is closed under cokernels of monomorphisms, it follows from (a) that $0 \to FC \to FI_0 \to FI_1 \to \cdots$ is exact. By using (b) we see that $H^i(\mathrm{Hom}_\Lambda(FB, F(I^\bullet)) = \mathrm{Ext}^i_\Lambda(FB, FC)$. Since F induces an isomorphism of complexes $\mathrm{Hom}_\Lambda(B, I^\bullet) \xrightarrow{\sim} \mathrm{Hom}_\Lambda(FB, F(I^\bullet))$, we deduce that $\mathrm{Ext}^i_\Lambda(B, C) \to \mathrm{Ext}^i_\Lambda(FB, FC)$ is an isomorphism for $i > 0$.

(e) This is analogous to the proof of (d).

We now use Proposition 1.2 to prove the connection between dualizing modules and strong cotilting modules.

**Proposition 1.3.** For an artin algebra $\Lambda$ , ${}_\Lambda\omega_\Lambda$ is a dualizing module if and only if ${}_\Lambda\omega$ and $\omega_\Lambda$ are strong cotilting modules and the natural algebra morphism $\Lambda \to \mathrm{End}({}_\Lambda\omega)^{op}$ is an isomorphism.

**Proof.** Assume first that ${}_\Lambda\omega_\Lambda$ is a dualizing module, and let $F = \mathrm{Hom}_\Lambda(\omega\ ,)$ and $G = \omega\otimes_\Lambda$ . Then F and G induce inverse equivalences of categories between $\mathcal{I}^\infty$ ($\Lambda$) and $\mathcal{P}^\infty$ ($\Lambda$). $\mathcal{P}^\infty$ ($\Lambda$) is a resolving subcategory of mod$\Lambda$ closed under cokernels of monomorphisms, and $\mathcal{I}^\infty$ ($\Lambda$) is a coresolving subcategory of mod$\Lambda$ closed under kernels of epimorphisms. Then $G(\Lambda) = {}_\Lambda\omega$ has finite injective dimension, and by Proposition 1.2 (e) we get $\mathrm{Ext}^i_\Lambda(\omega, \omega) \simeq \mathrm{Ext}^i_\Lambda(\Lambda, \Lambda) = 0$ for $i > 0$. If $id_\Lambda C < \infty$, we have $pd_\Lambda FC < \infty$, so that FC is in $\widehat{\mathrm{add}}\ \Lambda$. Since G is exact on sequences whose terms have finite projective dimension, it follows that C is in $\widehat{\mathrm{add}}\ \omega$. Hence we have $\mathcal{I}^\infty$ ($\Lambda$) $=\widehat{\mathrm{add}}\ (\omega)$, so that ${}_\Lambda\omega$ is a strong cotilting module.

We next show that $(\omega, D\Lambda) = \mathrm{Hom}_\Lambda(\omega, D\Lambda\ )$ is a strong tilting module. Since $\mathrm{Ext}^i_\Lambda(D\Lambda, D\Lambda) = 0$ for $i > 0$, we get $\mathrm{Ext}^i_\Lambda((\omega, D\Lambda), (\omega, D\Lambda)) = 0$ for $i > 0$ by Proposition 1.2 (d). Since $id_\Lambda D\Lambda = 0 < \infty$, we have $pd_\Lambda(\omega, D\Lambda) < \infty$. For a subcategory $\mathcal{X}$ of mod$\Lambda$ denote by $\check{\mathcal{X}}$ the category whose objects are the C for which there is an exact sequence $0 \to C \to X_0 \to \cdots X_n \to 0$ with $X_i$ in $\check{\mathcal{X}}$. If $pd_\Lambda C < \infty$, then GC is in $\mathcal{I}^\infty(\Lambda) = \text{add}\, D\Lambda$, so that C is in $\text{add}\,(\omega, D\Lambda)$, using the exactness property of F. This shows that $(\omega, D\Lambda) = {}_\Lambda D(\omega)$ is a strong tilting module, and consequently $\omega_\Lambda$ is a strong cotilting module.

Using that $G : \mathcal{P}^\infty(\Lambda) \to \mathcal{I}^\infty(\Lambda)$ is an equivalence of categories, we see that the natural algebra map $\Lambda \simeq \text{End}_\Lambda(\Lambda)^{op} \to \text{End}_\Lambda({}_\Lambda\omega)^{op}$ is an isomorphism.

Assume conversely that the bimodule ${}_\Lambda\omega_\Lambda$ has the property that ${}_\Lambda\omega$ and $\omega_\Lambda$ are strong cotilting modules and the natural algebra map $\Lambda \to \text{End}_\Lambda({}_\Lambda\omega)^{op}$ is an isomorphism. Then it was shown in [6] that $(\omega, ) : \text{mod}\Lambda \to \text{mod}\Lambda$ induces an equivalence of categories between $\mathcal{I}^\infty(\Lambda)$ and $\mathcal{P}^\infty(\Lambda)$, with $\omega\otimes$ giving an inverse equivalence. This shows that ${}_\Lambda\omega_\Lambda$ is a dualizing module.

We note that it follows from Proposition 1.3 that the definition of Cohen-Macaulay algebra and dualizing module given here coincides with the definition in [6]. We also point out that since for a dualizing module $\omega$, $\widehat{\text{add}}\,\omega = \mathcal{I}^\infty(\Lambda)$, we have $CM(\Lambda) = {}^\perp\omega$.

Recall that a cotilting module T is basic if no two modules in a direct sum decomposition of T into indecomposable modules are isomorphic. Since for a strong cotilting module T, add$T$ is the category of Extprojective modules in $\mathcal{I}^\infty(\Lambda)$, that is the set of C in $\mathcal{I}^\infty(\Lambda)$ with $\text{Ext}^1_\Lambda(C, \mathcal{I}^\infty(\Lambda)) = 0$, there is at most one basic strong cotilting module up to isomorphism. When ${}_\Lambda\omega_\Lambda$ is a dualizing module, it follows from the algebra isomorphisms $\Lambda \tilde{\to} \text{End}_\Lambda({}_\Lambda\omega)^{op}$ and $\Lambda \tilde{\to} \text{End}(\omega_\Lambda)$ that ${}_\Lambda\omega$ and $\omega_\Lambda$ are basic. Hence a dualizing module is uniquely determined up to isomorphism as a left and as a right module. However, it is not uniquely determined as a bimodule. For example, if $\Lambda$ is a Gorenstein algebra and we have an automorphism $\varphi : \Lambda \to \Lambda$ which is not inner, then the $\Lambda$-bimodule $\Lambda$ with the usual action of $\Lambda$ on the left and the right action given by $\lambda \cdot \lambda' = \lambda\varphi(\lambda')$, is a dualizing module not isomorphic to ${}_\Lambda\Lambda_\Lambda$ as a bimodule.

In [6] we gave analogues in tilting theory of the Cohen-Macaulay approximations studied in [3]. We here state the specializations of these results to Cohen-Macaulay modules for Cohen-Macaulay algebras. For this we recall that if $\mathcal{X}$ is a subcategory of mod$\Lambda$, then a map $g : X \longrightarrow C$ with X in $\mathcal{X}$ is said to be a right $\mathcal{X}$-approximation of C if for any map $h : Z \longrightarrow C$ with Z in $\mathcal{X}$ there is some $t : Z \longrightarrow X$ with $gt = h$. A map $g : X \longrightarrow C$ is right minimal if whenever there is a commutative diagram

$$\begin{array}{ccc} X & \xrightarrow{g} & C \\ \downarrow f & & \| \\ X & \xrightarrow{g} & C \end{array}$$

then $f$ is an isomorphism. A right minimal right $\mathcal{X}$ -approximation is said to be a minimal right $\mathcal{X}$- approximation. The concepts of left $\mathcal{X}$- approximations and left minimal maps are defined dually.

**Proposition 1.4** Let $\Lambda$ be a Cohen-Macaulay artin algebra, and let C be in mod$\Lambda$.

(a) There is a unique exact sequence $0 \longrightarrow Y_C \longrightarrow X_C \xrightarrow{g} C \longrightarrow 0$ with $X_C$ in $CM(\Lambda)$ and $Y_C$ in $\mathcal{I}^\infty(\Lambda)$, such that $g : X_C \longrightarrow C$ is a minimal right $\mathcal{X}$- approximation.

(b) There is a unique exact sequence $0 \longrightarrow C \xrightarrow{f} Y^C \longrightarrow X^C \longrightarrow 0$ with $X^C$ in $CM(\Lambda)$ and $Y^C$ in $\mathcal{I}^{\infty}(\Lambda)$, such that $f : C \longrightarrow X^C$ is a minimal left $\mathcal{I}^{\infty}(\Lambda)$-approximation.

We also state the following consequence of tilting theory (see [14]).

**Proposition 1.5.** Let $\Lambda$ be a Cohen-Macaulay artin algebra with dualizing module ${}_\Lambda\omega_\Lambda$. Then $(,\omega) : CM(\Lambda) \to CM(\Lambda^{op})$ is a duality.

We end this section with some information on the finitistic projective dimension of Cohen-Macaulay algebras. We recall that $fin.pd.(\Lambda) = \{sup(pd\ C); pd_\Lambda C < \infty\}$ is called the finitistic projective dimension of an artin algebra $\Lambda$. It is not true in general that $fin.pd.(\Lambda)$ is equal to $fin.pd.(\Lambda^{op})$, but we shall show that this is the case for Cohen-Macaulay algebras.

**Proposition 1.6.** Let $\Lambda$ be a Cohen-Macaulay artin algebra with dualizing module $\omega$. Then $id_\Lambda\omega = id\omega_\Lambda = fin.pd.(\Lambda) = fin.pd.(\Lambda^{op})$.

**Proof** Since ${}_\Lambda\omega$ is a strong cotilting module, we have for any C with $id_\Lambda C < \infty$ an exact sequence $0 \longrightarrow \omega_n \longrightarrow \ldots \longrightarrow \omega_0 \longrightarrow C \longrightarrow 0$ with $\omega_i$ in add$\omega$. This shows that $id_\Lambda C \leq id_\Lambda\omega$, and hence $id_\Lambda\omega = fin.pd.\Lambda^{op}$, and similarly $id\omega_\Lambda = fin.pd.\Lambda$. The rest follows from the following result in tilting theory, using that the natural algebra map $\Lambda \longrightarrow \mathrm{End}({}_\Lambda\omega)^{op}$ is an isomorphism. We include the proof for the sake of completeness.

**Lemma 1.7** Let $\Lambda$ be an artin algebra.

(a) Let T be a tilting module with $pd_\Lambda T = d$. Then the tilting module $T_{\mathrm{End}_\Lambda(T)^{op}}$ has projective dimension d.

(b) Let T be a cotilting module with $id_\Lambda T = d$. Then the cotilting module $T_{\mathrm{End}_\Lambda(T)^{op}}$ has injective dimension d.

**Proof** (a) Consider the exact sequence $0 \longrightarrow \Lambda \longrightarrow T_0 \longrightarrow \ldots \longrightarrow T_n \longrightarrow 0$ with $T_i$ in add$T$ and n smallest possible. Then $\mathrm{Ext}^n_\Lambda(T_n, \Lambda) \neq 0$, so that $pd_\Lambda T \geq pd_\Lambda T_n \geq n$. Consider further the sequence $0 \longrightarrow (T_n, T) \longrightarrow \ldots \longrightarrow (T_0, T) \longrightarrow (\Lambda, T) \simeq T \longrightarrow 0$, which is exact since $\mathrm{Ext}^i_\Lambda(T, T) = 0$ for $i > 0$. This is a projective resolution of T as a right $\mathrm{End}({}_\Lambda T)^{op}$- module. Hence we have $pdT_{\mathrm{End}_\Lambda(T)^{op}} \leq n \leq pd_\Lambda T$.
Similarly we get the opposite inequality, so that $pdT_{\mathrm{End}_\Lambda(T)^{op}} = pd_\Lambda T$.

(b) This is dual of (a).

## 2. Examples

In this section we give some examples of Cohen-Macaulay and Gorenstein artin algebras, including ring constructions preserving the properties. We also compare the notion of Gorenstein algebra with other possible definitions motivated by commutative Gorenstein rings.

**Proposition 2.1.** Let $\Lambda$ be an artin algebra.

(a) If $fin.pd.\Lambda = 0 = fin.pd.\Lambda^{op}$, then $\Lambda$ is Cohen-Macaulay with dualizing module $D\Lambda$, and $CM(\Lambda) = \text{mod}\Lambda$.

(b) If $\Lambda$ is local, then $\Lambda$ is Cohen-Macaulay with dualizing module $D\Lambda$.

(c) If $\Lambda$ is local, then $\Lambda$ is Gorenstein if and only if $\Lambda$ is selfinjective.

**Proof.** (a) When $fin.pd.\Lambda = 0 = fin.pd.\Lambda^{op}$, then $\mathcal{P}^\infty(\Lambda) = \text{add}\Lambda$ and $\mathcal{I}^\infty(\Lambda) = \text{add}\Lambda$. $\text{Hom}_\Lambda(D\Lambda, ) : \text{add}D\Lambda \longrightarrow \text{add}\Lambda$ is an equivalence of categories and $D\Lambda\otimes$ is an inverse equivalence. Hence $\Lambda$ is Cohen-Macaulay.

(b) This follows directly from (a), since it is easy to see that $fin.pd.\Lambda = fin.pd.\Lambda^{op} = 0$ when $\Lambda$ is local.

(c) If $\Lambda$ is local Gorenstein, then $\Lambda$ and $D\Lambda$ are both dualizing modules, and hence $\Lambda$ is an injective $\Lambda$-module.

It is further clear that the selfinjective algebras and the algebras of finite global dimension are Gorenstein, and hence Cohen-Macaulay. In the first case $CM(\Lambda) = \text{mod}\Lambda$, and in the second case $CM(\Lambda) = \text{add}\Lambda$. If the square of the radical of $\Lambda$ is 0, it is easy to verify that there are no other indecomposable Gorenstein algebras than the selfinjective ones and those of finite global dimension. In general there are however many other Gorenstein algebras, and the following result shows how to construct such examples.

**Proposition 2.2.** Let $\Lambda$ and $\Gamma$ be finite dimensional algebras over a field k.

(a) $\max(id_\Lambda\Lambda, id_\Gamma\Gamma) \leq id_{\Lambda\otimes_k\Gamma}(\Lambda\otimes_k\Gamma) \leq id_\Lambda\Lambda + id_\Gamma\Gamma$.

(b) $\Lambda\otimes_k\Gamma$ is Gorenstein if and only if $\Lambda$ and $\Gamma$ are Gorenstein.

(c) $\Lambda\otimes_k\Gamma$ is selfinjective if and only if $\Lambda$ and $\Gamma$ are selfinjective.

**Proof.** (a) $\Lambda\otimes_k\Gamma$ is a projective $\Lambda^{op}$- module, having $\Lambda$ as a direct summand as a left $\Lambda$-module. Hence the injective $\Lambda\otimes_k\Gamma$-modules are injective $\Lambda$-modules, so that $id_\Lambda\Lambda \leq id_\Lambda(\Lambda\otimes_k\Gamma) \leq id_{\Lambda\otimes_k\Gamma}(\Lambda\otimes_k\Gamma)$. Similarly $id_\Gamma\Gamma \leq id_{\Lambda\otimes_k\Gamma}(\Lambda\otimes_k\Gamma)$, so that we have the first inequality.
To prove the second inequality we can assume $id_\Lambda\Lambda = m < \infty$ and $id_\Gamma\Gamma = n < \infty$. It is then sufficient to show $pdD(\Lambda\otimes_k\Gamma)_{\Lambda\otimes_k\Gamma} \leq n+m$. For this we note that $D(\Lambda\otimes_k\Gamma) = \text{Hom}_k(\Lambda\otimes_k\Gamma, k) \simeq \text{Hom}_k(\Lambda, \text{Hom}_k(\Gamma, k)) = \text{Hom}_k(\Lambda, D\Gamma)$. We have an exact sequence $0 \longrightarrow Q_n \longrightarrow \ldots \longrightarrow Q_0 \longrightarrow D(\Gamma)_\Gamma \longrightarrow 0$ where the $Q_i$ are projective $\Gamma^{op}$-modules. This gives rise to the exact sequence
(*) $0 \longrightarrow \text{Hom}_k(\Lambda, Q_n) \longrightarrow \ldots \longrightarrow \text{Hom}_k(\Lambda, Q_0) \longrightarrow \text{Hom}_k(\Lambda, D\Gamma) \longrightarrow 0$ of right $\Lambda\otimes_k\Gamma$-modules. The exact sequence $0 \longrightarrow P_m \longrightarrow \ldots \longrightarrow P_0 \longrightarrow (D\Lambda)_\Lambda \longrightarrow 0$, where the $P_i$ are projective $\Lambda^{op}$-modules, gives rise to the exact sequence

$0 \longrightarrow P_m \otimes_k \Gamma \longrightarrow \ldots \longrightarrow D\Lambda \otimes_k \Gamma \longrightarrow 0$. Since the $P_i \otimes_k \Gamma$ are projective $(\Lambda \otimes_k \Gamma)^{op}$-modules, we have $pd(D\Lambda \otimes_k \Gamma)_{\Lambda\otimes_k\Gamma} \leq m$. Since $\mathrm{Hom}_k(\Lambda, \Gamma) \simeq \mathrm{Hom}_k(\Lambda, k) \otimes_k \Gamma = D\Lambda \otimes_k \Gamma$, we get $pd\mathrm{Hom}_k(\Lambda,\Gamma)_{\Lambda\otimes_k\Gamma} \leq m$. By (*) we then get $pd\mathrm{Hom}_k(\Lambda,\Gamma)_{\Lambda\otimes_k\Gamma} \leq n+m$, so that $id_{\Lambda\otimes_k\Gamma}(\Lambda \otimes_k \Gamma) \leq n+m$.

(b) and (c) are direct consequences of (a).

Since it is known that if $k$ is a perfect field $\Lambda \otimes_k \Gamma$ has finite global dimension if and only if $\Lambda$ and $\Gamma$ have finite global dimension, the above result gives a procedure for constructing Gorenstein algebras of infinite global dimension which are not selfinjective. For example, let $\Lambda$ be selfinjective of infinite global dimension, and let $\Gamma$ be a non selfinjective algebra of finite global dimension. Then $\Lambda \otimes_k \Gamma$ is a Gorenstein algebra with the desired property.

We note that when $\Lambda$ is a k-algebra, then $T_2(\Lambda) = \begin{pmatrix} \Lambda & 0 \\ \Lambda & \Lambda \end{pmatrix} = \Lambda \otimes_k T_2(k)$. Since $T_2(k)$ is Gorenstein, it follows from Proposition 2.2(b) that $T_2(\Lambda)$ is Gorenstein if and only if $\Lambda$ is Gorenstein. Actually, this holds for any artin algebra $\Lambda$ [10]. It was shown in [6] that $\Lambda$ is Cohen-Macaulay if and only if $T_2(\Lambda)$ is Cohen-Macaulay. But we do not know for k-algebras $\Lambda$ and $\Gamma$, whether $\Lambda$ and $\Gamma$ being Cohen-Macaulay implies that $\Lambda \otimes_k \Gamma$ is Cohen-Macaulay.

If $\Lambda$ is a Gorenstein k-algebra, it follows from Proposition 2.2(b) that $\Lambda[x]/(x^2) = k[x]/(x^2) \otimes_k \Lambda$ is Gorenstein.

If k is a field and G is a finite group, then kG is selfinjective and hence Gorenstein. Hence if $\Lambda$ is Gorenstein, it follows from Proposition 2.2(b) that $\Lambda G = \Lambda \otimes_k kG$ is Gorenstein. In fact, we have the following more general result.

**Theorem 2.3** Let $\Lambda$ be an artin algebra, G a finite group of automorphisms of $\Lambda$, and $\Lambda G$ the skew group ring. Then $id_\Lambda \Lambda = id_{\Lambda G}\Lambda G$, so that $\Lambda$ is Gorenstein if and only if $\Lambda G$ is Gorenstein.

**Proof** Assume $id_{\Lambda G}\Lambda G = d < \infty$, and let $0 \longrightarrow \Lambda G \longrightarrow I_0 \longrightarrow \ldots \longrightarrow I_d \longrightarrow 0$ be a minimal injective resolution of $\Lambda G$ as a left $\Lambda G$-module. Since $\Lambda G$ is free as a right $\Lambda$-module, the $I_j$ are injective $\Lambda$-modules. Then $id_\Lambda \Lambda G \leq d$, and since $\Lambda G$ is a free left $\Lambda$-module, we also have $id_\Lambda \Lambda \leq id_{\Lambda G}\Lambda G = d$.

Assume $id_\Lambda \Lambda = t < \infty$, and let $0 \longrightarrow \Lambda \longrightarrow J_0 \longrightarrow \ldots \longrightarrow J_t \longrightarrow 0$ be a minimal injective resolution of $\Lambda$ as a left $\Lambda$-module. Now $\mathrm{Hom}_\Lambda(\Lambda G, \Lambda)$ is a left $\Lambda G$-module by means of the right action of $\Lambda G$ on $\Lambda G$. Let $\tau \in \mathrm{Hom}_\Lambda(\Lambda G, \Lambda)$ be given by $\tau(\sum_{i=1}^n \lambda_i g_i) = \lambda_1$, where $g_1, \cdots, g_n$ are the elements of $G$ and $g_1$ is the identity element of G. Then the map $\varphi : \Lambda G \longrightarrow \mathrm{Hom}_\Lambda(\Lambda G, \Lambda)$ given by $\varphi(\mu) = \mu\tau$ is a $\Lambda G$-isomorphism of left $\Lambda G$-modules.

Consider the exact sequence $0 \longrightarrow \mathrm{Hom}_\Lambda(\Lambda G, \Lambda) \longrightarrow \mathrm{Hom}_\Lambda(\Lambda G, J_0) \longrightarrow \ldots \longrightarrow \mathrm{Hom}_\Lambda(\Lambda G, J_t) \longrightarrow 0$ of left $\Lambda G$-modules. Since the $\mathrm{Hom}_\Lambda(\Lambda G, J_i)$ are injective $\Lambda G$-modules, we have $id_{\Lambda G}\mathrm{Hom}_\Lambda(\Lambda G, \Lambda) \leq t$, and hence $id_{\Lambda G}\Lambda G \leq t = id_\Lambda \Lambda$.

By duality, we have the same inequalities for right modules, and hence we are done.

For a commutative local complete noetherian ring R the following conditions are equivalent.

(a) R is the dualizing module.

(b) $id_R R < \infty$.

(c) If $0 \longrightarrow R \longrightarrow T_0 \longrightarrow I_1 \longrightarrow \ldots \longrightarrow I_j \longrightarrow \ldots$ is a minimal injective resolution of R, then the flat dimension of $I_{j-1}$ is at most $j$ for all $j > 0$.

(d) grade $\mathrm{Ext}^i_R(M,R) \geq i$ for all $M$ in mod$R$ and all $i > 0$, that is $\mathrm{Ext}^j_R(\mathrm{Ext}^i_R(M,R),R) = 0$ for $0 \leq j < i$.

We have used property (a) (and property (b)) as basis for the definition of Gorenstein artin algebras. Both (c) and (d) make sense for artin algebras in the formulation we have given. But they are not equivalent to the definition of Gorenstein artin algebra. One nice feature of Gorenstein algebras is that they contain the algebras of finite global dimension in addition to the selfinjective algebras. But whereas a selfinjective algebra satisfies both (c) and (d), a hereditary algebra usually does not have the property that its injective envelope is projective.

Nevertheless, both (c) and (d) give rise to interesting classes of artin algebras, and for both we can talk about the property being true for all integers up to a given i. In the general noetherian noncommutative case Auslander introduced the notion of a ring being i-Gorenstein if (c) holds for all $j \leq i$ (see [10]). For artin algebras this class has not yet been investigated much, except for the case of 1-Gorenstein algebras, which coincides with the class of QF-3 algebras. In other branches of noncommutative ring theory the i-Gorenstein algebras have played an important role. If a ring $\Lambda$ is i-Gorenstein for all i and $\Lambda$ has finite injective dimension on both sides it is called Auslander-Gorenstein, and if in addition $\Lambda$ has finite global dimension, it is called Auslander regular (see [8]). It should be interesting to investigate these classes more also for artin algebras.

## 3. Cohen-Macaulay modules and almost split sequences

In this section we first give some basic properties of Cohen- Macaulay modules over Gorenstein algebras, which are similar to properties in the commutative complete local case. As a special case of results on cotilting modules the category of Cohen-Macaulay modules over Cohen-Macaulay algebras has almost split sequences. We give a relationship between the end terms of an almost split sequence for Gorenstein algebras. In analogy with the commutative case we also investigate projective summands of minimal right $CM(\Lambda)$-approximations.

Before we state the first basic result we introduce some notation. For $i \geq 1$ denote by $\Omega^i(\mathrm{mod}\Lambda)$ the additive category generated by direct summands of $i^{th}$ syzygy modules $\Omega^i(M)$ for $M$ in mod$\Lambda$, and projective modules. Denote by $TrC$ the transpose of a $\Lambda$-module $C$ and by $\underline{\mathrm{mod}\Lambda}$ and $\underline{CM(\Lambda)}$ the categories mod$\Lambda$ and $CM(\Lambda)$ modulo projectives. We denote the image of a $\Lambda$-module $C$ in $\underline{\mathrm{mod}\Lambda}$ by $\underline{C}$. We say that a Cohen-Macaulay algebra $\Lambda$ has dimension $d$ if $id_\Lambda\omega = d$ for the dualizing module $\omega$.

**Proposition 3.1** Let $\Lambda$ be a Gorenstein algebra with $id_\Lambda\Lambda = d$.

(a) $\mathrm{Hom}_\Lambda(\,,\Lambda) : CM(\Lambda) \longrightarrow CM(\Lambda^{op})$ is a duality.

(b) $CM(\Lambda) = \Omega^d(\mathrm{mod}\Lambda)$.

(c) $\Omega^1 : \underline{CM(\Lambda)} \longrightarrow \underline{CM(\Lambda)}$ is an equivalence of categories.

(d) If $C$ is in $CM(\Lambda)$, then $TrC$ is in $CM(\Lambda^{op})$.

(e) $CM(\Lambda) \cap \mathcal{P}^\infty(\Lambda) = \mathrm{add}\Lambda$.

**Proof.**

(a) This is a special case of Proposition 1.1.

(b) $CM(\Lambda) \supset \Omega^d(\mathrm{mod}\Lambda)$ holds more generally for a Cohen-Macaulay algebra of dimension d with dualizing module $\omega$. For let $C$ be in $\mathrm{mod}\Lambda$. Then $\mathrm{Ext}^i_\Lambda(\Omega^d C, \omega) \simeq \mathrm{Ext}^{i+d}_\Lambda(C,\omega) = 0$ for $i > 0$ since $id_\Lambda \omega = d$, and hence $\Omega^d C$ is in $CM(\Lambda)$. In addition the projective modules are in $CM(\Lambda)$.
Assume conversely that $C$ is in $CM(\Lambda)$ and write $C^* = \mathrm{Hom}_\Lambda(C,\Lambda)$. Let $\ldots \to P_d \to P_{d-1} \to \ldots \to P_1 \to P_0 \to C^* \to 0$ be a minimal projective resolution in $\mathrm{mod}\Lambda^{op}$. Since $C^*$ is in $CM(\Lambda^{op})$ by (a), we have $\mathrm{Ext}^i_\Lambda(C^*,\Lambda) = 0$ for $i > 0$, so that $0 \to C^{**} \to P_0^* \to P_1^* \to \cdots \to P_d^* \to \ldots$ is exact. Since $\mathrm{Hom}_\Lambda(\,,\Lambda) : CM(\Lambda) \to CM(\Lambda^{op})$ is a duality, we see that $C$ is in $\Omega^d(\mathrm{mod}\Lambda)$.

(c) Since for $C$ in $CM(\Lambda), \mathrm{Ext}^1_\Lambda(C,\Lambda) = 0$, we have that $\Omega^1 : \mathrm{Hom}(\underline{C},\underline{X}) \to \mathrm{Hom}(\underline{\Omega^1 C, \Omega^1 X})$ is an isomorphism for all X in $\mathrm{mod}\Lambda$ (see [2]), so that $\Omega^1 : \underline{CM(\Lambda)} \to \underline{CM(\Lambda)}$ is fully faithful. Since a Cohen-Macaulay module with no nonzero projective summands is an arbitrary syzygy as shown above, $\Omega^1$ is also dense.

(d) Let $C$ be in $CM(\Lambda)$, and consider a minimal projective resolution $\cdots \to P_d \to \cdots \to P_1 \to P_0 \to C \to 0$. We have the exact sequences $0 \to C^* \to P_0^* \to P_1^* \to P_2^* \to \cdots \to P_d^* \to \cdots$ and $P_0^* \to P_1^* \to TrC \to 0$, showing that $TrC$ is in $\Omega^d(\mathrm{mod}\Lambda) = CM(\Lambda)$.

(e) Since $\Lambda$ is a strong cotilting module, we know from [6] that $CM(\Lambda)^\perp = \{C; \mathrm{Ext}^i_\Lambda(CM(\Lambda),C) = 0 \text{ for } i > 0\}$ is equal to $\mathcal{I}^\infty(\Lambda)$. Since $\Lambda$ is Gorenstein, we have $\mathcal{P}^\infty(\Lambda) = \mathcal{I}^\infty(\Lambda)$, and we know that $CM(\Lambda)^\perp \cap CM(\Lambda) = \mathrm{add}\Lambda$ since $\Lambda$ is a cotilting module [6].

In view of Proposition 3.1 (b) it would be interesting to know if $\Lambda$ is Gorenstein when $\Lambda$ is Cohen-Macaulay and $CM(\Lambda) = \Omega^d(\mathrm{mod}\Lambda)$.

Let $\Lambda$ be a Cohen-Macaulay algebra with dualizing module $\omega$. Since $\omega$ is a cotilting module, we know from [6] that $CM(\Lambda) = {}^\perp\omega$ is a functorially finite resolving subcategory of $\mathrm{mod}\Lambda$ , and hence has almost split sequences by [7]. If $0 \to A \to B \to C \to 0$ is almost split, we write $\tau C = A$, and we will now establish a formula for $\tau$ when $\Lambda$ is Gorenstein. This will be based on the following.

**Lemma 3.2** For an arbitrary artin algebra let $\mathcal{X}$ be a contravariantly finite resolving subcategory of $\mathrm{mod}\Lambda$, and let $C$ in $\mathcal{X}$ be indecomposable and nonprojective.

(a) $\tau C$ is a direct summand of the minimal right $\mathcal{X}$- approximation $X_{DTrC}$ of $DTrC$.

(b) If $X_{DTrC} = A \amalg B$ where $A$ is indecomposable and $B$ is Extinjective in $\mathcal{X}$, that is $\mathrm{Ext}^1_\Lambda(\mathcal{X}, B) = 0$, then $\tau C = A$.

(c) If $X_{DTrC}$ is indecomposable, then $\tau C = X_{DTrC}$.

**Proof.** Let $0 \to DTrC \to E \to C \to 0$ be an almost split sequence in mod$\Lambda$. Since $\mathcal{X}$ is resolving, we have from [6, Prop. 3.6] the exact commutative diagram

$$\begin{array}{ccccccccc} & & 0 & & 0 & & & & \\ & & \downarrow & & \downarrow & & & & \\ & & Y_{DTrC} & = & Y & & & & \\ & & \downarrow & & \downarrow & & & & \\ 0 & \longrightarrow & X_{DTrC} & \longrightarrow & X & \xrightarrow{h} & C & \longrightarrow & 0 \\ & & \downarrow g & & \downarrow f & & \| & & \\ 0 & \longrightarrow & DTrC & \longrightarrow & E & \longrightarrow & C & \longrightarrow & 0 \\ & & \downarrow & & \downarrow & & & & \\ & & 0 & & 0 & & & & \end{array}$$

where $f : X \to E$ is a right $\mathcal{X}$- approximation. The map $f : X \to E$ is also right minimal. For otherwise there would be a sequence of maps $Z \to Y \to X_{DTrC} \to X \to Z$ with $Z$ nonzero, whose composition is the identity. This would contradict the fact that $g : X_{DTrC} \to DTrC$ is right minimal. Then the composition $h : X \to E \to C$ is right almost split, but not necessarily minimal right almost split. Hence we get that $\tau C$ is a summand of $X_{DTrC}$. The rest follows by using that $\tau C$ is indecomposable, and clearly not Extinjective in $\mathcal{X}$ .

We do not know if in general the minimal right $\mathcal{X}$- approximation $X_{DTrC}$ is indecomposable when $\mathcal{X}= CM(\Lambda)$ for a Cohen-Macaulay algebra $\Lambda$. But we shall show that this is the case up to a projective summand when $\Lambda$ is Gorenstein, and we shall also find an explicit expression for $\tau C$. For this we need some preliminary results on right $CM(\Lambda)$-approximations which are also of independent interest, and which are analogues of results from the commutative theory [3].

**Lemma 3.3** Let $\Lambda$ be a Gorenstein algebra and $\mathcal{X}= CM(\Lambda)$. For $B$ in mod$\Lambda$ we have that the minimal right $\mathcal{X}$-approximation $X_B$ is projective if and only if $pd_\Lambda B < \infty$.

**Proof.** Consider the exact sequence $0 \to Y_B \to X_B \to B \to 0$, where $pd_\Lambda Y_B < \infty$ since $\Lambda$ is Gorenstein. We then see that $pd_\Lambda B < \infty$ if and only if $pd_\Lambda X_B < \infty$. Since $X_B$ is in $CM(\Lambda)$, it follows that $pd_\Lambda B < \infty$ if and only if $X_B$ is projective.

**Lemma 3.4.** Let $\Lambda$ be a Gorenstein algebra and $0 \to A \to B \to C \to 0$ an exact sequence in mod$\Lambda$ with $pd_\Lambda B < \infty$. Then we have the following, where $\mathcal{X}= CM(\Lambda)$.

(a) $\underline{\Omega^1 X_C} \simeq \underline{X_A}$, where $X_C$ and $X_A$ are minimal right $\mathcal{X}$-approximations.

(b) $\underline{X_A}$ is indecomposable if and only if $\underline{X_C}$ is indecomposable.

**Proof.** Since $CM(\Lambda)$ is resolving, we have by [6, Prop. 3.6] an exact commutative diagram

$$\begin{array}{ccccccccc} & & 0 & & 0 & & 0 & & \\ & & \downarrow & & \downarrow & & \downarrow & & \\ 0 & \longrightarrow & Y_A & \longrightarrow & Y & \longrightarrow & Y_C & \longrightarrow & 0 \\ & & \downarrow & & \downarrow & & \downarrow & & \\ 0 & \longrightarrow & X_A & \longrightarrow & X & \longrightarrow & X_C & \longrightarrow & 0 \\ & & \downarrow & & \downarrow & & \downarrow & & \\ 0 & \longrightarrow & A & \longrightarrow & B & \longrightarrow & C & \longrightarrow & 0 \\ & & \downarrow & & \downarrow & & \downarrow & & \\ & & 0 & & 0 & & 0 & & \end{array}$$

with $X$ in $\mathcal{X}= CM(\Lambda)$ and $Y$ in $\mathcal{Y}=\mathcal{P}^\infty$ $(\Lambda)$. The map $X \longrightarrow B$ is then a right $CM(\Lambda)$-approximation, and since $\mathcal{X} \cap \mathcal{Y}= \text{add}\Lambda$, we have $X = X_B \amalg P$ where $P$ is projective. Then $X$ is projective by Lemma 3.3, since $pd_\Lambda B < \infty$. This proves (a), and then (b) follows by Proposition 3.1.

**Lemma 3.5.** Let $\Lambda$ be a Gorenstein algebra of dimension d, and let $0 \to A \to I_0 \to \dots \to I_{d-1} \to \Omega^{-d}A \to 0$ be exact, where $0 \to A \to I_0 \to \dots \to I_d \to \dots$ is a minimal injective resolution of the $\Lambda$-module $A$. Then we have the following.

(a) $\underline{\Omega^{-d}(X_{\Omega^{-d}A})} \simeq \underline{X_A}$.

(b) $\underline{X_A}$ is indecomposable if and only if $\underline{X_{\Omega^{-d}A}}$ is indecomposable.

(c) If $A$ is indecomposable nonprojective in $CM(\Lambda)$, then $\underline{X_{\Omega^{-d}A}}$ and $\overline{\Omega^{-d}A}$ are indecomposable, where $\overline{\Omega^{-d}A}$ denotes the image of $\Omega^{-d}A$ in the module category modulo injectives.

**Proof.** Parts (a) and (b) follow by repeated application of Lemma 3.4 since the $I_j$ have finite projective dimension. If $A$ is indecomposable in $CM(\Lambda)$, we have $X_A = A$, and if $A$ is not projective, it follows from (b) that $\underline{X_{\Omega^{-d}A}}$ is indecomposable. Since $D(\Omega^{-d}A)$ is in $\Omega^d(\text{mod}\Lambda) = CM(\Lambda)$, any direct summand of $D(\Omega^{-d}A)$ of finite projective dimension must be projective. Hence any direct summand of $\Omega^{-d}A$ of finite projective dimension must be injective. Assume that we can write $\Omega^{-d}A = B \amalg C$, where $B$ and $C$ are nonzero and not injective. Then $\underline{X_B}$ and $\underline{X_C}$ would be nonzero by Lemma 3.3, contradicting the fact that $\underline{X_{\Omega^{-d}A}}$ is indecomposable. Hence we conclude that $\overline{\Omega^{-d}A}$ is indecomposable.

For $C$ in $CM(\Lambda)$ there is by Proposition 3.1 an exact sequence $0 \to C \to P_{d-1} \to \dots \to P_0 \to A \to 0$, where the $P_i$ are projective and $A$ is in $CM(\Lambda)$ and has no nonzero projective summands if $d \geq 1$. We denote this uniquely determined module (up to isomorphism) by $A = \Omega^{-d}_{CM}(C)$. The following result is proved in the same way as for the commutative case in [5].

**Lemma 3.6.** For a Gorenstein algebra $\Lambda$ of dimension $d \geq 1$ and $C$ in mod$\Lambda$ we have the following.

(a) The nonprojective part $X_C$ of the minimal right $CM(\Lambda)$- approximation of $C$ is $\Omega^{-d}_{CM}\Omega^d C$.

(b) $\underline{\Omega^d(X_C)} \simeq \underline{\Omega^d(C)}$.

We now have the following main result.

**Theorem 3.7.** Let $\Lambda$ be a Gorenstein algebra of dimension $d$ and $0 \to \tau C \to B \to C \to 0$ an almost split sequence in $CM(\Lambda)$. Then $\tau C \simeq \Omega^{-d}_{CM} D\Omega^{-d}(TrC) \simeq \Omega^{-d}_{CM}\Omega^d(DTrC)$ and $\tau C$ is isomorphic to the nonprojective part of $\Omega^d D\Omega^d TrC$.

**Proof.** If $d = 0, CM(\Lambda) = \text{mod}\Lambda$, so the claims are obvious. Hence we can assume $d \geq 1$. Since the Extinjective objects in $CM(\Lambda)$ are the projective $\Lambda$-modules, we know that $\tau C$ is indecomposable nonprojective. By Lemma 3.6 we have $X_{DTrC} \simeq \Omega^{-d}_{CM}\Omega^d(DTrC) \amalg Q$ with $Q$ projective, which is isomorphic to $\Omega^{-d}_{CM}D\Omega^{-d}(TrC) \amalg Q$ since clearly $\Omega^d DX \simeq D\Omega^{-d}X$. Since $TrC$ is a Cohen-Macaulay $\Lambda^{op}$-module by Proposition 3.1, $\overline{\Omega^{-d}(TrC)}$ is indecomposable by Lemma 3.5. Hence it follows that $D\Omega^{-d}(TrC) \simeq \Omega^d(DTrC)$ is indecomposable up to a projective summand and is Cohen-Macaulay, so that $\Omega^{-d}_{CM}D\Omega^{-d}(TrC)$ is also indecomposable since $d \geq 1$. Since $\tau C$ is a nonprojective indecomposable summand of $X_{DTrC}$, we get $\tau C \simeq \Omega^{-d}_{CM}D\Omega^{-d}(TrC)$.

Consider the exact sequence $0 \to \Omega^d TrC \to P_{d-1} \to \cdots \to P_0 \to TrC \to 0$, which gives rise to the exact sequence $0 \to DTrC \to DP_0 \to \cdots \to DP_{d-1} \to D(\Omega^d TrC) \to 0$. Since $DP_i$ is injective for all $i$, we have $\underline{X_{DTrC}} \simeq \underline{\Omega^d(X_{D\Omega^d TrC})}$ by Lemma 3.5. We further have $\underline{\Omega^d(X_{D\Omega^d TrC})} \simeq \underline{\Omega^d(D\Omega^d TrC)}$ by Lemma 3.6. We can then conclude that the nonprojective part $\tau C$ of $X_{DTrC}$ is isomorphic to the nonprojective part of $\Omega^d D\Omega^d TrC$.

We remark that we could also have proved Theorem 3.7 by imitating the proof for commutative d-dimensional Gorenstein isolated singularities in [4].

We have seen in Lemma 3.4 that for a Gorenstein algebra $\Lambda, pd_\Lambda C < \infty$ if and only if the minimal right $CM(\Lambda)$-approximation $X_C$ is projective. In the rest of this section we shall, in analogy with the commutative case [1][9], investigate the influence of the projective part of $X_C$ on $C$. We define $\delta(C)$ to be the number of projective summands in a decomposition of $X_C$ into indecomposable modules, and denote by $t(C)$ the submodule of $C$ generated by the images of the Cohen-Macaulay modules with no nonzero projective summands. We have the following interpretation of $\delta$.

**Lemma 3.8.** Let $\Lambda$ be a Gorenstein algebra, and let $C$ be in mod$\Lambda$. Then $\delta(C) = l(C/t(C)/\mathfrak{r}(C/t(C)))$, where $l$ denotes length and $\mathfrak{r}$ denotes the radical of $\Lambda$. In particular, $\delta(C) = 0$ if and only if $t(C) = C$.

**Proof.** Consider the exact sequence $0 \to Y_C \xrightarrow{i} X_C \xrightarrow{f} C \to 0$, where $f : X_C \to C$ is a minimal right $CM(\Lambda)$-approximation. We write $X_C = Z \amalg P$, where $P$ is projective and $Z$ has no nonzero projective summands, and consider the exact commutative diagram

$$\begin{array}{ccccccccc}
 & & 0 & & 0 & & 0 & & \\
 & & \downarrow & & \downarrow & & \downarrow & & \\
0 & \longrightarrow & V & \longrightarrow & Z & \xrightarrow{f} & f(Z) & \longrightarrow & 0 \\
 & & \downarrow & & \downarrow i & & \downarrow & & \\
0 & \longrightarrow & Y_C & \longrightarrow & Z \amalg P & \longrightarrow & C & \longrightarrow & 0 \\
 & & \downarrow & & \downarrow q & & \downarrow & & \\
0 & \longrightarrow & K & \longrightarrow & P & \xrightarrow{g} & C/f(Z) & \longrightarrow & 0 \\
 & & \downarrow & & \downarrow & & \downarrow & & \\
 & & 0 & & 0 & & 0 & &
\end{array}$$

Here $i$ and $q$ are the natural maps.

We first show that $g : P \to C/f(Z)$ is a projective cover. If it is not, then $K = L \amalg Q$ with $Q$ nonzero projective, so that there is a map $h : P \to Q$ with the composition $Q \to K \to P \xrightarrow{h} Q$ the identity. Since $Y_C \to K \to 0$ is exact, there is then a map $Q \to Y_C$ such that the composition $Q \to Y_C \to K \to P \to Q$ is the identity. Then it follows from the above commutative diagram that the compostition $Q \to Y_C \to Z \amalg P \to P \xrightarrow{h} Q$ is the identity. Hence we get a contradition to the fact that $f : X_C = Z \amalg P \to C$ is right minimal, so that we can conclude that $g : P \to C/f(Z)$ is a projective cover.

We clearly have $f(C) \subseteq t(C)$. Using that $Z \amalg P \to C$ is a right $CM(\Lambda)$-approximation, we see that $t(C) \subseteq f(C) + \mathfrak{r}C$. It now follows that $\delta(C) = l(C/t(C)/\mathfrak{r}(C/t(C))$.

We have the following consequence.

**Corollary 3.9.** Let $\Lambda$ be a Gorenstein algebra, let $X$ be in $CM(\Lambda)$, and assume that $X$ has no nonzero projective summand. Then any nonzero factor module $Y$ of $X$ has infinite projective dimension.

**Proof.** Since clearly $t(Y) = Y$, we have $\delta(Y) = 0$ by Lemma 5.8, and hence the minimal right $\mathcal{X}$-approximation $X_Y$ is not projective. This implies $pd_\Lambda Y = \infty$ by Lemma 3.3.

In the commutative case the nonregular Gorenstein rings are characterized by the fact that $t(m) = m$, where $m$ denotes the maximal ideal of the ring $R$. [1][9]. Analogues to $t(m) = m$, we now investigate when $t(\mathfrak{r}P) = \mathfrak{r}P$ for an indecomposable projective module $P$ over a Gorenstein algebra.

**Proposition 3.10.** Let $\Lambda$ be a Gorenstein algebra of dimension $d$, and let $P$ be an indecomposable projective $\Lambda$-module. Then the following are equivalent.

(a) $P$ occurs in the minimal projective resolution of some $\Omega^d T$ with $T$ simple.

(b) $t(\mathfrak{r}P) = \mathfrak{r}P$.

(c) $pd_\Lambda P/\mathfrak{r}P = \infty$.

**Proof.** (c) $\Rightarrow$ (a). Assume that $P$ does not occur in any minimal projective resolution of $\Omega^d T$ for $T$ simple. Let $\cdots \to Q_{d+1} \to Q_d \to \cdots \to Q_0 \to T \to 0$ be a minimal projective resolution of the simple $\Lambda$−module $T$. Writing $S = P/\mathfrak{r}P$ we then have $(Q_j, S) = 0$ for $j \geq d$, so that $\operatorname{Ext}^j_\Lambda(T,S) = 0$ for $j > d$. Hence we get $id_\Lambda S < \infty$, and consequently $pd_\Lambda S < \infty$, so that (c) does not hold.

(a) $\Rightarrow$ (b). Assume now that $P$ occurs in the minimal projective resolution of some $\Omega^d T$ with $T$ simple. Let $\cdots \to Q_{d+1} \to Q_d \to \cdots \to Q_0 \to \Lambda/\mathfrak{r} \to 0$ be a minimal projective resolution, and assume that $P$ is an indecomposable summand of $Q_j$ for some $j \geq d$. Let $\mathfrak{a} = t(\Lambda)$, and consider the induced complex $\cdots \to Q_{d+1}/\mathfrak{a}Q_{d+1} \to Q_d/\mathfrak{a}Q_d \to \cdots \to Q_0/\mathfrak{a}Q_0 \to \Lambda/\mathfrak{r} \to 0$. $\Omega^d(\Lambda/\mathfrak{r})$ is in $CM(\Lambda)$, so $\Omega^j(\Lambda/\mathfrak{r})$ has no nonzero projective summand for $j > d$. Hence $t(\Omega^j(\Lambda/\mathfrak{r})) = \mathfrak{a}\Omega^j(\Lambda/\mathfrak{r}) = \Omega^j(\Lambda/\mathfrak{r})$ for $j > d$. We then conclude that the maps $Q_j/\mathfrak{a}Q_j \to Q_{j-1}/\mathfrak{a}Q_{j-1}$ are zero for $j > d$. Hence $\operatorname{Tor}_j(\Lambda/\mathfrak{a}, \Lambda/\mathfrak{r}) = Q_j/\mathfrak{a}Q_j$ for $j > d$. Since $\mathfrak{r}\operatorname{Tor}_j(\Lambda/\mathfrak{a}, \Lambda/\mathfrak{r}) = 0$, we get $\mathfrak{r}Q_j \subseteq \mathfrak{a}Q_j$. Since clearly $\mathfrak{a}P \subseteq \mathfrak{r}P$, we then get $t(\mathfrak{r}P) = \mathfrak{a}P = \mathfrak{r}P$.

(b) $\Rightarrow$ (c) This is obvious by Corollary 3.9.

We point out that an indecomposable Gorenstein algebra $\Lambda$ of infinite global dimension may have simple modules of finite projective dimension. This is the case for the algebra given by the quiver $\bullet \xrightarrow{\alpha} \bullet \circlearrowleft \beta$ with relation $\beta^2 = 0$.

We do not know if $t(S) = S$ when $S$ is a simple module over a Gorenstein algebra $\Lambda$, with $pd_\Lambda S = \infty$. Another problem motivated by the commutative case is the following [15]: If $\Lambda$ is a Cohen-Macaulay artin algebra with dualizing module $\omega$, is then the trivial extension algebra $\Lambda \ltimes \omega$ Gorenstein? Note that if $\Lambda$ is a Gorenstein algebra over a field $k$, we have seen in section 2 that $\Lambda \ltimes \Lambda$ is Gorenstein.

## 4. Gorenstein dimension

The notion of Gorenstein dimension was introduced in [2] for a twosided noetherian ring $\Lambda$. A $\Lambda$-module $M$ has Gorenstein dimension zero, written G-dim $M = 0$, if $M$ is a reflexive $\Lambda$-module and $\operatorname{Ext}^i_\Lambda(M,\Lambda) = 0 = \operatorname{Ext}^i_\Lambda(M^*,\Lambda)$ for $i > 0$, where $M^* = \operatorname{Hom}_\Lambda(M,\Lambda)$. Further G-dim$M \leq n$ if and only if there is an exact sequence $0 \to L_n \to \ldots \to L_0 \to M \to 0$ with G-dim $L_i = 0$ for $0 \leq i \leq n$. If $R$ is a commutative local noetherian ring, one has the formula G-dim $M$ + depth $M$ = depth $R$ for a $\Lambda$-module $M$ of finite Gorenstein dimension. Also $R$ is a Gorenstein ring if and only if every $R$-module has finite Gorenstein dimension [2]. The corresponding result for artin algebras, which is also proved in [12], is a consequence of a characterization of cotilting modules given in [6]. This point of view suggests a generalization of the notion of Gorenstein dimension and some of the results involving this notion.

Denote for an artin algebra $\Lambda$ by $\mathcal{G}_\Lambda$ the category of $\Lambda$-modules of $G$-dimension zero. For a $\Lambda$-module $T$ with $\operatorname{Ext}^i_\Lambda(T,T) = 0$ for $i > 0$, denote as in [6] by $\mathcal{X}_T$ the category whose objects are the $C$ in ${}^{\perp}T = \{X; \mathrm{Ext}^i_\Lambda(X,\Lambda) = 0, i > 0\}$ such that there is an exact sequence $0 \to C \to T_0 \xrightarrow{f_1} T_1 \xrightarrow{f_2} T_2 \to \ldots$ with $T_i$ in add$T$ and Im$f_i$ in ${}^{\perp}T$ for all $i$. We have the following relationship between these categories.

**Proposition 4.1** Let $\Lambda$ be an artin algebra.

(a) $\mathcal{G}_\Lambda \subset \mathcal{X}_\Lambda$ .

(b) If $\Lambda$ is Gorenstein, then $\mathcal{G}_\Lambda = \mathcal{X}_\Lambda = CM(\Lambda)$.

**Proof**

(a) Let $C$ be in $\mathcal{G}_\Lambda$ and consider the projective resolution $\ldots \to P_1 \to P_0 \to C^* \to 0$ of $C^*$ in mod$\Lambda^{op}$. Since $C$ is reflexive and $\mathrm{Ext}^i_\Lambda(C^*,\Lambda) = 0$ for $i > 0$, we get an exact sequence $0 \to C \to P_0^* \xrightarrow{f_1} P_1^* \to \ldots$ with the $P_i^*$ in add$\Lambda$. The exact sequence $0 \to C \to P_0^* \to \mathrm{Im} f_1 \to 0$ induces the exact sequence $0 \to (\mathrm{Im} f_1)^* \to P_0^{**} \to C^* \to \mathrm{Ext}^1_\Lambda(\mathrm{Im} f_1, \Lambda) \to 0$, showing that $\mathrm{Ext}^1_\Lambda(\mathrm{Im} f_1, \Lambda) = 0$. Continuing this we get that the Im$f_i$ are in ${}^{\perp}\Lambda$, and hence we can conclude that $C$ is in $\mathcal{X}_\Lambda$ .

(b) For a Gorenstein algebra $\Lambda$ we clearly have $\mathcal{X}_\Lambda \subset CM(\Lambda) \subset \mathcal{G}_\Lambda$ using Proposition 3.1, so that $\mathcal{X}_\Lambda = CM(\Lambda) = \mathcal{G}_\Lambda$ in this case.

We now give the following characterization of Gorenstein algebras.

**Proposition 4.2.** For an artin algebra $\Lambda$ the following are equivalent.

(a) $\Lambda$ is Gorenstein.

(b) $\hat{\mathcal{X}}_\Lambda =$ mod$\Lambda$.

(c) Every $\Lambda$-module has finite Gorenstein dimension.

**Proof.** For a $\Lambda$-module $T$ with $\mathrm{Ext}^i_\Lambda(T,T) = 0$ for $i > 0$ we have shown in [6] that $T$ is a cotilting module if and only if $\hat{\mathcal{X}}_T =$ mod$\Lambda$, and the equivalence of (a) and (b) then follows from the fact that $\Lambda$ is a cotilting module if and only if $\Lambda$ is Gorenstein.

Note that the objects of $\hat{\mathcal{G}}$ are the $\Lambda$-modules of finite Gorenstein dimension. If $\hat{\mathcal{G}}_\Lambda =$ mod$\Lambda$, it follows from Proposition 4.1 that $\hat{\mathcal{X}}_\Lambda =$ mod$\Lambda$. If $\Lambda$ is Gorenstein, we have $\mathcal{X}_\Lambda = \mathcal{G}_\Lambda$ by Proposition 4.1, and consequently $\hat{\mathcal{G}}_\Lambda = \hat{\mathcal{X}}_\Lambda =$ mod$\Lambda$.

For a Gorenstein algebra $\Lambda$, $\mathrm{Hom}_\Lambda( ,\Lambda)$ induces a duality between $\mathcal{X}_\Lambda = \mathcal{G}_\Lambda$ and $\mathcal{X}_{\Lambda^{op}} = \mathcal{G}_{\Lambda^{op}}$ . For an artitrary artin algebra, $\mathrm{Hom}_\Lambda( ,\Lambda)$ does not induce a duality between $\mathcal{X}_\Lambda$ and $\mathcal{X}_{\Lambda^{op}}$, but it does between $\mathcal{G}_\Lambda$ and $\mathcal{G}_{\Lambda^{op}}$. In fact, $\mathcal{G}_\Lambda$ can be viewed as the largest subcategory of $\mathcal{X}_\Lambda$ such that we have such a duality.

Viewing $\Lambda$ as a cotilting module for a Gorenstein algebra, we have the following generalization of the above definitions and results.

Let $T$ be a $\Lambda$-module satisfying $\mathrm{Ext}^i_\Lambda(T,T) = 0$ for $i > 0$ and assume that the natural map $\Lambda \to \mathrm{End}(T_{\mathrm{End}(T)})$ is an isomorphism. In other words, we assume that $T$ is a generalized tilting module in the sense of Wakamatsu [16]. We then define $\text{G-dim}_T C = 0$ if $\mathrm{Ext}^i_\Lambda(C,T) = 0 = \mathrm{Ext}^i_{\mathrm{End}(T)}((C,T),T)$ for $i > 0$, and the natural map $C \to ((C,T),T)$ is an isomorphism. We denote the corresponding category by $\mathcal{G}_T$. Then we have the following generalization of Proposition 4.1.

**Proposition 4.3.** For an artin algebra $\Lambda$ and a generalized tilting module $T$ we have the following.

(a) $\mathcal{G}_T \subset \mathcal{X}_T$.

(b) If $T$ is a cotilting module, then $\mathcal{G}_T = \mathcal{X}_T = {}^\perp T$.

**Proof.**

(a) This is similar to the proof of Proposition 4.1 (a), where for $C$ in $\mathcal{G}_T$ we consider a projective resolution $\dots \to P_1 \to P_0 \to (C,T) \to 0$ of $\mathrm{End}(T)$-modules. We use that by our assumption on $T$ the natural map $P \to ((P,T),T)$ is an isomorphism for any projective $\Lambda$-module $P$.

(b) This follows from [17].

Using Proposition 4.3 we get the following generalization of Proposition 4.2.

**Theorem 4.4** Let $\Lambda$ be an artin algebra and $T$ a generalized tilting module. Then the following are equivalent.

(a) $T$ is cotilting module.

(b) $\hat{\mathcal{X}}_T = \mathrm{mod}\Lambda$.

(c) $\hat{\mathcal{G}}_T = \mathrm{mod}\Lambda$.

**Proof.** As already recalled, the equivalence of (a) and (b) is proved in [6], and the rest is a direct consequence of Proposition 4.3.

## 5. Equivalences with functorially finite subcategories

Let $\Lambda$ be an artin algebra and $\mathcal{X}$ a functorially finite subcategory of $\mathrm{mod}\Lambda$ closed under extensions. Associated with $\mathcal{X}$ we have the categories $\mathcal{Y} = \{C; \mathrm{Ext}^1_\Lambda(\mathcal{X}, C) = 0\}$ and $\mathcal{Z} = \{C; \mathrm{Ext}^1_\Lambda(C, \mathcal{X}) = 0\}$ [1]. We shall study the relationship between $\mathcal{Y}$ and $\mathcal{Z}$. An especially interesting example is $\mathcal{X} = CM(\Lambda)$ for a Gorenstein algebra $\Lambda$, which is functorially finite by [6]. This is analogous to the corresponding investigation of $CM(R)$ for a complete local commutative Gorenstein ring $R$ [3][1].

In this case $\mathcal{Y} = \mathcal{P}^\infty(\Lambda)$, and several characterizations of the objects in $\mathcal{Z}$ are given in [1]. For a Gorenstein artin algebra, $\mathcal{P}^\infty(\Lambda)$ is also an example of a functorially finite subcategory of $\mathrm{mod}\Lambda$ closed under extensions, to which we can apply our general results. Here there are no counterparts in the higher dimensional commutative case.

Before we state our main result in this section, we recall some useful facts about contravariantly and covariantly finite subcategories [6][7]. The first part of part (b) is Wakamatsu's lemma.

**Lemma 5.1** Let $\Lambda$ be an artin algebra and $\mathcal{X}$ a contravariantly finite subcategory of mod$\Lambda$ closed under extensions.

(a) Then $\mathcal{Y}= \{C; \mathrm{Ext}^1_\Lambda(\mathcal{X},C)=0\}$ is covariantly finite, and if $\mathcal{X}$ contains the projectives, then $\mathcal{X}= \{C; \mathrm{Ext}^1_\Lambda(C,\mathcal{Y})=0\}$.

(b) For $C$ in mod$\Lambda$ let $g : X_C \to C$ be a minimal right $\mathcal{X}$-approximation and $h : C \to Y^C$ a minimal left $\mathcal{Y}$-approximation. Then Ker$g$ is in $\mathcal{Y}$ and Coker $h$ is in $\mathcal{X}$ .

Recall that for a subcategory $\mathcal{X}$ of mod$\Lambda$ we denote by Fac $\mathcal{X}$ the full subcategory of mod$\Lambda$ whose objects are factors of direct sums of objects in $\mathcal{X}$, and by Sub $\mathcal{X}$ the full subcategory of mod$\Lambda$ whose objects are the submodules of direct sums of objects in $\mathcal{X}$. If $\mathcal{Y}$ is a subcategory of $\mathcal{X}$ we denote by $\mathcal{Y}/\mathcal{X}$ the category $\mathcal{X}$ modulo the maps factoring through $\mathcal{Y}$. Then we have the following.

**Theorem 5.2.** Let $\Lambda$ be an artin algebra, $\mathcal{X}$ a functorially finite subcategory of mod$\Lambda$ closed under extensions, and let $\mathcal{Y}= \{C; \mathrm{Ext}^1_\Lambda(\mathcal{X},C)=0\}$ and
$\mathcal{Z}= \{C; \mathrm{Ext}^1_\Lambda(C,\mathcal{X})=0\}$.

For Z in $\mathcal{Z}$, define $\alpha(Z) = \mathrm{Ker} f$ where $f : X \to Z$ is a minimal right $\mathcal{X}$- approximation. Then $\alpha$ induces an equivalence of categories.

$$\alpha : \mathcal{Z} \cap \mathrm{Fac}\, \mathcal{X} \;/\; \mathcal{X} \cap \mathcal{Z} \to \mathcal{Y} \cap \mathrm{Sub}\, \mathcal{X} \;/\; \mathcal{X} \cap \mathcal{Y}.$$

To see this, the following preliminary result is useful.

**Lemma 5.3.** With the assumptions as in Theorem 5.2 we have the following.

(a) $\mathrm{Hom}(\mathcal{X},\mathcal{Y})/\ \mathcal{X} \cap \mathcal{Y}= 0$.

(b) $\mathrm{Hom}(\mathcal{Z},\mathcal{X})/\ \mathcal{X} \cap \mathcal{Z}= 0$.

**Proof.** (a) Let $h : X \to Y$ be a map with $X$ in $\mathcal{X}$ and $Y$ in $\mathcal{Y}$. Since $\mathcal{Y}$ is covariantly finite, we have a minimal left $\mathcal{Y}$-approximation $g : X \to Y'$, which is a monomorphism since $\mathcal{Y}$ contains the injectives. Then Coker$g$ is in $\mathcal{X}$ by Lemma 5.1, and hence $Y'$ is in $\mathcal{X} \cap \mathcal{Y}$ since $\mathcal{X}$ is closed under extensions. We are now done since $h : X \to Y$ factors through $g : X \to Y'$.

(b) This follows from (a) by duality.

**Proof of Theorem 5.2.** For $Z$ and $Z'$ in $\mathcal{Z} \cap \mathrm{Fac}\, \mathcal{X}$, consider the exact sequences $0 \to \alpha(Z) \xrightarrow{g} X \xrightarrow{f} Z \to 0$ and $0 \to \alpha(Z') \xrightarrow{g'} X' \xrightarrow{f'} Z' \to 0$, where $f : X \to Z$ and $f' : X' \to Z'$ are minimal right $\mathcal{X}$ -approximations. We have that $\alpha(Z)$ and $\alpha(Z')$ are in $\mathcal{Y}$ by Lemma 5.1, and consequently in $\mathcal{Y} \cap \mathrm{Sub}\, \mathcal{X}$. For $t : Z \to Z'$ in $\mathcal{Z} \cap \mathrm{Fac}\, \mathcal{X}$ we want to define a map $\alpha(t) : \alpha(Z) \to \alpha(Z')$ in $\mathcal{Y} \cap \mathrm{Sub}\, \mathcal{X} \;/\; \mathcal{X} \cap \mathcal{Y}$. Using that $f' : X' \to Z'$ is a right $\mathcal{X}$-approximation, we have a commutative diagram

$$\begin{array}{ccccccccc} 0 & \to & \alpha Z & \xrightarrow{s} & X & \xrightarrow{f} & Z & \to & 0 \\ & & \downarrow^{h} & & \downarrow^{g} & & \downarrow^{t} & & \\ 0 & \to & \alpha Z' & \xrightarrow{s'} & X' & \xrightarrow{f'} & Z' & \to & 0 \end{array}$$

We define $\alpha(t) = h$. To show that this is well defined, consider another commutative diagram

$$\begin{array}{ccccccccc} 0 & \to & \alpha Z & \xrightarrow{s} & X & \xrightarrow{f} & Z & \to & 0 \\ & & \downarrow^{h'} & & \downarrow^{g'} & & \downarrow^{t} & & \\ 0 & \to & \alpha Z' & \xrightarrow{s'} & X' & \xrightarrow{f'} & Z' & \to & 0 \end{array}$$

It is then clear that $h - h' : \alpha Z \to \alpha Z'$ factors through $\mathcal{X}$, and consequently through $\mathcal{X} \cap \mathcal{Y}$ by Lemma 5.3.

We next want to show that the functor $\alpha : \mathcal{Z} \cap \text{Fac}\, \mathcal{X} \to \mathcal{Y} \cap \text{Sub}\, \mathcal{X} \,/\, \mathcal{X} \cap \mathcal{Y}$ induces a functor $\alpha : \mathcal{Z} \cap \text{Fac}\, \mathcal{X} \,/\, \mathcal{X} \cap \mathcal{Z} \to \mathcal{Y} \cap \text{Sub}\, \mathcal{X} \,/\, \mathcal{X} \cap \mathcal{Y}$. So assume that $t : Z \to Z'$ factors through $\mathcal{X} \cap \mathcal{Z}$. Since $f' : X' \to Z'$ is a right $\mathcal{X}$-approximation, $t$ factors through $f'$. Then $h : \alpha Z \to \alpha Z'$ factors through $s : \alpha Z \to X$, and hence through $\mathcal{X} \cap \mathcal{Y}$.

Similarly we see that $\alpha : \mathcal{Z} \cap \text{Fac}\, \mathcal{X} \,/\, \mathcal{X} \cap \mathcal{Z} \to \mathcal{Y} \cap \text{Sub}\, \mathcal{X} \,/\, \mathcal{X} \cap \mathcal{Y}$ is fully faithful. We finally show that $\alpha$ is dense. For $Y$ in $\mathcal{Y} \cap \text{Sub}\, \mathcal{X}$ we have an exact sequence $0 \to Y \xrightarrow{u} X \xrightarrow{f} Z \to 0$, where $u : Y \to X$ is a minimal left $\mathcal{X}$-approximation, and consequently $Z$ is in $\mathcal{Z}$ by the dual of Lemma 5.1. And since $Y$ is in $\mathcal{Y}$, we have $\text{Ext}^1(\mathcal{X}, Y) = 0$, so that $f : X \to Z$ is a right $\mathcal{X}$-approximation. We can assume that $Y$ has no nonzero summand in $\mathcal{X} \cap \mathcal{Y}$. Then $f : X \to Z$ must be right minimal, since otherwise a nonzero summand of $X$ would be a summand of $Y$. Hence $Y = \alpha(Z)$, and the proof is finished.

The previous theorem is connected with the following problem. Let $\mathcal{X}$ be a functorially finite subcategory of $\text{mod}\Lambda$ closed under extensions for an artin algebra $\Lambda$, and let $\mathcal{Y}$ and $\mathcal{Z}$ be as before. Analogous to the basic property of almost split sequences we ask for the existence of exact sequences $0 \to Y \xrightarrow{s} X \xrightarrow{t} Z \to 0$ such that $t : X \to Z$ is a minimal right $\mathcal{X}$-approximation and $s : Y \to X$ is a minimal left $\mathcal{X}$-approximation. By what we just proved there is a (unique) such exact sequence for $Z$ if and only if $Z$ is in $\mathcal{Z} \cap \text{Fac}\, \mathcal{X}$, and for $Y$ if and only if $Y$ is in $\mathcal{Y} \cap \text{Sub}\, \mathcal{X}$.

We state explicitly some special cases of Theorem 5.2.

**Corollary 5.4.** Let $\Lambda$ be an artin algebra and $\mathcal{X}$ a functorially finite subcategory of $\text{mod}\Lambda$ closed under extensions, and let $\mathcal{Y}$, $\mathcal{Z}$ and $\alpha$ be as before.

(a) If $\mathcal{X}$ contains all projectives, we have an equivalence
$\alpha : \mathcal{Z} \,/\text{add}\Lambda \to \mathcal{Y} \cap \text{Sub}\, \mathcal{X} \,/\, \mathcal{X} \cap \mathcal{Y}$.

(b) If $\mathcal{X}$ contains all injectives, we have an equivalence

$\alpha : \mathcal{Z} \cap \mathrm{Fac}\,\mathcal{X} \,/\, \mathcal{X} \cap \mathcal{Z} \to \mathcal{Y} \,/\mathrm{add}D\Lambda$.

(c) If $\mathcal{X}$ contains all projectives and all injectives, we have an equivalence $\alpha : \mathcal{Z} \,/\mathrm{add}\Lambda \to \mathcal{Y} \,/\mathrm{add}D\Lambda$.

(d) If $\mathcal{X}$ contains all projectives, and the projectives are the Extinjectives in $\mathcal{X}$, we have an equivalence $\alpha : \mathcal{Z} \,/\mathrm{add}\Lambda \to \mathcal{Y} \cap \mathrm{Sub}\,\mathcal{X} \,/\mathrm{add}\Lambda$.

(e) If $\mathcal{X}=\mathcal{X}_T$ for a cotilting module $T$ we have an equivalence $\alpha : \mathcal{Z} \,/\mathrm{add}\Lambda \to \mathcal{Y} \cap \mathrm{Sub}\,\mathcal{X} \,/\mathrm{add}T$, where $\mathcal{Y}=\widehat{\mathrm{add}}\,T$.

(f) If $\Lambda$ is Gorenstein and $\mathcal{X}= CM(\Lambda)$, we have an equivalence $\alpha : \mathcal{Z} \,/\mathrm{add}\Lambda \to \mathcal{Y} \cap \mathrm{Sub}\,\mathcal{X} \,/\mathrm{add}\Lambda$, where $\mathcal{Y}=\mathcal{P}^{\infty}(\Lambda)$.

It would be interesting to have a description of the category $\mathcal{Z}$ in the case (e) for a cotilting module $T$. In the special case (f) where $T = \Lambda$, there is the following description. The proof is exactly the same as for the case of commutative local complete Gorenstein rings in [1], so it is omitted. We note that in [1] more equivalent statements are given.

**Proposition 5.5.** Let $\Lambda$ be a Gorenstein artin algebra, $\mathcal{X}= CM(\Lambda)$ and $\mathcal{Z}={}^{\perp}\mathcal{X}$. Then $\mathcal{Z}= \{C \,;\, TrC$ is in $\mathrm{Sub}\Lambda^{op}$ and $pdTrC < \infty\}$.

For Gorenstein artin algebras we have another important subcategory which is functorially finite and closed under extensions, namely $\mathcal{P}^{\infty}(\Lambda)$. We note first that this is a special feature of artin algebras, as follows from the following general observation. Note that $\mathcal{P}^{\infty}(\Lambda)$ is covariantly finite for commutative complete local Gorenstein rings, as follows from [3].

**Proposition 5.6.** Let $R$ be a noetherian commutative complete local ring, and $\Lambda$ an $R$-algebra which is a finitely generated $R$-module. Assume that there is some $x \in R$ which is not a zerodivisor in $\Lambda$. Let $\mathcal{X}$ be a subcategory of $\mathrm{mod}\Lambda$ closed under extensions and containing $\Lambda$ and $\Lambda/x\Lambda$. Then any $C$ in $\mathrm{mod}\Lambda$ which has a right $\mathcal{X}$-approximation is already in $\mathcal{X}$.

**Proof.** Assume that $C$ in $\mathrm{mod}\Lambda$ has a right $\mathcal{X}$- approximation. Since $R$ is complete, we have a minimal right $\mathcal{X}$- approximation $h : X_C \to C$, which is onto since $\mathcal{X}$ contains $\Lambda$. Consider the exact sequence $0 \to \mathrm{Ker}h \to X_C \to C \to 0$, where $\mathrm{Ext}^1_{\Lambda}(\Lambda/x\Lambda, \mathrm{Ker}h) = 0$ by Lemma 5.1. Now we have the exact sequence $0 \to \Lambda \xrightarrow{x} \Lambda \to \Lambda/x\Lambda \to 0$, which, writing $A = \mathrm{Ker}h$, gives rise to the exact sequence $A \xrightarrow{x} A \to \mathrm{Ext}^1_{\Lambda}(\Lambda/x\Lambda, A) = 0$. Then $A = 0$ by Nakayama's lemma, so that $X_C = C \in \mathcal{X}$.

**Corollary 5.7.** Let $R$ be a complete local commutative Gorenstein ring of dimension at least 1, such that $R$ is not a regular local ring. Then the category $\mathcal{X}=\mathcal{P}^{\infty}(\Lambda)$ is not contravariantly finite in $\mathrm{mod}\Lambda$.

**Proof.** Since $\dim R \geq 1, R$ contains a nonzero divisor $x$, and $R$ and $R/(x)$ are clearly in $\mathcal{X}$, which is also obviously closed under extensions. Since $R$ is not regular local, $\mathcal{X}$ is not equal to mod$\Lambda$. Hence it follows that $\mathcal{X}$ is not contravariantly finite.

For a Gorenstein artin algebra, however, $\mathcal{X}=\mathcal{P}^{\infty}(\Lambda)$ is in addition to being covariantly finite, also contravariantly finite. This easily follows by duality since $\Lambda^{op}$ is also Gorenstein. We want to investigate this case more closely. Here we have $\mathcal{Z}={}^{\perp}\mathcal{X}= CM(\Lambda)$, and by duality $\mathcal{Y}=\mathcal{X}^{\perp}= D(CM(\Lambda^{op}))$, which we shall call the Co-Cohen-Macaulay modules, and denote by $CoCM(\Lambda)$. Then $\mathcal{X} \cap \mathcal{Z}=$ add$\Lambda$, and $\mathcal{X} \cap \mathcal{Y}=$ add$D\Lambda$. This gives the following.

**Proposition 5.8.** Let $\Lambda$ be a Gorenstein artin algebra, and $\mathcal{X}=\mathcal{P}^{\infty}(\Lambda)$. Then we have an equivalence $\alpha : CM(\Lambda)/\text{add}\Lambda \to CoCM(\Lambda)/\text{add}D\Lambda$.

We point out the following connection with right $CM(\Lambda)$-approximations. Let $B$ be in $CoCM(\Lambda)$ with no nonzero injective summand, and consider the exact sequence $0 \to B \xrightarrow{s} X \xrightarrow{t} C \to 0$, where $s : B \to X$ is a minimal left $\mathcal{X}$-approximation. We then have $\alpha C = B$. Choose a surjection $P \to X$ where $P$ is projective. This gives rise to the exact commutative diagram

$$\begin{array}{ccccccccc}
 & & 0 & & 0 & & & & \\
 & & \downarrow & & \downarrow & & & & \\
 & & K & = & K & & & & \\
 & & \downarrow & & \downarrow & & & & \\
0 & \to & \Omega C \amalg Q & \to & P & \to & C & \to & 0 \\
 & & \downarrow_f & & \downarrow & & \| & & \\
0 & \to & B & \to & X & \to & C & \to & 0 \\
 & & \downarrow & & \downarrow & & & & \\
 & & 0 & & 0 & & & &
\end{array}$$

where Q is projective. Since $pd_{\Lambda}K < \infty$ , $f : \Omega C \amalg Q \to B$ is a right $CM(\Lambda)$-approximation. Hence $\Omega\alpha^{-1} : CoCM(\Lambda)/\text{add}D\Lambda \to CM(\Lambda)/\text{add}\Lambda$ is an equivalence of categories, associating with an object in $CoCM(\Lambda)$ its minimal right $CM(\Lambda)$-approximation. In particular, if $B$ is indecomposable noninjective, then its minimal right $CM(\Lambda)$-approximation is indecomposable modulo projectives. Note that this gives an alternative proof of the fact established in section 3 that if $C$ is indecomposable nonprojective in $CM(\Lambda)$ for $\Lambda$ Gorenstein, then $\underline{X_{DTrC}}$ is indecomposable, where $X_{DTrC}$ is the minimal right $CM(\Lambda)$-approximation of $DTrC$. For here $TrC$ is in $CM(\Lambda^{op})$, and hence $DTrC$ is in $CoCM(\Lambda)$.

We now give some applications to investigating the existence of exact sequences with given end terms, when the middle term is in a prescribed category. We here generalize results from [1].

**Proposition 5.9.** Let $\mathcal{X}$ be a covariantly finite resolving subcategory of mod$\Lambda$, containing all injective modules, and let $\mathcal{Z}=\{C; \mathrm{Ext}^1_\Lambda(C,\mathcal{X})=0\}$ and $\mathcal{Y}=\{C; \mathrm{Ext}^1_\Lambda(\mathcal{X},C)=0\}$. Let $B$ and $C$ be in mod$\Lambda$.

(a) If there is some exact sequence $0\to B\to X\to C\to 0$ with $X$ in $\mathcal{X}$, then $\underline{Z_C}\simeq \underline{Z^B}$. Here the notation is given by the exact sequences $0\to X_A\to Z_A\xrightarrow{f} A\to 0$ and $0\to A\xrightarrow{g} X^A\to Z^A\to 0$ where $f: Z_A\to A$ is a minimal right $\mathcal{Z}$-approximation and $g: A\to X^A$ is a minimal left $\mathcal{X}$-approximation.

(b) If $\mathcal{X}$ is also coresolving and $B$ is in $\mathcal{Y}$, then there is some exact sequence $0\to B\to X\to C\to 0$ with $X$ in $\mathcal{X}$ if and only if $\underline{Z_C}\simeq\underline{Z^B}$.

**Proof.** (a) Assume that there is an exact sequence $0\to B\to X\to C\to 0$ with $X$ in $\mathcal{X}$. Since $\mathcal{X}$ is covariantly finite and contains the injectives, we have an exact sequence $0\to B\xrightarrow{g} X^B\to Z^B\to 0$ where $g$ is a minimal left $\mathcal{X}$-approximation and $Z^B$ is in $\mathcal{Z}$ by the dual of Lemma 5.1. This gives rise to the exact commutative diagram

$$\begin{array}{ccccccccc}
0 & \to & B & \to & X^B & \to & Z^B & \to & 0\\
 & & \| & & \downarrow & & \downarrow & & \\
0 & \to & B & \to & X & \to & C & \to & 0
\end{array}$$

From this we get the exact commutative diagram

$$\begin{array}{ccccccccc}
 & & & & 0 & & 0 & & \\
 & & & & \downarrow & & \downarrow & & \\
 & & & & K & = & K & & \\
 & & & & \downarrow & & \downarrow & & \\
0 & \to & B & \to & X^B\amalg P & \to & Z^B\amalg P & \to & 0\\
 & & \| & & \downarrow & & \downarrow & & \\
0 & \to & B & \to & X & \to & C & \to & 0\\
 & & & & \downarrow & & \downarrow & & \\
 & & & & 0 & & 0 & &
\end{array}$$

with $P$ projective. Since $\mathcal{X}$ contains the projectives, we have that $X$ and $X^B\amalg P$ are in $\mathcal{X}$, and since $\mathcal{X}$ is closed under kernels of epimorphisms, it follows that $K$ is in $\mathcal{X}$. Using that $\mathcal{X}\cap\mathcal{Z}=\mathrm{add}\Lambda$, we see that $\underline{Z_C}\simeq\underline{Z^B\amalg P}\simeq\underline{Z^B}$.

(b) Assume now that $B$ is in $\mathcal{Y}$. Consider the exact sequence $0\to B\xrightarrow{g} X^B\to Z^B\to 0$, where $g$ is a minimal left $\mathcal{X}$-approximation, and assume that $Z^B\simeq Z_C\amalg Q$, where $Q$ is projective. Consider the exact sequence $0\to X_C\to Z_C\xrightarrow{f} C\to 0$ where $f: Z_C\to C$ is a minimal right $\mathcal{Z}$-approximation. This gives rise to the exact sequence $0 \to (C,B) \to (Z_C,B) \to (X_C,B) \to \mathrm{Ext}^1_\Lambda(C,B) \to \mathrm{Ext}^1_\Lambda(Z_C,B) \to \mathrm{Ext}^1_\Lambda(X_C,B)$, where $\mathrm{Ext}^1_\Lambda(X_C,B) = 0$ since $B$ is in $\mathcal{Y}$. Hence we conclude that $\mathrm{Ext}^1_\Lambda(C,B) \to \mathrm{Ext}^1_\Lambda(Z_C,B)$ is surjective. It follows that we get an exact commutative diagram

$$\begin{array}{ccccccccc}
 & & & & 0 & & 0 & & \\
 & & & & \downarrow & & \downarrow & & \\
 & & & & X_C \amalg Q & = & X_C \amalg Q & & \\
 & & & & \downarrow & & \downarrow & & \\
0 & \to & B & \xrightarrow{g} & X^B & \to & Z^B \simeq Z_C \amalg Q & \to & 0 \\
 & & \| & & \downarrow & & \downarrow f & & \\
0 & \to & B & \to & X & \to & C & \to & 0 \\
 & & & & \downarrow & & & & \\
 & & & & 0 & & & &
\end{array}$$

Since $\mathcal{X}$ is closed under cokernels of monomorphisms, it follows that $X$ is in $\mathcal{X}$, and we have our desired exact sequence.

We have the following special cases.

**Corollary 5.10.** Let $\Lambda$ be a Gorenstein artin algebra.

(a) Let $X = \mathcal{P}^\infty(\Lambda), \mathcal{Z} = CM(\Lambda)$ and let $B$ and $C$ be in $\mathrm{mod}\Lambda$. If there is an exact sequence $0 \to B \to X \to C \to 0$ with $pd_\Lambda X < \infty$, then $\underline{Z}_C \simeq \underline{Z}^B$ in the above notation, and if $B$ is in $CoCM(\Lambda)$ and $\underline{Z}_C \simeq \underline{Z}^B$, then there is an exact sequence $0 \to B \to X \to C \to 0$ with $pd_\Lambda X < \infty$.

(b) Let $\mathcal{X} = CM(\Lambda), \mathcal{Z} = \{C; \mathrm{Ext}^1_\Lambda(C,\mathcal{X}) = 0\}$ and let $C$ and $B$ be in $\mathrm{mod}\Lambda$. If there is an exact sequence $0 \to B \to X \to C \to 0$ with $X$ in $CM(\Lambda)$, then we must have $\underline{Z}_C \simeq \underline{Z}^B$.

**Proof.** (a) If $\mathcal{X} = \mathcal{P}^\infty(\Lambda)$, we know that $CM(\Lambda) = \{C; \mathrm{Ext}^1_\Lambda(C,\mathcal{X}) = 0\}$ by Lemma 5.1. Since $\mathcal{P}^\infty(\Lambda)$ is covariantly finite and resolving and coresolving, we can apply Proposition 5.9.

(b) Since $CM(\Lambda)$ is covariantly finite and resolving, we can apply Proposition 5.9 (a).

We also state the dual of Proposition 5.9.

**Proposition 5.11.** Let $\mathcal{X}$ be a contravariantly finite coresolving subcategory of $\mathrm{mod}\Lambda$, containing all projective modules, and let $\mathcal{Y} = \{C; \mathrm{Ext}^1_\Lambda(\mathcal{X}, C) = 0\}$, $\mathcal{Z} = \{C; \mathrm{Ext}^1_\Lambda(C,\mathcal{X}) = 0\}$. Let $B$ and $C$ be in $\mathrm{mod}\Lambda$.

(a) If there is an exact sequence $0 \to B \to X \to C \to 0$ with $X$ in $\mathcal{X}$, then $\overline{Y}^B \simeq \overline{Y}_C$ in $\overline{\text{mod}\Lambda}$. Here the notation is given by the exact sequences $0 \to Y_A \to X_A \xrightarrow{f} A \to 0$ and $0 \to A \xrightarrow{g} Y^A \to X^A \to 0$ where $f$ is a minimal right $\mathcal{X}$- approximation and $g$ a minimal left $\mathcal{Y}$-approximation.

(b) If $\mathcal{X}$ is also resolving, and C is in $\mathcal{Z}$, there is an exact sequence $0 \to B \to X \to C \to 0$ with $X$ in $\mathcal{X}$ if and only if $\overline{Y}^B \simeq \overline{Y}_C$.

## REFERENCES


[1] M. Auslander, Cohen-Macaulay approximations for Gorenstein rings, in preparation.

[2] M. Auslander and M. Brigder, Stable Module Theory, Mem. of the AMS 94, AMS, Providence 1969.

[3] M. Auslander and R.O. Buchweitz, The Homological theory of maximal Cohen-Macaulay approximations, Soc. Math. de France, Mém. no 38, 1989, 5-37.

[4] M. Auslander and I.Reiten, Almost split sequences for Cohen-Macaulay modules, Math. Ann. 277 (1989), 345-349.

[5] M. Auslander and I. Reiten, The Cohen-Macaulay type of Cohen-Macaulay rings, Adv. in Math., Vol. 73, No 1 (1989) 1-23.

[6] M. Auslander and I. Reiten, Applications of contravariantly finite subcategories, preprint Trondheim 1989, Adv. in Math. (to appear).

[7] M. Auslander and S.O. Smalø, Almost split sequences in subcategories, J. Algebra 69 (1981) 426-454.

[8] J.E. Bjørk and E.K. Ekstrøm, Filtered Auslander-Gorenstein rings, Coll. à l'honeur de Jacques Dixmier (1989), to appear in Birkhäuser.

[9] S. Ding, Cohen-Macaulay approximations over a Gorenstein local ring, preprint 1990.

[10] R. Fossum, P. Griffith and I. Reiten, Trivial extensions of abelian categories, SLN 456 (1975).

[11] D. Happel, On Gorenstein algebras, this volume.

[12] M. Hoshino, Algebras of finite selfinjective dimension, Proc. AMS.

[13] Y. Iwanaga and H. Sato, Minimal injective resolutions of Gorenstein rings, preprint 1990.

[14] Y. Miyashita, Tilting modules of finite projective dimension, Math. Z. 193 (1986) 113-146.

[15] I. Reiten, The converse to a theorem of Sharp on Gorenstein modules, Proc. AMS 32 (1970) 417-420.

[16] T. Wakamatsu, On modules with trivial selfextensions, J. Alg. 114 (1988) 106-114.

[17] T. Wakamatsu, On constructing stably equivalent functors, preprint.

Progress in Mathematics, Vol. 95, © 1991 Birkhäuser Verlag Basel

# Classical Invariants and the General Linear Group

J. A. GREEN

## 1. Introduction

Invariant theory in its classical sense is a theory of *algebraic forms*, *i.e.* of polynomials such as

$$f(a,x) = \sum \frac{r!}{r_1! \dots r_n!} a_{r_1 \dots r_n} x_1^{r_1} \dots x_n^{r_n} \ , \qquad 1.1$$

homogeneous of some degree $r$ in a set of $n$ independent *variables* $x = (x_1, \dots, x_n)$. The sum in 1.1 is over all vectors $\vec{r} = (r_1, \dots, r_n)$ of non-negative integers satisfying $r_1 + \dots + r_n = r$, and the *coefficients* $a_{r_1 \dots r_n} = a_{\vec{r}}$ are, like the variables $x_\nu$, regarded as independent indeterminates over a ground field $K$ (usually the field of real or complex numbers).

A linear *substitution* is the operation which consists in replacing $x_1, \dots, x_n$ by new variables $X_1, \dots, X_n$ according to equations of the form

$$x_\mu = \sum_{\nu \in \underline{n}} s_{\mu\nu} X_\nu \ , \qquad 1.2$$

for all $\mu$ in the set $\underline{n} = \{1, \dots, n\}$. The matrix $s = (s_{\mu\nu})$ must be non-singular, so that $s \in \mathrm{GL}_n(K)$. When the $x_\mu$'s in $f(a,x)$ are replaced by the $X_\mu$'s according to 1.2, we get a new form $f(A,X)$ of degree $r$; the coefficients $A_{p_1 \dots p_n} = A_{\vec{p}}$ of this new form are determined by the equation

$$f(a,x) = f(A,X) \ , \qquad 1.3$$

and are related to the original coefficients by equations

$$A_{\vec{p}} = \sum_{\vec{r}} T_{\vec{r},\vec{p}}(s) a_{\vec{r}} \ , \qquad 1.4$$

in which the $T_{\vec{r},\vec{p}}(s)$ are certain homogeneous polynomial functions of degree $r$ in the $n^2$ coefficients $s_{\mu\nu}$ of the substitution 1.2. A polynomial $J(a)$ in the $a_{\vec{r}}$ is said to be *invariant* of *weight* $w$ (for the given form $f$) if

$$J(A) = |s|^w J(a) \qquad 1.5$$

holds identically, *i.e.* for all $s \in \mathrm{GL}_n(K)$. A polynomial $J(a,x)$ which involves both the coefficients $a_{\vec{r}}$ and the variables $x$, is called a *covariant* of weight $w$ if

$$J(A,X) = |s|^w J(a,x) \qquad 1.6$$

holds identically. And then we may complicate matters still further by bringing in several variable-sets $x = (x_1, \ldots, x_n)$, $y = (y_1, \ldots, y_n)$, $\ldots$, and perhaps several forms of various degrees $f(a,x)$, $g(b,x)$, $h(c,x,y)$, etc.

D. Hilbert's papers of 1890 and 1893 mark the high point of classical invariant theory. In these he gave a general procedure to construct all invariant functions $J(a,b,\ldots)$ in the coefficients of a given set of forms $f(a,x)$, $g(b,x)$, $\ldots$, and showed that these invariants are all expressible as polynomials in a finite set $J_1, \ldots, J_s$ of them. The methods which Hilbert introduced at this time inspired a new kind of algebra, which was to change the course of algebraic geometry and, ironically, to displace invariant theory from its position as the chief algebraic tool for geometers.

Very soon after this the first ideas of group representation theory began to form. A. Hurwitz (1894) saw very clearly that the idea of an invariant depends only on the linear transformation of the arguments of the invariant, which is induced by the (matrix $s$ of the) substitution 1.2 ([Hu], p. 381). For example the definition 1.5 of the invariant $J(a)$, depends only on the coefficients $T_{\bar{r},\bar{p}}(s)$ in 1.4; the form $f(a,x)$ is there only to provide the "transformation rule" 1.4. I. Schur, who determined all the irreducible representations of $\mathrm{GL}_n(K)$ in his dissertation (1901), knew and used the fact that $s \to (T_{\bar{r},\bar{p}}(s))$ is a representation of $\mathrm{GL}_n(K)$ ([S], p. 16), and there are other references to invariant theory in the early sections of [S]. But Schur describes his main discoveries in terms of matrices, and not in terms of invariants.

The purpose of this article is to describe in modern terms some of the memoir "Essai d'une théorie générale des formes algébriques" by J. Deruyts. In this remarkable work Deruyts anticipates by nearly a decade the main results of Schur's dissertation. His language is the language of invariant theory, and he makes little use of matrices. But we can now look back on Deruyts's work and find a wealth of methods which, to our eyes, are pure representation theory; some of these methods are still unfamiliar today. One can speculate that the representation theory of linear groups could have taken a different course, or at least have matured more rapidly, if Schur or Weyl had taken up the ideas which are lying just below the surface of Deruyts's memoir!

Deruyts makes a detailed study of the covariants of a system of linear forms $a^{(1)}, \ldots, a^{(\ell)}$, and variable-sets $x^{(1)}, \ldots, x^{(n)}$ (see section 2 for notation). It is easy enough to define actions of $G_n = \mathrm{GL}_n(K)$ on the algebra $K[a,x]$ of all polynomials in the components of the $a$'s and $x$'s, so that the action of a given element $s$ of $G_n$ on a given $J(a^{(1)}, \ldots, x^{(1)}, \ldots)$ of $K[a,x]$ turns the latter into its "transform" $J(A^{(1)}, \ldots, X^{(1)}, \ldots)$. Each element $J$ of $K[a,x]$ can be written in the form $J = \gamma_1\xi_1 + \ldots + \gamma_r\xi_r$, where the $\gamma_i$ all lie in $K[a]$, and the $\xi_i$ all lie in $K[x]$. Assuming that $r$ is minimal, among all such expressions of $J$, the $K$-spans $\langle\gamma_1, \ldots, \gamma_r\rangle = L(J)$ and $\langle\xi_1, \ldots, \xi_r\rangle = R(J)$ are uniquely defined by $J$. If $J$ is a covariant of weight zero, then $L(J)$ and $R(J)$ are $KG_n$-submodules, of $K[a]$ and $K[x]$ respectively — $L(J)$ and $R(J)$ are dual in a certain sense (see section 8). If $J$ is a "primary" covariant, then $L(J)$ and $R(J)$ are irreducible (sections 9, 10). So what is a primary covariant? There is a procedure, much used by Deruyts, by which any function $\gamma = \gamma(a^{(1)}, \ldots)$ in $K[a]$ is turned into a function $[\Gamma]$ in $K[a,x]$: first take the transform $\Gamma = \gamma(A^{(1)}, \ldots)$ of $\gamma$, and regard this as polynomial in the $a$'s and the coefficients $s_{\mu\nu}$ of the substitution matrix (1.2). Now simply replace $s_{\mu\nu}$ by $x_\mu^{(\nu)}$ and call the result $[\Gamma]$, or $\sigma(\gamma)$, if we want to describe this procedure by a map $\sigma : K[a] \to K[a,x]$. This map takes $K[a]$ bijectively onto the set of all covariants (in $K[a,x]$) of weight zero. We may define a primary covariant to be a covariant of the form $\varphi = \sigma(\gamma)$, where $\gamma \in K[a]$ is a *semiinvariant, i.e.* $\gamma$ is a non-zero element of $K[a]$ which has the invariant property, not for all substitutions, but only for those whose matrix $s$ lies in the upper triangular subgroup $B_n$ of $G_n$. With these ideas Deruyts proves statements which imply

(1) all polynomial representations of $\mathrm{GL}_n(K)$ are semisimple (char $K = 0$, of course) and

(2) the irreducible ones are parametrized by weights $\pi = (\pi_1, \dots, \pi_n) \in \mathbf{N}_0^n$ such that $\pi_1 \geq \dots \geq \pi_n$ (section 12).

I hope to describe, in a future article, Deruyts's remarkable formula ([De, p. 142]) which effectively determines the irreducible characters of $\mathrm{GL}_n(K)$.

Finally I should like to thank J. Towber, who first told me of this pearl of nineteenth century algebra.

## 2. Notation. Action of $G_n$ on $K[a]$, $K[x]$

Throughout this paper $K$ denotes a field, assume to be of characteristic zero unless the contrary is indicated. If $r$ is a positive integer, we denote by $\underline{r}$ the set $\{1, \dots, r\}$. Let $n$ be a fixed positive integer.

By the "symbolic method" (see for example [Tu, p. 174] or [De, pp. 8–14]) the study of invariants of algebraic forms is reduced to the study of (a sufficiently large number of) *linear forms*. We use notation like $(a|x) = a_1x_1 + \dots + a_nx_n$ to denote a linear form in the variables $x_1, \dots, x_n$. Under the substitution 1.2 we get a new linear form: $(a|x) = (A|X)$, whose coefficients $A_1, \dots, A_n$ are related to the old $a_1, \dots, a_n$ by

$$A_\nu = \sum_{\mu \in \underline{n}} s_{\mu\nu} a_\mu \ , \quad \text{for all } \nu \in \underline{n} \ . \tag{2.1}$$

We shall refer to $A = (A_1, \dots, A_n)$ and $a = (a_1, \dots, a_n)$ as *coefficient-sets*, to distinguish them from the *variable-sets* $X = (X_1, \dots, X_n)$ and $x = (x_1, \dots, x_n)$.

We need a large number, say $m$, of variable-sets $x^{(j)} = (x_{1j}, \dots, x_{nj}), j = 1, \dots, m$. We need also a large number, say $\ell$, of coefficient-sets $a_{(i)} = (a_{i1}, \dots, a_{in}), i = 1, \dots, \ell$. By saying that $\ell, m$ are large, we mean that $\ell \geq n, m \geq n$. It is often convenient to use unadorned letters $a, b, c, \dots$ to refer to arbitrary elements of the set $\{a_{(1)}, \dots, a_{(\ell)}\}$.

Let $K[a]$ resp. $K[x]$ denote the $K$-algebras of all polynomials in the $\ell n$ indeterminates $a_{i\nu}$ ($i \in \underline{\ell}, \nu \in \underline{n}$), resp. the $mn$ indeterminates $x_{\mu j}$($\mu \in \underline{n}$, $j \in \underline{m}$). The algebra $K[a, x]$ of all polynomials in both the $a_{i\nu}$ and the $x_{\mu j}$ is usually written $K[a] \otimes K[x]$ ($\otimes$ means $\otimes_K$), so that a polynomial such as $2a_{11}x_{23}^3 - a_{21}a_{13}x_{11} + x_{12}^2 - a_{23}a_{22}^4$ is written $2a_{11} \otimes x_{23}^3 - a_{21}a_{13} \otimes x_{11} + 1 \otimes x_{12}^2 - a_{23}a_{22}^4 \otimes 1$ .

If $r$ is a positive integer, we denote the general linear group $\mathrm{GL}_r(K)$ by $G_r$. The group algebra of $G_r$ over $K$ is written $KG_r$; we may think of this as a vector space over $K$, having the elements of $G_r$ as basis.

We make $G_n$ act on $K[a]$ as follows: each element $s$ of $G_n$ acts on a polynomial $\gamma = \gamma(\dots, a_{i\nu}, \dots)$ in $K[a]$, by replacing each coefficient $a_{i\nu}$ by its transform $A_{i\nu}$ under the substitution $s$. Thus (with 2.1 in mind) we define

$$s \circ a_{i\nu} = \sum_{\mu \in \underline{n}} s_{\mu\nu} a_{i\mu} \ , \quad \text{for } i \in \underline{\ell}, \nu \in \underline{n} \ , \tag{2.2}$$

and extend this action on the $K$-algebra generators $a_{i\nu}$ of $K[a]$ by the rule $s \circ \gamma = \gamma(\dots, s \circ a_{i\nu}, \dots)$, for $\gamma \in K[a]$. Notice that $s$ acts linearly and multiplicatively on $K[a]$, and that $s \circ c = c$, for any constant polynomial $c \in K[a]$.

We could, in the same spirit, make $G_n$ act on $K[x]$ so that $s \in G_n$ acts on a polynomial $f(\dots, x_{\mu j}, \dots)$ to give $f(\dots, X_{\mu j}, \dots)$. But the equations 1.2 do not give $X_\mu$ directly in terms of the $x_\nu$'s; we must first solve them, which gives the formulae

$$s \circ x_{\mu j}\ (= X_{\mu j}) = \sum_{\nu \in \underline{n}} \hat{s}_{\mu\nu} x_{\nu j}\ , \quad \text{for all } j \in \underline{m}, \nu \in \underline{n}\ , \qquad 2.3^*$$

where $\hat{s}_{\mu\nu}$ is the $(\mu, \nu)$-coefficient of $s^{-1}$. To avoid these tiresome $\hat{s}_{\mu\nu}$ we shall first replace $s$ by $s^{-1}$ in 2.3*, and then write $s^{-1}x_{\mu j}$ as $x_{\mu j} \circ s$. This gives a valid *right* action of $G_n$ on $K[x]$, whereby $s \in G_n$ acts on the generators of $K[x]$ by the formulae

$$x_{\mu j} \circ s = \sum_{\nu \in \underline{n}} s_{\mu\nu} x_{\nu j}\ , \quad \text{for all } j \in \underline{m}, \mu \in \underline{n}\ , \qquad 2.3$$

and this action is extended to $K[x]$ by the rule $f \circ s = f(\dots, x_{\mu j} \circ s, \dots)$, for all $f \in K[x]$. Notice that our manipulation has secured that $x_{\mu j} \circ s^{-1} = X_{\mu j}$ for all $\mu \in \underline{n}$, $j \in \underline{m}$.

*Notes.* 1. Deruyts's notation [De, p. 4] for the "transformed" variables and coefficients — for example $X_\mu$ and $A_\nu$ — does not bring the substitution $s$ into evidence. A notation which makes $s$ explicit, allows formulation of the important properties $(x \circ s) \circ s' = x \circ ss'$, $s \circ (s' \circ a) = ss' \circ a$. See [GY, p. 5] or [Tu, pp. 160, 161].
2. It is clear that there is a natural *right* action of $G_\ell$ on $K[a]$, which commutes with the left action of $G_n$ on $K[a]$ just defined. Similarly there is a *left* action of $G_m$ on $K[x]$, which commutes with the right action of $G_n$ on $K[x]$ just defined. These actions correspond to "polar operations", and we shall discuss them in section 6.

## 3. Invariants and Covariants. First Fundamental Theorem

The actions of $G_n$ on $K[a]$ and $K[x]$ may be combined to make $K[a] \otimes K[x]$ into a $G_n$-$G_n$-bimodule by the rule:

$$s \circ (\alpha \otimes \xi) \circ s' = (s \circ \alpha) \otimes (\xi \circ s') \qquad 3.1$$

for all $s, s' \in G_n, \alpha \in K[a], \xi \in K[x]$.

Let $\varphi = \varphi(a, x)$ be an element of $K[a] \otimes K[x]$; we shall write $\varphi$ as a polynomial in the $a_{i\nu}$, $x_{\mu j}$. It follows from 2.2, 2.3 and 3.1 that

$$s \circ \varphi(\dots, a_{i\nu}, \dots, x_{\mu j}, \dots) \circ s^{-1} = \varphi(\dots, A_{i\nu}, \dots, X_{\mu j}, \dots)\ ,$$

where $A_{i\nu}$ and $X_{\mu j}$ are the transforms of $a_{i\nu}$ and $x_{\mu j}$, respectively, under the substitution 1.2 with matrix $s$. So we may restate definitions 1.5, 1.6 as follows:

*Definition 3.2.* Let $w$ be an integer. Then an element $\gamma$ of $K[a]$ is called an *invariant of weight* $w$ if

$$s \circ \gamma = |s|^w \gamma\ , \quad \text{for all } s \in G_n\ . \qquad 3.3$$

An element $\gamma$ of $K[a] \otimes K[x]$ is called a *covariant of weight* $w$ if

$$s \circ \varphi \circ s^{-1} = |s|^w \varphi \ , \ \text{for all } s \in G_n \ . \tag{3.4}$$

*Remark.* By identifying $\gamma \in K[a]$ with the element $\gamma \otimes 1 \in K[a] \otimes K[x]$, we see that 3.3 is a special case of 3.4. For $1 \circ s^{-1} = 1$, for all $s \in G_n$.

*Examples 3.5.* (1) Any linear form $\varphi = (a|x) = a_1x_1 + \ldots + a_nx_n$ is a covariant of weight zero, because $s \circ \varphi \circ s^{-1} = (s \circ a|x \circ s^{-1}) = (A|X) = (a|x)$, for all $s \in G_n$.
(2) Let $i_1, \ldots, i_n$ be $n$ integers drawn from $\underline{\ell} = \{1, \ldots, \ell\}$. Denote by $(a_{(i_1)}, \ldots, a_{(i_n)})$ the $n \times n$ determinant $\gamma = \det(a_{i_\rho,\nu})$, where $\rho$ and $\nu$ each run from 1 to $n$. Then $\gamma$ is an element of $K[a]$, and $s \circ \gamma = \det(s \circ a_{i_\rho,\nu}) = \det(\sum_{\mu \in \underline{n}} s_{\mu\nu} a_{i_\rho,\mu}) = |s|\gamma$, for all $s \in G_n$. Therefore $\gamma$ is an invariant of weight 1 (or a covariant, if we identify $\gamma$ with $\gamma \otimes 1$).
(3) Let $j_1, \ldots, j_n$ be $n$ integers drawn from $\underline{m} = \{1, \ldots, m\}$. Then the $n \times n$ determinant $\xi = (x^{(j_1)}, \ldots, x^{(j_n)}) = \det(x_{\mu,j_\rho})$ is an element of $K[x]$, and $\xi \circ s^{-1} = \det(x_{\mu,j_\rho}) \circ s^{-1}) = \det(\sum_{\nu \in \underline{n}} \hat{s}_{\mu\nu} x_{\nu,j_\rho}) = |s|^{-1}\xi$, for all $s \in G_n$. Thus $\xi$ $(= 1 \otimes \xi)$ is a covariant of weight -1.

The next theorem is sometimes called the "first fundamental theorem of invariant theory".

**Theorem 3.6.** *Every covariant $\varphi \in K[a] \otimes K[x]$ is a linear combination of products, each of whose factors is of one of the following types:*
(1) $(a_{(i)}|x^{(j)})$, *where* $i \in \underline{\ell}, j \in \underline{m}$,
(2) $(a_{(i_1)}, \ldots, a_{(i_n)})$, *where* $i_1, \ldots, i_n \in \underline{l}$ *or*
(3) $(x^{(j_1)}, \ldots, x^{(j_n)})$, *where* $j_1, \ldots, j_n \in \underline{m}$.
In particular, every invariant of the linear forms $(a_{(1)}|x), \ldots, (a_{(\ell)}|x)$ is a linear combination of products of type (2).

*Notes.* 1. Deruyts uses the term "invariant function" for what we have called a covariant; he uses the term "invariant" in the same sense as we do, *i.e.* for an invariant function which is independent of the variables; he has a term "identical covariant" for an invariant function independent of the coefficients. For example the polynomial $\xi$ (Example 3.5(3)) is an identical covariant ([De, pp. 4, 5]). See also [Tu, p. 206].
2. For a classical proof of theorem 3.6, based on the method of Hilbert, see [Tu, pp. 187–189 and pp. 210–211].
3. Deruyts derives theorem 3.6 very easily from a theorem on *semiinvariants* which occupies a central position in his theory (see 5.10, section 5 below, and [De, p. 60]).
4. Theorem 3.6 holds over infinite fields $K$ of arbitrary characteristic, see [Ro, p. 31].
5. Both theorem 3.6 and its generalizations for fields of finite characteristic should properly be called the "first fundamental theorem for a system of linear forms $(a_{(1)}|x), \ldots, (a_{(\ell)}|x)$". In the classical case ($\operatorname{char} K = 0$) the symbolic method allows one to extend the theorem to forms of arbitrary degree. But if $\operatorname{char} K$ is finite, the symbolic method breaks down — the reason is that the process for "reconstituting" a form $f(a, x)$ from its "symbolic expressions" involves rational coefficients which may be non-integral.

## 4. Isobaric Functions. Weights

We shall write $K^\times$ for the multiplicative group $K \setminus \{0\}$ of $K$, and $\mathbf{Z}^n$ for the additive group of all $n$-vectors $\pi = (\pi_1, \ldots, \pi_n)$ with integral coefficients.

*Definition 4.1.* A function $\varphi \in K[a] \otimes K[x]$ is said to be *isobaric* of weight $\pi \in \mathbf{Z}^n$ if

$$s \circ \varphi \circ s^{-1} = e_1^{\pi_1} \ldots e_n^{\pi_n} \cdot \varphi \qquad 4.2$$

holds for all diagonal matrices $s = d(e_1, \ldots, e_n)$, $e_1, \ldots, e_n \in K^\times$. Here

$$d(e_1, \ldots, e_n) = \begin{pmatrix} e_1 & & \\ & \ddots & \\ & & e_n \end{pmatrix} . \qquad 4.3$$

*Remark.* This definition is applied to elements $\gamma \in K[a]$ and elements $\xi \in K[x]$, by our convention which identifies these with $\gamma \otimes 1$ and $1 \otimes \xi$, respectively. So $\gamma$ is isobaric if $s \circ \gamma = e_1^{\pi_1} \ldots e_n^{\pi_n} \cdot \gamma$, and $\xi$ is isobaric (both $\gamma$ and $\xi$ being of weight $\pi$) if $\xi \circ s^{-1} = e_1^{\pi_1} \ldots e_n^{\pi_n} \cdot \xi$, for all $s = d(e_1, \ldots, e_n)$, $e_1, \ldots, e_n \in K^\times$.

*Examples 4.4.* (1) From 2.2 we see that $a_{i\nu}$ is isobaric of weight $(0, \ldots, 0, \underset{(\nu)}{1}, 0, \ldots, 0)$, for all $i \in \underline{\ell}, \nu \in \underline{n}$. Similarly 2.3 shows that $x_{\mu j}$ is isobaric of weight $(0, \ldots, 0, \underset{(\mu)}{-1}, 0, \ldots, 0)$, for all $\mu \in \underline{n}$, $j \in \underline{m}$.
(2) If $\varphi, \varphi'$ are isobaric, of weights $\pi, \pi'$ respectively, then clearly $\varphi\varphi'$ is isobaric of weight $\pi + \pi'$. The set of all $\varphi \in K[a] \otimes K[x]$ of given weight $\pi$, is a $K$-subspace of $K[a] \otimes K[x]$.
(3) Every monomial in the $a$'s and $x$'s is isobaric, *e.g.* if $n = 5$, then $a_{13}a_{25}^2x_{11}^3x_{15}x_{35}$ has weight $(-4, 0, 0, 0, 2)$. Consequently every $\varphi \in K[a] \otimes K[x]$ has a unique expression $\varphi = \sum_{\pi \in \mathbf{Z}^n} \varphi_\pi$, where $\varphi_\pi$ is isobaric of weight $\pi$.
(4) Comparing 3.4 with 4.2, we see that any covariant $\varphi$ is isobaric, and if $w$ is the weight of $\varphi$ as covariant, then $\varphi$ has weight $(w, \ldots, w)$ as isobaric function.
(5) Let $r$ be a positive integer, $r \leq n$ . Choose integers $h_1, \ldots, h_r$ from the set $\underline{n}$, and integers $i_1, \ldots, i_r$ from $\underline{\ell}$. Denote by $(a_{(i_1)}, \ldots, a_{(i_r)})_{h_1 \ldots h_r}$ the $r \times r$ determinant

$$\gamma = \det(a_{i_\rho h_\sigma}) = \begin{vmatrix} a_{i_1 h_1} & \cdots & a_{i_1 h_r} \\ \vdots & \ddots & \vdots \\ a_{i_r h_1} & \cdots & a_{i_r h_r} \end{vmatrix} . \qquad 4.5$$

By an easy calculation we find that $\gamma$ is isobaric of weight $(0, \ldots, 1, 0, \ldots, 1, 0, \ldots, 1, 0, \ldots, 0)$, where 1's stand in the places $h_1, \ldots, h_r$.
(6) Choose $r$ and $h_1, \ldots, h_r$ as in the last example, and let $j_1, \ldots, j_r$ be chosen from $\underline{m}$. Denote by $(x^{(j_1)}, \ldots, x^{(j_r)})_{h_1, \ldots, h_r}$ the $r \times r$ determinant

$$\xi = \det(x_{h_\rho j_\sigma}) = \begin{vmatrix} x_{h_1 j_1} & \cdots & x_{h_1 j_r} \\ \vdots & \ddots & \vdots \\ x_{h_r j_1} & \cdots & x_{h_r j_r} \end{vmatrix} . \qquad 4.6$$

Then $\xi$ is isobaric, and its weight is the negative of the weight shown in the last example.

*Notes.* 1. Deruyts defines weights for functions $\varphi$ involving coefficients of forms of any degree. For given $\nu \in \underline{n}$, he refers to $\pi_\nu$ as the "weight of $\varphi$ for the index $\nu$" (assuming that $\varphi$ is isobaric of weight $\pi$ in our sense). See [De, p. 32].
2. The $n \times n$ determinant denoted $(a_{(i_1)}, \dots, a_{(i_n)})$ in Example 3.5(2) can be described in the more detailed notation of Example 4.4(5) as $(a_{(i_1)}, \dots, a_{(i_n)})_{1\dots n}$. Similarly the determinant $(x^{(j_1)}, \dots, x^{(j_n)})$ of Example 3.5(3) has the alternate notation $(x^{(j_1)}, \dots, x^{(j_n)})_{1\dots n}$ under Example 4.4(6).

## 5. Semiinvariants

According to the definition in section 3, a function $\varphi \in K[a] \otimes K[x]$ is a *covariant* if the one-dimensional space $K\varphi$ generated by $\varphi$ is closed to the action $\varphi \to s \circ \varphi \circ s^{-1}$, for all $s$ in $G_n$. If we ask only that $K\varphi$ should be closed to this action for all $s$ in the "diagonal" subgroup

$$T_n = \left\{ d(e_1, \dots, e_n) \middle| e_1, \dots, e_n \in K^\times \right\}$$

of $G_n$, then $\varphi$ is *isobaric* (section 4). Between $T_n$ and $G_n$ lies the (now called) "Borel" or "upper triangular" subgroup of $G_n$, which consists of all $n \times n$ matrices

$$\begin{pmatrix} e_1 & s_{12} & \dots & s_{1n} \\ 0 & e_2 & \dots & s_{2n} \\ \vdots & \vdots & \ddots & \vdots \\ 0 & 0 & \dots & e_n \end{pmatrix} \tag{5.1}$$

with $e_\lambda \in K^\times, s_{\mu\nu} \in K$ for $\lambda, \mu, \nu \in \underline{n}$, $\mu < \nu$. Functions $\varphi$ such that $K\varphi$ is closed to the action of this group are called *semiinvariant*.

*Definition 5.2.* A function $\varphi \in K[a] \otimes K[x]$ is said to be a *semicovariant* of weight $\pi \in \mathbf{Z}^n$ if

$$s \circ \varphi \circ s^{-1} = e_1^{\pi_1} \dots e_n^{\pi_n} \cdot \varphi \tag{5.3}$$

holds for all matrices $s$ of form 5.1.

As usual, we may apply this definition to functions $\gamma \in K[a]$ and $\xi \in K[x]$; we adapt slightly Deruyts's terminology, as follows.

*Definition 5.4.* A function $\gamma \in K[a]$ is *semiinvariant* if $\gamma \otimes 1$ is a semicovariant, *i.e.* if $s \circ \gamma = e_1^{\pi_1} \dots e_n^{\pi_n} \cdot \gamma$, for all $s$ of form 5.1. A function $\xi \in K[x]$ is a *semiinvariant of the second kind* if $1 \otimes \xi$ is a semicovariant, *i.e.* if $\xi \circ s^{-1} = e_1^{\pi_1} \dots e_n^{\pi_n} \cdot \xi$, for all $s$ of form 5.1.

*Examples 5.5.* (1) Let $i_1, i_2, \dots, i_r$ be $r (\leq n)$ integers drawn from the set $\underline{\ell}$. For ease of notation we shall write $a = a_{(i_1)}$, $b = a_{(i_2)}$, $\dots h = a_{(i_r)}$. Then all the functions

$$a_1, (ab)_{12} = \begin{vmatrix} a_1 & a_2 \\ b_1 & b_2 \end{vmatrix}, \dots, (ab \dots h)_{12\dots r}$$

are semiinvariant. Their weights are $(1,0,0,\dots,0)$, $(1,1,0,\dots,0)$, $\dots$, $(\underbrace{1,1,1,\dots,1}_{r}, 0,\dots,0)$, respectively. (See Note 3, below.)

(2) Let $j_1, j_2, \ldots, j_r$ be $r(\leq n)$ integers drawn from the set $\underline{m}$. Write $u = x^{(j_1)}$, $v = x^{(j_2)}$, ..., $z = x^{(j_r)}$. Then all the functions

$$u_n, (uv)_{n-1,n} = \begin{vmatrix} u_{n-1} & v_{n-1} \\ u_n & v_n \end{vmatrix}, \ldots, (uv\ldots z)_{n-r+1\ldots n-1,n}$$

are semiinvariants (of the second kind). Their weights are $(0,\ldots,0,0,-1)$, $(0,\ldots,0,-1,-1)$ and $(0,\ldots,0,\underbrace{-1,\ldots,-1,-1}_{r})$, respectively. (See Note 3, below.)

**Theorem 5.6.** *Every semicovariant $\varphi \in K[a] \otimes K[x]$ is a linear combination of products, each of whose factors is of one of the following types:*
(1) $(a_{(i)}|x^{(j)})$, *where* $i \in \underline{\ell}, j \in \underline{m}$,
(2) $(a_{(i_1)}, \ldots, a_{(i_r)})_{1\ldots r}$, *where* $r \in \underline{n}$ *and* $i_1, \ldots, i_r \in \underline{\ell}$, *or*
(3) $(x^{(j_{r+1})}, \ldots, x^{(j_n)})_{r+1\ldots n}$, *where* $0 \leq r \leq n-1$ *and* $j_{r+1}, \ldots, j_n \in \underline{m}$.

*Remark 5.7.* We consider the weight of a product $P$ of factors of the above three types. A factor of type (1), since it is a covariant of weight 0, has weight $(0,0,\ldots,0)$ as isobaric function. A factor of type (2) has weight $(1,1,\ldots,1,0,\ldots,0)$, with $r$ 1's. A factor of type (3) has weight $(0,\ldots,0,-1,\ldots,-1)$, with $r$ 0's. So if $P$ contains $\alpha$ factors of type (1), $\beta_r$ factors of type (2) for each $r$ in the range $1 \leq r \leq n$, and $\gamma_r$ factors of type (3) for each $r$ in the range $0 \leq r \leq n-1$, then the weight of $P$ is

$$\pi = \sum_{r=0}^{n} (\beta_r, \ldots, \beta_r, -\gamma_r, \ldots, -\gamma_r) ,$$

where the term shown has $r$ entries $\beta_r$. From this we deduce: if $P$ has weight $\pi = (\pi_1, \ldots, \pi_n)$ then

$$\begin{aligned} \pi_r - \pi_{r+1} &= \beta_r + \gamma_r \ (1 \leq r \leq n-1), \text{ and} \\ \pi_n &= \beta_n - (\gamma_0 + \gamma_1 + \ldots + \gamma_{n-1}) . \end{aligned} \qquad 5.8$$

Since $\beta_r$, $\gamma_r$ are by definition non-negative integers, we have as corollary to theorem 5.6 the following

**Proposition 5.9.** *The weight $\pi = (\pi_1, \ldots, \pi_n)$ of any semi-covariant satisfies $\pi_1 \geq \pi_2 \geq \ldots \geq \pi_n$. If $\pi_1 = \pi_2 = \ldots = \pi_n$, then the semicovariant is a linear combination of products $P$ as above, for which $\beta_r = \gamma_r = 0$, for $r = 1, \ldots, n-1$. In this case we have $\pi_n = \beta_n - \gamma_0$.*

5.10. We may now deduce Theorem 3.6 from Theorem 5.6. Any covariant $\varphi \in K[a] \otimes K[x]$ of weight $w$ is a semicovariant, and the weight of $\varphi$ as isobaric function is $(w, w, \ldots, w)$ (see 4.4(4)). By 5.6 and 5.9 we see that $\varphi$ is a linear combination of products $P$, each of whose factors is of type 5.6(1), or of type 5.6(2) with $r = n$, or of type 5.6(3) with $r = 0$. But this is the statement of Theorem 3.6.

*Notes.* 1. Theorem 5.6 is proved in [De, pp. 57–60]. It has been shown to hold for infinite fields $K$ of arbitrary characteristic by C. de Concini, D. Eisenbud and C. Procesi ([CEP, p. 147]); the essential combinatorial argument of that proof is anticipated by D. G. Mead ([Me, pp. 169, 170]). See also [G", Theorem II'].
2. Deruyts uses "symbolic semi-invariant" ([De, p. 60]) for a "semiinvariant" in the sense of Definition 5.4.

3. To verify the semiinvariant property of the examples in 5.5, it is useful to notice that rules 2.2 and 2.3 can be written in matrix form; for example if $a \in \{a_{(1)}, \dots, a_{(\ell)}\}$ and $s \in \mathrm{GL}_n(K)$, then

$$(s \circ a_1, \dots, s \circ a_n) = (a_1, \dots, a_n)s \ . \tag{5.11}$$

So if $a, b \in \{a_{(1)}, \dots, a_{(\ell)}\}$ and if $s$ has form 5.1, then

$$\begin{pmatrix} s \circ a_1 & s \circ a_2 & \dots & s \circ a_n \\ s \circ b_1 & s \circ b_2 & \dots & s \circ b_n \end{pmatrix} = \begin{pmatrix} a_1 & a_2 & \dots & a_n \\ b_1 & b_2 & \dots & b_n \end{pmatrix} \begin{pmatrix} e_1 & s_{12} & \dots & s_{1n} \\ 0 & e_2 & \dots & s_{2n} \\ \vdots & \vdots & \ddots & \vdots \\ 0 & 0 & \dots & e_n \end{pmatrix},$$

hence

$$\begin{pmatrix} s \circ a_1 & s \circ a_2 \\ s \circ b_1 & s \circ b_2 \end{pmatrix} = \begin{pmatrix} a_1 & a_2 \\ b_1 & b_2 \end{pmatrix} \begin{pmatrix} e_1 & s_{12} \\ 0 & e_2 \end{pmatrix},$$

and taking determinants we get $s \circ (ab)_{12} = e_1 e_2 (ab)_{12}$.

In the same way, if $x \in \{x^{(1)}, \dots, x^{(m)}\}$ then the rule 2.3 shows that any $s \in \mathrm{GL}_n(K)$ acts on the column vector $(x_1, \dots, x_n)^T$ by *left* matrix multiplication, *i.e.*

$$(x_1 \circ s, \dots, x_n \circ s)^T = s(x_1, \dots, x_n)^T \ . \tag{5.12}$$

We may use this to verify example 5.5(2), for example that $(uv)_{n-1,n} \circ s^{-1} = e_{n-1}^{-1} e_n^{-1} \cdot (uv)_{n-1,n}$, when $s$ is the matrix 5.1.

4. *The operators $(h, l)$.* Our definition of semiinvariance (5.2, 5.3) is not the definition given by Deruyts in [De, p. 51], although he shows on pp. 52, 53 that the two are equivalent. In fact Deruyts's definition, and much of his discussion of semiinvariance, use certain operators $(h, l)$ which we now recognize as coming from the Lie algebras of the upper and lower unipotent subgroups of $G_n$. Let $U_n = U_n^+$ be the "upper unipotent" subgroup of $G_n$, which consists of all $n \times n$ matrices

$$\begin{pmatrix} 1 & s_{12} & \dots & s_{1n} \\ 0 & 1 & \dots & s_{2n} \\ \vdots & \vdots & \ddots & \vdots \\ 0 & 0 & \dots & 1 \end{pmatrix}, \tag{5.13}$$

with $s_{\mu\nu} \in K$ for all $\mu, \nu \in \underline{n}$, $\mu < \nu$. Each element $s$ of $B_n$ has a unique factorization $s = du$, with $d \in T_n$ and $u \in U_n$. So the condition 5.3 for a function $\varphi$ to be semiinvariant (*i.e.* to be a semicovariant) is equivalent to the two conditions

$$d \circ \varphi \circ d^{-1} = e_1^{\pi_1} \dots e_n^{\pi_n} \cdot \varphi, \text{ for all } d = d(e_1, \dots, e_n) \text{ in } T_n, \tag{1}$$

and

$$u \circ \varphi \circ u^{-1} = \varphi, \text{ for all } u \in U_n. \tag{2}$$

For any $\alpha, \beta \in \underline{n}$ with $\alpha \neq \beta$, and for any $t \in K$, let $u_{\alpha\beta}(t)$ be the $n \times n$ matrix whose $(\alpha, \beta)$-coefficient is $t$, and which coincides with $I_n \in \mathrm{GL}_n(K)$ at all its other coefficients. Then $U_{\alpha\beta} := \{u_{\alpha\beta}(t) : t \in K\}$ is a "root subgroup" of $G_n$. $U_n$

is generated by the set of all $U_{\alpha\beta}$ with $\alpha < \beta (\alpha, \beta \in \underline{n})$, and even by the set of all $U_{i,i+1}$ $(1 \leq i \leq n-1)$; Deruyts proves statements equivalent to these in [De, p. 52 and pp. 41, 42] (Deruyts uses $S_{hl}$ for a substitution of matrix $u_{lh}(\varepsilon)$, p. 36). Hence $\varphi$ is a semicovariant if and only if (1) holds (this just says that $\varphi$ is isobaric) and also

$$u \circ \varphi \circ u^{-1} = \varphi, \text{ for all } u \in U_{i,i+1}, i = 1, \dots, n-1 . \tag{2'}$$

Fix $\alpha, \beta \in \underline{n}$, $\alpha \neq \beta$ and let $u = u_{\alpha\beta}(t)$. We get $u \circ \varphi \circ u^{-1}$ from $\varphi = \varphi(\dots, a_{i\nu}, \dots, x_{\mu j}, \dots)$ by replacing each $a_{i\nu}$ by $u \circ a_{i\nu}$, and each $x_{\mu j}$ by $x_{\mu j} \circ u^{-1}$. By rules 2.2 and 2.3, $u \circ a_{i\nu} = a_{i\nu}$ if $\nu \neq \beta$, while $u \circ a_{i\beta} = a_{i\beta} + t a_{i\alpha}$ (all $i \in \underline{\ell}$); and $x_{\mu j} \circ u^{-1} = x_{\mu j}$ if $\mu \neq \alpha$, while $x_{\alpha j} \circ u^{-1} = x_{\alpha j} - t x_{\beta j}$ (all $j \in \underline{m}$). Now expand $u \circ \varphi \circ u^{-1} = \varphi(\dots, a_{i\beta} + t a_{i\alpha}, \dots, x_{\alpha j} - t x_{\beta j}, \dots)$ in powers of $t$, by Taylor's theorem. We get

$$u \circ \varphi \circ u^{-1} = \varphi + t \cdot \Delta_{\alpha\beta}\varphi + t^2 \cdot \frac{1}{2!}\Delta^2_{\alpha\beta}\varphi + \dots , \tag{5.14}$$

where $\Delta_{\alpha\beta}$ is the differential operator

$$\Delta_{\alpha\beta} = \sum_{i=1}^{\ell} a_{i\alpha} \frac{\partial}{\partial a_{i\beta}} - \sum_{j=1}^{m} x_{\beta j} \frac{\partial}{\partial x_{\alpha j}} . \tag{5.15}$$

In Deruyts's notation $\Delta_{\alpha\beta} = (\beta, \alpha)$ [De, p. 35]. By 5.14, $u \circ \varphi \circ u^{-1} = \varphi$ for all $t \in K$ (remember $u = u_{\alpha\beta}(t)$) if and only if $\Delta_{\alpha\beta}\varphi = 0$. Putting this into (2') we arrive at Deruyts's definition: an isobaric $\varphi$ is semiinvariant if and only if $\Delta_{i,i+1}\varphi = (i+1, i)\varphi = 0$, for $i = 1, \dots, n-1$.

## 6. Polar operators. Action of $G_l$, $G_m$

Deruyts makes much use of the standard invariant-theoretic technique of "polar operations". These can be derived from group actions exactly analogous to the actions of $G_n$ given in section 2.

Each element $g$ of $G_\ell = \mathrm{GL}_\ell(K)$ acts on $K[a] = K[a_{i\nu} : i \in \underline{\ell}, \nu \in \underline{n}]$ *on the right*, as follows. Let

$$a_{i\nu} \circ g = \sum_{h \in \underline{\ell}} g_{ih} a_{h\nu} , \text{ for all } i \in \underline{\ell}, \nu \in \underline{n} , \tag{6.1}$$

and extend this action to arbitrary $\gamma = \gamma(\dots, a_{i\nu}, \dots)$ by the rule $\gamma \circ g = \gamma(\dots, a_{i\nu} \circ g, \dots)$. Notice that $g$ acts linearly and multiplicatively on $K[a]$, and that $c \circ g = c$, for any constant polynomial $c \in K[a]$.

Similarly each $f$ in $G_m = \mathrm{GL}_m(K)$ acts on $K[x] = K[x_{\mu j} : \mu \in \underline{n}, j \in \underline{m}]$ *on the left*: we first let

$$f \circ x_{\mu j} = \sum_{k \in \underline{m}} f_{kj} x_{\mu k} , \text{ for all } \mu \in \underline{n}, j \in \underline{m} , \tag{6.2}$$

then extend this action to arbitrary $\xi = \xi(\dots, x_{\mu j}, \dots)$ by the rule $f \circ \xi = \xi(\dots, f \circ x_{\mu j}, \dots)$. Again, $f$ acts linearly and multiplicatively, and $f \circ c = c$, for any constant polynomial $c \in K[x]$.

The most important property of these actions is that they commute with appropriate actions of $G_n$, i.e. we have

$$(s \circ \gamma) \circ g = s \circ (\gamma \circ g), \ (f \circ \xi) \circ s = f \circ (\xi \circ s) \ , \qquad 6.3$$

for all $s \in G_n$, $g \in G_\ell$, $f \in G_m$, $\gamma \in K[a]$, $\xi \in K[x]$. The group actions 6.1, 6.2 extend, linearly, to the respective group algebras $KG_\ell$, $KG_m$; it is clear that 6.3 remains true if $g$, $f$ are replaced by any elements of $KG_\ell$, $KG_m$, respectively. Combining this last remark with the definitions of sections 2, 3, 4 we get the next proposition.

**Proposition 6.4.**

(1) *Let $\Omega \in KG_\ell$ and $\gamma \in K[a]$. If $\gamma$ is invariant of weight $w$, or isobaric of weight $\pi$, or semiinvariant of weight $\pi$, then $\gamma \circ \Omega$ has the corresponding property.*

(2) *Let $\Lambda \in KG_m$ and $\xi \in K[x]$. If $\xi$ is invariant of weight $w$, or isobaric of weight $\pi$, or semiinvariant (of the second kind) of weight $\pi$, then $\Lambda \circ \xi$ has the corresponding property.*

Now fix $x, y \in \underline{\ell}\ (x \neq y)$ and $t \in K$. Let $u_{xy}(t)$ be the matrix in $G_\ell$ whose $(x,y)$-coefficient is $t$, and which coincides with $I_\ell$ at all other coefficients. An argument like that in section 5, note 4, shows that if $\gamma = \gamma(\ldots, a_{i\nu}, \ldots)$ is any element of $K[a]$, then $\gamma \circ u_{xy}(t)$ is obtained by replacing $a_{x\nu}$ by $a_{x\nu} + t a_{y\nu}$ in $\gamma$, and leaving all the $a_{i\nu} (i \neq x)$ unchanged. Expanding $\gamma \circ u_{xy}(t)$ in powers of $t$ by means of Taylor's theorem we find

$$\gamma \circ u_{xy}(t) = \gamma_0 + t\gamma_1 + t^2\gamma_2 + \ldots + t^r\gamma_r \ , \qquad 6.5$$

where $\gamma_\rho = \frac{1}{\rho!}\Delta^\rho_{yx}\gamma \ (\rho = 0, 1, \ldots, r)$, and $\Delta_{yx}$ is *the polar operator in the coefficients*

$$\left(a_{(y)} \middle| \frac{\partial}{\partial a_{(x)}}\right) = \sum_{\nu \in \underline{n}} a_{y\nu} \frac{\partial}{\partial a_{x\nu}} \ . \qquad 6.6$$

We need to know that "polar operations" of this kind can be realized by the right action of suitable elements of $KG_\ell$.

**Proposition 6.7.** *Let $x, y \in \underline{\ell}\ (x \neq y)$ and let $\Delta = \left(a_{(y)} \middle| \frac{\partial}{\partial a_{(x)}}\right)$. Then for any finite-dimensional subspace $V$ of $K[a]$, there is an element $\Omega \in KG_\ell$ such that $\Delta\gamma = \gamma \circ \Omega$, for all $\gamma \in V$.*

**Proof.** The argument is standard. For each $\gamma \in V$ we have an equation 6.5. The "$r$" in 6.5 depends on $\gamma$, but since $\gamma \circ u_{xy}(t)$ is linear in $\gamma$ and $\dim V < \infty$, we can choose an $r$ so large that it will suffice for all $\gamma \in V$. From 6.5 we make a system of $r+1$ linear equations for $\gamma_0, \gamma_1, \ldots, \gamma_r$ by giving $t$ any $r+1$ distinct values $t_0, t_1, \ldots, t_r$ in $K$. The matrix of this system is the "Vandermonde" matrix $(t_v^\rho)\ (v, \rho = 0, 1, \ldots, r)$, which is non-singular. Therefore we may solve the system; in particular there exist $c_0, c_1, \ldots, c_r \in K$ such that $\Delta\gamma = \gamma_1 = \sum c_\rho(\gamma \circ u_{xy}(t_\rho)) = \gamma \circ \Omega$, where $\Omega = \sum c_\rho u_{xy}(t_\rho) \in KG_\ell$. Notice that $c_0, c_1, \ldots, c_r$ are made from the matrix $(t_v^\rho)$, which is the same for all $\gamma \in V$, hence $\Delta\gamma = \gamma \circ \Omega$ *for all* $\gamma \in V$.Q.E.D.

A general "polar operator on the coefficients", $O_c$, is a polynomial (over $K$) in the operators $\left(a_{(y)} \middle| \frac{\partial}{\partial a_{(x)}}\right)$ $(x, y \in \underline{\ell}, x \neq y)$; these operators do not commute, in general. Such a polar operator $O_c$ maps $K[a]$ into itself; by 6.7 its action on a given finite-dimensional subspace $V$ of $K[a]$ can be realized by the right action of some element $\Omega$ of $KG_\ell$. It is this last property which will be useful to us, and we shall not use the differential operators $\left(a_{(y)} \middle| \frac{\partial}{\partial a_{(x)}}\right)$ explicitly.

In the same way, a general "polar operator on the variables", $O_v$, is a polynomial in operators

$$\left(x^{(j)} \middle| \frac{\partial}{\partial x^{(k)}}\right) = \sum_{\mu \in \underline{n}} x_{\mu j} \frac{\partial}{\partial x_{\mu k}} \quad (j, k \in \underline{m}, j \neq k) ;$$

it maps $K[x]$ into itself, and its action on a given finite-dimensional subspace $V$ of $K[x]$ can be realized by the left action of some element $\Lambda$ of $KG_m$.

## 7. The map $\sigma : K[a] \to K[a] \otimes K[x]$

From now on we assume that $m = n$, so that we have exactly $n$ variable sets $x^{(1)}, \ldots, x^{(n)}$. Their components $x_{\mu j}$ $(\mu, j \in \underline{n})$ may be regarded as the coefficients of a "generic" $n \times n$ matrix $\mathcal{X}$, and the map

$$\sigma : K[a] \longrightarrow K[a] \otimes K[x] \qquad 7.1$$

which we next define is just the "generic" version of the left action on $K[a]$ of $G_n$ which we defined in 2.2. So we first define

$$\sigma(a_{i\nu}) = \sum_{\mu \in \underline{n}} a_{i\mu} \otimes x_{\mu\nu} \ , \ \text{for all } i \in \underline{\ell},\ \nu \in \underline{n} \ , \qquad 7.2$$

and then extend this to an arbitrary $\gamma = \gamma(\ldots, a_{i\nu}, \ldots)$ in $K[a]$ by the rule $\sigma(\gamma) = \gamma(\ldots, \sigma(a_{i\nu}), \ldots)$. Clearly $\sigma$ is a $K$-algebra map, *i.e.* it is linear, multiplicative and maps $1 \in K[a]$ to $1 \otimes 1 \in K[a] \otimes K[x]$ ($1 \otimes 1$ is just the constant polynomial 1 in the $a_{i\nu}$'s and $x_{\mu j}$'s).

This map (in a different notation, of course — see note 1 below) is fundamental in Deruyts's work. We give some of its properties.

**Proposition 7.3.** *If $s \in G_n$, let $\alpha_s : K[x] \to K$ be the "evaluation at $s$" map, i.e. the $K$-algebra map which takes $x_{\mu j}$ to $s_{\mu j}$ for all $\mu, j \in \underline{n}$. Then $(1' \otimes \alpha_s)\sigma(\gamma) = s \circ \gamma$, for all $\gamma \in K[a]$. ($1'$ is the identity map on $K[a]$.)*

This is immediate from the comparison of 7.2, 2.2. Now take $s = 1$, and write $\tau = 1' \otimes \alpha_1 : K[a] \otimes K[x] \to K[a]$, so that $\tau$ is the $K$-algebra map taking each $x_{\mu j}$ to $\delta_{\mu j}$, and each $a_{i\nu}$ to itself. From 7.3 we get

$$\tau\sigma = 1' \ . \ \text{Hence } \sigma \text{ is an injective map.} \qquad 7.4$$

**Theorem 7.5.** Im $\sigma$ *is the set of all $\varphi \in K[a] \otimes K[x]$ which satisfy*

$$s \circ \varphi = \varphi \circ s \ , \ \textit{for all } s \in G_n \ . \qquad 7.6$$

*In other words,* Im $\sigma$ *is the set of all covariants of weight zero in $K[a] \otimes K[x]$ (see 3.4).*

For proof of 7.5, see [De, p. 67], or note 4 below.

*Definition 7.7.* An element $\varphi \in K[a] \otimes K[x]$ will be said to be *doubly isobaric* of *double weight* $(\alpha, \beta)$ if it satisfies

$$\begin{aligned} d(e_1, \ldots, e_n) \circ \varphi &= e_1^{\alpha_1} \ldots e_n^{\alpha_n} \cdot \varphi \text{ and} \\ \varphi \circ d(e_1, \ldots, e_n) &= e_1^{\beta_1} \ldots e_n^{\beta_n} \cdot \varphi \end{aligned}, \qquad 7.8$$

for all $e_1, \ldots, e_n \in K^{\times}$.

It is clear (see 4.1) that such a $\varphi$ is isobaric in our earlier sense, of weight $\alpha - \beta$. Notice that the integers $\alpha_\nu, \beta_\nu$ which appear in 7.8 must be non-negative. Let $\mathbf{N}_0$ denote the set of all non-negative integers.

Every monomial in $K[a] \otimes K[x]$ is doubly isobaric; for example $a_{13}a_{25}^2 \cdot x_{11}^3 x_{15} x_{35}$ has double weight $(\alpha, \beta)$ with $\alpha = (0,0,1,0,2)$, $\beta = (4,0,1,0,0)$ (cf example 4.4(3)). It follows that every element $\varphi \in K[a] \otimes K[x]$ has a unique decomposition $\varphi = \sum_{\alpha,\beta} \varphi_{\alpha,\beta}$ where each $\varphi_{\alpha,\beta}$ is a doubly isobaric element of $K[a] \otimes K[x]$ of double weight $(\alpha, \beta)$; the sum is over all $\alpha, \beta \in \mathbf{N}_0^n$. We call $\varphi_{\alpha,\beta}$ the $(\alpha, \beta)$-component of $\varphi$.

**Proposition 7.9.** *Let $\gamma \in K[a]$, and let $\varphi = \sigma(\gamma)$. Then*
(1) *$\varphi$ is a sum of doubly isobaric functions, each having double weight of the form $(\alpha, \alpha)$, $\alpha \in \mathbf{N}_0^n$, and*
(2) *If $\gamma$ is isobaric of weight $\pi = (\pi_1, \ldots, \pi_n)$, then, for each $\nu \in \underline{n}$, $\varphi$ is homogeneous of degree $\pi_\nu$ in the variable-set $x^{(\nu)} = (x_{1\nu}, \ldots, x_{n\nu})$.*

**Proof.** (1) is true in the case $\gamma = a_{i\nu}$, since in that case $\varphi = a_{i1} \otimes x_{1\nu} + a_{i2} \otimes x_{2\nu} + \ldots + a_{in} \otimes x_{n\nu}$ (see 7.2). The general case follows from this.
(2) also comes easily from the case $\gamma = a_{i\nu}$.

*Notes.* 1. If $\gamma = \gamma(\ldots, a_{i\nu}, \ldots) \in K[a]$, the transform $\Gamma = \gamma(\ldots, A_{i\nu}, \ldots)$ of $\gamma$ under the substitution with matrix $s = (s_{\alpha\beta})$ may be regarded as polynomial in the $a_{i\nu}$'s and the coefficients $s_{\alpha\beta}$ of $s$. Deruyts writes $[\Gamma]$ for the result of replacing the $s_{\alpha\beta}$ by variables $x_{\alpha\beta}$; thus (apart from the $\otimes$ sign which we have introduced) $[\Gamma]$ is our $\sigma(\gamma)$ ([De, pp. 68, 69, 70]). Hilbert [Hi, p.248] also treats the "transformed" function $\Gamma$ as a polynomial in the coefficients of the substitution matrix.
2. We may think of $A = K[x]$ as the algebra of all polynomial functions on $G_n$, with $x_{\alpha\beta} : G_n \to K$ as the function which takes each $s \in G_n$ to its $(\alpha, \beta)$-coefficient. This is a *coalgebra* whose structure maps $\Delta : A \to A \otimes A$, $\varepsilon : A \to K$ are given by $\Delta(x_{\alpha\beta}) = \sum x_{\alpha\gamma} \otimes x_{\gamma\beta}$, $\varepsilon(x_{\alpha\beta}) = \delta_{\alpha\beta}$ (see [G', p. 19]). The action 2.2 makes $K[a]$ into a left $KG_n$-module which belongs to the category $M_K(n)$ (see [G', p. 8], notice that $A_K(n)$ in G' is our $A = K[x]$). It follows by a general procedure (see [G', p. 9]) that *$K[a]$ can be regarded as a right $A$-comodule* with structure map which is precisely the map $\sigma : K[a] \to K[a] \otimes A = K[a] \otimes K[x]$.
3. $K[a] \otimes K[x]$ can be made into a left $KG_n$-module, in which each $s \in G_n$ acts by the rule: $s(\gamma \otimes \xi) = \gamma \otimes s \circ \xi$ (here $s \circ \xi$ is given by 6.2; notice we assume $m = n$ ). It is now easy to check that *$\sigma : K[a] \to K[a] \otimes K[x]$ is a left $KG_n$-map*, in fact a $KG_n$-monomorphism (see 7.4). Moreover $K[a] \otimes K[x]$ (with the above $KG_n$-action) is an injective object of the category $M_K(n)$ (see [G, p. 150]), so that *$\sigma$ is an injective embedding of $K[a]$*.
4. **Proof of 7.5.** Let $\varphi \in K[a] \otimes K[x]$. The condition 7.6 for $\varphi$ be a covariant of weight zero can be written

$$(\sigma \otimes \mathbf{1})(\varphi) = (\mathbf{1}' \otimes \Delta)(\varphi) . \tag{7.10}$$

The comodule axioms for $K[a]$ give

$$(\sigma \otimes \mathbf{1})\sigma = (\mathbf{1}' \otimes \Delta)\sigma \text{ and} (\mathbf{1}' \otimes \varepsilon)\sigma = \mathbf{1}' . \tag{7.11}$$

In these equations we have written $\mathbf{1}$, $\mathbf{1}'$ for the identity maps on $K[x]$, $K[a]$, respectively. See [G, p. 138].

7.11 shows that $\varphi = \sigma(\gamma)$ satisfies 7.10, for all $\gamma \in K[a]$. Conversely let $\varphi$ be any element of $K[a] \otimes K[x]$ which satisfies 7.10. We shall prove that $\varphi = \sigma(\tau(\varphi))$ ($\tau$ is the map featured in 7.5), hence $\varphi \in \operatorname{Im} \sigma$.

Operate on both sides of 7.10 with the map $\mathbf{1}' \otimes \mathbf{1} \otimes \varepsilon$. On the left we get $(\sigma \otimes \varepsilon)(\varphi) = \sigma(\mathbf{1}' \otimes \varepsilon)(\varphi) = \sigma(\tau(\varphi))$; on the right is $(\mathbf{1}' \otimes (\mathbf{1} \otimes \varepsilon)\Delta)(\varphi) = (\mathbf{1}' \otimes \mathbf{1})(\varphi) = \varphi$, using the coalgebra axiom $(\mathbf{1} \otimes \varepsilon)\Delta = \mathbf{1}$ ([G, p. 137]), and remembering that $\varepsilon = \alpha_1$ in the notation of 7.3. This completes the proof of 7.5 — and the proof shows that 7.5 can be stated and proved for any right comodule (in place of $K[a]$) over any coalgebra (in place of $K[x]$).

## 8. Left and right spans of an element of $K[a] \otimes K[x]$

Deruyts had an effective method for expressing properties of a function $\gamma \in K[a]$ in terms of its image under $\sigma$. One feature of his technique can be expressed as follows. Let $X$, $Y$ be vector spaces over $K$ (whose dimensions need not be finite), and let $u$ be an element of the tensor product $X \otimes Y$ ($\otimes$ means $\otimes_K$, as usual). We may write

$$u = x_1 \otimes y_1 + \ldots + x_r \otimes y_r , \tag{8.1}$$

where $x_1, \ldots, x_r \in X$ and $y_1, \ldots, y_r \in Y$. Any such equation 8.1 is called an *expansion* of $u$, and $r$ is the *length* of this expansion.

*Definition 8.2.* Let $L(u)$ be the intersection of all subspaces $X'$ of $X$ such that $u \in X' \otimes Y$, and let $R(u)$ be the intersection of all subspaces $Y'$ of $Y$ such that $u \in X \otimes Y'$. These are subspaces of $X$ and $Y$, which will be called the *left* and *right spans*, respectively, of $u$.

**Proposition 8.3.** *Let $u$ be an element of $X \otimes Y$, and let 8.1 be an expansion of $u$. Then*

(1) *If $\{x_1, \ldots, x_r\}$ is a linearly independent set, then $R(u) = \langle y_1, \ldots, y_r \rangle$, the $K$-span of $\{y_1, \ldots, y_r\}$. Notice that it may happen that $\{y_1, \ldots, y_r\}$ is not independent.*
(2) *If $\{y_1, \ldots, y_r\}$ is linearly independent, then $L(u) = \langle x_1, \ldots, x_r \rangle$.*
(3) *Let $r(u)$ be the minimal length of all expansions 8.1 for $u$. If $r = r(u)$ we say that 8.1 is a* minimal *expansion of $u$. Then 8.1 is minimal if and only if both $\{x_1, \ldots, x_r\}$ and $\{y_1, \ldots, y_r\}$ are linearly independent sets. Minimal expansions of $u$ always exist (if $u = 0$, take $u = 0 \otimes 0$ as an expansion of length 0).*
(4) $\dim L(u) = r(u) = \dim R(u)$.
(5) *There is a non-singular bilinear form $\beta_u : L(u) \times R(u) \to K$ such that, if 8.1 is any minimal expansion of $u$, then $\beta_u(x_i, y_j) = \delta_{ij}$, for all $i, j \in \underline{r}$.*

These facts are quite easy to prove (proofs of (1) through (4) can be found in [G, pp. 141, 142]).

Now let $\varphi \in K[a] \otimes K[x]$, and let

$$\varphi = \gamma_1 \otimes \xi_1 + \ldots + \gamma_r \otimes \xi_r \qquad 8.4$$

be a minimal expansion of $\varphi$. If $\varphi$ is a covariant of weight zero, so that $s \circ \varphi = \varphi \circ s$ for all $s \in G_n$, it is easy to see from 8.3(3) that $L(\varphi)$ and $R(\varphi)$ are left and right $KG_n$-submodules of $K[a]$ and $K[x]$, respectively. Moreover the non-singular bilinear form $\beta = \beta_\varphi : L(\varphi) \times R(\varphi) \to K$ is $KG_n$-invariant, *i.e.* $\beta(s \circ \gamma, \xi) = \beta(\gamma, \xi \circ s)$ for all $s \in G_n$, $\gamma \in L(\varphi)$, $\xi \in R(\varphi)$. From this last fact it follows that $R(\varphi)$ is isomorphic to the dual $(L(\varphi))^*$ of $L(\varphi)$.

From 7.5 we know that $\varphi = \sigma(\gamma)$ for some $\gamma \in K[a]$; from 7.4 this $\gamma$ is unique and is given by $\gamma = \tau(\varphi)$. Deruyts calls $\gamma$ the *source* of $\varphi$ ([De, p. 67], see note 1., below).

**Proposition 8.5.** *If $\varphi = \sigma(\gamma)$, $\gamma \in K[a]$, then $L(\varphi)$ is the $KG_n$-submodule $KG_n \circ \gamma$ of $K[a]$ generated by $\gamma$.*

For a proof, see note 2, below. The next example shows how near we have come to a construction of the irreducible representations of $G_n$.

*Example 8.6.* Take $\gamma = a_{11}^r$, $r \in \mathbf{N}_0$. We calculate $\sigma(\gamma)$ from 7.2: $\sigma(\gamma) = \sigma(a_{11})^r = (a_{11} \otimes x_{11} + \ldots + a_{1n} \otimes x_{n1})^r = \sum_{\vec{r}} \kappa(\vec{r})(a_{11} \otimes x_{11})^{r_1} \ldots (a_{1n} \otimes x_{n1})^{r_n}$, the sum being over all $\vec{r} = (r_1, \ldots, r_n) \in \mathbf{N}_0^n$ such that $r_1 + \ldots + r_n = r$, and $\kappa(\vec{r}) = r!/r_1! \ldots r_n!$. Hence

$$\sigma(\gamma) = \sum_{\vec{r}} \kappa(\vec{r}) a_{11}^{r_1} \ldots a_{1n}^{r_n} \otimes x_{11}^{r_1} \ldots x_{n1}^{r_n} ,$$

and this is a *minimal* expansion for $\sigma(\gamma)$, because both sets $\{\kappa(\vec{r}) a_{11}^{r_1} \ldots a_{1n}^{r_n}\}$ and $\{x_{11}^{r_1} \ldots x_{n1}^{r_n}\}$ are linearly independent, in $K[a]$ and $K[x]$ respectively. So these sets are bases of the $KG_n$-modules $L(\sigma(\gamma))$ and $R(\sigma(\gamma))$, which may be identified with the Weyl modules (left and right $KG_n$-modules, respectively) of highest weight $(r, 0, \ldots, 0)$ (see [G', p. 67]).

*Notes.* 1. Let $\varphi \in K[a] \otimes K[x]$ be a covariant of weight zero, and homogeneous of degrees $\pi_1, \ldots, \pi_n$ in the variable-sets $x^{(1)}, \ldots, x^{(n)}$. The only monomial in these variables and of these degrees which is not annihilated by $\tau = \mathbf{1} \otimes \varepsilon_1$ (see 7.3, 7.4) is $x_{11}^{\pi_1} \ldots x_{nn}^{\pi_n}$. So we may describe the source $\tau(\varphi)$ of $\varphi$ as the coefficient, in $\varphi$, of $x_{11}^{\pi_1} \ldots x_{nn}^{\pi_n}$. This is Deruyts's definition of source [De, p. 67] (we take the case where Deruyts's variables $y_1, y_2, \ldots$ are absent).

2. **Proof of 8.5.** Since $\varphi = \sigma(\gamma)$ is a covariant, $L(\varphi)$ is a $KG_n$-submodule of $K[a]$; we must prove $L(\varphi) = KG_n \circ \gamma$. Take any minimal expansion 8.4 of $\varphi$. We know that $L(\varphi)$ is the $K$-span of $\gamma_1, \ldots, \gamma_r$. Also 7.3 gives us

$$s \circ \gamma = \gamma_1 \cdot \xi_1(s) + \ldots + \gamma_r \cdot \xi_r(s) \ , \ \text{all } s \in G_n \qquad 8.7$$

(notice that $\alpha_s(\xi) = \xi(s)$, for all $\xi \in K[x]$).

Since $\{\xi_1, \ldots, \xi_r\}$ is a linearly independent set we can find $s_1, \ldots, s_r \in G_n$ such that the matrix $(\xi_i(s_j))$ $(i, j \in \underline{r})$ is non-singular. Therefore we may solve for $\gamma_1, \ldots, \gamma_r$ the system of equations got by putting $s = s_1, \ldots, s_r$ in 8.7. Hence each

$\gamma_i$, and consequently the whole of $L(\varphi)$, lies in $KG_n \circ \gamma$. But if we put $s = 1$ in 8.7 we find that $\gamma = \gamma_1\xi_1(1) + \ldots + \gamma_r\xi_r(1) \in L(\varphi)$; therefore $KG_n \circ \gamma \subseteq L(\varphi)$, and 8.5 is proved.

3. Deruyts defines [De, p. 15] a set of functions $p_1, \ldots, p_r$ (we will assume that these lie in $K[a]$, to be specific), such that their $K$-span $\langle p_1, \ldots, p_r\rangle$ contains all their "transforms" $P_1, \ldots, P_r$ under any substitution, to be a "transformable system". In modern language, this says that $\langle p_1, \ldots, p_r\rangle$ is a $KG_n$-submodule of $K[a]$. The idea of a "minimal expansion" of a function $\varphi \in K[a] \otimes K[x]$ is explicit in [De, pp. 20, 21], where it is shown, if $\varphi$ is a covariant, that (in modern terms) $L(\varphi)$, $R(\varphi)$ are mutually "contragredient" (dual) $KG_n$-modules. Deruyts speaks of "coefficients" of a covariant $\varphi$, for example [De, p. 101], to mean (non-zero) elements of $L(\varphi)$. On pp. 104, 105 the term "irreducible expression" is used for a "minimal expansion" in our sense.

## 9. Primary covariants

*Definition 9.1.* A non-zero function $\varphi \in K[a] \otimes K[x]$ is a *primary covariant* if $\varphi$ is covariant of weight zero and its source $\gamma = \tau(\varphi)$ is semiinvariant.

*Remarks 9.2.* (1) This departs slightly from Deruyts's definition, see note 1, below.

(2) Every semiinvariant is isobaric (see 5.3, 5.4), and so $\gamma$ has a weight, $\pi = (\pi_1, \ldots, \pi_n)$, say. By 7.9(2), $\sigma(\gamma)$ is homogeneous of degree $\pi_\nu$ in the variable-set $x^{(\nu)}$, for each $\nu \in \underline{n}$.

*Examples 9.3.* (1) For any $r \in \mathbf{N}_0$, $\gamma = a_1^r$ is semiinvariant (cf Example 5.5(1)); hence the function $\varphi = \sigma(a_1^r)$ of Example 8.6 is a primary covariant.

(2) If $a, b, \ldots, h$ are any $r$ coefficient-sets drawn from $\{a_{(1)}, \ldots, a_{(\ell)}\}$, then $\gamma = (ab \ldots h)_{12\ldots r}$ is a semiinvariant (Example 5.5(1)), hence $\sigma(\gamma)$ is a primary covariant.

Primary covariants have a fundamental property, namely if $\varphi \in K[a] \otimes K[x]$ is a primary covariant, then both $L(\varphi)$ and $R(\varphi)$ are irreducible (= simple) $KG_n$-modules (left and right, respectively). This will be proved in the next section, but as a preliminary exercise, we shall try to calculate $L(\varphi)$ and $R(\varphi)$ for the primary covariant $\varphi = \sigma((ab \ldots h)_{12\ldots r})$ of example (2), above.

We introduce a convenient notation: if $i \in \underline{\ell}$, $\nu \in \underline{n}$, we shall write $(a_{(i)} \mid x^{(\nu)})$ for the element $\sigma(a_{i\nu}) = a_{i1} \otimes x_{1\nu} + \ldots + a_{in} \otimes x_{n\nu}$ (see 7.2); this looks more natural if we drop the anachronistic $\otimes$ signs and write $(a_{(i)} \mid x^{(\nu)}) = a_{i1}x_{1\nu} + \ldots + a_{in}x_{n\nu}$. Then

$$\sigma((a \ldots h)_{1\ldots r}) = \begin{vmatrix} (a \mid x^{(1)}) & \ldots & (a \mid x^{(r)}) \\ \vdots & \ddots & \vdots \\ (h \mid x^{(1)}) & \ldots & (h \mid x^{(r)}) \end{vmatrix}$$

The determinant on the right may be expanded by a well-known identity (see [Tu, p. 81]) as

$$\sum_{j_1 \ldots j_r} (a \ldots h)_{j_1 \ldots j_r} (x^{(1)} \ldots x^{(r)})_{j_1 \ldots j_r}\ , \tag{9.4}$$

where the sum extends over all $r$-element subsets $\{j_1, \dots, j_r\}$ of $\underline{n} = \{1, \dots, n\}$; we may introduce $\otimes$ between $(a \dots h)_{j_1 \dots j_r}$ and $(x^{(1)} \dots x^{(r)})_{j_1 \dots j_r}$ if we like.

It is quite easy to see that the summand displayed in 9.4 is doubly isobaric (see section 7) of double weight $(\alpha, \alpha)$, where $\alpha = \alpha_{j_1 \dots j_r} = (0, \dots, 1, 0, \dots, 1, 0, \dots, 1, 0, \dots, 0)$ is the element of $\mathbf{N}_0^n$ which has 1's in places $j_1, \dots, j_r$, and zeros elsewhere. But these weights $\alpha_{j_1 \dots j_r}$, as $\{j_1, \dots, j_r\}$ ranges over the $\binom{n}{r}$ $r$-element subsets of $\underline{n}$, are all distinct. Therefore each set $\{(a \dots h)_{j_1 \dots j_r}\}$ and $\{(x^{(1)} \dots x^{(r)})_{j_1 \dots j_r}\}$ is linearly independent, and so these sets are bases of $L(\varphi)$, and $R(\varphi)$, respectively. In case $r = 1$, $L(\varphi)$ is the $K$-span $E = \langle a_1, \dots, a_n \rangle$, and $G_n$ acts "naturally" *i.e.* $s \circ a_\nu = \sum_{\mu \in \underline{n}} s_{\mu\nu} a_\mu$ (see 2.2; remember $a = a_{(i)}$ for some $i \in \underline{\ell}$). For $1 \leq r \leq n$, $L(\varphi) = \bigwedge^r E$, the $r$th exterior power of $E$.

Now suppose $\gamma \in K[a]$ is an arbitrary (non-zero) semiinvariant of weight $\pi = (\pi_1, \dots, \pi_n)$. By 5.6 we may write $\gamma$ as a linear combination of products $P$, and since $\gamma$ does not involve the $x^{(\nu)}$'s, each $P$ is a product of determinants like $(ab \dots h)_{12 \dots r}$. Such a product may be represented by a "Young tableau", *e.g.* the product $P = (abcd)_{1234}(bce)_{123}(acd)_{123}(eb)_{12}$ is represented by

<table>
<tr><td>a</td><td>b</td><td>a</td><td>e</td></tr>
<tr><td>b</td><td>c</td><td>c</td><td>b</td></tr>
<tr><td>c</td><td>e</td><td>d</td><td></td></tr>
<tr><td>d</td><td></td><td></td><td></td></tr>
</table>

; 9.5

each column of the tableau 9.5 corresponds to a determinantal factor of $P$, and the length of the column is the order of this determinant. From 5.7, 5.8 (remember that $P$ has only factors of type 5.6(2)) we see that $P$ must have $\pi_r - \pi_{r+1}$ factors of order $r$ $(1 \leq r \leq n-1)$ and $\pi_n$ factors of order $n$; it is then easy to see that $\pi_1, \pi_2, \dots, \pi_n$ are the lengths of the rows of the tableau of $P$ (counting from the top). Thus the weight of the product $P$ in the example above is $(4, 4, 3, 1, 0, \dots, 0)$.

Giving the usual "lexicographic" order to the set $\mathbf{N}_0^n$ (so that if $\alpha, \beta \in \mathbf{N}_0^n$, $\alpha > \beta$ if there is some $k \in \underline{n}$ such that $\alpha_1 = \beta_1, \dots, \alpha_{k-1} = \beta_{k-1}$, $\alpha_k > \beta_k$), it is clear that the first or "highest" among the weights $\alpha_{j_1 \dots j_r}$ featuring in 9.4 is $\alpha_{1 \dots r} = (1, 1, \dots, 1, 0, \dots, 0)$ (there are $r$ 1's and $n - r$ zeros). The corresponding term in 9.4 is

$$(a \dots h)_{1 \dots r} \otimes (x^{(1)}, \dots, x^{(r)})_{1 \dots r} \, . \qquad 9.6$$

Now let $P = P_1 P_2 \dots P_t$ be a product of factors

$$P_i = (a[i], \dots, h[i])_{1, \dots, r[i]} \, ,$$

where for each $i \in \underline{t}$ the $a[i], \dots, h[i]$ are elements of the set $\{a_{(1)}, \dots, a_{(\ell)}\}$, and $r[i] \in \underline{n}$. Each $\sigma(P_i)$ can be expressed as a sum like 9.4. Multiplying out $\sigma(P) = \sigma(P_1) \dots \sigma(P_t)$ we arrive at an expansion

$$\sigma(P) = P \otimes x(\pi) + A_1 \otimes X_1 + A_2 \otimes X_2 + \dots \qquad 9.7$$

where

$$x(\pi) = \prod_{r \in \underline{n}} (x^{(1)}, \dots, x^{(r)})_{1 \dots r}^{\pi_r - \pi_{r+1}} \, . \qquad 9.8$$

The only summand in 9.7 with double weight $(\pi,\pi)$ is $P\otimes x(\pi)$; each of the other summands $A_\lambda\otimes X_\lambda$ has double weight $(\alpha,\alpha)$ for some $\alpha<\pi$. $P\otimes x(\pi)$ is the product of the highest terms 9.6 of the $\sigma(P_i)$, for $i\in \underline{t}$. Notice that the right factor in 9.6 depends only on $r$, hence $x(\pi)$ depends only on $\pi$; the exponent $\pi_r-\pi_{r+1}$ in 9.8 is the number of $P_i$ with $r[i]=r$ (we make the convention $\pi_{n+1}=0$ to cover the case $r=n$ ).

An arbitrary semiinvariant of weight $\pi$ is a linear combination of products like $P$ (see 5.6). So we have

**Theorem 9.9.** *If $\gamma\in K[a]$ is a non-zero semiinvariant of weight $\pi$, then the primary covariant $\varphi=\sigma(\gamma)$ has an expansion*

$$\sigma(\gamma)=\gamma\otimes x(\pi)+A_1\otimes X_1+A_2\otimes X_2+\dots\ , \qquad 9.10$$

*where $x(\pi)$ is given by 9.8. All the summands in 9.10 are doubly isobaric, and $\gamma\otimes x(\pi)$ is the only one whose double weight is $(\pi,\pi)$. Each of the other summands $A_\lambda\otimes X_\lambda$ has double weight $(\alpha,\alpha)$ for some $\alpha<\pi$.*

In general, 9.10 is not a *minimal* expansion of $\sigma(\gamma)$. But because $\gamma\otimes x(\pi)$ is the only summand in 9.10 with double weight $(\pi,\pi)$, then

$$\sigma(\gamma)=\gamma\otimes x(\pi)+B_1\otimes Y_1+B_2\otimes Y_2+\dots \qquad 9.11$$

is a minimal expansion of $\sigma(\gamma)$, for any minimal expansion $B_1\otimes Y_1+B_2\otimes Y_2+\dots$ of $A_1\otimes X_1+A_2\otimes X_2+\dots$. Moreover we may assume that each $B_\mu\otimes Y_\mu$ is doubly isobaric of double weight $(\alpha,\alpha)$, $\alpha<\pi$. Hence

**Theorem 9.12.** *If $\gamma\in K[a]$ is a non-zero semiinvariant of weight $\pi$, then $L(\sigma(\gamma))=KG_n\circ\gamma$ has a basis $\{\gamma,B_1,B_2,\dots\}$ where each $B_\mu$ is an isobaric element of $K[a]$ of weight $\alpha<\pi$.*

*Notes.* 1. If $r=n$, *all* the terms in 9.4 are divisible by $\delta=(x^{(1)},\dots,x^{(n)})_{1\dots n}$. It follows easily that $\sigma(\gamma)$ is divisible by $\delta^{\pi_n}$, for any semiinvariant $\gamma$ of weight $\pi$. Therefore $\chi=\sigma(\gamma)\delta^{-\pi_n}$ is a covariant of weight $\pi_n$ in $K[a]\otimes K[x]$, and moreover $\chi$ does not involve the variable-set $x^{(n)}$. Deruyts defines $\chi$ to be the primary covariant of source $\gamma$ ([De, p. 72]), whereas we have $\sigma(\gamma)$ to be the primary covariant of source $\gamma$.
2. Theorem 9.12 is proved in [De, p. 102].

## 10. Irreducibility of $L(\varphi)$, $\varphi$ a primary covariant

**Theorem 10.1.** *If $\varphi\in K[a]\otimes K[x]$ is a primary covariant, then $L(\varphi)$ is an irreducible left $KG_n$-submodule of $K[a]$.*

By definition 9.1 and proposition 8.5, this theorem is equivalent to

**Theorem 10.1'.** *If $\gamma\in K[a]$ is a non-zero semiinvariant, then the left $KG_n$-module $KG_n\circ\gamma$ is irreducible.*

We shall prove theorem 10.1 first in a special case, namely when $\varphi=\sigma(a(\pi))$, where

$$a(\pi) = \prod_{r \in \underline{n}} (a_{(1)}, \ldots, a_{(r)})_{1\ldots r}^{\pi_r - \pi_{r+1}} . \qquad 10.2$$

Here $\pi$ is any weight $(\pi_1, \ldots, \pi_n) \in \mathbf{N}_0^n$ satisfying $\pi_1 \geq \ldots \geq \pi_n$; it is clear that $a(\pi)$ has weight $\pi$ and is a non-zero semiinvariant; it is an analogue to $x(\pi)$ (9.8). In the notation of the last section, $a(\pi)$ is the product $P = P_1 \ldots P_t$ where, for each $i \in \underline{t}$ such that $r[i] = r$, $P_i = (a_{(1)}, \ldots, a_{(r)})_{1\ldots r}$. So 9.7 gives us

$$\sigma(a(\pi)) = a(\pi) \otimes x(\pi) + A_1 \otimes X_1 + A_2 \otimes X_2 + \ldots , \qquad 10.3$$

the first summand on the right is the only one which has double-weight $(\pi, \pi)$. The argument which leads up to 9.7 shows that we may take each $A_\lambda$ to be a product $A_\lambda = Q_{\lambda 1} \ldots Q_{\lambda t}$, where for each $i \in \underline{t}$ such that $r[i] = r$, $Q_{\lambda i} = (a_{(1)}, \ldots a_{(r)})_{k[i]}$, and $k[i]$ is short for an $r$-element subset $k[i]_1, \ldots, k[i]_r$ of $\underline{n}$. We may calculate $\sigma(A_\lambda) = \sigma(Q_{\lambda 1}) \ldots \sigma(Q_{\lambda t})$ in very much the same way as we did for $\sigma(P) = \sigma(P_1) \ldots \sigma(P_t)$, and get

$$\sigma(A_\lambda) = a(\pi) \otimes X_\lambda(\pi) + A_{\lambda 1} \otimes X_{\lambda 1} + A_{\lambda 2} \otimes X_{\lambda 2} + \ldots , \qquad 10.4$$

where the first summand on the right is the only summand of double-weight $(\pi, \pi)$, and

$$X_\lambda(\pi) = \prod_{i \in \underline{t}} (x^{(k_1)} \ldots x^{(k_r)})_{1\ldots r} , \qquad 10.5$$

where $k_1 = k[i]_1, \ldots, k_r = k[i]_r$. Let $q$ be any non-zero element of $L(a(\pi))$. Then 10.3 shows that $q = c_0 a(\pi) + c_1 A_1 + c_2 A_2 + \ldots$ for some $c_0, c_1, c_2, \ldots \in K$ (for this, it is not necessary that 10.3 be a *minimal* expansion). Use 10.3, 10.4 to calculate the $(\pi, \pi)$-component of $\sigma(q)$; it is $a(\pi) \otimes X$, where $X = c_0 x(\pi) + c_1 X_1(\pi) + c_2 X_2(\pi) + \ldots$. If $X = 0$ we have the relation

$$c_0 x(\pi) + c_1 X_1(\pi) + c_2 X_2(\pi) + \ldots = 0 . \qquad 10.6$$

But if we change each indeterminate $x_{\mu j}$ to $a_{j\mu}$ $(\mu, j \in \underline{n})$, $x(\pi)$ becomes $a(\pi)$, and $X_\lambda(\pi)$ becomes

$$\prod_{i \in \underline{t}} (a_{(1)}, \ldots, a_{(r)})_{k[i]} = A_\lambda .$$

Therefore 10.6 would give that $q = 0$, contradicting our hypothesis. It follows $X \neq 0$, so that the $(\pi, \pi)$ component of $a(\pi) \otimes X$ of $\sigma(q)$ is not zero. The argument leading up to theorem 9.12 now shows that $\sigma(q)$ has a minimal expansion containing $a(\pi) \otimes X$ as one of its summands. Therefore $KG_n \circ q = L(\sigma(q))$ contains $a(\pi)$. Since this holds for every non-zero element $q$ of $L(a(\pi)) = KG_n \circ a(\pi)$, it follows that $(a(\pi)) = KG_n \circ a(\pi)$ is irreducible.

The proof of theorem 10.1 is completed with the help of the following theorem, which will be proved in the next section.

**Theorem 10.7.** *Given any semiinvariant $\gamma \in K[a]$ of weight $\pi$, there exists an element $\Omega$ of $KG_\ell$ such that $\gamma = a(\pi) \circ \Omega$.*

Assuming this we may now prove 10.1'. Let $\gamma \in K[a]$ be any non-zero semiinvariant, and let $\Omega \in KG_\ell$ be such that $\gamma = a(\pi) \circ \Omega$. The map $u \to u \circ \Omega$ ($u \in K[a]$) is a $KG_n$-endomorphism of $K[a]$, and it induces a $KG_n$-epimorphism $\eta : KG_n \circ a(\pi) \to KG_n \circ \gamma$. But $KG_n \circ a(\pi)$ is irreducible, hence Ker $\eta = 0$, *i.e.* $\eta$ is a $KG_n$-isomorphism. So $KG_n \circ \gamma$, like $KG_n \circ a(\pi)$, is irreducible. This proves 10.1', and gives a little more, namely

**Theorem 10.8.** *Every non-zero semiinvariant $\gamma \in K[a]$ of weight $\pi$ generates a left $KG_n$-module $KG_n \circ \gamma$ which is isomorphic to $KG_n \circ a(\pi)$.*

*Notes.* 1. The proof that $L(\sigma(q))$ contains $a(\pi)$, for any non-zero $q \in KG_n \circ a(\pi)$, is given in [De, p. 103].
2. The proof that $KG_n \circ a(\pi)$ is simple is valid even for an infinite field $K$ of finite characteristic. However $KG_n \circ \gamma$ is not always irreducible in this case.

## 11. Proof of Theorem 10.7

Suppose that $\pi = (\pi_1, \ldots, \pi_n) \in \mathbf{N}_0^n$ is a given weight satisfying $\pi_1 \geq \pi_2 \geq \ldots \geq \pi_n$. The set SB($\pi$) of all $\gamma \in K[a]$ which are semiinvariant of weight $\pi$, is clearly a subspace of $K[a]$, and by 6.4 it is even a right $KG_\ell$-submodule of $K[a]$. Theorem 10.7 says that SB($\pi$) is a *cyclic* right $KG_\ell$-module, with $a(\pi)$ as generator.

Before we prove 10.7 we shall prove a proposition 11.1, below. This shows that the set ST($\pi$) of all $\gamma \in K[a]$ which are isobaric of weight $\pi$ is, like SB($\pi$), a cyclic right $KG_\ell$-submodule of $K[a]$. For this, we may take any $\pi \in \mathbf{N}_0^n$ — the condition $\pi_1 \geq \ldots \geq \pi_n$ is not needed.

**Proposition 11.1** *Let $\gamma \in K[a]$ be isobaric of weight $\pi$. Then there exists an element $\Omega$ of $KG_\ell$ such that $\gamma = (a_{11}^{\pi_1} a_{22}^{\pi_2} \ldots a_{nn}^{\pi_n}) \circ \Omega$.*

**Proof.** The space ST($\pi$) has as $K$-basis the set of all monomials

$$a(M) = \prod_{i \in \underline{\ell}} \prod_{\nu \in \underline{n}} a_{i\nu}^{m(i,\nu)} , \qquad 11.2$$

where $M = (m(i, \nu))$ runs over the set $\mathbf{M}(\pi)$ of all $\ell \times n$ matrices with non-negative integral coefficients satisfying

$$m(1, \nu) + \ldots + m(\ell, \nu) = \pi_\nu , \text{ for all } \nu \in \underline{n} . \qquad 11.3$$

Among these monomials is $a(J) = a_{11}^{\pi_1} a_{22}^{\pi_2} \ldots a_{nn}^{\pi_n}$, whose matrix $J$ is given by $m(i, \nu) = \delta_{i\nu}$. It will be enough to show that $a(J) \circ KG_\ell$ contains $a(M)$, for all $M \in \mathbf{M}(\pi)$.

Let $K[t]$ denote the $K$-algebra of all polynomials over $K$ in $\ell^2$ indeterminates $t_{hi}$ ($h, i \in \underline{\ell}$). Define a $K$-algebra map $\sigma' : K[a] \to K[t] \otimes K[a]$ as follows. First let

$$\sigma'(a_{i\nu}) = \prod_{h \in \underline{\ell}} t_{ih} \otimes a_{h\nu} , \qquad 11.4$$

for all $i \in \underline{\ell}$, $\nu \in \underline{n}$. Then extend this to arbitrary $\gamma = \gamma(\ldots, a_{i\nu}, \ldots)$ by the rule $\sigma'(\gamma) = \gamma(\ldots, \sigma'(a_{i\nu}), \ldots)$. Thus $\sigma'$ is a "generic" version of the right action of $G_\ell$ on $K[a]$ defined in 6.1; it is the right-hand analogue of the map $\sigma : K[a] \to K[a] \otimes K[x]$ (section 7). There holds therefore the following analogue of proposition 8.5.

**Proposition 11.5** *If* $\psi = \sigma'(\gamma)$, $\gamma \in K[a]$, *then* $R(\psi)$ *is the right* $KG_\ell$-*submodule* $\gamma \circ KG_\ell$ *of* $K[a]$.

Put $\gamma = a(J) = \prod_{\nu \in \underline{n}} a_{\nu\nu}^{\pi_\nu}$. Then by 11.4

$$\sigma'(\gamma) = \prod_{\nu \in \underline{n}} (t_{\nu 1} \otimes a_{1\nu} + t_{\nu 2} \otimes a_{2\nu} + \ldots + t_{\nu \ell} \otimes a_{\ell \nu})^{\pi_\nu} .$$

Expand the term $(\ldots)^{\pi_\nu}$ by the monomial theorem, and then form the product over $\nu \in \underline{n}$. Routine calculation gives

$$\sigma'(a(J)) = \sum_{M \in \mathbf{M}(\pi)} \kappa(M).t(M) \otimes a(M) , \qquad 11.6$$

where $t(M) = \prod_{i,\nu} t_{\nu i}^{m(i,\nu)}$ and $\kappa(M) = \pi_1! \ldots \pi_n! / \prod_{i,\nu} m(i,\nu)!$, for each $M \in \mathbf{M}(\pi)$. Both sets $\{\kappa(M).t(M)\}$ and $\{a(M)\}$ are linearly independent. Hence 11.6 is a minimal expansion of $\sigma'(a(J))$, and therefore $\{a(M)\}$ is a basis of $R(\sigma'(a(J)))$. By 11.5, each $a(M)$ lies in $a(J) \circ KG_\ell$, and proposition 11.1 is proved.

**Proof of 10.7.** Let $\pi \in N_0^n$ satisfy $\pi_1 \geq \ldots \geq \pi_n$. Let $\mathbf{N}(\pi)$ be the set of all $\ell \times n$ matrices with non-negative integral coefficients $m(i,\nu)$ which satisfy

$$m(i,1) + \ldots + m(i,n) = \pi_i \ , \ \text{for all } i \in \underline{n} \ , \qquad 11.7$$

and also $m(i,\nu) = 0$ for all $i \in \{n+1, \ldots, \ell\}$ and all $\nu \in \underline{n}$. Define $a(M)$ as in 11.2, and define

$$x(M) = \prod_{\nu \in \underline{n}} \prod_{i \in \underline{n}} x_{\nu i}^{m(i,\nu)} , \qquad 11.8$$

for all $M \in \mathbf{N}(\pi)$. Then $\{a(M) \mid M \in \mathbf{N}(\pi)\}$, resp. $\{x(M) \mid M \in \mathbf{N}(\pi)\}$, are the sets of all monomials of degrees $\pi_1, \ldots, \pi_n$ in $a_{(1)}, \ldots, a_{(n)}$, resp. $x^{(1)}, \ldots, x^{(n)}$. Taking $a(J)$ as before, a calculation like that which gave us 11.6 leads to

$$\sigma(a(J)) = \sum_{M \in \mathbf{N}(\pi)} \kappa(M).a(M) \otimes x(M) . \qquad 11.9$$

So the $(\pi, \pi)$-component of $\sigma(a(J))$ is

$$\sum_{M \in \mathbf{P}(\pi)} \kappa(M).a(M) \otimes x(M) \ , \ \text{where } \mathbf{P}(\pi) = \mathbf{M}(\pi) \cap \mathbf{N}(\pi) . \qquad 11.10$$

Let $\gamma \in K[a]$ be semiinvariant of weight $\pi$. By 11.1 there is some $\Omega \in KG_\ell$ such that $\gamma = a(J) \circ \Omega$. It follows easily from 11.10 (see note 4., below) that the $(\pi, \pi)$-component of $\sigma(\gamma) = \sigma(a(J) \circ \Omega)$ is

$$\sum_{M \in \mathbf{N}(\pi)} \kappa(M).(a(M) \circ \Omega) \otimes x(M) . \qquad 11.11$$

But 9.9 tells us that the $(\pi,\pi)$-component of $\sigma(\gamma)$ is $\gamma\otimes x(\pi)$; hence *11.11 is equal to* $\gamma\otimes x(\pi)$.

From their definitions (9.8 and 10.2) $a(\pi)$ and $x(\pi)$ can be written

$$a(\pi)=\sum_{M\in\mathbf{P}(\pi)}\zeta(M)a(M)\text{ and }x(\pi)=\sum_{M\in\mathbf{P}(\pi)}\zeta(M)x(M)\ , \qquad 11.12$$

respectively; the coefficients $\zeta(M)$ are the same for both. Comparison of 11.11 and $\gamma\otimes x(M)$ shows that $\kappa(M)(a(M)\circ\Omega)=\zeta(M)\gamma$, for all $M\in\mathbf{P}(\pi)$. Multiply both sides of this equation by $\zeta(M)\kappa(M)^{-1}$, and then sum over all $M\in\mathbf{P}(\pi)$. The sum on the left is $a(\pi)\circ\Omega$, by 11.12. So we find $a(\pi)\circ\Omega=\zeta'\gamma$, where $\zeta'$ is the positive rational number $\zeta(M)^2\ \kappa(M)^{-1}$. This completes the proof of 10.7.

*Notes.* 1. For proposition 11.1 see [De, pp. 61, 62]. Deruyts uses a polar operator on the coefficients in place of our $\Omega$ (see section 6), and the result he gives (top of p. 62) is a statement about covariants $\sigma(\gamma)$, and can be deduced from $\gamma=(a_{11}^{\pi_1}\dots a_{nn}^{\pi_n})\circ\Omega$ by applying $\sigma$.
2. The proof of 11.1 goes through without problem when $K$ is an infinite field of finite characteristic $p$, until we come to 11.6. Here some of the integers $\kappa(M)$ may become zero, and for such $M$ the monomial $a(M)$ will not lie in $a(J)\circ KG_\ell$.
3. We have followed Deruyts's elegant proof of 10.7 very closely ([De, pp. 62–64]).
4. To derive 11.11 from 11.10 one needs the following fact: if $f\in K[a]$ and if $\sigma(f)=\sum f_i\otimes\xi_i$ for certain $f_i\in K[a]$ and $\xi_i\in K[x]$, then

$$\sigma(f\circ\Omega)=\sum(f_i\circ\Omega)\otimes\xi_i\ , \qquad 11.13$$

for any $\Omega\in KG_\ell$. This is easily reduced to the case $f=a_{i\nu}$ (for some $i\in\underline{\ell}$, $\nu\in\underline{n}$) and $\Omega=g\in G_\ell$; for that case 11.13 comes at once from definitions 6.1, 7.2.

## 12. The Deruyts decomposition. Application: polynomial representations of $G_n$

We end our discussion with an impressive theorem of Deruyts [De, p. 81] which implies (see the end of this section) that all polynomial representations of $\mathrm{GL}_n(K)$ are completely reducible. As usual, we assume that $\operatorname{char} K=0$.

**Theorem 12.1** *Let $\gamma$ be an element of $K[a]$. Then there exist semiinvariants $\gamma^{(1)},\dots,\gamma^{(q)}$ and elements $\Lambda_1,\dots,\Lambda_q$ of $KG_n$ such that $\gamma=\Lambda_1\circ\gamma^{(1)}+\dots+\Lambda_q\circ\gamma^{(q)}$.* (See note 1, below).

**Proof.** Let

$$\sigma(\gamma)=\gamma_1\otimes\xi_1+\dots+\gamma_r\otimes\xi_r \qquad 12.2$$

be a minimal expansion of $\sigma(\gamma)$ ($\gamma_1,\dots,\gamma_r\in K[a]$ and $\xi_1,\dots,\xi_r\in K[x]$). We can arrange that $\{\gamma_1,\dots,\gamma_r\}$ is any given basis of $L(\sigma(\gamma))$, and if all the $\gamma_\rho$ are isobaric then all the $\xi_\rho$ will automatically be isobaric, and each $\gamma_\rho\otimes\xi_\rho$ will have double weight $(\alpha,\alpha)$ for some $\alpha\in\mathbf{N}_0^n$ (see note 2, below). Assume that the $\gamma_\rho$ are all isobaric, and let $\gamma'$ be one of them, which has maximal (lexicographic) weight. *Then $\gamma'$ is semiinvariant* (see note 3, below). Since $L(\sigma(\gamma'))=KG_n\circ\gamma'\subseteq KG_n\circ\gamma=L(\sigma(\gamma))$ (see 8.5) we may choose our basis $\{\gamma_1,\dots,\gamma_r\}$ of $L(\sigma(\gamma))$ so that it includes any given basis of $L(\sigma(\gamma'))$ as subset. Because $\gamma'$ is a non-zero semiinvariant (of weight $\pi$, say) there exists a minimal expansion $\sigma(\gamma')$ which is like 9.11,

$$\sigma(\gamma') = \gamma_1 \otimes \eta_1 + \ldots + \gamma_s \otimes \eta_s \;, \tag{12.3}$$

$\gamma_1 = \gamma'$ and $\eta_1 = x(\pi)$, and all $\gamma_\sigma \otimes \eta_\sigma$ doubly isobaric of weight $(\alpha, \alpha)$, $\alpha \in \mathbf{N}_0^n$. So we start with the basis $\gamma_1, \ldots, \gamma_s$ of $L(\sigma(\gamma'))$ provided by 12.3, extend this to a basis $\gamma_1, \ldots, \gamma_r$ (of isobaric elements) of $L(\sigma(\gamma))$, and we arrange that these $\gamma_\rho$ are the left factors appearing in 12.2.

If $\alpha$ is the weight of some $\gamma_\rho$ ($\rho \in \underline{r}$), then the weight of $\xi_\rho$ is $-\alpha$; hence, since $\gamma_1 = \gamma'$ has the greatest of the weights of $\gamma_1, \ldots, \gamma_r$, so must $\xi_1$ have least weight among the weights of $\xi_1, \ldots, \xi_r$, namely $-\pi$. By the "right" analogue of the fact proved in note 3., below, $\xi_1$ is a semiinvariant "of the second kind" (see 5.4). And by the right analogue of theorem 10.7, there must exist some $\Lambda \in KG_n$ (see note 4., below) such that

$$\xi_1 = \Lambda \circ x(\pi) = \Lambda \circ \eta_1 \;. \tag{12.4}$$

The minimal expansions 12.2 and 12.3 determine two $KG_n$-invariant, non-singular bilinear forms $\beta = \beta_{\sigma(\gamma)} : L(\sigma(\gamma)) \times R(\sigma(\gamma)) \to K$ and $\beta' = \beta_{\sigma(\gamma')} : L(\sigma(\gamma')) \times R(\sigma(\gamma')) \to K$, which satisfy the equations $\beta(\gamma_\rho, \xi_\sigma) = \delta_{\rho\sigma}$ ($\rho, \sigma \in \underline{r}$) and $\beta'(\gamma_\rho, \eta_\sigma) = \delta_{\rho\sigma}$ ($\rho, \sigma \in \underline{s}$) (see 8.3, 8.4). The map

$$h : R(\sigma(\gamma)) \longrightarrow R(\sigma(\gamma')) \tag{12.5}$$

which is adjoint, with respect to $\beta$ and $\beta'$, to the inclusion $i : L(\sigma(\gamma')) = KG_n \circ \gamma' \to KG_n \circ \gamma = L(\sigma(\gamma))$, is a $KG_n$-epimorphism (since $i$ is a $KG_n$-monomorphism), and may be calculated from the equations $\beta'(u', h(v)) = \beta(i(u'), v)$, for all $u' \in L(\sigma(\gamma'))$, $v \in R(\sigma(\gamma))$. From these we find at once

$$h(\xi_\rho) = \eta_\rho \; (\rho = 1, \ldots, s) \;, \; h(\xi_\rho) = 0 \; (\rho = s+1, \ldots, r) \;. \tag{12.6}$$

Theorem 10.1 tells us that $L(\sigma(\gamma'))$ is an irreducible left $KG_n$-module, hence its dual $R(\sigma(\gamma'))$ is an irreducible right $KG_n$-module. Therefore $R(\sigma(\gamma')) = \eta_1 \circ KG_n$, because $\eta_1$ is a non-zero element of $R(\sigma(\gamma'))$ (see 12.3).

We shall prove that the map $h$ (see 12.5) splits, *i.e.* that there exists a right $KG_n$-map $f : R(\sigma(\gamma')) \to R(\sigma(\gamma))$ such that $hf$ is the identity map on $R(\sigma(\gamma'))$. In fact this map is given by $f(\eta) = \Lambda \circ \eta$, for any $\eta \in R(\sigma(\gamma'))$. We know that $\eta = \eta_1 \circ \Omega$ for some $\Omega \in KG_n$. Hence $\Lambda \circ \eta = \Lambda \circ \eta_1 \circ \Omega = \xi_1 \circ \Omega$, by 12.4; this shows that $\Lambda \circ \eta \in R(\sigma(\gamma))$, and so $f$ is well-defined. It is clearly a right $KG_n$-map (see 6.3). Moreover $h(f(\eta_1)) = h(\Lambda \circ \eta_1) = h(\xi_1) = \eta_1$ (12.4, 12.6), and hence $h(f(\eta)) = \eta$ holds for all $\eta \in R(\sigma(\gamma')) = \eta_1 \circ KG_n$.

From 12.3 (see also section 7, note 3)

$$\sigma(\Lambda \circ \gamma') = \gamma_1 \otimes \Lambda \circ \eta_1 + \ldots + \gamma_s \otimes \Lambda \circ \eta_s \;. \tag{12.7}$$

By what has just been proved, $h(\xi_\sigma - \Lambda \circ \eta_\sigma) = \eta_\sigma - \eta_\sigma = 0$, hence $\xi_\sigma - \Lambda \circ \eta_\sigma \in \operatorname{Ker} h = K\xi_{s+1} + \ldots + K\xi_r$, for all $\sigma \in s$. So the right span of

$$\sigma(\gamma - \Lambda \circ \gamma') = \sum_{\sigma=1}^{s} \gamma_\sigma \otimes (\xi_\sigma - \Lambda \circ \eta_\sigma) + \sum_{\rho=s+1}^{n} \gamma_\rho \otimes \xi_\rho$$

lies in $K\xi_{s+1} + \ldots + K\xi_r$, and so has dimension $\leq r - s$. Thus the left span of $\sigma(\gamma - \Lambda \circ \gamma')$, which is $KG_n \circ (\gamma - \Lambda \circ \gamma')$, has dimension $\leq r - s$. Now $KG_n \circ \gamma$ and $KG_n \circ \gamma'$ have dimensions $r$, $s$ respectively, and clearly $KG_n \circ \gamma = KG_n \circ \gamma' + KG_n \circ (\gamma - \Lambda \circ \gamma')$. Therefore consideration of dimensions proves that this sum is direct:

$$KG \circ \gamma = KG \circ \gamma' \oplus KG \circ (\gamma - \Lambda \circ \gamma') . \qquad 12.8$$

An induction argument now completes the proof of theorem 12.1.

## Application of Deruyts theory: polynomial representations of $GL_n(K)$

In the notation of [G'], the objects of the category $M_K(n)$ are those locally finite $KG_n$-modules $V$ whose coefficient spaces cf$(V)$ lie in the coalgebra $A = K[c_{\mu\nu} : \mu, \nu \in \underline{n}]$ of all polynomial functions of $G_n$ into $K$.

To apply Deruyts's theory, take $\ell = n$ and identify $a_{i\nu}$ with $c_{i\nu}$ $(i, \nu \in \underline{n})$. This identifies $A$ with $K[a]$, and the left $G_n$-action on $K[a]$ (see 2.2) with the left $G_n$-action on $A$ given in [G', p. 34, (2.8a)].

Theorem 10.1 shows that the submodule $KG_n \circ \gamma$ generated by a non-zero semiinvariant $\gamma \in A$, is simple (= irreducible), and if $\gamma$ has weight $\pi$, then $KG_n \circ \gamma \cong KG_n \circ a(\pi)$. It is clear that if $\gamma$, $\gamma'$ are semiinvariants of distinct weights $\pi$, $\pi'$ then $KG_n \circ a(\pi) \ncong KG_n \circ (\pi')$, since these two modules have distinct "highest weights" (theorem 9.12). Every simple module in $M_K(n)$ is isomorphic to some left submodule $V'$ of $A$ ([G, p. 146, (1.3b)]). Take any non-zero element $\psi$ of $V'$. Since $V'$ is simple, $V' = KG_n \circ \psi$; now theorem 12.1 shows that $V' = KG_n \circ \gamma$ for some semiinvariant $\gamma \in A$. Therefore the set $KG_n \circ a(\pi)$, $\pi$ running over all elements of $\mathbb{N}_0^n$ which satisfy $\pi_1 \geq \ldots \geq \pi_n$, is a full set of simple objects of $M_K(n)$. Moreover 12.1 shows that every element of $A$ lies in a sum of simple submodules of $A$, hence $A$ is a semisimple coalgebra (see [G, p. 147]). It follows ([G, p. 147, (1.3c)]) every object $V$ of $M_K(n)$ is completely reducible.

*Notes.* 1. On p.81 of [De] is given the statement: "Every covariant of weight zero, in the variables $x^{(1)}, \ldots, x^{(n)}$, is a sum of powers of the determinant $(x^{(1)}, \ldots, x^{(n)})$, multiplied by polars [relative to the variables] of primary covariants." We may forget the powers of $(x^{(1)}, \ldots, x^{(n)})$, since these are built in to our definition of primary covariant (section 9, note 1). A polar operation relative to the variables, can be realized by the left action of an element $\Lambda$ of $KG_m = KG_n$ on $K[x]$ (see end of section 6). Then we may paraphrase Deruyts's statement as: "Every covariant $\varphi \in K[a] \otimes K[x]$ can be decomposed as a sum $\varphi = \Lambda_1\varphi_1 + \ldots + \Lambda_q\varphi_q$, where $\varphi_1, \ldots, \varphi_q$ are primary covariants, and $\Lambda_1, \ldots, \Lambda_q$ are elements of $KG_n$." Here the left action of $KG_n$ on $K[a] \otimes K[x]$ is by the rule $\Lambda(\gamma \otimes \xi) = \gamma \otimes \Lambda \circ \xi$. The map $\sigma$ induces a left $KG_n$-isomorphism from $K[a]$ onto the space of all covariants of weight zero (7.5, and note 3, section 7), and *primary* covariants correspond to semiinvariants in $K[a]$. So Deruyts's theorem (as paraphrased above) is equivalent to the statement that every element $\gamma$ of $K[a]$ can be written as a sum $\Lambda_1 \circ \gamma^{(1)} + \ldots + \Lambda_q \circ \gamma^{(q)}$, where $\gamma^{(1)}, \ldots, \gamma^{(q)}$ are semiinvariants, and $\Lambda_1, \ldots, \Lambda_q$ are elements of $KG_n$.
2. For any minimal expansion 12.2 of $\sigma(\gamma)$, the set $\{\gamma_1, \ldots, \gamma_r\}$ is a basis of $L(\sigma(\gamma))$. If we write each $\gamma_\rho$ as linear combination of the elements of some other basis $\{\gamma'_1, \ldots, \gamma'_r\}$ of $L(\sigma(\gamma))$, we can make a new minimal expansion of $\sigma(\gamma)$ in which $\gamma'_1, \ldots, \gamma'_r$ take the place of $\gamma_1, \ldots, \gamma_r$. Suppose now that each $\gamma_\rho$ (in 12.2) is isobaric, of weight $\alpha(\rho) = (\alpha_{\rho 1}, \ldots, \alpha_{\rho n})$. Putting $s = d(e_1, \ldots, e_n)$ in the equations

$\sum(s\circ\gamma_\rho)\otimes\xi_\rho = \sum\gamma_\rho\otimes(\xi_\rho\circ s)$ (see 7.6) we get $\sum\gamma_\rho\otimes(e_1^{\alpha_{\rho 1}}\dots e_n^{\alpha_{\rho n}})\xi_\rho = \sum\gamma_\rho\otimes(\xi_\rho\circ s)$. Comparing coefficients of $\gamma_\rho$, we find $\xi_\rho\circ s = (e_1^{\alpha_{\rho 1}}\dots e_n^{\alpha_{\rho n}})\xi_\rho$ for all $e_1,\dots,e_n$ in $K^\times$; hence $\xi_\rho$ is isobaric of weight $-\alpha(\rho)$, or equivalently, $\gamma_\rho\otimes\zeta_\rho$ is doubly isobaric of double weight $(\alpha(\rho),\alpha(\rho))$.

3. Deruyts proves that $\gamma'$ is semiinvariant on pp. 76, 77 of [De], using the operators $(h,l)$. We sketch a proof in terms of the group elements $u_{\alpha\beta}(t)$ (section 5, note 4.). Take any $\nu\in\{1,\dots,n-1\}$. We have equations

$$u_{\nu,\nu+1}(t)\circ\gamma' = \gamma' + t\gamma^{(1)} + t^2\gamma^{(2)} + \dots\,, \qquad 12.9$$

for all $t\in K$, where $\gamma^{(1)}$, $\gamma^{(2)}$, ... are elements of $K[a]$, independent of $t$. Equations 12.9 are exactly parallel to equations 6.5, and by the argument which proved proposition 6.7, we find that $\gamma^{(1)}$, $\gamma^{(2)}$, ... all lie in $KG_n\circ\gamma'\subseteq KG_n\circ\gamma = L(\sigma(\gamma))$. Apply $d = d(e_1,\dots,e_n)$ to both sides of 12.9, and use the fact that $d\cdot u_{\nu,\nu+1}(t)\cdot d^{-1} = u_{\nu,\nu+1}(e_\nu e_{\nu+1}^{-1}\cdot t)$. We find that, for each $m = 1,2,\dots$, $\gamma^{(m)}$ is isobaric of weight $\pi + (0,\dots,0,m,-m,0,\dots,0)$ ($m$ in place $\nu$). Now this weight is strictly larger than $\pi$, and $\pi$ is maximal among weights of non-zero elements of $KG_n\circ\gamma$. It follows $\gamma^{(m)} = 0$ for all $m\geq 1$, hence by 12.9, $\gamma'$ is invariant to the root subgroup $U_{\nu,\nu+1}$ ($\nu = 1,\dots,n-1$). But this shows that $\gamma'$ is invariant to $U_n^+$ and so is semiinvariant (section 5, note 4.).

4. If $\xi\in K[x]$ is semiinvariant of the second kind, of weight $-\pi$, then there exists an element $\Lambda\in KG_m$ such that $\xi = \Lambda\circ x(\pi)$. The proof is exactly parallel to that of theorem 10.7.

## REFERENCES


[CEP] C. de Concini, D. Eisenbud, C. Procesi, "Young diagrams and determinantal varieties", Invent. Math. 56 (1980), 129–165.

[De] J. Deruyts, "Essai d'une théorie générale des formes algébriques", Mém. Soc. Roy. Sci. Liège 17 (1892), 1–156.

[GY] J.H. Grace, A. Young, "The algebra of invariants", Cambridge University Press (1903).

[G] J.A. Green, "Locally finite representations", J. of Algebra 41 (1976), 137–171.

[G'] J.A. Green, "Polynomial representations of $GL_n$", Lect. Notes Math. 830, Springer, Berlin Heidelberg New York (1980).

[G"] J.A. Green, "Schur algebras and general linear groups", in Groups St. Andrews 1989, Cambridge University Press (1991).

[Hi] D. Hilbert, "Über die Theorie der algebraischen Formen", Math. Annalen 36 (1890), 473–534.

[Hi'] D. Hilbert, "Über die vollen Invariantensysteme", Math. Annalen 42 (1893), 313–373.

[Hu] A. Hurwitz, "Zur Invariantentheorie", Math. Annalen 45 (1894), 381–404.

[M] D. G. Mead, "Determinantal ideals, identities, and the Wronskian", Pacific J. Math. 42 (1972), 165–175.

[Ro] G.-C. Rota, "Théorie combinatoire des invariants classiques", IRMA 1/S-01, Université Louis Pasteur, Strasbourg (1977).

[S] I. Schur, "Über eine Klasse von Matrizen, die sich einer gegebenen Matrix zuordnen lassen", in I. Schur, Gesammelte Abhandlungen I, Springer, Berlin (1973), 1–70.

[Tu] H. W. Turnbull, "The theory of determinants, matrices and invariants", Blackie and Son, London Glasgow (1929).

[W] H. Weyl, "The classical groups", Princeton University Press, Princeton N. J. (1953).

Progress in Mathematics, Vol. 95, © 1991 Birkhäuser Verlag Basel

# Partial characters of $\pi$-separable groups

I. M. ISAACS*

## 1. Introduction.

Our concern in this expository paper is the character theory of a finite group $G$ as seen from the perspective of a set $\pi$ of prime numbers. There are no new results here and few new ideas. Our purpose is to present in as accessible a manner as possible, the proofs of some theorems in the character theory of $\pi$-separable groups, and to explain the significance of these results.

Our principal results are Theorems A and B, stated in the next section. These were first obtained in [2] as consequences of a more complete theory, and they were reproved with a simpler and more direct argument by Robinson and Staszewski in [5], where module theoretic interpretations were introduced.

The arguments of this paper are very closely related to (and are inspired by) [5]. We have removed the module theory, however, and we have (we believe) simplified the proofs. In order to make this work more accessible to readers unfamiliar with the methods of [2] or [5], we have also included (in Sections 3-5) some additional preliminary material.

## 2. The main results.

We fix a set $\pi$ of prime numbers and write $G^\circ$ to denote the set of $\pi$-elements of $G$. As usual, $cf(G)$ is the complex vector space of class functions on $G$ and we write $cf(G^\circ)$ to denote the space of class functions on $G^\circ$. For $\varphi \in cf(G)$, we use the notation $\varphi^\circ$ for the restriction of $\varphi$ to $G^\circ$. Of course, $\varphi^\circ \in cf(G^\circ)$.

The set $\mathrm{Irr}(G)$ of ordinary irreducible characters of $G$ is a basis for $cf(G)$ and in the case $\pi = p'$, the complement of a single prime, the set $\mathrm{IBr}_p(G)$ of irreducible $p$-Brauer characters of $G$ is a basis for $cf(G^\circ)$. (Note that in general, the set $\mathrm{IBr}_p(G)$ is not uniquely determined.)

A basis $\mathcal{U}$ of $cf(G^\circ)$ may happen to satisfy the following "decomposition property".

> (DP) If $\chi \in \mathrm{Irr}(G)$, then $\chi^\circ$ is a nonnegative integer linear combination of $\mathcal{U}$.

For example, $\mathrm{IBr}_p(G)$ satisfies (DP) if $\pi = p'$. Another property enjoyed by $\mathrm{IBr}_p(G)$ if $G$ happens to be $p$-solvable, is that if $\varphi \in \mathrm{IBr}_p(G)$, then there exists $\chi \in \mathrm{Irr}(G)$ with $\chi^\circ = \varphi$. (This is the Fong-Swan theorem.) In general, we consider bases $\mathcal{U}$ for $cf(G^\circ)$ (with $\pi$ arbitrary) which satisfy the following "Fong-Swan" property.

---

* Research partially supported by a grant from the National Science Foundation.

(FS) If $\varphi \in \mathcal{U}$, then there exists $\chi \in \mathrm{Irr}(G)$ with $\chi^\circ = \varphi$.

In the general case, where $\pi$ is arbitrary, we seek bases $\mathcal{U}$ for $cf(G^\circ)$ such that $\mathcal{U}$ satisfies both (DP) and (FS). Such bases do not always exist, but if $G$ is $\pi$-separable (i.e. each composition factor is either a $\pi$-group of a $\pi'$-group), the existence is guaranteed.

**THEOREM A.** *Let $G$ be $\pi$-separable. Then there exists a unique basis for $cf(G^\circ)$ which satisfies both (DP) and (FS).*

Since the uniqueness assertion in Theorem A is practically a triviality, we dispose of it immediately.

**Proof of uniqueness in Theorem A.** Suppose $\mathcal{U}_1$ and $\mathcal{U}_2$ are two bases for $cf(G^\circ)$, each satisfying (DP) and (FS). If $\varphi \in \mathcal{U}_1 \cup \mathcal{U}_2$, then by (FS), we have $\varphi = \chi^\circ$ for some $\chi \in \mathrm{Irr}(G)$, and thus $\varphi(1) = \chi(1)$ is a nonnegative integer, the *degree* of $\varphi$.

If $\mathcal{U}_1 \neq \mathcal{U}_2$, choose $\varphi \in \mathcal{U}_1 \cup \mathcal{U}_2$ of minimal degree such that $\varphi$ lies in just one of the bases (say $\mathcal{U}_1$). Since $\varphi = \chi^\circ$ for $\chi \in \mathrm{Irr}(G)$, we can use (DP) to write

$$\varphi = \sum_{\theta \in \mathcal{U}_2} a_\theta \theta$$

where the coefficients $a_\theta$ are nonnegative integers. Since $\varphi \notin \mathcal{U}_2$, this expresses $\varphi$ as a sum (with possible repeats) of members $\theta \in \mathcal{U}_2$ with properly smaller degree. By the choice of $\varphi$, all of these summands $\theta$ lie in $\mathcal{U}_1$ as well as in $\mathcal{U}_2$, and this contradicts the linear independence of $\mathcal{U}_1$. □

We shall henceforth write $\mathrm{I}_\pi(G)$ to denote the basis $\mathcal{U}$ of Theorem A. This set satisfies some useful properties with respect to a Hall $\pi$-subgroup $H$ of $G$. (Note that since $G$ is $\pi$-separable, $H$ exists and is unique up to conjugacy.)

If $\varphi \in \mathrm{I}_\pi(G)$, then the restriction $\varphi_H$ is defined on all of $H$ and, in fact, it is a character of $H$. This is so since we have $\varphi = \chi^\circ$ for some $\chi \in \mathrm{Irr}(G)$ by (FS), and hence $\varphi_H = \chi_H$.

**THEOREM B.** *Let $H$ be a Hall $\pi$-subgroup of the $\pi$-separable group $G$. Let $\varphi \in \mathrm{I}_\pi(G)$ and choose an irreducible constitutent $\alpha$ of $\varphi_H$ of smallest possible degree. Then*

a) $\alpha(1) = \varphi(1)_\pi$, *the $\pi$-part of* $\varphi(1)$.

b) $[\varphi_H, \alpha] = 1$.

c) $[\mu_H, \alpha] = 0$ *if* $\varphi \neq \mu \in \mathrm{I}_\pi(G)$.

It should be stressed that the character $\alpha \in \mathrm{Irr}(H)$ is not, in general, uniquely determined by $\varphi$; it need not even be unique up to conjugacy by $\mathbf{N}_G(H)$. Elsewhere, we have referred to $\alpha$ as a *Fong character* associated with $\varphi$ and we have characterized the Fong characters among the irreducible characters of $H$. (See [2] and [4].)

The original proofs of Theorems A and B in [2] depended on the construction of a certain canonically defined set $\mathrm{B}_\pi(G)$ of irreducible characters of $G$. While this approach enables one to prove certain additional properties of $\mathrm{I}_\pi(G)$ (see the discussion in Section 7), it is considerably more difficult than the method of Robinson and Staszewski in [5] which is essentially the content of Theorem C, below.

In order to state Theorem C, we introduce the vector subspace $vcf(G) \subseteq cf(G)$, consisting of functions with $\pi$-support. (In other words, $\varphi \in cf(G)$ lies in $vcf(G)$ provided that $\varphi(x) = 0$ whenever $x \in G$ is not a $\pi$-element.) Note that $vcf(G)$ is naturally isomorphic to $cf(G^\circ)$ via the restriction map $\varphi \mapsto \varphi^\circ$.

**THEOREM C.** *Let $G$ be $\pi$-separable and choose a Hall $\pi$-subgroup $H \subseteq G$. Then there exist subsets $\mathcal{A} \subseteq \mathrm{Irr(H)}$ and $\mathcal{B} \subseteq \mathrm{Irr(G)}$ and a bijection $\sigma : \mathcal{A} \to \mathcal{B}$ such that*

a) $\{\alpha^G \mid \alpha \in \mathcal{A}\}$ *spans* $vcf(G)$.

b) *If* $\alpha \in \mathcal{A}$ *and* $\beta \in \mathcal{B}$, *then*

$$[\alpha^G, \beta] = \begin{cases} 1 & \text{if } \beta = \sigma(\alpha) \\ 0 & \text{otherwise} . \end{cases}$$

c) $\sigma(\alpha)(1)_\pi = \alpha(1)$ *for* $\alpha \in \mathcal{A}$.

The proof of this result occupies most of the rest of this paper. We show now, however, how to derive Theorems A and B from it.

**Proof of Theorem A.** Let $H$ be a Hall $\pi$-subgroup of $G$ and apply Thoerem C to obtain subsets $\mathcal{A} \subseteq \mathrm{Irr}(H)$ and $\mathcal{B} \subseteq \mathrm{Irr}(G)$ and a bijection $\sigma : \mathcal{A} \to \mathcal{B}$ such that $\mathcal{A}^G$ spans $vcf(G)$ and also $[\alpha^G, \beta]$ is 1 or 0 according to whether or not $\sigma(\alpha) = \beta$.

We observe first that the functions $\beta^\circ \in cf(G^\circ)$ are distinct and linearly independent as $\beta$ runs over $\mathcal{B}$. To see this, suppose that

$$\sum_{\beta \in \mathcal{B}} a_\beta \beta^\circ = 0$$

with complex coefficients $a_\beta$. Restriction to $H$ yields

$$\sum_{\beta \in \mathcal{B}} a_\beta \beta_H = 0 .$$

Now fix $\alpha \in \mathcal{A}$. Then $[\alpha, \beta_H] = 0$ unless $\beta = \sigma(\alpha)$, in which case $[\alpha, \beta_H] = 1$. This gives

$$a_{\sigma(\alpha)} = [\alpha, \sum a_\beta \beta_H] = [\alpha, 0] = 0 ,$$

as required. It follows that

$$|\mathcal{B}| \leq \dim(cf(G^\circ)) .$$

Also, since $\mathcal{A}^G$ spans $vcf(G)$, we have

$$|\mathcal{A}| \geq |\mathcal{A}^G| \geq \dim(vcf(G)) .$$

Since $|\mathcal{A}| = |\mathcal{B}|$ and $\dim(cf(G^\circ)) = \dim(vcf(G))$, equality holds throughout and it follows that $\{\beta^\circ \mid \beta \in \mathcal{B}\}$ is a basis for $cf(G^\circ)$. Property (FS) is obvious for this basis and we proceed to check (DP).

Let $\chi \in \mathrm{Irr}(G)$ and express $\chi^\circ$ in terms of our basis, so that

$$\chi^\circ = \sum_{\beta \in \mathcal{B}} b_\beta \beta^\circ$$

with complex $b_\beta$. Then

$$\chi_H = \sum_{\beta \in \mathcal{B}} b_\beta \beta_H$$

and so, if $\alpha \in \mathcal{A}$ we have

$$b_{\sigma(\alpha)} = [\alpha, \sum b_\beta \beta_H] = [\alpha, \chi_H],$$

a nonnegative integer. This establishes (DP) and proves the existence of the desired basis. We have already proved uniqueness. □

We separate a part of the proof of Theorem B as a lemma.

**2.1 LEMMA.** *Assume the situation and notation of Theorem C. If $\chi \in vcf(G)$ is a character of $G$, then $\chi$ is a linear combination with nonnegative integer coefficients of $\{\alpha^G \mid \alpha \in \mathcal{A}\}$.*

**Proof.** By Theorem C(a), we can write

$$\chi = \sum_{\alpha \in \mathcal{A}} a_\alpha \alpha^G,$$

with complex $a_\alpha$. Then for $\beta \in \mathcal{B}$, we have

$$a_{\sigma^{-1}(\beta)} = \left[\sum a_\alpha \alpha^G, \beta\right] = [\chi, \beta],$$

which is a nonnegative integer since $\chi$ is a character and $\beta \in \mathrm{Irr}(G)$. □

**Proof of Theorem B.** Since we know (in the language of Theorem C) that $\mathrm{I}_\pi(G) = \{\beta^\circ \mid \beta \in \mathcal{B}\}$, choose $\beta \in \mathcal{B}$ with $\beta^\circ = \varphi$, and let $\alpha$ be any irreducible constituent of $\varphi_H = \beta_H$ with smallest degree. Write $\gamma = \sigma^{-1}(\beta)$.

Now $H$ consists entirely of $\pi$-elements and so $\alpha^G \in vcf(G)$ and by the lemma, $\alpha^G$ is a sum of some $\eta^G$ for $\eta \in \mathcal{A}$. We have $[\alpha^G, \beta] \neq 0$ and yet $[\eta^G, \beta] = 0$ for all $\eta \in \mathcal{A}$ except $\eta = \gamma$. It follows that $\gamma^G$ is a summand of $\alpha^G$, and in particular, $\alpha(1) \geq \gamma(1)$.

Since $\gamma$ is a constituent of $\beta_H = \varphi_H$, it follows from the choice of $\alpha$ that $\alpha(1) = \gamma(1)$. Thus

$$\alpha(1) = \gamma(1) = \beta(1)_\pi = \varphi(1)_\pi$$

by Theorem C(c), and (a) is established.

Since $\alpha(1) = \gamma(1)$ and $\gamma^G$ is a summand of $\alpha^G$, we deduce that $\alpha^G = \gamma^G$ and hence

$$[\varphi_H, \alpha] = [\beta_H, \alpha] = [\beta, \alpha^G] = [\beta, \gamma^G] = 1,$$

proving (b). If $\varphi \neq \mu \in \mathrm{I}_\pi(G)$, then $\mu = \delta^\circ$ for some $\delta \in \mathcal{B}$ with $\delta \neq \beta$. A similar calculation then yields

$$[\mu_H, \alpha] = [\delta, \gamma^G] = 0,$$

and (c) is proved. □

In order to justify the publication of a new proof of a known result, a necessary condition is that the result should have some importance. In fact, Theorems A and B appear to contribute significantly to our understanding of the representation theory of solvable groups. The following result, for example, is an easy consequence, and yet it appears inaccessible by other methods.

**THEOREM D.** *Let $H$ be a Hall subgroup of the solvable group $G$, and suppose that $\psi$ is a character of $H$. Then a necessary and sufficient condition that $\psi$ is the restriction $\chi_H$ for some character $\chi$ of $G$, is that $\psi(x) = \psi(y)$ whenever $x, y \in H$ are conjugate in $G$.*

Although the proof of Theorem D has appeared elsewhere [3], it is short enough to reproduce here so that we can illustrate one way that Theorems A and B can be used.

**Proof of Theorem D.** The necessity of the condition is clear and so we assume that $\psi \in \mathrm{Char}(H)$ satisfies the condition, and we produce $\chi \in \mathrm{Char}(G)$ with $\chi_H = \psi$.

Let $\pi$ be the set of prime divisors of $H$ and note that since $G$ is solvable, it is $\pi$-separable. If $g \in G^\circ$, then by the P. Hall $D_\pi$ property, there exists $x \in H$, conjugate to $g$. We define the function $\xi$ on $G^\circ$ by setting $\xi(g) = \psi(x)$ and we note that this is well defined since if $g$ is also conjugate to $y \in H$, then $\psi(x) = \psi(y)$ by the assumption on $\psi$.

Clearly $\xi \in cf(G^\circ)$ and so we can write

$$\xi = \sum_{\varphi \in \mathrm{I}_\pi(G)} a_\varphi \varphi$$

for some complex numbers $a_\varphi$. By (FS), we can choose $\chi_\varphi \in \mathrm{Irr}(G)$ such that $(\chi_\varphi)^\circ = \varphi$ and we write

$$\chi = \sum_\varphi a_\varphi \chi_\varphi .$$

Since

$$(\chi_\varphi)_H = ((\chi_\varphi)^\circ)_H = \varphi_H ,$$

we have

$$\chi_H = \sum_\varphi a_\varphi \varphi_H = \xi_H = \psi$$

and it suffices to show that $\chi$ is a character.

We show, in fact, that $a_\varphi$ is a nonnegative integer for $\varphi \in \mathrm{I}_\pi(G)$. Let $\alpha \in \mathrm{Irr(H)}$ be as in Theorem B, an irreducible constitutent of $\varphi_H$ of least degree. Then

$$a_\varphi = [\sum_{\mu \in \mathrm{I}_\pi(G)} a_\mu \mu_H, \alpha] = [\xi_H, \alpha] = [\psi, \alpha]$$

and this is a nonnegative integer since $\psi$ and $\alpha$ are characters of $H$. □

## 3. Functions with $\pi$-support.

Recall that we have defined $vcf(G)$ to be the subspace of $cf(G)$ consisting of class functions vanishing on $G - G^\circ$. We derive some useful but technical information about $vcf(G)$ in this section.

If $H \subseteq G$ and $X \subseteq cf(H)$ is any subset, we write $X^G$ to denote $\{\xi^G \mid \xi \in X\}$. Note that $X^G \subseteq cf(G)$ and that if $X$ is a subspace of $cf(H)$, then $X^G$ is a subspace of $cf(G)$.

**3.1 LEMMA.** *Let $G$ be $\pi$-separable and suppose $H \subseteq G$ is a Hall $\pi$-subgroup. Then*

$$cf(H)^G = vcf(G).$$

**Proof.** Let $C$ be a conjugacy class of $H$ and let $\varphi$ be its characteristic function on $H$. Then $\varphi^G$ is a nonzero class function on $G$ which vanishes outside of the class $K$ of $G$ containing $C$. It follows that $\varphi^G = a\theta$ where $a \neq 0$ and $\theta$ is the characteristic function of $K$ on $G$.

Since $cf(H)$ is spanned by the characteristic functions of the classes of $H$, we deduce that $cf(H)^G$ is the span of the characteristic functions of those classes of $G$ which meet $H$ nontrivially. Since $H$ is a $\pi$-group, these consist only of $\pi$-classes; since $H$ is Hall and $G$ is $\pi$-separable, these are all of the $\pi$-classes of $G$ and the result follows. □

As usual, we write $\mathrm{Irr}(G \mid \theta)$ to denote the set of irreducible constituents of $\theta^G$, where $\theta$ is an irreducible character of some subgroup. Similarly, we write $cf(G \mid \theta)$ to denote the span of $\mathrm{Irr}(G \mid \theta)$ and $vcf(G \mid \theta) = cf(G \mid \theta) \cap vcf(G)$. If $N \triangleleft G$ and $\theta_1, \theta_2, \ldots, \theta_t \in \mathrm{Irr}(N)$ are representatives for the orbits of the action of $G$ on $\mathrm{Irr}(N)$, then $\mathrm{Irr}(G)$ is partitioned by the sets $\mathrm{Irr}(G \mid \theta_i)$ and consequently, we have the direct sum decomposition

$$cf(G) = \sum_{i=1}^{t}{}^{\bullet}\, cf(G \mid \theta_i).$$

**3.2 LEMMA.** *Suppose $G$ is $\pi$-separable and that $N \triangleleft G$ is a $\pi$-group. As above, we let $\theta_1, \theta_2, \ldots, \theta_t$ be representatives for the $G$-orbits of $\mathrm{Irr}(N)$. Then*

$$vcf(G) = \sum_{i=1}^{t}{}^{\bullet}\, vcf(G \mid \theta_i).$$

**Proof.** It is clear that the sum on the right is direct and is contained in $vcf(G)$. Now let $H \subseteq G$ be a Hall $\pi$-subgroup. Since $vcf(G) = cf(H)^G$ is spanned by $\mathrm{Irr}(H)^G$, it suffices to show that

$$\alpha^G \in \sum_i vcf(G \mid \theta_i)$$

for each $\alpha \in \mathrm{Irr}(H)$.

Now $N \subseteq H$ and we choose an irreducible constitutent $\theta$ of $\alpha_N$. Then $\theta^g = \theta_i$ for some $g \in G$ and some $i$ with $1 \leq i \leq t$ and we see that $H^g$ is a Hall $\pi$-subgroup of $G$ and $\alpha^g \in \mathrm{Irr}(H^g)$ satisfies $(\alpha^g)^G \in vcf(G)$. Since $\alpha^g \in \mathrm{Irr}(H^g \mid \theta_i)$, we have $\alpha^G = (\alpha^g)^G \in \mathrm{Irr}(G \mid \theta_i)$ and so $\alpha^G \in vcf(G \mid \theta_i)$ and the result follows. □

**3.3 COROLLARY.** *In the situation of 3.2, let $\varphi \in vcf(G)$ and write*

$$\varphi = \varphi_1 + \cdots + \varphi_t$$

*with $\varphi_i \in cf(G \mid \theta_i)$. Then each $\varphi_i \in vcf(G)$.* □

Our next result is a refinement of 3.1.

**3.4 LEMMA.** *Let $G$ be $\pi$-separable and suppose $N \triangleleft G$ is a $\pi$-group. Let $\theta \in \mathrm{Irr}(N)$ with inertia group $T = I_G(\theta)$ and suppose $H$ is a Hall $\pi$-subgroup of $G$ chosen so that $H \cap T$ is a Hall $\pi$-subgroup of $T$. Then*

$$cf(H \mid \theta)^G = vcf(G \mid \theta).$$

**Proof.** By Lemma 3.1, $cf(H \mid \theta)^G \subseteq vcf(G)$ and since it is also clear that $cf(H \mid \theta)^G \subseteq cf(G \mid \theta)$, we have that $cf(H \mid \theta)^G \subseteq vcf(G \mid \theta)$ and it suffices to prove the reverse inclusion.

Let $\varphi \in vcf(G \mid \theta)$ and write $\varphi = \alpha^G$ with $\alpha \in cf(H)$ (using 3.1). Decompose $\alpha = \alpha_1 + \alpha_2$ where $\alpha_1$ is a linear combination of irreducible characters of $H$ lying over $G$-conjugates of $\theta$ and no irreducible "constituent" of $\alpha_2$ lies over a $G$-conjugate of $\theta$. Then no irreducible constituent of $(\alpha_2)^G$ lies over $\theta$ and yet

$$(\alpha_2)^G = \varphi - (\alpha_1)^G \in cf(G \mid \theta).$$

This forces $(\alpha_2)^G = 0$ and $\varphi = (\alpha_1)^G$.

To complete the proof, it suffices to show that if $\beta \in \mathrm{Irr}(H)$ lies over a $G$-conjugate of $\theta$, then $\beta^G \in cf(H \mid \theta)^G$. We can choose a conjugate $K$ of $H$ and a character $\gamma \in \mathrm{Irr}(K \mid \theta)$ such that $\beta^G = \gamma^G$. By the Clifford correspondence, we can write $\gamma = \psi^K$ for some character $\psi \in \mathrm{Irr}(K \cap T \mid \theta)$. However, $K \cap T$ is a $\pi$-subgroup of $T$ and thus is conjugate in $T$ to a subgroup $U$ of the Hall subgroup $H \cap T$ of $T$. Therefore, we can write $\psi^t \in \mathrm{Irr}(U)$ for some $t \in T$, and since $\theta^t = \theta$, we see that $\psi^t$ lies over $\theta$. Now

$$\beta^G = \gamma^G = \psi^G = (\psi^t)^G = ((\psi^t)^H)^G$$

and $(\psi^t)^H \in cf(H \mid \theta)$, as required. □

Our final result of this section is a "Clifford correspondence" for $vcf$.

**3.5 COROLLARY.** *Assume that $G$ is $\pi$-separable and that $N \triangleleft G$ is a $\pi$-group. Let $\theta \in Irr(N)$ and $T = I_G(\theta)$. Then*

$$vcf(T \mid \theta)^G = vcf(G \mid \theta).$$

**Proof.** Let $K$ be a Hall $\pi$-subgroup of $T$ and choose a Hall $\pi$-subgroup $H$ of $G$ with $H \cap T = K$. By 3.4, we have

$$cf(K \mid \theta)^T = vcf(T \mid \theta) \qquad \text{and} \qquad cf(H \mid \theta)^G = vcf(G \mid \theta)$$

and the Clifford correspondence gives

$$cf(K \mid \theta)^H = cf(H \mid \theta).$$

Therefore,

$$vcf(T \mid \theta)^G = cf(K \mid \theta)^G = cf(H \mid \theta)^G = vcf(G \mid \theta),$$

as required. □

## 4. Overbars.

In this section we hold $K \triangleleft G$ fixed and we use overbars to denote the canonical homomorphism $G \to G/K$. We thus write $\overline{g} = Kg$ and $\overline{H} = KH/K$ for elements $g \in G$ and subgroups $H \subseteq G$.

We extend the bar notation to class functions as follows. If $H \subseteq G$ and $\varphi \in cf(H \mid 1_{K\cap H})$, then $\varphi$ is constant on cosets of $K \cap H$ in $H$ and so we can define $\overline{\varphi} \in cf(\overline{H})$ by $\overline{\varphi}(\overline{h}) = \varphi(h)$. This defines overbar as a linear transformation

$$cf(H \mid 1_{K\cap H}) \to cf(\overline{H})$$

and in fact, this map is easily seen to be injective. It is routine to check that

$$\overline{\mathrm{Irr}(H \mid 1_{K\cap H})} = \mathrm{Irr}(\overline{H})$$

and from this it follows that the map is surjective too, and that it preserves inner products.

If $H \supseteq K$ and $\varphi \in cf(H \mid 1_K)$, it is well known that $\varphi^G \in cf(G \mid 1_K)$ and $\overline{\varphi^G} = \overline{\varphi}^{\overline{G}}$. In order to extend this to the case where $K \not\subseteq H$, we introduce the projection map

$$\tau : cf(G) \to cf(G \mid 1_K),$$

defined so that $\tau(\chi) = \chi$ if $\chi \in \mathrm{Irr}(G \mid 1_K)$ and $\tau(\chi) = 0$, otherwise.

**4.1 LEMMA.** *Let $H \subseteq G$ and $\varphi \in cf(H \mid 1_{K\cap H})$. Then*

$$\overline{\varphi}^{\overline{G}} = \overline{\tau(\varphi^G)}.$$

**Proof.** By linearity, we may assume that $\varphi \in \mathrm{Irr}(H)$. Write $\varphi^{HK} = \xi + \eta$ where $K \subseteq \ker\xi$ and no irreducible constituent of $\eta$ lies in $\mathrm{Irr}(HK \mid 1_K)$. Now

$$\xi(1) = [(\varphi^{HK})_K, 1_K] = [(\varphi_{K\cap H})^K, 1_K] = [\varphi_{K\cap H}, 1_{K\cap H}] = \varphi(1),$$

and since $\varphi$ is a constitutent of $\xi_H$, we deduce that $\xi_H = \varphi$ and so $\overline{\xi} = \overline{\varphi}$. It follows that

$$\overline{\varphi}^{\overline{G}} = \overline{\xi}^{\overline{G}} = \overline{\xi^G},$$

where the second equality holds since $K \subseteq HK$.

Since $\varphi^G = \xi^G + \eta^G$, we have $\tau(\varphi^G) = \xi^G$, and the result follows. □

**4.2 LEMMA.** *Suppose $G$ is $\pi$-separable and that $K$ is a $\pi'$-group. Then the map $\varphi \mapsto \overline{\tau(\varphi)}$ is a vector space isomorphism $vcf(G) \to vcf(\overline{G})$.*

**Proof.** If $\varphi \in vcf(G)$, write $\varphi = \alpha^G$, where $\alpha \in cf(H)$ for some Hall $\pi$-subgroup $H$. Since $K \cap H = 1$, Lemma 4.1 yields

$$\overline{\tau(\varphi)} = \overline{\tau(\alpha^G)} = \overline{\alpha}^{\overline{G}}.$$

This lies in $vcf(\overline{G})$ since $\overline{\alpha} \in cf(\overline{H})$ and $\overline{H}$ is a Hall $\pi$-subgroup of $\overline{G}$.

To show that the map is surjective, let $\theta \in vcf(\overline{G})$. Write

$$\theta = \beta^{\overline{G}}$$

for some $\beta \in cf(\overline{H})$. We can choose (uniquely) $\alpha \in cf(H)$ with $\overline{\alpha} = \beta$ and thus

$$\theta = \overline{\alpha}^G = \overline{\tau(\alpha^G)}.$$

Since $\alpha^G \in vcf(G)$, the map is surjective, as desired.

Since the number of $\pi$-classes of $G$ is the same as that for $\overline{G}$, we have

$$\dim vcf(G) = \dim vcf(\overline{G})$$

and we deduce that our map must be injective. □

## 5. Character-triple isomorphism.

Suppose that $N \triangleleft G$ and $\theta \in \mathrm{Irr}(N)$ is invariant in $G$; in other words $(G, N, \theta)$ is a *character-triple.* If we wish to study $\mathrm{Irr}(G \mid \theta)$ in this case, it is usually no loss to assume that $\theta$ is a linear character. More precisely, by Theorem 11.28 of [1], one can find another character triple, $(G^*, N^*, \theta^*)$ isomorphic to $(G, N, \theta)$, but with the additional property that $\theta^*$ is linear.

By definition (see 11.23 of [1]), if $(G, N, \theta)$ is isomorphic to $(G^*, N^*, \theta^*)$, then in particular, $G/N \cong G^*/N^*$ and we fix a particular isomorphism of these groups. If $N \subseteq H \subseteq G$, we write $H^*$ to denote the subgroup, $N^* \subseteq H^* \subseteq G^*$, such that $H^*/N^*$ is the image of $H/N$ under our fixed isomorphism. Also by the definition of character-triple isomorphism, we have a bijection $*: \mathrm{Irr}(H \mid \theta) \to \mathrm{Irr}(H^* \mid \theta^*)$ and we shall use the same symbol $*$ to denote its linear extension to a vector space isomorphism $cf(H \mid \theta) \to cf(H^* \mid \theta^*)$ for each subgroup $H$ with $N \subseteq H \subseteq G$.

If $N \subseteq H \subseteq K \subseteq G$ and $\chi \in \mathrm{Irr}(K \mid \theta)$ and $\psi \in \mathrm{Irr}(H \mid \theta)$, then the definition of character-triple isomorphism requires that

$$[\chi_H, \psi] = [(\chi^*)_{H^*}, \psi^*].$$

This implies that the vector space isomorphisms $*$ discussed in the previous paragraph are compatible with restriction. In other words, if $\varphi \in cf(K \mid \theta)$, then

$$(\varphi_H)^* = (\varphi^*)_{H^*},$$

and by Frobenius reciprocity, we also get

$$(\mu^K)^* = (\mu^*)^{K^*}$$

for $\mu \in cf(H \mid \theta)$.

Finally, we observe that the $*$ maps respect degrees in the sense that

$$\varphi^*(1) = \frac{\theta^*(1)}{\theta(1)}\varphi(1)$$

for $\varphi \in cf(H \mid \theta)$. To see this, note that

$$\varphi_N = \frac{\varphi(1)}{\theta(1)}\theta \quad \text{and} \quad (\varphi^*)_{N^*} = \frac{\varphi^*(1)}{\theta^*(1)}\theta^*$$

and thus

$$\varphi(1)/\theta(1) = \varphi^*(1)/\theta^*(1).$$

**5.1 LEMMA.** *Let $(G,N,\theta)$ and $(G^*,N^*,\theta^*)$ be isomorphic character-triples and assume that $G$ is $\pi$-separable and that both $N$ and $N^*$ are $\pi$-groups. Then the map $*: cf(G \mid \theta) \to cf(G^* \mid \theta^*)$ carries $vcf(G \mid \theta)$ onto $vcf(G^* \mid \theta^*)$.*

**Proof.** First, we note that $G^*$ is $\pi$-separable since $G^*/N^* \cong G/N$ is $\pi$-separable and $N^*$ is a $\pi$-group. Now let $H \subseteq G$ be a Hall $\pi$-subgroup. Then $H \supseteq N$ and so $H^*$ is defined and in fact, $H^*$ is a Hall $\pi$-subgroup of $GH^*$.

From the properties of character-triple isomorphisms, and using Lemma 3.4, we have

$$vcf(G \mid \theta)^* = (cf(H \mid \theta)^G)^* = cf(H^* \mid \theta^*)^{G^*} = vcf(G^* \mid \theta^*),$$

as required. □

Unfortunately, Theorem 11.28 of [1], which asserts the existence of a character-triple $(G^*,N^*,\theta^*)$ isomorphic to $(G,N,\theta)$ and with $\theta^*$ linear, does not guarantee that $N^*$ will be a $\pi$-group if $N$ is a $\pi$-group. It is pleasant that we can remedy this defect using known results and without the necessity of redoing the proof of the theorem.

**5.2 THEOREM.** *Let $(G,N,\theta)$ be a character-triple, where $N$ is a $\pi$-group. Then there exists an isomorphic triple $(G^*,N^*,\theta^*)$ where $N^*$ is a $\pi$-group contained in $\mathbf{Z}(G^*)$.*

**Proof.** Let $(G^*,N^*,\theta^*)$ be any isomorphic triple where $\theta^*$ is linear, and factor $\theta^* = \alpha\beta$ where the order $o(\alpha)$ (in the group of linear characters of $N^*$) is a $\pi$-number and $o(\beta)$ is a $\pi'$-number. Note that $\alpha$ and $\beta$ are uniquely determined by $\theta^*$.

Suppose $\beta$ is extendible to some linear character $\gamma$ of $G^*$. Then multiplication of all members of $\mathrm{Irr}(H^* \mid \theta^*)$ by $(\gamma^{-1})_{H^*}$, for all $H^*$ with $N^* \subseteq H^* \subseteq G^*$, defines a character-triple isomorphism $(G^*,N^*,\theta^*) \to (G^*,N^*,\alpha)$ (since $\theta^*\beta^{-1} = \alpha$). Next, factoring out $\ker(\alpha)$ constructs an isomorphic triple $(\overline{G^*},\overline{N^*},\overline{\alpha})$ in which $\overline{\alpha}$ is faithful. Then $\overline{N^*} \subseteq \mathbf{Z}(\overline{G^*})$ (since $\overline{\alpha}$ is invariant) and $|\overline{N^*}| = o(\alpha)$, a $\pi$-number.

We will show now that $\beta$ does extend to $G^*$. For this purpose, it suffices to extend $\beta$ to $P^*$, where $P^*/N^* \in \mathrm{Syl}_p(G^*/N^*)$ for an arbitrary prime $p$.

There are two cases. If $p \in \pi$, then since $o(\beta)$ is $\pi'$, extendibility follows by Gallagher's theorem (Corollary 8.16 of [1]). If $p \notin \pi$, consider the group $P$, corresponding to $P^*$, with $N \subseteq P \subseteq G$. Since $N$ is a $\pi$-group, Gallagher's theorem guarantees that $\theta$ extends to some $\psi \in \mathrm{Irr}(P \mid \theta)$. Writing $\lambda = \psi^* \in \mathrm{Irr}(P^* \mid \theta^*)$, we see that $\lambda(1) = \theta^*(1)$ and thus $\lambda$ is linear and extends $\theta^*$. Now $\lambda$ factors as $\mu\nu$ with $o(\mu)$ a $\pi$-number and $o(\nu)$ a $\pi'$-number and we have

$$\alpha\beta = \theta^* = (\lambda_{N^*})(\mu_{N^*}).$$

It follows that $\mu_{N^*} = \beta$ and $\beta$ does extend to $P^*$, as required. □

## 6. Theorem C.

Our goal now is to prove Theorem C. Actually, we can add an additional conclusion to the statement of Theorem C with negligible additional work in the proof.

**6.1 THEOREM.** *In the situation of Theorem C, the set $\mathcal{B}$ can be chosen so that every character $\beta \in \mathcal{B}$ has the form $\gamma^G$, where $\gamma$ is a character of some subgroup, and $\gamma(1)$ is a $\pi$-number.*

To prove Theorems C and 6.1, we use an inductive argument which necessitates a more general form of the result, as follows.

**6.2 THEOREM.** *Let $N \triangleleft G$ where $N$ is a $\pi$-group and $G$ is $\pi$-separable, and let $\theta \in \mathrm{Irr(N)}$. Choose a Hall $\pi$-subgroup $H$ of $G$ such that $H \cap I_G(\theta)$ is a Hall $\pi$-subgroup of the inertia group $I_G(\theta)$. Then there exist subsets $\mathcal{A} \subseteq \mathrm{Irr(H} \mid \theta)$ and $\mathcal{B} \subseteq \mathrm{Irr(G} \mid \theta)$ and a bijection $\sigma : \mathcal{A} \to \mathcal{B}$ such that*

a) *$\mathcal{A}^G$ spans $vcf(G \mid \theta)$.*

b) *If $\alpha \in \mathcal{A}$ and $\beta \in \mathcal{B}$, then*

$$[\alpha^G, \beta] = \begin{cases} 1 & \text{if } \beta = \sigma(\alpha) \\ 0 & \text{otherwise.} \end{cases}$$

c) *$\alpha(1) = \sigma(\alpha)(1)_\pi$ for $\alpha \in \mathcal{A}$.*

d) *Every $\beta \in \mathcal{B}$ is induced from a character with $\pi$-degree.*

Of course, Theorems C and 6.1 constitute the special case where $N = 1$ (and $\theta = 1_1$).

**Proof of Theorem** 6.2. We assume that $G, N$ and $\theta$ provide a counterexample to the theorem with the index $|G : N|$ minimal, and subject to this condition, we take $N \subseteq \mathbf{Z}(G)$ if we possibly can. Note that the theorem holds when $N = G$, since in that case $H = G$ and the sets $\mathcal{A} = \{\theta\} = \mathcal{B}$ satisfy conclusions $(a, b, c, d)$ (with $\sigma(\theta) = \theta$, of course). We thus have $N < G$.

STEP 1: $\theta$ is $G$-invariant.

**Proof**. Let $T = I_G(\theta)$ and suppose that $T < G$. Then $|T : N| < |G : N|$ and so the theorem holds for $T, N$ and $\theta$ and we can choose $\mathcal{A}_0 \subseteq \mathrm{Irr}(H \cap T \mid \theta)$ and $\mathcal{B}_0 \subset \mathrm{Irr}(T \mid \theta)$ and a bijection $\sigma_0 : \mathcal{A}_0 \to \mathcal{B}_0$ such that $(a, b, c, d)$ hold.

We define $\mathcal{A} = (\mathcal{A}_0)^H$ and $\mathcal{B} = (\mathcal{B}_0)^G$. By the Clifford correspondence, $\mathcal{A} \subseteq \mathrm{Irr}(H \mid \theta)$, $\mathcal{B} \subseteq \mathrm{Irr}(T \mid \theta)$ and the map $\sigma : \mathcal{A} \to \mathcal{B}$ defined by $\sigma(\alpha^H) = \sigma_0(\alpha)^G$ is a well-defined bijection. We proceed to show that $\mathcal{A}, \mathcal{B}$ and $\sigma$ satisfy $(a, b, c, d)$. Since $G, N$ and $\theta$ supposedly form a counterexample, this will be the contradiction we seek.

Now $(\mathcal{A}_0)^T$ spans $vcf(T \mid \theta)$ and therefore $\mathcal{A}^G = ((\mathcal{A}_0)^H)^G = ((\mathcal{A}_0)^T)^G$ spans $vcf(T \mid \theta)^G$, which equals $vcf(G \mid \theta)$ by Lemma 3.5. This establishes (a). For (b), let $\alpha \in \mathcal{A}_0$ and $\beta \in \mathcal{B}_0$ and compute

$$[(\alpha^H)^G, \beta^G] = [(\alpha^T)^G, \beta^G] = [\alpha^T, \beta],$$

where the second inequality follows by the Clifford correspondence since $\alpha^T, \beta \in cf(T \mid \theta)$. We know that $[\alpha^T, \beta]$ is 1 or 0 according to whether or not $\sigma_0(\alpha) = \beta$. Statement (b) now follows.

Now let $\alpha \in \mathcal{A}_0$ and compute

$$\sigma(\alpha^H)(1)_\pi = \sigma_0(\alpha)^G(1)_\pi = |G : T|_\pi \sigma_0(\alpha)(1)_\pi = |H : H \cap T|\alpha(1) = \alpha^H(1),$$

proving (c). Statement (d) is immediate since $\mathcal{B} = (\mathcal{B}_0)^G$, and we have our desired contradiction.

STEP 2. $N \subseteq \mathbf{Z}(G)$.

**Proof.** By Theorem 5.2, construct a character-triple $(G^*, N^*, \theta^*)$ isomorphic to $(G, N, \theta)$ and with $N^*$ a $\pi$-group central in $G^*$. If, in fact, $N \not\subseteq \mathbf{Z}(G)$, then by our choice of $G$, we see that $G^*, N^*$ and $\theta^*$ do not form a counterexample and we have $\mathcal{A}^* \subseteq \mathrm{Irr}(H^* \mid \theta^*)$ and $\mathcal{B}^* \subseteq \mathrm{Irr}(G^* \mid \theta^*)$ and a bijection $\sigma^* : \mathcal{A}^* \to \mathcal{B}^*$ such that $(a, b, c, d)$ hold. (Note that $G^*$ is $\pi$-separable and $H^*$ is a Hall subgroup.) Let $\mathcal{A} \subseteq \mathrm{Irr}(H \mid \theta)$ and $\mathcal{B} \subseteq \mathrm{Irr}(G \mid \theta)$ be the preimages of $\mathcal{A}^*, \mathcal{B}^*$ under the appropriate $*$ maps. Define $\sigma : \mathcal{A} \to \mathcal{B}$ by $\sigma(\alpha)^* = \sigma^*(\alpha^*)$, so that $\sigma$ is a bijection. We will verify $(a, b, c, d)$ for $\mathcal{A}, \mathcal{B}, \sigma$, and this will provide the desired contradiction.

By Lemma 5.1, we have

$$vcf(G \mid \theta)^* = vcf(G^* \mid \theta^*),$$

and since $(\mathcal{A}^*)^{G^*} = (\mathcal{A}^G)^*$ spans $vcf(G^* \mid \theta^*)$, we see that $\mathcal{A}^G$ must span $vcf(G \mid \theta)$ and (a) holds. Assertion (b) is an immediate consequence of the fact that

$$[\alpha^G, \beta] = [(\alpha^G)^*, \beta^*] = [(\alpha^*)^{G^*}, \beta^*].$$

Writing $r = \theta(1)$, we have

$$\sigma(\alpha)(1)_\pi = r\sigma(\alpha)^*(1)_\pi = r\sigma^*(\alpha^*)(1)_\pi = r\alpha^*(1) = \alpha(1)$$

and (c) follows. Statement (d) follows since $*$ respects induction and degrees.

STEP 3. $\mathbf{O}_{\pi'}(G) = 1$.

**Proof.** Let $K = \mathbf{O}_{\pi'}(G)$ and assume that $K > 1$. We write $\overline{G} = G/K$ and we use the notation of Section 4.

Note that $\overline{H}$ is a Hall $\pi$-subgroup of $\overline{G}$. Since $|\overline{G} : \overline{N}| < |G : N|$, we have $\overline{\mathcal{A}} \subseteq \mathrm{Irr}(\overline{H} \mid \overline{\theta})$ and $\overline{\mathcal{B}} \subseteq \mathrm{Irr}(\overline{G} \mid \overline{\theta})$ and a bijection $\overline{\sigma} : \overline{\mathcal{A}} \to \overline{\mathcal{B}}$ satisfying $(a, b, c, d)$.

Since overbar defines bijections $\mathrm{Irr}(H) \to \mathrm{Irr}(\overline{H})$ and $\mathrm{Irr}(G \mid 1_K) \to \mathrm{Irr}(\overline{G})$, this uniquely defines subsets $\mathcal{A} \subseteq \mathrm{Irr}(H)$ and $\mathcal{B} \subseteq \mathrm{Irr}(G \mid 1_K)$ and a bijection $\sigma : \mathcal{A} \to \mathcal{B}$ such that $\overline{\sigma(\alpha)} = \overline{\sigma}(\overline{\alpha})$. Also, if $\eta \in \mathcal{A} \cup \mathcal{B}$, then $\overline{\eta_N} = \overline{\eta}_{\overline{N}}$ is a multiple of $\overline{\theta}$ and thus $\eta_N$ is a multiple of $\theta$. This shows that $\mathcal{A} \subseteq \mathrm{Irr}(H \mid \theta)$ and $\mathcal{B} \subseteq \mathrm{Irr}(G \mid \theta)$. We verify $(a, b, c, d)$.

For (a), let $\varphi \in vcf(G \mid \theta)$. By 3.4, we have $\varphi = \mu^G$ with $\mu \in cf(H \mid \theta)$ and thus by 4.1,

$$\overline{\tau(\varphi)} = \overline{\mu}^{\overline{G}} \in vcf(\overline{G} \mid \overline{\theta}).$$

By property (a) for $\overline{G}$, we can write

$$\overline{\mu}^{\overline{G}} = \overline{\nu}^{\overline{G}}$$

where $\overline{\nu}$ is a linear combination of $\overline{\mathcal{A}}$. This defines $\nu \in cf(H \mid \theta)$, a linear combination of $\mathcal{A}$ and by 4.1, we have

$$\overline{\tau(\varphi)} = \overline{\tau(\nu^G)}$$

and thus $\varphi = \nu^G$ by 4.2. This establishes (a).

For (b), let $\alpha \in \mathcal{A}$ and $\beta \in \mathcal{B}$. Then

$$[\alpha^G, \beta] = [\tau(\alpha^G), \beta] = [\overline{\tau(\alpha^G)}, \overline{\beta}] = [\overline{\alpha}^{\overline{G}}, \overline{\beta}]$$

where the first equality follows since $\beta \in \mathrm{Irr}(G \mid 1_K)$. Thus $[\alpha^G, \beta]$ is 1 or 0 as required.

Part (c) follows since the bar map preserves degrees and finally, if $\beta \in \mathcal{B}$, then $\overline{\beta} = \overline{\gamma}^{\overline{G}}$ where $\overline{\gamma} \in \mathrm{Irr}(\overline{U})$ for some subgroup $\overline{U} \subseteq \overline{G}$ and $\overline{\gamma}(1)$ is a $\pi'$-number. We may take $U \supseteq K$. Then

$$\overline{\beta} = \overline{\gamma}^{\overline{G}} = \overline{\gamma^G}$$

and so $\beta = \gamma^G$, proving (d).

STEP 4. We have a contradiction.

**Proof.** We know that $N \subseteq \mathbf{Z}(G)$ and $N < G$. Since $\mathbf{O}_{\pi'}(G) = 1$, it follows that $\mathbf{O}_\pi(G) > N$ and we write $M = \mathbf{O}_\pi(G)$.

Now $G$ acts on $\mathrm{Irr}(M \mid \theta)$ and we choose orbit representatives $\xi_1, \xi_2, \ldots, \xi_t$ such that $I_G(\xi_i) \cap H$ is a Hall subgroup of $I_G(\xi_i)$.

Since $|G : M| < |G : N|$, we have subsets $\mathcal{A}_i \subseteq \mathrm{Irr}(H \mid \xi_i)$ and $\mathcal{B}_i \subseteq \mathrm{Irr}(G \mid \xi_i)$ and bijections $\sigma_i : \mathcal{A}_i \to \mathcal{B}_i$ such that $(a, b, c, d)$ hold. Note that the sets $\mathcal{A}_i$ are pairwise disjoint as are the $\mathcal{B}_i$, and so if $\mathcal{A} = \bigcup \mathcal{A}_i$ and $\mathcal{B} = \bigcup \mathcal{B}_i$, we have a bijection $\sigma : \mathcal{A} \to \mathcal{B}$ obtained by piecing together the $\sigma_i$.

We obtain our contradiction by showing that $\mathcal{A}, \mathcal{B}, \sigma$ satisfy $(a, b, c, d)$. First, note that $cf(G \mid \theta)$ is the direct sum of the subspaces $cf(G \mid \xi_i)$. By Corollary 3.3, it follows that

$$vcf(G \mid \theta) = \sum_{i=1}^{t} vcf(G \mid \xi_i)$$

and since $(\mathcal{A}_i)^G$ spans $vcf(G \mid \theta_i)$, we see that $\mathcal{A}^G$ spans $vcf(G \mid \theta)$.

If $\alpha \in \mathcal{A}_i$ and $\beta \in \mathcal{B}_j$ with $i \neq j$, then $[\alpha^G, \beta] = 0$ since $\alpha^G \in cf(G \mid \xi_i)$ and $\beta \in \mathrm{Irr}(G \mid \xi_j)$. Statement (b) is now immediate. In this situation (c) and (d) are trivially true and the proof is complete. □

## 7. Further remarks.

All groups considered in this section will be assumed to be $\pi$-separable. We shall say that a function $\varphi \in cf(G^\circ)$ is a *$\pi$-partial character* of $G$ if $\varphi$ is nonzero and is a nonnegative integer linear combination of $\mathrm{I}_\pi(G)$. It follows from the linear independence of $\mathrm{I}_\pi(G)$ that the members of $\mathrm{I}_\pi(G)$ are precisely those $\pi$-partial characters which are irreducible (i.e. which cannot be written as a proper sum of $\pi$-partial characters).

**7.1 LEMMA.** *The $\pi$-partial characters of $G$ are precisely the functions $\chi^\circ$ for characters $\chi$ of $G$.*

**Proof.** This is immediate via (DP) and (FS). □

Note that Lemma 7.1 suggests an alternative approach to Theorem A: simply define the $\pi$-partial characters of $G$ as the functions $\chi^\circ$ for $\chi \in \mathrm{Char}(G)$, and construct $\mathrm{I}_\pi(G)$ as the set of irreducible $\pi$-partial characters. (This approach is even algorithmic; it allows $I_\pi(G)$ to be read off from the character table of $G$.) It is easy to see that with this definition, $I_\pi(G)$ spans $cf(G^\circ)$. What is not obvious is how to establish the linear independence from this point of view.

We can induce and restrict $\pi$-partial characters. If $H \subseteq G$ and $\varphi \in cf(G^\circ)$, we write $\varphi_H$ to denote the restriction of $\varphi$ to $H^\circ$ and we observe that $\varphi_H \in cf(H^\circ)$. Also, if $\theta \in cf(H^\circ)$, we define

$$\theta^G(g) = \frac{1}{|H|} \sum_{\substack{x \in G \\ xgx^{-1} \in H}} \varphi(xgx^{-1})$$

for $g \in G^\circ$. Then clearly $\theta^G \in cf(G^\circ)$. Note that if $\xi \in cf(G)$ and $\eta \in cf(H)$, then

$$(\xi^\circ)_H = (\xi_H)^\circ \quad \text{and} \quad (\eta^\circ)^G = (\eta^G)^\circ .$$

**7.2 LEMMA.** *Let $H \subseteq G$ and suppose $\varphi$ and $\theta$ are $\pi$-partial characters of $G$ and $H$ respectively. Then $\varphi_H$ and $\theta^G$ are $\pi$-partial characters of $H$ and $G$.*

**Proof.** Write $\varphi = \chi^\circ$ for some character $\chi$ of $G$. Then

$$\varphi_H = (\chi^\circ)_H = (\chi_H)^\circ$$

is a $\pi$-partial character of $H$ since $\chi_H$ is a character of $H$. The proof for induction is similar. □

One might wonder whether or not other familiar properties of characters work also for the $\pi$-partial characters. For example, if $N \triangleleft G$ and $\varphi$ is an irreducible $\pi$-partial character of $G$ (i.e. $\varphi \in I_\pi(G)$), then can we say that all irreducible constitutents of $\varphi_N$ are $G$-conjugate? The answer is "yes", but the author has been unable to find a proof using only the methods of this paper. (The deeper results in [2] prove this assertion easily.)

It is not generally true that if $H \subseteq G$ and $\theta \in I_\pi(H)$ and $\varphi \in I_\pi(G)$, then $\theta$ is a constituent of $\varphi_H$ iff $\varphi$ is a constituent of $\theta^G$. This does hold if $H \triangleleft G$, but here too, it is not clear how to prove this without resorting to the deeper theory of [2].

There is one result about induction of $\pi$-partial characters which we can prove and for which there is no analog for ordinary characters.

**7.3 THEOREM.** *Let $\varphi \in I_\pi(G)$. Then $\varphi = \theta^G$ where $\theta \in I_\pi(U)$ for some subgroup $U \subseteq G$, and where $\theta(1)$ is a $\pi$-number.*

**Proof.** We can write $\varphi = \beta^\circ$ where $\beta \in \mathrm{Irr}(G)$ and in fact, we can take $\beta \in \mathcal{B}$ as in Theorem C. By Theorem 6.1, we have $\beta = \gamma^G$ with $\gamma \in \mathrm{Irr}(U)$, where $\gamma(1)$ is a $\pi$-number and $U \subseteq G$. Writing $\theta = \gamma^\circ$, we have

$$\theta^G = (\gamma^\circ)^G = (\gamma^G)^\circ = \beta^\circ = \varphi .$$

Also, $\theta$ is a $\pi$-partial character and if $\theta$ can be written as a sum of two $\pi$-partial characters, then by 7.2, so can $\varphi$. Since $\varphi \in I_\pi(G)$, we deduce that $\theta \in I_\pi(U)$. □

## REFERENCES

[1] I. M. Isaacs, *Character theory of finite groups*, Academic Press, New York, 1976.

[2] I. M. Isaacs, Characters of $\pi$-separable groups, *J. of Algebra* **86** (1984) 98-128.

[3] I. M. Isaacs, Extensions of characters from Hall $\pi$-subgroups, *Proc. Edinburgh Math. Soc.* **38** (1985) 313-317.

[4] I. M. Isaacs, Fong characters in $\pi$-separable groups, *J. of Algebra* **99** (1986) 89-107.

[5] G. R. Robinson and R. Staszewski, On the representation theory of $\pi$-separable groups, *J. of Algebra* **119** (1988) 226-232.

Progress in Mathematics, Vol. 95, © 1991 Birkhäuser Verlag Basel

# First Cohomology Groups for Classical Lie Algebras

JENS C. JANTZEN *

## Introduction

The classical Lie algebras in the title are the Lie algebras of semisimple algebraic groups in prime characteristic. This includes the groups of exceptional type. The word classical was chosen to distinguish from the simple Lie algebras of Cartan type.

We want to look at the first cohomology groups of these modules with values in simple modules. The simple modules that we study will be denoted by $L(\lambda)$ and arise as simple socles of certain induced modules $H^0(\lambda)$ for the algebraic group, cf. 1.4 for a more precise statement. It is rather straightforward — given the ideas from [2] — to compute the first cohomology for the $H^0(\lambda)$, cf. 4.1, 5.1, 5.2. Unfortunately, some special considerations in characteristic 2 (and 3 for groups of type $G_2$) are needed.

The transition from $H^0(\lambda)$ to $L(\lambda)$ is difficult. The most pleasant case is the one where we look at the Lie algebra of $SL_{n+1}$ and where the characteristic does not divide $n+1$. Then there are exactly $n$ simple modules $L(\lambda_i)$ for which the corresponding $H^0(\lambda_i)$ has a nonvanishing first cohomology. For these $n$ modules the first cohomology groups of $L(\lambda_i)$ and $H^0(\lambda_i)$ coincide, cf. 4.10. For all other $L(\lambda)$ the first cohomology can be described in terms of the module structure of $H^0(\lambda)$. (This is rather similar to the description of the cohomology of $L(\lambda)$ for the algebraic group, cf. 1.7.) However, this leads to the deepest unsolved problems in the representation theory of these groups.

If the characteristic divides $n+1$ or for other groups the results are weaker. In those cases where $H^0(\lambda)$ has a nonzero cohomology, we usually get just a nonzero map from the first cohomology of $L(\lambda)$ to that of $H^0(\lambda)$, cf. 4.5, 5.4, 5.5 — no longer an isomorphism — and there are a few cases, where even that is false, mainly in characteristic 2, cf. 5.9.

However, for small representations (see 6.1 for the meaning of small) we can compute the cohomology completely. Questions from Dietrich Burde involving these small cases inspired this work. So did some open problems in John Sullivan's paper [21].

## 1. Generalities

**1.1.** Let $G$ be a connected semisimple algebraic group over an algebraically closed field $k$ of prime characteristic $p > 0$. Assume that $G$ is almost simple and simply connected. Let $\mathfrak{g}$ be the Lie algebra of $G$. In this paper I want to discuss the first cohomology groups of simple $\mathfrak{g}$–modules.

It is enough to look at those modules that have the same central character as the trivial one dimensional module. This means that we have to look only at restricted modules for the restricted Lie algebra $\mathfrak{g}$. By a formula of Hochschild

* Supported in part by the NSF.

(cf. [15],I.9.19), the first Lie algebra cohomology coincides with the first restricted Lie algebra cohomology for simple modules, except possibly for the trivial simple module. Therefore I shall look at restricted cohomology only.

Let $G_1$ be the first Frobenius kernel of $G$. The representation theory of the group scheme $G_1$ and of $\mathfrak{g}$ regarded as a restricted Lie algebra can be identified. The Hochschild cohomology groups $H^i(G_1, L)$ of a $G_1$–module $L$ and the restricted cohomology of the corresponding $\mathfrak{g}$–module are equal. We shall usually express the results in terms of the $H^i(G_1, L)$.

**1.2.** Let $T$ be a maximal torus in $G$. Denote the group of characters of $T$ by $X(T)$ and the root system by $R \subset X(T)$. For any root $\alpha$ denote the dual root by $\alpha^\vee$. Choose a set of positive roots $R^+$ in $R$. Denote the simple roots by $\alpha_1, \alpha_2, \ldots, \alpha_n$ using the same numbering as in Bourbaki's tables in [3]. Note, however, that we denote the rank by $n$. Let $\omega_1, \omega_2, \ldots, \omega_n$ be the fundamental weights, let $X(T)_+$ be the set of dominant weights and $X_1(T)$ the set of restricted dominant weights, i.e., the set of all $\sum_{i=1}^n r_i\omega_i$ with $r_i \in \mathbf{Z}$ and $0 \le r_i < p$ for all $i$.

We consider on $X(T)$ the usual order relation where $\lambda \le \mu$ holds if and only if there are integers $r_i \ge 0$ with $\mu - \lambda = \sum_{i=1}^n r_i\alpha_i$.

For any $T$–module $M$ and any $\mu \in X(T)$ denote the weight space of $M$ for the character $\mu$ by $M_\mu$.

**1.3.** There is a Lie algebra $\mathfrak{g}_\mathbf{Z}$ over the integers with $\mathfrak{g}_\mathbf{Z} \otimes_\mathbf{Z} k \simeq \mathfrak{g}$. It is the span of a Chevalley basis in a semisimple complex Lie algebra of the same type. So there is for each root $\alpha$ a root vector $X_\alpha$ in $\mathfrak{g}_\mathbf{Z}$. If $\alpha, \beta$ are roots with $\alpha + \beta \in R$ then $[X_\alpha, X_\beta] = N_{\alpha,\beta}X_{\alpha+\beta}$ for some integer $N_{\alpha,\beta}$. Up to sign it can be read off easily from the root system.

We shall use the notation $X_\alpha$ also for the element $X_\alpha \otimes 1$ of $\mathfrak{g}$. The $p$-th power map $X \mapsto X^{[p]}$ in $\mathfrak{g}$ has the property $X_\alpha^{[p]} = 0$ for all $\alpha$.

**1.4.** Let $B \supset T$ be the Borel subgroup of $G$ corresponding to the negative roots and let $U$ be the unipotent radical of $B$. So the Lie algebra $\mathfrak{u}$ of $U$ is spanned by all $X_{-\alpha}$ with $\alpha \in R^+$.

Each $\lambda \in X(T)$ defines a one dimensional module $k_\lambda$ of $B$ via the isomorphism $B/U \simeq T$. The induced module $H^0(\lambda) = \operatorname{ind}_B^G k_\lambda$ is nonzero if and only if $\lambda \in X(T)_+$. If so, the socle $L(\lambda)$ of $H^0(\lambda)$ is the simple $G$–module with highest weight $\lambda$. It can also be constructed as the unique simple image of the Weyl module $V(\lambda)$. The $L(\lambda)$ with $\lambda \in X_1(T)$ remain simple under restriction to $G_1$. Any simple $G_1$–module is isomorphic to exactly one $L(\lambda)$ with $\lambda \in X_1(T)$.

**1.5.** We shall assume that $G$ and $T$ are defined and split over the prime field $\mathbf{F}_p$. The Frobenius morphism $F$ on $G$ is then an endomorphism that maps any $t \in T$ to its $p$-th power $t^p$. It maps $B$ and $U$ to themselves. Its kernels on these subgroups will be denoted by $T_1$, $B_1$, and $U_1$.

If one composes a representation of $G$ on a vector space $M$ with $F$ one gets a new representation where $G_1$ (and therefore also $\mathfrak{g}$) acts trivially. The corresponding new module will be denoted by $M^{(1)}$. For any $\mu \in X(T)$ the $\mu$ weight space of $M$ is the $p\mu$ weight space of $M^{(1)}$.

On the other hand, if $V$ is a $G$–module on which $G_1$ (or $\mathfrak{g}$) acts trivially, then there is a unique $G$–module $M$ with $V = M^{(1)}$. We shall denote this $G$–module $M$

by $V^{(-1)}$. (One has to use that $F$ induces an isomorphism of $G/G_1$ with $G$.) For example, if $L$ is a $G$-module, then each cohomology group $H^i(G_1, L)$ is a $G$-module with $G_1$ acting trivially. So $H^i(G_1, L)^{(-1)}$ makes sense.

The situation is similar for the groups $B$, $T$, and $U$, and similar notations will be used for these groups.

**1.6.** The $G$-cohomology and the $G_1$-cohomology of a $G$-module $V$ are related by the Lyndon-Hochschild-Serre spectral sequence, cf. [15], I.6.6(3), with $E_2$-term

$$E_2^{r,s} = H^r(G/G_1, H^s(G_1, V)) = H^r(G, H^s(G_1, V)^{(-1)}).$$

If $V$ is simple and non trivial as a $G_1$-module, then $H^0(G_1, V) = 0$. Then the five term exact sequence of the spectral sequence yields isomorphisms (for all $\lambda \in X_1(T)$ with $\lambda \neq 0$)

$$H^1(G, L(\lambda)) \simeq H^1(G_1, L(\lambda))^G. \tag{1}$$

This is a very special case of results by Donkin and Andersen relating extension groups for $G$ and its Frobenius kernels, cf. [15], II.10.17.

**1.7.** One has for all $\lambda \in X(T)_+$ an isomorphism, cf. [15], II.2.14(4),

$$H^1(G, L(\lambda)) \simeq \mathrm{Hom}_G(L(0), H^0(\lambda)/L(\lambda)). \tag{1}$$

For $\lambda \in X_1(T)$ there is a similar formula for $H^1(G_1, L(\lambda))$ where $H^0(\lambda)$ is replaced by an induced module for $G_1$, cf. [15], II.9.16(4). However, in the cases to be studied, that induced module seems to have a more complicated structure than $H^0(\lambda)$. So, to get at the $G_1$-cohomology of the $L(\lambda)$, we use a different method. One has for all $\lambda \in X(T)_+$ an isomorphism

$$H^1(G_1, H^0(\lambda))^{(-1)} \simeq \mathrm{ind}_B^G(H^1(B_1, k_\lambda)^{(-1)}), \tag{2}$$

cf. [15], II.12.2(2). Using this, we can compute the first cohomology of the $H^0(\lambda)$ and then look at the obvious long exact sequence to get some information on the cohomology of the $L(\lambda)$.

## 2. The $U_1$-cohomology

**2.1.** The group scheme $U_1$ is normal not only in $U$, but also in $B$. Therefore $B$ acts on the cohomology groups $H^i(U_1, k)$. The isomorphisms below are compatible with the action of $B$.

**Lemma:** *One has*

$$H^1(U_1, k) \simeq H^1(\mathfrak{u}, k) \simeq (\mathfrak{u}/[\mathfrak{u}, \mathfrak{u}])^*.$$

**Proof**: The second isomorphism is a well known fact that holds for any Lie algebra. A 1-cocycle (for the standard complex computing Lie algebra cohomology) is just a linear form on the Lie algebra annihilating all commutators, and all 1-coboundaries are zero (for the trivial module). Any 1-cocycle $\varphi$ as above defines a semilinear functional $\hat{\varphi}$ on the Lie algebra via $\hat{\varphi}(X) = \varphi(X^{[p]})$. The kernel of the map $\varphi \mapsto \hat{\varphi}$

is the first restricted cohomology group of the trivial module. In our case $X_\alpha^{[p]} = 0$ for all roots $\alpha$ implies $\hat\varphi(X_\alpha) = 0$ for all negative roots and all $\varphi$ as above. This implies $\hat\varphi = 0$, because these $X_\alpha$ span $\mathfrak{u}$ and because $\hat\varphi$ is semilinear. So in this case the restricted Lie algebra cohomology $H^1(U_1, k)$ is equal to the ordinary Lie algebra cohomology $H^1(\mathfrak{u}, k)$.

**2.2.** We shall call $p$ *special* (or special for $R$) if

(*A*) $$p = 2 \quad \text{and } R \text{ has two root lengths,}$$

or if

(*B*) $$p = 3 \quad \text{and } R \text{ is of type } G_2.$$

**Proposition:** *If p is not special, then*

$$H^1(U_1, k) \simeq \bigoplus_{i=1}^{n} k_{\alpha_i}.$$

**Proof**: Obviously $[\mathfrak{u}, \mathfrak{u}]$ is spanned by all $[X_\alpha, X_\beta] = N_{\alpha,\beta} X_{\alpha+\beta}$ for all negative roots $\alpha, \beta$ with $\alpha + \beta \in R$. Therefore it is spanned by all $X_\gamma$ where $\gamma$ is a negative root that can be written as a sum of two negative roots $\gamma = \alpha + \beta$ with $N_{\alpha,\beta} \neq 0$. The $N_{\alpha,\beta}$ here are the reduction modulo $p$ of those in 1.3. The only possible values of these structure constants (over $\mathbf{Z}$) are $1, 2, 3, -1, -2, -3$, cf. [12], 25.2 and 9.4. If $\alpha, \beta, \alpha + \beta$ have the same length, then $N_{\alpha,\beta}$ is 1 or $-1$. The values 3 and $-3$ occur only if $R$ is of type $G_2$. Having excluded the cases (A) and (B) ensures therefore that all these $N_{\alpha,\beta}$ are nonzero in $k$.

This shows that $[\mathfrak{u}, \mathfrak{u}]$ is spanned by all $X_\gamma$ with $\gamma$ negative and $\gamma$ a sum of two negative roots. The last condition is equivalent to $-\gamma$ not simple. Therefore $\mathfrak{u}/[\mathfrak{u}, \mathfrak{u}]$ has the classes of the $X_{-\alpha_i}$ as a basis. These are weight vectors for $T$ for the weights $-\alpha_i$. There are no order relations between these weights, so $U$ has to act trivially. As a $B$–module $\mathfrak{u}/[\mathfrak{u}, \mathfrak{u}]$ is a direct sum of $n$ one dimensional modules with $U$ acting trivially and $T$ via the different $-\alpha_i$. Dualizing yields the claim.

**2.3.** We shall use in the following proposition as well as later on the description of the root systems as in Bourbaki's tables. Especially, the $\varepsilon_i$ have the same meaning as there.

The filtrations described below are filtrations as $B$–modules. The factors occur in the given order from top to bottom. The same remark holds for the filtrations in 2.4, 3.6, 3.7.

**Proposition:** *Suppose that $p = 2$ and that $R$ is of type $B_n$, $C_n$, or $F_4$. Then one has a decomposition*

$$H^1(U_1, k) \simeq \bigoplus_{i=1}^{n} M_i$$

*where:*

*(a) If $R$ is of type $B_n$, then $M_i = k_{\alpha_i}$ for all $i \neq n-1$, and $M_{n-1}$ has a filtration with factors $k_{\alpha_{n-1}+2\alpha_n}$ and $k_{\alpha_{n-1}}$.*

*(b) If $R$ is of type $C_n$, then $M_i = k_{\alpha_i}$ for all $i \neq n$, and $M_n$ has a filtration with factors $k_{2\varepsilon_1}, k_{2\varepsilon_2}, \ldots, k_{2\varepsilon_{n-1}}, k_{2\varepsilon_n} = k_{\alpha_n}$.*

*(c) If $R$ is of type $F_4$, then $M_i = k_{\alpha_i}$ for all $i \neq 2$, and $M_2$ has a filtration with factors $k_{\alpha_2+2\alpha_3+2\alpha_4}$, $k_{\alpha_2+2\alpha_3}$, and $k_{\alpha_2}$.*

**Proof**: If $\alpha, \beta$ are roots with $\alpha + \beta \in R$, then the coefficient $N_{\alpha,\beta}$ (over $\mathbf{Z}$) is 2 or $-2$ if and only if $\alpha + \beta$ is long and $\alpha$ or $\beta$ is short. (If so, then both $\alpha$ and $\beta$ have to be short.) In all other cases the coefficient is 1 or $-1$.

If $\gamma$ is a negative root such that $-\gamma$ is not simple, then there are negative roots $\alpha, \beta$ with $\gamma = \alpha + \beta$. If $\gamma$ is short, then the discussion above implies that $X_\gamma$ is a commutator. This holds for $\gamma$ long if and only if we can find long negative roots with $\gamma = \alpha + \beta$, i.e., if and only if $-\gamma$ is not simple in the root subsystem of all long roots.

Therefore $\mathfrak{u}/[\mathfrak{u}, \mathfrak{u}]$ has as a basis all $X_{-\beta}$ with $\beta$ simple in $R$ or simple in the subsystem of the long roots. For $R$ of type $C_n$ that subsystem has type $(A_1)^n$ and all positive long roots $2\varepsilon_i$ with $1 \leq i \leq n$ occur. For $R$ of type $B_n$ it is of type $D_n$ with simple roots $\varepsilon_i - \varepsilon_{i+1} = \alpha_i$ for $1 \leq i < n$ and $\varepsilon_{n-1} + \varepsilon_n = \alpha_{n-1} + 2\alpha_n$. For $R$ of type $F_4$ the subsystem has type $D_4$ with simple roots $\alpha_1$, $\alpha_2$, $\alpha_2 + 2\alpha_3 + 2\alpha_4$, and $\alpha_2 + 2\alpha_3$.

The discussion so far shows that $H^1(U_1, k)$ has exactly the weights as described in (a), (b), or (c). It remains to discuss the operation of $U$. Because $U_1$ acts trivially, any root subgroup $U_\alpha$ has to shift weights by multiples of $p\alpha = 2\alpha$. So the weights in an indecomposable summand of $H^1(U_1, k)$ have to belong to the same coset modulo $p\mathbf{Z}R$. Collecting all weight spaces for weights congruent to $\alpha_i$ modulo $p\mathbf{Z}R$ yields the summand $M_i$ in the claim. Because $U$ corresponds to the negative roots, one gets a composition series starting with the largest weight and ending with the smallest one.

*Remark*: One can show that the $M_i$ are indecomposable. The point is that suitable operators in the distribution algebra of $U$ act like $\mathrm{ad}(X_\alpha)^2/2!$ and map suitable $X_\beta$ to $X_{\beta+2\alpha}$. A similar remark holds for type $G_2$ below.

**2.4. Proposition:** *Let $R$ be of type $G_2$ and $p = 2$ or $p = 3$. Then*

$$H^1(U_1, k) \simeq k_{\alpha_1} \oplus M_2$$

*where $M_2$ has a filtration with factors $k_{2\alpha_1+\alpha_2}$ and $k_{\alpha_2}$ for $p = 2$ resp. $k_{3\alpha_1+\alpha_2}$ and $k_{\alpha_2}$ for $p = 3$.*

**Proof**: For some choice of signs the commutators of negative root vectors in $\mathfrak{g}_{\mathbf{Z}}$ look like

$$[X_{-\alpha_1}, X_{-\alpha_2}] = X_{-(\alpha_1+\alpha_2)}, \quad [X_{-\alpha_1}, X_{-(\alpha_1+\alpha_2)}] = 2X_{-(2\alpha_1+\alpha_2)},$$

$$[X_{-\alpha_1}, X_{-(2\alpha_1+\alpha_2)}] = 3X_{-(3\alpha_1+\alpha_2)}, \quad [X_{-\alpha_2}, X_{-(3\alpha_1+\alpha_2)}] = X_{-(3\alpha_1+2\alpha_2)},$$

$$[X_{-(\alpha_1+\alpha_2)}, X_{-(2\alpha_1+\alpha_2)}] = -3X_{-(3\alpha_1+2\alpha_2)},$$

cf. [9]. It is now easy to find which $X_\gamma$ belong to $[\mathfrak{u}, \mathfrak{u}]$. Then one uses the same arguments as in 2.3.

## 3. The $B_1$–cohomology

**3.1.** We shall write $\lambda$ instead of $k_\lambda$ whenever no confusion is likely. For example, we shall generally use notations like $H^i(B_1, \lambda)$.

Because $B_1/U_1 \simeq T_1$ is a diagonalizable group scheme, one has

$$H^1(B_1, \lambda) \simeq H^1(U_1, \lambda)^{T_1} \simeq (H^1(U_1, k) \otimes \lambda)^{T_1}, \tag{1}$$

cf. [15], I.6.9(3). One has obviously

$$H^1(B_1, \lambda + p\nu) \simeq H^1(B_1, \lambda) \otimes p\nu \tag{2}$$

for all $\lambda, \nu \in X(T)$. So it is enough to compute $H^1(B_1, \lambda)$ for all $\lambda \in X_1(T)$.

**3.2. Lemma:** *Let $\lambda \in X(T)$. Then $H^1(B_1, \lambda) \neq 0$ if and only if there is an $i$ $(1 \leq i \leq n)$ with $\lambda \equiv -\alpha_i \pmod{pX(T)}$.*

**Proof**: Considered as a $T$–module $H^1(U_1, k) \otimes \lambda$ is the direct sum of certain $k_{\alpha+\lambda}$ with $\alpha \in R^+$. Such a summand yields a nonzero contribution to $H^1(B_1, \lambda)$ if and only if $\alpha + \lambda \in pX(T)$, i.e., if and only $\lambda \equiv -\alpha \pmod{pX(T)}$. The $\alpha$ that can occur are always the simple roots $\alpha_i, 1 \leq i \leq n$ and for special $p$ some other roots (cf. 2.3/2.4) that are congruent to one of the $\alpha_i$ modulo $pX(T)$. The lemma follows.

**3.3.** For each simple root $\alpha_i$ let $\lambda_i$ be the unique weight in $X_1(T)$ with $\lambda_i \equiv -\alpha_i \pmod{pX(T)}$. By Lemma 3.2 the $\lambda_i$ are exactly the weights $\lambda \in X_1(T)$ with $H^1(B_1, \lambda) \neq 0$.

**Lemma:** *One has $\lambda_i = p\omega_i - \alpha_i$ except in the following cases:*

*(a) $R$ is of type $C_n$ with $n \geq 2$, one has $p = 2$ and $i = n$. Then $\lambda_n = 0$.*

*(b) $R$ is of type $B_n$ with $n \geq 3$, one has $p = 2$ and $i = n - 1$. Then $\lambda_{n-1} = \omega_{n-2}$.*

*(c) $R$ is of type $F_4$, one has $p = 2$ and $i = 2$. Then $\lambda_2 = \omega_1$.*

*(d) $R$ is of type $G_2$, one has $p \in \{2, 3\}$ and $i = 2$. Then $\lambda_2 = \omega_1$ for $p = 2$, and $\lambda_2 = \omega_2$ for $p = 3$.*

**Proof**: We have to show that $p\omega_i - \alpha_i \in X_1(T)$ except for those cases mentioned. Expressed in terms of the fundamental weights $p\omega_i - \alpha_i$ is the sum of $(p-2)\omega_i$ and all $-\langle \alpha_i, \alpha_j^\vee \rangle \omega_j$ with $\langle \alpha_i, \alpha_j^\vee \rangle < 0$. So $p\omega_i - \alpha_i \in X_1(T)$ is equivalent to $-\langle \alpha_i, \alpha_j^\vee \rangle < p$ for all these $j$. If there is only one root length one has $\langle \alpha_i, \alpha_j^\vee \rangle \geq -1$ for all $i, j$. If $R$ has two root lengths, there is exactly one pair $(i, j)$ for which this inequality is violated: One has to have $\alpha_i$ long, $\alpha_j$ short, and $i, j$ joined in the Dynkin diagram. This leads exactly to the four cases for $R$ and $i$ above. One checks easily that only the primes mentioned occur and that $\lambda_i$ is the weight mentioned.

**3.4. Lemma:** *The $\lambda_i$ with $1 \leq i \leq n$ are distinct except for the following cases:*

*(a) $R$ is of type $A_2$ and $p = 3$. In this case $\lambda_1 = \lambda_2 = \omega_1 + \omega_2$.*

*(b) $R$ is of type $A_3$ and $p = 2$. In this case $\lambda_1 = \lambda_3 = \omega_2$.*

*(c) $R$ is of type $B_3$ and $p = 2$. In this case $\lambda_1 = \lambda_3 = \omega_2$.*

*(d) R is of type $B_4$ and $p = 2$. In this case $\lambda_1 = \lambda_3 = \omega_2$.*

*(e) R is of type $D_4$ and $p = 2$. In this case $\lambda_1 = \lambda_3 = \lambda_4 = \omega_2$.*

*(f) R is of type $D_n$ with $n \geq 5$ and $p = 2$. In this case $\lambda_{n-1} = \lambda_n = \omega_{n-2}$.*

**Proof**: Consider first the case where all roots are equally long. Then $\lambda_i = p\omega_i - \alpha_i$ is the sum of $(p-2)\omega_i$ and of those $\omega_j$ with $j$ linked to $i$ in the Dynkin diagram. For $p > 3$ the coefficient $p - 2$ is the only one greater than 1 and thus determines $i$. For $p = 3$ we can have $\lambda_i = \lambda_j$ if and only if $i$ and $j$ are linked in the Dynkin diagram and if they are not linked to any other vertex. (Observe that the diagram does not contain any cycle.) This can happen only in type $A_2$ where we get case (a). For $p = 2$ we can have $\lambda_i = \lambda_j$ if and only if $i$ and $j$ are not linked in the Dynkin diagram and if they are both linked to the same vertices. The absence of cycles implies that there is exactly one vertex linked to $i$ and $j$. This shows easily that (b), (e) and (f) are the only possibilities.

Before dealing with the remaining cases observe first that $\lambda_i = \lambda_j$ implies $\alpha_i \equiv \alpha_j \pmod{pX(T)}$. As $\alpha_i$ and $\alpha_j$ are not congruent modulo $p\mathbf{Z}R$ (assuming $i \neq j$ all the time), we must have that $p$ divides the index of connection $(X(T) : \mathbf{Z}R)$. This shows that there are no possible cases for $R$ of type $F_4$ or $G_2$. For $R$ of type $C_n$ or $B_n$ we have to deal with $p = 2$ only. For $C_n$ and $p = 2$ one has

$$\lambda_i = \begin{cases} \omega_2, & \text{if } i = 1; \\ \omega_{i-1} + \omega_{i+1}, & \text{if } 1 < i < n; \\ 0, & \text{if } i = n. \end{cases}$$

These weights are obviously distinct. For $R$ of type $B_n$ with $n \geq 3$ and $p = 2$ one gets

$$\lambda_i = \begin{cases} \omega_2, & \text{if } i = 1; \\ \omega_{i-1} + \omega_{i+1}, & \text{if } 1 < i < n-1; \\ \omega_{n-2}, & \text{if } i = n-1; \\ \omega_{n-1}, & \text{if } i = n. \end{cases}$$

The only equalities occur when $\omega_2$ is either equal to $\omega_{n-2}$ or to $\omega_{n-1}$. This leads to the cases (c) and (d).

*Remarks*: 1)For $R$ of type $A_n$ the result goes back to Lemma 2.3(a) in [21] where the case (b) is overlooked.

2) For $p = 2$ the congruence $\alpha_i \equiv \alpha_j \pmod{pX(T)}$ implies that $\alpha_i + \alpha_j$ is a degenerate sum in the sense of [16], 2.4. Indeed, all the special cases in 3.3 and 3.4 correspond to one of those degenerate sums, cf. the table in [16], 2.8.

**3.5. Proposition:** *Suppose that $p$ is not special. One has $H^1(B_1, \lambda) = 0$ for all $\lambda \in X_1(T)$ with $\lambda \neq p\omega_i - \alpha_i$ for all $i$. One has (for each $i$)*

$$H^1(B_1, p\omega_i - \alpha_i)^{(-1)} \simeq k_{\omega_i}$$

*except in the following cases:*

*(a) R of type $A_2$ and $p = 3$ where $H^1(B_1, \omega_1 + \omega_2)^{(-1)} \simeq k_{\omega_1} \oplus k_{\omega_2}$,*

*(b) R of type $A_3$ and $p = 2$ where $H^1(B_1, \omega_2)^{(-1)} \simeq k_{\omega_1} \oplus k_{\omega_3}$,*

*(c) R of type $D_4$ and $p = 2$ where $H^1(B_1, \omega_2)^{(-1)} \simeq k_{\omega_1} \oplus k_{\omega_3} \oplus k_{\omega_4}$,*

*(d) R of type $D_n$ with $n \geq 5$ and $p = 2$ where $H^1(B_1, \omega_{n-2})^{(-1)} \simeq k_{\omega_{n-1}} \oplus k_{\omega_n}$.*

**Proof**: Under our assumption $H^1(U_1, \lambda)$ is the direct sum of all $k_{\alpha_i+\lambda}$. The $i$-th summand yields a contribution to $H^1(B_1, \lambda) = H^1(U_1, \lambda)^{T_1}$ if and only if $\alpha_i + \lambda$ is divisible by $p$ in $X(T)$, i.e., for $\lambda = \lambda_i = p\omega_i - \alpha_i$ in which case we get a contribution $k_{p\omega_i}$. If $\lambda_i \neq \lambda_j$ for all $j \neq i$ this is the only contribution and we get the general formula. The cases (determined in 3.4) where distinct $i$ yield the same $\lambda_i$ lead to the exceptions in (a) to (d). Finally, because we have applied the untwist $(?)^{(-1)}$ to the cohomology groups, we have to divide all weights by $p$.

**3.6. Proposition:** *Suppose that $p = 2$ and that $R$ is of type $B_n$, $C_n$, or $F_4$. The only weights $\lambda \in X_1(T)$ with $H^1(B_1, \lambda) \neq 0$ are those occurring in (a) to (f) below.*

*(a) One has $H^1(B_1, p\omega_i - \alpha_i)^{(-1)} = k_{\omega_i}$ for all $i$ with $p\omega_i - \alpha_i \in X_1(T)$ except for the two cases in (b) and (c).*

*(b) If $R$ is of type $B_3$, then $H^1(B_1, \omega_2)^{(-1)} = k_{\omega_1} \oplus k_{\omega_3}$.*

*(c) If $R$ is of type $B_4$, then $H^1(B_1, \omega_2)^{(-1)} = k_{\omega_1} \oplus M$ where $M$ has a filtration with factors $k_{\omega_4}$ and $k_{\omega_3 - \omega_4}$.*

*(d) If $R$ is of type $B_n$ with $n \geq 3$ and $n \neq 4$, then $H^1(B_1, \omega_{n-2})^{(-1)}$ has a filtration with factors $k_{\omega_n}$ and $k_{\omega_{n-1} - \omega_n}$.*

*(e) If $R$ is of type $C_n$ with $n \geq 2$, then $H^1(B_1, k)^{(-1)}$ has a filtration with factors $k_{\omega_1}$ and $k_{\omega_i - \omega_{i-1}}$ with $i = 2, 3, \ldots, n$.*

*(f) If $R$ is of type $F_4$, then $H^1(B_1, \omega_1)^{(-1)}$ has a filtration with factors $k_{\omega_4}$ and $k_{\omega_3 - \omega_4}$ and $k_{\omega_2 - \omega_3}$.*

**Proof**: All weights of any $M_i$ as in 2.3 are congruent modulo $pX(T)$. So either all weights of $M_i \otimes \lambda$ are divisible by $p$ in $X(T)$ or none. The first case occurs if and only if $\lambda = \lambda_i$. So $H^1(B_1, \lambda)$ is the direct sum of all $M_i \otimes \lambda_i$ with $\lambda = \lambda_i$. The only cases with two summands arise from 3.4(c),(d) leading to (b),(c) above. In most cases one has $M_i = k_{\alpha_i}$; then $\lambda_i = p\omega_i - \alpha_i$ yields $M_i \otimes \lambda_i = k_{p\omega_i}$. In the other cases it is left to the reader to add $\lambda_i$ from 3.3 to the weights of $M_i$ from 2.3 and to express these sums in terms of fundamental weights. As in 3.5 we have to divide all weights by $p$ at the end.

**3.7. Proposition:** *Suppose that $p \in \{2, 3\}$ and that $R$ is of type $G_2$. The only weights $\lambda \in X_1(T)$ with $H^1(B_1, \lambda) \neq 0$ are those occurring in (a) resp. (b) below.*

*(a) If $p = 2$, then $H^1(B_1, \omega_2)^{(-1)} = k_{\omega_1}$, and $H^1(B_1, \omega_1)^{(-1)}$ has a filtration with factors $k_{\omega_1}$ and $k_{\omega_2 - \omega_1}$.*

*(b) If $p = 3$, then $H^1(B_1, \omega_1 + \omega_2)^{(-1)} = k_{\omega_1}$, and $H^1(B_1, \omega_2)^{(-1)}$ has a filtration with factors $k_{\omega_1}$ and $k_{\omega_2 - \omega_1}$.*

**Proof**: One uses the same arguments as in 3.6. Details are left to the reader.

## 4. The $G_1$–cohomology: Good Cases

**4.1. Proposition:** *Suppose that p is not special. One has*

$$H^1(G_1, H^0(\lambda)) = 0$$

*for all* $\lambda \in X_1(T)$ *with* $\lambda \neq p\omega_i - \alpha_i$ *for all i, and*

$$H^1(G_1, H^0(p\omega_i - \alpha_i))^{(-1)} \simeq H^0(\omega_i)$$

*for all i except in the following cases:*

*(a) R of type* $A_2$ *and* $p = 3$ *where*

$$H^1(G_1, H^0(\omega_1 + \omega_2))^{(-1)} \simeq H^0(\omega_1) \oplus H^0(\omega_2),$$

*(b) R of type* $A_3$ *and* $p = 2$ *where*

$$H^1(G_1, H^0(\omega_2))^{(-1)} \simeq H^0(\omega_1) \oplus H^0(\omega_3),$$

*(c) R of type* $D_4$ *and* $p = 2$ *where*

$$H^1(G_1, H^0(\omega_2))^{(-1)} \simeq H^0(\omega_1) \oplus H^0(\omega_3) \oplus H^0(\omega_4),$$

*(d) R of type* $D_n$ *with* $n \geq 5$ *and* $p = 2$ *where*

$$H^1(G_1, H^0(\omega_{n-2}))^{(-1)} \simeq H^0(\omega_{n-1}) \oplus H^0(\omega_n).$$

**Proof**: This is an immediate consequence of 1.7(2) and Proposition 3.5.

*Remarks*: 1) It is easy to compute $H^1(G_1, H^0(\mu))$ for all $\mu \in X(T)_+$ using 3.1(2). This is left to the interested reader.

2) For $p$ greater than the Coxeter number of $R$ this proposition is a special case of the general formulas for $H^\bullet(G_1, H^0(\lambda))$ in [2], 3.7 and 5.5 (with open problems for the exceptional types). One gets a contribution to $H^1$ in that formula if the Weyl group element there has length 1, i.e., is one of the simple reflections $s_1, s_2, \ldots, s_n$. One has $s_i \bullet 0 = -\alpha_i$ so the formula in [2] yields $H^1(G_1, H^0(p\nu - \alpha_i))^{(-1)} \simeq H^0(\nu)$ when it can be applied.

**4.2.** We have for all $\lambda \in X(T)_+$ an exact sequence

$$0 \longrightarrow L(\lambda) \longrightarrow H^0(\lambda) \longrightarrow H^0(\lambda)/L(\lambda) \longrightarrow 0. \tag{1}$$

We shall consider the corresponding long exact sequence of $G_1$–cohomology. One term is $H^0(G_1, H^0(\lambda))$, the space of fixed points for $G_1$ in $H^0(\lambda)$. This is a $G$–submodule. If it is nonzero, it has to contain the simple socle $L(\lambda)$ of $H^0(\lambda)$ as a $G$–module. Then $G_1$ has to act trivially on $L(\lambda)$ which happens if and only if $\lambda \in pX(T)$. If we restrict to $\lambda \in X_1(T)$ this occurs only for $\lambda = 0$. This shows

$$H^0(\lambda)^{G_1} = 0 \qquad \text{for all } \lambda \in X_1(T), \lambda \neq 0. \tag{2}$$

We get therefore an exact sequence

$$\begin{aligned} 0 \to (H^0(\lambda)/L(\lambda))^{G_1} &\to H^1(G_1, L(\lambda)) \\ &\to H^1(G_1, H^0(\lambda)) \to H^1(G_1, H^0(\lambda)/L(\lambda)) \end{aligned} \tag{3}$$

for all $\lambda \in X_1(T)$. (It works also for $\lambda = 0$ because of $H^0(0) = L(0) = k$.)

**4.3. Proposition:** *Suppose that p is not special. One has for all $\lambda \in X_1(T)$ with $\lambda \neq p\omega_i - \alpha_i$ for all i:*

$$H^1(G_1, L(\lambda)) \simeq (H^0(\lambda)/L(\lambda))^{G_1}.$$

**Proof:** This is an obvious consequence of 4.2(3) and 4.1.

**4.4.** We have

$$H^1(G_1, H^0(\lambda)) \simeq H^1(\mathfrak{g}, H^0(\lambda)) \tag{1}$$

for all $\lambda \in X_1(T), \lambda \neq 0$. This follows from a formula of Hochschild, cf. [15], I.9.19, and from 4.2(2).

One can compute $H^1(\mathfrak{g}, M)$ for any $\mathfrak{g}$–module $M$ using the standard complex. The space $Z^1(\mathfrak{g}, M)$ of 1–cocycles consists of all linear maps $\varphi : \mathfrak{g} \to M$ with

$$\varphi([X,Y]) = X\varphi(Y) - Y\varphi(X) \tag{2}$$

for all $X, Y \in \mathfrak{g}$. The 1–cocycle $\varphi$ is a coboundary if and only if there is an $m \in M$ with $\varphi(X) = Xm$ for all $X \in \mathfrak{g}$.

Suppose now that we get $M$ from a $G$–module by differentiating the $G$–module structure. Then $Z^1(\mathfrak{g}, M)$ is a submodule of the $G$–module $\mathrm{Hom}(\mathfrak{g}, M)$ where we use the adjoint representation on $\mathfrak{g}$. The coboundaries form a $G$–submodule of $Z^1(\mathfrak{g}, M)$. We get thus a $G$–module structure on $H^1(\mathfrak{g}, M)$ inducing the one on $H^1(G_1, M)$ which we have been using. We can find for each cohomology class of weight say $\mu \in X(T)$ a representing cocycle $\varphi \in Z^1(\mathfrak{g}, M)$ that has weight $\mu$, i.e., with $\varphi(\mathfrak{g}_\alpha) \subset M_{\alpha+\mu}$ for all $\alpha \in R \cup \{0\}$.

**4.5. Proposition:** *Suppose that p is not special. Let $\lambda = p\omega_i - \alpha_i$ for some i with $1 \leq i \leq n$. Then $H^1(G_1, H^0(\lambda))^{(-1)}$ has a direct summand isomorphic to $H^0(\omega_i)$. The socle $L(\omega_i)$ of this summand is contained in the image of the canonical map*

$$H^1(G_1, L(\lambda))^{(-1)} \to H^1(G_1, H^0(\lambda))^{(-1)}.$$

**Proof:** The first statement is just a repetition of part of 4.1. We have to prove the second one. Consider a nonzero element of weight $p\omega_i$ in $H^1(G_1, \lambda)$. As explained in 4.4 it can be represented by a 1–cocycle $\varphi \in Z^1(\mathfrak{g}, H^0(\lambda))$ of weight $p\omega_i$. We shall show that $\varphi$ takes values in $L(\lambda)$. That will imply the proposition.

Any $\mathfrak{g}_\alpha$ is mapped under $\varphi$ into the $p\omega_i + \alpha$ weight space of $H^0(\lambda) = H^0(p\omega_i - \alpha_i)$. If this weight space is nonzero, then $p\omega_i + \alpha \leq p\omega_i - \alpha_i$, i.e., $\alpha \leq -\alpha_i$. So we have to show that $\varphi(X_\alpha) \in L(\lambda)$ for all negative roots $\alpha$ with $\alpha \leq -\alpha_i$. We use induction on $\alpha$ from above for the usual order $\leq$.

To start with, for $\alpha = -\alpha_i$ the image $\varphi(X_\alpha)$ has weight $p\omega_i - \alpha_i = \lambda$ and is therefore contained in $H^0(\lambda)_\lambda = L(\lambda)_\lambda$. Suppose now that $\alpha < -\alpha_i$. Then $-\alpha$ is not simple and there are negative roots $\beta, \gamma$ with $\alpha = \beta + \gamma$. By induction we have $\varphi(X_\beta), \varphi(X_\gamma) \in L(\lambda)$. (Note that if, e.g., $\beta \not\leq -\alpha_i$, then $\varphi(X_\beta) = 0 \in L(\lambda)$.). The cocycle condition 4.4(2) implies that also $\varphi([X_\beta, X_\gamma]) \in L(\lambda)$. But $[X_\beta, X_\gamma] = N_{\beta,\gamma} X_\alpha$ for some coefficient $N_{\beta,\gamma}$. Because we assume that $p$ is not special, this coefficient is not zero in $k$, cf. 2.2. We get therefore that also $\varphi(X_\alpha) \in L(\lambda)$.

*Remark*: The proposition implies especially that $H^1(G_1, L(\lambda)) \neq 0$. This was first shown in [21], Thm. 2.2.

**4.6.** The last proposition implies (for $p$ not special) that the natural map from $H^1(G_1, L(\lambda))$ to $H^1(G_1, H^0(\lambda))$ is onto if all $H^0(\omega_i)$ with $\lambda = p\omega_i - \alpha_i$ are irreducible. Let me therefore report what is known about that point.

In general, one can decide for any $\mu \in X(T)_+$ whether $H^0(\mu)$ is irreducible by evaluating the sum formula as in [15], II.8.19. For the fundamental weights one can often use easier approaches. For example, if $\omega_i$ is minuscule, then $H^0(\omega_i)$ is irreducible for any characteristic. (Recall that $\omega_i$ is called minuscule, if there is not a dominant weight $\mu$ with $\mu < \omega_i$. See [3], ch. VI, §1, exerc. 24 for a more thorough discussion of this notion, and ibid. §4, exerc. 15 for a list of all minuscule weights.)

For $R$ of type $A_n$ with $n \geq 1$ all $\omega_i$ are minuscule, hence all $H^0(\omega_i)$ are irreducible in any characteristic. For $R$ of type $B_n$ with $n \geq 2$ the weight $\omega_n$ is minuscule. For $R$ of type $D_n$ with $n \geq 4$ the weight $\omega_i$ is minuscule for $i \in \{1, n-1, n\}$. So these $H^0(\omega_i)$ are irreducible irreducible in any characteristic. Any other $H^0(\omega_i)$ is irreducible if and only if $p \neq 2$ (for both types). For $R$ of type $C_n$ with $n \geq 2$ the weight $\omega_1$ is minuscule, hence $H^0(\omega_1)$ is irreducible in any characteristic. An $H^0(\omega_i)$ with $1 < i \leq n$ is irreducible if and only if $p$ does not divide any binomial coefficient

$$\binom{n+1-(i+j)/2}{(i-j)/2}$$

with $0 \leq j < i$ and $j \equiv i \pmod 2$, cf. [18].

For the exceptional types I shall write down the results in a table. The third column contains the primes for which the $H^0(\omega_i)$ with $\omega_i$ in the second column is not irreducible. In many cases one can find more detailed information in [8].

| | | |
|---|---|---|
| Type $E_6$ | $\omega_1, \omega_6$ | none |
| | $\omega_3, \omega_5$ | 2 |
| | $\omega_2$ | 3 |
| | $\omega_4$ | 2,3 |
| Type $E_7$ | $\omega_7$ | none |
| | $\omega_1$ | 2 |
| | $\omega_2$ | 3 |
| | $\omega_3, \omega_5$ | 2,3 |
| | $\omega_6$ | 2,7 |
| | $\omega_4$ | 2,3,13 |
| Type $E_8$ | $\omega_8$ | none |
| | $\omega_1$ | 2 |
| | $\omega_7$ | 2,3,5 |
| | $\omega_2$ | 2,3,7 |
| | $\omega_6$ | 2,3,5,7 |
| | $\omega_3$ | 2,3,19 |
| | $\omega_5$ | 2,3,5 |
| | $\omega_4$ | 2,3,5,13,19 |
| Type $F_4$ | $\omega_4$ | 3 |
| | $\omega_1$ | 2 |
| | $\omega_3, \omega_2$ | 2,3 |
| Type $G_2$ | $\omega_1$ | 2 |
| | $\omega_2$ | 3 |

**4.7. Proposition:** *Suppose that $\omega_i$ is minuscule. Set $\lambda = p\omega_i - \alpha_i$.*

*(a) If $p\mu$ with $\mu \in X(T)_+$ is a weight of $H^0(\lambda)$, then $\mu = 0$.*

*(b) If $p$ does not divide the index of connection $(X(T) : \mathbf{Z}R)$, then there is no $\mu \in X(T)_+$ such that $p\mu$ is a weight of $H^0(\lambda)$.*

**Proof**: Suppose that $p\mu$ is a weight of $H^0(\lambda)$. Then we have necessarily:

$$p\mu \leq p\omega_i - \alpha_i. \tag{1}$$

(b): The inequality (1) implies especially $p\mu \equiv p\omega_i \pmod{\mathbf{Z}R}$. Because $p$ does not divide the index of connection, this implies $\mu \equiv \omega_i \pmod{\mathbf{Z}R}$. Then $\omega_i$ is the smallest dominant weight of the Weyl module $V(\mu)$ (by the representation theory in characteristic 0). This shows that $\omega_i \leq \mu$, hence $p\omega_i \leq p\mu \leq p\omega_i - \alpha_i$, a contradiction.

(a): Let $\alpha_0$ be the largest short root in $R$. Then $\alpha_0^\vee$ is the largest root in the dual root system. Each $\mu' \in X(T)_+$ with $\mu' \neq 0$ satisfies $\langle \mu', \alpha_0^\vee \rangle \geq 1$, with equality if and only if $\mu'$ is minuscule, cf. [3], ch.VI, §1, exerc.24. The inequality (1) implies $p\langle \mu, \alpha_0^\vee \rangle \leq p\langle \omega_i, \alpha_0^\vee \rangle = p$, hence $\langle \mu, \alpha_0^\vee \rangle \leq 1$. Therefore $\mu$ is either 0 (as desired) or minuscule. We have to show that the second case does not arise.

Let us look first at the case where $R$ is of type $A_n$. Suppose that $p\omega_j \leq p\omega_i - \alpha_i < p\omega_i$ for some $j$. Express both sides in terms of the simple roots — for example, by looking at Bourbaki's tables — and compare the coefficients of $\alpha_1$ and $\alpha_n$. One gets $n+1-j \leq n+1-i$ and $j \leq i$, hence $i = j$. But $p\omega_i \leq p\omega_i - \alpha_i$ is clearly impossible.

When dealing with the other types, we can assume that $p$ divides the index of connection. For $R$ of type $B_n$, $C_n$, or $D_n$ this means $p = 2$. Then $\lambda$ is a fundamental weight (it is $\omega_{n-1}$ for $B_n$, $\omega_2$ for $C_n$, $\omega_2$ or $\omega_{n-2}$ for $D_n$). The only dominant weights of such $H^0(\lambda)$ are certain $\omega_j$ and possibly 0. The only weight divisible by $p = 2$ is 0 (if it occurs). For $R$ of type $E_6$ we have $p = 3$ and $\lambda \in \{\omega_1 + \omega_3, \omega_5 + \omega_6\}$. For $R$ of type $E_7$ we have $p = 2$ and $\lambda = \omega_6$. In both cases it is easy to check that 0 is the only weight of $H^0(\lambda)$ divisible by $p$.

*Remarks*: 1) It can happen (for $R$ of type $A_n$ or $D_n$) that $p$ divides the index of connection and that $p\omega_i \notin \mathbf{Z}R$. Then 0 cannot be a weight of $H^0(\lambda)$, so (a) implies that there cannot be any $\mu \in X(T)_+$ such that $p\mu$ is a weight of $H^0(\lambda)$. In the case where $R$ is of type $A_n$ and $i = 1$ this was observed by Ballard, cf. [22].

2)If $R$ is not of type $A_n$, then there is an index $i$ with $\omega_i = \alpha_0$, the largest short root as in the proof. Using the same arguments as for (b), one can show: If $p$ does not divide the index of connection, then 0 is the only dominant weight of $H^0(p\omega_i - \alpha_i)$ divisible by $p$. The point is that $p\mu < p\alpha_0$ implies $\mu \in \mathbf{Z}R$ under our assumption on $p$. On the other hand, any dominant $\mu \neq 0$ in $\mathbf{Z}R$ satisfies $\alpha_0 \leq \mu$.

**4.8. Corollary:** *Suppose that $\omega_i$ is minuscule. Set $\lambda = p\omega_i - \alpha_i$.*

*(a) One has*

$$(H^0(\lambda)/L(\lambda))^{G_1} = (H^0(\lambda)/L(\lambda))^G = H^1(G, L(\lambda)). \tag{1}$$

*(b) If $p$ does not divide the index of connection, then*

$$(H^0(\lambda)/L(\lambda))^{G_1} = 0. \tag{2}$$

**Proof**: Any composition factor of the $G$–module $(H^0(\lambda)/L(\lambda))^{G_1}$ has the form $L(p\mu)$ for some $\mu \in X(T)_+$. Obviously $p\mu$ has to be a weight of $H^0(\lambda)$. Proposition 4.7 implies that $\mu = 0$. So $G$ acts trivially on $(H^0(\lambda)/L(\lambda))^{G_1}$ and we get (a). Part (b) follows immediately from part (b) in 4.7.

*Remark*: The remarks in 4.7 yield similar results in the cases studied there. The same is true for the next proposition. We leave the formulations to the reader.

**4.9. Proposition:** *Assume that p is not special. Suppose that $\omega_i$ is minuscule. Set $\lambda = p\omega_i - \alpha_i$.*

*(a) There is an exact sequence*

$$0 \to (H^0(\lambda)/L(\lambda))^G \to H^1(G_1, L(\lambda)) \to H^1(G_1, H^0(\lambda)) \to 0. \tag{1}$$

*(b) If p does not divide the index of connection, then*

$$H^1(G_1, L(p\omega_i - \alpha_i))^{(-1)} \simeq H^0(\omega_i). \tag{2}$$

**Proof**: We use the exact sequence 4.2(3). Proposition 4.5 shows that the map from the first cohomology of $L(\lambda)$ to that of $H^0(\lambda)$ is surjective, because $H^0(\omega_i)$ is simple. (In the exceptional cases 4.1(a)–(d) one has to observe that $\omega_j$ is minuscule for each direct summand $H^0(\omega_j)$ of $H^1(G_1, H^0(\lambda))^{(-1)}$.) For the kernel of this map we use 4.8 to get the claim. (Note that the cases 4.1(a)–(d) do not show up in part (b) here.)

*Remark*: Let me mention explicitly one of the generalisations that we get from the remarks in 4.7. Take $R$ of type $E_8$. Then the adjoint representation $H^0(\omega_8)$ is irreducible in each characteristic. So the remark 2 in 4.7 shows that (1) holds always for $E_8$ and $i = 8$.

**4.10.** Part (b) of the last proposition answers the questions 3.6(a),(b) in [21] for $R$ of type $A_n$. Recall that the index of connection is equal to $n+1$ in that case.

**Proposition:** *Suppose that $R$ is of type $A_n$. Let $\lambda = p\omega_i - \alpha_i$ for some $i$.*

*(a) If $1 < i < n$, then there is an exact sequence*

$$0 \to (H^0(\lambda)/L(\lambda))^G \to H^1(G_1, L(\lambda))^{(-1)} \to H^0(\omega_i) \to 0. \tag{1}$$

*If the first term is not 0, then p divides $n+1$ and $i$ is an integral multiple of $(n+1)/p$.*

*(b) If $i \in \{1, n\}$, then*

$$H^1(G_1, L(\lambda))^{(-1)} \simeq H^0(\omega_i), \tag{2}$$

*except for the following two cases: if $n = 2$ and $p = 3$, then*

$$H^1(G_1, L(\omega_1 + \omega_2))^{(-1)} \simeq H^0(0) \oplus H^0(\omega_1) \oplus H^0(\omega_2), \tag{3}$$

*and if $n = 3$ and $p = 2$, then*

$$H^1(G_1, L(\omega_2))^{(-1)} \simeq H^0(\omega_1) \oplus H^0(\omega_3). \tag{4}$$

**Proof**: Part (a) is more or less a restatement of 4.9 for $R$ of type $A_n$. By excluding $i = 1, n$ we can write $H^0(\omega_i)$ instead of the first cohomology of $H^0(\lambda)$: The exceptions in 4.1(a),(b) lead to the exceptions in (3) and (4) of part (b). Also, following Remark 1 in 4.7, we have replaced the condition from 4.9(b) by the condition: $p\omega_i \notin \mathbf{Z}R$.

In order to get (b) we can assume by symmetry that $i = 1$. We have to show that $(H^0(\lambda)/L(\lambda))^G = 0$ except for the case in (3) where it is equal to $H^0(0) = L(0) = k$. Then 4.9(1) yields the claim. In the situation of (3) one has to use in addition that the three simple modules on the right hand side do not extend each other in order to get a direct sum decomposition. This is clear, however, because the highest weights belong to different cosets modulo $\mathbf{Z}R$. (In (3) and (4) I have replaced $\lambda$ by its actual value.)

Now $\lambda$ is equal to $(p-2)\omega_1 + \omega_2$ for $n > 1$ resp. to $(p-2)\omega_1$ for $n = 1$. One has obviously $H^0(\lambda) = L(\lambda)$ if $n = 1$ or $p = 2$ (where $\lambda = \omega_2$). Let us therefore assume $n > 1$ and $p \geq 3$. Set $\mu_j = (p-j)\omega_1 + \omega_j$ for all $j$ with $1 \leq j \leq \min(p, n+1)$ using the convention $\omega_{n+1} = 0$. We want to show that $H^0(\mu_j)/L(\mu_j) \simeq L(\mu_{j+1})$ for all $j < \min(p, n+1)$ and $H^0(\mu_j) = L(\mu_j)$ for $j = \min(p, n+1)$. Because of $\lambda = \mu_2$ this implies that $H^0(\lambda)/L(\lambda) \simeq L(\mu_3) = L((p-3)\omega_1 + \omega_3)$ does not have a composition factor $L(0)$ except in the case $p = 3$ and $n = 2$ as desired.

In order to prove the claim in the last paragraph we first evaluate the sum formula as in [15], II.8.19. We use the notation $\chi(\mu)$ for the formal character of the Weyl module $V(\mu)$. An elementary computation — left to the reader — shows that the sum formula for $V(\mu_j)$ evaluates to $\sum_{r=j+1}^{\min(p,n+1)} (-1)^{r-j+1} \chi(\mu_r)$. For $j = \min(p, n+1)$ this sum is empty proving the irreducibility of $V(\mu_j)$, hence of $H^0(\mu_j)$. For smaller $j$ induction implies that the sum is equal to the formal character of $L(\mu_{j+1})$. This implies that $L(\mu_{j+1})$ is the unique largest proper submodule of $L(\mu_j)$. By duality, the claim on $H^0(\mu_j)$ follows.

## 5. The $G_1$–cohomology: The Special Cases

**5.1. Proposition:** *Suppose that $p = 2$ and that $R$ is of type $B_n$, $C_n$ or $F_4$. One has $H^1(G_1, H^0(\lambda)) = 0$ for all $\lambda \in X_1(T)$ not mentioned explicitly below.*

*(a) If $R$ is of type $B_n$ with $n \geq 3$, then*

$$H^1(G_1, H^0(\omega_2))^{(-1)} \simeq H^0(\omega_1), \tag{1}$$

*and*

$$H^1(G_1, H^0(\omega_{i-1} + \omega_{i+1}))^{(-1)} \simeq H^0(\omega_i) \qquad \text{for } 1 < i < n-1 \tag{2}$$

*and*

$$H^1(G_1, H^0(\omega_{n-2}))^{(-1)} \simeq H^1(G_1, H^0(\omega_{n-1}))^{(-1)} \simeq H^0(\omega_n), \tag{3}$$

*except in the following cases: For $n = 3$ one has*

$$H^1(G_1, H^0(\omega_2))^{(-1)} \simeq H^0(\omega_1) \oplus H^0(\omega_3), \tag{4}$$

*and for $n = 4$ one has*

$$H^1(G_1, H^0(\omega_2))^{(-1)} \simeq H^0(\omega_1) \oplus H^0(\omega_4). \tag{5}$$

*(b) If $R$ is of type $C_n$ with $n \geq 2$, then*

$$H^1(G_1, H^0(0))^{(-1)} \simeq H^1(G_1, H^0(\omega_2))^{(-1)} \simeq H^0(\omega_1), \tag{6}$$

*and*

$$H^1(G_1, H^0(\omega_{i-1} + \omega_{i+1}))^{(-1)} \simeq H^0(\omega_i) \qquad \text{for } 1 < i < n. \tag{7}$$

*(c) If $R$ is of type $F_4$, then*

$$H^1(G_1, H^0(\omega_2))^{(-1)} \simeq H^0(\omega_1), \tag{8}$$

*and*

$$H^1(G_1, H^0(\omega_2 + \omega_4))^{(-1)} \simeq H^0(\omega_3), \tag{9}$$

*and*

$$H^1(G_1, H^0(\omega_1))^{(-1)} \simeq H^1(G_1, H^0(\omega_3))^{(-1)} \simeq H^0(\omega_4). \tag{10}$$

**Proof**: This follows from 3.6 more or less in the same way as 4.1 follows from 3.5. (Well, after computing the explicit values of all $p\omega_i - \alpha_i$.) Additional thought is required only for the direct summands of $H^1(B_1, \lambda)^{(-1)}$ that have dimension greater than one. Denote one of these summands by $M$. We have to determine $\text{ind}_B^G M$. There is exactly one dominant weight of $M$, say $\mu$. All other weights $\mu'$ of $M$ have one coefficient equal to $-1$ when expressed as a linear combination of the fundamental weights. The claim of the proposition is now that $\text{ind}_B^G M = \text{ind}_B^G \mu = H^0(\mu)$. This follows, because by Kempf's vanishing theorem all higher derived functors $R^j\text{ind}_B^G$ with $j > 0$ vanish on $k_\mu$, cf. [15], II.4.5., and because on the other weights $\mu'$ all $R^j\text{ind}_B^G$ with $j \geq 0$ disappear, cf. [15], II.5.4.a.

*Remark*: Again, it easy to compute $H^1(G_1, H^0(\mu))$ for all $\mu \in X(T)_+$ using 3.1(2). Note that for $M$ as in the last part of the proof each weight of any $M \otimes p\nu$ with $\nu \in X(T)_+$ either is dominant or has a coefficient equal to $-1$. One can therefore argue as before. However, in general $\text{ind}_B^G(M \otimes p\nu)$ will have a filtration with more than one factor of the form $H^0(\lambda')$.

**5.2. Proposition:** *Suppose that $R$ is of type $G_2$ and that $p \in \{2, 3\}$. One has $H^1(G_1, H^0(\lambda)) = 0$ for all $\lambda \in X_1(T)$ not mentioned explicitly below.*

*(a) If $p = 2$, then*

$$H^1(G_1, H^0(\omega_1))^{(-1)} \simeq H^1(G_1, H^0(\omega_2))^{(-1)} \simeq H^0(\omega_1). \tag{1}$$

*(b) If $p = 3$, then*

$$H^1(G_1, H^0(\omega_1 + \omega_2))^{(-1)} \simeq H^1(G_1, H^0(\omega_2))^{(-1)} \simeq H^0(\omega_1). \tag{2}$$

**Proof**: This follows from 3.7 in the same way as 5.1 follows from 3.6.

**5.3.** The formulas in 4.2 and the discussion in 4.4 did not require any restriction on $p$. They are still valid in the cases looked at in 5.1 and 5.2. Before dealing with the transition from $H^0(\lambda)$ to $L(\lambda)$ in general, let us get rid of one trivial case.

**Proposition:** *Suppose that $R$ is of type $G_2$ and $p = 2$. Then*

$$H^1(G_1, L(0)) = H^1(G_1, L(\omega_1 + \omega_2)) = 0 \tag{1}$$

*and*

$$H^1(G_1, L(\omega_2))^{(-1)} \simeq H^0(\omega_1). \tag{2}$$

*There is an exact sequence*

$$0 \to L(0) \to H^1(G_1, L(\omega_1))^{(-1)} \to H^0(\omega_1) \to 0. \tag{3}$$

**Proof**: One has $H^0(\omega_1)/L(\omega_1) \simeq L(0)$ and $H^0(\lambda) = L(\lambda)$ for all other $\lambda \in X_1(T)$, i.e., for $\lambda = 0, \omega_2, \omega_1 + \omega_2$. So for these three weights the claim follows immediately from 5.2. For $\omega_1$ we combine the exact sequence 4.2(3) with 5.2 and the result for $L(0)$.

**5.4.** In the cases considered in 5.1/2 there is exactly one simple root $\alpha_i$ with $p\omega_i - \alpha_i \notin X_1(T)$, cf. 3.3. In order to express the next result in a uniform way, let us denote by $\lambda_0$ the unique weight in $X_1(T)$ with $\lambda_0 \equiv -\alpha_i \pmod{pX(T)}$ for this $i$ (determined in 3.3) and denote by $\mu_0$ the dominant weight with (usually) $H^0(G_1, H^0(\lambda_0))^{(-1)} \simeq H^0(\mu_0)$. The one exception that forces me to write *usually* is the case $B_4$ where $H^0(\mu_0)$ is only one of two direct summands of the cohomology group. Let me list the different cases for $\lambda_0$ and $\mu_0$:

| Type | $\lambda_0$ | $\mu_0$ |
|---|---|---|
| $B_n, n \geq 3$ | $\omega_{n-2}$ | $\omega_n$ |
| $C_n, n \geq 2$ | $0$ | $\omega_1$ |
| $F_4$ | $\omega_1$ | $\omega_4$ |
| $G_2, p = 2$ | $\omega_1$ | $\omega_1$ |
| $G_2, p = 3$ | $\omega_2$ | $\omega_1$ |

**Proposition:** *Except for $R$ of type $B_4$, one has in all these cases an exact sequence*

$$0 \to (H^0(\lambda_0)/L(\lambda_0))^G \to H^1(G_1, L(\lambda_0)) \to H^0(\mu_0)^{(1)} \to 0.$$

**Proof**: We use the same ideas that lead to 4.9(1). One checks easily that 0 is the only weight of $H^0(\lambda_0)$ divisible by $p$. We can therefore replace the $G_1$ invariants in 4.2(3) by the $G$ invariants. It remains to show (by 5.1/2) that the natural map from $H^1(G_1, L(\lambda_0))^{(-1)}$ to $H^0(G_1, H^0(\lambda_0))^{(-1)} \simeq H^0(\mu_0)$ is surjective. This follows from 5.3(3) in the case $R$ of type $G_2$ and $p = 2$. Let us exclude this case from now on. In all other cases $H^0(\mu_0)$ is irreducible, cf. 4.6. Therefore it is enough to show that the map is not zero.

We proceed as in the proof of Proposition 4.5. Let $\varphi \in Z^1(\mathfrak{g}, H^0(\lambda_0))$ be a cocycle of weight $p\mu_0$. We have to show that $\varphi$ takes values in $L(\lambda_0)$. There is a negative root $\beta$ with $\beta + p\mu_0 = \lambda_0$. It is always the negative of the largest weight of the $M_i$ in 2.3/4. In the five cases of the table one has $\beta = -\alpha_{n-1} - 2\alpha_n$, $-2\varepsilon_1$, $-\alpha_2 - 2\alpha_3 - 2\alpha_4$, $-2\alpha_1 - \alpha_2$, $-3\alpha_1 - \alpha_2$.

As in 4.5 the formula

$$\varphi(\mathfrak{g}_\alpha) \subset H^0(\lambda_0)_{\alpha + p\mu_0}$$

(together with $\lambda_0 = p\mu_o + \beta$) implies that we have to look only at the $\varphi(X_\alpha)$ for negative roots $\alpha \leq \beta$. This image is in $L(\lambda_0)$ for $\alpha = \beta$, because it has then weight $\lambda_0$. For $\alpha < \beta$ we can use induction and the cocycle condition because $X_\alpha$ is in $[\mathfrak{u}, \mathfrak{u}]$. Indeed, the computations in 2.3/4 show: If $\gamma$ is a negative root with $X_\gamma \notin [\mathfrak{u}, \mathfrak{u}]$, then $\gamma \geq \beta$ or $-\gamma$ is a simple root.

*Remarks*: 1)For $R$ of type $B_4$ the proof shows that the image of the natural map from $H^1(G_1, L(\lambda_0))^{(-1)}$ to $H^0(G_1, H^0(\lambda_0))^{(-1)}$ contains the direct summand $H^0(\mu_0)$. We shall get more complete information on this case in 6.9(c).

2) For $R$ of type $F_4$ one may compare the discussion of this case in 6.9(d).

**5.5. Proposition:** *Suppose that $p = 2$. In the following cases one has $\lambda = p\omega_i - \alpha_i \in X_1(T)$ and the canonical map $H^1(G_1, L(\lambda))^{(-1)} \to H^1(G_1, H^0(\lambda))^{(-1)}$; has $L(\omega_i)$ in its image:*

*(a) $R$ is of type $B_n$ with $n \geq 3$ and $1 \leq i < n-1$.*

*(b) $R$ is of type $C_n$ with $n \geq 2$ and $1 \leq i < n$ with $i \equiv n \pmod 2$.*

*(c) $R$ is of type $F_4$ and $i \in \{1, 4\}$.*

**Proof:** We have to show in these cases (as in the proof of Proposition 4.5) that any 1–cocycle $\varphi \in Z^1(\mathfrak{g}, H^0(\lambda))$ of weight $p\omega_i$ takes values in $L(\lambda)$. Again, we have to look only at the $\varphi(X_\alpha)$ with $\alpha \leq -\alpha_i$, and the case $\alpha = -\alpha_i$ is trivial. For the $X_\alpha \in [\mathfrak{u}, \mathfrak{u}]$ we can use the same induction argument as in 4.5.

So we have to deal with all negative roots $\alpha < -\alpha_i$ such that $X_\alpha \notin [\mathfrak{u}, \mathfrak{u}]$. For $R$ of type $B_n$ the only possible candidate is $-\alpha_{n-1} - 2\alpha_n$. It is not less $-\alpha_i$ for $i < n-1$. This yields (a).

For $R$ of type $F_4$ the only roots to be looked at are $\beta = -\alpha_2 - 2\alpha_3$ and $\gamma = -\alpha_2 - 2\alpha_3 - 2\alpha_4$. None of these is less $-\alpha_1$. This yields our claim for $i = 1$. For $i = 4$ we have to deal with $\varphi(X_\gamma)$. This is a vector of weight $2\omega_4 + \gamma = 2\omega_4 + (\omega_1 - 2\omega_4) = \omega_1$ in $H^0(\omega_3)$. According to [23] the weight spaces for the weight $\omega_1$ in $H^0(\omega_3)$ and $L(\omega_3)$ coincide. This implies $\varphi(X_\gamma) \in L(\lambda)$ and settles (c).

For $R$ of type $C_n$ we have to look at the roots $-2\varepsilon_j = -2\sum_{r=j}^{n-1} \alpha_r - \alpha_n$ with $1 \leq j \leq i$. The arguments are more complicated in this case, and the proof is carried out in the next three sections that yield the other direction to (b) at the same time.

*Remark*: For $R$ of type $B_n$ and $i = 1$ we shall see (in 6.9) that the image contains all of $H^0(\omega_i)$.

**5.6.** We look first at the special case $i = 1$ of 5.5(b):

**Lemma:** *Suppose that $p = 2$ and that $R$ is of type $C_n$ with $n \geq 2$. Let $\varphi \in Z^1(\mathfrak{g}, H^0(\omega_2))$ be a nonzero 1–cocycle of weight $p\omega_1$. Then $\varphi$ takes values in $L(\omega_2)$ if and only $n$ is odd.*

**Proof:** The weights of $H^0(\omega_2)$ are the short roots and 0. The weight spaces corresponding to the short roots have dimension 1, both in $H^0(\omega_2)$ and in $L(\omega_2)$. The zero weight space has dimension $n-1$ in $H^0(\omega_2)$. Its dimension in $L(\omega_2)$ (in characteristic 2) is $n-1$ for $n$ odd, and $n-2$ for $n$ even, cf. 6.6 below. We see especially: If $n$ is odd, then $H^0(\omega_2)$ is irreducible, so the lemma is obvious in that case.

Assume from now on that $n$ is even. The map $\varphi$ has weight $p\omega_1 = 2\varepsilon_1$, it sends any $X_\alpha$ to a vector of weight $\alpha + 2\varepsilon_1$ that will be automatically in $L(\omega_2)$ unless this weight is 0, i.e., unless $\alpha = -2\varepsilon_1$. So we have to show (for $n$ even!):

$$\varphi(X_{-2\varepsilon_1}) \notin L(\omega_2). \tag{1}$$

We know in general that $\varphi(X_\alpha) \neq 0$ implies $\alpha \leq -\alpha_1$, i.e.,

$$\alpha \in \{-2\varepsilon_1, -(\varepsilon_1 - \varepsilon_j), -(\varepsilon_1 + \varepsilon_j) \mid 1 < j \leq n\}.$$

We have especially $\varphi(X_\beta) = 0$ for all positive roots $\beta$, hence, by the cocycle condition 4.4(2)

$$\varphi([X_\beta, X_{-2\varepsilon_1}]) = X_\beta \varphi(X_{-2\varepsilon_1}). \tag{2}$$

The commutator formulas (without sign problems in characteristic 2)

$$[X_{\varepsilon_1-\varepsilon_j}, X_{-2\varepsilon_1}] = X_{-(\varepsilon_1+\varepsilon_j)}, \qquad [X_{\varepsilon_1+\varepsilon_j}, X_{-2\varepsilon_1}] = X_{-(\varepsilon_1-\varepsilon_j)} \tag{3}$$

for all $j > 1$ imply therefore: If $\varphi(X_{-2\varepsilon_1}) = 0$, then $\varphi(X_\alpha) = 0$ for all roots $\alpha \leq -\alpha_1$, hence $\varphi = 0$, a contradiction. We conclude that

$$\varphi(X_{-2\varepsilon_1}) \neq 0. \tag{4}$$

For all simple roots $\alpha_j$ with $j > 1$ the sum $\alpha_j - 2\varepsilon_1$ is not a root. So in that case (2) yields

$$X_{\alpha_j}\varphi(X_{-2\varepsilon_1}) = 0 \qquad \text{for all } j > 1. \tag{5}$$

Set

$$R' = R \cap \sum_{j>1} \mathbf{Z}\alpha_j,$$

and let $G'$ be the subgroup of $G$ generated by the root subgroups $U_\beta$ with $\beta \in R'$. Then $G'$ is semisimple of type $C_{n-1}$. The Lie algebra $\mathfrak{g}'$ of $G'$ is the Lie subalgebra of $\mathfrak{g}$ generated by all $X_\beta$ with $\beta \in R'$. Consider the sum $L'$ of all weights spaces in $L(\omega_2)$ with weights in $\mathbf{Z}R'$. Then $L'$ is a submodule for $G'$ and $\mathfrak{g}'$. Its weights are 0 and the short roots in $R'$. Therefore $L'$ has the analogue of $L(\omega_2)$ for $G'$ as a composition factor. The zero weight space of that analogue has dimension $n-2$, because $n-1$ is odd. That number is equal to the dimension of $L(\omega_2)_0$. Therefore $L'$ is equal to that simple module for $\mathfrak{g}'$.

Suppose now that (1) is false, i.e., that $\varphi(X_{-2\varepsilon_1})$ is contained in $L(\omega_2)$. Because it has weight 0, it would belong to $L'$. By (5) it would be a highest weight vector of weight 0 in the simple $\mathfrak{g}'$-module $L'$, a contradiction. So (1) and the lemma follow.

**5.7. Lemma:** *Let $M$ be a $G$-module and $\mu \in X(T)$ such that the weight space $Z^1(\mathfrak{g}, M)_{p\mu}$ has dimension 1. Denote a generator of this weight space by $\varphi$. Then one has for all $w$ in the Weyl group of $G$ with $w(\mu) = \mu$, for all $G$-submodules $N$ of $M$, and for all roots $\alpha$:*

$$\varphi(X_\alpha) \in N \iff \varphi(X_{w(\alpha)}) \in N. \tag{1}$$

**Proof**: Let $g$ be a representative of $w$ in the normalizer of $T$ in $G$. Then $g$ permutes the weight spaces of a $G$-module as $w$ permutes the weight. Therefore $gX_\alpha$ has weight $w(\alpha)$, and $g\varphi$ has weight $w(p\mu) = p\mu$. Because the weight spaces in question have dimension 1, there are constants $c_1, c_2 \in k$ with $gX_\alpha = c_1 X_{w(\alpha)}$ and $g\varphi = c_2\varphi$. These constants are nonzero because $g$ acts bijectively. We get now

$$c_2\varphi(X_{w(\alpha)}) = (g\varphi)(X_{w(\alpha)}) = g(\varphi(g^{-1}X_{w(\alpha)})) = c_1^{-1}g(\varphi(X_\alpha)),$$

hence

$$\varphi(X_\alpha) = c_1c_2g^{-1}(\varphi(X_{w(\alpha)})).$$

This yields the claim.

*Remark*: If we take $N = 0$ in the lemma, we get

$$\varphi(X_\alpha) = 0 \iff \varphi(X_{w(\alpha)}) = 0, \tag{2}$$

for all $\alpha \in R$ and all $w$ with $w(\mu) = \mu$.

**5.8. Lemma:** *Suppose that $p = 2$ and that $R$ is of type $C_n$ with $n \geq 2$. Let $1 \leq i < n$ and let $\varphi \in Z^1(\mathfrak{g}, H^0(p\omega_i - \alpha_i))$ be a nonzero 1-cocycle of weight $p\omega_i$. Then $\varphi$ takes values in $L(p\omega_i - \alpha_i)$ if and only $n + 1 - i$ is odd.*

**Proof**: Set $\lambda = p\omega_i - \alpha_i$.

We claim first of all that there exists an $r$ with $1 \leq r \leq i$ and

$$\varphi(X_{-2\varepsilon_r}) \neq 0. \tag{1}$$

Indeed, because we have again $\varphi(X_\beta) = 0$ for all positive roots $\beta$, we can extend 5.6(2) from $r = 1$ to all $r$. The commutator formulas 5.6(3) hold still, if we replace $\varepsilon_1$ by any $\varepsilon_r$ with $r < j$. So, if (1) were not true, we would get $\varphi = 0$, a contradiction.

For each $r$ with $1 \leq r < i$ we can find an element $w$ of the Weyl group with $w(\omega_i) = \omega_i$ and $w(-2\varepsilon_i) = -2\varepsilon_r$. For example, the reflection with respect to the root $\varepsilon_r - \varepsilon_i$ has this property. Now Lemma 5.7 implies that we can replace (1) by:

$$\varphi(X_{-2\varepsilon_i}) \neq 0. \tag{2}$$

On the other hand, we know that $\varphi$ takes values in $L(\lambda)$ if and only if it maps all $X_{-2\varepsilon_r}$ with $1 \leq r \leq i$ to $L(\lambda)$. Using 5.7 we get now:

$$\varphi(\mathfrak{g}) \subset L(\lambda) \iff \varphi(X_{-2\varepsilon_i}) \in L(\lambda). \tag{3}$$

Set

$$R' = R \cap \sum_{j \geq i} \mathbf{Z}\alpha_j,$$

let $G'$ be the subgroup of $G$ generated by the root subgroups $U_\beta$ with $\beta \in R'$, and let $\mathfrak{g}'$ be the Lie algebra of $G'$. Then $G'$ is semisimple with root system $R'$ of type $C_{n+1-i}$, and $\mathfrak{g}'$ is the subalgebra of $\mathfrak{g}$ generated by all $X_\beta$ with $\beta \in R'$. Set $H'$ resp. $L'$ equal to the sum of all weight spaces in $H^0(\lambda)$ resp. in $L(\lambda)$ with weights in $\lambda + \mathbf{Z}R'$. Then $H'$ resp. $L'$ is the analogue for $G'$ of $H^0(\omega_2)$ resp. $L(\omega_2)$. The restriction $\varphi'$ of $\varphi$ to $\mathfrak{g}'$ is an element of $Z^1(\mathfrak{g}', H')$ and its weight for $G'$ is the analogue of $p\omega_1$. We have $\varphi' \neq 0$ by (2). We can now apply 5.6 to $\varphi'$ and see that $\varphi(X_{-2\varepsilon_i}) \in L'$ if and only if $n + 1 - i$ is odd. This implies the lemma by (3).

**5.9. Proposition:** *Suppose that $p = 2$. In the following cases one has $\lambda = p\omega_i - \alpha_i \in X_1(T)$ and the canonical map $H^1(G_1, L(\lambda))^{(-1)} \to H^1(G_1, H^0(\lambda))^{(-1)}$ is 0. One has*

$$H^1(G_1, L(\lambda)) \simeq (H^0(\lambda)/L(\lambda))^{G_1}.$$

*(a) $R$ is of type $B_n$ with $n \geq 4$ and $i = n$.*

*(b) $R$ is of type $C_n$ with $n \geq 2$ and $1 \leq i < n$ with $i \equiv n + 1 \pmod 2$.*

*(c) $R$ is of type $F_4$ and $i = 3$.*

**Proof**: By 4.2(3), we have to show that the canonical map mentioned is 0. In all these cases, we have $H^1(G_1, H^0(\lambda))^{(-1)} \simeq H^0(\omega_i)$. It is enough to show that the socle $L(\omega_i)$ of $H^0(\omega_i)$ is not contained in the image. So we have to take a nonzero 1–cocycle $\varphi \in Z^1(\mathfrak{g}, H^0(\lambda))$ of weight $p\omega_i$ and show that it does not take values in $L(\lambda)$. For $R$ of type $C_n$ this follows from 5.8.

We have to show, cf. the proof of 5.5: For $R$ of type $B_n$ one has $\varphi(X_\alpha) \notin L(\lambda)$ where $\alpha = -(\alpha_{n-1} + 2\alpha_n)$. For $R$ of type $F_4$ one has $\varphi(X_\beta) \notin L(\lambda)$ or $\varphi(X_\gamma) \notin L(\lambda)$ where $\beta = -\alpha_2 - 2\alpha_3$ and $\gamma = -\alpha_2 - 2\alpha_3 - 2\alpha_4$. The weights of these images are $2\omega_n + \alpha = \omega_{n-2}$ resp. $\omega_1 + 2\omega_4$ resp. $\omega_1 + 2\omega_3 - 2\omega_4$.

For $R$ of type $B_2$ one knows in characteristic 2 that $0 = \omega_1 - (\alpha_1 + \alpha_2)$ is not a weight of $L(\omega_1)$. If we apply this to the subsystem spanned by $\alpha_{n-1}, \alpha_n$ in $B_n$ resp. by $\alpha_2, \alpha_3$ in $F_4$, we see that $\omega_{n-1} - (\alpha_{n-1} + \alpha_n) = \omega_{n-2}$ is not a weight of $L(\omega_{n-1}) = L(\lambda)$ resp. that $\omega_2 + \omega_4 - (\alpha_2 + \alpha_3) = \omega_1 + 2\omega_4$ is not a weight of $L(\omega_2 + \omega_4) = L(\lambda)$. In the second case, also $\omega_1 + 2\omega_3 - 2\omega_4$ is not a weight of $L(\lambda)$ being conjugate to $\omega_1 + 2\omega_4$ under the Weyl group.

This implies now: If $\varphi$ takes values in $L(\lambda)$, then $\varphi(X_\alpha) = 0$ resp. $\varphi(X_\beta) = \varphi(X_\gamma) = 0$. In the $B_n$ case we have

$$[X_{\alpha_1+\alpha_2}, X_\alpha] = X_{-\alpha_n},$$

hence

$$\varphi(X_{-\alpha_n}) = X_{\alpha_1+\alpha_2}\varphi(X_\alpha) = 0.$$

This implies $\varphi = 0$, i.e., a contradiction. In the $F_4$ case we get similarly first $\varphi(X_{-\alpha_3}) = 0$ from $\varphi(X_{-\beta}) = 0$ and then reach the same contradiction.

*Remark*: For $R$ of type $B_3$ and $i = 3$ we have $H^1(G_1, H^0(\lambda))^{(-1)} \simeq H^0(\omega_1) \oplus H^0(\omega_3)$. The proof shows that the image of $H^1(G_1, L(\lambda))^{(-1)}$ is contained in $H^0(\omega_1)$. We shall see (in 6.9) that it is equal to $H^0(\omega_1)$.

**5.10. Proposition:** *Suppose that $R$ is of type $G_2$ and $p = 3$. Then*

$$H^1(G_1, L(\omega_2))^{(-1)} \simeq H^0(\omega_1). \tag{1}$$

*One has for all $\lambda \in X_1(T)$ with $\lambda \neq \omega_2$:*

$$H^1(G_1, L(\lambda)) \simeq (H^0(\lambda)/L(\lambda))^{G_1}. \tag{2}$$

**Proof**: The first claim follows from 5.2(b) and 4.2(3) because of

$$H^0(\omega_2)/L(\omega_2) \simeq L(\omega_1)$$

and $L(\omega_1) = H^0(\omega_1)$. The second claim follows from the same results except in the case $\lambda = \omega_1 + \omega_2$. In that case we have to prove that the canonical map from $H^1(G_1,, L(\lambda))$ to $H^1(G_1, H^0(\lambda)) \simeq H^0(\omega_1)^{(1)}$ is zero. As in the last proof, it is enough to show that a 1–cocyle $\varphi \in Z^1(\mathfrak{g}, L(\lambda))$ of weight $p\omega_1$ is zero.

The module $L(\lambda)$ can be identified with $L(\omega_1) \otimes L(\omega_2)$. The weights of these factors are 0 and the short resp. the long roots. All weight spaces have dimension

1. Let $v_0$ be a basis for the 0 weight space in $L(\omega_1)$, and let $v^+$ be a basis for the $\omega_2$ weight space in $L(\omega_2)$. Then $v_0 \otimes v^+$ is a basis for the $\omega_2$ weight space in $L(\lambda)$, cf. [19]. Because $\alpha_2$ is not a weight of $L(\omega_1)$ we have $X_{\alpha_2}v_0 = 0$, therefore $v_1 = X_{\alpha_1}v_0$ cannot be 0. So $v_1$ is a basis of the $\alpha_1$ weight space in $L(\omega_1)$. Because $2\alpha_1$ is not a weight of $L(\omega_1)$ we have $X_{\alpha_1}v_1 = 0$.

We have $\varphi(X_\alpha) = 0$ unless $\alpha \leq -\alpha_1$. For $\alpha = -(3\alpha_1 + \alpha_2)$ the image has weight $\omega_2$, so there is $a \in k$ with

$$\varphi(X_{-(3\alpha_1+\alpha_2)}) = a\, v_0 \otimes v^+.$$

Using the cocycle condition 4.4(2) and commutator formulas where we choose the signs as in [9] we compute more values of $\varphi$. We get from $[X_{\alpha_1}, X_{-(3\alpha_1+\alpha_2)}] = X_{-(2\alpha_1+\alpha_2)}$ that

$$\varphi(X_{-(2\alpha_1+\alpha_2)}) = a\, v_1 \otimes v^+.$$

We get then from

$$[X_{\alpha_1}, X_{-(2\alpha_1+\alpha_2)}] = 2X_{-(\alpha_1+\alpha_2)}, \quad [X_{\alpha_2}, X_{-(\alpha_1+\alpha_2)}] = -X_{-\alpha_1}$$

that

$$\varphi(X_{-(\alpha_1+\alpha_2)}) = \varphi(X_{-\alpha_1}) = 0.$$

Now $[X_{-\alpha_1}, X_{-(\alpha_1+\alpha_2)}] = 2X_{-(2\alpha_1+\alpha_2)}$ implies $\varphi(X_{-(2\alpha_1+\alpha_2)}) = 0$, hence $a = 0$ and $\varphi(X_{-(3\alpha_1+\alpha_2)}) = 0$. Finally $[X_{-\alpha_2}, X_{-(3\alpha_1+\alpha_2)}] = X_{-(3\alpha_1+2\alpha_2)}$ gives $\varphi(X_{-(3\alpha_1+\alpha_2)}) = 0$, hence $\varphi = 0$.

## 6. Small Representations

**6.1.** In this part we shall compute $H^1(G_1, L(\lambda))$ for simple modules $L(\lambda)$ that have a small dimension in some sense. More precisely, we shall look at the cases where $\lambda$ is 0 or a minuscule weight or a root. For $R$ of type $A_n$ we shall look also at the symmetric powers of the natural representation. In all these cases the structure of $H^0(\lambda)$ is well known and we can use 4.2(3) to get complete information.

**6.2.** The trivial one dimensional module $k = L(0)$ is equal to $H^0(0)$. So its $G_1$-cohomology can be read off the propositions 4.1 and 5.1/2. We cannot have $0 = p\omega_i - \alpha_i$ for some $i$ unless $R$ is of type $A_1 = C_1$ and $p = 2$. Combining this with the cases in 5.1/2 we get the following well known result, cf. [15], II.12.2:

**Proposition:** *If $R$ is of type $C_n$ with $n \geq 1$ and if $p = 2$, then*

$$H^1(G_1, k)^{(-1)} \simeq H^0(\omega_1).$$

*In all other cases $H^1(G_1, k) = 0$.*

**6.3.** If we are to get a contribution to the cohomology of $L(\lambda)$ from the cohomology of $H^0(\lambda)$, then $\lambda$ has to be one of the $\lambda_i$ from Lemma 3.1. Most of the highest weights to be considered in this part are fundamental weights. Let us first determine which $\lambda_i$ are fundamental weights or multiples of fundamental weights

For $R$ of type $A_1$ we have $\lambda_1 = (p-2)\omega_1$ for all $p$. This is always a multiple of a fundamental weight; it is equal to one if and only if $p = 3$.

Assume from now on that the rank is at least 2. For $R$ of type $G_2$ and $p = 3$ we have $\lambda_2 = \omega_2$. Let us show that this is the only case for $p \geq 3$. Indeed, each $p\omega_i - \alpha_i$ has at least two nonzero coefficients: that of $\omega_i$ and that of some $\omega_j$ with $j$ joined to $i$ in the Dynkin diagram. Therefore $p\omega_i - \alpha_i$ is not a multiple of a fundamental weight.

Consider now $p = 2$. Then $\lambda_i = p\omega_i - \alpha_i$ has only one nonzero coefficient if and only if $i$ is an endpoint of the Dynkin diagram. Together with the special cases from 3.3 (b) – (d), we see that exactly the following $\lambda_i$ are fundamental weights (for $p = 2$):

| Type | $i$ | $\lambda_i$ |
|---|---|---|
| $A_n,\ n \geq 2$ | 1 | $\omega_2$ |
| | $n$ | $\omega_{n-1}$ |
| $B_n,\ n \geq 3$ | 1 | $\omega_2$ |
| | $n-1$ | $\omega_{n-2}$ |
| | $n$ | $\omega_{n-1}$ |
| $C_n,\ n \geq 2$ | 1 | $\omega_2$ |
| $D_n,\ n \geq 4$ | 1 | $\omega_2$ |
| | $n-1$ | $\omega_{n-2}$ |
| | $n$ | $\omega_{n-2}$ |
| $E_n,\ n = 6,7,8$ | 1 | $\omega_3$ |
| | 2 | $\omega_4$ |
| | $n$ | $\omega_{n-1}$ |
| $F_4$ | 1 | $\omega_2$ |
| | 2 | $\omega_1$ |
| | 4 | $\omega_3$ |
| $G_2$ | 1 | $\omega_2$ |
| | 2 | $\omega_1$ |

**6.4. Proposition:** *If $\lambda$ is a minuscule weight, then $H^1(G_1, L(\lambda)) = 0$ except in the following cases:*

*(a) One has for $R$ of type $A_1$ and $p = 3$:*

$$H^1(G_1, L(\omega_1))^{(-1)} \simeq H^0(\omega_1).$$

*(b) One has for $R$ of type $A_n$ with $n \geq 2, n \neq 3$ and $p = 2$:*

$$H^1(G_1, L(\omega_2))^{(-1)} \simeq H^0(\omega_1) \quad \text{and} \quad H^1(G_1, L(\omega_{n-1}))^{(-1)} \simeq H^0(\omega_n).$$

*(c) One has for $R$ of type $A_3$ and $p = 2$:*

$$H^1(G_1, L(\omega_2))^{(-1)} \simeq H^0(\omega_1) \oplus H^0(\omega_3).$$

**Proof**: A look at a table of the minuscule weights — for example, in [3], ch.VI, §4, exerc. 15 — shows that only those mentioned occur also in 6.3. We have $H^0(\lambda) = L(\lambda)$ for any minuscule $\lambda$. Therefore the claim follows from 4.1.

**6.5.** For $R$ of type $A_n$ the $r$-th symmetric power of the natural representation (resp. of its dual) of $G = SL_{n+1}(k)$ is isomorphic to $H^0(r\omega_1)$ (resp. to $H^0(\omega_n)$), cf. [15], II.2.16. It is irreducible for $r < p$.

**Proposition:** *(a) If $R$ is of type $A_1$ and $0 \leq r < p$, then*

$$H^1(G_1, L(r\omega_1))^{(-1)} \simeq \begin{cases} H^0(\omega_1), & \text{if } r = p-2; \\ 0, & \text{otherwise.} \end{cases}$$

*(b) If $R$ is of type $A_n$ with $n \geq 2$, then*

$$H^1(G_1, L(r\omega_1)) = H^1(G_1, L(r\omega_n)) = 0$$

*for all $r$ with $0 \leq r < p$ except in the case $p = 2$, $n = 2$, $r = 1$ occurring in 6.4(b).*

**Proof**: Because of the irreducibility of the $H^0(r\omega_1)$ with $r < p$, this follows easily from 4.1 and 6.3.

*Remark*: For $R$ of type $A_1$, the result is part of Theorem 4 in [7].

**6.6.** Suppose in this section that $R$ has two root lengths. Let $\alpha_0$ be the largest short root. It is equal to $\omega_1$ for $R$ of type $B_n$, to $\omega_2$ for $R$ of type $C_n$, to $\omega_4$ for $R$ of type $F_4$, and to $\omega_1$ for $R$ of type $G_2$. The weights of $H^0(\alpha_0)$ are the short roots — all conjugate to $\alpha_0$ under the Weyl group — and 0. The dimension of the 0 weight space in $H^0(\alpha_0)$ is equal to the number of short simple roots. Denote this number for the moment by $r$. If $p$ divides $r+1$, then $H^0(\alpha_0)/L(\alpha_0) \simeq L(0)$, otherwise $H^0(\alpha_0)$ is irreducible, cf. [13], p. 20.

**Proposition:** *One has $H^1(G_1, L(\alpha_0)) = 0$ except in the following cases:*

*(a) If $R$ is of type $B_3$ and $p = 2$, then $H^1(G_1, L(\alpha_0))^{(-1)} \simeq k \oplus H^0(\omega_3)$.*

*(b) If $R$ is of type $B_n$ with $n \geq 4$ and $p = 2$, then $H^1(G_1, L(\alpha_0)) \simeq k$.*

*(c) If $R$ is of type $C_n$ with $n \geq 3$ and if $p$ is odd and divides $n$, then $H^1(G_1, L(\alpha_0)) \simeq k$.*

*(d) If $R$ is of type $C_n$ with $n \geq 2$ and if $p = 2$, then*

$$H^1(G_1, L(\alpha_0))^{(-1)} \simeq \begin{cases} k, & \text{for } n \text{ even;} \\ H^0(\omega_1), & \text{for } n \text{ odd.} \end{cases}$$

*(e) If $R$ is of type $F_4$ and $p = 3$, then $H^1(G_1, L(\alpha_0)) \simeq k$.*

*(f) If $R$ is of type $G_2$ and $p = 2$, then there is an exact sequence*

$$0 \to k \to H^1(G_1, L(\alpha_0))^{(-1)} \to H^0(\alpha_0) \to 0.$$

**Proof**: Assume at first that $p \neq 2$. Then the discussion in 6.3 shows that $\alpha_0$ is not among the $\lambda_i$. Therefore the cohomology wanted is equal to $H^0(\alpha_0)/L(\alpha_0)$. It is nonzero only in the cases (c) and (e).

Assume now that $p = 2$. The $G_2$ case is just a restatement of 5.3(3). If $R$ is of type $F_4$, then $H^0(\alpha_0)$ is irreducible and $\alpha_0$ is not among the $\lambda_i$, cf. 6.3, so the cohomology is 0. If $R$ is of type $B_n$ with $n \geq 3$, then $H^0(\alpha_0)/L(\alpha_0) \simeq k$. Now 6.2

implies that the last term in the exact sequence 4.2(3) is 0, and 6.3 shows that $\alpha_0$ is one of the $\lambda_i$ only for $n = 3$. This yields (b). In order to get (a) we have to observe that any extension of $k$ and $H^0(\omega_3)$ splits because $\omega_3 \notin \mathbf{Z}R$ for $B_3$.

Finally we have to look at $C_n$ with $n \geq 2$ for $p = 2$. In this case $\alpha_0 = \lambda_1$. If $n$ is odd, then $H^0(\alpha_0)$ is irreducible and the result follows from 5.1(6). If $n$ is even, we have to use Proposition 5.9(b).

**6.7.** Let $\tilde{\alpha}$ be the largest root in $R$. For $R$ of type $C_n$ with $n \geq 1$ (including the case $C_1 = A_1$) one has $\tilde{\alpha} = 2\omega_1$. In that case $\tilde{\alpha} \notin X_1(T)$ for $p = 2$ and we shall not discuss $L(\tilde{\alpha})$ for $p = 2$ in this case.

**Proposition:** *Suppose that $p \neq 2$. Then $H^1(G_1, L(\tilde{\alpha})) = 0$ except in the following cases:*

*(a) One has $H^1(G_1, L(\tilde{\alpha}))^{(-1)} \simeq k \oplus H^0(\omega_1) \oplus H^0(\omega_2)$, if $R$ is of type $A_2$ with $p = 3$.*

*(b) One has $H^1(G_1, L(\tilde{\alpha}))^{(-1)} \simeq k$, if $R$ is of type $A_n$ with $n \geq 3$ and $p \mid n+1$.*

*(c)One has $H^1(G_1, L(\tilde{\alpha}))^{(-1)} \simeq k$, if $R$ is of type $E_6$ with $p = 3$.*

*(d) One has $H^1(G_1, L(\tilde{\alpha}))^{(-1)} \simeq H^0(\omega_1)$, if $R$ is of type $G_2$ with $p = 3$.*

**Proof:** Unless $R$ is of type $A_n$ with $n \geq 2$ the largest root is a multiple of a fundamental weight. Now 6.3 shows that is does not occur among the $\lambda_i$ except for the case $R$ of type $G_2$ and $p = 3$. That case leads to (d) which is just a restatement of 5.10(1). In the $A_n$ case one has $\tilde{\alpha} = \omega_1 + \omega_n$. It can occur as an $\lambda_i = p\omega_i - \alpha_i$ only if $p = 3$ and $n = 2$. This leads to (a) which is just 4.10(3).

In the remaining cases we have to determine the $G_1$ invariants in $H^0(\tilde{\alpha})/L(\tilde{\alpha})$. This space is equal to $L(0) = k$ for $R$ of type $A_n$ for $p$ dividing $n+1$ and for $R$ of type $E_6$ for $p = 3$. For $R$ of type $G_2$ and $p = 3$ it is isomorphic to $L(\omega_1)$. In all other cases $H^0(\tilde{\alpha})$ is irreducible for $p \geq 3$.

**6.8. Proposition:** *Suppose that $p = 2$ and that $R$ has only one root length.*

*(a) One has $H^1(G_1, L(\tilde{\alpha})) = 0$, if $R$ is of type $A_n$ with $n$ even or $E_6$ or $E_8$.*

*(b) One has $H^1(G_1, L(\tilde{\alpha})) = k$, if $R$ is of type $A_n$ with $n \geq 5$ odd or $E_7$.*

*(c) One has $H^1(G_1, L(\tilde{\alpha}))^{(-1)} = k \oplus H^0(\omega_2)$, if $R$ is of type $A_3$.*

*(d) One has $H^1(G_1, L(\tilde{\alpha}))^{(-1)} = k \oplus H^0(\omega_1)$, if $R$ is of type $D_n$ with $n \geq 5$ odd.*

*(e) One has $H^1(G_1, L(\tilde{\alpha}))^{(-1)} = k \oplus k \oplus H^0(\omega_1)$, if $R$ is of type $D_n$ with $n \geq 6$ even.*

*(f) One has $H^1(G_1, L(\tilde{\alpha}))^{(-1)} = k \oplus k \oplus H^0(\omega_1) \oplus H^0(\omega_3) \oplus H^0(\omega_4)$, if $R$ is of type $D_4$.*

**Proof:** The only possible composition factor of $H^0(\tilde{\alpha})/L(\tilde{\alpha})$ is $L(0)$. The first cohomology of this quotient is 0, because we have excluded the case $A_1$. So 4.2(3) yields in this case an exact sequence

$$0 \to H^0(\tilde{\alpha})/L(\tilde{\alpha}) \to H^1(G_1, L(\tilde{\alpha})) \to H^1(G_1, H^0(\tilde{\alpha})) \to 0. \tag{1}$$

On one side, the quotient $H^0(\tilde{\alpha})/L(\tilde{\alpha})$ is isomorphic to $L(0) \oplus L(0)$ if the type is $D_n$ with $n$ even,it is 0 if the type is $E_6$, $E_8$ or $A_n$ with $n$ even, and it is isomorphic to $L(0)$ in the remaining cases, i.e., for $E_7$, for $A_n$ with $n$ odd, and for $D_n$ with $n$ odd.

On the other side, $H^0(\tilde{\alpha})$ has nonzero $G_1$–cohomology only if $\tilde{\alpha}$ is one of the $\lambda_i$. In the $E_n$ cases $\tilde{\alpha}$ is a fundamental weight not in the table in 6.2, hence not a $\lambda_i$. For $R$ of type $D_n$ we have always $\tilde{\alpha} = \omega_2 = \lambda_1$ and for $n = 4$ it also equal to $\lambda_3 = \lambda_4$. Finally, $\tilde{\alpha} = \omega_1 + \omega_n$ cannot be an $\lambda_i = 2\omega_i - \alpha_i$ for $R$ of type $A_n$ unless $n = 3$.

The weights of $H^1(G_1, H^0(\tilde{\alpha}))^{(-1)}$ are not in $\mathbf{Z}R$. Therefore the exact sequence (1) splits in the cases (c) – (f).

**6.9. Proposition:** *Suppose that* $p = 2$.

*(a) If $R$ is of type $B_n$ with $n \geq 3$ odd, then* $H^1(G_1, L(\tilde{\alpha}))^{(-1)} \simeq H^0(\omega_1)$.

*(b) If $R$ is of type $B_n$ with $n \geq 6$ even, then there is an exact sequence*

$$0 \to k \to H^1(G_1, L(\tilde{\alpha}))^{(-1)} \to H^0(\omega_1) \to 0.$$

*(c) If $R$ is of type $B_4$, then there is an exact sequence*

$$0 \to k \to H^1(G_1, L(\tilde{\alpha}))^{(-1)} \to H^0(\omega_1) \oplus H^0(\omega_4) \to 0.$$

*(d) If $R$ is of type $F_4$, then* $H^1(G_1, L(\tilde{\alpha}))^{(-1)} \simeq H^0(\omega_4)$.

*(e) If $R$ is of type $G_2$, then* $H^1(G_1, L(\tilde{\alpha}))^{(-1)} \simeq H^0(\omega_1)$.

**Proof**: The $G_2$ claim is just a restatement of 5.3(2). If $R$ is of type $F_4$, then $H^0(\tilde{\alpha})/L(\tilde{\alpha}) \simeq L(\alpha_0)$. This quotient has no invariants under $G_1$ and its first cohomology is zero by 6.6. So the $F_4$ claim follows from 5.1(10).

Suppose from now on that $R$ is of type $B_n$ with $n \geq 3$. The (untwisted) first cohomology group of $H^0(\tilde{\alpha})$ is isomorphic to $H^0(\omega_1)$, if $n \geq 5$, whereas there is an additional summand $H^0(\omega_n)$ for $n = 3, 4$, cf. 5.1(a).

The Weyl module $V(\tilde{\alpha})$ is just $\mathfrak{g}$ with the adjoint representation, its submodules are just the ideals. The explicit description of all ideals in [11] shows: A root vector of weight $\alpha_0$ generates a submodule isomorphic to $V(\alpha_0)$. This is the whole radical of $V(\tilde{\alpha})$ for $n$ odd; for $n$ even there is one more factor $L(0)$ on top of this $L(\alpha_0)$. Dualising yields the structure of $H^0(\tilde{\alpha})$: One has $H^0(\tilde{\alpha})/L(\tilde{\alpha}) \simeq H^0(\alpha_0)$ for $n$ odd and there is an exact sequence

$$0 \to L(0) \to H^0(\tilde{\alpha})/L(\tilde{\alpha}) \to H^0(\alpha_0) \to 0$$

for $n$ even.

This shows on one hand that $(H^0(\tilde{\alpha})/L(\tilde{\alpha}))^{G_1}$ is 0 for $n$ odd and $k$ for $n$ even. On the other hand the first cohomology groups of $L(0)$ and (except for $n = 3$) for $H^0(\alpha_0)$ vanish. Then the same is true for $H^0(\tilde{\alpha})/L(\tilde{\alpha})$. Now everything follows using 4.2(3) unless $n = 3$. In that case 4.2(3) yields an exact sequence

$$0 \to H^1(G_1, L(\tilde{\alpha}))^{(-1)} \to H^0(\omega_1) \oplus H^0(\omega_3) \to H^0(\omega_3).$$

The remark in 5.9 shows that the image of $H^1(G_1, L(\tilde{\alpha}))^{(-1)}$ is contained in the summand $H^0(\omega_1)$. The exactness yields now equality and establishes (a) also in this case.

**6.10.** Let us finally consider a few cases not of small representations, but of small groups. First, take $R$ of type $A_2$. For $p = 2$ all $\lambda \in X_1(T)$ have already been considered, cf. 6.2, 6.4(b), 6.8(a). Assume now $p \geq 3$. All composition factors of an $H^0(\lambda)$ with $\lambda \in X_1(T)$ have a highest weight in $X_1(T)$. The trivial module $L(0)$ occurs as a composition factor of $H^0(\lambda)$ only for $\lambda = 0$ and $\lambda = (p-2)(\omega_1 + \omega_2)$. One has $H^0(\lambda)/L(\lambda) = L(0)$ in the second case. This implies that for $p > 3$ exactly the following three cohomology groups do not vanish:

$$\begin{aligned} H^1(G_1, L((p-2)(\omega_1+\omega_2))) &\simeq L(0), \\ H^1(G_1, L((p-2)\omega_1+\omega_2)) &\simeq L(\omega_1)^{(1)}, \\ H^1(G_1, L(\omega_1+(p-2)\omega_2)) &\simeq L(\omega_2)^{(1)}. \end{aligned}$$

For $p = 3$ there is a collapse of the cases leading to the situation in Proposition 6.7(a).

Consider now $R$ of type $A_3$. We shall write $(a,b,c)$ for $a\omega_1 + b\omega_2 + c\omega_3$. For $p = 2$ one gets the full information easily by checking that all $H^0(\lambda)$ with $\lambda \in X_1(T)$ are irreducible with the only exception of $\lambda = (1,0,1) = \tilde{\alpha}$. Assume now $p \geq 3$. The examples in [14] yield (with some additional work for $p = 3$) that $L(0)$ occurs as a factor of $H^0(\lambda)/L(\lambda)$ exactly for $\lambda = (p-3,0,p-3)$ (except for $p = 3$) and for $\lambda = (p-3,2,p-3)$. Furthermore $L(p\omega_1)$ resp. $L(p\omega_3)$ occurs exactly for $\lambda = (2,p-2,p-2)$ resp. $\lambda = (p-2,p-2,2)$. No other $L(p\mu)$ occurs. One checks using the sum formula that the $L(p\mu)$ mentioned appear in the socle of $H^0(\lambda)/L(\lambda)$. Combining this with 4.3 and 4.10 we see that exactly the following groups do not vanish:

$$\begin{aligned} H^1(G_1, L(p-3,0,p-3)) \simeq H^1(G_1, L(p-3,2,p-3)) &\simeq L(0), \\ H^1(G_1, L(2,p-2,p-2)) \simeq H^1(G_1, L(p-2,1,0)) &\simeq L(\omega_1)^{(1)}, \\ H^1(G_1, L(p-2,p-2,2)) \simeq H^1(G_1, L(0,1,p-2)) &\simeq L(\omega_3)^{(1)}, \\ H^1(G_1, L(1,p-2,1)) &\simeq L(\omega_2)^{(1)}. \end{aligned}$$

For $p = 3$ the first term has to be dropped.

## REFERENCES


[1] H. H. Andersen: Extensions of modules for algebraic groups, *Amer. J. Math.* **106** (1984), 489–504

[2] H. H. Andersen, J. C. Jantzen: Cohomology of induced representations for algebraic groups, *Math. Ann.* **269** (1984), 487–525

[3] N. Bourbaki: *Groupes et algèbres de Lie, ch. IV, V et VI*, Paris 1968 (Hermann)

[4] Chiu Sen, Shen Guangyu: Cohomology of graded Lie algebras of Cartan type of characteristic *p*, *Abh. Math. Sem. Univ. Hamburg* **57** (1987), 139–156

[5] E. Cline, B. Parshall, L. Scott: Cohomology, hyperalgebras, and representations, *J. Algebra* **63** (1980), 98–123

[6] S. Donkin: On $\mathrm{Ext}^1$ for semisimple groups and infinitesimal subgroups, *Math. Proc. Camb. Phil. Soc.* **92** (1982), 231–238

[7] A. Džumadil'daev: On the cohomology of modular Lie algebras (russ.), *Matem. Sbornik* **119** (1982), 132–149 (engl. transl.: *Math. USSR Sbornik* **47** (1984), 127–143)

[8] P. Gilkey, G. Seitz: Some representations of exceptional Lie algebras, *Geometriae Dedicata* **25** (1988), 407–416

[9] G. Grélaud, C. Quitté, P. Tauvel: Bases de Chevalley et $\mathfrak{sl}(2)$-triplets des algèbres de Lie simples exceptionelles, preprint, Univ. de Poitiers, June 1990

[10] G. Hochschild: Cohomology of restricted Lie algebras, *Amer. J. Math.* **76** (1954), 555–580

[11] G. M. D. Hogeweij: Almost classical Lie algebras I, *Indag. math.* **44** (1982), 441–452

[12] J. Humphreys: *Introduction to Lie Algebras and Representation Theory*, New York, etc., 1972 (Springer)

[13] J. C. Jantzen: Darstellungen halbeinfacher algebraischer Gruppen und zugeordnete kontravariante Formen, *Bonner math. Schriften* **67**(1973)

[14] J. C. Jantzen: Weyl modules for groups of Lie type, pp.291–300 in: M. Collins (ed.), *Finite Simple Groups II*, Proc. Durham 1978, London etc. 1980 (Academic Press)

[15] J. C. Jantzen: *Representations of Algebraic Groups*, Pure and Applied Mathematics, vol. 131, Boston, etc., 1987 (Academic)

[16] W. van der Kallen: *Infinitesimally central extensions of Chevalley groups*, Lecture Notes in Mathematics **356**, Berlin, etc.,1973(Springer)

[17] Z. Lin: Extensions between simple modules for Frobenius kernels, to appear

[18] A. A. Premet, I. D. Suprunenko: The Weyl modules and the irreducible representations of the symplectic group with the fundamental highest weights, *Commun. Algebra* **11** (1983), 1309–1342

[19] T. A. Springer: Weyl's character formula for algebraic groups, *Invent. math.* **5** (1968), 85–105

[20] J. B. Sullivan: Relations between the cohomology of an algebraic group and its infinitesimal subgroups, *Amer. J. Math.* **100**(1978) , 995–1014

[21] J. B. Sullivan: Lie algebra cohomology at irreducible modules, *Ill. J. Math.* **23**(1979) , 363–373

[22] J. B. Sullivan: The second Lie algebra cohomology group and Weyl modules, *Pacific J. Math.* **86** (1980), 321–326

[23] F. Veldkamp: Representations of algebraic groups of type $F_4$ in characteristic 2, *J. Algebra* **16** (1970), 326–339

Progress in Mathematics, Vol. 95, © 1991 Birkhäuser Verlag Basel

# Endotrivial modules and the Auslander-Reiten quiver

CHRISTINE BESSENRODT*

## Introduction

As Dade stated it in [8]:

"There are just too many modules over $p$-groups!"

More precisely, if $P$ is a $p$-group and $R$ a suitable commutative valuation ring, then almost always the group algebra $RP$ is of wild representation type and there is no classification of all its indecomposable modules. Searching for a useful family of modules that could still be classified Dade was led to study *endopermutation* $RP$-modules, i.e. $RP$-lattices whose $R$-endomorphisms form a permutation $RP$-module. These modules play an important rôle for example in the study of sources of simple modules. The isomorphism classes of indecomposable endopermutation $RP$-modules with vertex $P$ form an abelian group under a multiplication induced by tensor product. For abelian $P$, Dade determined the structure of this group [8]; for non-abelian $P$ Puig [11] proved at least that this group is finitely generated.

On the way to his result Dade classified the indecomposable *endotrivial* $FP$-modules where $F$ is a field of characteristic $p$. These modules satisfy the stronger property that their $F$-endomorphism ring is isomorphic to the direct sum of the trivial module $F$ and a projective module. It turns out that for abelian $P$ only the Heller modules $\Omega^n(F)$, for $n \in \mathbb{Z}$, are endotrivial. For a while, endotrivial modules which are not of this form were only known for the generalised quaternion, the dihedral and the semidihedral 2-groups; so it was conjectured that only for these 2-groups there were such exceptions [8]. But then, Okuyama found endotrivial modules which were not Heller translates of the trivial module also for other 2-groups and for the extraspecial groups of order $p^3$ and exponent $p$, for odd primes $p$ [10].

In this article we want to provide some explanation for the exceptions in the dihedral and semidihedral case which includes information about the classification in these cases by locating the endotrivial modules in their components of the Auslander-Reiten quiver for $FP$.

We start by proving some results on the Auslander-Reiten component of an endotrivial module and its position in it. Then we deduce the classification of the endotrivial modules for the dihedral 2-groups from the known classification of all the indecomposable modules; they are just the non-projective modules in the components of $F$ and $\Omega(F)$. At the end we make some remarks on the relation to the earlier conjecture.

*This work has been supported by the Deutsche Forschungsgemeinschaft

For the following we want to fix some notation. By $G$ we will always denote a finite group, and by $F$ a field of characteristic $p$, where $p$ is a prime dividing the order of $G$. Furthermore, all $FG$-modules are finitely generated left modules. For further standard terminology we refer the reader to [3].

## 1. Preliminaries on endotrivial modules

An $FG$-module $V$ is called an *endotrivial* module if $V \otimes V^* \simeq F \oplus Q$, where $F$ denotes the trivial $FG$-module and $Q$ is a projective $FG$-module. So obviously, for any endotrivial module $V$ also the modules $\Omega^n(V)$ are endotrivial for all $n \in \mathbb{Z}$. In particular, for any group $G$ all the modules $\Omega^n(F)$ are examples of endotrivial modules.

As we will often work modulo projectives, we will abbreviate the projective-free part of a module $M$ by $core(M)$. So $core(M)$ is the module defined (up to isomorphism) by $M \simeq core(M) \oplus Q$, where $Q$ is projective and $core(M)$ has no projective summand. Moreover, if $M$ and $N$ are $FG$-modules, we will write $M \equiv N$ *(mod projectives)* if $core(M) \simeq core(N)$.

First we collect some known results on endotrivial modules. We start with a consequence of a theorem of Alperin and Evens [1] which implies that projectivity can be tested on elementary abelian subgroups (see [3]).

**Proposition 1.1.** *Let $V$ be an $FG$-module. Then $V$ is endotrivial if and only if $V_E$ is endotrivial for any elementary abelian $p$-subgroup $E$ of $G$.*

It is easy to see that tensoring with an endotrivial module preserves the number of non-projective indecomposable summands. In fact, this property characterises endotrivial modules (see [6]):

**Proposition 1.2.** *Let $V$ be a non-projective $FG$-module. Then the module $V$ is endotrivial if and only if $core(V \otimes W)$ is indecomposable for all indecomposable $FG$-modules $W$.*

The endotrivial modules form an abelian group with respect to tensor products, modulo projectives. This group is finitely generated; more generally, such a result holds even for endopermutation modules (see Puig [11]).

Despite of this result, there is no classification of the indecomposable endotrivial modules for general $p$-groups. The best theorem in this direction was obtained by Dade [8]:

**Theorem 1.3.** *Let $G$ be an abelian $p$-group. Then the indecomposable endotrivial $FG$-modules are exactly the modules $\Omega^n(F)$, where $n \in \mathbb{Z}$.*

## 2. Components of endotrivial modules in the Auslander-Reiten quiver

For the definition of the Auslander-Reiten sequence of an $FG$-module and the Auslander-Reiten quiver for the group algebra $FG$ we refer the reader to [3].

First we want to prove that tensoring an Auslander-Reiten sequence with an endotrivial module gives again an Auslander-Reiten sequence.

**Proposition 2.1.** *Let $V$ be an endotrivial $FG$-module, let $M$ be an indecomposable non-projective $FG$-module and let*

$$0 \to \Omega(M) \to E \to \Omega^{-1}(M) \to 0$$

*be the Auslander-Reiten sequence for $\Omega^{-1}(M)$. Then*

$$0 \to \Omega(M) \otimes V \to E \otimes V \to \Omega^{-1}(M) \otimes V \to 0$$

*is the Auslander-Reiten sequence for $\Omega^{-1}(M \otimes V)$ plus injective resp. projective split sequences.*

**Proof.** Let $\alpha = [E] - [\Omega(M)] - [\Omega^{-1}(M)]$ as an element of the Green ring $A(G)$. By Proposition 1.2, $W = \Omega^{-1}(M \otimes V)$ is indecomposable. Now for any indecomposable $FG$-module $N$ we have

$$< \alpha \cdot [V], [N] > = < \alpha, [V^* \otimes N] > = \begin{cases} d_M & \text{if } M \simeq core(V^* \otimes N) \\ 0 & \text{otherwise} \end{cases}$$

by [3, 2.18.4]; here we use that $core(V^* \otimes N)$ is indecomposable. As $M \simeq core(V^* \otimes N)$ is equivalent to $N \simeq core(M \otimes V)$ and $d_M = d_{\Omega(W)}$, this proves that $\alpha \cdot [V]$ is the element in the Green ring corresponding to the Auslander-Reiten sequence for $W$. Hence the assertion follows. □

For convenience we let $X_V$ denote the core of the middle term in the Auslander-Reiten sequence ending with an indecomposable $FG$-module $V$. Furthermore, we write $\beta(V)$ for the number of indecomposable summands of $X_V$. Moreover, for any indecomposable module $V$ let $AR(V)$ denote the component of the stable Auslander-Reiten quiver containing $V$.

**Corollary 2.2.** *Let $V$ be an indecomposable endotrivial $FG$-module. Then*

$$X_F \otimes V \equiv X_V .$$

*In particular, we have $\beta(V) = \beta(F)$ and the set of multiplicities of the indecomposable summands in $X_V$ and $X_F$ is the same.*

From the proposition above we get now the following important fact:

**Theorem 2.3.** *Let $V$ be an endotrivial $FG$-module and $M$ any indecomposable $FG$-module. Then tensoring with $V$ induces an isomorphism between the components $AR(M)$ and $AR(core(M \otimes V))$ of the stable Auslander-Reiten quiver of $FG$.*

*In particular, $AR(V)$ is isomorphic to $AR(F)$.*

We say that a module $V$ lies *at the end* of its component in the Auslander-Reiten quiver if $X_V$ is indecomposable. From the above we see immediately that if $F$ lies at the end then so does every endotrivial module.

The Auslander-Reiten sequence for $F$ is known for any finite group. Let us write down instead the sequence for $\Omega(F)$ from which it can be immediately deduced. Let

$P_F$ denote the projective cover of the trivial module, and let $ht\, P_F = rad\, P_F/soc\, P_F$ be the heart of $P_F$.

Then for any finite group $G$ we always have the almost split sequence

$$0 \to \Omega(F) \to P_F \oplus ht\, P_F \to \Omega^{-1}(F) \to 0$$

The behaviour of $ht\, P_F$ is well understood:

**Theorem 2.4.** *(Webb [13]; Auslander-Carlson [2]) The module $ht\, P_F$ is indecomposable except in the following cases:*

*(i) $p = 2$ and the Sylow 2-subgroups of $G$ are dihedral of order $\geq 8$.*

*(ii) $p = 2$, $Q = \mathbb{Z}_2 \times \mathbb{Z}_2$ is a Sylow 2-subgroup of $G$, and if $N_G(Q) \neq C_G(Q)$ then $F$ contains a primitive cube root of unity.*

*In these cases, $ht\, P_F$ decomposes into two indecomposable endotrivial summands.*

Together with Corollary 3.2 the theorem above implies:

**Corollary 2.5.** *In the exceptional cases of Theorem 3.4 the endotrivial modules form a union of components in the stable Auslander-Reiten quiver on which $\beta \equiv 2$. In all other cases they lie at the end of their components.*

More precisely, using [9] and [13] we obtain:

**Theorem 2.6.** *Let $V$ be an indecomposable endotrivial $FG$-module and let $\Gamma$ be the tree class of $\Delta = AR(V)$. Let $P$ be a Sylow $p$-subgroup of $G$.*

*(i) If $P$ is cyclic, then $\Gamma = A_n$ for some $n$. The endotrivial modules in $\Delta$ are exactly the modules in the two end orbits.*

*(ii) If $P = C_2 \times C_2$ and $N_G(P) = C_G(P)$, then $\Gamma = \tilde{A}_{1,2}$. All modules in $\Delta$ are endotrivial.*

*(iii) If $P = C_2 \times C_2$ and $N_G(P) \neq C_G(P)$ but $F$ does not contain a primitive cube root of unity then $\Gamma = \tilde{B}_3$. The endotrivial modules in $\Delta$ are exactly the modules in the two end orbits.*

*(iv) If $P$ is a dihedral 2-group and neither (ii) nor (iii) holds, then $\Gamma = A_\infty^\infty$. All modules in $\Delta$ are endotrivial.*

*(v) If $P$ is a semidihedral 2-group, then $\Gamma = D_\infty$. The endotrivial modules in $\Delta$ are exactly the modules in the two end orbits.*

*(vi) In all other cases, $\Gamma = A_\infty$, and the endotrivial modules in $\Delta$ form the unique end orbit.*

We may also extend Webb's Theorem C in [13] from $AR(F)$ and $AR(\Omega(F))$ to the components of arbitrary endotrivial modules. For this, we need the following lemma.

**Lemma 2.7.** *Let $V$ be an endotrivial $FG$-module and $M$ an indecomposable non-projective $FG$-module. Then $M$ and $W = core(M \otimes V)$ have the same vertices. More precisely, if $D$ is a vertex of $M$ and $U$ a $D$-source for $M$, then $core(U \otimes V_D)$ is a source for $W$.*

**Proof.** Let $D$ be a vertex of $M$ and let $Y$ be a vertex of $W$. Clearly, $Y \leq_G D$. Since $M = core(W \otimes V^*)$, we also get $D \leq_G Y$ and hence $D =_G Y$. Now $W$ is a direct summand of $U^G \otimes V \simeq (U \otimes V_D)^G$ and $core(U \otimes V_D)$ is indecomposable as $V_D$ is endotrivial. This proves the assertion. □

Using Webb's Theorem C this gives now:

**Theorem 2.8.** *Let $\Delta$ be a component of the stable Auslander-Reiten quiver for $FG$ containing an endotrivial module, and suppose that the Sylow p-subgroups of $G$ are non-cyclic. Then all modules in $\Delta$ have a Sylow p-subgroup of $G$ as a vertex.*

*Remark.* In fact, a suitable modification of Webb's proof shows that the result holds for any component containing a module $V$ with the property that $V_P$ decomposes into a non-zero sum of indecomposables with vertex $P$ and a projective module, for all $p$-subgroups $P$ in $G$.

## 3. Classification of the endotrivial modules for the dihedral 2-groups

The classification of all the indecomposable modules for the modular group algebra over a dihedral 2-group was obtained independently by Bondarenko [4] and Ringel [12]. Here we will use the terminology introduced by Ringel (see also [3]). Let us first note that an endotrivial module is always of odd dimension and hence such an indecomposable module has to be a string module.

First we have to fix some notation. Let $n \geq 3$ and set

$$D = D_{2^n} = \langle x, y \mid x^2 = 1 = y^2 = (xy)^{2^{n-1}} \rangle$$

and

$$z = (xy)^{2^{n-2}}.$$

Furthermore, we write $\tilde{x} = 1 + x$, $\tilde{y} = 1 + y$ and abbreviate

$$\overline{(\tilde{x}\tilde{y})^j} = (\tilde{x}\tilde{y})^j + (\tilde{y}\tilde{x})^j \quad \text{and} \quad \overline{(\tilde{x}\tilde{y})^j\tilde{x}} = (\tilde{x}\tilde{y})^j\tilde{x} + \tilde{y}(\tilde{x}\tilde{y})^j.$$

Then it is straightforward to check:

**Lemma 3.1.**

$$1 + z = (\tilde{x}\tilde{y})^{2^{n-2}} + \overline{(\tilde{x}\tilde{y})^{2^{n-2}-1}\tilde{x}} + \sum_{j=0}^{n-3} \overline{(\tilde{x}\tilde{y})^{2^{n-2}-2^j}}.$$

Now set $N = 2^{n-2} - 1$ and let us first consider the string module $M = M((ab)^N)$ on $E = <x, z>$. From the equation for $1 + z$ above, the following properties can easily be checked:

(i) $ann_M(1+z) \simeq M((ab)^{2^{n-3}-1}a)$.

(ii) $(1+z)M \simeq M((ab)^{2^{n-3}-1})$.

(iii) $\tilde{x}\, ann_M(1+z) \subseteq (1+z)M$.

From (i) we get $dim\, soc(M_E) = 2^{n-3}$. Since

$$dim\, M = 2^{n-1} - 1 = 4 \cdot 2^{n-3} - 1\,,$$

we could already deduce from this:

$$M_E \simeq (FE)^{2^{n-3}-1} \oplus \Omega(F_E)\,.$$

Alternatively, we have from (ii):

$$dim\, \tilde{x}(1+z)M = 2^{n-3} - 1 = |projective\ indecomposables\ in\ M_E|,$$

since $\tilde{x}(1+z)$ annihilates any non-projective module.

Similarly, we have on $E = <y, z>$:

$$M_E \simeq (FE)^{2^{n-3}-1} \oplus \Omega^{-1}(F_E)\,.$$

There are only two longer uniserial string modules, namely those to the words $(ab)^N a$ resp. $b(ab)^N$. These are projective on $<x, z>$ resp. on $<y, z>$.

If we have a shorter uniserial string module $M(a) \leq M(C) \leq M$, then there is a ("standard basis") vector $v \in ann_{M(C)}(1+z)$ such that $\tilde{x}v$ is not contained in $(1+z)M(C)$. Then it is easy to see that $M(C)_{<x,z>}$ has a two-dimensional indecomposable summand $FE \cdot v$, so it is not endotrivial.

In the following, $M = M(C)$ will always be a string module, say $C = C_1 C_2^{-1} \cdots C_{2g-1} C_{2g}^{-1}$, where the letters in all the words $C_i$ are direct and only $C_1$ may be trivial.

If $M$ is endotrivial, then it has to have odd dimension, so $C$ is a word of even length. Since the string modules associated with $C$ and $C^{-1}$ are isomorphic we can (and will) always assume that $C$ starts with $a^{\pm}$.

For an endotrivial module $M$ the same arguments as above give immediately the following restrictions for the length of the $C_i$ :

**Lemma 3.2.** *If $M$ is endotrivial, then we have for all $i$:*

$$2^{n-1} - 2 \leq |C_i| \leq 2^{n-1} - 1 \quad or \quad |C_i| \leq 2.$$

*Moreover, the end terms are not of length 2.*

If some $|C_i| = 1$, then depending on whether $C_i = a$ or $b$, $M$ decomposes on $E = <y,z>$ or on $<x,z>$ into:

$$M_E \simeq M(C_1 \cdots C_{i-1}^{\pm})_E \oplus M(C_{i+1}^{\pm} \cdots C_{2g}^{-1})_E$$

If $M$ is endotrivial, then one of the summands is projective and the other is endotrivial.

In this case we can determine a large part of $C$ :

**Lemma 3.3.** *Suppose $M(C)$ is projective on $<x,z>$. Then we have:*

$$C = (a(ba)^N b^{-1})^m a(ba)^N \quad or \quad C = (a^{-1}(b^{-1}a^{-1})^N b)^m a^{-1}(b^{-1}a^{-1})^N$$

*for some $m \in \mathbb{N}_0$.*

The words associated with endotrivial modules are composed mainly of such projective pieces as above. For $m \in \mathbb{Z}$ set

$$w_a(m) = \begin{cases} (a(ba)^N b^{-1})^m & \text{if } m > 0 \\ 1 & \text{if } m = 0 \\ (a^{-1}(b^{-1}a^{-1})^N b)^{-m} & \text{if } m < 0 \end{cases}$$

$$w_b(m) = \begin{cases} (a^{-1}(ba)^N b)^m & \text{if } m > 0 \\ 1 & \text{if } m = 0 \\ (a(b^{-1}a^{-1})^N b^{-1})^{-m} & \text{if } m < 0 \end{cases}$$

**Lemma 3.4.** *If $M = M(C)$ is endotrivial and $|C_i| = 2$ for some $i$, then $M \simeq M(w_a(s)w_b(t))$ for some $s,t > 0$ or some $s,t < 0$. Conversely, all these modules are endotrivial.*

**Proof.** Set $C_{(i)} = C_1 C_2^{-1} \cdots C_i^{\pm}$ and $C^{(i)} = C_i^{\pm} \cdots C_{2g-1} C_{2g}^{-1}$. Here $C$ need not necessarily start with $a^{\pm}$. Suppose $i$ is odd and $C_i = ab$ (the other cases can be treated similarly).

Then we have:

$$M_{<x,z>} \simeq M(C_{(i-1)}a)_{<x,z>} \oplus M(C^{(i+1)})_{<x,z>}.$$

This implies that $M(C^{(i+1)})$ is projective on $<x,z>$, so by Lemma 4.3 $C^{(i+1)} = w_a(s)a^{-1}(b^{-1}a^{-1})^N$, for some $s \leq 0$. Similarly, the restriction on $<y,z>$ shows that $M(C_{(i-1)})$ is projective on this subgroup, so $C_{(i-1)} = b^{-1}(a^{-1}b^{-1})^N w_b(t)$ for some $t \leq 0$. Gluing these pieces together, we get $C^{-1} = w_a(1-s)w_b(1-t)$. As $M \simeq M(C^{-1})$, this proves the assertion. □

**Lemma 3.5.** *If $M = M(C)$ is endotrivial and $|C_i| = 2N$ for some $i$, then $M \simeq w_a(s)(ab)^N w_b(t)$ for some $s,t \geq 0$, or $M \simeq w_a(s)(a^{-1}b^{-1})^N w_b(t)$ for some $s,t \leq 0$. Conversely, all these modules are endotrivial.*

**Proof.** Suppose $i$ is odd, and assume that $C_i = (ab)^N$. Let us look at the restriction of $M$ to $<x,z>$. If $|C_{i+1}| \geq 2N$, then we split off the projectives and obtain a two-dimensional summand coming from $C_i$ - contradicting the endotriviality. Similarly, the restriction onto $<y,z>$ shows that $C_{i-1}$ cannot be long. By the previous lemma we already know that no $C_j$ is of length 2. Hence if $C_{i-1}$ resp. $C_{i+1}$ are non-trivial, then they are of length 1. Now the result follows from Lemma 4.3 and the remarks preceding it. □

The last case where some $C_i$ is of length $2N+1$ can be handled similarly, i.e. by splitting off projectives on the Klein four groups.

Putting all the pieces together, we obtain now:

**Theorem 3.6.** *The indecomposable endotrivial modules for the dihedral 2-group of order $2^n$ are the string modules associated with the following words:*

*(i)* $w_a(s)\, w_b(t)$, *where* $s,t \in \mathbb{Z}$;

*(ii)* $w_a(s)\,(ab)^N\, w_b(t)$, *where* $N = 2^{n-2}-1$ *and* $s,t$ *are not both negative;*

*(iii)* $w_a(s)\,(a^{-1}b^{-1})^N\, w_b(t)$, *where* $s,t \le 0$.

Butler and Shahzamanian [5] have computed the Auslander-Reiten quiver for the dihedral 2-groups (see also [3]). Comparing the list above with their results shows that the endotrivial modules of type (i) above form the component of the trivial module in the Auslander-Reiten quiver, whereas the modules of type (ii) and (iii) are exactly the non-projective modules in the component of the module $\Omega(F)$. So we can state:

**Corollary 3.7.** *The indecomposable endotrivial modules for the dihedral 2-groups are exactly the non-projective modules in the components containing the trivial module $F$ resp. the module $\Omega(F)$.*

As we have seen before, the Heller translates and the tensor powers of an endotrivial module are again endotrivial. For the dihedral 2-groups all the endotrivial modules can be obtained from a single endotrivial module using these operations:

**Proposition 3.8.** *As before, let $N = 2^{n-2}-1$ and set $M = M((ab)^N)$. Then all the endotrivial modules for the dihedral group of order $2^n$ are of the form $\Omega^m(M^{\otimes i})$ for suitable $m,i \in \mathbb{Z}$.*

*More precisely, we have:*

$$\Omega^0(M^{\otimes i}) = \begin{cases} M(w_a(\frac{i}{2})w_b(\frac{i}{2})) & \text{if } i \text{ is even} \\ M(w_a(\frac{i-1}{2})(ab)^N w_b(\frac{i-1}{2})) & \text{if } i \text{ is odd and positive} \end{cases}$$

**Proof.** We already know that

$$\Omega^0(M_{<x,z>}) \simeq \Omega(F_{<x,z>})$$

and

$$\Omega^0(M_{<y,z>}) \simeq \Omega^{-1}(F_{<y,z>})\,.$$

Thus

$$\Omega^0(M^{\otimes i}{<}x,z{>}) \simeq \Omega^i(F{<}x,z{>})$$

and

$$\Omega^0(M^{\otimes i}{<}y,z{>}) \simeq \Omega^{-i}(F{<}y,z{>})\,.$$

Now $\Omega^0(M^{\otimes i})$ is an indecomposable endotrivial module, so it is associated to one of the words classified in our theorem. It is easy to compute the restrictions of the corresponding modules to the subgroups $<x,z>$ resp. $<y,z>$. One finds that the endotrivial modules can be distinguished by these restrictions, and thus we arrive at a unique word from our list.

For example, $M(w_a(s)w_b(t))$ restricts to $\Omega^{2t}(F_E)$ for $E = <x,z>$, and to $\Omega^{-2s}(F_E)$ for $E = <y,z>$. □

From the result above it is straightforward (but somewhat tedious) to write down the formula for all the $\Omega^j(M^{\otimes i})$ explicitly, i.e. by giving the associated word.

## 4. Remarks

For abelian $p$-groups we know by Dade's theorem that the indecomposable endotrivial modules are just the modules $\Omega^n(F)$, $n \in \mathbb{Z}$. So all endotrivial modules lie in $AR(F) \cup AR(\Omega(F))$, the components of the Auslander-Reiten quiver containing the trivial module and its Heller translate. In Section 4 we have seen that the endotrivial modules for the dihedral 2-groups are exactly all the non-projective modules in $AR(F) \cup AR(\Omega(F))$. In the semidihedral case, the component of $F$ is of type $D_\infty$, so we already have endotrivial modules which are not Heller modules $\Omega^n(F)$ in the other end orbits of the components $AR(F)$ and $AR(\Omega(F))$. These should be all exceptions. Of course, in the case of a generalised quaternion 2-group, there are endotrivials outside these components; these groups are the only non-cyclic $p$-groups where we have periodic endotrivials. In fact, there are exactly eight endotrivial indecomposable modules, lying at the end of four tubes [7].

Now we want to restate Dade's conjecture as a question in the language of $AR$-quivers:

**Question.** *Let $G$ be a $p$-group. When do all the indecomposable endotrivial $FG$-modules lie in the components $AR(F)$ and $AR(\Omega(F))$ ?*

Here is the connection to the earlier conjecture:

**Proposition 4.1.** *If $G$ is a $p$-group which is neither dihedral, nor semidihedral, nor generalised quaternion and for which all indecomposable endotrivial $FG$-modules lie in $AR(F) \cup AR(\Omega(F))$, then all its indecomposable endotrivial modules are of the form $\Omega^n(F)$, $n \in \mathbb{Z}$.*

**Proof.** By Dade's theorem we may assume that $G$ is not abelian. Now by Theorem 2.6 the components $AR(F)$ and $AR(\Omega(F))$ are of type $A_\infty$ and we know that the endotrivials are the modules at the end of these components; these are exactly the modules $\Omega^n(F)$, $n \in \mathbb{Z}$. □

Note that apart from giving a good "explanation" for the occurrence of the exceptions in the dihedral and semidihedral cases a positive answer to the question above does also contain a precise classification of the endotrivials in these cases. By Section 3 the endotrivials for the semidihedral 2-group should be just the modules $\Omega^n(F)$ and $\Omega^n(L)$, $n \in \mathbb{Z}$, where $L$ is the module with $ht\,FG$ as the middle term of its Auslander-Reiten sequence.

The obstacle in answering the question above is that we would need some criterio to decide whether a given module is in the same component as $F$ or $\Omega(F)$. Hopefully any good sufficient criterion should be applicable to endotrivial modules. There are a few useful necessary conditions known, for example the vertex and the variety should be maximal, but these are already satisfied by any module of dimension not divisible by $p$.

## REFERENCES

[1] J.L. Alperin and L. Evens: Representations, resolutions, and Quillen's dimension theorem, *J. Pure Appl. Algebra* **22** (1981), 1-9

[2] M. Auslander and J.F. Carlson: Almost split sequences and group rings, *J. Algebra* **103** (1986), 122-140

[3] D. Benson: Modular representation theory: New trends and methods, *Lecture Notes in Math.* **1081**, Springer 1984

[4] V.M. Bondarenko: Representations of dihedral groups over a field of characteristic 2, *Math. USSR Sbornik* **25** (1975), 58-68

[5] M.C.R. Butler and M. Shahzamanian: The construction of almost split sequences, III: Modules over two classes of tame local algebras, *Math. Ann.* **247** (1980), 111-122

[6] J.F. Carlson: The variety of an indecomposable module is connected, *Invent. math.* **77** (1984), 291-299

[7] J.F. Carlson: personal communication

[8] E.C. Dade: Endo-permutation modules over $p$-groups, I, II, *Annals Math.* **107** (1978), 459-494; **108** (1978), 317-346

[9] T. Okuyama: On the Auslander-Reiten quiver of a finite group, *J. Algebra* **110** (1987), 425-430

[10] T. Okuyama: personal communication

[11] L. Puig: The source algebra of a nilpotent block, Preprint 1981

[12] C.M. Ringel: The indecomposable representations of the dihedral 2-groups, *Math. Ann.* **214** (1975), 19-34

[13] P.J. Webb: The Auslander-Reiten quiver of a finite group, *Math. Z.* **179** (1982), 97-121

Progress in Mathematics, Vol. 95, © 1991 Birkhäuser Verlag Basel

# Tame curve singularities with large conductor

ERNST DIETERICH

In this article I intend to introduce the reader to my thesis [Di 90]. Accordingly, here I emphasize problems and results. Only in section 2 I try to sketch the main line of arguments. For proofs see [Di 90].

In the sequel, the notions "summand", "isoclass", "epivalence" are understood to mean "direct summand", "isomorphism class", "representation equivalence".

## 1. Generalities

**1.1.** Throughout, let $k$ be an algebraically closed field, and let $R = k[[\pi]]$ be the ring of formal power series over $k$ in one indeterminate $\pi$. These are the ground field and the ground ring for the categories to be considered.

Recall that an *R-order* is an $R$-algebra which is finitely generated free as an $R$-module. A *curve singularity* (over $k$) is, by definition, an $R$-order which is commutative, local and isolated singular.

Here, the term "isolated singular" has two possible interpretations. Namely, a commutative local $R$-order $\Lambda$ is called *isolated singular* in the sense of commutative algebra, if $\Lambda$ is not regular but $\Lambda_{\mathcal{P}}$ is regular for each nonmaximal prime ideal $\mathcal{P}$ of $\Lambda$, respectively in the sense of noncommutative algebra, if $\Lambda$ is not hereditary and $K \otimes_R \Lambda$ is semisimple, where $K = \text{fract}(R)$.

In fact, it is easy to see that both interpretations are equivalent.

**1.2.** Curve singularities, in the sense defined above, can be characterized among a wider class of $k$-algebras as follows.

**Proposition 1.** *Let $\Lambda$ be a $k$-algebra which is commutative, noetherian and complete local. Let $\mathcal{M}$ be the unique maximal ideal of $\Lambda$. Then the following assertions are equivalent:*

*(i) There exists a ring homomorphism $R \to \Lambda$ such that $\Lambda$ is a curve singularity over $k$.*

*(ii) $\Lambda$ is not regular, reduced, onedimensional, and the canonical monomorphism $k \to \Lambda/\mathcal{M}$ is an isomorphism.*

*(iii) There exists an affine-algebraic curve $C \subset \mathbb{A}^n k$, with singular point $O \in C$, such that $\Lambda$ and $\hat{\mathcal{O}}_{C,O}$ are analytically isomorphic.*

Here, $\hat{\mathcal{O}}_{C,O}$ denotes the *complete local ring* of $(C, O)$. It may be defined as $\hat{\mathcal{O}}_{C,O} := k[[X_1, \ldots, X_n]]/(\{f \in k[X_1, \ldots, X_n] | f(C) = 0\})$. It contains the algebraic information of the geometric object $(C, O)$, viewed locally. "Analytical isomorphism" means $k$-algebra isomorphism. Hence, curve singularities in our sense are, up to $k$-algebra isomorphism, precisely the complete local rings of affine-algebraic curve singularities in the classical sense. This justifies our terminology.

Also note that the ground ring $R$ has disappeared in (ii) and (iii). It can be reconstructed by means of a ring homomorphism $R \to \Lambda$, $\pi \mapsto \rho$, with $\rho$ being any nonzerodivisor in $\mathcal{M}$. Hence, there are various possible choices for the ground ring $R$ of a curve singularity $\Lambda$. They correspond to the possible choices of regular elements in $\Lambda$.

**1.3.** Let $\Lambda$ be a curve singularity over $k$. In this context, we are not primarily interested in $\Lambda$ itself, but in the category latt $\Lambda$ of $\Lambda$-*lattices* attached to it. Recall that latt $\Lambda$ is the full subcategory of mod $\Lambda$, the category of all finitely generated left $\Lambda$-modules, with object class $\{M \in \operatorname{mod}\Lambda \mid M \text{ is finitely generated free as } R\text{-module}\}$.

By Auslander-Buchsbaum-Serre's Theorem, latt $\Lambda$ coincides with cm $\Lambda$, the category of *Cohen-Macaulay* $\Lambda$-*modules.* This provides an alternative way of describing latt $\Lambda$, without reference to the ground ring $R$.

Since $\Lambda$ is complete, latt $\Lambda$ is a *Krull-Schmidt category* (i.e. every $\Lambda$-lattice is a finite direct sum of indecomposable $\Lambda$-lattices, and in this decomposition, the isoclasses of indecomposable summands and the multiplicities to which they occur are uniquely determined).

Therefore, the problem of classifying all indecomposable $\Lambda$-lattices, up to isomorphism, arises naturally. This problem we call briefly the *classification problem* of $\Lambda$. The attempt to solve the classification problem, for a class of curve singularities as large as possible, is the main source of motivation for all of the following material.

**1.4.** Denote by Is(ind $\Lambda$) the set of all isoclasses of indecomposable $\Lambda$-lattices. A curve singularity $\Lambda$ is called *representation-finite*, if Is(ind $\Lambda$) is a finite set. Representation-finite curve singularities certainly show the simplest behaviour with respect to solution of the classification problem. I proceed to recall the main result concerning them, as an illustration and motivation for the type of questions to be asked later in the tame case.

**Theorem 2 ([Dr/Ro 67], [Ja 67], [Gr/Kn 85]).** *Let $\Lambda$ be a curve singularity over $k$, with* $\operatorname{char} k = 0$. *Then the following assertions are equivalent:*

*(i)* $\Lambda$ *is representation-finite.*
*(ii)* $\mu_\Lambda(\Omega/\Lambda) \leq 2$ *and* $\mu_\Lambda(\operatorname{rad}(\Omega/\Lambda)) \leq 1$, *where* $\Omega$ *is the unique maximal order in* $K \otimes_R \Lambda$.
*(iii)* $\Lambda$ *dominates* $\hat{O}_{C_s,O}$ *for some simple plane curve singularity* $(C_s, O)$.

Here, $\mu_\Lambda(M)$ is the minimal number of generators for $M$, where $M \in \operatorname{mod}\Lambda$. A curve singularity $\Lambda$ is said to dominate a curve singularity $\Sigma$, if $\Sigma \subset \Lambda \subset K \otimes_R \Sigma$. The simple plane curve singularities $(C_s, O)$ are classified up to analytical isomorphism and their list, in bijection to the list of all Dynkin diagrams $\mathsf{A}_n$, $\mathsf{D}_n$, $\mathsf{E}_6$, $\mathsf{E}_7$, $\mathsf{E}_8$, is well-known (see e.g. [Gr/Kn 85]).

Interpreting Theorem 2, we conclude with two remarks.

(a) The equivalence (i) $\Leftrightarrow$ (ii) provides an effective criterion, in terms of the simply accessible numerical invariant $(\mu_\Lambda(\Omega/\Lambda), \mu_\Lambda(\mathrm{rad}(\Omega/\Lambda)))$, for deciding finite representation type, for any given curve singularity $\Lambda$.

(b) The equivalence (i) $\Leftrightarrow$ (iii) implies the classification of all representation-finite curve singularities, up to analytical isomorphism.

**1.5.** A curve singularity $\Lambda$ is called *tame*, if $\Lambda$ is representation-infinite and for each $d \in \mathbb{N}$ there are finitely many one-parameter series of indecomposable $\Lambda$-lattices of rank $d$, representing almost all isoclasses of indecomposable $\Lambda$-lattices of rank $d$.

Being interested in the solution of the classification problem, and leaving the representation-finite case, we have to study tame curve singularities. Here, we are faced with the following key problems.

(A) Find an effective criterion, in terms of some simply accessible invariant, for deciding tame representation type, for any given curve singularity $\Lambda$.
(B) Classify all tame curve singularities, up to analytical isomorphism.
(C) Find an effective general strategy for classifying $\mathrm{Is}(\mathrm{ind}\,\Lambda)$, for any given tame curve singularity $\Lambda$.

We reflect, for a moment, the analogues of these problems in the representation-finite case. As already mentioned, Theorem 2 solves the analogues of problems (A) and (B). Also, there are effective algorithms for constructing the *AR*-quiver of $\Lambda$, or alternatively, for solving the matrix problem to which the classification problem of $\Lambda$ can be reduced. These provide solutions to the analogue of problem (C).

The equivalence (i) $\Leftrightarrow$ (iii) of Theorem 2 has the following analogue for tame curve singularities, which we state as a conjecture.

**Conjecture.** *Let $\Lambda$ be a curve singularity over $k$. Then the following assertions are equivalent:*

*(i) $\Lambda$ is tame.*
*(ii) $\Lambda$ dominates $\hat{\mathcal{O}}_{C_u,O}$ for some unimodular plane curve singularity $(C_u, O)$.*

The unimodular plane curve singularities $(C_u, O)$ are classified, up to analytical isomorphism [Ar 74]. They fall into three types: there are two series of *parabolic* singularities, $\mathsf{T}_{4,4,2,a}$ and $\mathsf{T}_{6,3,2,a}$, with $a \in k \setminus \{0,1\}$; there is one series of *hyperbolic* singularities, $\mathsf{T}_{p,q,2}$, with $p, q \in \mathbb{N} \setminus \{0\}$ such that $\frac{1}{p} + \frac{1}{q} + \frac{1}{2} < 1$; and there are 16 *exceptional* singularities. For normal forms see [Ar 74], [Scha 85a].

So far, tame representation type for the parabolic singularities $\mathsf{T}_{6,3,2,a}$ and $\mathsf{T}_{4,4,2,a}$ is known to be true [Di 85], [Ka 87], [Ka 89], [Di 90]. On the other hand, tame representation type for the hyperbolic and exceptional unimodular plane curve singularities is still conjectural. For a historical account of these and related problems we refer to [Di 88].

Finally we observe that the validity of the conjecture, together with the full classifications of $\mathrm{Is}(\mathrm{ind}\,\hat{\mathcal{O}}_{C_u,O})$ for all unimodular plane curve singularities $(C_u, O)$, would imply the solution of problem (B).

## 2. Curve singularities with large conductor

**2.1.** With any given curve singularity $\Lambda$ over $k$ we associate the following data: $\mathcal{M}$ is the unique maximal ideal of $\Lambda$; $K = \text{fract}(R) = k((\pi))$ is the field of Laurent series over $k$ in one indeterminate $\pi$; $K \otimes_R \Lambda$ is the finite-dimensional semisimple $K$-algebra generated by $\Lambda$; $\Omega$ is the unique maximal $R$-order in $K \otimes_R \Lambda$, with $\mathcal{J} := \text{rad}\,\Omega$.

There is an isomorphism of $k$-algebras $K \otimes_R \Lambda \xrightarrow{\sim} \prod_{i=1}^n k((\pi_i))$, inducing an isomorphism of $k$-algebras $\Omega \xrightarrow{\sim} \prod_{i=1}^n k[[\pi_i]]$. ¿From now on we assume, without loss of generality, that $\Omega = \prod_{i=1}^n k[[\pi_i]]$. We call $n$ the *number of branches* of $\Lambda$. It coincides with the number of nonmaximal prime ideals of $\Lambda$.

We say that $\Lambda$ has *large conductor*, in case $\mathcal{J}^2 \subset \mathcal{M}$. We denote by $C_n^l(k)$ the class of all curve singularities over $k$ with $n$ branches and large conductor.

**2.2.** Consider the quiver $Q_n$ : $\begin{array}{c} 1 \circ \searrow \\ \vdots \quad \vdots \quad \circ\, 0 \\ n \circ \nearrow \end{array}$, and the category $\text{rep}_k\, Q_n$ of $k$-representations of $Q_n$. We define a map $\chi_n : C_n^l(k) \to \text{rep}_k\, Q_n$, $\chi_n(\Lambda) = Z = (Z_0, Z_i, \zeta_i)$, by $Z_0 := \mathcal{M}/\mathcal{J}^2$, $Z_i := k\,\pi_i$, and by commutativity of the triangle

$$\begin{array}{ccc} \mathcal{M}/\mathcal{J}^2 & \xrightarrow{(\zeta_i)} & \prod_{i=1}^n k\,\pi_i \\ {\scriptstyle \text{incl}} \searrow & & \nearrow {\scriptstyle \text{can}} \\ & \mathcal{J}/\mathcal{J}^2 & \end{array}$$

We call $Z = \chi_n(\Lambda)$ the *characteristic representation* of $\Lambda$. Moreover, we denote by $\text{rep}_k^\bullet\, Q_n$ the full subcategory of $\text{rep}_k\, Q_n$ with object class $\{V \in \text{rep}_k\, Q_n \mid V \text{ has no simple projective summand}\}$.

**2.3.** Let $\Lambda$ be a curve singularity over $k$, with $n$ branches and large conductor. Let $Z = \chi_n(\Lambda)$ be the characteristic representation of $\Lambda$. We define the restricted endofunctor $\mathcal{Z} : \text{rep}_k^\bullet\, Q_n \to \text{rep}_k\, Q_n$ on objects $X = (X_0, X_i, \xi_i)$ by $\mathcal{Z}(X) := Z \otimes X := (Z_0 \otimes_k X_0, Z_i \otimes_k X_i, \zeta_i \otimes_k \xi_i)$, and on morphisms in the obvious way. Denote by $E$ the bifunctor $\text{Ext}^1_{Q_n}(?,?) : (\text{rep}_k^\bullet\, Q_n)^{\text{op}} \times \text{rep}_k\, Q_n \to \text{mod}\, k$. Then we define the category $\text{rep}_{\mathcal{Z}}\, E$ as follows. Objects in $\text{rep}_{\mathcal{Z}}\, E$ are pairs $(X, e)$, where $X \in \text{rep}_k^\bullet\, Q_n$ and $e \in E(X, \mathcal{Z}(X))$ . Morphisms $\mu : (X, e) \to (X', e')$ in $\text{rep}_{\mathcal{Z}}\, E$ are those morphisms $\mu : X \to X'$ in $\text{rep}_k\, Q_n$ which satisfy the equation $\mathcal{Z}(\mu)e = e'\mu$, in $E(X, \mathcal{Z}(X'))$. We call $\text{rep}_{\mathcal{Z}}\, E$ the category of *representations of* $E$, *over the graph of* $\mathcal{Z}$. Finally, let $\text{rep}_{\mathcal{Z}}^\vee\, E$ be the full subcategory of $\text{rep}_{\mathcal{Z}}\, E$ with object class $\{(X, e) \in \text{rep}_{\mathcal{Z}}\, E \mid (I_0, 0) \text{ is not a summand of } (X, e)\}$, where $I_0$ is the simple injective representation.

**2.4.** In terms of the above notation we can now formulate our main theoretical result. It describes a reduction for lattice categories over curve singularities with large conductor.

**Theorem 3 ([Di 90]).** *Let $\Lambda$ be a curve singularity over $k$, with $n$ branches and large conductor. Then there exists an epivalence*

$$\mathcal{E} : \text{latt}\,\Lambda \to \text{rep}_{\mathcal{Z}}^\vee\, E\,,$$

*with kernel* $(\{\psi \in \mathrm{Hom}_\Lambda(M,M') \mid \psi(M) \subset \mathcal{M}M'\})_{(M,M')\in \mathrm{latt}\,\Lambda\times\mathrm{latt}\,\Lambda}$.

Consider the $n$-ad $\Gamma := \mathcal{J} \oplus k \cdot 1_\Omega$. It is the unique maximal element in $C_n^l(k)$. Then $(\mathcal{M}, \Lambda, \Gamma)$ is an *admissible triple* (in the sense of [Di 88]) and therefore determines a $k$-category $\mathcal{K}$ together with an epivalence $\mathcal{F} : \mathrm{latt}\,\Lambda \to \mathcal{K}$, according to [Di 88, §3]. This is the initial and trivial step in the proof of Theorem 3. The substantial part of the proof exhibits an equivalence of categories $\mathcal{K} \to \mathrm{rep}_{\mathcal{Z}}^{\vee} E$. The construction of this equivalence is not straightforward.

Let $(X, e)$ be an object in $\mathrm{rep}_{\mathcal{Z}} E$, and let $\delta : X \xrightarrow{\sim} \oplus X_i^{n_i}$ be a decomposition of $X$ into indecomposable summands $X_i$, such that $X_i \not\cong X_j$ for $i \neq j$. Consider the sequence of isomorphisms

$$\mathrm{Ext}^1_{Q_n}(X, \mathcal{Z}(X)) \underset{\iota_1}{\xrightarrow{\sim}} \mathrm{Hom}_{Q_n}(\tau^-\mathcal{Z}(X), X)^* \underset{\iota_2}{\xrightarrow{\sim}} \prod_{i,j} [\mathrm{Hom}_{Q_n}(\tau^-\mathcal{Z}(X_i), X_j)^*]^{n_i\times n_j},$$

where $\iota_1$ is the well-known AR-isomorphism, and $\iota_2$ is canonically induced by $\delta$. Then $\iota_2\iota_1(e)$ is a block-matrix, with block $(i,j)$ having size $n_i \times n_j$ and entries in the $k$-vectorspace $\mathrm{Hom}_{Q_n}(\tau^-\mathcal{Z}(X_i), X_j)^*$, for all $(i,j)$. This indicates roughly, how $\mathrm{rep}_{\mathcal{Z}} E$ can be viewed as a matrix category.

It is well-known that each curve singularity with more than 4 branches is of wild representation type. Being interested in tame curve singularities we may therefore assume that $n \leq 4$. In this situation, the categories $\mathrm{rep}_k Q_n$ are well-understood and it is not very difficult to describe the endofunctors $\tau^-\mathcal{Z} : \mathrm{rep}_k^\bullet Q_n \circlearrowleft$, for all possible characteristic representations $Z = \chi_n(\Lambda)$. This opens the door to derive explicit descriptions of $\mathrm{rep}_{\mathcal{Z}} E$ as matrix categories, to relate them to categories which have been objects of study in classical representation theory of finite-dimensional algebras, and to apply the theories which have been developed there.

For the substantial work that has to be done in carrying out this general strategy, we refer to [Di 90]. Here we are content with the presentation of main results which have been obtained along this line. To this we turn in the following sections.

## 3. Tame curve singularities with four branches and large conductor

**3.1.** Let $\Lambda$ be a curve singularity over $k$ and assume that $\Lambda$ is tame. For each $d \in \mathbb{N}$, let $n(d)$ be the minimal number of one-parameter series of indecomposable $\Lambda$-lattices of rank $d$ representing almost all isoclasses of indecomposable $\Lambda$-lattices of rank $d$. Recall the following refinement of the notion "tame".

a) $\Lambda$ is called *domestic tame* if there exists $N \in \mathbb{N}$ such that $n(d) \leq N$, for all $d \in \mathbb{N}$.
b) $\Lambda$ is called *nondomestic tame of finite growth* if $\Lambda$ is not domestic tame and there exists $N \in \mathbb{N}$ such that $n(d) \leq Nd$, for all $d \in \mathbb{N}$.
c) $\Lambda$ is called *tame of infinite growth* if there exists a real number $N > 1$ and an infinite subset $D \subset \mathbb{N}$ such that $n(d) \geq N^d$, for all $d \in D$.

**3.2.** From now on, we consider only curve singularities with large conductor and — excepting the general remarks in 3.3 — with at most 4 branches.

**Proposition 4.** *Let $\Lambda$ be a curve singularity over $k$, with 4 branches and large conductor. Let $Z = \chi_4(\Lambda)$ be the characteristic representation of $\Lambda$ in $\operatorname{rep}_k Q_4$. Then the following assertions hold:*

*(i) If $Z$ is the direct sum of indecomposable preinjective representations, then $\Lambda$ is domestic tame.*
*(ii) If $Z$ is indecomposable homgeneous regular, then $\Lambda$ is nondomestic tame of finite growth.*
*(iii) If $Z$ is the direct sum of indecomposable nonhomogeneous regular and indecomposable preinjective representations, with at least one indecomposable regular summand, then $\Lambda$ is tame of infinite growth.*
*(iv) If $Z$ contains an indecomposable preprojective summand, then $\Lambda$ is wild.*

Observe that the assumptions on $Z$ in (i)–(iv) cover all possibilities, since $\dim_k Z_i = 1$, for all $i = 1, \ldots, 4$.

Forgetting some of the information contained in Proposition 4, we immediately obtain the following criterion for tame representation type of curve singularities in $C_4^l(k)$.

**Proposition 5.** *For all curve singularities $\Lambda$ in $C_4^l(k)$, the following assertions are equivalent:*

*(i) $\Lambda$ is tame.*
*(ii) $\chi_4(\Lambda)$ contains no indecomposable preprojective summand.*

**3.3.** Let $\mathcal{V}_n$ be the class of all objects $V = (V_0, V_i, \alpha_i) \in \operatorname{rep}_k Q_n$, such that $I_0$ is not a summand of $V$ and $\dim_k V_i = 1$ for all $i = 1, \ldots, n$. Then evidently $\chi_n(\Lambda) \in \mathcal{V}_n$, for all $\Lambda \in C_n^l(k)$.

Conversely, there is the following construction $\Sigma_n : \mathcal{V}_n \to C_n^l(k)$. For any given $V \in \mathcal{V}_n$, set $\alpha := (\alpha_i) : V \hookrightarrow \oplus_{i=1}^n V_i$; choose basis elements $v_i \in V_i$, for all $i = 1, \ldots, n$, and define the $k$-linear isomorphism $\beta : \oplus_{i=1}^n V_i \xrightarrow{\sim} \oplus_{i=1}^n k\pi_i$ by $\beta(v_i) = \pi_i$; set $U := \beta\alpha(V_0)$, and finally set $\Sigma_n(V) := \mathcal{J}^2 \oplus U \oplus k \cdot 1_\Omega$. Then in fact $\Sigma_n(V)$ is in $C_n^l(k)$.

Let $\sim$ be the equivalence relation on $\mathcal{V}_n$ which is generated by isomorphisms in $\operatorname{rep}_k Q_n$ and by permutations of the $k$-linear maps $\alpha_i$ in $V = (V_0, V_i, \alpha_i) \in \mathcal{V}_n$. Then it is easy to see that $\chi_n$ and $\Sigma_n$ induce a bijection between the set of analytical isoclasses of $C_n^l(k)$ and $\mathcal{V}_n/\sim$.

For $n \leq 4$, the classification of $\mathcal{V}_n/\sim$ is an easy exercise. For $n = 4$ this yields, in combination with Proposition 4, the following classification of all tame curve singularities in $C_4^l(k)$, up to analytical isomorphism.

**Proposition 6.** *A complete system of representatives for the analytical isoclasses of tame curve singularities with 4 branches and large conductor is given by the following list (where $a \in k \backslash \{0, 1\}$ ):*

| $\Lambda$ | $Z$ | $U$ |
|---|---|---|
| $\Lambda_4^1$ | $\oplus_{i=1}^4 I_i$ | 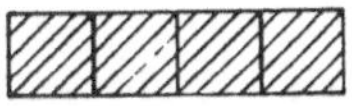 |
| $\Lambda_4^2$ | $\tau I_0$ | 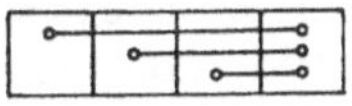 |
| $\Lambda_4^3$ | $I_1 \oplus \tau I_1$ | 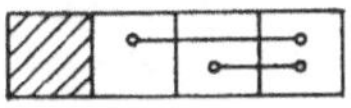 |
| $\Lambda_4^a$ | $R_1(a)$ | 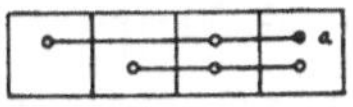 |
| $\Lambda_4^4$ | $I_1 \oplus I_2 \oplus R_1(1)$ | 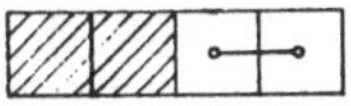 |
| $\Lambda_4^5$ | $R_1(1) \oplus \tau R_1(1)$ | 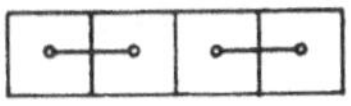 |
| $\Lambda_4^6$ | $R_2(0)$ | 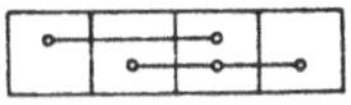 |

In this table, we have listed the curve singularities $\Lambda$ in the first column, their characteristic representation $Z = \chi_4(\Lambda)$ in the second column, and the subspaces $U \subset \oplus_{i=1}^4 k\pi_i$ in the third column. In the second column we have used the following notation: $I_i$ is the indecomposable injective representation corresponding to $i \in \{0, \dots, 4\}$, and $R_l(\lambda)$ is the indecomposable regular representation of regular length $l$ and belonging to the AR-component of the indecomposable representation defined by $\left(\begin{smallmatrix}1\\0\end{smallmatrix} \middle| \begin{smallmatrix}0\\1\end{smallmatrix} \middle| \begin{smallmatrix}1\\1\end{smallmatrix} \middle| \begin{smallmatrix}\lambda\\1\end{smallmatrix}\right)$.

Combining Proposition 4 and Proposition 6, we immediately obtain the following Corollary.

**Corollary 7.** *The analytical isoclasses of tame curve singularities in $C_4^l(k)$ fall into three types:*

1) *there are 3 isoclasses of domestic tame curve singularities, represented by $\Lambda_4^1$, $\Lambda_4^2$, $\Lambda_4^3$;*
2) *there is one $(k\backslash\{0,1\})$-series of isoclasses of nondomestic tame curve singularities of finite growth, represented by $\Lambda_4^a$;*
3) *there are 3 isoclasses of tame curve singularities of infinite growth, represented by $\Lambda_4^4$, $\Lambda_4^5, \Lambda_4^6$.*

**3.4.** Now we turn to the classification of $\mathrm{Is}(\mathrm{ind}\,\Lambda)$, for $\Lambda \in \{\Lambda_4^1, \Lambda_4^2, \Lambda_4^3, \Lambda_4^a \mid a \in k\backslash\{0,1\}\}$. First, we fix some notation and terminology.

For any $\Lambda$-lattice $M$, we denote by $[M]$ the isoclass of $M$ and we set $M^* := \mathrm{Hom}_R(M, R)$. We define the *dimension type* of $M$ to be the vector $\underline{\dim}\, M := (\dim_k(M/\mathcal{M}M), \dim_k(e_i\Omega M/\mathcal{J}M))_{i=1,\dots,4}$ in $\prod_{i=0}^4 \mathbf{Z}$, where $e_1 = (1,0,0,0)$, $\dots$, $e_4 = (0,0,0,1)$ are the primitive orthogonal idempotents of $\Omega$.

We denote by $\mathcal{A}(\Lambda)$ the AR-quiver of $\Lambda$, by $\mathcal{P}(\Lambda)$ the connected component of $\mathcal{A}(\Lambda)$ which contains the unique projective point $[\Lambda]$, and by $\mathcal{A}_s(\Lambda)$ the stable AR-quiver of $\Lambda$.

A translation quiver $\mathcal{T}$ is called a $\mathsf{P}^1k$*-tubular family* if $\mathcal{T} = \dot{\bigcup}_{\lambda \in \mathsf{P}^1 k} \mathcal{T}(\lambda)$ and for each $\lambda \in \mathsf{P}^1 k$ there is an isomorphism of translation quivers $\mathcal{T}(\lambda) \xrightarrow{\sim} \mathbb{Z}A_\infty/\tau^{n(\lambda)}$. Let $\mathcal{T}$ be a $\mathsf{P}^1k$-tubular family and set $\mathcal{E} := \mathsf{P}^1 k \backslash n^{-1}(1)$, for some parametrization $\mathcal{T} = \dot{\bigcup}_{\lambda \in \mathsf{P}^1 k} \mathcal{T}(\lambda)$. Then $\mathcal{T}$ is said to be of *tubular type* $\mathsf{D}_4$, respectively $\tilde{\mathsf{D}}_4$, in case $n(\varepsilon) = 2$ for all $\varepsilon \in \mathcal{E}$ and $\operatorname{card} \mathcal{E} = 3$, respectively $\operatorname{card} \mathcal{E} = 4$. (See [Ri 84].)

Suppose $\mathcal{A}(\Lambda)$ contains a translation subquiver $\mathcal{T}$ which is a $\mathsf{P}^1k$-tubular family of tubular type $\mathsf{D}_4$ or $\tilde{\mathsf{D}}_4$. Choose a parametrization $\mathcal{T} = \dot{\bigcup}_{\lambda \in \mathsf{P}^1 k} \mathcal{T}(\lambda)$ and set $\mathfrak{M} := \{M \in \operatorname{ind} \Lambda \mid [M] \text{ is at the mouth of } \mathcal{T}(\lambda), \text{ for some } \lambda \in n^{-1}(1)\} \dot{\cup} \{M \in \operatorname{ind} \Lambda \mid [M] \text{ is direct successor of a point at the mouth of } \mathcal{T}(\varepsilon), \text{ for some } \varepsilon \in n^{-1}(2)\}$. If there exists a vector $d \in \prod_{i=0}^4 \mathbb{Z}$ such that $\dim M = d$ for all $M \in \mathfrak{M}$, then $d$ is called the *dimension type of* $\mathcal{T}$.

**Proposition 8.** *Let* $i \in \{1,2,3\}$. *Then* $\mathcal{A}(\Lambda_4^i) = \mathcal{T} \dot{\cup} \mathcal{P}(\Lambda_4^i)$, *where* $\mathcal{T}$ *is a* $\mathsf{P}^1k$*-tubular family of tubular type* $\mathsf{D}_4$ *and of dimension type* $(2,1,1,1,1)$, *and where* $\mathcal{P}(\Lambda_4^i)$ *is of the following form:*

$\mathcal{P}(\Lambda_4^1)$ :

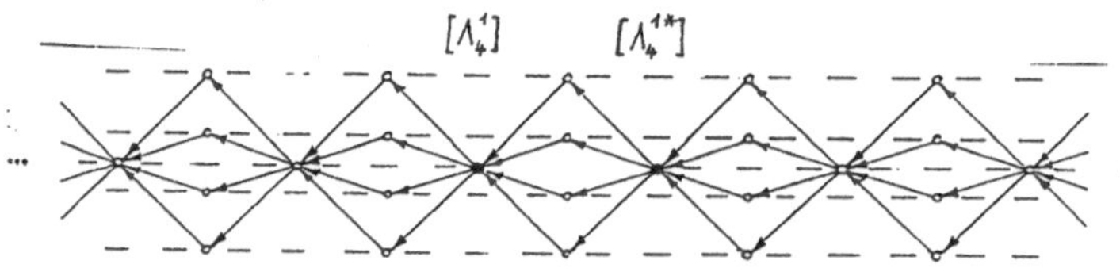


$\mathcal{P}(\Lambda_4^2)$ :

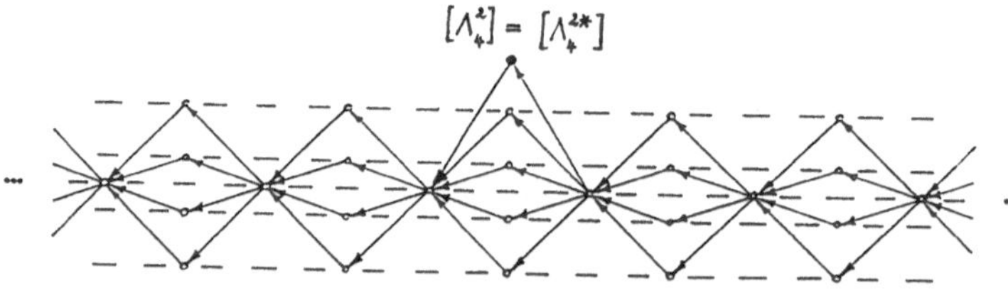


$\mathcal{P}(\Lambda_4^3)$ :

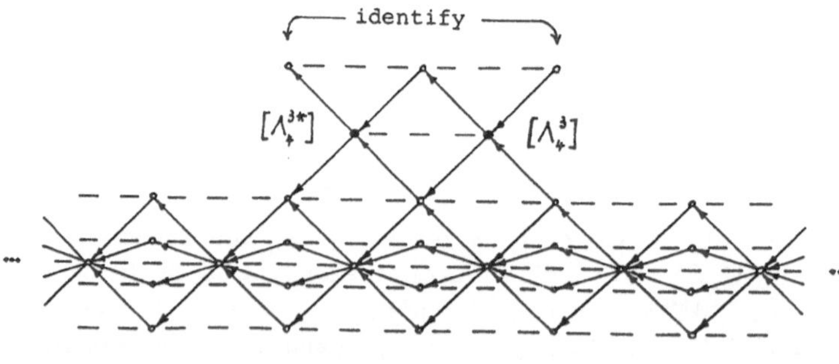

**Proposition 9.** *Let $a \in k\backslash\{0,1\}$. Then $\mathcal{A}_s(\Lambda_4^a) = \dot{\bigcup}_{\beta:\alpha\in \mathbf{P}^1\mathbf{Q}} \mathcal{T}_{\beta:\alpha}$, where $\mathcal{T}_{\beta:\alpha}$ is a $\mathbf{P}^1k$-tubular family of tubular type $\tilde{\mathrm{D}}_4$ and of dimension type $\frac{2|\alpha|+|\beta|}{\gcd(\alpha,\beta)} \cdot (2,1,1,1,1)$, for all $\beta : \alpha \in \mathbf{P}^1\mathbf{Q}$.*

**3.5.** Finally we turn to the classification of $\mathrm{Is}(\mathrm{ind}\,\Lambda)$, for $\Lambda \in \{\Lambda_4^4, \Lambda_4^5, \Lambda_4^6\}$. Again we have to fix some notation first.

Let $Q = (Q_0, Q_1, s, t)$ be any quiver, and let $L := \{\alpha \in Q_1 \mid s(\alpha) = t(\alpha)\}$ be the set of all loops of $Q$. With $Q$ we associate its *alphabet* $A := A^+ \dot{\cup} A^- \dot{\cup} A^*$, built by the set of *direct letters* $A^+ := \{\alpha^+ \mid \alpha \in Q_1 \setminus L\}$, the set of *inverse letters* $A^- := \{\alpha^- \mid \alpha \in Q_1 \setminus L\}$, and the set of *special letters* $A^* := \{\alpha^* \mid \alpha \in L\}$. Define the starting point function $s_A : A \to Q_0$ and the terminal point function $t_A : A \to Q_0$ in the obvious way. We set $I(x) := t_A^{-1}(x)$ for all $x \in Q_0$, and $I := (I(x))_{x\in Q_0}$. Now suppose that each set $I(x)$ carries the structure of a poset which is either a chain or the disjoint union of two chains. Then the pair $(Q, I)$ can be completed to a *clan* $C$ in the sense of [Cr 88], by setting $C := (k, Q, L, q, I)$. Here, after a choice of elements $b, c \in k$ subject to $bc(b-c) \neq 0$, we define $q : L \to k[X]$ to be the constant function assigning to each $\alpha \in L$ the quadratic polynomial $q_\alpha(X) := (X-b)(X-c)$.

Given any pair $(Q, I)$ as above, we set $\mathrm{rep}_k(Q, I) := \mathrm{rep}\, C$, where $\mathrm{rep}\, C$ is the category of representations of the clan $C = (k, Q, L, q, I)$, as defined in [Cr 88].

Let $\Lambda$ be any curve singularity in $C_4^l(k)$. Observe that $\Lambda_4^1$ is the *tetrad*, the unique maximal element in $C_4^l(k)$. Hence there is the restriction functor $\mathrm{latt}\,\Lambda_4^1 \to \mathrm{latt}\,\Lambda$, and in fact this is an embedding. Thus we may view $\mathrm{latt}\,\Lambda_4^1$ as a full subcategory of $\mathrm{latt}\,\Lambda$. Moreover, $\Lambda_4^1$ is a Bäckström order. Therefore it admits the classical epivalence $\mathcal{R} : \mathrm{latt}\,\Lambda_4^1 \to \mathrm{rep}_k^\circ\, Q_4$, where $\mathrm{rep}_k^\circ\, Q_4$ is the full subcategory of $\mathrm{rep}_k\, Q_4$ with object class $\{V \in \mathrm{rep}_k\, Q_4 \mid V \text{ has no simple summand }\}$ (see e.g. [Ri/Ro 79], [Di 88]).

Now the classification of $\mathrm{Is}(\mathrm{ind}\,\Lambda_4^4)$ can be described in the following way. Let $\mathcal{K}_4$ be the full subcategory of $\mathrm{ind}\,\Lambda_4^4$ with object class $\{N \in \mathrm{ind}\,\Lambda_4^1 \mid \mathrm{Hom}_{Q_4}(R_1(1), \mathcal{R}(N)) = 0 \text{ and } \mathrm{Hom}_{Q_4}(\mathcal{R}(N), R_1(1)) = 0\}$, and let $\mathcal{L}_4$ be the full subcategory of $\mathrm{latt}\,\Lambda_4^4$ with object class $\{M \in \mathrm{latt}\,\Lambda_4^4 \mid \Lambda_4^1 M \text{ has no summand in } \mathcal{K}_4\}$.

**Proposition 10.** *(i)* $\mathrm{Is}(\mathrm{ind}\,\Lambda_4^4) = \mathrm{Is}(\mathcal{K}_4) \dot{\cup} \mathrm{Is}(\mathrm{ind}\,\mathcal{L}_4)$.

*(ii) There exists an epivalence $\mathcal{F} : \mathcal{L}_4 \to \mathrm{rep}_k(Q, I)$, where the pair $(Q, I)$ is defined as follows:*

$Q$:

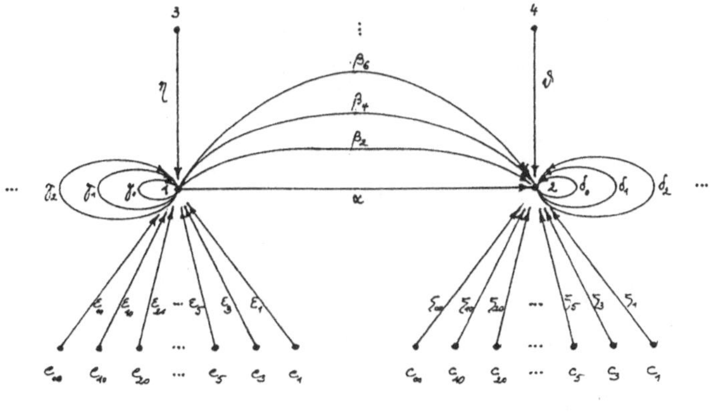

$I(1)$:

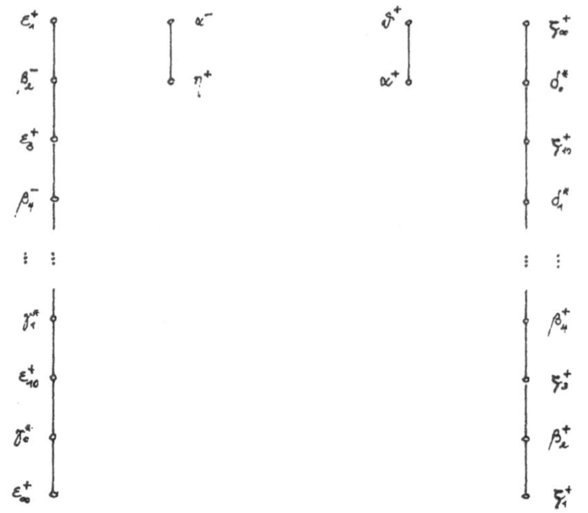

$:I(2)$

In view of Crawley-Boevey's classification of $\mathrm{Is}(\mathrm{ind}\, C)$ [Cr 88] it is clear that Proposition 10 implies the classification of $\mathrm{Is}(\mathrm{ind}\, \Lambda_4^4)$.

The classification of $\mathrm{Is}(\mathrm{ind}\, \Lambda_4^5)$ admits a similar description. Namely, let $Z = R_1(1) \oplus \tau R_1(1)$, let $\mathcal{K}_5$ be the full subcategory of $\mathrm{ind}\, \Lambda_4^5$ with object class $\{N \in \mathrm{ind}\, \Lambda_4^1 \mid \mathrm{Hom}_{Q_4}(Z, \mathcal{R}(N)) = 0$ and $\mathrm{Hom}_{Q_4}(\mathcal{R}(N), Z) = 0\}$, and let $\mathcal{L}_5$ be the full subcategory of $\mathrm{latt}\, \Lambda_4^5$ with object class $\{M \in \mathrm{latt}\, \Lambda_4^5 \mid \Lambda_4^1 M$ has no summand in $\mathcal{K}_5\}$.

**Proposition 11.** *(i)* $\mathrm{Is}(\mathrm{ind}\, \Lambda_4^5) = \mathrm{Is}(\mathcal{K}_5)\, \dot{\cup}\, \mathrm{Is}(\mathrm{ind}\, \mathcal{L}_5)$.
*(ii) There exists an epivalence $\mathcal{F} : \mathcal{L}_5 \to \mathrm{rep}_k(Q, I)$, where the pair $(Q, I)$ is defined as follows:*

$Q$:

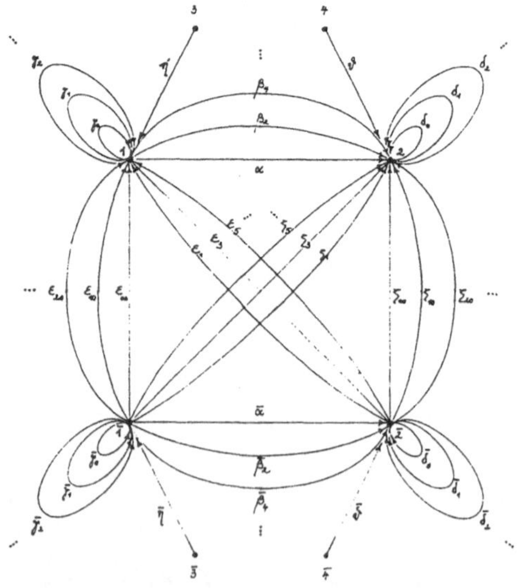

$I(1)$:

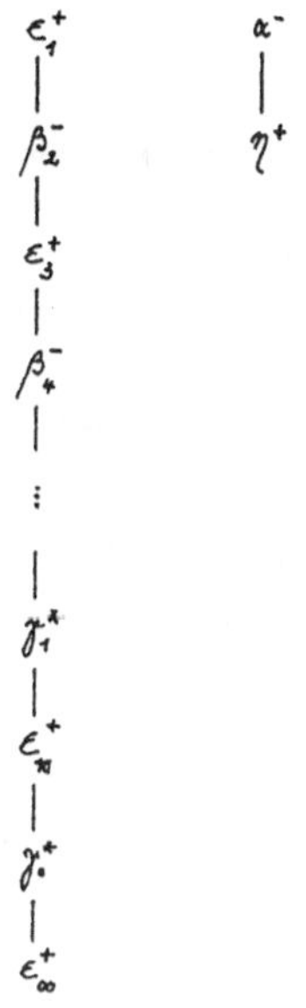

:$I(2)$

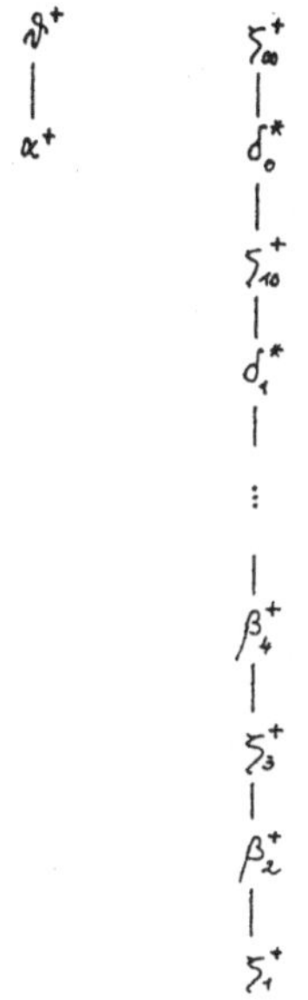

$I(\bar{1})$:

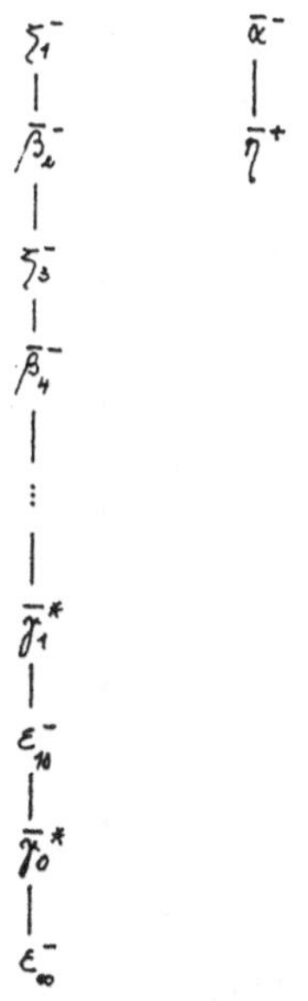

:$I(\bar{2})$

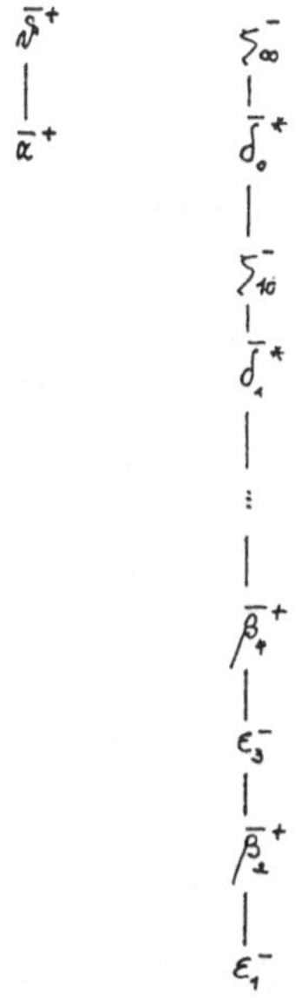

The classification of $\mathrm{Is}(\mathrm{ind}\,\Lambda_4^6)$ should also admit a similar description. However, my analysis of this case is not yet complete.

## 4. Tame curve singularities with $n$ branches, $n < 4$, and large conductor

As with tame curve singularities in $C_4^l(k)$, it is also possible to derive complete information on tame curve singularities in $C_n^l(k)$, $n < 4$, on the basis of Theorem 3. Here we shall only present the analogues to Proposition 5 and Proposition 6 (which will solve Problems (A) and (B) for $C_n^l(k)$, $n < 4$), whereas we shall not describe the classification of Is(ind $\Lambda$), for $\Lambda$ a tame curve singularity in $C_n^l(k)$, $n < 4$.

**Proposition 12.** *Let $\Lambda$ be a curve singularity over $k$ with $n$ branches, $n < 4$, and large conductor.*

*(i) Suppose $n = 3$. Let $\nu(\Lambda)$ be the number of indecomposable projective summands of $\chi_3(\Lambda)$. Then the following holds:*
  *If $\nu(\Lambda) = 0$, then $\Lambda$ is representation-finite.*
  *If $\nu(\Lambda) = 1$, then $\Lambda$ is tame.*
  *If $\nu(\Lambda) \geq 2$, then $\Lambda$ is wild.*

*(ii) Suppose $n = 2$. Then the following holds:*
  *If $\chi_2(\Lambda)$ is not semisimple projective, then $\Lambda$ is representation-finite.*
  *If $\chi_2(\Lambda)$ is semisimple projective, then $\Lambda$ is tame.*

*(iii) Suppose $n = 1$. Then $\Lambda$ is representation-finite.*

Combining Proposition 12 with the remarks of section 3.3, it is easy to derive the full classification of all tame curve singularities in $C_n^l(k)$, with $n < 4$. The respective result is the following.

**Proposition 13.** *A complete system of representatives for the analytical iso-classes of tame curve singularities with $n$ branches, $n < 4$, and large conductor is given by the following list:*

| $\Lambda$ | $Z$ | $U$ |
|---|---|---|
| $\Lambda_3^1$ | $P_0$ | 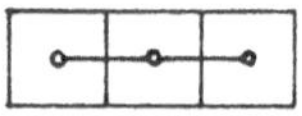 |
| $\Lambda_3^2$ | $P_1 \oplus I_2 \oplus I_3$ | 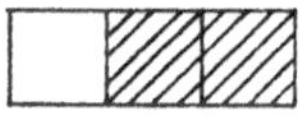 |
| $\Lambda_3^3$ | $P_1 \oplus \tau^- P_1$ | 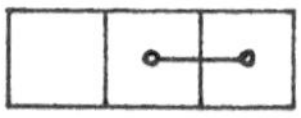 |
| $\Lambda_2^1$ | $P_1 \oplus P_2$ | 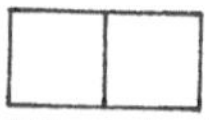 |

(For reading this table, see conventions subsequent to Proposition 6.)

## 5. Complete list of all tame curve singularities with large conductor

Proposition 6 together with Proposition 13 provide a complete classification for all tame curve singularities with large conductor, up to analytical isomorphism. The property which distinguishes them from other curve singularities with large conductor is that they are *tame*. By definition, this is a property of their lattice category. And moreover, by definition, this property implies that Is(ind $\Lambda$) can be classified. Which in turn means that, in some good sense, the lattice category latt $\Lambda$ can be "made visible". Hence the substanticial interest in tame curve singularities is to "see" their lattice category. (In this respect, compare sections 3.4 and 3.5.)

A question of secondary interest is to "see" the tame curve singularities themselves. Indeed, their classification given in Proposition 6 and Proposition 13 is not just of a qualitative nature, but provides an entirely constructive description of all tame curve singularities with large conductor, up to analytical isomorphism, in view of $\Sigma_n: \mathcal{V}_n \to C_n^l(k)$ (see section 3.3). From this description it is straightforward to read off geometric information such as multiplicity, $\delta$-invariant, embedding dimension, or parametrization of the respective singularity. Since this may not be evident for the reader who is not familiar with representation theory, we include below a table which summarizes this information.

Here, we denote by $\Lambda = \Lambda_n^i$ a tame curve singularity with $n$ branches and index $i$, where $i$ runs through an index set which is in bijection to the set of analytical isoclasses of all tame curve singularities in $C_n^l(k)$. We denote by $\delta$ the $\delta$-invariant of $\Lambda$ and by $e$ the embedding dimension of $\Lambda$. (Recall that $\delta = \dim_k(\Omega/\Lambda)$, and $e = \dim_k(\mathcal{M}/\mathcal{M}^2)$.) In the last column we indicate for each $\Lambda = \Lambda_n^i$ a chosen affine-algebraic curve singularity $(C, O)$, subject to $\hat{\mathcal{O}}_{C,O} \xrightarrow{\sim} \Lambda$, by giving a parametrization $\varphi_i: k \to k^e$, $i = 1, \ldots, n$, of its branches.

| $\Lambda$ | $\delta$ | $e$ | $\varphi_i: k \mapsto k^e$ |
|---|---|---|---|
| $\Lambda_2^1$ | 3 | 4 | $\varphi_1: t \mapsto (t^2, t^3, 0, 0)$<br>$\varphi_2: t \mapsto (0, 0, t^2, t^3)$ |
| $\Lambda_3^1$ | 4 | 3 | $\varphi_1: t \mapsto (t, 0, 0)$<br>$\varphi_2: t \mapsto (t, t^2, 0)$<br>$\varphi_3: t \mapsto (t, 0, t^2)$ |
| $\Lambda_3^2$ | 3 | 4 | $\varphi_1: t \mapsto (t, 0, 0, 0)$<br>$\varphi_2: t \mapsto (0, t, 0, 0)$<br>$\varphi_3: t \mapsto (0, 0, t^2, t^3)$ |
| $\Lambda_3^3$ | 4 | 4 | $\varphi_1: t \mapsto (t, 0, 0, 0)$<br>$\varphi_2: t \mapsto (t, t^2, 0, 0)$<br>$\varphi_3: t \mapsto (0, 0, t^2, t^3)$ |

| $\Lambda$ | $\delta$ | $e$ | $\varphi_i: k \to k^e$ |
|---|---|---|---|
| $\Lambda_4^1$ | 3 | 4 | $\varphi_1: t \mapsto (t,0,0,0)$<br>$\varphi_2: t \mapsto (0,t,0,0)$<br>$\varphi_3: t \mapsto (0,0,t,0)$<br>$\varphi_4: t \mapsto (0,0,0,t)$ |
| $\Lambda_4^2$ | 4 | 3 | $\varphi_1: t \mapsto (t,0,0)$<br>$\varphi_2: t \mapsto (0,t,0)$<br>$\varphi_3: t \mapsto (0,0,t)$<br>$\varphi_4: t \mapsto (t,t,t)$ |
| $\Lambda_4^3$ | 4 | 3 | $\varphi_1: t \mapsto (t,0,0)$<br>$\varphi_2: t \mapsto (0,t,0)$<br>$\varphi_3: t \mapsto (0,0,t)$<br>$\varphi_4: t \mapsto (t,t,0)$ |
| $\Lambda_4^a$<br>$a \in k \setminus \{0,1\}$ | 5 | 3 | $\varphi_1: t \mapsto (t,0,0)$<br>$\varphi_2: t \mapsto (0,t,0)$<br>$\varphi_3: t \mapsto (t,t,0)$<br>$\varphi_4: t \mapsto (at,t,t^2)$ |
| $\Lambda_4^4$ | 4 | 4 | $\varphi_1: t \mapsto (t,0,0,0)$<br>$\varphi_2: t \mapsto (0,t,0,0)$<br>$\varphi_3: t \mapsto (0,0,t,0)$<br>$\varphi_4: t \mapsto (0,0,t,t^2)$ |
| $\Lambda_4^5$ | 5 | 4 | $\varphi_1: t \mapsto (t,0,0,0)$<br>$\varphi_2: t \mapsto (t,t^2,0,0)$<br>$\varphi_3: t \mapsto (0,0,t,0)$<br>$\varphi_4: t \mapsto (0,0,t,t^2)$ |
| $\Lambda_4^6$ | 5 | 3 | $\varphi_1: t \mapsto (t,0,0)$<br>$\varphi_2: t \mapsto (0,t,0)$<br>$\varphi_3: t \mapsto (t,t,0)$<br>$\varphi_4: t \mapsto (0,t,t^2)$ |

## 6. Epilogue

In conclusion, let us reflect the presented results in view of the problems (A), (B), (C) posed in 1.5. If we restrict ourselves to the class of all curve singularities with large conductor, then Proposition 5 and Proposition 12 solve problem (A), with the characteristic representation as the desired invariant, and Propostion 6 together with Proposition 13 solve problem (B). Moreover, Theorem 3 is the first important step to a solution of problem (C). As evidence for this claim we quote Propositions 8-11 which are derived on the basis of Theorem 3.

Altogether, problems (A), (B), (C) are solved for the class of all curve singularities with large conductor, and the presented results are consistent with Conjecture 1.5. But still we are very far away from a general solution of these problems.

## REFERENCES

[Ar 74] V.I. Arnol'd: Critical points on smooth functions. Proc. Intern. Congr. Math. Vancouver 1974, Vol.1, 19-39.

[Cr 88] W.W. Crawley-Boevey: Functorial filtrations II: Clans and the Gelfand problem. Preprint, Liverpool 1988.

[Di 85] E. Dieterich: Solution of a nondomestic tame classification problem from integral representation theory of finite groups ($\Lambda = RC_3$, $v(3) = 4$). Memoirs of the Amer. Math. Soc. (to appear).

[Di 88] E. Dieterich: Tame orders. Banach Center Publications, Vol.26, part 1 (to appear).

[Di 90] E. Dieterich: Gitterkategorien über Kurvensingularitäten mit großem Führer. Habilitationsschrift, Zürich 1990.

[Dr/Ro 67] J.A. Drozd, A.V. Roiter: Commutative rings with a finite number of indecomposable integral representations. Izv. Akad. Nauk. SSSR Ser. Mat. 31(1967), 783-798. Math. USSR Izv.1 (1967), 757-772.

[Gr/Kn 85] G.-M. Greuel, H. Knörrer: Einfache Kurvensingularitäten und torsionsfreie Moduln. Math. Ann. 270 (1985), 417-425.

[Ja 67] H. Jacobinski: Sur les ordres commutatifs avec une nombre fini de réseaux indecomposables. Acta Math. 118 (1967), 1-31.

[Ka 87] C. Kahn: Reflexive Moduln auf einfach-elliptischen Flächensingularitäten. Dissertation, Bonn 1987. MPI Preprint 87-25; Bonner Math. Schriften 188 (1988).

[Ka 89] C. Kahn: Reflexive modules on minimally elliptic singularities. Math. Ann. 285 (1989), 141-160.

[Ri 84] C.-M. Ringel: Tame algebras and integral quadratic forms. Lecture Notes in Math., Springer 1099 (1984).

[Ri/Ro 79] C.-M. Ringel, K.W. Roggenkamp: Diagrammatic methods in the representation theory of orders. Journal of Alg. 60 (1979), 11-42.

[Scha 85] A. Schappert: Die Klassifikation aller torsionsfreien Moduln vom Rang 1 über den unimodularen Kurvensingularitäten. Diplomarbeit, Kaiserslautern 1985.

Progress in Mathematics, Vol. 95, © 1991 Birkhäuser Verlag Basel

# Polynomial representations of finite general linear groups in non-describing characteristic

RICHARD DIPPER*

## §1 Introduction

Let $\Gamma$ be a connected reductive group over an algebraically closed field $F$ of positive characteristic and let $G$ be the set of fixed points of $\Gamma$ under some Frobenius map of $\Gamma$. Then $G$ is a finite group of Lie type. The $p$-modular representations of $G$ for $p = \operatorname{char} F$ (the "describing" characteristic case) are closely related to rational representations of $\Gamma$, and thus results from the theory of reductive algebraic groups can be used to develop the representation theory of finite groups of Lie type in the describing characteristic case. One of the outstanding open questions in this area is to determine the dimensions of the irreducible representations and the decomposition matrix whose entries record multiplicities of irreducible modules as composition factors of Weyl modules. One way to define Weyl modules is by reducing modulo $p$ ($p = \operatorname{char} F$) the irreducible modules of the complex algebraic group of the same type. Here reduction modulo $p$ happens in both, in the field underlying the group and in the field over which the representing matrices are defined, simultaneously. One of the main tools to study these questions is derived from the Weyl group of $\Gamma$. If, for example, $\Gamma$ is a general linear group, the decomposition matrix of the Weyl modules can be calculated from decomposition numbers of Schur algebras, which in turn can be calculated in terms of symmetric groups. We shall discuss this phenomenon in more detail below.

From the viewpoint of finite group theory the describing characteristic for a finite group $G$ of Lie type is just one among many primes $p$ dividing the order of $G$. Until the beginning to the eighties little was known in general about representations of $G$ in positive non-describing characteristic. Around 1979, Alperin suggested that the $p$-block theory of the general linear group $GL_n(q)$ for primes $p$ not dividing $q$ should resemble the block theory of symmetric groups. This was demonstrated (and extended to unitary groups) by Fong and Srinivasan in [23] about 1982. Subsequent results ([6], [7], [8], [15], [16], [31], [32], [33], [34]), showed that there are many more analogies and similarities between the describing and the non-describing characteristic case for finite general linear groups.

Indeed there is a reason for the similarities between the describing and the non-describing characteristic case (for general linear groups). Both are special cases of the representation theory of a certain algebra $R_q(G)$, which is a $q$-deformation of the coordinate ring $R(G)$ of the general linear group $G = GL_n(R)$ (where $R$ is a commutative ring with identity) and is what Drinfeld called a quantum group in his

* The author gratefully acknowledges support received from the Deutsche Forschungsgemeinschaft, from NATO under Grant 0222/87 and from NSF under Grant 9002606.
AMS subject classification: 20C20,20C30,20G05,20G40.

JCM talk in Berkeley, 1986 [18]. Associated with $R_q(G)$ are certain algebras, $q$-Schur algebras, whose module category embeds into the category of $R$-representations of $GL_n(q)$, if $q$ is specialized to a prime power. The representations of $GL_n(q)$ which come from $q$-Schur algebras are what we call polynomial representations of $GL_n(q)$.

The aim of this survey article is to explain how representations of quantum general linear groups and finite general linear groups in the non-describing characteristic cases are related.

The starting point is the diagram below giving the classical situation of polynomial (rational) representations of algebraic general linear groups on the left and the $q$-analogue on the right hand side, with the following notation:

$R$ is a commutative ring with 1

$\mathfrak{S}_r$ is the symmetric group on $r$ letters

$\mathcal{H}_{R,q}(\mathfrak{S}_r)$ is the Hecke algebra

$\mathcal{S}(n,r)$ is the classical Schur algebra and $\mathcal{S}_{R,q}(n,r)$ is the $q$-Schur algebra defined in [16].

We shall define quantization $A_{R,q}(n)$ and $R_q(G)$ of the coordinate rings $A_R(n)$ of $n \times n$-matrices and $R(G)$ of $GL_n(R)$ respectively and discuss their relation to quantum hyper-algebra $\mathcal{U}_q$, recently introduced by Lusztig [37], [38], a $q$-deformation of the hyper-algebra $\mathcal{U}$.

1.1

| Classical | Quantized version | non-describing characteristic case |
|---|---|---|
| $\bigoplus_{1\leq r\in\mathbb{Z}} \operatorname{mod} R\mathfrak{S}_r$ | $\bigoplus_{1\leq r\in\mathbb{Z}} \operatorname{mod} \mathcal{H}_{R,q}(\mathfrak{S}_r)$ | |
| $\downarrow$ | $\downarrow$ | |
| $\bigoplus_{1\leq r\in\mathbb{Z}} \operatorname{mod} \mathcal{S}(n,r)$ | $\bigoplus_{1\leq r\in\mathbb{Z}} \operatorname{mod} \mathcal{S}_{R,q}(n,r)$ | $\operatorname{mod} RGL_n(q)$ |
| $\cong$ | $\cong$ | |
| comod $A(n)$ | comod $A_q(n)$ | |
| $\cong$ | | for a domain $R$ of char $p$ and a prime power $q$ such that $(p,q)=1$ |
| polynomial representations of $GL_n$ | | |
| $\cap$ | $\cap$ | |
| rational representations of $GL_n$ | | |
| $\cong$ | | |
| comod $R(G)$ | comod $R_q(G)$ | |
| $\cong$ | $\cong$ | |
| mod $R(G)^*$ | mod $R_q(G)^*$ | |
| $\cap$ | $\cap$ | |
| mod $\mathcal{U}$ | mod $\mathcal{U}_q$ | |

The polynomial representations of general linear groups are precisely the modules of the various Schur algebras and direct sums of those. If $R$ is a field of characteristic 0, the Weyl modules are the irreducible $\mathcal{S}(n,r)$-modules (labelled by weights, which can be identified with partitions of $r$ into at most $n$ parts). If $R$ is a field of positive characteristic, the Weyl modules are not irreducible anymore, but have unique maximal submodules and the resulting irreducible factor modules give a complete set of irreducible non-isomorphic $\mathcal{S}(n,r)$-modules. The matrix recording the composition multiplicities of Weyl modules is unitriangular, for a suitable ordering of the partitions. It appears here as decomposition matrix of Schur algebras in the classical meaning, recording composition multiplicities of modular reductions of irreducible characteristic zero Schur algebra modules. For general linear groups the modular reduction means reducing the field of scalars simultaneously with the field over which the group is defined.

Schur algebras act on tensor space centralizing the action of the symmetric group given by place permutations of tensors. As a consequence one can calculate the decomposition numbers of Schur algebras, hence the composition factors of Weyl modules of general linear groups, entirely in terms of symmetric groups [30].

We shall show that analogous results hold for the corresponding $q$-deformations and how these relate to representations of finite general linear groups in the non-describing characteristic case. Here the resulting decomposition matrices arise from reduction of the scalar alone, the groups remain unchanged.

The quantum group $R_q(G)$ covers both the describing and non-describing case. If we take $q$ to 1 we get the classical situation describing rational representations of $GL_n(R)$. Taking $q$ to be a prime power, and $R$ to be a field of characteristic $p$ not dividing $q$ we have the non-describing characteristic case for $GL_n(q)$.

Thus we have the following general method for the representation theory of finite general linear groups in non-describing characteristic: We start with a known results on rational representations of general linear groups defined over an algebraically closed field $R$. We find a $q$-analogue for $R_q(G)$ and then specialize $q$ to a prime power not divisible by the characteristic of $R$ to get a result for $GL_n(q)$ in non-describing characteristic. We shall demonstrate this technique in the last section, formulating "Steinberg's tensor product theorem" for the non-describing characteristic case.

There is however another application. If $\operatorname{char} R = p$ and if $q$ is a prime power relatively prime to $p$, then $q$ is a root of unity modulo $p$, say a primitive $e$-$th$ root of unity. Instead considering $R_q(G)$ we may investigate $\mathbb{C}_\omega(G)$, where $\omega$ is a primitive $e$-$th$ complex root of unity. It turns out that $\mathbb{C}_\omega(G)$ behaves in many respects similar to $R_q(G)$, in fact it appears that we may consider $\mathbb{C}_\omega(G)$ as a first approximation to $R_q(G)$. In the last section we shall explain why this should be true and why the representation theory of $\mathbb{C}_\omega(G)$ is simpler than that of $R_q(G)$.

This paper evolved from a series of lectures which I gave at the meeting of the DFG-Schwerpunkt "Darstellungstheorie endlicher Guppen und endlich-dimensionaler Algebren" in Bad Honnef, June 1988, and of talks in Manchester in July 1988 and again in Bad Honnef in January 1989. It reports on joint work with Gordon James (representations of Hecke algebras, $q$-Schur algebras and finite general linear groups) and on joint work with Steve Donkin on quantum general linear groups. In the meantime the theory of quantum groups took a stormy development. We shall try to take some of the newer results into account. In particular, in the fall 1988, Manin's book [39] appeared introducing a different quantum $GL_n$ and subsequently Parshall and Wang [40] made a deep investigation into it. We shall explain how these results relate to ours.

Moreover we include outlines of two more recent investigations: First (in section two) of joint work with Peter Fleischmann on modular Harish-Chandra theory [12], which is an important step toward generalization of the main results to other families

of finite groups of Lie type. Secondly, in the last section we shall outline some
results from joint work with Jie Du on Green correspondence for Hecke algebras
and a Steinberg tensor product theorem for general linear groups in non-describing
characteristic, [11].

This paper should help the reader to understand the big picture rather than to
see the technical details. So we frequently omit the details of the proofs and refer to
the original papers.

I would like to thank the Deutsche Forschungsgemeinschaft and G. Michler for
the hospitality during my visits in Essen.

## §2 Decomposition numbers and quotients of Hom-functors

In this section we shall establish the link between $q$-Schur algebras and general
linear groups. We exhibit a commutative triangle of functors

(2.1)

$$
\begin{array}{c}
\operatorname{mod} RGL_n(q) \\
\nearrow \qquad \nwarrow \\
\operatorname{mod} \mathcal{H}_{R,q}(\mathfrak{S}_n) \xrightarrow{\qquad} \operatorname{mod} \mathcal{S}_{R,q}(n,n)
\end{array}
$$

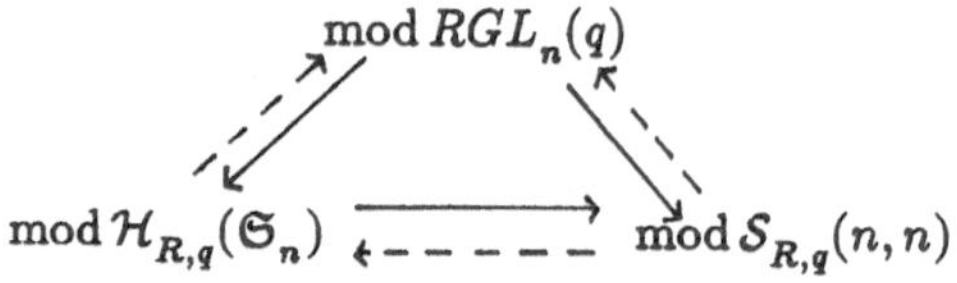

where $R$ is a domain whose characteristic does not divide $q$. The solid line arrows
are functors which we are going to define. It turns out that these have right inverses,
denoted by dotted arrows. The functor from $\operatorname{mod} \mathcal{S}_{R,q}(n,n)$ to $\operatorname{mod} \mathcal{H}_{R,q}(\mathfrak{S}_n)$ is a $q$-
analogue of the Schur functors (c.f. [26]). We begin with some general considerations,
of which the functors in (2.1) are special cases. In general the proofs are omitted
and can be found in [8].

Let $\mathcal{O}$ be a discrete complete valuation ring with quotient field $K$ and residue
field $F$. Let $T_K$ be a semisimple finite dimensional $K$-algebra, and let $T_{\mathcal{O}}$ be an
$\mathcal{O}$-order in $T_K$, that is $T_{\mathcal{O}}$ is a $\mathcal{O}$-subalgebra of finite rank of $T_K$ such that $T_{\mathcal{O}}$ is free
as $\mathcal{O}$-module, and a $\mathcal{O}$-basis of $T_{\mathcal{O}}$ is also a $K$-basis of $T_K$. Throughout $R$ means
"one of the rings $F, \mathcal{O}$ or $K$". We set $T_R = R \otimes_{\mathcal{O}} T_{\mathcal{O}}$.

Let $M_{\mathcal{O}}$ be a right $T_{\mathcal{O}}$-lattice, so $M_{\mathcal{O}}$ is free of finite rank as an $\mathcal{O}$-module. We
set $M_R = R \otimes_{\mathcal{O}} M_{\mathcal{O}}$, and $\mathcal{H}_R = \operatorname{End}_{T_R}(M_R)$ acting on $M_R$ from the left. Note that
$\mathcal{H}_K \cong K \otimes_{\mathcal{O}} \mathcal{H}_{\mathcal{O}}$ and $\mathcal{H}_F \supseteq F \otimes_{\mathcal{O}} \mathcal{H}_{\mathcal{O}}$ canonically, but in general we do not have
equality in the second case.

Let $\{M_{\mathcal{O}}^{(1)}, \dots, M_{\mathcal{O}}^{(t)}\}$ be a complete set of non-isomorphic indecomposable direct summands of $M_{\mathcal{O}}$, and let $\{S^{(1)}, \dots, S^{(k)}\}$ be a complete set of irreducible
$T_K$-modules occurring as constituents in $M_K$. Note that the assumptions on $\mathcal{O}$ imply that the Krull-Schmidt theorem holds for all algebras involved, [5, 6.17]. The
matrix $D_M = (d_{ij})$, where $1 \le i \le t$, $1 \le j \le k$ and $d_{ij}$ is the multiplicity of $S_j$
as composition factor of $K \otimes_{\mathcal{O}} M_{\mathcal{O}}^{(i)}$ is the **decomposition matrix** of $M_{\mathcal{O}}$. The
decomposition matrix $D_{T_{\mathcal{O}}}$ respectively $D_{\mathcal{H}_{\mathcal{O}}}$ of the regular module $T_{\mathcal{O}}$ respectively
$\mathcal{H}_{\mathcal{O}}$ is the classical decomposition matrix of the ring $T_{\mathcal{O}}$ respectively $\mathcal{H}_{\mathcal{O}}$.

Fitting's lemma implies immediately:

**2.2 Lemma**  *There is a bijection between the indecomposable direct summands*
*of $M_{\mathcal{O}}$ and the projective indecomposable $\mathcal{H}_{\mathcal{O}}$-modules and similarly between the*

*irreducible constituents of $M_K$ and the irreducible representations of $\mathcal{H}_K$. Ordering rows and columns of $D_M$ and $D_{\mathcal{H}_\mathcal{O}}$ accordinly we have $D_M = D_{\mathcal{H}_\mathcal{O}}$.*

We want to relate $D_M$ (hence $D_{\mathcal{H}_\mathcal{O}}$) with $D_{T_\mathcal{O}}$. So we consider a projective cover of $M_\mathcal{O}$, that is a projective $T_\mathcal{O}$-lattice $P_\mathcal{O}$ together with a $T_\mathcal{O}$-epimorphism $\beta : P_\mathcal{O} \to M_\mathcal{O}$.

We do not assume that $P_\mathcal{O}$ is a minimal projective cover, that is, that $\ker \beta_\mathcal{O}$ has no projective direct summand.

**2.3 Hypothesis** *Let $\beta_K = 1 \otimes \beta : P_K \to M_K$. Suppose that $M_K$ and $\ker(\beta_K)$ have no irreducible composition factor in common.*

An immediate consequence of (2.3) is the following observation:

**2.4 Lemma** *Assume (2.3). Let $\beta_R = 1 \otimes \beta : P_R \to \mathcal{H}_R$. Then the kernel $\ker \beta_R$ is invariant under the endomorphism ring $\mathrm{End}_{T_R}(P_R)$ of $P_R$ for every choice of $R$ as $F, \mathcal{O}$ or $K$.*

We denote the endomorphism ring of $P_R$ by $\mathcal{E}_R$. Since $P_\mathcal{O}$ is projective, $\mathcal{E}_R \cong R \otimes_\mathcal{O} \mathcal{E}_\mathcal{O}$ canonically not only for $R = K$, but for $R = F$ as well. We consider $P_\mathcal{O}$ as $\mathcal{E}_\mathcal{O} - T_\mathcal{O}$-bimodule. So (2.4) says that $\ker \beta_R$ is a subbimodule of $P_R$, provided hypothesis (2.3) holds. If so, let $J_R$ be the set of endomorphisms $\varphi$ of $P_R$ such that $\varphi(P_R) \subseteq \ker \beta_R$. So $J_R$ is the right annihilator of $\beta_R$ in $\mathcal{E}_R$, considering $\mathrm{Hom}_{T_R}(P_R, M_R)$ as right $\mathcal{E}_R$-module. In particular $J_R$ is an ideal in $\mathcal{E}_R$.

**2.5 Lemma** *Assume hypothesis (2.3). Then $\mathcal{E}_R/J_R \cong \mathcal{H}_R$ canonically.*

Note that $J_R = \mathrm{Hom}_{T_R}(P_R, \ker \beta_R)$, so $J_R \cong R \otimes_\mathcal{O} J_\mathcal{O}$, since $P_\mathcal{O}$ is projective. A dimension argument yields now immediately that $F \otimes_\mathcal{O} \mathcal{H}_\mathcal{O}$ is indeed the endomorphism ring $\mathcal{H}_F$ of $M_F$, that is every endomorphism of $M_F$ is liftable to an endomorphism of $M_\mathcal{O}$. From this we see that the isomorphism in (2.5) is compatible with change of rings, that is if $\sigma_R : \mathcal{E}_R/J_R \to \mathcal{H}_R$ is the canonical isomorphism of (2.5), then $\sigma_R = 1 \otimes_\mathcal{O} \sigma_\mathcal{O}$.

Now the following results can be seen easily:

**2.6 Consequences** *Suppose hypothesis (2.3) holds.*

*i)* $\mathcal{H}_F \cong F \otimes_\mathcal{O} \mathcal{H}_\mathcal{O}$.

*ii)* *$\beta_R$ induces a bijection between the indecomposable direct summands of $P_R$ not contained in $\ker \beta_R$ and the indecomposable direct summands of $M_R$. So the minimal projective cover of an indecomposable direct summand $U$ of $M_R$ is indecomposable, and therefore $U$ has a simple head, that is $U$ has a unique* *maximal submodule.*

*iii) The decomposition matrix $D_{\mathcal{H}}$ of $\mathcal{H}_{\mathcal{O}}$ is a submatrix of $D_{T_{\mathcal{O}}}$, (rows and columns suitably ordered).*

We interrupt the general theory here to give first application of the results so far. So let $G$ be a finite group of Lie type. Let $B \leq G$ be a Borel subgroup with split torus $T$ and unipotent radical $U$. So $B$ is the semidirect product of $U$ by $T$. Note that the order of $U$ is a power of $q$, so it is invertible in $\mathcal{O}$ and $F$. We assume that $F$ and $K$ are splitting fields for all subgroups of $G$.

We denote the trivial $RB$-module by $I_R$. Since $B$ is the semidirect product of the normal $p$-regular subgroup $U$ and the torus $T$, the regular representation $P_{\mathcal{O}}$ of $T$ considered as $B$-module with trivial $U$-action is a projective cover of $I_{\mathcal{O}}$. The corresponding idempotent in $\mathcal{O}B$ is $\widehat{U} = |U|^{-1}\sum_{u\in U} u$, and the kernel of the canonical projection $\beta_{\mathcal{O}}$ of $P_{\mathcal{O}}$ onto $I_{\mathcal{O}}$ is the augmentation ideal of $T$ lifted to $B$ (e.g. by multiplying it by $\widehat{U}$). Since $T$ is abelian, the $KT$-module $P_K$ decomposes into the direct sum of one dimensional $KT$-modules each occurring with multiplicity one, and we may lift this decomposition to the group $B$ by letting $U$ act trivially. Let $\mathcal{R}$ be this set, then

$$P_K = I_K \oplus \sum_{\substack{\rho \neq I_K \\ \rho\in\mathcal{R}}} \rho .$$

For $\rho, \tau \in \mathcal{R}$ we have by the Mackey decomposition and the Bruhat decomposition of $G$:

$$\text{(2.7)} \qquad \begin{aligned} \operatorname{Hom}_{KG}(\rho^G, \tau^G) &= \bigoplus_{w\in W} \operatorname{Hom}_{K({}^{\dot w}B\cap B)}(\rho_{\,{}^{\dot w}B\cap B}, \tau_{\,{}^{\dot w}B\cap B}) \\ &= \begin{cases} K & \text{if } \rho = \tau \\ (0) & \text{otherwise}, \end{cases} \end{aligned}$$

since ${}^{\dot w}B \cap B \supseteq T$ for all $w \in W$. Here $W$ is the Weyl group of $G$, so $W = N_G(T)/T$, and for $w \in W$ we choose a representative $\dot w \in N_G(T)$. In particular for $\tau \neq I_K$ we have $\operatorname{Hom}_{KG}(I_{K_1}^G, \tau^G)$ is zero, since $T \leq_{\dot w} B \cap B$.

If $\beta_{\mathcal{O}}$ is an epimorphism from $P_{\mathcal{O}}$ onto $\mathcal{O}_B$, we may induce the exact sequence

$$0 \to \ker\beta_R \to P_R \to I_R \to 0$$

to $G$, to get

$$\text{(2.8)} \qquad 0 \to \ker\beta_R^G \to P_R^G \to I_R^G \to 0 .$$

Obviously $P_R^G$ is projective, so (2.7) implies that $\beta_R^G : P_R^G \to I_R^G$ satisfies hypothesis (2.3). The endomorphism ring of $I_R^G$ is the Hecke algebra $\mathcal{H}_{R,q}(W)$, where $GF(q)$ is the field of definition for $G$. Generators and relations for $\mathcal{H}_{R,q}(W)$ have been given by Iwahori [28]. We shall give them below. By (2.6) we now have the following result

**2.9 Theorem** *Let $G$ be a finite group of Lie type defined over $GF(q)$. Let $p$ be a prime not dividing $q$ and let $\mathcal{H}_{R,q}(W)$ be the Hecke algebra associated with the Weyl group $W$ of $G$. Then the decomposition matrix of $\mathcal{H}_{\mathcal{O},q}(W)$ is a submatrix of the $p$-decomposition matrix of $G$.*

Here are the generators and relations for Hecke algebras associated with Weyl groups (or more general associated with Coxeter groups): As an $R$-module $\mathcal{H}_{R,q}(W)$ is free with basis $\{T_w \mid w \in W\}$. The multiplication is given by the rule

$$T_vT_w = \begin{cases} T_{vs} & \text{if } \ell(vw) > \ell(w) \\ qT_{vw} + (q-1)T_w & \text{if } \ell(vw) < \ell(w) \end{cases}$$

for a basic reflection $v \in W$ and arbitrary $w \in W$, while $\ell(w)$ is the usual length of $w$.

There is a remarkable extension of result (2.9). A representation of $G$ is called cuspidal if its restriction to the unipotent radical of every proper parabolic subgroup does not contain a trivial subrepresentation. Let $P = L \cdot U$ be a parabolic subgroup of $G$ with Levi complement $L$ and unipotent radical $U$. Then $L$ is again a finite group of Lie type. Let $C_R$ be a cuspidal representation of $L$ such that $C_K$ is irreducible. We take the $G$-module $M_R$ to be $C_R \otimes_{RP} RG$, where we consider $C_R$ as $P$-module letting $U$ act trivially on $C_R$, (the Harish-Chandra induced module). Howlett and Lehrer showed in [27] that the endomorphism ring $\mathrm{End}_{KG}(M_K)$ is always a Hecke algebra $\mathcal{H}_K(M_K)$ associated with a certain group $W_M$, which contains a large normal reflection group with abelian factor. More precisely $W_{M_K}$ is the inertial group of $M_K$ in the stabilizer of $L$ in the Weyl group $W$ of $G$, so $W_{M_K} = \{w \in \mathrm{Stab}_W(L) \mid {}^w M_K \cong M_K\}$, where $W$ acts on the Levi subgroups of $G$ by conjugation. $\mathcal{H}_K(M_K)$ has a certain $K$-basis $\{T_w \mid w \in W_{M_K}\}$ and one might ask if the $\mathcal{O}$-order generated by this basis is the endomorphism ring $\mathcal{H}_{\mathcal{O}}(M_{\mathcal{O}})$ of $M_{\mathcal{O}}$ for a suitable choice of a lattice $C_{\mathcal{O}}$ in $C_K$, and if then $\mathrm{End}_{FG}(M_F) \cong F \otimes_{\mathcal{O}} \mathcal{H}_{\mathcal{O}}(M_{\mathcal{O}})$. This is a hard problem in general, one does not even know if $\mathrm{End}_{FL}(C_F) = F$ always, certainly a necessary condition for the statement above. But there is a certain class of cuspidal irreducible $KL$-modules having lattices with this property. They are the so called regular cuspidal representations, defined by the property that their character is constituent of the Gelfand-Graev character. This is a character induced from a linear character in general position of the unipotent radical of a Borel subgroup of $L$. In particular inducing a lattice affording this linear character to $L$ gives a projective $\mathcal{O}L$-lattice $V_{\mathcal{O}}$. It is known that $V_K$ is multiplicity free, since its endomorphism ring is commutative, (see e.g. [4]). Let $X_{\mathcal{O}}$ be the indecomposable direct summand of $V_{\mathcal{O}}$ such that $C_K$ is constituent of $X_K$. Let $\beta_K : X_K \to C_K$ be the canonical projection. Let $C_{\mathcal{O}}$ be the image of the restriction $\beta_{\mathcal{O}}$ of $\beta_K$ to $X_{\mathcal{O}}$. Then $C_{\mathcal{O}}$is a $\mathcal{O}L$-lattice in $C_K$ and $\beta_{\mathcal{O}} : X_{\mathcal{O}} \to C_{\mathcal{O}}$ is the minimal projective cover of $C_{\mathcal{O}}$. This lifts to the parabolic group $P = L \cdot U$ and we may apply Harish-Chandra induction to get a projective cover of $M_{\mathcal{O}}$. The composition factors of $X_K$ come with multiplicity 1. The corresponding column in the $p$-decomposition matrix of $L$ has therefore only ones and zeros. Consequently the simple head of $X_F$ can occur only with multiplicity one in reductions modulo $p$ of the irreducible constituents of $X_K$, in particular in $C_F$. But this is also the head of $C_F$ and from this it follows easily that every endomorphism of $C_F$ is an automorphism. Since $F$ is a splitting field for $L$ we see that $\mathrm{End}_{FL}^{(}C_F) = \mathrm{End}_{FP}(G) = F$. In [12] it is shown that indeed the endomorphism ring $\mathcal{H}_{\mathcal{O}}(M_{\mathcal{O}})$ of $M_{\mathcal{O}}$ has $\mathcal{O}$-basis $\{T_w \mid w \in W_{M_K}\}$ and hypothesis (2.3) is satisfied, provided $W_{M_F} = \{w \in \mathrm{Stab}_W(L) \mid {}^w M_F \cong M_F\} = W_{M_K}$, and a sufficient condition for this to happen is given in terms of the associated Deligne-Lusztig characters. So the theory applies in this case as well.

If $G = GL_n(q)$, all cuspidal characters are regular. In this case $C_F$ is always irreducible and $\mathrm{End}_{FG}(M_F)$ is always a Hecke algebra associated with $W_{M_F}$. A proof of this fact and consequences for the representations of $GL_n(q)$ can be found in [6], [7], [15], and [33].

We return now to our general setting where $\mathcal{H}_{\mathcal{O}}$ is the endomorphism ring of the $T_{\mathcal{O}}$-lattice $M_{\mathcal{O}}$. Throughout we assume hypothesis (2.3). In (2.6) we have seen that $\mathcal{H}_R$ contains a lot of structural information about representations of $T_R$. In fact, there is a functor carrying this information. Recall that $\mathcal{E}_R = \mathrm{End}_{T_R}(P_R)$, and $J_R = \{\varphi \in \mathcal{E} \mid \ \mathrm{im}\ \varphi \subseteq \ker \beta_R\}$. We define the functor $H_R : \mathrm{mod}\, T_R \to \mathrm{mod}\, \mathcal{H}_R$ setting

$$(2.10) \qquad H_R(V) = \mathrm{Hom}_{T_R}(T, V)/\mathrm{Hom}_{T_R}(P_R, V)J_R$$

for $V \in \mathrm{mod}\, T_R$.

Note that $H_{T_R}(P_R, V)$ is a right $\mathcal{E}_R$-module, since $P_R$ is a left $\mathcal{E}_R$-module. So $H_R$ is the Hom-functor $\mathrm{Hom}_{T_R}(P_R, -) : \mathrm{mod}\, T_R \to \mathrm{mod}\, \mathcal{E}_R$, followed by factoring out the ideal $J_R$.

**2.11 Theorem** *The functor $\widehat{H}_R = - \otimes_{\mathcal{H}_R} M_R$ is a right inverse of $H_R$. Moreover, if $R$ is a field, $\widehat{H}_R$ maps irreducible $\mathcal{H}_R$-modules to irreducible $T_R$-modules, and $H_R(V)$ is the zero module or irreducible, if $V \in \mathrm{mod}\, T_R$ is irreducible. So we have a bijection between the irreducible $\mathcal{H}_R$-modules and the irreducible $T_R$-modules which are not mapped to the zero module under $H_R$.*

Note that the irreducible $T_R$-modules ($R$ a field) which are not mapped to (0) under $H_R$ are precisely the irreducible $T_R$-modules in the head of $M_R$.

We need now an additional condition:

**2.12 Hypothesis** *Assume (2.3) and in addition suppose that $Hom_{T_R}(P_R, U) \neq (0)$ for every submodule $U$ of $M_R$.*

**2.13 Theorem** *Assume (2.12) and let $X = \sum_i^{\oplus} X_i$ be a direct sum of right ideals $X_i$ of $\mathcal{H}_R$. Let $N_R = \sum_i^{\oplus} X_i M_R$ and $\widetilde{N}_R = \sum_i^{\oplus} X_i \otimes_{\mathcal{H}_R} M_R$. Then $H_R(N) = H_R(\widetilde{N}) \cong X$ and $H_R$ induces an $R$-algebra isomorphism $H_R : \mathrm{End}_{T_R} N \to \mathrm{End}_{\mathcal{H}_R} X$.*

We need a further refinement of (2.13). It can happen that even if $X_i$ is a pure right ideal of $\mathcal{H}_{\mathcal{O}}$, that is $\mathcal{H}_{\mathcal{O}}/X_i$ has no $\mathcal{O}$-torsion, $X_i M_{\mathcal{O}}$ is not pure in $M_{\mathcal{O}}$. We denote the purification of $X_i M_{\mathcal{O}}$ in $M\mathcal{O}$, that is the unique minimal pure sublattice of $M_{\mathcal{O}}$ containing $X_i M_{\mathcal{O}}$, by $\sqrt{X_i M_{\mathcal{O}}}$.

**2.14 Theorem** *Assume (2.12) and let $X_{\mathcal{O}} = \sum_i^{\oplus} X_i$ be a direct sum of pure right ideals $X_i$ of $\mathcal{H}_{\mathcal{O}}$. Let $N_R = \sum_i^{\oplus} R \otimes_{\mathcal{O}} \sqrt{X_i M_{\mathcal{O}}}$. Then $H_R(N_R) = X_R = R \otimes_{\mathcal{O}} X_{\mathcal{O}}$ and $H_R$ induces an isomorphism $H_R : \mathrm{End}_{T_R}(N_R) \to \mathrm{End}_{\mathcal{H}_R}(X_R)$. Moreover $H_R = 1_R \otimes H_{\mathcal{O}}$.*

We turn now again to applications. As we have seen (2.9) for a finite group $G$ of Lie type we have a pair of functors $H_R$ and $\widehat{H}_R$:

$$\operatorname{mod} RG \underset{\widehat{H}_R}{\overset{H_R}{\rightleftarrows}} \operatorname{mod} \mathcal{H}_{R,q}(W)$$

giving for $G = GL_n(q)$ one side of the triangle (2.1). In fact from the discussion subsequent to (2.9) we see that we have many such functors arising from irreducible regular cuspidal modules for various Levi subgroups satisfying certain conditions.

As a consequence we can produce many irreducible $FG$-modules using these functors by (2.11) and the calculation of a part of the decomposition matrix of $G$ is reduced to calculating decomposition matrices of Hecke algebras. It is a well known theorem ([3] and [36]) that $\mathcal{H}_{K,q}(W)$, where $W$ is a Weyl group, is isomorphic to the group algebra $KW$.

The representation theory of $\mathcal{H}_{R,q}(W)$ for Weyl groups $W$ of type $A_{n-1}$ (that is for $W = \mathfrak{S}_n$), has been investigated in [13] and [14]. Many classical results for symmetric groups turned out to have extensions to $\mathcal{H}_{R,q}(W)$. In general, if $R$ is a field and $1 \neq q \in R$ is a primitive *e-th* root of unity, the representations of $\mathcal{H}_{R,q}(W)$ behave like "e-modular" representations of symmetric groups. So we have Specht modules $S^\lambda$ parametrized by partitions of $n$, and for $e$-regular partitions, these have unique maximal submodules and their irreducible top factors $D^\lambda$ provide a complete set of non-isomorphic irreducible $\mathcal{H}_{R,q}(\mathfrak{S}_n)$-modules. In particular $\mathcal{H}_{R,q}(\mathfrak{S}_n)$ is semisimple if and only if $e > n$ or $q$ is not a root of unity. Moreover the Nakayama conjecture on $p$-blocks of symmetric groups generalizes as well: The blocks of $\mathcal{H}_{R,q}(\mathfrak{S}_n)$ are given by $e$-cores of partitions of $n$.

If $q = 1$ we set $e = \operatorname{char} R$. So $e$ is the smallest positive integer such that $1 + q + q^2 + \cdots + q^{e-1} = 0$ in $R$ and $e = \infty$, if no such integer exists.

We are going to apply (2.14) now. A composition $\lambda$ of $n$ into $k$ parts is a sequence of $k$ non-negative integers whose sum is $n$. We denote the set of compositions of $n$ into $k$ parts by $\Lambda(k,n)$. If $\lambda \in \Lambda(k,n)$ for some $k$ we write often $\lambda \models n$. Let $\lambda \models n$. The standard Young subgroup (or standard parabolic subgroup) $W_\lambda$ of $\mathfrak{S}_n$ is the group $\mathfrak{S}_{\{1,\dots,\lambda_1\}} \times \mathfrak{S}_{\{\lambda_1+1,\dots,\lambda_1+\lambda_2\}} \times \cdots$ canonically embedded into $\mathfrak{S}_n$, where $\lambda = (\lambda_1, \lambda_2, \dots)$. It is obvious from the definition of the multiplication in $\mathcal{H} = \mathcal{H}_{R,q}(\mathfrak{S}_n)$ that $\{T_w \mid w \in W_\lambda\}$ spans a subalgebra $\mathcal{H}_\lambda$ of $\mathcal{H}$.

We define $x_\lambda = \sum_{w \in \mathcal{H}_\lambda} T_w$ and $y_\lambda = \sum_{w \in W_\lambda} (-q)^{-\ell(w)} T_w$. So $x_\lambda, y_\lambda \in \mathcal{H}_\lambda$ and it can be seen easily that $T_w x_\lambda = q^{\ell(w)} x_\lambda$ and $y_\lambda T_w = (-1)^{\ell(w)} y_\lambda$ for $w \in W_\lambda$, so $Rx_\lambda$ and $Ry_\lambda$ are one dimensional right ideals of $\mathcal{H}_\lambda$. For $q = 1$ we get the trivial respectively the alternating representation of $W_\lambda$, and thus we call the representation afforded by $Rx_\lambda$ the trivial and by $Ry_\lambda$ the alternating representation of $\mathcal{H}_\lambda$. So $M^\lambda = M^\lambda_{R,q} = x_\lambda \mathcal{H}_{R,q}(\mathfrak{S}_n)$ is a $q$-deformation of the permutation representation of $\mathfrak{S}_n$ on the cosets of $W_\lambda$ and similarly $\widetilde{M}^\lambda = \widetilde{M}^\lambda_{R,q} = y_\lambda \mathcal{H}$ of the sign representation of $\mathfrak{S}_n$ on the cosets of $W_\lambda$. Each coset $W_\lambda w$ $(w \in \mathfrak{S}_n)$ contains a unique element $d$ of minimal length and then $\ell(ud) = \ell(u) + \ell(d)$ for all $u \in W_\lambda$. Such a coset representative is called distinguished, and the set of distinguished coset representative of $W_\lambda$ in $\mathfrak{S}_n$ is denoted by $\mathcal{D}_\lambda$ (compare e.g. [12]). It is easy to see that $\mathcal{H}$ is free as $\mathcal{H}_\lambda$-module with basis $\{T_d \mid d \in \mathcal{D}_\lambda\}$. In particular $M^\lambda$ is free as $R$-module with basis $\{x_\lambda T_d \mid d \in \mathcal{D}_\lambda\}$. Note that $x_\lambda T_d = \sum_{w \in W_\lambda d} T_w$, and therefore $M^\lambda_{\mathcal{O},q}$ is a pure right ideal of $\mathcal{H}_{\mathcal{O},q}(\mathfrak{S}_n)$. Similarly $\widetilde{M}^\lambda_{\mathcal{O},q}$ is pure in $\mathcal{H}_{\mathcal{O},q}(\mathfrak{S}_n)$.

It is common in the theory of polynomial representation of general linear groups to use the letter $n$ to denote the degree of the group $G = GL_n(q)$, and to use $r$ for the rank of the involved symmetric group. We shall follow this tradition. However, we shall see that the case $n = r$ is relevant for the non-describing characteristic case.

**2.15 Definition** *Let* $0 \leq n, r \in \mathbb{Z}$. *The* $q$**-Schur algebra** $\mathcal{S}_{R,q}(n,r)$ *is the endomorphism ring* $\operatorname{End}_{\mathcal{H}}(\sum^{\oplus}_{\lambda \in \Lambda(n,r)} M^\lambda_{R,q})$, *where* $\mathcal{H} = \mathcal{H}_{R,q}(\mathfrak{S}_r)$.

Obviously Definition (2.15) works for arbitrary commutative rings $R$, where $q \in R$ is a unit.

It is well known that the classical Schur algebra $\mathcal{S}(n,r)$ is the endomorphism ring of the corresponding permutation module for $\mathfrak{S}_r$. So $\mathcal{S}_{R,q}(n,r)$ is indeed a $q$-analogue of $\mathcal{S}(n,r)$. We shall demonstrate this fact in more detail in the next section using $q$-tensor space.

Note that for $r = n$ we may choose $\lambda_0 = (1^n)$, so $x_{\lambda_0} = 1 \in \mathcal{H}$ and $M^{\lambda_0} = \mathcal{H}$. The projection $\zeta$ of $\sum^{\oplus}_{\lambda \in \Lambda(n,n)} M^\lambda$ onto $M^{\lambda_0}$ is an idempotent of $\mathcal{S}_{R,q}(n,n)$. Fitting's lemma implies that $\zeta \mathcal{S}_{R,q}(n,n)\zeta \cong \operatorname{End}_{\mathcal{S}}(M^{\lambda_0}) \cong \mathcal{H}_{R,q}(\mathfrak{S}_n)$. Moreover the functor $V \mapsto V\zeta$ from $\bmod \mathcal{S}_{R,q}(n,n)$ to $\bmod \mathcal{H}_{R,q}(\mathfrak{S}_n)$ is a special case of the functor of (2.10) with $M_R = P_R = \zeta\mathcal{S}_{R,q}(n,n)$. It has an inverse $- \otimes_{\mathcal{H}} \zeta\mathcal{S}_{R,q}(n,n)$ (identifying $\mathcal{H}$ and $\zeta\mathcal{S}_{R,q}(n,n)\zeta$ and $\operatorname{End}_{\mathcal{S}_{R,q}(n,n)}(\zeta\mathcal{S}_{R,q}(n,n))$ canonically). For $q = 1$ we get the classical Schur functor (compare e.g. [26, §6.3]) and its right inverse. Consequently we call the functor $V \to V\zeta$ from $\bmod \mathcal{S}$ to $\bmod \mathcal{H}$ $q$**-Schur functor**.

This establishes the horizontal functors in the triangle (2.1).

Applying (2.14) we shall construct now the third side of (2.1). For details and proofs the reader is referred to [16].

Again let $G$ be the general linear group $GL_n(q)$ and let $(F, \mathcal{O}, K)$ be a split $p$-modular system for $G$ where $p$ does not divide $q$. Our module $M_R$ is again $I_R^G$, where $I_R$ is the trivial $RB$-module for a Borel subgroup $B$ of $G$. So the endomorphism ring of $I_R^G$ is $\mathcal{H}_{R,q}(\mathfrak{S}_n) = \mathcal{H}_R$. In (2.8) we saw that for a certain projective cover $\beta_{\mathcal{O}} = P_{\mathcal{O}} \to I_{\mathcal{O}}$ the induced diagram: $\beta_{\mathcal{O}}^G : P_{\mathcal{O}}^G \to I_{\mathcal{O}}^G$ satisfies (2.3).

In fact $P_{\mathcal{O}}^G = \widehat{U}\mathcal{O}G$. Now, as a right ideal $I_R^G$ is $\overline{B}RG$, where $\overline{B} = (\sum_{t \in T} t)\widehat{U}$, and it is free as $R$-module with basis $\{\overline{B}\pi u \mid \pi \in W, u \in U_\pi^-\}$ by the Bruhat decomposition. Here $W$ is the Weyl group of $G$ (identified with the set of permutation matrices), and $U_\pi^- = \{u \in U \mid \pi u \pi^{-1} \in w_0 U w_0^{-1}\}$, where $w_0$ is the longest element of $W$. Note that for $\pi = 1$ we have $U_\pi^- = (1)$ and that the subspace $Z$ spanned by $\{\overline{B}\pi u \mid 1 \neq \pi \in W, u \in U_\pi^-\}$ is $U$-invariant. Now it can be seen easily that every non-zero submodule of $I_R^G$ contains an element of the form $x = \lambda\overline{B} + z$ for some $z \in Z$, and $0 \neq \lambda \in R$. Thus $x\hat{U} = \lambda\overline{B}\hat{U} + z\hat{U} = \lambda\overline{B} + \tilde{z}$ for $q^k\lambda\overline{B} + \tilde{z}$ some integer $k$ and $\tilde{z} = z\hat{U} \in Z$, so in particular $x\hat{U} \neq 0$. For an $RG$-module $V$ we have $\operatorname{Hom}_{RG}(P_R^G, V) \cong V\hat{U}$ (as $\mathcal{E}_R$-modules, $\mathcal{E}_R = \operatorname{End}_{RG}(P_R^G) \cong \hat{U}RG\hat{U}$ ) so we have:

**2.16 Lemma**  *Let $V$ be a submodule of $I_R^G$. Then $\operatorname{Hom}_{RG}(P_R^G, V) \neq (0)$.*

In particular $\beta_R^G : P_R^G \to I_R^G$ satisfies hypothesis (2.12). So we may apply (2.14).

We consider now the $\mathcal{H}_R$-$RG$-bimodule $I_R^G$. Obviously, for $\lambda \in \Lambda(n,n)$ we have $x_\lambda\mathcal{H}_R I_R^G = x_\lambda I_R^G$, and $y_\lambda\mathcal{H}_R I_R^G = y_\lambda I_R^G$. Whereas it is true that $x_\lambda I_R^G$ is pure in $I_R^G$, this is in general false for $y_\lambda I_R^G$. We define the $RG$-modules $N_R, \tilde{N}_R$ by:

(2.17)
$$a) \qquad N_R = \bigoplus_{\lambda \in \Lambda(n,n)} x_\lambda I_R^G$$

$$b) \qquad \tilde{N}_R = \bigoplus_{\lambda \in \Lambda(n,n)} R \otimes_\mathcal{O} \sqrt{y_\lambda I_\mathcal{O}^G}$$

From (2.14) and (2.15) we conclude:

**2.18 Theorem**  *The functor $H_R = \operatorname{Hom}_{RG}(P_R^G, -)/\operatorname{Hom}_{RG}(P_R^G, -)J_R$ from $\operatorname{mod} RG$ to $\operatorname{mod}\mathcal{H}_R$ induces isomorphisms $H_R : \operatorname{End}_{RG}(N_R) \to S_{R,q}(n,n)$ and $H_R : \operatorname{End}_{RG}(\tilde{N}_R) \to \operatorname{End}_{\mathcal{H}_R}(\sum_{\lambda \in \Lambda(n,n)}^{\oplus} y_\lambda\mathcal{H}_R).$*

There is an outer automorphism $\#$ of $\mathcal{H}_{R,q}(\mathfrak{S}_r)$ induced by $T_{(i,i+1)}^\# = (q-1) - T_{(i,i+1)}$ for $1 \leq i \leq r-1$. If $\lambda \models r$ then $\#$ interchanges $M_{R,q}^\lambda$ and $\widetilde{M}_{R,q}^\lambda$. Consequently, $\operatorname{End}_{\mathcal{H}_R}(\sum_{\lambda \in \Lambda(n,r)}^{\oplus} y_\lambda\mathcal{H}_R) \cong S_{R,q}(n,r)$ canonically and we have:

**2.19 Corollary**  *$H_R$ induces an algebra isomorphism $H_R : \operatorname{End}_{RG}(\tilde{N}_R) \to S_{R,q}(n,n)$.*

Next we want to verify hypothesis (2.3) for one of these modules (in fact for $\tilde{N}_\mathcal{O}$) to apply (2.6) again, now to $RG$ and $S_{R,q}(n,n)$. For this we need to understand the modules $x_\lambda I_R^G$ and $R \otimes_\mathcal{O} \sqrt{y_\lambda I_\mathcal{O}^G}$. Associated with $\lambda = (\lambda_1, \lambda_2, \ldots)$ we have a certain parabolic subgroup of $G$, namely

$$P_\lambda = \left\{ \begin{pmatrix} \boxed{\begin{smallmatrix} \lambda_1 \end{smallmatrix}} & & & \\ & \boxed{\begin{smallmatrix} \lambda_2 \end{smallmatrix}} & & \\ & & \ddots & \\ 0 & & & \end{pmatrix} \in G \right\}$$

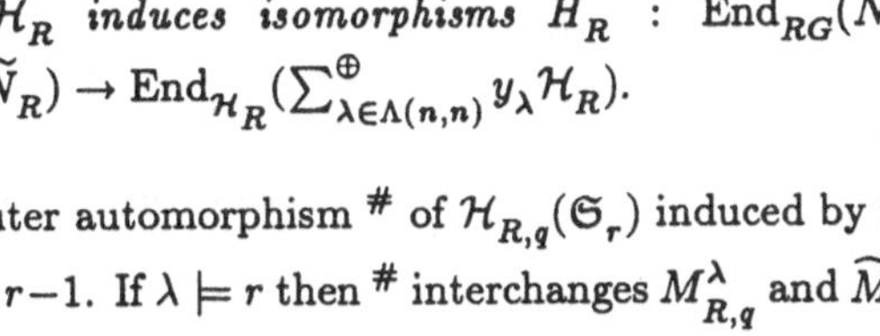

We define the Levi complement $L_\lambda = GL_{\lambda_1}(q) \times GL_{\lambda_2}(q) \times \cdots \leq G$ and the Levi kernel $U_\lambda$ in the obvious way. For $w \in W_\lambda\mathcal{S}_n$, the subspace $T_w I_R^{P_\lambda}$ of $I_R^G$ is $P_\lambda$-invariant. Indeed $x_\lambda I_R^G \cong (x_\lambda I_R^{P_\lambda})^G$ and $R \otimes \sqrt{y_\lambda I_\mathcal{O}^G} \cong (R \otimes \sqrt{y_\lambda I_\mathcal{O}^{P_\lambda}})^G$ canonically.

**2.20 Lemma**

*i) The $RP_\lambda$-module $x_\lambda I_R^{P_\lambda}$ is the trivial $P_\lambda$-module. So $x_\lambda I_R^G$ is the permutation representation of $G$ on the cosets of $P_\lambda$.*

*ii) The $KP_\lambda$-module $K \otimes_{\mathcal{O}} \sqrt{y_\lambda I_{\mathcal{O}}^{P_\lambda}} = y_\lambda I_K^{P_\lambda}$ is the Steinberg module of $L_\lambda$ with trivial $U_\lambda$-action. Moreover the irreducible character of $L_\lambda$ afforded by $y_\lambda I_K^{P_\lambda}$ (considered as $L_\lambda$-module) is constituent of the Gelfand-Graev character of $L_\lambda$. Since the Gelfand-Graev character is afforded by a projective module, there is a unique $\mathcal{O}L_\lambda$-lattice $Q_{\lambda,\mathcal{O}}$ affording it. Since $Q_{\lambda,K}$ is multiplying free, there is a unique epimorphism $\psi_K : Q_{\lambda,K} \to y_\lambda I_K^{P_\lambda}$. The lattice $\sqrt{y_\lambda I_{\mathcal{O}}^{P_\lambda}}$ in $y_\lambda I_K^{P_\lambda}$ is the image of the restriction $\psi_{\mathcal{O}}$ of $\psi_K$ to the subset $Q_{\lambda,\mathcal{O}}$ of $Q_{\lambda,K}$.*

We define the **Steinberg module** $St_{\lambda,R}$ of $P_\lambda$ over $R$ to be $RP_\lambda$-module $R \otimes \sqrt{y_\lambda I_{\mathcal{O}}^{P_\lambda}}$. So $\widetilde{N}_R = \sum_{\lambda\in\Lambda(n,n)}^{\oplus} {St_{\lambda,R}}^G$. We note that "$\lambda \in \Lambda(n,n)$" contains some redundancy: $P_\lambda$ and the modules involved are not changed by removing from $\lambda$ all zeros. So we could have replaced "$\lambda \in \Lambda(n,n)$" by "all compositions of $n$" getting a Morita equivalent $q$-Schur algebra and the same results. Moreover, ordering $\lambda$ to get a partition $\mu$, rather than a composition of $n$ yields isomorphic modules, that is $P_\lambda$ and $P_\mu$ are not conjugate, but the $\mathcal{O}G$-lattices $\sqrt{y_\lambda I_{\mathcal{O}}^G}$ and $\sqrt{y_\mu I_{\mathcal{O}}^G}$ are isomorphic. So we could replace "$\lambda \in \Lambda(n,n)$" by "$\lambda \vdash n$".

We know which characters occur in the Gelfand-Gaev character of $L_\lambda$, when we inflate it to $P_\lambda$ and then induce it to $G$ from $P_\lambda$ to $G$. So we induce $\psi_K : Q_{\lambda,K} \to St_{\lambda,K}$ from $P_\lambda$ to $G$. Then some character arguments enable us to verify (2.3) for $Q_{\mathcal{O}} = \sum_{\lambda\in\Lambda(n,n)}^{\oplus} Q_{\lambda,\mathcal{O}}^G \to \widetilde{N}_{\mathcal{O}}$.

**2.21 Theorem** *Let the projective cover $Q_{\mathcal{O}} \to \widetilde{N}_{\mathcal{O}}$ be given by inducing $\psi_{\mathcal{O}} : Q_{\lambda,\mathcal{O}} \to St_{\lambda,\mathcal{O}}$ to $G$ and summing over $\lambda \in \Lambda(n,n)$. Then $Q_{\mathcal{O}} \to \widetilde{N}_{\mathcal{O}}$ satisfies hypothesis (2.3).*

We remark that (2.21) was shown in [16, 3.7]. Its proof involved some special results on irreducible representations of $G$. The proof outlined here uses only some elementary facts on the Gelfand-Gaev character, which can be entirely formulated and shown using characteristic zero representations. The details of the proof will appear in [9].

We remark also that the $\mathcal{O}G$-module $N_{\mathcal{O}}$ does not satisfy hypothesis (2.3) in general.

We may apply now (2.10) to relate $\operatorname{mod} RG$ and $\operatorname{mod} S_{R,q}(n,n)$ by the functor $\mathcal{S}_{R,q} : \operatorname{mod} RG \to \operatorname{mod} S_{R,q}(n,n)$ given by $\mathcal{S}_{R,q} = \operatorname{Hom}_{RG}(Q_R, -)/\operatorname{Hom}_{RG}(Q_R, -)J_R$ and its right inverse $\widehat{\mathcal{S}}_{R,q} = - \otimes_S \widetilde{N}_R$. This completes the construction of triangle (2.1).

## §3 Decomposition matrices

In the previous section we constructed a functor $\mathcal{S}_{R,q} : \operatorname{mod} RG \to \operatorname{mod} S_{R,q}(n,n)$ and its right inverse $\widehat{\mathcal{S}}_{R,q}$. By (2.6) the decomposition matrix of $S_{\mathcal{O},q}(n,n)$ is a submatrix of the $p$-decomposition matrix of $G = GL_n(q)$. We turn now to some consequences and further results. The main references for details and proofs in this section are [13], [15] and [16].

We need some ingredients: First the Jordan decomposition of characters and blocks of $\mathcal{O}G$. In [25] Green constructed the irreducible characters of $G$. They are parametrized by pairs $(s, \lambda)$ where $s$ is taken from a set of representatives of the semisimple conjugacy classes of $G$ (the semisimple part) and $\lambda$ is a multi-partition labelling an irreducible character of the Weyl group of the centralizer $C_G(s)$ of $s$ in $G$ (which is a direct product of groups of the form $GL_d(q^k)$). This label $\lambda$ is the unipotent part of the character. If $t \in G$ is semisimple and conjugate to $s$, the set $(t^G)$ of all irreducible characters with semisimple part $s$ is the **geometric conjugacy class** associated with $t$. Similarly by Fong's and Srinivasan's result [23] the $p$-blocks of $G$ are parametrized by pairs $(s, \rho)$, where $s$ is again a representative of a semisimple conjugacy class of $G$ and is in addition $p$-regular and $\rho$ arises from taking $\tilde{e}$-cores of the unipotent parts of the characters. Here $\tilde{e}$ is a variable denoting the multiplicative order of $q^k$ modulo $p$ corresponding to the factor $GL_d(q^k)$ of $C_G(s)$.

We remark that for $s = 1 \in G$, obviously $G = C_G(s)$, and therefore $\lambda$ is a partition of $n$. Then $\tilde{e}$ equals $e$, as defined in the second section, if $p$ does not divide $q-1$, that is $q \neq 1$ modulo $p$. If $q = 1$ modulo $p$, then $e = p$ but $\tilde{e} = 1$.

The main result of [23] is now that the set of irreducible characters in $b_s$, the collection of all blocks with semisimple parts, is precisely the union of all geometric conjugacy classes of the form $(t^G)$ where $t = sy$ and $y$ is a $p$-element of $C_G(s)$.

We turn to the case $s = 1$. We denote the collection of blocks with semisimple part 1 by $b$. If convenient we consider $b$ as ideal of $\mathcal{O}G$ and denote $R \otimes_{\mathcal{O}} b$ by $b_R$.

The geometric conjugacy class $(1^G)$ is the set of irreducible characters afforded by the constituents of $I_R^G$ (the trivial $B$-module induced to $G$ for a Borel subgroup $B$ of $G$).

The triangle (2.1) connects the decomposition matrices of $\mathcal{H}_{\mathcal{O},q}(\mathfrak{S}_n) \cong \mathrm{End}_{\mathcal{O}G}(I_{\mathcal{O}}^G)$, of $\mathcal{S}_{\mathcal{O},q}(n,n)$ and of $\mathcal{O}G$. From the construction it is clear that the decomposition matrix of $\mathcal{S}_{\mathcal{O},q}(n,n)$ describes how the indecomposable direct summands of $y_\lambda I_{\mathcal{O}}^G$ decompose when tensored by $K$. Using the dual description of decomposition numbers as composition multiplicities of irreducible $KG$-module reduced modulo $p$, we see that the decomposition numbers of $\mathcal{S}_{\mathcal{O},q}(n,n)$ describe how the characters in $(1^G)$ decompose into irreducible $p$-modular Brauer characters. So the decomposition matrix $D_1$ of $\mathcal{S}_{\mathcal{O},q}(n,n)$ is part of the decomposition matrix $\widetilde{D}$ of the block ideal $b$ of $\mathcal{O}G$.

There is another interpretation of the decomposition numbers of $\mathcal{S}_{\mathcal{O},q}(n,n)$: We have seen in (2.18) that the $q$-Schur algebra is the endomorphism ring of the $G$-module $\sum^{\oplus}_{\lambda \in \Lambda(n,n)} x_\lambda I_R^G$ as well. So $D_1$ also records how the indecomposable direct summands of $x_\lambda I_{\mathcal{O}}^G$ decompose, when they are tensored by $K$. Using the $q$-Schur functor we see that these numbers also record how the indecomposable direct summands of $x_\lambda \mathcal{H}_{\mathcal{O}} = M_{\mathcal{O}}^\lambda$ (so called Young modules) decompose, when they are tensored by $K$. The composition multiplicities of the "permutation" modules $M_K^\lambda$ are known (the so called Kostka numbers), so we may deduce $D_1$ from the numbers which record the multiplicities of indecomposable direct summands (Young modules) of $x_\lambda \mathcal{H}_{\mathcal{O}}$. Applying the functor $\widehat{H}_{\mathcal{O}}$ those multiplicities are precisely the multiplicities of indecomposable direct summands of $x_\lambda I_{\mathcal{O}}^G$, that is of the permutation representations of $G$ on the cosets of parabolic subgroups.

It turns out that $D_1$ is a square matrix, its columns and rows labelled by partitions of $n$. If these are ordered lexicographically downward, (or any order compatible with the dominance order of partitions), then $D_1$ is lower unitriangular, that is it is lower triangular with ones on the diagonal.

There are similar results for other geometric conjugacy classes. More precisely, with $(s^G)$ we may associate a certain parabolic subgroup $P_s = L_s U_s$, with Levi complement $L_s$ and Levi kernel $U_s$, and a certain irreducible cuspidal $L_s$-module $C_s$ (compare the discussion following (2.9) and [12]). It can be shown that $C_s$ remains irreducible modulo $p$, so there is a unique $\mathcal{O}L_s$-lattice $C_{\mathcal{O}}$ in $C_s$. Consider the module $M_R$ which we get if we turn $C_R$ into a $P_s$-module with trivial $U_s$-action and induce it to $G$. The endomorphism ring of $M_R$ is a Hecke algebra. More precisely it is a tensor product of Hecke algebras of the form $\mathcal{H}_{R,q^d}(\mathfrak{S}_k)$, with various $d$ and $k$. It may happen that $R \otimes_{\mathcal{O}} \mathcal{H}_{\mathcal{O}}$ is a proper subalgebra of $\mathcal{H}_F$, but we can still construct the functors $H_R$ using a weaker hypothesis than (2.3). We can also construct associated $q$-Schur algebras (as tensor products of factors of the form $\mathcal{S}_{R,q^d}(k,k)$) and relate their decomposition matrices to the decomposition matrix of $G$.

Counting the irreducible modules which come from the $q$-Schur algebras corresponding to $p$-regular geometric conjugacy class one sees that one has constructed a complete set of irreducible $FG$-modules. In particular $\mathcal{S}_{R,q}(n,n)$ gives all irreducible modules in $b_F$. So $\widetilde{D}$ looks now as follows:

**3.1.**

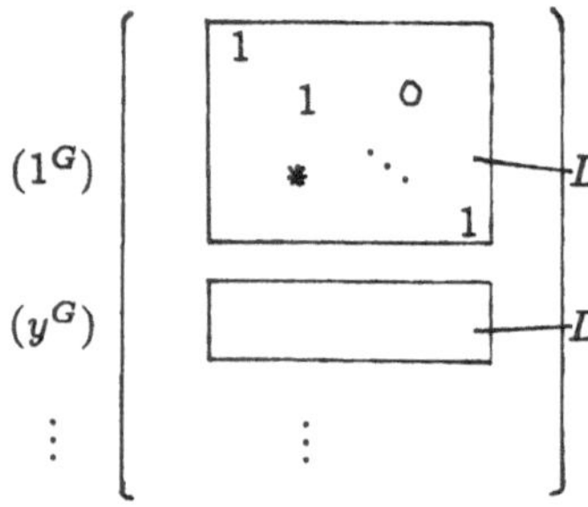


$(y^G)$ is a geometric conjugacy class for the $p$-element $y$.

Part of the matrices $D_y$ can be described as decomposition matrix of a certain $q$-Schur algebra. However the columns are labelled differently, namely by multi-partitions rather than partitions of $n$. The problem here is to match the labels of the columns in $D_y$ with those of the columns of $D_1$. There is an algorithm, described in [7] and [15] which does precisely that matching of labels. It appears as "$e$-$p$-padic" decomposition of partitions of $n$ and looks like a $q$-analog for the $p$-adic decomposition of weights appearing in Steinberg's tensor product theorem. In the last section we shall explain that the reason behind this is indeed a "tensor product theorem in non-describing characteristic".

There is (for $y \neq 1$) in general a part of $D_y$, whose column labels do not correspond to multi-partitions associated with the geometric class $(y^G)$ (so there are more columns than rows in $D_y$ in general). There is a combinatorial algorithm involving the Littlewood-Richardson rule (see e.g. [29]), which computes the entries in these additional columns. For details the reader is referred to [16].

So we have the result that the decomposition matrix $\widetilde{D}$ of $b$ is completely determined by the decomposition matrices of various $q^k$-Schur algebras. Combined with the remark above on permutation modules of $G$ on the cosets of parabolic subgroups this leads to the remarkable fact that we may deduce the decomposition numbers of general linear groups from the multiplicities of indecomposable direct summands of trivial $P$-modules induced to $G$ for general linear groups $G$ and parabolic subgroups $P$.

## §4 $q$ stands for quantum

Our main reference for this section is [10]. Let $R$ be a commutative ring and let $q$ be a unit of $R$. In section 2 we introduced the $q$-Schur algebra $\mathcal{S}_{R,q}(n,r)$ as endomorphism ring of the $\mathcal{H}_{R,q}(\mathfrak{S}_r)$-right module $\sum^{\oplus}_{\lambda\in\Lambda(n,r)} x_\lambda\mathcal{H}$. Classically the Schur algebra can be defined as centralizing algebra of the $\mathfrak{S}_r$-action on tensor space $V^{\otimes r}$ by place permutation, where $V$ is the natural module for $GL_n$, (compare e.g. [26]). Our strategy is now to give an $R$-linear isomorphism between $V^{\otimes r}$ and $\sum^{\oplus}_{\lambda\in\Lambda(n,r)} x_\lambda\mathcal{H}$, and use the $\mathcal{H}$-module structure of the latter to define an $\mathcal{H}$-structure of $V^{\otimes r}$. Now $V^{\otimes r}$ has a basis $\{e_{\mathbf{i}} \mid \mathbf{i} \in I(n,r)\}$, where $I(n,r)$ denotes the set of functions from $\{1,\dots r\}$ to $\{1,\dots,n\}$. So $\mathbf{i} \in I(n,r)$ is a multi-index $(i_1,\dots,i_r)$, with $i_\nu \in \{1,\dots,n\}$ for $1 \le \nu \le r$, and $e_{\mathbf{i}} = e_{i_1}\otimes\cdots\otimes e_{i_r}$, where the basis of $V$ is denoted by $\{e_i \mid 1 \le i \le n\}$. The symmetric group $\mathfrak{S}_r$ acts on $I(n,r)$ from the right by $\mathbf{i}\pi = (i_{1\pi^{-1}}, i_{2\pi^{-1}},\dots,i_{r\pi^{-1}})$, hence on $V^{\otimes r}$ by $e_{\mathbf{i}}\pi = e_{\mathbf{i}\pi}$.

Let $\lambda = (\lambda_1,\dots,\lambda_n) \in \Lambda(n,r)$. Define $e_\lambda \in V^{\otimes r}$ setting $e_\lambda = e_1^{\otimes\lambda_1}\otimes e_2^{\otimes\lambda_2}\otimes\cdots\otimes e_n^{\otimes\lambda_n}$, where $v^{\otimes s}$ means the $s$-fold tensor product of the vector $v \in V$ with itself, for $s \in \mathbb{N}$. For $\mathbf{i} \in I(n,r)$ define its **content** $\lambda_{\mathbf{i}} \in \Lambda(n,r)$ to be the composition $(\lambda_1,\dots,\lambda_n)$ of $r$ where $\lambda_j$ is the number of indices $i_\nu$ equal to $j$ in $\mathbf{i}$ (for $1 \le j \le n$). Obviously we find a unique $d \in \mathcal{D}_\lambda$ such that $e_{\mathbf{i}}d = e_\lambda$ for $\lambda = \lambda_{\mathbf{i}}$. In fact this establishes a bijection between the basis $\{x_\lambda T_d \mid \lambda \in \Lambda(n,r),\ d \in \mathcal{D}_\lambda\}$ of $\sum^{\oplus}_{\lambda\in\Lambda(n,r)} x_\lambda\mathcal{H}$ and the basis $\{e_{\mathbf{i}} \mid \mathbf{i} \in I(n,r)\}$ given by $x_\lambda T_d \to q^{\ell(d)}e_\lambda d$, hence an $R$-isomorphism between these two $R$-modules.

In [13,3.2] the action of a generator $T_v$ of $\mathcal{H}$ is given, where $v = (a, a+1) \in \mathfrak{S}_r$ for some $1 \le a \le r-1$. The corresponding action of $T_v$ on tensor space is

$$(4.1) \qquad e_{\mathbf{i}}T_v = \begin{cases} qe_{\mathbf{i}v} & , \quad \text{if } i_a \le i_{a+1} \\ e_{\mathbf{i}v} + (q-1)e_{\mathbf{i}} & , \quad \text{if } i_a > i_{a+1}, \end{cases}$$

where $\mathbf{i} = (i_1,\dots,i_r) \in I(n,r)$.

So $V^{\otimes r}$ is isomorphic to $\sum^{\oplus}_{\lambda\in\Lambda(n,r)} x_\lambda\mathcal{H}$ as $\mathcal{H}$-module.

**4.2 Theorem** *Let $0 \le r \in \mathbb{Z}$, and $\mathcal{H}_{R,q} = \mathcal{H}_{R,q}(\mathfrak{S}_r)$. Then:*

*i)* $\mathrm{End}_{\mathcal{H}_{R,q}}(V^{\otimes r}) = \mathcal{S}_{R,q}(n,r)$

*ii)* $\mathrm{End}_{\mathcal{S}_{R,q}(n,r)}(V^{\otimes r}) = \mathcal{H}_{R,q}$

*So the $\mathcal{S}_{R,q}(n,r) - \mathcal{H}_{R,q}$ bimodule $V^{\otimes r}$ has the double centralizer property.*

**Proof** i) is the definition (2.15). For ii) see [17, 6.6].

The coordinate ring of the variety of $n \times n$-matrices (over $R$) is the polynomial ring $A_R(n) = R[x_{ij}]$ in the $n^2$ variables $x_{ij}, 1 \leq i,j \leq n$. This is a bialgebra with comultiplication $\Delta : x_{ij} \to \sum_k x_{ik} \otimes x_{kj}$ and augmentation $\epsilon : x_{ij} \to \delta_{i,j}$. $A_R(n)$ is graded by the total degree of polynomials, so $A_R(n) = \sum_{0\leq r\in\mathbb{Z}}^{\oplus} A_R(n,r)$, where $A_R(n,r)$ is spanned by the monomials of total degree $r$. Obviously $A_R(n,r)$ is a subcoalgebra of $A_R(n)$, hence the dual space $A_R^*(n,r) = \mathcal{S}_R(n,r)$ is an associative algebra. The free $R$-module of dimension $n$ over $R$ is naturally a left $A_R(n)$-comodule, and so the tensor space $V^{\otimes r}$ as well. In fact $\tau(V^{\otimes r}) \subseteq V^{\otimes r} \otimes A_R(n,r)$, where $\tau$ is the structure map, so $V^{\otimes r}$ is an $A_R(n,r)$-comodule. The right action of $\mathfrak{S}_r$ on $V^{\otimes r}$ is a comodule map. Therefore $A_R^*(n,r)$ is contained in the centralizing algebra of $\mathfrak{S}_r$ which is in the Schur algebra $\mathcal{S}_R(n,r)$, in fact it can be shown that $A_R(n,r)^* = \mathcal{S}_R(n,r)$.

We want to find a $q$-deformation of this situation. More precisely we want to find a bialgebra involving $n^2$ (non commuting) variables and a parameter $q$. It should be graded by subcoalgebras, have the sam Hilbert series as $A_R(n)$, and $V^{\otimes r}$ should be a comodule such that multiplication by elements of $\mathcal{H}_{R,q}(\mathfrak{S}_r)$ is a comodule map.

So we start with the free $R$-bialgebra in $n^2$ noncommuting variables: $\mathcal{F}_R(n) = R\langle x_{ij}\rangle$, where the comultiplication is given by $\Delta : x_{ij} \to \sum_k x_{ik} \otimes x_{kj}$ and the augmentation by $\epsilon : x_{ij} \to \delta_{i,j}$. Again the total degree of monomials induces a grading on $\mathcal{F}_R(n)$ by subcoalgebras $\mathcal{F}_R(n,r)$ $(0 \leq r \in \mathbb{Z})$.

$V^{\otimes r}$ is a $\mathcal{F}_R(n,r)$-comodule with structural map (written on the right) $\tau_r$ : $V^{\otimes r} \to V^{\otimes r} \otimes \mathcal{F}_R(n,r)$ given by $e_{\mathrm{i}}\tau_r = \sum_{\mathrm{j}\in I(n,r)} e_{\mathrm{j}} \otimes x_{\mathrm{ji}}$, where $x_{\mathrm{ji}} = x_{j_1 i_1} \cdots x_{j_r i_r}$ for $\mathrm{j} = (j_1, \ldots, j_r), \mathrm{i} = (i_1, \ldots, i_r)$ in $I(n,r)$. We determine now relations generating an ideal $I$ of $\mathcal{F}_R(n)$, which we have to factor out to make multiplication by elements of $\mathcal{H}$ on $V^{\otimes r}$ a comodule map. That is $\tilde{\tau}(T \otimes 1) = T\tilde{\tau}$ for $T \in \mathcal{H}$, where the induced structure map $\tilde{\tau} : V^{\otimes r} \to V^{\otimes r} \otimes \mathcal{F}_R(n)/I$ is given by $\tilde{\tau} = \tau(1 \otimes \pi)$ and $\pi : \mathcal{F}_R(n) \to \mathcal{F}_R(n)/I$ is the canonical projection.

It suffices to consider the case $r = 2$. Write $\tau = \tau_2$. The method is to evaluate $(e_k \otimes e_\ell)\tau(T \otimes 1)$ and $(e_k \otimes e_\ell)T\tau$ and compare coefficients. We demonstrate this in the special case where $k \leq \ell$. So by (4.1)

$$
\begin{aligned}
(e_k \otimes e_\ell)\tau(T \otimes 1) &= (\sum_{r,s} e_r \otimes e_s \otimes x_{rk}x_{s\ell})(T \otimes 1) \\
&= q\sum_{r\leq s} e_s \otimes e_r \otimes x_{rk}x_{s\ell} \\
&+ \sum_{r>s} e_s \otimes e_r \otimes x_{rk}x_{s\ell} + (q-1)\sum_{r>s} e_r \otimes e_s \otimes x_{rk}x_{s\ell} \\
&= \sum_{r<s} e_s \otimes e_r \otimes (qx_{rk}x_{s\ell} + (q-1)x_{sk}x_{r\ell}) \\
&+ q\sum_{s} e_s \otimes e_s \otimes x_{sk}x_{s\ell} + \sum_{r>s} e_s \otimes e_r \otimes x_{rk}x_{s\ell}.
\end{aligned}
$$

On the other hand

$$e_k \otimes e_\ell T\tau = qe_\ell \otimes e_k\tau = q\sum_{r,s} e_s \otimes e_r \otimes x_{s\ell}x_{rk}\,.$$

Comparing coefficients of $e_s \otimes e_r$ for $s < r$ forces the element $x_{rk}x_{s\ell} - qx_{s\ell}x_{rk}$ to be in $I$, and comparing those of $e_s \otimes e_s$ shows that $x_{sk}x_{s\ell} - x_{s\ell}x_{sk}$ is in $I$. Considering the other case $r < s$ and treating $k > \ell$ similarly we get the following defining relations for $I$.

4.3 The ideal $I$ of $\mathcal{F}_R(n)$ is generated by:

| | | |
|---|---|---|
| A) | $x_{ik}x_{j\ell} - qx_{j\ell}x_{ik}$ | for $i > j$ and $k \leq \ell$ |
| B) | $x_{ik}x_{j\ell} - x_{j\ell}x_{ik} - (q-1)x_{jk}x_{i\ell}$ | for $i > j$ and $k > \ell$ |
| C) | $x_{ik}x_{i\ell} - x_{i\ell}x_{ik}$ | for all $i, \ell, k$ |

We set $A_{R,q}(n) = \mathcal{F}_R(n)/I$ and $c_{ij} = x_{ij} + I \in A_{r,q}(n)$.

**4.4 Theorem** *$I$ is a homogeneous biideal of $\mathcal{F}_R(n)$. so $\mathcal{F}_R(n)/I = A_{r,q}(n)$ is a graded bialgebra: $A_{R,q}(n) = \sum^{\oplus}_{0\leq 1\in\mathbf{Z}} A_{R,q}(n,r)$. The $A_{R,q}(n,r)$ are subcoalgebras. Moreover $A_{r,q}(n)$ is free as $R$-module with basis $\{c_{11}^{\nu_{11}}c_{10}^{\nu_{12}}\cdots c_{nn}^{\nu_{nn}} \mid 0 \leq \nu_{ij} \in \mathbf{Z},\ 1 \leq i,j \leq n\}$.*

By construction the associative $R$-algebra $A_{R,q}(n,r)^*$ $(0 \leq r \in \mathbf{Z})$ centralizes the action of $\mathcal{H}_{R,q}(\mathfrak{S}_r)$ on $V^{\otimes r}$, hence is contained in the $q$-Schur algebra $\mathcal{S}_{R,q}(n,r)$ by (4.2), (it is easy to see that the derived action of $A_{R,q}(n,r)^*$ on $V^{\otimes r}$ is faithful). In fact, we have:

**4.5 Theorem** *$A_{R,q}(n,r)^*$ is the $q$-Schur algebra $\mathcal{S}_{R,q}(n,r)$.*

The symmetric and exterior powers of the $R$-space $V$ have $q$-analogues as well: The tensor algebra $\mathcal{T}(V)$ is a right $A_{R,q}(n)$-comodule with structure map $\tau : \mathcal{T}(V) \to \mathcal{T}(V) \otimes A_{R,q}(n)$ sending $e_{\mathbf{j}}$ to $\sum_{\mathbf{i}\in I(n,r)} e_{\mathbf{i}} \otimes c_{\mathbf{ij}}$ for $\mathbf{j} \in I(n,r)$. This extends the $A_{R,q}(n,r)$-comodule structure on $V^{\otimes r}$ given above. We consider the ideals $J^{(1)}$ and $J^{(2)}$ of $\mathcal{T}(V)$ generated by $\{e_i \otimes e_i, e_k \otimes e_\ell + qe_\ell \otimes e_k \mid 1 \leq i \leq n,\ 1 \leq k < \ell \leq n\}$ respectively by $\{e_i \otimes e_j - e_j \otimes e_i \mid 1 \leq i < j \leq n\}$. Then $J^{(1)}$ and $J^{(2)}$ are subcomodules of $\mathcal{T}(V)$. We define the **$q$-exterior algebra** $\Lambda_q(V)$ of $V$ to be the $A_{R,q}(n)$-comodule $\mathcal{T}(V)/J^{(1)}$ and the $q$-symmetric algebra $S_q(V)$ to be the $A_{R,q}(n)$-comodule $\mathcal{T}(V)/J^{(2)}$. Abusing notation we write $e_i$ for the image of $e_i \in V$ in $\Lambda_q(V)$ and in $S_q(V)$, while multiplication is denoted in $\Lambda_q(V)$ by "$\wedge$" and in $S_q(V)$ by composition.

**4.6 Theorem** *i) $\Lambda_q(V)$ is graded by subspaces $\Lambda_q^r(V)$, $0 \leq r \in \mathbf{Z}$. Moreover $\Lambda_q^r(V) = (0)$ for $r > n$. For $r \leq n$, $\Lambda_q^r(V)$ is free as $R$-module with basis $\{e_{i_1} \wedge \ldots \wedge e_{i_r} \mid 1 \leq i_1 < i_2 < \cdots < i_r \leq n\}$.*

*ii) $S(V)$ is graded by subspaces $S^r_q(V)$, $0 \le r \in \mathbf{Z}$. For $0 \le r \in \mathbf{Z}$, $S^r_q(V)$ is free as $R$-module with basis $\{e_1^{r_1} \cdots e_n^{r_n} \mid 1 \le r_\nu \le n,\ r_1 + \cdots + r_n = r\}$.*

There is also a left comodule structure on $\mathcal{T}(V)$. Here we have to factor out $\tilde{\mathcal{J}}^{(1)}, \tilde{\mathcal{J}}^{(2)}$, which are the ideals generated by $\{e_i \otimes e_i, e_j \otimes e_k + e_k \otimes e_j \mid 1 \le i \le n, 1 \le j < k \le n\}$ respectively by $\{qe_i \otimes e_j - e_j \otimes e_i \mid 1 \le i < j \le n\}$.

The elements of $\Lambda(n,r), (0 \le r \in \mathbf{Z})$, are called **weights**. For $\lambda \in \Lambda(n,r), \lambda = (\lambda_1, \ldots, \lambda_n)$, we define the **dual weight** $\lambda' = (\lambda'_1, \lambda'_2, \ldots)$ by setting $\lambda'_i$ to be the number of parts $\lambda_j$ of $\lambda$ with $\lambda_j \ge i$. We write $\lambda' = (1^{a_1}, \ldots, n^{a_n})$ where $a_i$ is the number of parts $\lambda'_j = i$ of $\lambda'$. As in the classical case $A_{R,q}(n)$-comodules which are free of finite rank as $R$-modules admit a weight space decomposition. For details we refer to [10, §2.2]. We give here the classification of the irreducible finite dimensional $A_{(n)}$-comodules for fields $R$. For $\lambda \in \Lambda(n,r)$, let $\lambda' = (1^{a_1}, \ldots, n^{a_n})$. Then define the comodule $M(\lambda)$ to be $V^{\otimes a_1} \otimes (\Lambda^2_q V)^{\otimes a_2} \otimes \cdots \otimes (\Lambda^n_q V)^{\otimes a_n}$, where the tensor product of comodules is turned into a comodule in the usual way.

A weight $\lambda \in \Lambda(n,r)$ is called **dominant** if $\lambda$ is a partition, that is if $\lambda = \lambda''$. The set of dominant weights in $\Lambda(n,r)$ is denoted by $\Lambda^+(n,r)$. It turns out that for a dominant weight $\lambda$, the weight space of $M(\lambda)$ to weight $\lambda$ is one dimensional. As a consequence $M(\lambda)$ has a unique composition factor $L(\lambda) = L_{R,q}(\lambda)$ such that the $\lambda$-weight space of $L(\lambda)$ is one dimensional.

**4.7 Theorem** *Let $R$ be a field. Then $\{L(\lambda) \mid \lambda \in \Lambda^+(n,r), 0 \le r \in \mathbf{Z}\}$ is a complete set of non-isomorphic irreducible $A_{R,q}(n)$-comodules.*

For the proof of (4.7) the reader is referred to [10, 3.3.2] and [17, 8.8]. We remark that (4.7) provides in particular a classification of the irreducible representations of $q$-Schur algebras, hence of finite general linear groups in non-describing characteristic in view of the results kin sections 2 and 3.

So far we constructed a quantization of the coordinate ring of $n \times n$-matrices. To quantize the coordinate ring of general linear groups we have to find a quantized version of the determinant function $d = \sum_{\pi \in \mathfrak{S}_n} (-1)^{\ell(\pi)} x_{1,1\pi} \cdots x_{n,n\pi}$. Classically $d$ appears as coefficient in the one dimensional comodule $\Lambda^n_1(V)$. Theorem (4.6) implies that $\Lambda^n_q(V)$ is one dimensional. An easy calculation yields $\tau : e_1 \wedge e_2 \wedge \cdots e_n \to e_1 \wedge \cdots \wedge e_n \otimes d$, where $d = d_{R,q} \in A_{R,q}(n,n)$ is $\sum_{\pi \in \mathfrak{S}_n} (-1)^{\ell(\pi)} c_{1,1\pi} \cdots c_{n,n\pi}$. Using the basis element $e_n \wedge e_{n-1} \wedge \cdots \wedge e_1$ of $\Lambda^n_q(V)$ (which is a multiple of $e_1 \wedge \cdots \wedge e_n$) or the left comodule structure we get:

**4.8 Lemma** *The $q$-determinant $d = d_{R,q}$ is given as*

$$d = \sum_{\pi \in \mathfrak{S}_n} (-1)^{\ell(\pi)} c_{1,1\pi} \cdots c_{n,n\pi} = \sum_{\pi \in \mathfrak{S}_n} (-q)^{\ell\pi)} c_{n\pi,n} \cdots c_{1\pi,1}$$
$$= \sum_{\pi \in \mathfrak{S}_n} (-q)^{-\ell(\pi)} c_{1\pi,1} \cdots c_{n\pi,n}\,.$$

If the multiplicative order of $q$ is greater than $n$, then $d$ is the only group like element of $A_{R,q}(n)$, where group like means that $\Delta(d) = d \otimes d$. We have seen in (4.8) that $d$ is normalizing, that is $dA_{R,q}(n) = A_{R,q}(n)d$. Thus the set $\{d^i \mid 0 \le i \in \mathbb{Z}\}$ satisfies the right and left Ore condition and we can localize at this set, that is we can form the $R$-algebra $A_{R,q}(n)(d^{-1})$. The diagonal $\Delta$ extends to this extension by $\Delta(d^{-i}) = d^{-i} \otimes d^{-i} (1 \le i \in \mathbb{Z})$ and similarly the augmentation by $\epsilon(d^{-i}) = 1$.

**4.9 Lemma** *The algebra $A_{R,q}(n)(d^{-1})$ is a bialgebra with comultiplication $\Delta$ and counit $\epsilon$. It is called the quantum $GL_n$ and denoted by $R_q(G)$.*

Classically the coordinate ring of general linear groups is a Hopf algebra, where the antipode is given by dualizing the map "taking inverses" in $GL_n$. More precisely the antipode maps the variable $x_{ij}$, $(0 \le i, j \le n)$ to $h_{ij}$, where $h_{ij}$ is the $ij$-th entry of the adjunct matrix of the generic matrix $(x_{k\ell})$, that is $h_{ij}$ is the determinant of the $(n-1) \times (n-1)$-matrix $X_{ji}$ which we get by removing row $j$ and column $i$ from the generic matrix $(x_{k\ell})$ multiplied by the inverse of the determinant function. It is easy to see that these elements occur precisely as cocoefficients of $\tau(e_{i_1} \wedge \ldots \wedge e_{i_{n-1}})$ in the $(n-1)$th exterior power $\Lambda^{n-1}(V)$ considered as comodule for the coordinate ring of $n \times n$-matrices.

Obviously the quantum determinant gives a rule how to evaluate determinants of $n \times n$-matrices with entries in $A_{R,q}(n)$, and we may extend this to square matrices of arbitrary size. It is easy to see that the $q$-determinant $f_{ij}$ of the $(n-1) \times (n-1)$-minor $C_{ij}$ of the generic $n \times n$-matrix $(c_{ij})$ over $A_{R,q}(n)$ is the cocoefficient of

$$\tau(e_1 \wedge \ldots \wedge e_{i-1} \wedge e_{i+1} \wedge \ldots \wedge e_n) \in \Lambda_q^{n-1}(V) \otimes A_{R,q}(n).$$

Setting $h_{ij} = (-1)^{i+j} f_{ji} d^{-1} \in R_q(G)$ we define the map $S : R_q(G) \to R_q(G)$ by $S(c_{ij}) = h_{ij}$ and $S(d^r) = d^{-r}$. This is an algebra anti-endomorphism, and we have:

**4.10 Theorem** *The quantum group $R_q(G)$ is a Hopf algebra with antipode $S$.*

So far we discussed all objects and their relations from (1.1) except the quantum hyper-algebra $\mathcal{U}_q$. This Hopf algebra was introduced by Lusztig [37], [38]. Classically the hyper-algebra and the coordinate ring of semisimple algebraic groups are in Hopf duality. So one should suspect that $R_q(G)$ is related to $\mathcal{U}_q$ as well by Hopf algebra duality. That is not the case. The reason for this is that there are several in fact many ways to quantize the coordinate ring of general linear groups.

First there is Manin's group introduced (for $n = 2$) in [39]. Parshall and Wang investigated this quantization $\widetilde{R}_q(G)$ in [40]. Remarkably, $R_q(G)$ and $\widetilde{R}_q(G)$ cannot be isomorphic: The quantum determinant (characterized by the property to be the only generically group like element of the quantum coordinate ring of $n \times n$-matrices) is not central in $R_q(G)$ but is so in $\widetilde{R}_q(G)$. Takeuchi defined recently a two parameter quantum linear group $G_{\alpha,\beta}$ in [45], such that $G_{(1,q)} = R_q(G)$ and $G_{(q,q)} = \widetilde{R}_q(G)$. He also constructed Hopf algebra duals, that is quantum enveloping algebras corresponding to $G_{\alpha,\beta}$. A family of deformations of the coordinate ring of $GL_n$ depending on $1+\binom{n}{2}$ parameter, of which Takeuchi is a special case has been constructed independently by Sudbery [44], Reshetikhin [41] and Artin, Schelter and Tate [2]. For this multiparameter groups it is not clear if they are nonisomorphic as algebras. They are all graded algebras with the same Hilbert series and it is a remarkable fact that their coalgebra structure depends only on one of the parameters, which comes up in a special way. So in particular taking the dual of the homogeneous subcoalgebras results in $q$-Schur algebra as for $R_q(G)$, $q$ being the value of the special parameter. This was shown for Manin's group $\widetilde{R}_{q^{\frac{1}{2}}}(G)$ for fields $R$ in [10, 4.4.8], for Takeuchi's two parametric groups in [21, 2.1] and in general in [2, thm. 3].

For Manin's group $\widetilde{R}_q(G)$ the quantum determinant $\tilde{d}_{R,q}$ is central. So one might factor out the ideal generated by $1-\tilde{d}_{R,q}$ to get a quantization of the coordinate ring of special linear groups.

Takeuchi constructed in [45] a two-parameter quantized enveloping algebra $U_{\alpha,\beta}$ associated with $G_{(\alpha,\beta)}$. Taking $q=\alpha=\beta$, one gets the (by now classical) quantum enveloping algebra $U_q$ (of $\ell(n)$) as quotient of $U_{q,q}$, which has been defined by Jimbo [35] and Drinfeld [18] and has been studied by many authors in the past few years. Lusztig defined the quantum hyper algebra $\mathcal{U}_{R,q}$ by taking certain quantization of Kostant's $\mathbb{Z}$-form in $U_q(s\ell(n))$ [37], [38]. Anderson, Polo and Wen showed recently [1] that $\mathcal{U}_{R,q}$ is in Hopf algebra duality with Manin's quantum special linear group.

## §5 Steinberg's tensor product theorem

As pointed out in the introduction, there are two ways quantum groups may be utilized for the representation theory of general linear groups: First the subcategory of polynomial representations of general linear groups $G$ (that is the image of the module categories of $q$-Schur algebras in mod $FGL_n(q)$ under the functors defined in the second section) is essentially the comodule category of $A_{F,q}(n)$, and can therefore be described as representations of the quantum hyper-algebra $\mathcal{U}_q$. So we may describe this part of mod $FGL_n(q)$ in terms of the classical representation theory of algebraic groups, (or rather $q$-deformations of it). Secondly, replacing $F$ by the complex field $\mathbb{C}$ and $q$ by a primitive *eth* root of unity we get an approximation of $p$-modular representation theory of $GL_n(q)$, where $e$ is the multiplicative order of $q$ modulo $p$, ($p = \operatorname{char} F$). In this last section we want to demonstrate this through an example. More results and detailed proofs will appear in a forthcoming paper [11].

It is clear that for our purposes quantum groups at roots of unity (that is where $q$ is a root of unity) are the most important ones. In fact if $q$ is not a root of unity, then the quantized objects (Hecke algebras, $q$-Schur algebras, $A_{R,q}(n), R_q(G)$) behave like the algebraic groups in characteristic zero. In particular the $q$-Weyl modules are precisely the irreducible representations and Hecke and $q$-Schur algebras are semisimple.

So let $q$ be a primitive *eth* root of unity.

For the following we let $A_{R,q}(n)$ be defined as in the previous section. However everything we say is working in the same way for Manin's groups $\widetilde{A}_{R,q}(n)$, in fact for every quantization of $n\times n$-matrices, as long as the graded duals are $q$-Schur algebras (which is true for all versions in Section 4).

Recall that $A_{R,q}(n)$ has $n^2$ generators $c_{ij} (1\le i,j\le n)$. The following results hae been shown in [10, §1.3]. (For the analogous statements for $\widetilde{A}_{R,q}(n)$ see [40]).

**5.1 Lemma** *Let $1 \leq i,j \leq n$. Then $c_{ij}^e$ is central in $A_{R,q}(n)$. The subalgebra $Z_{R,q}(n)$ of $A_{R,q}(n)$ generated by $\{c_{ij}^e \mid 1 \leq i,j \leq n\}$ is the polynomial ring $R[c_{ij}^e \mid 1 \leq i,j \leq n]$ in the commuting variables $y_{ij} = c_{ij}^e$. Moreover $A_{R,q}(n)$ is free as $Z_{R,q}(n)$-module with basis $\{c_{11}^{\nu_{11}} c_{12}^{\nu_{12}} \cdots c_{nn}^{\nu_{nn}} \mid 1 \leq \nu_{ij} \leq e-1, 1 \leq i,j \leq n\}$.*

We define now Frobenius homomorphisms for $A_{R,q}(n)$ and $R_q(G)$ as follows (compare [40, 7.2]). If $R$ is a domain of characteristic $p > 0$ and $k$ is a non-negative integer, let $e_k = ep^k$ and note that $q^{e_k} = 1$. The subalgebra of $A_{R,q}(n)$ generated by $\{c_{ij}^{e_k} \mid 1 \leq i,j \leq n\}$ is contained in $Z_{R,q}(n)$ and is canonically isomorphic to $A_{R,q^{e_k}}(n) = A_{R,1}(n)$, the classical coordinate ring of the variety of $n \times n$-matrices. Indeed the canonical embedding $\mathcal{F}_k : A_{R,q^{e_k}}(n) \to A_{R,q}(n)$ taking a generator $y_{ij}$ to $c_{ij}^{e_k} (1 \leq i,j \leq n)$ is a bialgebra map. This map is called the **$k$-th Frobenius homomorphism** of $A_{R,q}(n)$. If the characteristic of $R$ is 0, we still have $q^e = 1$, so the analogous definition for $\mathcal{F}_0 = \mathcal{F}$ embedding $A_{R,q^e}(n)$ into $A_{R,q}(n)$ (with image $Z_{R,q}(n)$) yields the **Frobenius homomorphism** in this case. If in addition $e = 1$, that is $q = 1$, then $\mathcal{F}$ is the identity map and everything is trivial. We shall therefore exclude this case in the following discussion to avoid lengthy special hypotheses.

We remark that for positive characteristic we may interpret $\mathcal{F}_k$ as $\mathcal{F}_0$ composed with the $k$-th power of the classical Frobenius homomorphism ($p$-th power map).

For any characteristic it can be seen easily that $\mathcal{F}_k$ takes the determinant function $d = \sum_{\pi \in \mathfrak{S}_n} (-1)^{\ell(\pi)} y_{1,1\pi} \cdots y_{n,n\pi}$ to the $e_k$-th power $\det_q^{ep^k}$ of the quantum determinant. So $\mathcal{F}_k$ be extended to a Frobenius homomorphism (also denoted by $\mathcal{F}_k$) from $R(G) = R_{q^{e_k}}(G)$ into $R_q(G)$.

Let $0 \leq r \in \mathbb{Z}$. An **$e$-$p$-adic decomposition** of $r$ is a decomposition $r = r_{-1} + r_0 + r_1 ep + \cdots + r_k ep^k$ for some non-negative integers $r_{-1}, r_0, \cdots, r_k$, and $k$. Note that we do not assume that the coefficients $r_i (-1 \leq i \leq k)$ are less than $e$ or $p$. If $p = 0$, we have always $k = 0$, and we get $e$-0-adic decompositions of $r$ determined by two numbers $r_{-1}$ and $r_0$. If $e = 1$ (that is $q = 1$) we always assume that $r_{-1} = 0$ to avoid ambiguities, and we get 1-$p$-adic decompositions of $r$.

Consider the map $\mu(1 \otimes \mathcal{F}_0 \otimes \mathcal{F}_1 \otimes \cdots \otimes \mathcal{F}_k)$ from $A_{R,q}(n) \otimes A_{R,q^e}(n) \otimes \cdots \otimes A_{R,q^{e_k}}(n)$ into $A_{R,q}(n)$, where $\mu$ is the multiplication map from tensor powers of $A_{R,q}(n)$ into $A_{R,q}(n)$. Obviously this map takes $A_{R,q}(n, r_{-1}) \otimes A_{R,q^e}(n, r_0) \otimes \cdots \otimes A_{R,q^{e_k}}(n, r_k)$ into $A_{R,q}(n, r)$, where $r = r_{-1} + r_0 e + \cdots + r_k ep^k$ is an $e$-$p$-adic decomposition of $r$. Denoting $(r_{-1}, r_0, r_1, \cdots, r_k) \in \mathbb{Z}_{\leq 0}^{k+2}$ by $\alpha$, we let $\varphi_\alpha$ be this restriction of $\mu(1 \otimes \mathcal{F}_0 \otimes \cdots \otimes \mathcal{F}_k)$ to $A_{R,q}(n, r_{-1}) \otimes A_{R,q^e}(n, r_0) \otimes \cdots \otimes A_{R,q^{e_k}}(n, r_k)$. The dual map $\varphi_\alpha^*$ maps the $q$-Schur algebra $\mathcal{S}_{R,q}(n,r)$ to $\mathcal{S}_{R,q}(n, r_{-1}) \otimes \mathcal{S}_{R,q^e}(n, r_0) \otimes \cdots \otimes \mathcal{S}_{R,q^{e_k}}(n, r_k) = \mathcal{S}_{R,q,\alpha}$. Note that for $i > 0$, $\mathcal{S}_{R,q^{e_i}}(n,r) = \mathcal{S}_R(n,r)$ is the classical Schur algebra. So $\mathcal{S}_{R,q,\alpha}$ is a tensor product of the $q$-Schur algebra $\mathcal{S}_{R,q}(n, r_{-1})$ and classical Schur algebras.

In [42] Scott defined a Brauer homomorphism for endomorphism rings of permutation modules. In (2.15) $\mathcal{S}_{R,q}(n,r)$ was defined as endomorphism ring of a certain $q$-analogue of a permutation module, and it turns out that $\varphi_\alpha$ defined above is the

$q$-analogue of Scott's construction.

**5.2 Definition** *The homomorphism $\varphi_\alpha^*$ constructed above is the* **Brauer homomorphism** *associated with the $\alpha = (r_{-1}, r_0, r_1, \dots, r_k) \in \mathbb{Z}_{\geq 0}^{r+2}$.*

The Brauer homomorphism $\varphi_\alpha^*$ is an algebra homomorphism and therefore induces an inflation functor $\phi_\alpha$ from $\operatorname{mod} \mathcal{S}_{R,q,\alpha}$ to $\operatorname{mod} \mathcal{S}_{R,q}(n,r)$. The irreducible $\mathcal{S}_{R,q,\alpha}$-modules are parametrized by multi-partitions $\boldsymbol{\lambda} = (\lambda^{(-1)}, \lambda^{(0)}, \lambda^{(1)}, \dots, \lambda^{(k)})$, where $\lambda^{(i)} \in \Lambda^+(n, r_i)$. We define the dominant weight $\lambda = \lambda_\alpha \in \Lambda^+(n,r)$ to be $\lambda = \lambda^{(-1)} + e\lambda^{(0)} + ep\lambda^{(1)} + \cdots + ep^k\lambda^{(k)})$, where we define $\mu + \rho \in \Lambda(n,r)$ to be $(\mu_1 + \rho_1, \mu_2 + \rho_2, \dots)$ for $\mu = (\mu_1, \mu_2, \dots)$, $\rho = (\rho_1, \rho_2, \dots) \in \Lambda(n,r)$ and similarly multiples of weights.

The irreducible $\mathcal{S}_{R,q,\alpha}$-module $L(\boldsymbol{\lambda})$ associated with $\boldsymbol{\lambda}$ is the (outer) tensor product of the irreducible $\mathcal{S}_{R,q^{e_i}}(n, r_i)$-modules $L(\lambda^{(i)})$ where $i = -1, 0, 1, \dots, k$, and $L(\lambda^{(i)})$ is defined in (4.7).

**5.3 Theorem** (Steinberg's tensor product theorem) *Let $R$ be a field and let $r = r_{-1} + r_0 e + r_r ep + \cdots + r_k ep^k$, $\alpha = (r_{-1}, r_0, \dots, r_k) \in \mathbb{Z}_{\geq 0}^{k+2}$. Let $\lambda^{(i)} \in \Lambda^+(n, r_i)$, $\boldsymbol{\lambda} = (\lambda^{(-1)}, \lambda^{(0)}, \dots, \lambda^{(k)})$ and let $L(\boldsymbol{\lambda})$ be the corresponding irreducible $\mathcal{S}_{R,q,\alpha}$-module. Then $\phi_\alpha(L(\boldsymbol{\lambda})) = L(\lambda_\alpha)$, where $\lambda_\alpha = \lambda^{(-1)} + e\lambda^{(0)} + \cdots + ep^k\lambda^{(k)} \in \Lambda^+(n,r)$.*

For $p > 0$, $q = 1$ this is Steinberg's tensor product theorem for general linear groups [43]. For $\operatorname{char} R = 0$, Du defined the Brauer homomorphism $\varphi_\alpha$ by completely different methods [20] and he and Scott gave a proof of the tensor product theorem (still for $\operatorname{char} R = 0$) in [22]. Independently, (5.3) was shown by Parshall and Wang in [40, 11.7.1], (as well for $\operatorname{char} R = 0$). Following Parshall and Wang, the basic idea for the proof of (5.3) is to compare weight space decompositions of $L(\boldsymbol{\lambda})$ and $L(\lambda_\alpha)$. Details will appear in [11].

So far we finished step one of our program by finding a quantized version of Steinberg's tensor product theorem. If $R = F$ is a field of characteristic $p > 0$ and $q$ is a prime power of a prime different from $p$, we let $e$ be the multiplicative order of $q$ modulo $p$ and write $q$ again for $q \cdot 1_F \in F$. Theorem (5.3) may be specialized now to this case. We now use the triangle (2.1) to reinterpret (5.3) for the $p$-modular representations of $GL_n(q)$.

We need some notation. First we now have $r = n$, and the $q$-Schur algebra we are considering is $\mathcal{S}_q(n,n)$. So we have an $e$-$p$-adic decomposition $n = r_{-1} + r_0 e + \cdots + r_k ep^k$ with $\alpha = (r_{-1}, r_0, \dots, r_k) \in \mathbb{Z}_{\geq 0}^{k+2}$ as above. Since $r_i \leq n$, the $F$-algebras $\mathcal{S}_{F,q^{e_i}}(n, r_i)$ and $\mathcal{S}_{F,q^{e_i}}(r_i, r_i)$ are Morita equivalent and their module categories may be identified. With other words we replace $\mathcal{S}_{F,q,\alpha}$ by the $F$-algebra $\widetilde{\mathcal{S}}_{F,q,\alpha} = \mathcal{S}_{F,q}(r_{-1}, r_{-1}) \otimes \mathcal{S}_{F,1}(r_0, r_0) \otimes \cdots \otimes \mathcal{S}_{F,1}(r_k, r_k)$. The Brauer homomorphism (5.2) together with Morita equivalence induces a functor $Br_{F,q,\alpha}$ from $\operatorname{mod} \widetilde{\mathcal{S}}_{F,q,\alpha}$ to $\operatorname{mod} \mathcal{S}_{F,q}(n,n)$.

For $0 \leq i \leq k$ we let $\tilde{y}_i$ be a $p$-element of maximal order in the finite field

$GF(q^{ep^i})$ and $y_i \in GL_{ep^i}(q)$ the companion matrix of its minimum polynomial over $GF(q)$. We set $L_i = GL_{ep^i}(q)^{r_i} \leq GL_{ep^i k_i}(q)$, $L_{-1} = GL_1(q)^{r_{-1}} \leq GL_{r_{-1}}(q)$ and $L_\alpha = L_{-1} \times L_0 \times L_1 \times \cdots \times L_k$. We define the Levi subgroup $\widetilde{L}_\alpha$ of $GL_n(q)$ to be $GL_{r_{-1}}(q) \times GL_{er_0}(q) \times \cdots \times GL_{ep^k r_k}(q)$. So $L_\alpha$ is a Levi subgroup of $\widetilde{L}_\alpha$ (and of $G$).

We define $P_\alpha, \widetilde{P}_\alpha$ to be the corresponding standard parabolic subgroups of $G$. Let $y_{-1} = 1 \in GL_1(q)$, and define $y$ to be the matrix direct sum $y_{-1}^{r_{-1}} \times y_0^{r_0} \times \cdots \times y_k^{r_k}$. So $y$ is a $p$-element of $L_\alpha$ (hence of $P_\alpha, \widetilde{L}_\alpha, \widetilde{P}_\alpha$ and $G$ as well). The geometric conjugacy class $(y^{L_\alpha})$ of $L_\alpha$ consists of precisely one irreducible cuspidal character. Let $C_\mathcal{O}$ be the $\mathcal{O}L_\alpha$-lattice affording it, where again $(F, \mathcal{O}, K)$ is a split $p$-modular system for $G$. Note that $C_\mathcal{O}$ is unique since $C_F$ is an irreducible $FL_\alpha$-module. We may inflate $C_R$ to the parabolic subgroup $P_\alpha \cap \widetilde{L}_\alpha$ of $\widetilde{L}_\alpha$ and induce it to $\widetilde{L}_\alpha$. The resulting $R\widetilde{L}_\alpha$-module is denoted by $M_\mathcal{O}^{(\alpha)}$. Its endomorphism ring is $\mathcal{H}_{R,\alpha} = \mathcal{H}_{R,q}(\mathfrak{S}_{r_{-1}}) \otimes \mathcal{H}_{R,q^e}(\mathfrak{S}_{r_0}) \otimes \mathcal{H}_{R,q^{ep^k}}(\mathfrak{S}_{r_k})$, which is isomorphic to $\mathcal{H}_{R,q}(\mathfrak{S}_{r_{-1}}) \otimes R\mathfrak{S}_{r_0} \otimes \cdots R\mathfrak{S}_{r_k}$ if $R = F$, and we have a functor $H_{R,\alpha}$ from $\operatorname{mod} RL_\alpha$ to $\operatorname{mod} \mathcal{H}_{R,\alpha}$. Denoting the set of multi-partitions $\boldsymbol{\lambda} = (\lambda^{(-1)}, \lambda^{(0)}, \ldots, \lambda^{(k)})$ with $\lambda^{(i)} \in \Lambda(r_i, r_i)$ $(-1 \leq i \leq n)$ by $\Lambda_\alpha(n,n)$ we define $\widetilde{N}_R^{(\alpha)} = \sum^{\oplus}_{\boldsymbol{\lambda} \in \Lambda_\alpha(n,n)} R \otimes_\mathcal{O} \sqrt{y_{\boldsymbol{\lambda}} M_\mathcal{O}^{(\alpha)}}$. Then $H_{R,\alpha}$ induces an isomorphism $H_{R,\alpha} : \operatorname{End}_{R\widetilde{L}_\alpha}(\widetilde{N}_R^{(\alpha)}) \to \widetilde{\mathcal{S}}_{R,q,\alpha} = \mathcal{S}_{R,q}(r_{-1}, r_{-1}) \otimes \mathcal{S}_{R,q^e}(r_0, r_0) \otimes \cdots \otimes \mathcal{S}_{R,q^{ep^k}}(r_k, r_k)$. As in section 2 we have functors $S_{R,q,\alpha} : \operatorname{mod} RG \to \operatorname{mod} \widetilde{\mathcal{S}}_{R,q,\alpha}$ with right inverses $- \otimes_{\widetilde{\mathcal{S}}_{R,q,\alpha}} \widetilde{N}_R^{(\alpha)}$. So we have a rectangle of functors:

$$
(5.4)\qquad
\begin{array}{ccc}
\operatorname{mod} \mathcal{S}_{F,q}(n,n) & \xleftarrow{\ \mathcal{B}r_{F,q,\alpha}\ } & \operatorname{mod} \widetilde{\mathcal{S}}_{F,q,\alpha} \\
\downarrow S_{F,q} & & \uparrow -\otimes \widetilde{N}_F^{(\alpha)} \\
\operatorname{mod} b_F & \xleftarrow{\ St_{1,\alpha}\ } & \operatorname{mod} F\widetilde{L}_\alpha ,
\end{array}
$$

where $\mathcal{B}r_{F,q,\alpha}$ is the Brauer functor defined above, $S_{F,q}$ is the functor defined in section 2, (in particular it maps into the module category of the sum $b_F$ of blocks with semisimple part 1, compare section 3), and the Steinberg functor $St_{1,\alpha} : \operatorname{mod} F\widetilde{L}_\alpha \to \operatorname{mod} b_F \subseteq \operatorname{mod} FG$ is defined as composite functor $V \mapsto S_{F,q}\mathcal{B}r_{F,q,\alpha}(V \otimes \widetilde{N}_F^{(\alpha)})$ for $F\widetilde{L}_\alpha$-modules $V$. For sums $b_s$ of blocks with semisimple part $s$ we have similar functors $St_{s,\boldsymbol{\alpha}}$ (where $\alpha$ has to be replaced by multi-vectors $\boldsymbol{\alpha}$). We remark that the construction of the functors $S_{R,q}$ and $- \otimes \widetilde{N}_F^{(\alpha)}$ (using Steinberg modules of parabolic subgroups rather than trivial ones) forces to dualize the involved compositions. So we deduce from (5.3):

**5.5 Theorem** *$St_{1,\alpha}$ takes the irreducible $F\widetilde{L}_\alpha$-modules in the sum of blocks of $\widetilde{L}_\alpha$ with semisimple part 1 to irreducible modules in $b_1$. More precisely if $\boldsymbol{\lambda} = (\lambda^{(-1)}, \lambda^{(0)}, \ldots, \lambda^{(k)})$ is a multi-partition where $\lambda^{(i)}$ is a partition of $r_i$ $(-1 \leq i \leq$*

*k), we let $D(\boldsymbol{\lambda})$ be the corresponding irreducible $F\widetilde{L}_\alpha$-module. Let $\mu^{(i)}$ be the dual partition of $\lambda^{(i)}$ and set $\mu = \mu^{(-1)} + e\mu^{(0)} + ep\mu^{(1)} + \cdots + ep^k\mu^{(k)}$. Let $\lambda = \mu'$. Then $\lambda$ is a partition of $n$ and $St_{1,\alpha}(D(\boldsymbol{\lambda})) = D(\lambda)$ is the irreducible $FGL_n(q)$-module in $b_1$ associated with $\lambda$.*

From [15,5.3] one concludes:

**5.6 Corollary** *$D(\lambda)$ is the Harish-Chandra induction of $D(\boldsymbol{\lambda})$, that is $D(\lambda)$ is considered as $\widetilde{P}_\alpha$-module by the natural epimorphism $\widetilde{P}_\alpha \to \widetilde{L}_\alpha$ and then induced to $G$. So on irreducible modules $St_{1,\alpha}$ is Harish-Chandra induction.*

We remark that (5.5) shows that the matching algorithm of the column labels in $D_1$ and $D_y$ in (3.1) comes indeed from a $q$-version of Steinberg's tensor product theorem.

There is another interpretation of (5.3) for $GL_n(q)$. If $y \in G$ is defined as above let $L = C_G(y) \cong GL_{r_{-1}}(q) \times GL_{r_0}(q^e) \times GL_{r_1}(q^{ep}) \times \cdots \times GL_{r_k}(q^{ep^k})$. Note that $L$ is a regular subgroup of $\widetilde{L}_\alpha$ and of $G$, and that $y$ is contained in a maximal split torus of $L$. Let $\beta_1$ be the sum of $p$-blocks of $L$ with semisimple part 1. Then we have functors $S_{R,q,\alpha} : \operatorname{mod} \widetilde{S}_{R,q,\alpha} \to \operatorname{mod} \beta_1$ and get therefore a Steinberg functor $St_{1,\alpha} : \operatorname{mod} \beta_1 \to \operatorname{mod} b_1$, which takes irreducible $\beta_1$-modules to irreducible $b_1$-modules, where the weights involved are the same as in (5.5). This functor splits into the composite of a functor from $\operatorname{mod} \beta_1$ into $\operatorname{mod} \tilde{\beta}_1$, and our previous Steinberg functor from $\operatorname{mod} \tilde{\beta}_1$ into $\operatorname{mod} \beta_1$, where $\tilde{\beta}_1$ is the sum of $p$-blocks of $\widetilde{L}_\alpha$ with semisimple part 1.

The Frobenius twist of the classical tensor product theorem corresponds in (5.5) to the involvement of the $p$-element $y$ of $GL_n(q)$.

So far we demonstrated how the triangle in (2.1) can be used to turn classical results on polynomial representations of $GL_n$ into results on representations of finite general linear groups in non-describing characteristic.

Let now $R$ be a field of characteristic 0 and let $q \in R$ be a primitive $e$-th root of unity. We have seen that the quantum group $R_q(G)$ has only one Frobenius morphism (raising $c_{ij}$ to the $e$-th power). Corresponding in Steinberg's tensor product theorem we have to consider only $e$-0-adic decompositions of weights, that is $\lambda = \lambda^{(-1)} + e\lambda^{(0)}$ where $\lambda^{(-1)}$ is $e$-restricted (that is all parts are less than $e$). Going over to a field of characteristic $p$ then forces to use in addition the $p$-adic decomposition of $\lambda^{(0)}$. So further decomposition of irreducible modules can be expected. This idea is supported by some more results on Young modules in [11], concerning Green correspondence for Hecke algebras, which generalize work of Grabmeier [24].

A Young subgroup $W_\lambda$ of $\mathfrak{S}_n$ is said to be $e$-0-parabolic if $\lambda = (1^{r_{-1}}, e^{r_0})$ and $e$-$p$-parabolic, if $\lambda = (1^{r_{-1}}, e_0^{r_0}, e_i^{r_1}, \ldots, e_k^{r_k})$, where again $e_i = ep^i$ and $\alpha = (r_{-1}, r_0, \ldots, r_k) \in \mathbb{Z}_{\geq 0}^{k+2}$ such that $r_{-1} + \sum_{r=0}^{k} r_i e_i = n$. We define $\mathcal{H}(\alpha)$ to be $\mathcal{H}_{R,q}(\mathfrak{S}_{r_{-1}}) \otimes R\mathfrak{S}_{r_0} \otimes \cdots \otimes R\mathfrak{S}_{r_k}$, (where $R\mathfrak{S}_{r_i}$ is considered as $\mathcal{H}_{R,q^{e_i}}(\mathfrak{S}_{r_i})$, $0 \leq i \leq k$). We may consider $\mathcal{H}(\alpha)$ as $q$-analogue of the group algebra of the normalizer $N_{\mathfrak{S}_n}(W_\lambda)$ of $W_\lambda$ in $\mathfrak{S}_n$ modulo $W_\lambda$. Let $R$ be a field of characteristic $p \geq 0$, let $q \in R$ be a primitive $e$-th root of unity, and write $\mathcal{H} = \mathcal{H}_{R,q}(\mathfrak{S}_n)$. Let $V$ be an indecomposable $\mathcal{H}$-module. A **vertex** $vx(V)$ of $V$ is a minimal standard Young subgroup $W_\lambda(\lambda \models n)$ such that $V$ is $\mathcal{H}_\lambda$-projective ($\mathcal{H}_\lambda = \mathcal{H}_{R,q}(W_\lambda) \leq \mathcal{H}$), that is $V$ is direct summand of $V \otimes_{\mathcal{H}_\lambda} \mathcal{H}$. It can be shown that $vx(V)$ is unique up to conjugation. We concentrate on indecomposable direct summands of $x_\mu\mathcal{H}(\mu \models n)$, so called Young modules.

**5.7 Theorem** *Let $Y$ be a Young module. Then $vx(Y)$ is e-p-parabolic.*

This was shown in [20] for $p = 0$. A proof of the general result will appear in [11]. This is also the reference for the following:

**5.8 Theorem** *Let $\lambda \mapsto n$ be e-p-parabolic corresponding to $\alpha \in \mathbb{Z}^{k+2}_{\geq 2}$, and let $E = \mathrm{End}_\mathcal{H}(x_\lambda\mathcal{H})$. Then $E$ is a subalgebra of the q-Schur algebra $S_{R,q}(n,n)$ and the Brauer homomorphism $\varphi^*_\alpha$ maps $E$ onto $\mathcal{H}(\alpha)$. Moreover if $f \in E$ is a primitive idempotent of $E$ then $\varphi^*_\alpha(f) \neq 0$ if and only if the vertex of the indecomposable direct summand $fx_\lambda\mathcal{H}$ of $x_\lambda\mathcal{H}$ is $W_\lambda$. So $\varphi^*_\alpha$ induces so one by one correspondence between the primitive idempotents $f$ of $E$ such that $W_\lambda$ is a vertex of $fx_\lambda\mathcal{H}$ and the primitive idempotents of $\mathcal{H}(\alpha)$. In particular if $\varphi^*_\alpha(f) \neq 0$ then the dimension of the irreducible $\mathcal{H}(\alpha)$-module whose minimal projective cover is $\varphi^*_\alpha(f)\mathcal{H}(\alpha)$ gives the multiplicity of the indecomposable $\mathcal{H}$-module of $fx_\lambda\mathcal{H}$ as direct summand in $x_\lambda\mathcal{H}$.*

Theorem (5.8) has various consequences. It compares to the groups situation: Let $G$ be a finite group, $H$ a $p$-subgroups and $N$ the normalizer of $H$ in $G$. If $F$ is field of characteristic $p$ we may consider $V = F^G$, where $F$ is the trivial $H$-module (or any indecomposable $FH$-module with vertex $H$, endomorphism ring $F$ and inertial group $N$). Now $F^G = (F^N)^G$, thus the Green correspondence gives a one by one correspondence between the indecomposable direct summands of $F^N$ and the indecomposable direct summands of $F^G$ with vertex $H$. Clifford theory gives a one by one correspondence between the indecomposable direct summands of $F^N$ and those of $F(N/H)$, that is the $PIM$'s of $F(N/H)$, and multiplicities are preserved. This is Fitting's lemma, since $\mathrm{End}_{FN}(F^N) \cong F(N/H)$ canonically. Although we do not have a $q$-analogue for the normalizer of $W_\lambda$ in $\mathfrak{S}_n$, there is one for $N_{\mathfrak{S}_n}(W_\lambda)/W_\lambda$ namely $\mathcal{H}(\alpha)$, and theorem (5.8) is Green correspondence for trivial source modules (that is Young modules) for Hecke algebras. Analogeous results hold if the trivial $\mathcal{H}_\lambda$-module $Fx_\lambda$ is replaced by the alternating $\mathcal{H}_\lambda$-module $Fy_\lambda$, resulting in Green correspondence for alternating source modules. As in [24] Green correspondence can be used to derive a classification of the trivial respectively alternating source modules providing so another proof for the fact that the irreducible $q$-Schur algebra modules (hence to $A_{R,q}(n)$-comodules) are parametrized by dominant weights $\lambda \in \Lambda^+(n,r)$, [17,8.8]. Moreover, similar results hold of $S_{F,q}(n,n)$ is replaced by $S_{F,q}(n,r)$, [11].

It should be also true that the vertex of any indecomposable $\mathcal{H}$-module is $e$-$p$-parabolic. For $p = 0$ this was proved in [19].

Note that every $e$-0-parabolic subgroup is $e$-$p$-parabolic but not vice versa. So there are more vertices in characteristic $p$ than there are in characteristic 0. The $H(\alpha)$'s in characteristic $p > 0$ are subalgebras of those in characteristic 0. This means for the dimension of the associated irreducible $q$-Schur algebra modules that (for fixed weight) transition from characteristic 0 to characteristic $p$ results in further decomposition.

As we have seen in section 3 we need to determine the decomposition numbers of $q$-Schur algebras in order to find the $p$-decomposition matrices of finite general linear groups. Thus we need to determine composition multiplicities of $q$-Weyl modules, which can be considered as modules for the $q$-hyper algebra.

The Lusztig conjectures connect these multiplicities with the Kazhdan-Lusztig polynomials, (actually for all families of groups of Lie type): The old Lusztig conjecture deals with the case $\operatorname{char} F = p > 0$ and $q = 1$, the new one with $\mathcal{U}_q$ over a field $F$ of characteristic 0, where $q$ is a primitive $e$-th root of unity. The new conjecture should apply to all Weyl modules, the old one only for Weyl modules, whose highest weight satisfies certain conditions which amount basically to be in the lowest $p$-alcove far away from the walls.

The reason for this distinction is easy to explain: The existence of only one Frobenius map in the $e$-0-case and the classification of the vertices above indicates the $\mathcal{U}_q$ has only one (bottom) $e$-0-alcove, so all weights satisfy the condition above.

The connection with the vertex theory (in the special type $A$ case) is that if a weight in the $e$-$p$-case satisfies the restrictions of the old Lusztig conjecture, than the vertex of the corresponding Young module is an $e$-0-parabolic subgroup and the same as the vertex of the Young module in characteristic 0 (with the same $e$) corresponding to the same weight. It seems likely that the corresponding irreducible module in characteristic 0 remains irreducible if reduced modulo $p$; this would be indeed an important step for proving that the new Lusztig conjecture implies the old one (in the type $A$ case).

The Green correspondence (5.8) gives us tools to calculate multiplicities of Young modules as direct summands of permutation type modules $x_\lambda \mathcal{H}$ in terms of "smaller" Hecke algebras $\mathcal{H}(\alpha)$ by comparing (for a fixed weight) the $e$-0 and the $e$-$p$-case. This could be valuable for post Lusztig conjecture times, where the $e$-0-case should be given by the Lusztig conjecture and the Kostka numbers (compare section 3), and so Green correspondence could be used to get decomposition numbers of $q$-Schur algebras (hence of general linear groups in arbitrary characteristic) even for $q$-Weyl modules whose weight is in higher $e$-$p$-alcoves.

So (5.8) does not tell us that the $e$-0-case is a first approximation to the $p$-modular $e$-$p$-case, but although provides tools for actually deriving results in the $e$-$p$-case from the $e$-0-case.

## REFERENCES


[1] H.H. Anderson, P. Polo, K. Wen , Representations of quantum algebras, Aarhus Universitet, preprint,1990.

[2] M. Artin, W. Schelter, J. Tate, Quantum deformations of $GL_n$, preprint.

[3] C.T. Benson, C.W. Curtis, On the degrees and rationality of certain characters of finite Chevalley groups, Trans. Amer. Math. Soc. vol 165, 1972, 251–273.

[4] R. Carter, Finite groups of Lie type: Conjugacy classes and complex characters, John Wiley and Sons, New York, 1985.

[5] C.W. Curtis, J. Reiner, Methods of representation theory with applications to finite groups and orders, vol. I, John Wiley and Sons, New York, 1981.

[6] R. Dipper, On the decomposition numbers of the finite general linear groups, Trans. Amer. Math. Soc. 290, 1985, 315–344.
[7] R. Dipper, On the decomposition numbers of the finite general linear groups II, Trans. Amer. Math. Soc. 292, 1985, 123–133.
[8] R. Dipper, On quotients of Hom-functors and representations of finite general linear groups I, J. of Algebra 130, 1990, 235–259.
[9] R. Dipper, On quotients of Hom-functors and representations of finite general linear groups II, in preparation.
[10] R. Dipper, S. Donkin, Quantum $GL_n$, to appear in Proc. London Math. Soc..
[11] R. Dipper, J. Du, Trivial and alternating source modules of Hecke algebras of type $A$, in preparation.
[12] R. Dipper, P. Fleischmann, Modular Harish-Chandra theory, in preparation.
[13] R. Dipper, G.D. James, Representations of Hecke algebras of general linear groups, Proc. London Math. Soc. (3)52, 1986, 20–52.
[14] R. Dipper, G.D.James, Blocks and idempotents of Hecke algebras of general linear groups, Proc. London Math. Soc. (3)54, 1987, 57–82.
[15] R. Dipper, G.D.James, Identification of the irreducible modular representations of $GL_n(q)$, J. Algebra 104, 1986, 266–288.
[16] R.Dipper, G.D.James, The $q$-Schur algebra, Proc. London Math. Soc. (3)59, 1989, 23-50.
[17] R. Dipper, G.D. James $q$-tensor space and $q$-Weyl modules, Trans. Amer. Math. Soc., to appear.
[18] V.G. Drinfeld, Quantum groups, Proc. ICM 1986, Berkeley, California, 798–820.
[19] J. Du, The Green correspondence for the representations of Hecke algebras of type $A_{r-1}$, Trans. Amer. Math. Soc., to appear.
[20] J. Du, The modular representation theory of $q$-Schur algebras, Trans. Amer. Math. Soc., to appear.
[21] J. Du, B. Parshall, J.P. Wang, Two-parameter quantum linear groups and the hyperbolic invariance of $q$-Schur algebras, preprint.
[22] J. Du, L.L. Scott, Brauer homomorphisms and tensor product theorems, in preparation.
[23] P. Fong, B. Srinivasan, The blocks of finite general linear and unitary groups, Invent. Math. 69, 1982, 105–153.
[24] J. Grabmeier, Unzerlegbare Moduln mit trivialer Youngquelle und Darstellungstheorie der Schuralgebra, Doctorial Thesis, Bayreuth, (1985).
[25] J.A. Green, The characters of the finite general linear groups, Trans. Amer. Math. Soc. 80, 1955, 402–447.
[26] J.A. Green, Polynomial representations of $GL_n$, Lect. Notes in Math. 830 , Springer, New York, 1980.
[27] R.B. Howlett, G.J. Lehrer, Induced cuspidal representations and generalized Hecke rings, Invent. Math. 58, 1980, 37–64.
[28] N. Iwahori, On the structure of the Hecke ring of a Chevalley group over a finite field, J. Fac. Sci. Univ. Tokyo Sect. IA Math. 10, 1964, 215–236.
[29] G.D. James, The representation theory of the symmetric groups, Lect. Notes in Math. 830, Springer, New York, 1980.
[30] G.D. James, The decomposition of tensors over fields of prime characteristic, Math. Zeitschrift 172, 1980, 161–178.

[31] G.D. James, Unipotent representations of the finite general linear groups, J. Algebra 74, 1982, 443–465.
[32] G.D. James, Representations of general linear groups, London Math. Soc. Lect. Notes 94, Cambridge, 1984.
[33] G.D. James, The irreducible representations of the finite general linear groups, Proc. London Math. Soc. (3)52, 1986, 236–268.
[34] G.D. James, The decomposition matrices of $GL_n(q)$ for $n \leq 10$, Proc. London Math. Soc. (3)60, 1990, 225–265.
[35] M. Jimbo, A $q$-analogue of $U(gl(n+1))$, Hecke algebra and the Yang-Baxter equation, Letters Math. Phys.11, 1986, 247–252.
[36] G. Lusztig, On a theorem of Benson and Curtis, J. Algebra 71, 1981, 490–498.
[37] G. Lusztig, Finite dimensional Hopf algebras arising from quantized universal enveloping algebras, J. Amer. Math. Soc. 3, 1990, 257–296.
[38] G. Lusztig, Quantum groups at root of 1, Geom. Ded., to appear, 1990.
[39] Yu. I. Manin, Quantum groups and non-commutative geometry, CRM Université de Montréal, (1988).
[40] B. Parshall, J.P. Wang, Quantum linear groups, Memoirs Amer. Math. Soc., to appear.
[41] N. Reshetikhin, Multiparameter quantum groups and twisted quasitriangular Hopf algebras.
[42] L.L. Scott, Modular permutation representations, Trans. Amer. Math. Soc. 175, 1973, 101–121.
[43] R. Steinberg, Representations of algebraic groups, Najoya Math. J. 22, 1963, 33–56.
[44] A. Sudbery, Consistent Multiparameter Quantization of $GL(n)$, preprint.
[45] M. Takeuchi, Two parameter quantization of $GL(n)$, (summary) preprint.

Progress in Mathematics, Vol. 95, © 1991 Birkhäuser Verlag Basel

# On Auslander-Reiten components for wild blocks

KARIN ERDMANN

## Introduction

Let $B$ be a block of wild representation type over an algebraically closed field. Then the stable Auslander-Reiten quiver of $B$ has infinitely many components of the form $\mathbf{Z}A_\infty$.

Suppose $G$ is a finite group and $K$ is a field of characteristic $p > 0$. We are interested in the graph structure of components of the stable Auslander-Reiten quiver $\Gamma_s(KG)$ , or of $\Gamma_s(B)$ where $B$ is a block of KG ; the motivation is that this graph has been used as a powerful tool in the study of blocks of finite and of tame representation type.

Suppose $\Delta$ is a component of $\Gamma_s(KG)$ , then by Webb's theorem the tree class of $\Delta$ is either one of the infinite trees $A_\infty$, $B_\infty$, $C_\infty$, $D_\infty$ or $A_\infty^\infty$, or else it is $A_n$ or one of a few Euclidean diagrams; see [$W$]. We are interested in the converse, that is we would like to know which of these occur for which types of group algebras or blocks; in particular it would be of great importance to know whether there are finiteness conditions.

We assume that the field $K$ is algebraically closed. By [BD], the representation type of a block is finite if the defect groups are cyclic, and tame if $p = 2$ and the defect groups are dihedral or semidihedral or quaternion, and wild otherwise. The following is known:

If $B$ is a block of finite type then $\Gamma_s(B)$ has only one component, of the form $\mathbf{Z}A_n/ < \tau^e >$ . Here $e$ is the number of simple modules, and $n+1$ is the order of a defect group. For a tame block $B$, the components of $\Gamma_s(B)$ have also been determined [E 2]: Non-periodic components have tree class $A_\infty^\infty$ (or $\tilde{A}_5$ or $\tilde{A}_{1,2}$) and $D_\infty$. The tubes have rank $\leq 3$, and the number of 3-tubes is finite. In general, Euclidean components occur only for blocks whose defect groups are Klein 4-groups, by results in [$O$, $L$, Bes, ES]. If $B$ is a block of wild representation type over an algebraically closed field then the possible tree classes of a component of $\Gamma_s(B)$ are $A_\infty$ or $A_\infty^\infty$ or $D_\infty$. For wild group algebras the modules $K$ and $\Omega K$ lie in components $\cong \mathbf{Z}A_\infty$, see [$W, L$]. There are no known examples of components for wild blocks whose tree class is $A_\infty^\infty$ or $D_\infty$; however, not many explicit components have been investigated.

Here we study components of wild local group algebras $\Lambda$ which are isomorphic to KD for $D$ a $p$-group. We investigate cyclic submodules of $\Lambda$; the reason for this is that any such module lies either at the end of a component with tree class $A_\infty$; or else it must lie in a component of tree class $A_\infty^\infty$ or $D_\infty$, see 1.4.

We show that any cyclic submodule of $\Lambda$ on which the commutator subgroup of $D$ acts trivially must lie at the end of a component with tree class $A_\infty$. In particular, if $\Lambda$ is commutative then this accounts for all cyclic submodules of $\Lambda$ . In general, these components may be tubes. To establish that the algebra has infinitely many $\mathbf{Z}A_\infty$-components, we study cyclic submodules of $\Lambda$ of small dimensions. The result is that for $D \neq C_2 \times C_2r$ or $M_n(2)$ , defined below, the algebra $\Lambda$ has infinitely many 2-dimensional cyclic submodules of $\Lambda$ which lie at ends of $ZA_\infty$-components. Moreover this gives rise to infinitely many such components since distinct modules lie in different components. As far as the groups $C_2 \times C_2r$ $(r \geq 2)$ and $M_n(2)(n \geq 3)$ are concerned, we show that uniserial modules of length 3 lie at ends of $\mathbf{Z}A_\infty$-components; and as before one obtains infinitely many such components.

We use these results to show that an arbitrary wild block $B$ has infinitely many components of the form $\mathbf{Z}A_\infty$ . Moreover, each of the components obtained contains a module of dimension $< dimB$.

We assume throughout this paper that $\Lambda = KD$ where $D$ is a $p$-group and $K$ is an algebraically closed field of characteristic $p$ , and we assume that $D$ is not cyclic or dihedral, semidihedral or quaternion. The radical of $\Lambda$ is denoted by $J$ , and $Z(\Lambda)$ is the centre of $\Lambda$ . If $M$ is a module then soc $M$ is the largest semisimple submodule of $M$ ; and we denote by top $M$ the largest semisimple factor module of $M$ . We write $|M|$ for $\dim_k M$. The group $M_n(2)$ has the presentation

$$< x, y : x^{2^n} = y^2 = 1,\ x_y = x^{1+2^{n-1}} > .$$

It has order $2^{n+1}$ , and its commutator subgroup has order 2 and the commutator factor group is isomorphic to $C_2 \times C_2n-1$ . We denote by $\phi(D)$ the Frattini subgroup of the group $D$ . As far as background on representations of algebras is concerned, we refer to [AR, *G*, *R*]. For basic facts on groups and group representations, see for example [*H*, HB, Go, *B*].

## 1. On Auslander-Reiten components

1.1 We will first summarize some basic concepts and definitions from Auslander - Reiten theory.

(a) Suppose that A is an arbitrary finite-dimensional $K$-algebra and that $X$, $Y$ are indecomposable $A$-modules. A homomorphism $f : X \to Y$ is said to be irreducible provided $f$ is not an isomorphism, and given any factorization

$$\begin{array}{ccc} X & \xrightarrow{f} & Y \\ & {\scriptstyle g}\searrow \quad \nearrow{\scriptstyle h} & \\ & Z & \end{array}$$

then either $h$ is a split epimorphism or $g$ is a split monomorphism. Thus $f$ is irreducible if and only if $f \in rad\ Hom_a(X,Y)$ but $f \notin rad^2 Hom_a(X,Y)$.
Set

$$a_{X,Y} := \dim rad\ Hom(X,Y)/rad^2 Hom(X,Y).$$

(b) The Auslander-Reiten quiver of A , denoted by $\Gamma(A)$, is the directed graph whose vertices are the isomorphism classes of indecomposable A-modules, and the number of edges $[X] \to [Y]$ is $a_{X,Y}$. We usually identify a module and the corresponding vertex in $\Gamma$.

(c) An Auslander-Reiten sequence (AR-sequence, or almost split sequence) is a non-split short exact sequence of A-modules

$$(*)\ 0 \to M' \xrightarrow{\mu} E \xrightarrow{\lambda} M \to 0$$

where $M$ and $M'$ are indecomposable satisfying the following (equivalent) conditions:
(AR) If $g : X \to M$ does not split then $g$ factors through $\lambda$.
(AR*) If $h : M' \to Y$ does not split then $h$ factors through $\mu$.

The following important result is due to Auslander and Reiten [AR]:

**Theorem** *Suppose $M$ is indecomposable and nonprojective, then there is a (unique) almost split sequence ending with $M$. Dually, suppose $M'$ is indecomposable and not injective then there is a (unique) almost split sequence starting with $M'$. Moreover, for any indecomposable module $X$ we have $a_{X,M} = a_{M',X} =$ the multiplicity of $X$ as a direct summand of $E$.*

For $M$ indecomposable and non-projective, the module $M'$ in (*) is denoted by $\tau M$ and is called the Auslander-translate of $M$ . Dually, for $M'$ indecomposable and not injective, the module $M$ satisfying (*) is $\tau^{-1}M'$ . In the special case when $\Lambda$ is symmetric then $\tau \cong \Omega^2$ , see for example [$G$].

Suppose $P$ is indecomposable projective, then it is easy to see that the irreducible maps ending at $P$ are precisely the inclusion maps $X \mapsto P$ where $X$ is a summand of rad $P$ . Dually, if I is indecomposable injective then the irreducible maps starting at I are the epimorphism $I \to Y$ where $Y$ is a summand of $I/soc\ I$.

(d) Suppose that $P$ is an indecomposable injective- projective A-module. Then there is always an AR- sequence of the form $0 \to rad\ P \to rad\ P/soc\ P \oplus P \to P/soc\ P \to 0$ .
Moreover, this is the only AR-sequence where $P$ occurs.

(e) The *stable Auslander-Reiten quiver* of A, denoted by $\Gamma_s(A)$ , is obtained from $\Gamma(A)$ by removing all vertices of the form $[\tau^{-k}P]$ , $P$ projective, and $[\tau^k I]$ , I injective with $k \geq 0$ . In the case when A is self-injective one loses only finitely many vertices, namely all $[P]$ where $P$ is projective. We say that a connected component of $\Gamma_s(A)$ is regular if it is also a component of $\Gamma(A)$ . There are general theorems describing the graph structure of connected components of $\Gamma_s(A)$ in [Ri, HPR] and in [$W$] for the case when A is a group algebra.

1.2 Here we will study components of tree class $A_\infty$, $A_\infty^\infty$ , $D_\infty$. We say that a module $M$ lies "at the end" of its component if $M$ has either one or three predecessors in $\Gamma_s(A)$.

Suppose $\theta \cong \mathbf{Z}A_\infty$ or $\mathbf{Z}A_\infty/ < \tau^k >$ ($a$ tube). Then $\theta$ has an end which consists of a single $\tau$-orbit. We say that a module $M$ in $\theta$ lies in the $k-th$ row if there are irreducible maps $M = M_k \to M_{k+1} \to \ldots \to M_1$ where $M_1$ lies at the end, and where $M_{i-2} \neq \tau^{-1} M_i$ for $k \geq i \geq 3$ .

A component $\cong \mathbf{Z}\, A_\infty^\infty$ does not have an end; and the end of a component which is of the form $\mathbf{Z}D_\infty$ consists of three $\tau$-orbits.

The fact that AR-sequences are exact implies that the composition lengths of modules are additive. For example, if $\theta$ is a regular component of $\Gamma_s(A)$ then for each subgraph of the form

$$\begin{array}{ccccc} & & X & & \\ & \nearrow & & \searrow & \\ \tau M & & & & M \\ & \searrow & & \nearrow & \\ & & Y & & \end{array}$$

(where $M$ has two predecessors), we have $|M| + |\tau M| = |X| + |Y|$ . This gives local conditions for distinguishing tree classes.

For symmetric algebras, $\tau$-periodicity and $\Omega$-periodicity is the same. Moreover, for a block of infinite type, periodic modules form tubes, by [HPR,W].

1.3 (i) Suppose that $\theta \cong \mathbf{Z}A_\infty$ or $\mathbf{Z}A_\infty / < \tau^k >$, and that $\theta$ is regular. Then the lengths of modules in $\theta$ are strictly increasing with growing row number. In particular, given $M$ in $\theta$ , then either $M$ lies at the end, or else, if $M$ lies in the $k-th$ row where $k \geq 2$ then there is a unique irreducible monomorphism $X \to M$ with $X$ in row $k-1$ , and there is a unique irreducible epimorphism $Y \to M$ with $Y$ in row $k+1$ .

(ii) Suppose $\theta \cong \mathbf{Z}\, A_\infty^\infty$ and $\theta$ is regular, so that $\theta$ does not have an end. Take a module $M$ in $\theta$ whose length is minimal, then both irreducible maps ending at $M$ are onto, and both irreducible maps starting at $M$ are one-to-one.

Now assume that $\Lambda = KD$, as defined in the introduction. The following motivates to study components containing modules with simple socles and tops.

**Lemma 1.4** *Suppose $M$ is a cyclic submodule of $\Lambda$ which is not simple or projective, and let $\theta$ be the component of $\Gamma_s(\Lambda)$ containing $M$ . Then one of the following holds:*
(i) $\theta \cong \mathbf{Z}A_\infty$ *or* $\mathbf{Z}A_\infty / < \tau^k >$, *and $M$ lies at the end.*
(ii) $\theta \cong \mathbf{Z}\, A_\infty^\infty$ *or* $\mathbf{Z}D_\infty$ .

**Proof.** We note first that $\theta$ is not Euclidean: If $\Lambda = KD$ has an Euclidean component then by [*L*, *O*, Bes, ES], $D$ is a Klein 4-group which we exclude here.

(1) $\theta$ is regular and not the component containing $K$ : Let $\Delta$ be the component of $\Lambda$; we have to show that $M$ does not lie in $\Delta \cup \Omega\Delta$. We excluded dihedral, semidihedral and quaternion groups. Therefore, by [W], the stable part of *Delta* is isomorphic to $\mathbf{Z}A_\infty$ with $\Lambda$ attached at the end, and moreover $\Omega\Delta \neq \Delta$. Consider first $\Omega\Delta$, this is a regular component. A module $X$ at the end of $\Omega\Delta$ is isomorphic to $\Omega^n(K)$ for $n$ even and has therefore dimension $equiv 1 (mod |D|)$. Hence if $dim X < |D|$ then $dim X = 1$ and $X = k$. Now suppose $X$ in $\Omega\Delta$is a module not at the end; then by 1.3(i) we have $dim X > |D|$, in particular $X$ does not have a simple socle or top. It follows that $M$ does not lie in $\Omega\Delta$.

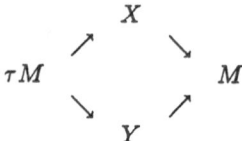

Now suppose there is a module $X$ in $\Delta$ which has a simple top. Then $\Omega X$ lies in $\Omega\Delta$ and is of dimension $< |D|$. By the first part of the argument, $\Omega X \cong K$ and hence $X \cong \Omega^{-1}K$, which does not have a simple socle. Therefore $M$ does not lie in $\Delta$.

(2) <u>If there is an irreducible monomorphism $X \xrightarrow{\alpha} M$ then (ii) holds:</u>

Since $\alpha$ is not onto and since $M$ has a unique maximal submodule, there is a factorization

$$\begin{array}{ccc} X & \xrightarrow{\alpha} & M \\ & {\scriptstyle\tilde{\alpha}}\searrow \quad \nearrow{\scriptstyle j} & \\ & radM & \end{array}$$

Now $\alpha$ is irreducible and $j$ does not split, hence $\tilde{\alpha}$ is a split monomorphism. Moreover, rad $M$ is indecomposable and hence $X \cong rad\ M$ . In particular $|M| - |X| = 1$.

Suppose that the tree class of $\theta$ is $A_\infty$. Since $\theta$ is regular, we can therefore exploit the exactness of Auslander-Reiten sequences. In each row there is a module $Z$ and an irreducible map $W \to Z$ and $|Z| - |W| = 1$ , except at the end where we must have a module of dimension 1. That is $\theta$ contains $K$ , contrary to (1).

It follows now that (ii) is true. Dually one shows that if there is an irreducible epimorphism $M \to Z$ then also (ii) holds.

Therefore we may assume that there is no irreducible monomorphism ending at $M$ (and no irreducible epimorphism starting at $M$). Suppose (ii) does not hold, then the tree class of $\theta$ is $A_\infty$ . Moreover, $\theta$ is regular and therefore by 1.3 there is no module in $\theta$ at which two irreducible epimorphisms end. Consequently there is only one irreducible map ending at $M$; that is, $M$ lies at the end of $\theta$ .

1.5 Incidentally, there is also a more general result by Butler and Ringel which gives a necessary condition for certain cyclic modules to have a unique predecessor in $\Gamma(\Lambda)$ . Suppose A is an arbitrary basic finite - dimensional $K$-algebra, with radical $J$, and let $e$, $f$ be primitive idempotents of $\Lambda$ . Following [BR], a non-zero element $a \in eJf$ is *non-supportive* if for any indecomposable summand $C$ of $eJ/aA$,, the intersection $C \cap [soc(eA/aJ)]\pi$ is non-zero; where $\pi : eA/aJ \to eA/aA$ is the canonical projection. The result is as follows:

**Theorem [BR]** *Suppose $a \in eJf$ is non-supportive, and let* $0 \to \tau(eA/aA) \to E \to eA/aJ \to 0$ *be the AR-sequence. Then $E$ is indecomposable.*

Butler and Ringel proved that if $E$ is decomposable then $E \cong C \oplus N$ where $C$ is a summand of $eJ/aA$ having zero intersection with (soc $eA/aJ)\pi$.

1.6 It has been proved in [BR] that any element $a \in eJf - J^2$ is non-supportive. Moreover, let $0 \neq a \in eJf$ . Suppose $eJ/aA$ is indecomposable and soc $eA/aJ$ is not simple; then a is non-supportive: In this case consider $C = eJ/aA$ ; we have that $\ker\pi$ is simple and hence $0 \neq [soc(eA/aJ)]\pi \subset soc\ C$ . For local algebras, the condition $a \in eJf$ is trivially satisfied by all elements of $J$ .

Returning to group algebras $\Lambda$, let $\omega\Lambda \subset \Lambda$. We will now study whether $\omega$ is non-supportive.

1.7 Suppose we know that $\omega\Lambda$ has a unique predecessor in $\Gamma_s(\Lambda)$. Then either $\omega\Lambda$ satisfies 1.4(i), or else $\omega\Lambda$ could lie at the end of a $\mathbf{Z}D_\infty$-component. The second possibility is most often excluded by the following:

**Lemma** *Suppose $\Lambda$ is a local p-group algebra and $K \neq \omega\Lambda \subset \Lambda$. Assume that $\omega\Lambda$ has a unique predecessor in $\Gamma(\Lambda)$; and assume that either $\omega\Lambda$ is self-dual; or else that $\Omega(\omega\Lambda)/K$ and rad $\Omega^{-1}(\omega\Lambda)$ both have no proper direct summands whose dimensions are $\equiv (-1) \bmod p$. Then $\omega\Lambda$ does not lie at the end of a $\mathbf{Z}D_\infty$-component.*

**Proof.** Let $\theta$ be the component of $M := \Lambda/\omega\Lambda = \Omega^{-1}(\omega\Lambda)$. Since $\Omega^{-1}$ induces a graph isomorphism of $\Gamma_s(\Lambda)$. $M$ has also one predecessor, and we have to show that $\theta \neq \mathbf{Z}D_\infty$. Suppose this is not so. Then there are AR-sequences $0 \to \tau M \xrightarrow{\lambda} X \xrightarrow{\mu} M \to 0$ and also $0 \to \tau R \xrightarrow{\lambda'} X \xrightarrow{\mu'} R \to 0$, and by the hypothesis, $X$ is indecomposable. The maps $\lambda$, $\lambda'$ are one-to-one, and we consider $\tau M$ and $\tau R$ as submodules of $X$. We have that soc $\tau M$ is simple, since soc $\tau M \cong soc\ \Omega^2 M \cong top\ \Omega M \cong top\ \omega\Lambda$.

Assume first that $\tau M \cap \tau R \neq 0$, then $K \cong soc\ \tau M \subset \tau R$. Both sequences are almost split, therefore $X/K \cong \tau M/K \oplus M$ but also $X/K \cong \tau R/K \oplus R$. Now $R$ is indecomposable *and* $\neq M$ since $M$ and $R$ represent distinct vertices of the AR-quiver; consequently $R$ is isomorphic to a summand of $\tau M/K$; similarly $M$ is a summand of $\tau R/K$. Hence

$$(*)\ \tau M/K \cong R \oplus C \text{ and } \tau R/K \cong \tau M \oplus C.$$

Assume first that $\omega\Lambda$ is self-dual; then $M^* \cong [\Omega^{-1}(\omega\Lambda)]^* \cong \Omega(\omega\Lambda^*) \cong \Omega(\omega\Lambda) \cong \tau M$. We deduce that $*$ fixes $\theta$ and induces a graph isomorphism of this component, a "reflection"; therefore $X \cong X^*$ and then $R^* \cong \tau R$. It follows that $\dim R = (\dim\ X)/2 = \dim M = \dim \tau M$; on the other hand, by (*), $\dim R < \dim \tau M$, a contradiction.

In general, $\dim M = \dim \tau M = m$, say. By $[E]$, the vertex of modules is constant in a $\mathbf{Z}D_\infty$-component. Let $V$ be such a vertex; then $V$ is non-cyclic; otherwise $M$ would be periodic, and hence there is a subgroup $< x > \subset V$ of order $p$. The Auslander-Reiten sequences in $\theta$ split on restriction *to* $< x >$, see [*B* 2.17.10]. Moreover $M_{<x>} \cong \tau M_{<x>}$ modulo projective summands (since all $K < x >$ *−modules* have $\Omega$-period $\leq 2$), and similarly $R_{<x>} \cong \tau R_{<x>}$ modulo projective summands. Hence $X_{<x>} \cong 2M_{<x>} \cong 2R_{<x>}$, and therefore $M_{<x>} \cong R_{<x>}$ modulo projective summands. In particular, $\dim M \equiv \dim R \bmod p$. Let $\dim R = r$; then $m = r + kp$. Let $c = \dim C$ where $C$ is as in (*); then it follows that $m - 1 = r + c$ and $c = pk - 1$, a contradiction to the hypothesis.

Now assume $\tau M \cap \tau R = 0$; this is dual to the first case exploiting the fact that top $M$ is simple.

1.8 The following simple observation is useful to show that very often modules lie in different components: Suppose $\Lambda$ is a group algebra and $M$ is an indecomposable $\Lambda$-module which is not projective or periodic. Assume also that $M \cong M^*$; then $M$ is the unique module in its $\Omega$-orbit which is self-dual. This is clear since $(\Omega^n M)^* \cong \Omega^{-n}(M^*) \cong \Omega^{-n}(M) \neq \Omega^n(M)$ for $n \neq 0$. In particular, if $M'$ is also a self-dual module and $M' \neq M$ then $M'$ does not lie in the same $\tau$-orbit of $M$. If in addition

$M$ and $M'$ lie at ends of components of the form $\mathbf{Z}A_\infty$ then they must lie in different components.

## 2. Group algebras of $p$-groups, indecomposable subquotients

2.1 We summarize a number of well-known elementary facts which will be used frequently.

(1) Suppose A is a local algebra, and for $\zeta \in A$ , let $R_\zeta = \{x \in A : \zeta x = 0\}$. Then $R_\zeta$ is a right ideal of A , and $\zeta A \cong A/R_\zeta$ , as A-modules, and in the case when $\zeta$ is central also as $\bar{A}$-modules where $\bar{A} = A/R_\zeta$.

(2) Let $\Lambda = KD$ where $D$ is a $p$-group and char $K = p$. Take $z \in Z(D)$ of order $p$ , and put $\xi = (1-z)$ and $\rho = (1-z)^{p-1}$ . Then (1) applies to $\zeta = \xi$ and $\zeta = \rho$. Here we have that $R_\rho = \xi\Lambda$ and $R_\xi = \rho\Lambda$.
Moreover, let $\bar{D} = D/ < z >$ . Then $\Lambda/R_\rho = \Lambda/\xi\Lambda \cong K\bar{D}$ where an isomorphism is induced by the group epimorphism $d \to \bar{d}$ .

(3) Suppose I is a right ideal of $\Lambda = KD$ and assume that I contains the ideal $J(KD')KD$ . Then with the identification $\rho\Lambda \cong \Lambda/R_\rho \cong K\bar{D}$ , we have that $\rho I$ is a right ideal of $K\bar{D}$, and it contains $(1-\bar{d})$, $\bar{d} \in \bar{D}'$.

(4) For $\Lambda = KD$, let $\Lambda^{d'} = \{m \in \Lambda : mg = m \text{ for } g \in D'\}$. Then $\Lambda^{d'} \cong K(D/D') \cong K^d_{d'}$.

2.2 Let $\Lambda = KD$ where $D$ is a $p$-group and char $K = p$ . Then there is an explicit description of the Loewy series of $\Lambda$ , due to Jennings; see [HB, $p$.255], or [$B$, p187]. In particular, this gives a basis for $J/J^3$ which we describe in detail.

Let $H_1 = D$, $H_2 = \phi(D)$ and

$$H_3 = \begin{cases} < [\phi(D), D], \phi(D)^2 > & p = 2 \\ < [\phi(D), D], D^p > & p > 2 \end{cases}$$

Then $H_i$ is a normal subgroup of $D$ and $H_i/H_{i+1}$ is elementary abelian.

Choose $x_1, ..., x_{d_1} \in D$ such that the elements $x_i\phi(D)$ form a basis of $D/\phi(D)$; and choose also $y_1, \ldots, y_{d_2} \in \phi(D)$ such that $\{y_iH_3\}$ is a basis for $\phi(D)/H_3$. Then we have that

(a) The cosets of $\{(x_i - 1) : 1 \leq i \leq d_1\}$ form a $K$-basis for $J/J^2$.

(b) If $p = 2$ then the cosets of $\{(x_i-1)(x_j-1) : 1 \leq i < j \leq d_1\} \cup \{(y_i - 1) : 1 \leq i \leq d_2\}$ form a basis for $J^2/J^3$ . In case $p > 2$ , a basis for $J^2/J^3$ is given by the cosets of $\{(x_i-1)(x_j-1) : 1 \leq i \leq j \leq d_1\} \cup \{(y_i - 1) : 1 \leq i \leq d_2\}$. In particular, $\dim J/J^2 = d_1$ and

$$\dim\ J^2/J^3 = \begin{cases} d_1(d_1-1)/2 + d_2 & p = 2 \\ d_1(d_1+1)/2 + d_2 & p > 2 \end{cases}$$

Moreover, the radical series and the socle series of KD coincide.

2.3 Assume that $D$ is elementary abelian, say $D = (C_p)^n$ ; Then the group algebra KD is isomorphic to $K[X_1, X_2, ..., X_n]/(X_i^p)$. Moreover, whenever $v_1, v_2, ..., v_n \in$

$\Lambda$ such that the cosets modulo $J^2$ form a $K$-basis of $J/J^2$ then the map $X_i \to v_i$ induces an algebra isomorphism of KD .

More generally, let $D$ be an abelian $p$-group and $\Lambda = KD$ . Then $\Lambda \cong K[X_1, ..., X_n]/(X_i^{r_i})$ where each $r_i$ is some $p$-power $> 1$ . Then it is not true any more that arbitrary generators of $J$ satisfy the same relation. We have still some information as far as $J^2$ modulo $J^3$ is concerned.

**Lemma** *Suppose $w_1, w_2, ..., w_n \in J(\Lambda)$ are independent modulo $J^2$ .*
(a) *If $r_i > 2$ for all $i$ then the set $\{w_i w_j : i \leq j\}$ is a basis for $J^2$ modulo $J^3$ .*
(b) *Suppose $p = 2$ . Then the set $B = \{w_i w_j : i < j\}$ is independent modulo $J^3$ . Moreover, if $w_i{}^2 \neq 0$ then $B \cup \{w_i{}^2\}$ is also independent modulo $J^3$.*

**Proof.** A basis of $J^2$ modulo $J^3$ is given by $\{X_i X_j : i < j\} \cup \{X_i{}^2 : X_i{}^2 \neq 0\}$ .
(a) Assume $r_i > 2$ for all $i$ ; then dim $J^2/J^3 = |B|$ , and since $B$ is clearly a spanning set, it must be a basis.
(b) Suppose $p = 2$ . We express the $w_i$ in terms of the basis $\{X_i\}$ modulo $J^2$ ; let $C$ be the corresponding matrix. Then (*)

$$w_i w_j = \sum_{k<l} (c_{ki} c_{lj} + c_{li} c_{kj}) X_k X_l + \sum_k c_{ki} c_{kj} X_k{}^2$$

The non-zero squares $X_k{}^2$ are independent of $\{X_k X_l : k < l\}$ . Hence it suffices to show that the matrix of coefficients for $X_k X_l$ in (*) is non-singular. This is true since for char $K = 2$ this is the matrix of all 2-minors of $C$ .

2.4 Suppose $D$ is a $p$-group such that $z \in D' \cap Z(D)$ has order $p$ , and $\bar{D} = D/<z>$ *is* abelian. Put $\rho = (1-z)^{p-1}$ . Let $X_1, X_2, ..., X_r$ be as in 2.2; and assume $x_i D'$ has order $a_i$ . Then $soc_2 KD$ has a basis (modulo soc $\Lambda$) given by $\rho Z_1$, $\rho Z_2$, ..., $\rho Z_r$ where $Z_i = X_1^{a_1-1} ... X_i^{a_i-2} ... X_r^{a_r-1}$ :

To see this, let $Z = Z_1 X_1 (= Z_i X_i$ for $i = 1, ..., r)$. Then $<Z> = soc\ \Lambda$ since $\rho\Lambda \cong K\bar{D}$ and $Z$ corresponds to a generator of soc $K\bar{D}$ . Moreover, this correspondence takes $\rho Z_i$ to a basis of $soc_2 K\bar{D}$ .

2.5 We will now study indecomposability of a number of subquotients of $J/soc\ \Lambda$. For the commutative situation, we have the following observation:

**Lemma** *Let $\Lambda$ be a local algebra such that $J/J^3$ is commutative and moreover for any two elements $\alpha$ and $\beta$ with $\alpha$, $\beta$ independent modulo $J^2$ the product $\alpha\beta$ does not lie in $J^3$ . Then $J/J^3$ is indecomposable.*

**Proof.** Let $J/J^3 = A + B$ where $A \cap B = 0$ , and suppose A and $B$ are both non-zero. Then there exist $\alpha \in A - J^2$ and $\beta \in B - J^2$ ; and $\alpha$ and $\beta$ are independent modulo $J^2$ . Therefore, by the hypothesis, $0 \neq \alpha\beta$; but $\alpha\beta = \beta\alpha \in A \cap B$ , a contradiction.

We deduce, using 2.3 and duality:

**Corollary** *Let $\Lambda = KD$ where $D$ is an abelian $p$-group; and let char $K = p$ . Let $J$ be the radical of $\Lambda = KD$ . Then $J/J^3$ and $soc_3\Lambda/soc\ \Lambda$ are indecomposable.*

It is not always true for wild group algebras of $p$-groups that $J/J^3$ is indecomposable. For example, let $D$ be extraspecial of order $p^3$ and exponent $p$ ; then we know that $\dim J/J^2 = 2$ and $\dim J^2/J^3 = 4$, by 2.2. Consequently $\Lambda/J^3$ is the local algebra with two generators, $\alpha$ and $\beta$ say, where no relations hold except $J^3 = 0$; and hence $J/J^3 = <\alpha> \oplus <\beta>$. However, the following holds which applies for example to extraspecial groups of exponent $p^2$ when $p > 2$.

**Lemma** *Let $\Lambda = KD$ and let $H_2$ and $H_3$ be the Jennings groups, as in 2.2. If $H_2 = H_3$ then $J/J^3$ and $soc_3\Lambda/soc\ \Lambda$ are indecomposable.*

**Proof.** The canonical epimorphism $D \to D/\phi(D) = \bar{D}$ gives rise to an algebra homomorphism $\psi : KD \to K\bar{D}$. If $H_2 = H_3$ then $\psi$ induces an isomorphism $J/J^3 \to J(K\bar{D})/J^3(K\bar{D})$ by 2.2, and by the above Corollary $J/J^3$ is indecomposable. The other part is dual.

We shall now study more general subquotients of $\Lambda$.

**Proposition 2.6** *Suppose $\Lambda = KD$ where $D$ is an abelian p-group which is not cyclic and not the Klein 4-group, and let $\omega \in J$. Then $J/\omega\Lambda$ is indecomposable, except possibly when $D = C_{p^r} \times C_{p^s}$ and $\omega = XY + \rho$ for $X$, $Y$ in $J$ independent modulo $J^2$ and $\rho \in J^3$.*

**Proof.** Let $I = J/\omega\Lambda$. Assume first that $\omega$ does not belong to $J^2$. Let $J = A + B$ with $A \cap B \subset \omega\Lambda$. Choose generators $\omega = w_1, w_2, ..., w_n$ of $J$ such that $w_2, ..., w_r$ generate A and $w_{r+1}, ..., w_n$ generate $B$. We consider now $J^2$ modulo $J^3$. Suppose A and $B$ are both non-zero (modulo $\omega\Lambda$) : that is $2 \leq r < n$. The element $w_2w_n$ lies in $A \cap B$, hence it lies in $\omega\Lambda$. However, by 2.3, $w_2w_n$ is independent of $B = \{w_1w_j : j > 1\}$ modulo $J^3$ (or of $B \cup \{{w_1}^2\}$). Hence $w_2w_n \in J^3$, a contradiction to 2.3.

Now suppose $\omega$ lies in $J^3$, then $\omega\Lambda \subset J^3$ and $I/(J^3/\omega\Lambda) \cong J/J^3$ which is indecomposable, by 2.5. On the other hand, this is also $I/rad^2 I$, and it follows that I is indecomposable.

It remains to study the case when $\omega \in J^2$ *but* $\notin J^3$. Suppose that I is decomposable. Let $J = A + B$ where $A \cap B \subset \omega\Lambda$; and choose generators $w_1, ..., w_n$ of $J$ modulo $J^2$ such that $w_1, ..., w_r$ generate A and $w_{r+1}, ..., w_n$ generate $B$. Then $w_1w_n \in A \cap B - J^3$, therefore $\omega = c(w_1w_n) + \rho$ for $0 \neq c \in K$ and $\rho \in J^3$. Suppose $2 \leq r$, then also $w_2w_n \in A \cap B - J^3$, and hence $w_2w_n$ and $w_1w_n$ are dependent modulo $J^3$, a contradiction. Therefore $r = 1$ and similarly $r + 1 = n$. Hence $J$ has two independent generators $X, Y$ say, and $\omega$ is of the stated form.

2.7 Now let $\Lambda = KD$ where $D$ is an arbitrary $p$-group, and suppose that I is a right ideal of $\Lambda$. We wish to study possible decompositions of $I/soc\ \Lambda$; assume therefore that $I = A + B$ such that $A \cap B = soc\ \Lambda$ where A, $B$ are right ideals of $\Lambda$.

The following holds for an arbitrary local self-injective algebra.

**Lemma** *Suppose $0 \neq \rho \in Z(\Lambda) \cap I$ and $\rho = a + b$ with $a \in A$ and $b \in B$. Then $\rho I = aI + bI$ and $aI \cap bI \subset soc\ \Lambda$.*

**Proof.** It is clear that $aI \cap bI \subset A \cap B \subset soc\ \Lambda$; and that $\rho I \subset aI + bI$. For the converse, observe that $aB \subset soc\ \Lambda$ and $bA \subset soc\ \Lambda$ : For example, if $x \in B$ then $ax = (\rho - b)x = x\rho - bx \in A \cap B$. Suppose $m \in I$, say $m = a' + b'$ with $a' \in A$ and $b' \in B$. Then $am = aa' + ab' \equiv aa'$ (modulo soc $\Lambda$) $\equiv aa' + ba' = \rho a' \in \rho I$ and $aI \subset \rho I$; similarly $bI \subset \rho I$.

**Proposition 2.8** *Let $\Lambda = KD$ where $D$ is a non-abelian p-group, and assume that $D$ is not dihedral or semidihedral or quaternion, and suppose that $I$ is a right ideal of $\Lambda$. If the commutator $D'$ acts trivially on $\Lambda/I$ then $I/soc\ \Lambda$ is indecomposable. Dually, if $D'$ acts trivially on $I$ then $rad\ \Lambda/I$ is indecomposable.*

**Proof.** It suffices to prove the first statement. Note that $D'$ acts trivially on $\Lambda/I$ if and only if I contains the ideal $J_1 := J(KD')KD$.

Let $I = A + B$ where A and $B$ are right ideals which are not simple, such that $A \cap B = soc\ \Lambda$. Since $D$ is not abelian, the group $D' \cap Z(D)$ is not trivial; choose $z \in D' \cap Z(D)$ of order $p$. Let $\xi := (1 - z)$, then $\xi$ belongs to $J_1$; and hence $\xi \in I$ and also $\rho$ where $\rho = (1 - z)^{p-1}$.

We will proceed by induction on the group order. Let $\bar{D} = D/ < z >$, and suppose first that $\bar{D}$ does not satisfy the hypothesis. Then

(1) <u>$\bar{D}$ must be abelian:</u> Assume not; then $p = 2$ and $\bar{D}$ is dihedral or semidihedral or quaternion. We have $< z > \subset D' \subset D$ and therefore $D/D' \cong \bar{D}/\bar{D}'$, which is a Klein 4-group. By a Theorem of Tausski [$H$, p.339], the group $D$ is dihedral or semidihedral or quaternion, this contradicts the hypothesis. Hence $\bar{D} = D/D'$ and therefore $\Lambda/\xi\Lambda \cong K\bar{D} = K(D/D')$.

(2) <u>$\xi$ does not lie in $J^3$:</u> Suppose $\xi \in J^3$, then by 2.2, $soc_3\Lambda \cong soc_3 K\bar{D}$, and hence $soc_3\Lambda/soc\ \Lambda$ is indecomposable, by (1) and 2.5. On the other hand, $soc_3 K\bar{D} \cong soc_3 \rho\Lambda \subset I$ and hence $soc_3\Lambda \subset I(2.1(3))$. Since $soc_3\Lambda/soc\ \Lambda$ is indecomposable, one of $soc_3 A/soc\ A$ and $soc_3 B/soc\ B$ is zero, and then one of A and $B$ is simple, contrary to the hypothesis.

(3) <u>$\{x \in \Lambda : x\xi \in soc\ \Lambda\} = \rho\Lambda + \omega K$ where $\omega \in soc_3\Lambda$:</u> We have $\rho\Lambda = R_\xi \subset \{x : x\xi \in soc\ \Lambda\}$, and the quotient has dimension 1. Therefore (1) follows if we show that $soc_3\Lambda$ is not contained in $\rho\Lambda$. Recall that $|soc_3\Lambda/soc_2\Lambda| = |J^2/J^3|$, see 2.2; and by (2), we have $|J^2/J^3| = 1 + |J^2(K\bar{D})/J^3(K\bar{D})|$ and hence $|soc_3\Lambda/soc_2\Lambda| = 1 + |soc_3 K\bar{D}/soc_2 K\bar{D}| = 1 + |soc_3(\rho\Lambda)/soc_2(\rho\Lambda)|$. We have that $\rho \in I$, so we may write $\rho = a + b$ for $a \in A$ and $b \in B$.

(4) <u>$a\xi \in soc\ \Lambda$ and $b\xi \in soc\ \Lambda$:</u> First observe that a and $b$ commute: $ab = (\rho - b)b = b\rho - b^2 = ba$. Now $(a + b)\xi = \rho\xi = 0$ and $a\xi = -b\xi \in A \cap B = soc\ \Lambda$. By (3), there are $u, v \in \Lambda$ and $c, d \in K$ such that $a = \rho u + c\omega$ and $b = \rho v + d\omega$. Since $a + b = \rho$ (and $\omega \notin \rho\Lambda$), we deduce $\rho(u + v - 1) = 0$ and hence $u + v - 1 \in J$. Therefore at least one of $u$, $v$ does not lie in $J$ and is then a unit, since $\Lambda$ is local. Say $u$ is a unit.

(5) <u>$soc_2\Lambda \subset A$:</u> We apply 2.4. With the notation there, suppose, say, $\omega Z_1$ does not lie in soc $\Lambda$; then $Z_1 \notin J^2$. Consequently $a_1 - 2 + \Sigma\ a_i - 1 = 1$. Now, $r \geq 2$ and hence $r = 2$ and $Z_1 = X_2$ and $X_1{}^2$ and $X_2{}^2 \equiv 0 \bmod \xi\Lambda$. This implies that $\Lambda/\xi\Lambda$ has dimension 4 and hence $D/D'$ is a Klein 4-group. By Tausski's Theorem [$H$ p.339], $D$ is dihedral or semidihedral or quaternion, a contradiction. Hence $\omega Z_1, \omega Z_2, ..., \omega Z_r$ lie in soc $\Lambda$, and therefore $aZ_i \equiv u\rho Z_i \bmod soc\ \Lambda$, and it follows that $u\rho Z_i \in A$ for $1 \leq i \leq r$. Now, $soc_2\Lambda$ is central in $\Lambda$ (since $\Lambda$ is local symmetric), and hence also $\rho Z_i u$ and $\rho Z_i$ belong to A ; that is, A contains $soc_2\Lambda$.

This implies now that $B \subset soc\ \Lambda$, contrary to the hypothesis.

Now suppose that $\bar{D}$ also satisfies the hypothesis; and by induction assume the statement holds for $\bar{D}$ . Let $\bar{\Lambda} = \Lambda/R_\rho$; by 2.1 we have that $\bar{\Lambda} \cong K\bar{D}$; and moreover from 2.7 and 2.1(3) we have that $J(K\bar{D}') \subset \rho I = aI + bI$ , and $aI \cap bI \subset soc\ K\bar{D}$ . By the induction hypothesis, this decomposition must be trivial; say $aI \subset soc\ \Lambda$ and $bI = \rho I$ .

(6) $\underline{soc_2\Lambda \subset \rho\Lambda}$: Suppose $m \in \Lambda$ does not belong to $\rho\Lambda$ ; we have to show that $J^2 m \neq 0$ . By the hypothesis, $z \in \phi(D)$ , and hence, $\xi \in J^2$ by 2.2. Suppose $\xi m = 0$ , then $m \in R_\xi = \rho\Lambda$ .

(7) $\underline{\xi \notin A}$: Suppose $\xi \in A$ ; then $\rho\Lambda \subset A$ and $soc_2\Lambda \subset A$ ; and then $soc_2\Lambda \cap B \subset soc\ \Lambda$ and $B \subset soc\ \Lambda$ , a contradiction.

(8) $\underline{\text{rad } A = A \cap \xi\Lambda}$: By 2.7 we have that aB and bA are contained in soc $\Lambda$ ; therefore modulo the socle, $aI = aA = (a+b)A = \rho A \subset soc\ \Lambda$; consequently $\rho(rad\ A) = 0$ and rad $A \subset R_\rho = \xi\Lambda$.
Suppose $m = \xi v \in A \cap \xi\Lambda$ , for some $v \in \Lambda$ . Since $\xi \notin A$ , we know that $v$ is not a unit and then $v \in J$. Write $\xi = a_1 + b_1$ with $a_1 \in A$ and $b_1 \in B$ . Then $m - a_1 v = b_1 v \in A \cap B \subset soc\ \Lambda \subset rad\ A$ ; and since also $a_1 v \in AJ = rad\ A$ , we deduce $m \in rad\ A$ .
This shows that $0 \neq A/rad\ A = A/A \cap \xi\Lambda \cong A + \xi\Lambda/\xi\Lambda \subset \Lambda/\xi\Lambda$ . Now by 2.1, $\Lambda/\xi\Lambda \cong K\bar{D}$ , hence has a simple socle. On the other hand, $A/rad\ A$ is semisimple and non-zero, and therefore $A/rad\ A$ is simple.
Write $A = \alpha\Lambda$ where $\alpha \in \Lambda$ and $\alpha + \xi\Lambda$ spans the socle of $\Lambda/\xi\Lambda$. We have here that $< z > \not\subseteq D'$ (since $\bar{D}$ is not abelian), and hence $\xi\Lambda \not\subseteq J_1 = J(KD')KD$ and $A \subset \alpha\Lambda + \xi\Lambda \subset J_1$. On the other hand, $J_1 \subset I$, and $\alpha \in I - rad\ I$ which implies that $\alpha \in J_1 - rad\ J_1$. Since $\Lambda/\xi\Lambda$ has a simple socle, this means that $J_1/\xi\Lambda$ must be simple. However, then $\dim \Lambda/J_1 = \dim \Lambda/\xi\Lambda - 1$; on the other hand, $\dim \Lambda/J_1 = |D/D'|$ and this divides $|\bar{D}| = \dim \Lambda/\xi\Lambda$. It follows that $\dim \Lambda/J_1 = 1$ and $D = D' = 1$, a contradiction.

## 3. Components contaning cyclic submodules of $\Lambda$

We will now study Auslander-Reiten components of cyclic submodules for $\Lambda$ when $\Lambda$ is a commutative wild $p$-group algebra.

**Lemma 3.1** *Let $\Lambda = KD$ where $D$ is an abelian non-cyclic p-group. Suppose $\omega\Lambda \subset \Lambda$ such that $\Lambda/\omega J$ has a simple socle, then $\omega\Lambda = K$ or $\Lambda$.*

**Proof.** Assume $\omega J \neq 0$; then we have soc $\Lambda \subset \omega J$; let $k \geq 1$ such that $soc_k\Lambda \subset \omega J$ but $soc_{k+1}$ is not contained in $\omega J$. Take $x \in soc_{k+}{}^1\Lambda - \omega J$, then $xJ \subset soc_k\Lambda \subset \omega J$ and hence the coset of $x$ belongs to soc $\Lambda/\omega J$. The hypothesis implies that soc $\Lambda/\omega J$ is spanned by the coset of $\omega$; consequently $x = c\omega$ mod $soc_k\Lambda$, for some $c \in K*$; in particular $\omega \in soc_{k+1}\Lambda$, and moreover $soc_{k+1}\Lambda/soc_k\Lambda$ is spanned by the coset of $\omega$.

For an abelian $p$-group algebra the only socle factors which are simple are $K$ and $\Lambda/J$ : From 2.3 one sees that the socle factors have bases consisting of cosets of homogenion polynomials of the appropriate degree. We assumed $\omega J \neq 0$ and therefore $\omega\Lambda = \Lambda$.

**Proposition 3.2** *Let $\Lambda = KD$ be commutative of wild type where $D$ is a $p$-group and char $K = p$. Suppose $\omega\Lambda \subset \Lambda$. Then $\omega\Lambda$ lies at the end of a component of tree class $A_\infty$ .*

**Proof.** First, we will show that $\omega\Lambda$ has one predecessor in $\Gamma_s(\Lambda)$ . This follows from 1.5 if we prove that $\omega$ is non-supportive. If $\omega \in J - J^2$ then this is true by 1.6. Suppose $\omega \in J^2$ ; and assume first that $I = J/\omega\Lambda$ is indecomposable. By 2.6, this is always the case for $\omega \in J^3$ . Now $\Lambda/\omega J$ does not have a simple socle 3.1); hence by 1.6, $\omega\Lambda$ is non-supportive.

It remains to consider the case that $\omega \in J^2 - J^3$ where $I = J/\omega\Lambda$ is decomposable; by 2.6 then $D$ has rank 2 and $\omega = XY + \rho$ for $\rho \in J^3$ where $X$, $Y$ are independent generators of $\Lambda$ . Here one checks directly that $\omega$ is non-supportive.

Now we will show that $\omega\Lambda$ does not lie at the end of a $\mathbf{Z}D_\infty$-component. Let $\theta$ be the component of $\omega\Lambda$ ; we are done by 1.7 in the case when $J/\omega\Lambda$ and $\Omega(\omega\Lambda)/K$ are indecomposable. Suppose now that $J/\omega\Lambda$ is decomposable, then we have the situation as given in 2.6. In this case, actually, $J/\omega\Lambda$ is a direct sum of two uniserial modules. Assume for contradiction that $\theta \cong \mathbf{Z}D_\infty$; then we have, from the proof of 1.7, that the module $R$ appearing in the component of $\Omega^{-1}(\omega\Lambda)$ is one of the summands of $J/\omega\Lambda$. Then $R$ is uniserial, consequently $R \cong \alpha\Lambda$ for some $\alpha \in \Lambda$ , and by the lengths, $R \subset J^3$. Hence by 2.6, $J/\alpha\Lambda$ and $\Omega(\alpha\Lambda)/K$ are indecomposable. Therefore by the first part of the proof, $R$ should belong to an $A_\infty$-component, a contradiction.

The case when $\Omega(\omega\Lambda)/K$ is decomposable is dual.

**Proposition 3.3** *Let $\Lambda = KD$ where $D$ is a $p$-group such that $\Lambda$ is wild. Suppose $M$ is a cyclic submodule of $\Lambda$ on which the commutator subgroup $D'$ of $D$ acts trivially. Let $\theta$ be the component of $M$ ; then the tree class of $\theta$ is $A_\infty$ and $M$ lies at the end.*

**Proof.** Let $M = \omega\Lambda$ for $\omega \in \Lambda$ . By 3.2, we may assume that $D$ is not abelian. Also, without loss of generality $M$ is not of the form $K_u^d$ for a subgroup $U$ of $D$ ; in this case the statement follows from [W] or [E].

By 2.8, applied to $\omega\Lambda$ and also to $I = R_\omega$ we have that $J/\omega\Lambda$ and $R_\omega/K$ are indecomposable. We claim that the element $\omega$ is non-supportive: Using 1.6, it suffices to show that $\Lambda/\omega J$ does not have a simple socle.

By the hypothesis, $\omega\Lambda$ is contained in the module $\Lambda^{d'}$ of $D'$-fixed elements. Suppose $\Lambda/\omega J$ has a simple socle, then also $\Lambda^{d'}/\omega J$ has a simple socle. Now $\Lambda^{d'} \cong K(D/D')$ , see 2.1(4), and $D/D'$ is abelian; therefore we may apply 3.1 and it follows that $\omega\Lambda = \Lambda^{d'}$ or $\omega\Lambda = K$ . Both possibilities were excluded since $\Lambda^{d'} \cong (K_{d'})^d$ and $K \cong K_d^d$ .

Hence $\omega\Lambda$ has one predecessor in $\theta$ , by 1.5. Moreover, since $J/\omega\Lambda$ is indecomposable, it follows from 1.7 that the tree class of $\theta$ is not $D_\infty$ . This completes the proof.

## 4. $\mathbf{Z}A_\infty$ -Components of $\Lambda$

We will now show that a $p$-group algebra of wild type has infinitely many components of the form $\mathbf{Z}A_\infty$ .

4.1 Suppose $M$ is indecomposable of dimension 2. Then the group $D$ is represented on $M$ as a group of invertible unipotent upper triangular $2 \times 2$ matrices. Therefore $D/ker\ M$ is elementary abelian, and $\phi(D)$ and $D'$ are contained in $\ker M$. Let $D/\phi(D) \cong (C_p)^n$ ; fix generators $x_1, x_2, \ldots, x_n$ . Then with respect to a basis of $M$ containing a generator for soc $M$ , $x_i$ is represented by $\left[\begin{smallmatrix} 1 & \lambda_i \\ 0 & 1 \end{smallmatrix}\right]$ . Not all $\lambda_i$ are zero since $M$ is indecomposable. Hence $\underline{\lambda} = [(\lambda_1, \ldots, \lambda_n)] \in P^{n-1}(K)$; put $M = M_{\underline{\lambda}}$, then $M_{\underline{\mu}} \cong M_{\underline{\mu}}$ if and only if $\underline{\lambda} = \underline{\mu}$ in $P^{n-1}(K)$.

An easy calculation shows that indecomposable modules of length 2 are always self-dual, where $M^* = Hom_k(M, K)$ with KD-right action given by $(\varphi g)(m) = \varphi(mg^{-1})$ , for $d \in D$ and $\varphi \in M^*$, $m \in M$ .

**Proposition 4.2** *Suppose $D$ $C_2 \times C_2r$ or $M_n(2)$ and $\Lambda = KD$. Then there are infinitely many $\Lambda$-modules of dimension 2 lying at ends of $\mathbf{Z}A_\infty$-components. Moreover, non-isomorphic such modules belong to different components.*

**Proof.** Suppose $M$ is a 2-dimensional submodule of $\Lambda$ . Then $D'$ acts trivially on $M$, hence by 3.2 and 3.3, the module lies at the end of a component of tree class $A_\infty$. Moreover, by 1.8 and self-duality, if such modules are non-periodic then distinct modules belong to different components. By 4.1 there are infinitely many such modules, and we require infinitely many non-periodic ones.

By [AE], $M$ is periodic if and only if $M_e$ is periodic or projective for all elementary abelian subgroups $E$ of $D$. Moreover, if $E$ has order $p^n$ and $M_e$ is periodic then $p^{n-1}$ divides $\dim M[C]$; therefore if $p > 2$ then we are done.

Now let $p = 2$, and suppose $D$ has only finitely many non-periodic modules of dimension 2. We will show that then $D \cong C_2 \times C_2r$ or $M_n(2)$.

Then there are 2-dimensional modules which are periodic; therefore by the above dimension argument, a maximal elementary abelian subgroup of $D$ has order $\leq 4$ ; that is, the 2-rank of $D$ is at most 2. Actually, the 2-rank is 2 since the only 2-groups with rank 1 are cyclic or quaternion [$H$], which we exclude.

(1) The Frattini subgroup $\phi(D)$ is cyclic: By 4.1, $\phi(D)$ acts trivially on $M$. On the other hand, $M_{\phi(d)}$ must be periodic, consequently by [AE], the 2-rank of $\phi(D)$ is 1, and $\phi(D)$ is either cyclic or quaternion. Then the centre of $\phi(D)$ is cyclic, and consequently $\phi(D)$ must be cyclic [Su (4.21)].

(2) $|D/\phi(D)| = 4$ : Clearly $|D/\phi(D)| \geq 4$ since $D$ is not cyclic. Suppose $|D/\phi(D)| \geq 8$. Since the 2-rank of $D$ is 2, $D$ is not elementary abelian and $\phi(D) > 1$, and then $\phi(D)$ contains a central element of order 2, $z$ say. Take an elementary abelian subgroup $E \leq D$ of order 4; then $E \cap \phi(D)$ must contain $z$; otherwise the group $< E, z >$ *would* have rank 3. Now, if $M_e$ is non-periodic then $M$ is non-periodic, by [AE]. Choose $x \in E - \phi(D)$, and parametrize the 2-dimensional modules as in 4.1, taking $x_1 = x$. Then for $M = M_\lambda$ with $\lambda_1 = 0$, we have that $x$ acts trivially on $M$. Moreover, $z$ also acts trivially on $M$ since $z \in \phi(D)$. Hence $M_e \cong K \oplus K$ and is non-periodic. Since $n \geq 3$, there are infinitely many such modules, contrary to the assumption.

If $D$ is abelian then by (1) and (2) we have that $D \cong C_2 \times C_2r$ . Suppose $D$ is non-abelian. If $\phi(D) = D'$ then by the Theorem of Tausski [*H* p.339], $D$ would be dihedral or semidihedral. Hence we have that $D' \not\subseteq \phi(D)$; consequently

$D/D' \cong C_2 \times C_2r$ for $r > 1$ . Moreover, the image of $\phi(D)$ in $D/D'$ is cyclic of order $2^{r-1}$ , so there is an element $x \in D$ such that $< x^2, D' >= \phi(D)$ . Since $\phi(D)$ is cyclic and $D' \subset \phi(D)$ , we deduce $\phi(D) =< x^2 >$ . Then $H =< x >$ *is* a maximal subgroup of $D$ which is cyclic. Since $D$ is not abelian, $H = C_d(H)$ , and by [Go], $D \cong M_n(2)$ , as required.

4.3 We will now study the 3-dimensional cyclic submodules of $\Lambda$ for the remaining two types of groups. Suppose $p = 2$ .

(a) Assume first that $D = C_2 \times C_2r$ for $r \geq 2$. Then the group algebra $\Lambda = KD$ is isomorphic to the algebra $K[X,Y]/(X^2, Y^m)$ where $m = 2^r$. Let $M = \omega\Lambda \subset \Lambda$ be of dimension 3; then $M$ is uniserial and $\omega X = c\omega Y$ mod soc $\Lambda$ for some $c \in K$. Consequently $\omega X^2 = c^2\omega Y^2$, and since $X^2 = 0$ it follows that $c = 0$ and $\omega X$ belongs to soc $\Lambda$. Hence $M$ has a $K$-basis $\{\omega Y^2, \omega Y, \omega\}$, and with respect to this basis, $X$ and $Y$ are represented by

$$X \to \begin{pmatrix} 0 & 0 & s \\ 0 & 0 & 0 \\ 0 & 0 & 0 \end{pmatrix} \text{ and } Y \to \begin{pmatrix} 0 & 1 & 0 \\ 0 & 0 & 1 \\ 0 & 0 & 0 \end{pmatrix}$$

We denote this module by $M_s$. Suppose $M_s \cong M_t$ ; then $M_s$ and $M_t$ have equal annihilators. The element $X - sY^2$ annihilates $M_s$; and if it also annihilates $M_t$ then it follows that $s = t$. An easy calculation shows that $M_s \cong M_s^*$.

(b) Now let $D = M_n(2)$ , and let $M$ be a 3-dimensional $\Lambda$-module. Then $g^4$ acts trivially on $M$ , for all $g \in D$ . It follows that the central involution, $z$ , say, belongs to Ker $M$ . Hence $M$ may be considered as a $D/ < z >$ -module; and since this group is $C_2 \times C_2r$ , the 3-dimensional $\Lambda$-modules are parametrized as in (a).

**Proposition 4.4** *Let* $\Lambda = KD$ *where* $D = C_2 \times C_2r$ *with* $r \geq 2$ , *or* $M_n(2)$ *with* $n \geq 3$ , *and let* $M \leq \Lambda$ *be 3-dimensional cyclic. Then* $M$ *lies at the end of a component* $\cong \mathbf{Z}A_\infty$ , *and distinct such modules lie in different components.*

**Proof.** We see from 4.3 that the commutator of $D$ acts trivially on $M$. Moreover since $\dim M$ is odd, $M$ is non-periodic; and the first statement follows from 3.2 and 3.3. The last statement holds by 4.3.1(4) and 1.8.

4.5 Suppose $D$ is an arbitrary $p$-group and $\Lambda = KD$ . Consider 2 - dimensional indecomposable $\Lambda$-modules. We note that all but finitely many have vertex $D$ : Suppose $V = vx(M) \subset D$ and $\dim M = 2$ ; then there is an indecomposable $V$-module $S$ such that $M$ is a direct summand of the induced module $S \otimes_{kv} KD$ . On the other hand, by Green's theorem, the induced module is indecomposable, so it is isomorphic to $M$ . In particular, for $p > 2$ we have $V = D$ , and in case $p = 2$ and $V \neq D$ we deduce $\dim S = 1$ and $|D : V| = 2$ . Consequently $S \cong K$ , and these modules correspond to the maximal subgroups of D. Similarly if dim $M = 3$ but $p = 2$ then the vertex of $M$ is $D$.

We shall use that if $M$ lies at the end of a component $\theta$ and $vx(M) = D$ then all modules in $\theta$ have maximal vertex; this is proved in [$E$].

**Corallary 4.6** *Let* $D$ *be a p-group such that* $\Lambda = KD$ *is of wild type. Then* $\Lambda$ *has infinitely many components of the form* $\mathbf{Z}A_\infty$. *Moreover, there are infinitely* *many such components which contain a submodule of $KD$ and in which all modules have vertex $D$.*

## 5. Arbitrary wild blocks

The results on $p$-group algebras have the following consequence for arbitrary wild blocks:

**Theorem 5.1** *Suppose $B$ is a block of some group algebra such that $B$ is of wild representation type. Then the stable Auslander-Reiten quiver of $B$ has infinitely many components $\cong \mathbb{Z}A_\infty$. Moreover, there are infinitely many components which contain a module of dimension $< \dim B$.*

**Proof.** We will actually show that there are infinitely many such components in which all modules have maximal vertex. Let $D$ be a defect group of $B$ , then $D$ is not cyclic or dihedral, semidihedral or quaternion [BD]. If $D = G$ then $B = KD$, and the statement holds by 4.6.

Now we assume that $G = DC_g(D)$ . Then the block $B$ is Morita equivalent to KD ; this is well-known (see for example [E 2]). Moreover, the usual equivalence preserves vertices of modules. Hence $\Gamma_s(B)$ has infinitely many components $\cong \mathbb{Z}A_\infty$ on which all modules have vertex $D$ .

Next, we consider the case when $D$ is a normal subgroup of $G$ . Let $C = DC_g(D)$, then $C \trianglelefteq G$. Suppose $b$ is a block of $C$ which is covered by $B$ ; then $b$ has also defect group $D$ . By the previous stage, we know that $\Gamma_s(b)$ has infinitely many components $\cong \mathbb{Z}A_\infty$ in which all modules have maximal vertex.

It is enough to consider the case when $b$ is $G$-stable: In general, there is a unique block $b_1$ of the stabilizer $N_g(b)$ with $b_1^g = B$ , and the blocks $b_1$ and $B$ are Morita equivalent; and vertices are preserved under the usual Morita equivalence; this is again well-known. In particular, $B$ and $b_1$ have isomorphic AR-quivers; so we may replace $G$ by $N_g(b)$ and $B$ by $b_1$.

Then the index of $C$ in $G$ is not divisible by $p$ since $b$ and $B$ have the same defect group (see for example [A]). Therefore we can apply a result of [S], namely:

($*$) If $X$ is an indecomposable KC-module and $\mathcal{A}(X)$ is the AR-sequence of $X$ and if $X^g \cong \oplus X_i$ then $\mathcal{A}(X)^g \cong \oplus \mathcal{A}(X_i)$ .

Let $\theta$ be a component of $\Gamma_s(b)$ which is isomorphic to $\mathbb{Z}A_\infty$ in which all modules have vertex $D$, and let $M$ be a module at the end of $\theta$. Suppose $N := \{g \in G : M \otimes g \cong M\}$. then $N = \{g \in G : X \otimes g \cong X$ for all $X \in \theta\}$. This is true since $N$ must fix the modules at the end of $\theta$, and a component $\cong \mathbb{Z}A_\infty$ does not have an automorphism which is induced by $- \otimes g$ for $g \in G$ . This has as a consequence that for any $X \in \theta$ , the number of indecomposable direct summands of $X^g$ is the same, namely the number, $m$ say, of indecomposable direct summands of $K(N/C)$ in a fixed decomposition.

Now let $\mathcal{A}(M) : 0 \to \tau M \to E \to M \to 0$ be the AR-sequence ending in $M$; then $\mathcal{A}(M)^G \cong \bigoplus_{i=1}^m \mathcal{A}(M_i)$ where $M^G \cong \bigoplus_{i=1}^m M_i$. Since $M$ lies at the end, $E$ is indecomposable, and then $E^g$ is a direct sum of $m$ indecomposable modules. Now, there are also $m$ AR-sequences and each has at least one non-zero summand in the middle. It follows that for all $i$, the middle of $\mathcal{A}(M_i)$ must be indecomposable.

Let $E_i$ be the middle of $\mathcal{A}(M_i)$ and $\theta_i$ be the component of $M_i$ , then $\theta_i$ has an end. Consider the AR-sequence of $b$-modules starting with $E$ , the middle has two indecomposable summand, one of them is $M$ . Say the sequence is $\mathcal{A}(\tau^{-1}E) : 0 \to E \to M \oplus X \to \tau^{-1}E \to 0$. By $(*)$, we have that $\mathcal{A}(\tau^{-1}E)^G \cong \bigoplus_{i=1}^m \mathcal{A}(\tau^{-1}E_i)$ . For each $i$ , $M_i$ must occur in the middle of $\mathcal{A}(\tau^{-1}E_i)$ . Moreover, by the exactness, there is at least one other indecomposable summand. On the other hand, $X^g$ has exactly $m$ indecomposable summands which is also the number of sequences. Therefore the middle of $A(\tau^{-1}E_i)$ consists of two indecomposable direct summands, for each $i$ .

This shows that $\theta_i \neq \mathbf{Z}D_\infty$ , and therefore, using [W] and [ES], the only other possibility for $K$ algebraically closed is that $\theta_i \cong \mathbf{Z}A_\infty$ . Moreover, inducing from $C$ to $N$ preserves vertices since $|N : C|$ is not divisible by $p$ ; and inducing could merge at most finitely many components. The block $b$ has infinitely many such components $\theta$ , and hence $B$ has infinitely many components $\cong \mathbf{Z}A_\infty$ on which the modules have maximal vertex.

Now let $G$ be arbitrary. Let $b$ be the Brauer correspondent of $B$ in $N_g(D)$; then $D$ is the defect group of $b$ . By the previous part of the proof, $\Gamma_s(b)$ has infinitely many components $\cong \mathbf{Z}A_\infty$ on which the modules have maximal vertex. Let $\theta$ be such a component. The Green correspondents of $b$-modules belong to mod $B$ . Moreover, by [K], Green correspondence induces a graph isomorphism from $\theta$ to a subquiver of a unique component $\Delta$ of $\Gamma_s(B)$ . If now $\theta \cong \mathbf{Z}A_\infty$ then there is an unbounded additive function of $\theta$ , given by $d(M) = \dim Hom_{kd}(W, M)$ where $W$ is a fixed periodic KD-module, for $M$ in $\theta$ (see [O], [ES]). It follows from the Nakayama relations that $d$ also induces an additive function on $\Delta$, still unbounded. Therefore the tree class of $\Delta$ must be $A_\infty$, by [W], since $D_\infty$ and $A_\infty^\infty$ admit only bounded additive functions (see [HPR] or [B]). This completes the first part of the proof.

The block $B$ of the form $B = eKG$ where $e$ is a central idempotent of $KG$. Each of the component in question contains a module $M$ of the block $B$ which is a direct summand of $X^G$ for a submodule $X$ of $KD$, by 4.6. Inducing and multiplying with a block idempotent preserves inclusion; therefore $(X^G)e$ and $M = Me$ are submodules of $B$. Since $M$ is indecomposable and not projctive, we have $\dim M < \dot{\imath}mB$.

## REFERENCES

[A] J.L. Alperin, "Local Methods in Block Theory", Cambridge University Press (1986).

[AE] J.L. Alperin, L. Evens, "Representations, resolutions and Quillen's dimension theorem", J. Pure Appl. Algebra 22 (1981), 1-9.

[AR] M. Auslander, I. Reiten, "Representation theory of Artin algebras III, Almost split sequences", Commun. Algebra 3 (1975), 239-294.

[B] D. Benson, "Modular representation theory: new trends and methods", Springer Lect. Notes in Maths. 1081 (1984).

[BD] V.M. Bondarenko, J.A. Drozd, "The representation type of finite groups", Zap. Nauchn. Sem.

Leningrad. Otdel. Mat. Inst. Steklov. 71 (1977), 24-71.

[Bes] C. Bessenrodt, "The Auslander-Reiten quiver of a modular group algebra revisited", Preprint (1989).

[BR] M.C.R. Butler, C.M. Ringel, "Auslander-Reiten sequences with few middle terms and applications to string algebras", Comm. Algebra 15 (1987), 145-181.

[C] J. F. Carlson, "The dimensions of periodic modules over modular group" algebras. III", J. Math. 23 (1979), 295-306.

[E] K. Erdmann, "On the vertices of modules in the Auslander-Reiten quiver of $p$-groups", Math. Z. 203 (1990), 321-334.

[E 2] K. Erdmann, "Blocks of tame representation type and related algebras", Springer Lect. Notes in Mathss 1428 (1990).

[ES] K. Erdmann, A. Skowronski, "On Auslander-Reiten components of blocks and self-injective biserial algebras", To appear.

[G] P. Gabriel, "Auslander-Reiten sequences and representation-finite algebras, Representation theory I" Springer Lect. Notes in Maths. 831 (1980), 1-71.

[Go] D. Gorenstein, "Finite groups", Harper and Row (1968).

[H] B. Huppert, "Endliche Gruppen I", Grundlehren der mathematischen Wissenschaften. Springer 134 (1967).

[HB] B. Huppert, N. Blackburn, "Finite groups II" Grundlehren der mathematischen Wissenschaften. Springer 242 (1982).

[HPR] D. Happel, U. Preiser, C.M. Ringel, "Vinberg's caracerization of Dynkin diagrams using subadditive functions with applications to DTr-periodic modules", Representation Theory II", Springer Lect. Notes Maths. 832 (1980), 280-294.

[K] S. Kawata, "Module correspondence in Auslander-Reiten quivers for finite groups", Preprint 1988.

[L] P.A. Linnell, "The Auslander-Reiten quiver of a finite group", Arch. Math. 45 (1985), 289-295.

[O] T. Okuyama, "On the Auslander-Reiten quiver of a finite group", J. Alg. 110 (1987), 425-430.

[Ri] C. Riedtmann, "Algebren, Darstellungsköcher, Überlagerungen und zurück" Comment. Math. Helv. 55 (1980), 199-224.

[R] C.M. Ringel, "Tame algebras and integral quadratic forms", Springer Lect. Notes Maths. 1099 (1984).

[S] O. Solberg, "Strongly graded rings and almost split sequences", Trondheim, preprint (1987).

[Su] M. Suzuki, "Finite groups II", Grundlehren der mathematischen Wissenschaften, Springer 248 (1986).

[W] P. Webb, "The Auslander-Reiten quiver of a finite group" Math. Z. 179 (1982), 97-121.

Progress in Mathematics, Vol. 95, © 1991 Birkhäuser Verlag Basel

# On Gorenstein Algebras

DIETER HAPPEL

## Introduction

Let $A$ be a finite-dimensional $k$-algebra (associative, with unit) over some fixed algebraically closed field $k$. Let $\operatorname{mod} A$ be the category of finitely generated left $A$-modules. With $D = \operatorname{Hom}_k(-,k)$ we denote the standard duality with respect to the ground field. Then ${}_AD(A_A)$ is an injective cogenerator for $\operatorname{mod} A$. For an arbitrary $A$-module ${}_AX$ we denote by $\operatorname{proj.dim}_A X$ (resp. $\operatorname{inj.dim}_A X$) the projective dimension (resp. the injective dimension) of the module ${}_AX$.

Following Auslander and Reiten [AR2], we call a finite-dimensional algebra $A$ a Gorenstein algebra if $A$ satisfies $\operatorname{proj.dim}_A D(A_A) < \infty$ and $\operatorname{inj.dim}_A A < \infty$. We refer to [AR2],[AB],[Bu1],[Bu2] for the relationship of this notion to the theory of Gorenstein rings in commutative algebra. Note that selfinjective algebras as well as algebras of finite global dimension are Gorenstein algebras.

It is conjectured that $\operatorname{proj.dim}_A D(A_A) < \infty$ is equivalent to the condition that $\operatorname{inj.dim}_A A < \infty$. We will briefly discuss this in section one. It is easily seen that $\operatorname{proj.dim}_A(DA_A) = \operatorname{inj.dim}_A A$, if both are finite.

Let us now formulate some of the results established here. For this we need some more notation. Let ${}_A\mathcal{P} \subset \operatorname{mod} A$ be the full subcategory containing the finitely generated projective $A$-modules. Let $K^b({}_A\mathcal{P})$ be the homotopy category of bounded complexes over ${}_A\mathcal{P}$. Similarly, we denote by $K^b({}_A\mathcal{I})$ the homotopy category of bounded complexes over the full subcategory ${}_A\mathcal{I}$ containing the finitely generated injective $A$-modules. Let $D^b(A)$ be the derived category of bounded complexes over $\operatorname{mod} A$. We consider $K^b({}_A\mathcal{P})$ and $K^b({}_A\mathcal{I})$ as full subcategories of $D^b(A)$.

In section one we will show that an algebra $A$ is a Gorenstein algebra if and only if $K^b({}_A\mathcal{P}) = K^b({}_A\mathcal{I})$.

Also $K^b({}_A\mathcal{P})$ is an épaisse subcategory in the sense of Verdier [V]. So we may localize with respect to this subcategory. The resulting triangulated category is denoted by $\mathcal{D}_\mathcal{P}$. It seems that in general it is hard to describe this category explicitly.

In section four we will gather some information on $\mathcal{D}_\mathcal{P}$. Note, that $\operatorname{mod} A$ is embedded into $D^b(A)$. So we obtain a functor from $\operatorname{mod} A$ to $\mathcal{D}_\mathcal{P}$ We will show that this functor is dense, if $\operatorname{proj.dim}_A(DA_A) < \infty$. If $A$ is a Gorenstein algebra, then $\mathcal{D}_\mathcal{P}$ has a particularly nice structure. We denote by

$$\mathcal{X} = \{{}_AX \mid \operatorname{Ext}^i_A(D(A_A), X) = 0, \text{ for } i > 0\}$$

the category of Cohen-Macaulay modules. We refer to [AR2] and the articles quoted there for an intensive study of the dual of this category. It turns out that $\mathcal{D}_\mathcal{P} \simeq \mathcal{X}/D(A_A)$, where $\mathcal{X}/D(A_A)$ is the category obtained from $\mathcal{X}$ by factoring out those maps which factor over an injective $A$-module. Note that this generalizes a result of Rickard [R1]. In fact, if $A$ is a selfinjective algebra then $\mathcal{X} = \text{mod}\, A$. Thus $\mathcal{D}_{\mathcal{P}}$ is equivalent to the stable module category. Also note that for a Gorenstein algebra $A$ a module $X \in \mathcal{X}$ of finite projective dimension is an injective $A$-module. In particular this implies that $\mathcal{D}_{\mathcal{P}} = 0$ for an algebra of finite global dimension.

The main result in section three asserts that $K^b({}_A\mathcal{P})$ has Auslander-Reiten triangles if and only if $\text{proj.dim}_A D(A_A) < \infty$. In particular, if $A$ is a Gorenstein algebra we see that the quiver of $K^b({}_A\mathcal{P})$ has the structure of a stable translation quiver. We will include some computations of certain components in this section. We will also recall in section 3 the relevant background.

In section one we recall some easily established homological properties of Gorenstein algebras.

We refer to section two for a discussion of some examples of Gorenstein algebras.

We denote the composition of morphisms $f : X \to Y$ and $g : Y \to Z$ in a given category $\mathcal{K}$ by $fg$. For the unexplained representation-theoretic terminology we refer to [Ri].

## Acknowledgements

Some of the results presented here were obtained while the author was visiting the University of Trondheim. I am thankful to Maurice Auslander and Idun Reiten for numerous inspiring discussions.

## 1. Homological properties

1.1. In this paper $A$ will always denote a finite-dimensional $k$-algebra over some fixed algebraically closed field $k$. Unless stated otherwise all modules will be finitely generated left $A$-modules.

1.2. The following lemma collects some homological properties of Gorenstein algebras. The proofs are rather easy and will be omitted. For (ii) we refer to [AR2] and (i) is a special case of lemma 1.2 in [H1]. Recall that the finistic dimension fin.dim $(A)$ of an algebra $A$ is defined as

$$\text{fin.dim}(A) = \sup \{\text{proj.dim}_A X \mid \text{proj.dim}_A X < \infty\}.$$

It is conjectured (see [AR1] for a discussion) that $\text{fin.dim}(A) < \infty$.

**Lemma:** *Let $A$ be a finite-dimensional $k$-algebra with $\text{proj.dim}_A D(A_A) = r < \infty$. Then:*

(i) *If $\text{inj.dim}_A A < \infty$, then $\text{proj.dim}_A D(A_A) = \text{inj.dim}_A A$.*
(ii) *$\text{inj.dim}_A A < \infty$ if and only if $\text{fin.dim}(A) < \infty$.*
(iii) *If $\text{inj.dim}_A A < \infty$, then $\text{proj.dim}_A X < \infty$ if and only if $\text{inj.dim}_A X < \infty$.*

We remark that an algebra $A$ is a Gorenstein algebra if and only if $D(A_A)$ is a (generalized) tilting module. This will turn out to be useful in section 4. So in particular it follows from a result in [Bo] that $\text{proj.dim}_A D(A_A) \leq 1$ implies that $\text{inj.dim}_A A \leq 1$. For related results we refer to [RS].

1.3. Recall that $\mathcal{X} = \{{}_AX \mid \text{Ext}^i_A(D(A_A), X) = 0, \text{ for } i > 0\}$. If $\mathcal{X} = \text{mod}\, A$ we infer that $A$ is a selfinjective algebra.

Now assume that $A$ satisfies $\text{proj.dim}_A D(A_A) = r < \infty$.

Let ${}_AX \in \text{mod}\, A$ and let

$$0 \longrightarrow {}_AX \xrightarrow{d^0} I^0 \xrightarrow{d^1} I^1 \xrightarrow{d^2} I^2 \longrightarrow \cdots$$

be a minimal injective resolution of ${}_AX$. Set $W_i = \text{cok}\, d^i$. Then it is easily seen that $W_i \in \mathcal{X}$ if $i \geq r - 1$. This also trivially implies for an algebra of finite global dimension that ${}_AX$ is injective if ${}_AX \in \mathcal{X}$.

1.4. We will need some facts about the derived category $D^b(A)$ of bounded complexes over mod $A$ as well as some notation for complexes which we will introduce next. We also recall the definition of some triangulated categories which will be needed later.

Let $\mathfrak{a}$ be an arbitrary additive subcategory of mod$A$.
A complex $X^\bullet = (X^i, d^i_X)_{i \in \mathbb{Z}}$ over $\mathfrak{a}$ is a collection of objects $X^i$ from $\mathfrak{a}$ and morphisms $d^i = d^i_X : X^i \to X^{i+1}$ such that $d^i d^{i+1} = 0$. A complex $X^\bullet = (X^i, d^i_X)$ is bounded below if $X^i = 0$ for all but finitely many $i < 0$. It is called bounded above if $X^i = 0$ for all but finitely many $i > 0$. It is bounded if it is bounded below and bounded above. It is said to have bounded cohomology if $H^i(X^\bullet) = 0$ for all but finitely many $i \in \mathbb{Z}$, where by definition $H^i(X^\bullet) = \text{ker} d^i_X / \text{im} d^{i-1}_X$. Denote by $C(\mathfrak{a})$ the category of complexes over $\mathfrak{a}$, by $C^{-,b}(\mathfrak{a})$ (resp.$C^{+,b}(\mathfrak{a})$, resp. $C^b(\mathfrak{a})$) the full subcategories of complexes bounded above with bounded cohomology (resp. bounded below with bounded cohomology, resp. bounded above and below).

If $X^\bullet = (X^i, d^i_X)_{i \in \mathbb{Z}}$ and $Y^\bullet = (Y^i, d^i_Y)_{i \in \mathbb{Z}}$ are two complexes, a morphism $f^\bullet : X^\bullet \to Y^\bullet$ is a sequence of morphisms $f^i : X^i \to Y^i$ of $\mathfrak{a}$ such that

$$d^i_X f^{i+1} = f^i d^i_Y$$

for all $i \in \mathbb{Z}$. The translation functor is defined by

$$(X^\bullet[1])^i = X^{i+1} \quad , \quad (d_{X[1]})^i = -(d_X)^{i+1} \; .$$

The mapping cone $C_{f^\bullet}$ of a morphism $f^\bullet : X^\bullet \to Y^\bullet$ is the complex

$$C_{f^\bullet} = ((X^\bullet[1])^i \oplus Y^i, d^i_{C_f})$$

with 'differential'

$$d^i_{C_f} = \begin{pmatrix} -d^{i+1}_X & f^{i+1} \\ 0 & d^i_Y \end{pmatrix} .$$

We denote by $K^{-,b}(\mathfrak{a}), K^{+,b}(\mathfrak{a})$ and $K^b(\mathfrak{a})$ the homotopy categories of the categories of complexes introduced above. Note that all these categories are triangulated categories in the sense of [V].

Recall that two morphisms $f^\bullet, g^\bullet : X^\bullet \to Y^\bullet$ are called homotopic, if there exist morphisms $h^i : X^i \to Y^{i-1}$ such that $f^i - g^i = d_X^i h^{i+1} + h^i d_Y^{i-1}$ for all $i \in \mathbb{Z}$.

We have denoted by ${}_A\mathcal{P}$ (resp. ${}_A\mathcal{I}$) the full subcategory of mod $A$ formed by the projective (resp. injective) $A$-modules. Then we identify the derived category $D^b(A)$ of bounded complexes over mod $A$ with $K^{-,b}({}_A\mathcal{P})$ or with $K^{+,b}({}_A\mathcal{I})$. In case $A$ has finite global dimension this yields the identification of $D^b(A)$ with $K^b({}_A\mathcal{P})$ or with $K^b({}_A\mathcal{I})$, since the natural embedding of $K^b({}_A\mathcal{P})$ into $K^{-,b}({}_A\mathcal{P})$ is an equivalence in this case.

1.5. The following lemma is straightforward

**Lemma:** *Let $A$ be a finite-dimensional $k$-algebra. Then*
*(i) $\text{proj.dim}_A D(A_A) < \infty \Leftrightarrow K^b({}_A\mathcal{I}) \subseteq K^b({}_A\mathcal{P})$.*
*(ii) $\text{inj.dim}_A A < \infty \Leftrightarrow K^b({}_A\mathcal{P}) \subseteq K^b({}_A\mathcal{I})$.*
*(iii) $A$ is Gorenstein $\Leftrightarrow K^b({}_A\mathcal{I}) = K^b({}_A\mathcal{P})$.*

Also note the following for a Gorenstein algebra $A$.
Denote by $\nu = \text{Hom}_A(D(A_A), -)$ the functor from $K^b({}_A\mathcal{I})$ to $K^b({}_A\mathcal{P})$. Then there exists a commutative diagram of triangulated categories

$$\begin{array}{ccc} D^b(A) & \xrightarrow{F} & D^b(A) \\ \mu \uparrow & & \uparrow \mu' \\ K^b({}_A\mathcal{I}) & \xrightarrow{\nu} & K^b({}_A\mathcal{P}) \end{array}$$

where F is the right derived functor of the functor $\text{Hom}_A(D(A_A), -)$ considered as a functor from mod $A$ to mod $A$ and $\mu, \mu'$ the canonical inclusions. Note that both $F$ and $\nu$ are triangle equivalences. We will come back to this diagram in section 4.

1.6. Let $\mathcal{C}$ be a triangulated category and let $\mathcal{U} \subset \mathcal{C}$ be a full triangulated subcategory of $\mathcal{C}$. Following [V] we define

$$^{\perp}\mathcal{U} = \{X \in \mathcal{C} \mid \text{Hom}(X, U) = 0 \text{ for all } U \in \mathcal{U}\},$$

and

$$\mathcal{U}^{\perp} = \{X \in \mathcal{C} \mid \text{Hom}(U, X) = 0 \text{ for all } U \in \mathcal{U}\}.$$

Then trivially ${}^{\perp}\mathcal{U}$ and $\mathcal{U}^{\perp}$ are full triangulated subcategories of $\mathcal{C}$.

**Proposition:** *Let $A$ be a finite-dimensional $k$-algebra. Then*
*(i) If $\text{fin.dim}(A) < \infty$, then $K^b({}_A\mathcal{I})^{\perp} = 0$.*
*(ii) If ${}^{\perp}K^b({}_A\mathcal{P}) = 0$, then given $0 \neq {}_AX$ there exists $i \geq 0$ such that $\text{Ext}_A^i(X, {}_AA) \neq 0$.*
*(iii) If $K^b({}_A\mathcal{I})^{\perp} = 0$, then given $0 \neq {}_AX$ there exists $i \geq 0$ such that $\text{Ext}_A^i(D(A_A), X) \neq 0$.*

**Proof:** For (i) assume that $K^b({}_A\mathcal{I})^{\perp} \neq 0$. Let $I^\bullet \in K^b({}_A\mathcal{I})^{\perp}$. We assume that $I^\bullet = (I^i, d_I^i) \neq 0$. Applying the translation functor if necessary we may assume that $I^i = 0$ for $i < 0$. Next consider the complex

$$\cdots 0 \to \text{Hom}(D(A_A), I^0) \to \text{Hom}(D(A_A), I^1) \to \cdots$$

Observe that $\mathrm{Hom}(D(A_A), I^j) \in {}_A\mathcal{P}$. Since $I^\bullet \in K^b({}_A\mathcal{I})^\perp$ we infer that this complex is acyclic. Note that $\mathrm{proj.dim}\,\mathrm{cok}\,\mathrm{Hom}(D(A_A), d^i) < \infty$. But this contradicts $\mathrm{fin.dim}(A) < \infty$.

For (ii) assume that there exists $0 \neq {}_AX$ with $\mathrm{Ext}^i_A(X, {}_AA) = 0$ for all $i \geq 0$. Let $P^\bullet$ be a minimal projective resolution of ${}_AX$. We consider $P^\bullet$ as element of $D^b(A)$. We claim that $P^\bullet \in {}^\perp K^b({}_A\mathcal{P})$.

For this let $Q^\bullet = (Q^i, d^i_Q) \in K^b({}_A\mathcal{P})$. So there exists $r \leq s$ such that $Q^i = 0$ for $i < r$ and $i > s$. The width $w(Q^\bullet)$ of $Q^\bullet$ is by definition $s - r + 1$. The assertion $\mathrm{Hom}(P^\bullet, Q^\bullet) = 0$ now follows easily by induction on $w(Q^\bullet)$ by considering a triangle as in lemma 1.1 of [H1] and applying the cohomological functor $\mathrm{Hom}_{D^b(A)}(-, Q^\bullet)$ to this triangle. The start of the induction is just the assumption $\mathrm{Ext}^i_A(X, {}_AA) = 0$ for all $i \geq 0$.

The assertion (iii) is clearly dual to (ii).

If $\mathrm{proj.dim}_A D(A_A) < \infty$ we infer that ${}^\perp K^b({}_A\mathcal{P}) = 0$ by 1.5[(i) and the simple observation that ${}^\perp K^b({}_A\mathcal{I}) = 0$ is always zero. And dually, 1.5(ii) implies that $K^b({}_A\mathcal{I})^\perp = 0$ if $\mathrm{inj.dim}_A A < \infty$.

The proposition above shows that ${}^\perp K^b({}_A\mathcal{P}) = 0$ implies the generalized Nakayama conjecture. See [AR1] for a discussion. We do not know if the converses of the proposition above can be shown directly.

## 2. Examples

In this section we want to present several examples of Gorenstein algebras.

2.1. Important examples of Gorenstein algebras are the algebras of finite global dimension and the selfinjective algebras. Note that the later class contains the group algebras $kG$ of a finite group $G$ over the ground field $k$, or more generally the symmetric algebras. Recall that a finite-dimensional selfinjective $k$-algebra $A$ is called symmetric if the Nakayama functor $\nu : {}_A\mathcal{I} \to {}_A\mathcal{P}$ given by $\nu = \mathrm{Hom}_A(D(A_A), -)$ is naturally equivalent to the identity functor. We refer to section 3 for some particular results for symmetric algebras.

2.2. Let $A$ be a Gorenstein algebra with $\mathrm{proj.dim}_A D(A_A) = r$. Then it was shown in [FGR] that the lower triangular matrix algebra

$$B = T_2(A) = \begin{pmatrix} A & 0 \\ A & A \end{pmatrix}$$

is a Gorenstein algebra with $\mathrm{proj.dim}_B D(B_B) = r + 1$. Below we will sketch a proof.

For instance, if $A = k[x]/(x^2)$, then $T_2(A)$ is given as a bound quiver algebra by

$$\alpha \circlearrowright \underset{S}{\circ} \xrightarrow{\beta} \underset{T}{\circ} \circlearrowleft \gamma \quad \text{with } \alpha\beta = \beta\gamma, \alpha^2 = \gamma^2 = 0$$

The indecomposable projective $T_2(A)$-modules are given by their Loewy series:

$$P(S) = \begin{array}{cc} \multicolumn{2}{c}{S} \\ S & T \\ \multicolumn{2}{c}{S} \end{array} \quad ; \quad P(T) = \begin{array}{c} T \\ T \end{array}.$$

and the indecomposable injective $T_2(A)$-modules are given by their Loewy series:

$$I(T) = \begin{matrix} & S & \\ S & & T \\ & S & \end{matrix} \quad ; \quad I(S) = \begin{matrix} S \\ S \end{matrix} .$$

For the convenience of the reader we sketch part of the argument in [FGR].

We will show that $\text{inj.dim}_B B = r + 1$. The other part of the statement follows dually.

We can consider a $B$-module ${}_BX$ as a pair of $A$-modules ${}_AY, {}_AZ$ and a map $f : Y \to Z$. Such a module will be denoted by $\begin{pmatrix} Y \\ Z \end{pmatrix}_f$. So $B$ operates on ${}_BX = \begin{pmatrix} Y \\ Z \end{pmatrix}_f$ by

$$\begin{pmatrix} a & 0 \\ b & c \end{pmatrix} \begin{pmatrix} y \\ z \end{pmatrix} = \begin{pmatrix} ay \\ byf + cz \end{pmatrix}$$

for $a, b, c \in A$ and $y \in Y, z \in Z$.

It is easily seen that a $B$-module $\begin{pmatrix} Y \\ Z \end{pmatrix}_f$ is injective if and only if ${}_AZ$ is $A$-injective, $f$ is surjective and $\ker f$ is $A$-injective, if and only if ${}_AZ$ is isomorphic to $\begin{pmatrix} Z' \oplus Z \\ Z \end{pmatrix}_{\pi_2}$ with $Z, Z'$ being $A$-injective and $\pi_2$ the canonical projection onto the second summand.

Moreover, given $\begin{pmatrix} Y \\ Z \end{pmatrix}_f$ then the injective envelope is easily computed as $\begin{pmatrix} E \oplus F \\ F \end{pmatrix}_{\pi_2}$ where $E$ is the $A$-injective envelope of $\ker f$ and $F$ is the $A$-injective envelope of $Z$.

In the following we will not specify the maps, for they are clear from the context. Now let

$$0 \to {}_AA \xrightarrow{d_0} E_0 \xrightarrow{d_1} E_1 \xrightarrow{d_2} \cdots \xrightarrow{d_r} E_r \to 0$$

be a minimal injective resolution of ${}_AA$.

Clearly

$$0 \to \begin{pmatrix} A \\ A \end{pmatrix} \to \begin{pmatrix} E_0 \\ E_o \end{pmatrix} \to \cdots \to \begin{pmatrix} E_r \\ E_r \end{pmatrix} \to 0$$

is a minmal injective resolution of the projective $B$-module $\begin{pmatrix} A \\ A \end{pmatrix}$.

And

$$0 \to \begin{pmatrix} 0 \\ A \end{pmatrix} \to \begin{pmatrix} E_0 \\ E_0 \end{pmatrix} \to \begin{pmatrix} E_0 \oplus E_1 \\ E_1 \end{pmatrix} \to \begin{pmatrix} E_{r-1} \oplus E_r \\ E_r \end{pmatrix} \to \begin{pmatrix} E_r \\ 0 \end{pmatrix} \to 0$$

is a minimal injective resolution of the projective $B$-module $\begin{pmatrix} 0 \\ A \end{pmatrix}$. So we see that $\text{inj.dim}_B B = r + 1$.

2.3. Another example is obtained by considering the algebra $A$ given as a bound quiver algebra by

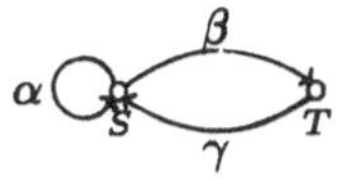


with $\alpha^2 = \beta\gamma, \gamma\beta = \gamma\alpha = \alpha\beta = 0.$

The indecomposable projective $A$-modules are given by their Loewy series:

$$P(S) \;=\; \begin{matrix} & S & \\ S & & T \\ & T & \end{matrix} \quad ; \quad P(T) \;=\; \begin{matrix} T \\ S \end{matrix} .$$

and the indecomposable injective $A$-modules are given by their Loewy series:

$$I(S) \quad \begin{matrix} & S & \\ S & & T \\ & T & \end{matrix} \quad ; \quad I(T) \;=\; \begin{matrix} S \\ T \end{matrix} .$$

Then it is easily seen that $\text{proj.dim}_A D(A_A) = 2 = \text{inj.dim}_A A$. So $A$ is in fact a Gorenstein algebra.

## 3. Auslander-Reiten triangles

In [H1] we introduced the notion of an Auslander-Reiten triangle in a triangulated category. We first recall the relevant definitions.

3.1. Let $\mathcal{C}$ be a triangulated category with $\text{Hom}_{\mathcal{C}}(X, Y)$ finite-dimensional as $k$-vector space for all $X, Y \in \mathcal{C}$ and assume that the endomorphism ring of any indecomposable object is local. This assumption ensures that $\mathcal{C}$ is a Krull-Schmidt category (compare 2.2 of [Ri]). We denote by $X[1]$ the value of the translation functor on the object $X$ of $\mathcal{C}$.

A triangle $X \xrightarrow{u} Y \xrightarrow{v} Z \xrightarrow{w} X[1]$ in $\mathcal{C}$ is called an Auslander-Reiten triangle (starting at the right) if the following conditions are satisfied:

(AR1) $X$ is indecomposable
(AR2) $w \neq 0$
(AR3) If $f : W \to Z$ is not a retraction, then there exists $f' : W \to Y$ such that $f'v = f$.

The object $X$ is called the translate of the object $Z$. Dually, we may define an Auslander-Reiten triangle starting at the left. It follows from [H1] that an Auslander-Reiten triangle starting at the right is an Auslander-Reiten triangle starting at the left. We will say that $\mathcal{C}$ has Auslander-Reiten triangles (starting at the right) if for all indecomposable objects $Z \in \mathcal{C}$ there exists a triangle satisfying the conditions above. We do not know whether in this case all indecomposable objects $X$ occur as a translate or equivalently, whether $\mathcal{C}$ has Auslander-Reiten triangles (starting at the left).

It is easily seen that in a triangle $X \xrightarrow{u} Y \xrightarrow{v} Z \xrightarrow{w} X[1]$ satisfying (AR1), (AR2) and (AR3) the object $Z$ is indecomposable. In fact, suppose that $Z = Z_1 \oplus Z_2$.

Then $w = \begin{pmatrix} w_1 \\ w_2 \end{pmatrix}$. So we may assume that $w_1 \neq 0$ and $Z_1$ is indecomposable. Let $\mu_1 : Z_1 \to Z$ be the canonical inclusion. Then $\mu_1 w = w_1 \neq 0$. So by the equivalent statement of (AR3) which is mentioned below we infer that $\mu_1$ is a retraction, hence $Z_1 = Z$

The following observations were obtained in [H1]. For more details and related results we refer to [H1] and [H2]. In particular we point out that in an Auslander-Reiten triangle $X \xrightarrow{u} Y \xrightarrow{v} Z \xrightarrow{w} X[1]$ the morphism $u$ is a source morphism and $v$ is a sink morphism (see for example [Ri] for a definition).

First note that the following are equivalent for a triangle as above:
(i) (AR2);
(ii) $u$ is not a section;
(iii) $v$ is not a retraction;

And if this is not the case we infer that $Y \simeq X \oplus Z$. Indeed, let $w = 0$. Then we obtain the following diagram of triangles in $\mathcal{C}$

$$\begin{array}{ccccccc} X & \xrightarrow{u} & Y & \xrightarrow{v} & Z & \xrightarrow{0} & X[1] \\ {\scriptstyle 1_X}\downarrow & & {\scriptstyle f}\downarrow & & {\scriptstyle 1_Z}\downarrow & & {\scriptstyle 1_{X[1]}}\downarrow \\ X & \xrightarrow{\mu} & X \oplus Z & \xrightarrow{\pi} & Z & \xrightarrow{0} & X[1] \end{array}$$

where $\mu$ and $\pi$ denote the canonical maps. Note that the second row is indeed a triangle, since it follows from the octahedral axiom that the direct sum of triangles is a triangle. Now we obtain by the third axiom of a triangulated category (see [V]) a morphism $f : Y \to X \oplus Z$, which is an isomorphism, by a well-known fact in triangulated categories, (see for example [V]).

And also the following are equivalent for a triangle as above:
(i) (AR3);
(ii) If $f : W \to Z$ is not a retraction, then $fw = 0$.

3.2. Let $\mathcal{C}$ be a triangulated category satisfying the general assumptions of 3.1. We recall the definition of the quiver $\vec{\Gamma}(\mathcal{C})$ of a triangulated category. Let $X, Y$ be two objects in $\mathcal{C}$. If $X, Y$ are indecomposable objects $\operatorname{rad}(X, Y)$ denotes the subspace of $\operatorname{Hom}_{\mathcal{C}}(X, Y)$ of non-invertible morphisms from $X$ to $Y$. If $X, Y$ are arbitrary, say with decompositions $X = \oplus X_i$ and $Y = \oplus Y_j$ then $\operatorname{rad}(X, Y) = \oplus \operatorname{rad}(X_i, Y_j)$. Then $\operatorname{rad}^2(X, Y)$ is given by the set of morphisms of the form $fg$ with $f \in \operatorname{rad}(X, M)$, $g \in \operatorname{rad}(M, Y)$ for some object $M$ in $\mathcal{C}$. We denote by

$$\operatorname{Irr}(X, Y) = \operatorname{rad}(X, Y)/\operatorname{rad}^2(X, Y)$$

and let $d_{XY} = \dim_k \operatorname{Irr}(X, Y)$.

By definition, the vertices of the quiver $\vec{\Gamma}(\mathcal{C})$ are the isomorphism classes $[X]$ of the indecomposable objects $X$ of $\mathcal{C}$. The quiver has $d_{XY}$ arrows from $[X]$ to $[Y]$.

The quivers we are interested in have an additional structure. Recall that a translation quiver $\vec{\Gamma} = (\Gamma_0, \Gamma_1, \tau)$ is given by a (locally finite) quiver $(\Gamma_0, \Gamma_1)$

($\Gamma_0$ denotes the vertex set, $\Gamma_1$ denotes the set of arrows) together with an injective map $\tau : \Gamma_0' \to \Gamma_0$ defined on a subset $\Gamma_0' \subseteq \Gamma_0$ such that for any $z \in \Gamma_0'$, and any $y \in \Gamma_0$ the number of arrows from $y$ to $z$ is equal to the number of arrows from $\tau z$ to $y$. The map $\tau$ is called the translation. If $\Gamma_0' = \Gamma_0$ and $\tau$ is a bijection we say that $\vec{\Gamma}$ is a stable translation quiver. A morphism of translation quivers is a morphism of quivers which commutes with the translations.

If the triangulated category $\mathcal{C}$ has Auslander-Reiten triangles then $\vec{\Gamma}(\mathcal{C})$ has the structure of a translation quiver (see [H2]).

3.3. Let $D^b(A)$ be the derived category of bounded complexes over $\operatorname{mod} A$. It was shown in [H1] that $D^b(A)$ has Auslander-Reiten triangles if $A$ has finite global dimension. The converse of this was established in [H3]. Also, in this case, the quiver $\vec{\Gamma}(D^b(A))$ is a stable translation quiver. A different kind of example is easily obtained. Let $A$ be a finite-dimensional selfinjective algebra. Then the stable category $\underline{\operatorname{mod}} A$ is a triangulated category with Auslander-Reiten triangles, and also in this case the associated quiver is a stable translation quiver.

This is rather simple. Indeed let $Z$ be an indecomposable object in $\underline{\operatorname{mod}} A$. We can assume that ${}_AZ$ is an indecomposable non-projective module. Let

$$0 \to X \xrightarrow{u} Y \xrightarrow{v} Z \to 0$$

be the Auslander-Reiten sequence ending in Z. Let

$$w \in \underline{\operatorname{Hom}}(Z, X[1]) \simeq \operatorname{Ext}_A^1(Z, X)$$

be the corresponding element. Then it is clear that

$$X \xrightarrow{\underline{u}} Y \xrightarrow{\underline{v}} Z \xrightarrow{w} X[1]$$

is an Auslander-Reiten triangle, where $\underline{u}$ denotes the residue class of $u$ in $\underline{\operatorname{mod}} A$.

For later purposes we also need the repetitive category. For the convenience of the reader we recall the definition.

Let $T(A)$ be the trivial extension algebra of $A$ by the injective cogenerator $Q = \operatorname{Hom}_k(A, k)$. This is by definition the following finite-dimensional $k$-algebra. The additive structure is $A \oplus Q$ and the multiplication is defined by

$$(a, \varphi) \cdot (b, \psi) = (ab, a\psi + \varphi b)$$

for $a, b \in A$ and $\varphi, \psi \in Q$.

The algebra $T(A)$ is a $\mathbb{Z}$-graded algebra, where the elements of $A \oplus 0$ are the elements of degree 0 and those of $0 \oplus Q$ the elements of degree 1. We denote by $\operatorname{mod}^{\mathbb{Z}} T(A)$ the category of finitely generated $\mathbb{Z}$-graded $T(A)$-modules with morphisms of degree zero. This is a Frobenius category in the sense of [H1] and is called the repetitive category for $A$. The stable category of $\operatorname{mod}^{\mathbb{Z}} T(A)$ is denoted by $\underline{\operatorname{mod}}^{\mathbb{Z}} T(A)$. It follows from [AR3] that also this triangulated category has Auslander-Reiten triangles with a stable translation quiver.

We also recall from [H2] or [H3] that we have a sequence of exact embeddings of triangulated categories

$$K^b({}_A\mathcal{P}) \hookrightarrow D^b(A) \hookrightarrow \underline{\mathrm{mod}}^{\mathbb{Z}}T(A)\,.$$

3.4. We can now formulate the main result in this section.

**Theorem:** *Let $A$ be a finite-dimensional $k$-algebra. Then the following are equivalent.*

*(i) $\mathrm{proj.dim}_A D(A_A) < \infty$*

*(ii) $K^b({}_A\mathcal{P})$ has Auslander-Reiten triangles.*

**Proof:** If $\mathrm{proj.dim}_A D(A_A) < \infty$, then the Auslander-Reiten triangles can be constructed as in the proof of 3.6 of [H1]. For the convenience of the reader we sketch the argument.

First observe that ${}_A\mathcal{P}$ and ${}_A\mathcal{I}$ are equivalent under the Nakayama functor $\nu = D\mathrm{Hom}_A(-, {}_AA)$, where $D$ denotes the duality on $\mathrm{mod}A$ with respect to the base field $k$. There is also an invertible natural transformation
$\alpha_P : D\mathrm{Hom}(P, -) \to \mathrm{Hom}(-, \nu P)$ for $P \in {}_A\mathcal{P}$.

This induces an equivalence of triangulated categories again denoted by $\nu$ between $K^b({}_A\mathcal{P})$ and $K^b({}_A\mathcal{I})$ and an invertible natural transformation
$\alpha_{P^\bullet} : D\mathrm{Hom}(P^\bullet, -) \to \mathrm{Hom}(-, \nu P^\bullet)$ for $P^\bullet \in K^b({}_A\mathcal{P})$. We have noted in section one that our assumption implies that $K^b({}_A\mathcal{I}) \subseteq K^b({}_A\mathcal{P})$.

If $Z^\bullet$ is indecomposable in $K^b({}_A\mathcal{P})$ we let $\varphi \in D\mathrm{Hom}(Z^\bullet, Z^\bullet)$ be a linear form on $\mathrm{End}Z^\bullet$ which vanishes on the radical $\mathrm{rad}\,\mathrm{End}Z^\bullet$ and satisfies $\varphi(id_{Z^\bullet}) = 1$. We consider the image $\alpha_{Z^\bullet}(\varphi)$; it is a non-zero linear map from $Z^\bullet$ to $\nu Z^\bullet$ such that $f\alpha_{Z^\bullet}(\varphi) = 0$ whenever the morphism $f$ of $K^b({}_A\mathcal{P})$ is not a retraction. Consider the triangle

$$\nu Z^\bullet[-1] \to Y^\bullet \to Z^\bullet \xrightarrow{\alpha_{Z^\bullet}(\varphi)} \nu Z^\bullet$$

having $\alpha_{Z^\bullet}(\varphi)$ as last morphism. We infer that this is an Auslander-Reiten triangle.

For the converse let $Z^\bullet$ be indecomposable in $K^b({}_A\mathcal{P})$. And let

$$X^\bullet \to Y^\bullet \to Z^\bullet \to X^\bullet[1]$$

be the Auslander-Reiten triangle. We will show below that this is even an Auslander-Reiten triangle in $D^b(A)$. Let $P(S)$ be an indecomposable projective $A$-module, which we consider as a complex concentrated in degree zero. The translate of $P(S)$ in $D^b(A)$ is isomorphic to $I(S)[-1]$, where $I(S) = \nu P(S)$. Thus $I(S) \in K^b({}_A\mathcal{P})$, but this is clearly equivalent to $\mathrm{proj.dim}_A D(A_A) < \infty$.

So it remains to show that the triangle above is an Auslander-Reiten triangle in $D^b(A)$. For this let $W^\bullet \in K^{-,b}({}_A\mathcal{P})$ and $f^\bullet : W^\bullet \to Z^\bullet$ a morphism which is not a retraction. We will show that $f^\bullet w^\bullet = 0$ (i.e. homotopic to zero in $K^{-,b}({}_A\mathcal{P})$, where $w^\bullet$ is the morphism from $Z^\bullet$ to $X^\bullet[1]$ in the Auslander-Reiten triangle.

Let $W^\bullet = (W^i, d^i_W)$ and $X^\bullet = (X^i, d^i_X)$. Applying the translation functor if necessary we may assume that $W^i = X^i = 0$ for $i > 0$. For $n < 0$ we denote by $W^\bullet_n$

the truncated complex $W_n^\bullet = (W_n^i, d_n^i)$ with $W_n^i = 0$ for $i < n$, $W_n^i = W^i$ for $i \geq n$ and $d_n^i = d_W^i$ for $i \geq n$ and zero otherwise. We obtain a morphism $\mu_n^\bullet : W_n^\bullet \to W^\bullet$ with $\mu_n^i = id_{W^i}$ for $i \geq n$ and zero otherwise. If $f^\bullet$ is not a retraction, then also $\mu_n^\bullet f^\bullet$ is not a retraction. Thus for $n < 0$ we infer that $\mu_n^\bullet f^\bullet w^\bullet = 0$ (i.e homotopic to zero in $K^b({}_A\mathcal{P})$), since $W_n^\bullet \in K^b({}_A\mathcal{P})$ and the triangle is an Auslander-Reiten triangle in $K^b({}_A\mathcal{P})$. Since $Z^\bullet = (Z^i, d_Z^i)$ and $X^\bullet[1]$ belong to $K^b({}_A\mathcal{P})$ there exists an $n_0$ such that $Z^i = (X^\bullet[1])^i = 0$ for $i < n_0$. By the observation above there exists $k^i : W_{n_0}^i \to (X^\bullet[1])^{i-1} = X^i$ such that $\mu_{n_0}^i f^i w^i = d_{n_0}^i k^{i+1} - k^i d_X^i$. Note that

$$k^{n_0} : W^{n_0} \to (X^\bullet[1])^{n_0-1} = X^{n_0} = 0$$

is the zero map.

Set $h^i = k^i$ for $i \geq n_0$ and $h^i = 0$ for $i < n_0$. Then it is straightforward to see that $f^i w^i = d_W^i h^{i+1} - h^i d_X^i$. Thus $f^\bullet w^\bullet = 0$ in $K^{-,b}({}_A\mathcal{P})$.

This finishes the proof of the theorem.

3.5. The following two corollaries follow immediately from this theorem and its dual. We refrain from stating the dual version explicitly.

**Corollary:** *Let $A$ be a finite-dimensional $k$-algebra. If $A$ is a Gorenstein algebra then the quiver of $K^b({}_A\mathcal{P})$ has the structure of a stable translation quiver.*

**Corollary:** *Suppose that $A$ satisfies $inj.dim_A A < \infty$. Then the following two statements are equivalent.*
*(i) $A$ is a Gorenstein algebra*
*(ii) $K^b({}_A\mathcal{P})$ has Auslander-Reiten triangles.*

3.6. We also note the following corollary.

**Corollary:** *Let $A$ be a finite-dimensional $k$-algebra. If $proj.dim_A D(A_A) < \infty$, then the embedding $K^b({}_A\mathcal{P}) \hookrightarrow \underline{mod}^{\mathbf{Z}} T(A)$ induces an embedding of the corresponding translation quivers.*

**Proof:** Let $Z^\bullet \in K^b({}_A\mathcal{P})$ be indecomposable. It is easily seen (compare [H2],or [H4]) that the computation of the translate of $Z^\bullet$ in both categories coincides. So the assertion follows from the fact that $K^b({}_A\mathcal{P})$ is a full triangulated subcategory of $\underline{mod}^{\mathbf{Z}} T(A)$.

So we see that if $A$ is a Gorenstein algebra a component of $\vec{\Gamma}(\underline{mod}^{\mathbf{Z}} T(A))$ containing vertices corresponding to modules contained in the image of $K^b({}_A\mathcal{P}) \hookrightarrow \underline{mod}^{\mathbf{Z}} T(A)$ gives rise to a full component of $\vec{\Gamma}(K^b({}_A\mathcal{P}))$.

3.7. We conclude this section by some examples.

Let $A$ be a Gorenstein algebra. We denote by $\tau_{K^b({}_A\mathcal{P})} Z^\bullet$ the translate of $Z^\bullet \in K^b({}_A\mathcal{P})$. Note that $\tau_{K^b({}_A\mathcal{P})}$ is an equivalence.

If $A$ is a finite-dimensional hereditary $k$-algebra we refer to [H1] for the description of $\vec{\Gamma}(K^b({}_A\mathcal{P}))$. Observe that in this case we know that $K^b({}_A\mathcal{P}) \simeq D^b(A)$, for gl.dim $A = 1$.

If $A$ is a symmetric algebra (see 2.1 for a definition), then it turns out that $\tau_{K^b({}_A\mathcal{P})} Z^\bullet = Z^\bullet[-1]$. For example, if $A = k[x]/(x^2)$, then it is easily seen that

$\vec{\Gamma}(K^b({}_A\mathcal{P})) = \mathbb{Z}A_\infty$.

More generally, for a symmetric algebra each indecomposable projective $A$-module $P(S)$ determines a component of $\vec{\Gamma}(K^b({}_A\mathcal{P}))$ of type $\mathbb{Z}A_\infty$, whose vertices correspond to complexes $P^\bullet_{r,s} = (P^i_{rs}, d^i_{rs})$, where $P^i_{rs} = P(S)$ for $-\infty < r \leq i \leq s < \infty$ and $d^i_{rs} = w$ for $r \leq i < s$ and zero otherwise and $w$ is the endomorphism of $P(S)$ sending the top of $P(S)$ to the socle of $P(S)$. The Auslander-Reiten triangles in this component are then given by

$$P^\bullet_{r+1,s+1} \to P^\bullet_{r,s+1} \oplus P^\bullet_{r,s-1} \to P^\bullet_{r,s} \to P^\bullet_{r,s}\,,$$

where $P^\bullet_{r,s-1} = 0$ if $s-r \leq 1$ and the last morphism is computed as $f^\bullet_{r,s}$ with $f^s_{r,s} = w$ and zero otherwise. Note that $P^\bullet_{r+1,s+1}[1] = P^\bullet_{r,s}$.

## 4. The description of the category $\mathcal{D}_\mathcal{P}$

4.1. Let $A$ be a finite-dimensional $k$-algebra. We consider $K^b({}_A\mathcal{P})$ as an épaisse subcategory of $D^b(A)$. Following [V] (see also [GZ], [G]) we can form the category of fractions. To be more precise, let $\mathcal{S}$ be the set of morphisms $f : X \to Y$ in $D^b(A)$ whose cone $C_f$ lies in $K^b({}_A\mathcal{P})$, where the cone of $f$ is the third term in a triangle in $D^b(A)$ whose first morphism is $f$. Then $\mathcal{S}$ is a saturated multiplicative system compatible with the triangulation. We denote by $\mathcal{D}_\mathcal{P}$ the category obtained from $D^b(A)$ by localizing with respect to $\mathcal{S}$. Similarly we denote by $\mathcal{D}_\mathcal{I}$ the category obtained from $D^b(A)$ by considering the épaisse subcategory $K^b({}_A\mathcal{I})$. If $A$ is a Gorenstein algebra it follows from 1.5 that $\mathcal{D}_\mathcal{P} = \mathcal{D}_\mathcal{I}$. Also $\mathcal{D}_\mathcal{P}$ is a triangulated category. We denote by $Q_\mathcal{S}$ the localization functor. Note that this is an exact functor and that $Q_\mathcal{S}(f)$ is an isomorphism if $f \in \mathcal{S}$ and that $Q_\mathcal{S}$ is universal with respect to these properties.

So we have a sequence of triangulated categories

$$K^b({}_A\mathcal{P}) \hookrightarrow D^b(A) \to \mathcal{D}_\mathcal{P}.$$

Following [G] (see also [Q] for related results) we may define the Grothendieck group $K_0(\mathcal{C})$ of a triangulated category $\mathcal{C}$. Applying $K_0$ to the sequence above gives an exact sequence of abelian groups

$$K_0(K^b({}_A\mathcal{P})) \to K_0(D^b(A)) \to K_0(\mathcal{D}_\mathcal{P}) \to 0.$$

Now $K_0(K^b({}_A\mathcal{P})) = K_0({}_A\mathcal{P})$ and $K_0(D^b(A)) = K_0(\operatorname{mod} A)$. It is easily seen that the induced map $K_0({}_A\mathcal{P}) \to K_0(\operatorname{mod} A)$ coincides with the Cartan map $C_A$ (see for example [Ba] for a definition). In particular this trivially implies that $K_0(\mathcal{D}_\mathcal{P}) \simeq \operatorname{cok} C_A$. If $C_A$ is an isomorphism we see that $K_0(\mathcal{D}_\mathcal{P}) = 0$.

We point out that this may happen even for non-trivial selfinjective algebras.

For example, let $A$ be a finite-dimensional $k$-algebra whose Cartan matrix is $\begin{pmatrix} 2 & 1 \\ 2 & 1 \end{pmatrix}$. Such an algebra is easily constructed: We may choose $A$ as the following bound quiver algebra:

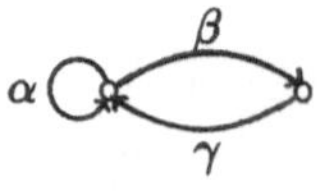


with $\alpha^2 = \beta\gamma = \gamma\alpha = \gamma\beta = 0$.

Then let $\Lambda$ be the trivial extension algebra of $A$ by the minimal injective cogenerator. Thus $\Lambda$ is selfinjective and the Cartan matrix for $\Lambda$ is computed as $\begin{pmatrix} 4 & 3 \\ 3 & 2 \end{pmatrix}$. And so $\det C_\Lambda = -1$, hence $C_\Lambda$ is an isomorphism.

In the applications given below this description of $K_0(\mathcal{D}_\mathcal{P})$ also follows from [AR4].

Let $A$ be a Gorenstein algebra. We apply $K_0$ to the commutative diagram of triangulated categories in 1.5 and obtain:

$$\begin{array}{ccc} K_0(A) & \xrightarrow{\tilde{\Phi}} & K_0(A) \\ {\scriptstyle C_A^t}\uparrow & & \uparrow{\scriptstyle C_A} \\ K_0(\mathcal{I}) & \xrightarrow{\Phi} & K_0(\mathcal{P}) \end{array}$$

for some maps $\Phi, \tilde{\Phi}$.

Using the description of the Coxeter map as contained for example in [Ri] we see that $\Phi$ is identified with the Coxeter map. In particular, we see that the translation $\tau_{K^b({}_A\mathcal{P})}$ for a Gorenstein algebra induces a map on the Grothendieck group which is identified with $-\Phi$. We thank H. Lenzing for this observation.

4.2. If $A$ has finite global dimension, then trivially $\mathcal{D}_\mathcal{P} = 0$. If $A$ is a selfinjective algebra, then Rickard [R1] has shown that $\mathcal{D}_\mathcal{P}$ is identified with the stable category $\underline{\mathrm{mod}}\, A$. We recall that Heller's loop space functor [He] serves as a translation functor in this case. Both results will be immediate consequences from the theorem below.

4.3. The following lemma is a first step towards a description of $\mathcal{D}_\mathcal{P}$.

Note that we have a natural functor from $\mathrm{mod}\, A \to \mathcal{D}_\mathcal{P}$ which is the composition of the embedding of $\mathrm{mod}\, A$ into $D^b(A)$ with $Q_S$ defined above.

**Lemma:** *If $\mathrm{proj.dim}_A D(A_A) < \infty$, then the natural functor $\mathrm{mod}\, A \to \mathcal{D}_\mathcal{P}$ is dense.*

**Proof:** Let $P^\bullet = (P^i, d^i_P) \in K^{-,b}({}_A\mathcal{P})$. So there exists $n_0$ such that $H^j(P^\bullet) = 0$ for all $j \leq n_0$. Let $X = \ker d^j_P$. Then $P^\bullet$ is isomorphic to $X[-j]$ in $\mathcal{D}_\mathcal{P}$ for all $j \leq n_0$. So we may assume that $P^\bullet$ is isomorphic to some $X[r]$ for some $r \geq 0$. Let

$$X \longrightarrow I^0 \xrightarrow{d^0} I^1 \xrightarrow{d^1} \cdots \xrightarrow{d^r} I^r \longrightarrow \cdots$$

be an injective resolution of $X$. Let $W = \mathrm{cok}\, d^r$.
Since $\mathrm{proj.dim}_A D(A_A) < \infty$ we infer that the following bounded complex

$$I^0 \xrightarrow{d^0} I^1 \xrightarrow{d^1} \cdots \xrightarrow{d^r} I^r$$

belongs to $K^b({}_A\mathcal{P})$.
Hence $X[r] \simeq W$ in $\mathcal{D}_\mathcal{P}$. In particular we see that $P^\bullet \simeq W$ in $\mathcal{D}_\mathcal{P}$. This shows the assertion.

There are also examples such that proj.dim$_A D(A_A) = \infty$ and that the functor mod $A \to \mathcal{D}_{\mathcal{P}}$ is dense. For example consider the algebra given as a bound quiver algebra by

$$\alpha \circlearrowleft \underset{S}{\circ} \xrightarrow{\beta} \underset{T}{\circ} \text{ with } \alpha\beta = \alpha^2 = 0.$$

It is not hard to show that the only indecomposable object in $\mathcal{D}_{\mathcal{P}}$ is the simple module $S$ in this case.

4.4. Let $M$ be an $A$-module. By add $M$ we denote the full subcategory of mod $A$ whose objects are the direct sums of direct summands of $M$. Also let

$$\mathcal{X} = \{X \mid \mathrm{Ext}^i_A(D(A_A), X) = 0, \text{ for } i > 0\}.$$

**Lemma:** *Suppose that proj.dim$_A(D(A_A) < \infty$. The following statements are equivalent.*

*(i) inj.dim$_A A < \infty$*

*(ii) For $X \in \mathcal{X}$ there is an exact sequence*

$$0 \to X' \to I \to X \to 0$$

*with $X' \in \mathcal{X}$ and $I \in \mathrm{add}\, D(A_A)$.*

*(iii) If $X \in \mathcal{X}$ satisfies proj.dim$_A X < \infty$, then $X \in \mathrm{add}\, D(A_A)$.*

**Proof:** First we show that (i) $\Rightarrow$ (ii). For this observe that $D(A_A)$ is a generalized tilting module. This implies that $X \in \mathcal{X}$ is generated by $D(A_A)$. Let $f_1, \ldots, f_r$ be a $k$-basis of $\mathrm{Hom}_A(D(A_A), X)$ and let

$$D(A_A)^r \xrightarrow{\begin{pmatrix} f_1 \\ \vdots \\ f_r \end{pmatrix}} X$$

be the corresponding map. Note that this map is surjective and that the kernel belongs to $\mathcal{X}$ by construction.

Next we show (ii) $\Rightarrow$ (iii). For this let ${}_AX$ be a module belonging to $\mathcal{X}$ and satisfying proj.dim$_A X = r < \infty$. Using (ii) we get for all $s$ an exact sequence:

$$0 \to X^s \longrightarrow I^s \xrightarrow{d^s} \cdots \xrightarrow{d^1} I^0 \xrightarrow{d^0} X \to 0$$

with $X^i = \ker d^i \in \mathcal{X}$ for all $i$ and $I^j \in \mathrm{add}\, D(A_A)$ for all $j$. So we infer that

$$\mathrm{Ext}^1_A(X, X^0) \simeq \mathrm{Ext}^{s+1}_A(X, X^s) = 0 \text{ for all } s \geq r,$$

hence the assertion.

The remaining implication (iii) $\Rightarrow$ (i) follows by applying the remark in 1.3 to a minimal injective resolution of ${}_AA$. Note that the syzygies of ${}_AA$ have finite projective dimension.

4.5. Now suppose that $A$ is a Gorenstein algebra. For $X \in \mathcal{X}$ we choose an exact sequence $0 \to X' \to I \to X \to 0$ as in 4.4. It is easily seen that this choice induces a functor $S : \mathcal{X}/D(A_A) \to \mathcal{X}/D(A_A)$, where $\mathcal{X}/D(A_A)$ is obtained from $\mathcal{X}$ by factoring out those maps which factor over an injective $A$-module. Also it is easily seen that any other choice will yield a functor naturally equivalent to $S$ and that $S$ serves as an inverse to Heller's loop space functor. The proofs are similar to those in [He] and therefore are omitted.

4.6. Let $A$ be a Gorenstein algebra. The functor $\operatorname{mod} A \to \mathcal{D}_{\mathcal{P}}$ induces a functor from $\mathcal{X}/D(A_A) \xrightarrow{\psi} \mathcal{D}_{\mathcal{P}}$.

**Theorem:** *The functor $\psi$ is an equivalence of categories.*

**Proof:** We will give the construction of an inverse $\varphi$ to $\psi$, but will leave out the tedious technicalities, for they are similar to those in [R1]. We will identify $\mathcal{D}_{\mathcal{P}}$ with $\mathcal{D}_{\mathcal{I}}$.

Let $P^\bullet = (P^i, d^i_P), Q^\bullet = (Q^i, Q^i_Q) \in K^{+,b}({}_A\mathcal{I})$ and $\alpha^\bullet$ a morphism from $P^\bullet$ to $Q^\bullet$. Using the remark in 1.3 there exists $n_0 \in \mathbb{N}$ such that $\operatorname{cok} d^j_P, \operatorname{cok} d^j_Q \in \mathcal{X}$ for all $j \geq n_0$. Let $\alpha_j : \operatorname{cok} d^j_P \to \operatorname{cok} d^j_Q$ be the map induced by $\alpha^\bullet$. We set $\varphi(P^\bullet) = S^{n_0}(\operatorname{cok} d^{n_0}_P), \varphi(Q^\bullet) = S^{n_0}(\operatorname{cok} d^{n_0}_Q)$ and $\varphi(\alpha^\bullet) = S^{n_0}(\alpha_{n_0})$.

It is easily seen that in $\mathcal{D}_{\mathcal{I}}$ this does not depend on the choice of $n_0$. This finishes the construction of $\varphi$.

4.7. Let $A$ be a Gorenstein algebra. We have noticed in section one that $\mathcal{X} = \operatorname{add} D(A_A)$ if and only if $\operatorname{gl.dim} A < \infty$, and also that $\mathcal{X} = \operatorname{mod} A$ if and only if $A$ is selfinjective. In the later case the category $\mathcal{X}/D(A_A)$ coincides with the stable category $\underline{\operatorname{mod}}\, A$.

It follows from [AR2] that the triangulated category $\mathcal{D}_{\mathcal{P}} \simeq \mathcal{X}/D(A_A)$ has Auslander-Reiten triangles.

4.8. We consider the specific example from 2.3. In this case it turns out that $\mathcal{D}_{\mathcal{P}} = \operatorname{add} Q_S N$, where $N$ is the module whose Loewy series is given by $\begin{smallmatrix} & S & \\ S & & T \end{smallmatrix}$. Note that $\operatorname{End} Q_S N = k$ in $\mathcal{D}_{\mathcal{P}}$ for the non-trivial endomorphism in $\operatorname{mod} A$ factors as follows

$$\begin{matrix} & S & \\ S & & T \end{matrix} \twoheadrightarrow \begin{matrix} S \\ T \end{matrix} \to \begin{matrix} & S & \\ S & & T \end{matrix},$$

and $\begin{matrix} S \\ T \end{matrix}$ is injective.

## REFERENCES

[AB] Auslander, M.; Bridger, M. Stable module theory, Mem. of the AMS 94, Providence 1969.

[AR1] Auslander, M.; Reiten, I. On a generalized version of the Nakayama conjecture, Proc. AMS. 52 (1972), 69-74.

[AR2] Auslander, M.; Reiten, I. Applications of contravariantly finite subcategories, preprint.

[AR3] Auslander, M.; Reiten, I. Stable equivalence of dualizing R-varieties, Adv. in Math. 12 (1974), 306-366.

[AR4] Auslander, M.; Reiten, I. Grothendieck groups of algebras and orders, J. Pure Appl. Alg. 39 (1986), 1-51.

[Ba] Bass, H. Algebraic K-Theory, Benjamin, New York 1968.

[Bo] Bongartz, K. Tilted algebras, Springer Lecture Notes 903, Heidelberg 1981, 26-38.

[Bu1] Buchweitz, R.O. Maximal Cohen-Macaulay modules and Tate-cohomology over Gorenstein rings, preprint.

[Bu2] Buchweitz, R.O. Appendix to Cohen-Macaulay modules on quadrics by Buchweitz-Eisenbud-Herzog, Proceedings Lambrecht Conference, Springer Lecture Notes 1273, Heidelberg 1987, 96-116.

[FGR] Fossum, R.M.; Griffith. Ph.A.; Reiten, I. Trivial Extensions of Abelian Categories, Springer Lecture Notes 456, Heidelberg 1975.

[GZ] Gabriel, P.; Zisman, M. Calculus of fractions and homotopy theory, Ergebnisse der Mathematik 35, Heidelberg 1967.

[G] Grothendieck, A. Groupes des classes des catégories abeliennes et triangulée, in SGA 5, Éxposé VIII, Springer Lecture Notes 589, Heidelberg 1977, 351-371.

[H1] Happel, D. On the derived category of a finite-dimensional algebra, Comment. Math. Helv. 62 (1987), 339-389.

[H2] Happel, D. Triangulated categories in the representation theory of finite-dimensional algebras, LMS Lecture Notes Vol.119, Cambridge University Press 1988.

[H3] Happel, D. Auslander-Reiten triangles in derived categories of finite-dimensional algebras, to appear Proc. AMS.

[H4] Happel, D. Repetitive categories, Proceedings Lambrecht Conference, Springer Lecture Notes 1273, Heidelberg 1987, 298-317.

[He] Heller, A. The loop space functor in homological algebra, Trans. AMS. 96 (1960), 382-394.

[Q] Quillen, D. Higher algebraic K-Theory I, Springer Lecture Notes 341, Heidelberg 1971, 85-147.

[R1] Rickard, J. Derived categories and stable equivalence, J. Pure Appl. Algebra 61(1989), 303-317.

[RS] Rickard, J.; Schofield, A. Cocovers and tilting modules, Math. Proc. Camb. Phil. Soc. 106 (1989), 1-5.

[Ri] Ringel, C.M. Tame algebras and integral quadratic forms, Springer Lecture Notes 1099, Heidelberg 1984.

[V] Verdier,J.L. Catégories dérivées, état 0, Springer Lecture Notes 569, Heidelberg 1977, 262-311.

Progress in Mathematics, Vol. 95, © 1991 Birkhäuser Verlag Basel

# Decomposition numbers of finite groups of Lie type in non-defining characteristic

GERHARD HISS

## Introduction

In 1982 P. Fong and B. Srinivasan published a paper [12] on the blocks of the finite general linear and unitary groups, which revealed the surprising fact that block theory and Deligne–Lusztig theory are highly compatible for these groups. Since then a lot of work has been done on the $\ell$-modular character theory of groups of Lie type where $\ell$ is a prime different from the underlying characteristic of the group. In this survey article I shall present some of these results. I shall concentrate on those which are general in the sense that they can be formulated and proved for all groups of Lie type and not just for a particular series such as the general linear groups.

The results I am going to summarize are due to various authors. Contributions come from Fong, Srinivasan, Broué, Dipper, James, Geck and myself. In particular, I shall present the recent results of my Habilitationsschrift and of Geck's Ph. D. thesis [23, 18].

The article is divided into three sections. The first summarizes the theory developed so far to obtain the blocks. The second section describes the first steps to get information about the number of irreducible Brauer characters and to simplify the problem of finding the decomposition numbers. In the final section I shall introduce some general procedures to obtain decomposition numbers. Most of the result are stated without proofs, but with appropriate references.

## 1. Blocks

Let $G$ be a connected reductive algebraic group defined over $\mathbb{F}_q$, the field with $q$ elements of characteristic $p$. The Frobenius map corresponding to the $\mathbb{F}_q$-structure of $G$ will be denoted by $F$ and the finite group of fixed points of $F$ by $G^F$. Thus we only consider those finite groups of Lie type which can be defined by a standard Frobenius map. We shall assume throughout this paper that the centre of $G$ is connected. An $F$-stable subgroup of $G$ is called regular if it is conjugate to some standard Levi subgroup.

We shall write $\mathrm{Irr}(G^F)$ for the set of (absolutely) irreducible ordinary characters of $G^F$. The inner product on the set of class functions on $G^F$ will be denoted by $\langle\ ,\ \rangle$. Let $G^*$ be a group dual to $G$, defined over $\mathbb{F}_q$. There is a natural bijection

$$\{(T',\theta)\} \bmod G^F\text{-conjugacy} \leftrightarrow \{(T,s)\} \bmod G^{*F}\text{-conjugacy}$$

where $T'$ (resp. $T$) runs over the $F$-stable maximal tori of $G$ (resp. $G^*$), $\theta$ is an irreducible character of $T'^F$ and $s$ is an element of $T^F$. If $(T', \theta)$ and $(T, s)$ correspond in this way, we denote the Deligne–Lusztig character $R^G_{T'}(\theta)$ by $R^G_T(s)$. For semisimple $s \in G^{*F}$ let

$$\mathcal{E}(G^F, s) = \{\chi \in \mathrm{Irr}(G^F) \mid \langle \chi, R^G_T(s)\rangle \neq 0 \text{ for some } T \text{ containing } s\}.$$

This set only depends on the $G^{*F}$-conjugacy class of $s$. It is called a geometric conjugacy class of characters.

The irreducible characters of $G^F$ can be parametrized via the so-called Jordan decomposition of characters. Let $s$ be a semisimple element in $G^{*F}$ and denote by $G(s)$ a group dual to $C_{G^*}(s)$. Then there is a bijection

$$\mathcal{L}^G_s : \mathcal{E}(G^F, s) \longrightarrow \mathcal{E}(G(s)^F, 1)$$

such that for any $\chi \in \mathcal{E}(G^F, s)$ and any $F$-stable maximal torus $T$ of $G^{*F}$ containing $s$,

$$\langle \chi, R^G_T(s)\rangle = \varepsilon_G \varepsilon_{G(s)} \langle \mathcal{L}^G_s(\chi), R^{G(s)}_T(1)\rangle,$$

(see [5, p. 199] for the definition of the signs $\varepsilon_G$, $\varepsilon_{G(s)}$). Furthermore,

$$\chi(1) = \mathcal{L}^G_s(\chi)(1) |G^{*F} : C_{G^*}(s)^F|_{p'}.$$

This map is called the Jordan decomposition of characters. If $G(s)$ is a regular subgroup of $G$, then $\mathcal{L}^G_s$ is the inverse of the map $\mathcal{E}(G(s)^F, 1) \to \mathcal{E}(G^F, s)$, $\lambda \mapsto R^G_{G(s)}(\theta_s \lambda)$, where $\theta_s$ is a linear character of $G(s)^F$ and $R^G_{G(s)}$ denotes the twisted induction from $G(s)^F$ to $G^F$. We shall write $\chi = \chi_{s,\lambda}$ for $\chi \in \mathcal{E}(G^F, s)$ with $\mathcal{L}^G_s(\chi) = \lambda$. The elements in $\mathcal{E}(G^F, 1)$ are called unipotent characters. We refer the reader to Section 4 of [13] for a nice description of the Jordan decomposition of characters and some of its properties. But note that in [13] the symbol $\mathcal{L}'_s$ is used for what is called $\mathcal{L}^G_s$ in the present paper.

From now on $\ell$ always denotes a prime number different from $p$. If $\chi$ is a class function of $G^F$, let $\hat{\chi}$ denote the restriction of $\chi$ to the $\ell$-regular classes of $G^F$. If $H$ is a subgroup of $G^F$, we write $H_\ell$ for the set of $\ell$-elements of $H$.

Our first question concerns the $\ell$-blocks of $G^F$. There are natural subsets of irreducible characters of $G^F$, which are also unions of $\ell$-blocks. Namely, if $s$ is a semisimple $\ell'$-element in $G^{*F}$, define

$$\mathcal{E}_\ell(G^F, s) = \bigcup_{t \in C_{G^*}(s)^F_\ell} \mathcal{E}(G^F, st).$$

This definition may be motivated by the following observation (see [22]), called the *basic relation.* If $s \in G^{*F}$ is a semisimple $\ell'$-element, $t \in C_{G^*}(s)^F$ an $\ell$-element and $T$ a maximal torus of $G^*$ containing $st$, then

$$\hat{R}^G_T(st) = \hat{R}^G_T(s).$$

**Theorem 1.1.** (Broué and Michel [3]): *$\mathcal{E}_\ell(G^F, s)$ is a union of $\ell$-blocks.*

This result is true even if the centre of $G$ is not connected. We call $\mathcal{E}_\ell(G^F, s)$ a geometric conjugacy class of $\ell$-blocks. The task now is to obtain information such as the number of irreducible Brauer characters and the decomposition numbers of these geometric conjugacy classes of $\ell$-blocks. For this purpose we first consider the Gelfand–Graev character. This is an ordinary, multiplicity free character of $G^F$ obtained by inducing a certain linear character from a Sylow $p$-subgroup up to $G$. Since $\ell \neq p$, the Gelfand–Graev character is the character of a projective module. We now have the following fundamental result.

**Theorem 1.2.** (Hiss [22]): *The restriction to $\mathcal{E}_\ell(G^F, s)$ of the Gelfand–Graev character is the character of a projective indecomposable module.*

The main ingredient of the proof is Theorem 1.1 and the basic relation. We mention some consequences of Theorem 1.2.

1.3. The sets $\mathcal{E}_\ell(G^F, s)$ are the smallest sets of characters, which are at the same time unions of geometric conjugacy classes and $\ell$-blocks.

This was observed by R. Carter in [6]. An $\ell$-block containing a unipotent character is called unipotent. A second consequence was observed by M. Geck.

1.4. An $\ell$-block is unipotent, if and only if it is contained in $\mathcal{E}_\ell(G^F, 1)$.

An irreducible character of $\mathcal{E}(G^F, s)$ is called *semisimple* if it is of the form $\chi_{s,1}$ and *regular* if it is of the form $\chi_{s,\mathrm{St}}$, where St denotes the Steinberg character of $G(s)^F$. It is known that the regular characters are exactly the constituents of the Gelfand–Graev character.

1.5. All regular characters of $\mathcal{E}_\ell(G^F, s)$ lie in a unique $\ell$-block.

Up to sign, the semisimple characters are the Curtis–Alvis duals of the regular characters. Since Curtis–Alvis duality preserves $\ell$-blocks (see [1]), we have:

1.6. All semisimple characters of $\mathcal{E}_\ell(G^F, s)$ lie in a unique $\ell$-block.

These two distinguished blocks may or may not coincide.

We finally mention a result which applies only when the regular block has a cyclic defect group (see [22, Theorem 3.5]).

1.7. If the regular block has a cyclic defect group and contains exceptional characters, then the unique regular character of $\mathcal{E}(G^F, s)$ is joined to the exceptional node on the Brauer tree.

## 2. Basic sets

We shall now investigate the number of irreducible Brauer characters of $\mathcal{E}_\ell(G^F, s) =: \mathcal{B}$. We shall use the following terminology:

2.1. A set of (proper) Brauer characters of $\mathcal{B}$ is called a *generating set of Brauer characters* if its integral linear span contains all irreducible Brauer characters.

2.2. A *basic set of Brauer characters* of $\mathcal{B}$ is a generating set which is linearly independent over the integers.

2.3. Let $S$ be a basic set of Brauer characters of $\mathcal{B}$. The *relations* of $\mathcal{B}$ corresponding to $S$ are the set of equations obtained by expressing the characters $\hat{\chi}$, $\chi \in \mathrm{Irr}(\mathcal{B})$, in terms of $S$.

We have to make some remarks concerning these definitions:

2.4. The definiton of a basic set of Brauer characters is usually given in a different, more general form: it is not required for the characters in the basic set to be proper Brauer characters.

2.5. Given a basic set $S$ and the corresponding relations, the knowledge of the decomposition numbers of the elements of $S$ is then equivalent to the knowledge of the entire decomposition matrix of $\mathcal{B}$.

2.6. As is well known, the set $\{\hat{\chi} \mid \chi \in \mathrm{Irr}(\mathcal{B})\}$ is a generating set of Brauer characters of $\mathcal{B}$.

2.7. If $S$ is a set of Brauer characters of $\mathcal{B}$ such that every $\hat{\chi}$, $\chi \in \mathrm{Irr}(\mathcal{B})$, can be expressed as an integral linear combination of elements of $S$, then $S$, by Remark 2.6, is a generating set.

By the basic relation we know that $\hat{R}_T^G(st) = \hat{R}_T^G(s)$, if $s$ is a semisimple $\ell'$-element of $G^{*F}$ and $t$ an $\ell$-element in $G^{*F}$ centralizing $s$. So it is a natural question to ask whether $\hat{\mathcal{E}}(G^F, s) = \{\hat{\chi} \mid \chi \in \mathcal{E}(G^F, s)\}$ is a generating set or even a basic set of Brauer characters for $\mathcal{E}_\ell(G^F, s)$.

The following examples show that in general $\hat{\mathcal{E}}(G^F, s)$ is not a generating set. And even if it is, it need not be a basic set. For the first example we have to consider a group with disconnected centre (see Theorem 2.8 below). Let $G^F = SU_3(q^2)$. Assume $3 \mid (q+1)$ and let $\ell = 3$. Then $\mathcal{E}(G^F, 1)$ has 3 elements, whereas $\mathcal{E}_\ell(G^F, 1)$ contains 5 irreducible Brauer characters (see [15]). So $\hat{\mathcal{E}}(G^F, 1)$ is not a generating set for $\mathcal{E}_\ell(G^F, 1)$.

In the next example, let $G^F = G_2(q)$, $q$ odd and $\ell = 2$. Then $\hat{\mathcal{E}}(G^F, 1)$ is a generating, but not a basic set for $\mathcal{E}_\ell(G^F, 1)$ (see [26]). The reason for this is the fact that every ordinary character in $\mathcal{E}(G^F, 1)$ can be expressed as a sum of Deligne–Lusztig characters and a class function whose support is on 2-singular elements (see [7]).

On the other hand, Fong and Srinivasan have shown in [12] that if $G^F$ is either a general linear or a unitary group, then for every $2 \neq \ell \neq p$ and every semisimple $\ell'$-element in $G^{*F}$, the set $\hat{\mathcal{E}}(G^F, s)$ is a basic set for $\mathcal{E}_\ell(G^F, s)$. Their proof consists of two steps. They first show that $\hat{\mathcal{E}}(G^F, s)$ is a generating set, by using Remark 2.6 above. A counting argument then shows that $\hat{\mathcal{E}}(G^F, s)$ is in fact a basic set.

A similar argument was used by M. Geck and the author in [20] to show that this statement is true in a more general situation.

**Theorem 2.8.** (Geck and Hiss [20]): *Assume that $\ell$ is a good prime for $G$. Then $\hat{\mathcal{E}}(G^F, s)$ is a basic set of Brauer characters of $\mathcal{E}_\ell(G^F, s)$, for all semisimple $\ell'$-elements $s \in G^{*F}$. In particular, the number of irreducible Brauer characters of any block $B$ in $\mathcal{E}_\ell(G^F, s)$ equals $|B \cap \mathcal{E}(G^F, s)|$.*

We conjecture that $\hat{\mathcal{E}}(G^F, s)$ is a generating set for $\mathcal{E}_\ell(G^F, s)$, even if $\ell$ is a bad prime for $G$. As the first of the above examples shows, this conjecture cannot be true without the assumption that $G$ has connected centre. In any case we have [22, Theorem 3.1]:

**Theorem 2.9.** *Every irreducible Brauer character in $\mathcal{E}_\ell(G^F, s)$ is a constituent of some character of $\hat{\mathcal{E}}(G^F, s)$.*

This result is true even if the centre of $G$ is not connected. As already mentioned above, the problem of finding the decomposition numbers for $\mathcal{E}_\ell(G^F, s)$ can be reduced to the corresponding problem for $\mathcal{E}(G^F, s)$, provided that $\hat{\mathcal{E}}(G^F, s)$ is a basic set and the relations of $\mathcal{E}_\ell(G^F, s)$ corresponding to $\hat{\mathcal{E}}(G^F, s)$ are known. Now if $\ell$ is as in Theorem 2.8, the relations can be obtained by decomposing certain characters $R_L^G(\mu)$ into its irreducible constituents in $\mathcal{E}(G^F, s)$ (see the proof of Theorem 3.1 in [20]). An important consequence of this is the fact that, roughly speaking, the relations are the same in $\mathcal{E}_\ell(G^F, s)$ and in $\mathcal{E}_\ell(C_{G^*}(s)^{*F}, 1)$. In particular, if $\mathcal{E}(G^F, s)$ and $\mathcal{E}(C_{G^*}(s)^{*F}, 1)$ have the same decomposition numbers, this will be true for $\mathcal{E}_\ell(G^F, s)$ and $\mathcal{E}_\ell(C_{G^*}(s)^{*F}, 1)$ also. We shall now make these statements more precise. As in Section 1, we shall write $G(s) := C_{G^*}(s)^{*F}$.

Let $L^*$ be a regular subgroup of $G^*$ and assume that the semisimple $\ell'$-element $s$ is contained in $L^*$. If $L \leq G$ is a regular subgroup of $G$ dual to $L^*$, then we have the following diagram:

$$\begin{array}{ccc} \mathcal{E}(L^F, s) & \xrightarrow{\mathcal{L}_s^L} & \mathcal{E}(L(s)^F, 1) \\ \Big\downarrow R_L^G & & \Big\downarrow R_{L(s)}^{G(s)} \\ \mathcal{E}(G^F, s) & \xrightarrow{\mathcal{L}_s^G} & \mathcal{E}(G(s)^F, 1) \end{array} \tag{1}$$

Now assume in addition that the centre of $G(s)$ is connected. For each $t \in C_{G^*}(s)_\ell^F$ we have bijections

$$\begin{aligned} \mathcal{L}_{st}^G &: \mathcal{E}(G^F, st) \longrightarrow \mathcal{E}(G(st)^F, 1) \\ \left(\mathcal{L}_t^{G(s)}\right)^{-1} &: \mathcal{E}(G(st)^F, 1) \longrightarrow \mathcal{E}(G(s)^F, t) \end{aligned} \tag{2}$$

Combining these, we obtain a bijection

$$\mathcal{L}_{s,\ell}^G : \mathcal{E}_\ell(G^F, s) \longrightarrow \mathcal{E}_\ell(G(s)^F, 1). \tag{3}$$

We can now state:

**Proposition 2.10.** (Geck and Hiss [20]): *Let $\ell$ be a good prime for $G$, and let $s$ be a semisimple $\ell'$-element in $G^{*F}$. If $t$ is an $\ell$-element in $C_{G^*}(s)^F$, then $C_{G^*}(t)$ and $C_{G^*}(st)$ are regular subgroups of $G^*$ respectively $C_{G^*}(s)$. Let $L$ and $L(s)$ denote regular subgroups dual to $C_{G^*}(t)$, respectively $C_{G^*}(st)$ in $G$ respectively $G(s)$. Suppose that for every such $t$ the diagram (1) is commutative. Then $\mathcal{E}_\ell(G^F, s)$ and $\mathcal{E}_\ell(G(s)^F, 1)$ have the same relations with respect to the basic sets $\hat{\mathcal{E}}(G^F, s)$ and $\hat{\mathcal{E}}(G(s)^F, 1)$.*

The diagram (1) is commutative if $G(s)$ is a regular subgroup of $G$. Also, in a classical group, it is always commutative (see the remarks in the appendix of [13]).

**Corollary 2.11.** (Dipper and James): *Let $G^F = GL_n(q)$. Then the decomposition numbers of $\mathcal{E}_\ell(G^F, s)$ and $\mathcal{E}_\ell(C_{G^*}(s)^{*F}, 1)$ are the same.*

**Proof.** By [11, Theorem 6.2], $\mathcal{E}(G^F, s)$ and $\mathcal{E}(C_{G^*}(s)^{*F}, 1)$ have the same decomposition numbers, if the characters in the two sets are ordered according to the Jordan decompositon of characters. Q.E.D.

The known examples suggest the following

**Conjecture 2.12.** *$\mathcal{E}_\ell(G^F, s)$ and $\mathcal{E}_\ell(C_{G^*}(s)^{*F}, 1)$ always have the same decomposition numbers. To be more precise: the maps in (2) can be chosen such that $\mathcal{L}^G_{s,\ell}$ maps the projective indecomposable characters of $\mathcal{E}_\ell(G^F, s)$ onto those of $\mathcal{E}_\ell(C_{G^*}(s)^{*F}, 1)$.*

Please note that by Theorem 2.8 the two sets have the same number of irreducible Brauer characters if $\ell$ is good. Broué conjectures that the equality of the decomposition numbers in Conjecture 2.12 is in fact a result of a Morita equivalence [1]. He proves his conjecture in the case when $C_{G^*}(s)$ is a maximal torus [2, Thérème 3.3]. Furthermore, as Broué remarks at the end of his paper, when $C_{G^*}(s)$ is the Levi complement of an $F$-stable parabolic subgroup of $G^*$ (i.e., it is a maximally split regular subgroup), the Morita equivalence is an immediate consequence of [2, Théorème 0.2].

If we consider very special primes, then Conjecture 2.12 is also true.

**Proposition 2.13.** (Puig [33], Hiss [23, Satz 6.5]): *If $\ell$ is a prime dividing $q-1$, but not the order of the Weyl group of $G$, then $\ell$ is good and the characters in $\mathcal{E}(G^F, s)$ remain irreducible on reduction modulo $\ell$. In particular, $\mathcal{E}_\ell(G^F, s)$ and $\mathcal{E}_\ell(C_{G^*}(s)^{*F}, 1)$ have the same decomposition numbers if $C_{G^*}(s)$ is a regular subgroup of $G^*$.*

Finally, Conjecture 2.12 is true in the following case.

**Proposition 2.14.** (Fong and Srinivasan [14, (9B)]): *Let $G$ be a classical group and suppose that the defect groups of the blocks in $\mathcal{E}_\ell(G^F, s)$ are cyclic. Then $\mathcal{E}_\ell(G^F, s)$ and $\mathcal{E}_\ell(C_{G^*}(s)^{*F}, 1)$ have the same decomposition numbers.*

The truth of Conjecture 2.12 implies the truth of the following well known

**Conjecture 2.15.** (Jordan decomposition of $\ell$-blocks): *The map $\mathcal{L}^G_{s,\ell}$ respects $\ell$-blocks.*

This Jordan decomposition of $\ell$-blocks is known to be true in case $C_{G^*}(s)$ is a regular subgroup (see [1]) and in case $G$ is a classical group and $\ell$ is odd (see [13]). It would follow in general if one could proof that $\mathcal{L}^G_{s,\ell}$ can be chosen to be a perfect isometry.

The truth of Conjecture 2.12 also implies the truth of

**Conjecture 2.16.** (Jordan decomposition of Brauer characters): *There is a bijection $\mathcal{L}'^G_{s,\ell}$ between the irreducible Brauer characters of $\mathcal{E}_\ell(G^F, s)$ and those of $\mathcal{E}_\ell(C_{G^*}(s)^{*F}, 1)$ such that $\varphi(1) = \mathcal{L}'^G_{s,\ell}(\varphi)|G^{*F}: C_{G^*}(s)^{*F}|_{p'}$ if $\varphi$ is an irreducible Brauer character of $\mathcal{E}_\ell(G^F, s)$.*

## 3. Decomposition numbers

Having obtained some information on the $\ell$-blocks, we are now interested in finding the decomposition numbers for the groups of Lie type (of course again as always for

$\ell \neq p$.) The decomposition numbers are completely known for the groups of type $A_1$, the Suzuki groups [4], the Ree groups ${}^2G_2(3^{2m+1})$ [31, 24] and $GL_n(q)$, all $q$ and all $n \leq 10$ (see James [28]). Whereas the problem for the first two series is easy since the defect groups are cyclic, it is already extremely difficult to find the 2-modular decomposition numbers for ${}^2G_2(3^{2m+1})$ (see [31]). The Brauer trees are known for all classical groups [13], for the Ree groups of type ${}^2F_4(2^{2m+1})$ [24]. (The Suzuki and Ree groups are mentioned here for the sake of completeness, although of course they cannot be defined by a standard Frobenius map.) There are partial results for the groups $Sp_4(q)$ [34, 35] and $G_2(q)$ [21, 26, 25].

The unitary groups $SU_3(q^2)$ in three dimension are the smallest example of a series where not all decomposition numbers are known. J. McCorkindale has recently succeeded in finding their 2-modular decomposition numbers (with some restrictions on $q$). Her methods seem to apply also to the other characteristics, so there is some hope to solve this problem in the near future.

In all these examples, the principal methods to determine the decomposition numbers is Brauer reciprocity. Instead of producing Brauer characters and trying to show they are irreducible, one produces projective characters which are close to beeing indecomposable. By the Brauer reciprocity law, the multiplicities of the ordinary characters in the PIM's (the projective indecomposable characters) give exactly the decomposition numbers. If the irreducible Brauer characters are close to beeing ordinary characters (as seems to be the case in groups of Lie type in non-defining characteristic) it is better to produce projectives. On the other hand, if the Brauer characters are very far away from the ordinaries (as is the case in groups of Lie type in defining characteristic), it is better to produce Brauer characters.

Now the question arises of how to obtain projective characters. There are various methods of which we only mention the two most important. The first is taking tensor products. The tensor product of an ordinary character with a projective is projective. To start with, one can use defect 0 characters. This approach can be applied if explicit calculation with the character table of $G^F$ is possible (e.g. in $Sp_4(q)$ or $G_2(q)$).

The second method is induction from subgroups. A character induced from a projective is again projective. This is particularly favourable if one induces from an $\ell'$-subgroup, since then every ordinary character is projective. We mention that in our case $\ell \neq p$ Harish-Chandra induction also takes projectives to projectives.

In view of the many sources of projective characters, a systematic approach seems to be necessary in order to obtain some general results. I shall now describe some systematic approaches which promise to be very useful.

*Generalized Gelfand–Graev characters.* The first approach was suggested by Geck in his Ph. D. thesis [18]. It uses ideas of Kawanaka to produce characters which generalize the Gelfand–Graev character and which are therefore called *generalized Gelfand–Graev characters.*

Geck first defines a total ordering on the set $\mathcal{E}(G^F, s)$ as follows. Let $\xi : \mathrm{Irr}(G^F) \to \mathcal{C}$ denote the map from the set of irreducible characters of $G^F$ to the set of $F$-stable unipotent conjugacy classes of $G$ defined by Lusztig in [32, Chapter 13]. Now define a total ordering on $\mathcal{C}$ which satisfies: If $C \subseteq \overline{C'}_j$, then $C \leq C'$ (here the bar denotes the closure of the conjugacy class). Then choose any total ordering on $\mathcal{E}(G^F, s)$ which is consistent via $\xi$ with the ordering on $\mathcal{C}$. This can of course be done by prescribing any ordering on the fibres of $\xi$.

Next Geck considers generalized Gelfand–Graev characters. These are induced characters $\gamma_u$, where $u$ is a unipotent element of $G^F$. Since $\gamma_u$ is induced from a unipotent subgroup of $G^F$, it is projective. In case $G$ is simple of type $A_n$, the above map $\xi$ is injective. In particular, considering only unipotent characters of $G^F$, we have the same number of generalized Gelfand–Graev characters and unipotent characters.

We need the following proposition, which is based upon results of Kawanaka [30].

**Proposition 3.1.** (Geck [16]): *Let $G$ be a group of type $A$. Let $\chi \in \mathrm{Irr}(G^F)$, $C = \xi(\chi)$ and $u \in C^F$. Then:*

(1) *$\langle \gamma_u, \chi \rangle$ is non-zero and independent of $q$.*

(2) *$\langle \gamma_v, \chi \rangle = 0$ for every unipotent $v \in G^F$ such that $\overline{C_v} \not\subseteq \overline{C}$, if $C_v$ denotes the conjugacy class containing $v$.*

From this follows easily:

**Theorem 3.2.** (Geck [16]): *Let $G$ be a simple group of adjoint type $A$. Then, with the above ordering of characters, the decomposition matrix of $\mathcal{E}_\ell(G^F, s)$ is lower unitriangular.*

It seems very likely that the method sketched above can also be applied for groups of type other than $A$ to show that the decomposition matrices are unitriangular (possibly with some restrictions on $p$, $q$ and $\ell$).

It should be noted that Theorem 3.2 for the general linear groups was first proved with different methods by Dipper. We shall sketch Dipper's method later in this article.

A disadvantage of the generalized Gelfand–Graev characters is the fact that they have many ordinary constituents, i.e., they are far from beeing indecomposable in general.

*Harish-Chandra induction.* The second systematic approach uses projectives obtained by Harish-Chandra induction of projective characters of Levi subgroups. This was first used by Dipper in his papers on the decomposition numbers of the general linear groups.

To ease notation, we shall assume $s = 1$ in the following. Thus we only consider the geometric conjugacy class of unipotent blocks. If Conjecture 2.12 is true, this means no restriction of generality.

We fix a standard Levi subgroup $L_J = L$ of $G$, where $J$ is an $F$-stable subset of the simple roots. To calculate the Harish-Chandra induction of a projective of $L^F$, we only need to induce unipotent characters, provided $\ell$ is good. First of all, by Theorem 2.8 the unipotent characters of $\mathcal{E}_\ell(G^F, 1)$ are a basic set. In order to find the decomposition numbers, it suffices therefore to find the unipotent contributions of the PIM's. But in the Harish-Chandra induction of a PIM of $L^F$, only the unipotent characters yield contributions to unipotent characters. It is easy to generalize these statements to arbitrary geometric conjugacy classes $\mathcal{E}_\ell(G^F, 1)$.

By the Howlett–Lehrer theory one knows how to decompose the Harish-Chandra induction of a cuspidal unipotent character. Non-cuspidal unipotents can be induced by using the formulae given in [27]. For example, consider the Gelfand–Graev character of $L^F$. Its unique unipotent constituent is the Steinberg character $\mathrm{St}_{L^F}$. Hence the Harish-Chandra induction of $\mathrm{St}_{L^F}$ to $G^F$ is the unipotent contribution to the Harish-Chandra induction of the Gelfand–Graev character of $L^F$ (which is a PIM by Theorem 1.2). By the theory of Hecke algebras, there is a bijection $\varphi \mapsto \chi_\varphi$ beetween the irreducible characters of the Weyl group and the unipotent characters of $G^F$ which lie in the principal series. Let $P^F$ denote the parabolic subgroup of $G^F$ containing $L^F$. In order to calculate the constituents of $\mathrm{St}_{P^F}^{G^F}$ we can make use of the following formulae:

**Proposition 3.3.** [8, Theorem (70.24)]

(1) $\langle \mathrm{St}_{P^F}^{G^F}, \chi_\varphi \rangle = \langle \varepsilon_{W_J}^W, \varphi \rangle$, *where $\varepsilon$ is the alternating character of the parabolic subgroup $W_J$ of $W$ and $\varphi$ is an irreducible character of $W$.*

(2) *This can also be expressed as* $\langle \mathrm{St}_{P^F}^{G^F}, \chi \rangle = \langle 1_{P^F}^{G^F}, D(\chi) \rangle$, *where $D$ denotes the Curtis–Alvis duality and $\chi$ a unipotent character* [17, Proposition 5.2].

Of course, the projectives considered above give only information on the decomposition numbers of unipotent characters in the principal series. On the other hand, they are universal in the sense that they are projective for every characteristic $\ell \neq p$.

We shall now apply the method to the special case of the general linear groups.

**Corollary 3.4.** (Dipper [9, Theorem 1.2]): *Let $G^F = GL_n(q)$, $\ell$ any prime not dividing $q$. We order the partitions of $n$ lexicographically as in* [29, 2.2.1]. *Let $\chi_\alpha$ denote the unipotent character and $L_\alpha^F$ the Levi subgroup of $G^F$ corresponding to the partition $\alpha$. Let $\Phi_\alpha$ denote the Harish-Chandra induction of the Steinberg character of $L_\alpha^F$. Then $\Phi_\alpha$ is the unipotent contribution of a projective character. By* Proposition 3.3 *and* [29, Sections 2.1 and 2.2], *the matrix $\langle \chi_\alpha, \Phi_\beta \rangle$, where $\alpha$ and $\beta$ run through the partitions of $n$, is lower unitriangular. Thus $\mathcal{E}_\ell(G^F, 1)$ has a lower unitriangular decomposition matrix.*

*Hecke algebras.* It would be nice to be able to decompose the projective characters obtained by Harish-Chandra induction into indecomposables. That this is possible in particular cases, using the theory of Hecke algebras, will now be discussed.

Having talked merely about characters so far, we have to bring modules into the play. For this purpose let $(K, R, k)$ denote a splitting $\ell$-modular system for $G^F$. If $X$ is an $RG$-lattice, we denote the extension of scalars $X \otimes_R K$ by $X^K$. The starting point is the following observation of Dipper.

**Proposition 3.5.** (Dipper, [10]): *Let $Y$, $X$ be $RG^F$-lattices such that:*

(1) *$Y$ is projective.*

(2) *There is an epimorphism $\beta : Y \to X$.*

(3) $(\mathrm{Ker}\,\beta)^K$ *and $X^K$ have no common constituent.*

*Set $E = \mathrm{End}_{RG}(X)$. Let $\{V_1, \dots, V_t\}$ denote a system of representatives of the simple $KG^F$-modules occuring in $X^K$. Let $D_E$ denote the decomposition matrix of $E$, whose rows are indexed by the simple $E^K$-modules corresponding to the $V^K$. If finally $D_{RG^F}$ is the decomposition matrix of $RG^F$ with rows indexed by $(V_1, \dots, V_t, V_{t+1}, \dots, V_n)$ then with a suitable numbering of the columns we have:*

$$D_{RG} = \left( \begin{array}{c|c} D_E & * \\ \hline * & \end{array} \right).$$

This result can also be stated as follows: The decomposition of the $RG^F$-lattice $X$ into indecomposables determines the decomposition of the projective $Y$ into PIM's. For every indecomposable direct summand $Z$ of $X$ there is an indecomposable direct summand of $Y$ covering $Z$. The pair $Y$ and $Z$ again satisfies condition (3) of the proposition. Thus the decomposition of $Z^K$ into ordinary characters of $G^F$ gives part of a column of the decomposition matrix of $G^F$.

Now two natural questions arise. First of all we have to find pairs of lattices for which the assumptions of Proposition 3.5 are satisfied. And secondly, suppose we are given such a lattice $X$, can we determine the decomposition matrix of its endomorphism ring?

We are now going to investigate the occurence of such lattice pairs. Again we only consider the special case of unipotent characters. The reader is referred to [23, Section 5.2] for more general statements.

**Proposition 3.6.** *Let $L^F$ be a standard Levi subgroup of $G^F$. Let $Y_0$ denote the* PIM *whose Brauer character is the Gelfand–Graev character $\gamma$. Then $Y_0$ has a well defined pure sublattice $Z_0$ corresponding to the non-unipotent characters in $\gamma$. Let $X_0$ denote the factor $Y_0/Z_0$. Then the Brauer character of $X_0$ is the Steinberg character. Let $X$ and $Y$ denote the Harish-Chandra inductions of $X_0$ and $Y_0$ respectively. Then $Y \to X$ is a lattice pair satisfying assumptions* (1)–(3) *of* Proposition 3.5.

**Proof.** We only have to check condition (3) of Proposition 3.5. Now $Z_0^K$ consists entirely of non-unipotent characters. Since Harish-Chandra induction respects geometric conjugacy classes, the Harish-Chandra induction of $Z_0^K$ has no unipotent constituent. Q.E.D.

We can apply this result to the general linear groups $G^F = GL_n(q)$ as follows. For every partition $\alpha$ of $n$ let $X_\alpha$ be the $RG^F$-lattice constructed as in Proposition 3.6 from the Gelfand–Graev character of the standard Levi subgroup corresponding to $\alpha$.

Define

$$S := \mathrm{End}_{RG^F}\left(\bigoplus_{\alpha \vdash n} X_\alpha\right).$$

The $R$-algebra $S$ introduced above is the $q$-Schur algebra (see [11, Theorem 2.24(iv) and Section 3]). With this notation we have:

**Corollary 3.7.** (Dipper and James [11]): *The decomposition matrix of $S$ gives that part of the decomposition matrix of $G^F$ which corresponds to the unipotent characters.*

We now give a second important example of a lattice pair as in 3.5, the proof of which is similar to that of Proposition 3.6.

**Proposition 3.8.** *Let $L^F$ be a standard Levi subgroup of $G^F$. Suppose $Y_0$ is a* PIM *such that the character of $Y_0^K$ contains the unipotent cuspidal irreducible character $\chi_0$ exactly once. Suppose furthermore that $\chi_0$ is invariant under conjugation by $N_{G^F}(L^F)$. Then $Y_0$ has a factor lattice $X_0$ with character $\chi_0$. Let $X$ and $Y$ denote the Harish-Chandra inductions of $X_0$ and $Y_0$ respectively. Then $Y \to X$ is a lattice pair satisfying assumptions* (1)–(3) *of* Proposition 3.5.

In order to apply this proposition in connection with Proposition 3.5, we have to be able to decompose the $RG^F$-lattice $X$ into indecomposable direct summands. Equivalently, we have to decompose the Hecke algebra $E = \mathrm{End}_{RG^F}(X)$ into PIM's. For this purpose, we have to know $E$ of course. Now $X$ is the Harish-Chandra induction of the lattice $X_0$ with a cuspidal character. By the Howlett–Lehrer theory we know the endomorphism ring of $X^K$. It can be described as follows. Let

$$W^{J,\phi} = \{w \in W \mid w(J) = J, {}^w\phi = \phi\}$$

denote the inertia subgroup of $\phi = \chi_0$ (remember that $L = L_J$). Then the Hecke algebra of $X^K$ has a $K$-basis $\{T_w \mid w \in W^{J,\phi}\}$, whose elements satisfy the multiplication rules given in [5, Theorem 10.8.5]. Now it is very natural to ask whether these elements $T_w$ leave invariant the $R$-form $X$ of $X^K$ and if so, whether their $R$-span of the $T_w$ is the endomorphism ring of $X$. This is indeed the case under suitable assumptions on the lattice $X_0$.

**Theorem 3.9.** (Geck [19]): *Let $L_J^F$ be a parabolic subgroup of $G^F$, and let $\phi$ be a cuspidal character of $L^F$. Set $W^{J,\phi} = \{w \in W \mid w(J) = J, {}^w\phi = \phi\}$. Then there is an $RL_J^F$-lattice $X_0$ with character $\phi$ such that the following holds: If $X$ is the Harish-Chandra induction of $X_0$, then* $\mathrm{End}_{RG^F}(X)$ *has an $R$-basis $\{T_w \mid w \in W^{J,\phi}\}$ which multiplies together as in* [5, Theorem 10.8.5].

Geck proves this theorem under a certain technical condition on the lattice for $\phi$. Such a lattice for $\phi$ can always be choosen if the centre of $G$ is connected as is shown in [19, Lemma 3.1]. However, it is not essential to assume the connectedness of the centre of $G$ to obtain the above result.

Another sufficient condition to obtain a lattice pair as in 3.5 from a lattice of a cupidal character $\phi$ of a Levi subgroup is given in [19]: The reduction of $\phi$ modulo $\ell$ should be an irreducible $\ell$-modular character and should have the same inertia subgroup as $\phi$. So the question arises of when a cuspidal character is irreducible on restriction to $\ell$-modular classes. Nothing is known in this direction in general, but the examples lead Geck to put forward a conjecture of which we state the following special case.

**Conjecture 3.10.** (Geck [19]): *If $\phi$ is a cuspidal unipotent character of $G^F$ then the reduction of $\phi$ modulo $\ell$ is irreducible and has the same inertia subgroup as $\phi$.*

It should be noted that it is essential in this conjecture to have a standard Frobenius map $F$. Namely, in the Ree groups of type ${}^2G_2$ there are cuspidal unipotent characters which do not remain irreducible on reduction modulo 2 (see the decomposition matrix in [31]).

We finally give an indication of how the theory developed so far can be used to prove a special case of Proposition 2.13.

**Proposition 3.11.** *Let $\ell$ be a prime dividing $q-1$ but not the order of $W$. Then every unipotent character of $G^F$ remains irreducible on reduction modulo $\ell$.*

**Proof.** Let $B^F$ denote the Borel subgroup of $G^F$ containing the maximal torus $T^F$. Let $E$ denote the endomorphism algebra of the Harish-Chandra induction of the trivial $RT^F$-lattice. Of course, in this special case it is not really necessary to appeal to Geck's result [19] to find the $R$-basis $\{T_w \mid w \in W\}$ of $E$ and their multiplication relations. Our assumptions on $\ell$ now readily imply that the decomposition matrix of $E$ is a unit matrix. The result is now clear with 3.5. Q.E.D.

## REFERENCES

[1] M. Broué, "Isométries parfaites, types de blocs, catégories dérivées", *in:* Représentations Linéaires de Groupes Finis, Astérisque 181–182, 1990, pp. 61–92.

[2] M. Broué, "Isométries de caractères et equivalences de Morita ou dérivées", preprint.

[3] M. Broué et J. Michel, "Blocs et séries de Lusztig dans un groupe réductif fini", *J. reine angew. Math.* **395** (1989), 56–67.

[4] R. Burkhardt, "Über die Zerlegungszahlen der Suzukigruppen", *J. Algebra* **59** (1979), 421–433.

[5] R. W. Carter, "Finite Groups of Lie Type: Conjugacy Classes and Complex Characters", Wiley, New York, 1985.

[6] R. W.Carter, "Deligne–Lusztig theory and block theory", preprint.

[7] B. Chang and R. Ree, "The characters of $G_2(q)$", *in:* Symposia Mathematica Vol. 13, pp. 395–413, Academic Press, London, 1974.

[8] C. W. Curtis and I. Reiner, Methods of Representation Theory Vol. II, Wiley, New York, 1987.

[9] R. Dipper, "On the decomposition numbers of the finite general linear groups II", *Trans. Amer. Math. Soc.* **292** (1985), 123–133.

[10] R. Dipper, "On quotients of Hom-functors and representations of finite general linear groups I", *J. Algebra* **130** (1990), 235–259.

[11] R. Dipper and G. James, "The $q$-Schur algebra", *Proc. London Math. Soc.* **59** (1989), 23–50.

[12] P. Fong and B. Srinivasan, "The blocks of finite general linear and unitary groups", *Invent. Math.* **69** (1982), 109–153.

[13] P. Fong and B. Srinivasan, "The blocks of finite classical groups", *J. reine angew. Math.* **396** (1989), 122–191.

[14] P. Fong and B. Srinivasan, "Brauer trees in classical groups", (1989), *J. Algebra* **131** (1990), 179–225.

[15] M. Geck, "Irreducible Brauer characters of the 3-dimensional special unitary groups in non-defining characteristic", *Comm. Algebra* **18** (1990), 563–584.

[16] M. Geck, "On the decomposition numbers of the finite unitary groups in non-defining characteristic", to appear in *Math. Z.*

[17] M. Geck, "Generalized Gelfand–Graev characters for Steinberg's triality groups and their applications", to appear in *Comm. Algebra.*

[18] M. Geck, "Verallgemeinerte Gelfand-Graev Charaktere und Zerlegungszahlen endlicher Gruppen vom Lie-Typ", Dissertation, Aachen 1990.

[19] M. Geck, "Hecke algebras over valuation rings", preprint.

[20] M. Geck and G. Hiss, "Basic sets of Brauer characters of finite groups of Lie type", to appear in *J. reine angew. Math.*

[21] G. Hiss, "On the decomposition numbers of $G_2(q)$", *J. Algebra* **120** (1989), 339–360.

[22] G. Hiss, "Regular and semisimple blocks in finite reductive groups", *J. London Math. Soc.* **41** (1990), 63–68.

[23] G. Hiss, "Zerlegungszahlen endlicher Gruppen vom Lie-Typ in nicht-definierender Charakteristik", Habilitationsschrift, Aachen, 1990.

[24] G. Hiss, "The Brauer trees of the Ree groups", to appear in *Comm. Algebra.*

[25] G. Hiss and J. Shamash, "3-blocks and 3-modular characters of $G_2(q)$", *J. Algebra* **131** (1990), 371–387.

[26] G. Hiss and J. Shamash, "2-blocks and 2-modular characters of $G_2(q)$", to appear in *Math. Comp.*

[27] R. B. Howlett and G. I. Lehrer, "Representations of generic algebras and finite groups of Lie type", *Trans. Amer. Math. Soc.* **280** (1983), 753–779.

[28] G. James, "The decomposition numbers of $GL_n(q)$ for $n \leq 10$", *Proc. London Math. Soc.* **60** (1990), 225–265.

[29] G. James and A. Kerber, "The Representation Theory of the Symmetric Group", *Encyclopedia Math.* **16** (1981).

[30] N. Kawanaka, "Shintani lifting and Gelfand–Graev representations", *in:* The Arcata Conference on Representations of Finite Groups, Symp. Pure Math. **47** 1987, pp. 147–163.

[31] P. Landrock and G. O. Michler, "Principal 2-blocks of the simple groups of Ree type", *Trans. Amer. Math. Soc.* **260** (1980), 83–111.

[32] G. Lusztig, "Charakters of Reductive Groups over a Finite Field", *Ann. Math. Studies* **107**, Princeton University Press, 1984.

[33] L. Puig, "Algèbres de source de certains blocs de groupes de Chevalley", *in*: Représentations Linéaires de Groupes Finis, Astérisque 181–182, 1990, pp. 221–236.

[34] D. White, "The 2-decomposition numbers of $Sp(4, q)$, $q$ odd", *J. Algebra* **131** (1990), 703–725.

[35] D. White, "Decomposition numbers of $Sp(4, q)$ for primes dividing $q \pm 1$", *J. Algebra* **132** (1990), 488–500.

Progress in Mathematics, Vol. 95, © 1991 Birkhäuser Verlag Basel

# Counting blocks of defect zero

**REINHARD KNÖRR**

## Introduction

Let $G$ be a finite group and $p$ a prime number. Then the number of $p$-blocks of defect zero in $G$ - *i.e.* the number of ordinary irreducible characters $\chi$ of $G$ with $p \nmid \chi(1)^{-1}|\overline{G}|$ - is clearly the dimension of a suitable ideal in the center of the group algebra $FG$, where $F$ is a field of characteristic $p$. It is well known that this number is zero, if $G$ has a non-trivial normal $p$-subgroup $Q$, say. In this situation, however, one may still ask for the number of blocks of the factor group $\overline{G} = G/Q$ with defect zero - *i.e.* the number of ordinary irreducible characters $\chi$ of $G$ with $Q \leq \operatorname{Ker}\chi$ and $p \nmid \chi(1)^{-1}|\overline{G}|$. It will turn out that this again is the dimension of a suitable ideal (to be described below) in $Z(FG)$. In fact, one can be slightly more precise: given $p$-blocks $B_1, ..., B_n$ of $G$, one can count the irreducible characters of defect zero in $\overline{G}$ which belong to one of the given blocks when viewed as characters of $G$.

The description of the relevant ideal requires a classical construction — briefly recalled below — applied to a particular module, namely the permutation module $F_S^G$, where $S$ is a Sylow $p$-subgroup of $G$.

## 1. Notation and simple remarks

The notation follows [1] and [2]; any undefined notion can be found there.

*Definition 1.1.* Let $M$ be a finitely generated right $FG$-module, *i.e.* there is an $F$-algebra homomorphism

$$\mu : FG \longrightarrow E = E(M) = \operatorname{End}_F(M).$$

Define
$$\tau = \tau_M : E \longrightarrow FG$$
by
$$\tau(\alpha) = \sum_g tr_M[\alpha\mu(g^{-1})]g \quad \text{for } \alpha \in E.$$

(As usual, the $\mu$ will often be omitted in the following.)

**Lemma 1.2.** *$\tau$ is $F(G \times G)$-linear.*

**Proof.** Clearly $\tau[f(\alpha + \beta)] = f[\tau(\alpha) + \tau(\beta)]$ for $f \in F$ and $\alpha, \beta \in E$. Let $x, y \in G$, so we have to show that $\tau(x\alpha y) = x\tau(\alpha)y$. For arbitrary $u \in FG$ and $g \in G$, write $u_g$ for the coefficient at $g$ of $u$, so $u = \sum_g u_g g$. Then

$$\begin{aligned}[x\tau(\alpha)y]_{xgy} &= \tau(\alpha)_g \\ &= tr_M(\alpha g^{-1}) \\ &= tr_M(x\alpha g^{-1}x^{-1}) \\ &= tr_M[x\alpha y(xgy)^{-1}] \\ &= \tau(x\alpha y)_{xgy}\end{aligned}$$

Since this holds for every $g \in G$, the assertion follows.

As usual, for a subgroup $H \leq G$, we define $T_H^G : E^H = \mathrm{End}_{FH}(M) \to E^G$ by

$$T_H^G(\alpha) = \sum_{G=\dot{\cup}Hg} \alpha^g .$$

**Corollary 1.3.** *If $\alpha \in E^H$, then $\tau(\alpha) \in C_{FG}(H)$ and $\tau[T_H^G(\alpha)] = T_H^G[\tau(\alpha)] \in Z(FG)$. In particular, $\tau(E^G) \subseteq Z(FG)$.*

*Remark 1.4.* If $\alpha \in \mathrm{End}_{FG}M$, then the last statement just means that $\tau(\alpha)$ is constant on conjugacy classes. In fact, more is true: If $g = su = us$ with a $p'$-element $s$ and a $p$-element $u$ and $\alpha \in E^G$, then $\alpha g = \alpha s + \alpha s(u-1)$. But $\alpha s(u-1)$ is nilpotent since $u-1$ is nilpotent and commutes with $\alpha s$. Hence $tr_M(\alpha g) = tr_M(\alpha s)$. This shows that $\tau(\alpha)$ is constant on $p'$-sections.

*Remark 1.5.* If $L = M \oplus N$ is a decomposition of $FG$-modules and $\varepsilon$ is the projection on $M$ and if $\alpha \in E(L)$ maps $M$ to $M$ [so that $\alpha_{|M} \in E(M)$], then $\tau_M(\alpha_{|M}) = \tau_L(\varepsilon\alpha)$; to see this, just pick a basis of $L$ respecting the decomposition when calculating the traces.

*Remark 1.6.* For the definition of $\tau$, one is free to replace $F$ by an arbitrary commutative ring $R$ as long as $M$ is finitely generated and free over $R$. The classical situation is $R = \mathbf{C}$ and $M$ an irreducible $\mathbf{C}G$-module affording the character $\chi$. This leads to an old acquaintance, namely

$$\tau(id_M) = \sum_g \chi(g^{-1})g = \chi(1)^{-1}|G|e_\chi ,$$

where $e_\chi$ is the primitive central idempotent associated with $\chi$.

More important for the following is the case $R = O$, a complete discrete rank one valuation ring with quotient field $K$, a splitting field of characteristic 0 for $G$, and with maximal ideal $(\pi)$ such that $O/(\pi) = F$.

In the more general context, 1.2, 1.3 and 1.5 remain valid, while 1.4 depends on the characteristic.

*Example 1.7.* Let $M = F_S^G$, where $S$ is a Sylow $p$-subgroup and $F$ is the trivial $FS$-module. Then

$$\tau(id_M) = \lambda\sigma ,$$

where $\lambda = |G : S| \cdot 1$ is a non-zero scalar in $F$ and $\sigma \in FG$ is the sum of all $p$-elements in $G$.

**Proof.** It is enough to show that

$$\gamma(g) \equiv \left\{ \begin{array}{cl} |G:S| & \text{if } g \text{ is a } p\text{-element} \\ 0 & \text{otherwise} \end{array} \right\} \bmod p \,,$$

where $\gamma$ denotes the permutation character of $G$ on the cosets of $S$.

If $\gamma(g) \neq 0$, then $g$ has fixed points, *i.e.* $g$ belongs to some point stabilizer. But this is a Sylow $p$-subgroup, so $g$ is a $p$-element. Therefore the non trivial orbits of $g$ habe $p$-power size, so that

$$\begin{aligned} |G:S| &= \text{number of all points} \\ &\equiv \text{number of fixed points of } g \bmod p \\ &= \gamma(g). \end{aligned}$$

[Alternatively, one can use 1.4.]

**Corollary 1.8.** *Let $\chi$ be an ordinary irreducible character of $G$ and let $D$ be the defect group of the p-block containing $\chi$. Let $\omega_\chi$ be the associated central homomorphism. If $D = 1$, then $\omega_\chi(\sigma) = 1$; if $D \neq 1$, then $\omega_\chi(\sigma) \in p\mathbf{Z}$; here $\sigma$ denotes the sum in $OG$ of all p-elements of $G$.*

**Proof.** If $D = 1$, then $\chi$ vanishes on $p$-singular elements, so $\chi(\sigma) = \chi(1)$ and $\omega_\chi(\sigma) = \chi(1)^{-1}\chi(\sigma) = 1$. So let $D \neq 1$.

If $1_S^G = \sum_{\psi \in \mathrm{Irr}\, G} a_\psi \psi$, then in $OG$ one has (for $M = O_S^G$ )

$$\begin{aligned} |G:S|\sigma &\equiv \tau(id_M) && \bmod p \text{ (by 1.7)} \\ &= \sum_{g,\psi} a_\psi \psi(g^{-1}) g \\ &= \sum_{\psi} a_\psi \psi(1)^{-1} |G| e_\psi \,, && \text{hence} \\ |G:S|\omega_\chi(\sigma) &\equiv \sum_{\psi} a_\psi \psi(1)^{-1} |G| \omega_\chi(e_\psi) && \bmod p \\ &= a_\chi \chi(1)^{-1} |G| \\ &\equiv 0 && \bmod p \text{ (since } D \neq 1). \end{aligned}$$

It is well known that $\omega_\chi$ has integral values. Since the set of $p$-elements is closed under powers, it follows that $\omega_\chi(\sigma)$ is rational (see [2], chapter 3 and exercises 2.12). Therefore $\omega_\chi(\sigma) \in \mathbf{Z}$ . Since $p \not| \, |G:S|$, the assertion follows.

*Remark 1.9.* An alternative proof of 1.8 can be deduced from [3], 2.8.

## 2. The result

*Notation/Remark 2.1.* As in the introduction, let $B_1, ..., B_2$ be $p$-blocks of $G$; denote $e = \sum_i e_i$, where $e_i \in OG$ is the block idempotent of $B_i$. Let $L = F_S^G$ and set $M = Le$ and $N = L(1-e)$, so $L = M \oplus N$ as in 1.5. Further, let $Q$ be a normal $p$-subgroup of $G$. For an irreducible character $\chi$ of $G$, we will write:

$$\chi \in B \text{ if } \chi(e) = \chi(1)$$
$$\chi \in \overline{B} \text{ if } \chi \in B \text{ and } Q \leq \operatorname{Ker}\chi \text{ and}$$
$$\chi \in \overline{B}^0 \text{ if } \chi \in \overline{B} \text{ and } p \nmid \chi(1)^{-1}|G:Q|.$$

Since $Q$ acts trivially on $M$, one has $E = E^Q$, so $T_Q^G(\alpha) \in E^G$ for any $\alpha \in E$. In particular, this holds for $\alpha = \mu(x)$ with $x \in FG$. Denote $I = (\tau_M T_Q^G \mu)(FG)$. Then $I \subseteq Z(FG) = Z$ by 1.3; in fact, I is an ideal of $Z$, since both $\tau_M$ (by 1.2) and $T_Q^G$ (trivially) are $Z$-linear.

**Theorem 2.2.** *Dim $I$ is the number of irreducible characters $\chi \in \overline{B}^0$ of $G$.*

**Proof.** Let $x \in FG$. Then

$$\begin{aligned}(\tau_M T_Q^G)(x) &= \tau_L[eT_Q^G(x)] && \text{by 1.5}\\ &= eT_Q^G[(x)\tau_L(id_L)] && \text{by 1.2}\\ &= \lambda e T_Q^G(x\sigma) && \text{by 1.7.}\end{aligned}$$

It is now more convenient to calculate in $OG$, since one can then use ordinary characters. Changing notation, we let $\sigma$ be the sum in $OG$ of all $p$-elements of $G$ and $x \in OG$. Moreover $\hat{M} = (O_S^G)e$. Then, by the above,

$$(\tau_{\hat{M}} T_Q^G)(x) \equiv |G:S| e T_Q^G(x\sigma) \qquad \operatorname{mod} \pi.$$

Let $\rho$ be the regular character of $G$. Taking the coefficient at $g$ gives (all congruences mod $\pi$)

$$\begin{aligned}[(\tau_{\hat{M}} T_Q^G)(x)]_g &\equiv |G:S||G|^{-1}\rho[eT_Q^G(x\sigma)g^{-1}]\\ &= |S|^{-1}\sum_{\chi\in B}\chi(1)\chi[T_Q^G(x\sigma)g^{-1}]\\ &= |S|^{-1}\sum_{\chi\in B}\chi(1)\omega_\chi[T_Q^G(x\sigma)]\chi(g^{-1})\\ &\qquad\qquad \text{since } T_Q^G(x\sigma)\in Z\\ &= |S|^{-1}\sum_{\chi\in B}\chi[T_Q^G(x\sigma)]\chi(g^{-1})\\ &= |G:S||Q|^{-1}\sum_{\chi\in B}\chi(x\sigma)\chi(g^{-1})\\ &= |G:S||Q|^{-1}\sum_{\chi\in B}\omega_\chi(\sigma)\chi(x)\chi(g^{-1})\\ &\qquad\qquad \text{since } \sigma\in Z.\end{aligned}$$

Since $Q$ is a normal $p$-subgroup, $\sigma$ is a sum over full cosets of $Q$. Therefore $\chi(\sigma) = 0$ (hence $\omega_\chi(\sigma) = 0$) if $Q \not\leq \operatorname{Ker}\chi$ (see [2], lemma 8.14).

If $\chi \in \operatorname{Irr}\overline{B}$, then $\chi(\sigma) = |Q|\chi(\overline{\sigma})$, where $\overline{\sigma}$ is the sum in $O\overline{G}$ of all the $p$-elements of $\overline{G}$, so $\omega_\chi(\sigma) = |Q|\overline{\omega}_\chi(\overline{\sigma})$, where $\overline{\omega}_\chi$ is the associated central character of $\overline{G}$. Therefore

$$\begin{aligned}[\tau_{\hat{M}} T_Q^G(x)]_g &\equiv |G:S| \sum_{\chi\in\overline{B}} \overline{\omega}_\chi(\overline{\sigma})\chi(x)\chi(g^{-1}) \\ &\equiv |G:S| \sum_{\chi\in\overline{B}^0} \chi(x)\chi(g^{-1}) ,\end{aligned}$$

where the last congruence holds by 1.8, used for $\overline{G}$.

For $\chi \in \overline{B}^0$, denote $\overline{\chi} : G \longrightarrow F$ the map $\overline{\chi}(g) = \chi(g) + (\pi)$ and $b_\chi = \sum_g \overline{\chi}(g^{-1})g \in FG$.

Reducing the above mod $\pi$ and recalling that $|G:S| = \lambda$ in $F$, we have

$$\begin{aligned}\tau_M T_Q^G(x) &= \lambda \sum_{\chi\in\overline{B}^0, g\in G} \overline{\chi}(x)\overline{\chi}(g^{-1})g \\ &= \lambda \sum_{\chi\in\overline{B}^0} \overline{\chi}(x) b_\chi ,\end{aligned}$$

so the $b_\chi$'s generate a space containing I. We show that they form in fact a basis of I; the assertion is then immediate.

Fix $\psi \in \overline{B}^0$ ; then $\psi$ belongs to a block of defect zero of $\overline{G}$ and

$$\varepsilon_\psi = |\overline{G}|^{-1}\psi(1) \sum_{\overline{g}\in\overline{G}} \overline{\psi}(\overline{g}^{-1})\overline{g} \in F\overline{G}$$

is the corresponding block idempotent. Choose $\overline{g} \in \overline{G}$ such that $\overline{\psi}(\overline{g}) \neq 0$ and let $u \in FG$ be a preimage of $\varepsilon_\psi\overline{g}$. For any $\chi \in \overline{B}^0$ then, $\overline{\chi}(u) = \overline{\chi}(\varepsilon_\psi\overline{g}) = \delta_{\chi\psi}\overline{\psi}(\overline{g})$, hence

$$(\tau_M T_Q^G)(u) = \lambda \sum_{\chi\in\overline{B}^0} \overline{\chi}(u) b_\chi = \lambda\overline{\psi}(\overline{g}) b_\psi \in I ,$$

so $b_\psi \in I$ — recall $\lambda \neq 0$.

If $\sum_{\chi\in\overline{B}^0} f_\chi b_\chi = 0$ for $f_\chi \in F$, then for every $g \in G$ , one has

$$0 = \sum_{\chi\in\overline{B}^0} f_\chi \overline{\chi}(g^{-1}) = \sum_{\chi\in\overline{B}^0} f_\chi \overline{\chi}(\overline{g}^{-1}).$$

Hence

$$\begin{aligned}0 &= \sum_{\chi\in\overline{B}^0, \overline{g}\in\overline{G}} f_\chi \overline{\chi}(\overline{g})^{-1}\overline{g} \\ &= \sum_{\chi\in\overline{B}^0} f_\chi \chi(1)^{-1}|\overline{G}|\varepsilon_\chi .\end{aligned}$$

Since the $\varepsilon_\chi$'s are orthogonal and $\chi(1)^{-1}|\overline{G}| \neq 0$ in $F$ for $\chi \in \overline{B}^0$, this forces $f_\chi = 0$ for all $\chi$, so the $b_\chi$'s are linearly independent.

*Remark 2.3.* Let $Q$ be a $p$-subgroup of $H$ and let $\beta \leftrightarrow \varepsilon$ be a $p$-block of $H$. Let $G = N_H(Q)$ and $B_i \leftrightarrow e_i, i = 1, ..., n$ be the blocks of $G$ with $B_i^H = \beta$. Then $e = \sum_i e_i = \mathrm{Br}(\varepsilon)$, where Br is the Brauer homomorphism (see [1], III 9). Therefore $|\overline{B}^0|$ is exactly the contribution comming from (the conjugacy class of) $Q$ in *Alperin's* conjectured count of $\ell(\beta)$. It may therefore be interesting to have other interpretations of that number. One is offered here.

## REFERENCES

[1] W. Feit, The representation theory of finite groups. North-Holland, Amsterdam (1982).
[2] I.M. Isaacs, Character theory of finite groups. Academic Press, New York (1976).
[3] R. Knörr, Virtually irreducible lattices. Proc. London Math. Soc. (3) 59 (1989), 99-132.

Progress in Mathematics, Vol. 95, © 1991 Birkhäuser Verlag Basel

# Group-theoretical descriptions of ring-theoretical invariants of group algebras

BURKHARD KÜLSHAMMER*

Let $F$ be a field, and let $G$ be a finite group. Then the elements of $G$ form a distinguished basis of the group algebra $FG$ of $G$ over $F$. In this paper we shall be concerned with the following two closely related questions:

- Which invariants of the ring $FG$ can be described in terms of the group basis $G$ of $FG$?
- Which invariants of the group $G$ are determined by the structure of the ring $FG$?

A well-known and trivial, but typical example of the results we have in mind is the description of the center $\mathbf{Z}FG$ of $FG$ as the linear span of all class sums.

We have chosen to concentrate here on facts which are valid for arbitrary groups and to neglect those valid for special classes of groups only (like $p$-groups, $p$-solvable groups or groups of Lie type). Most of the results to be presented are known to the specialists but usually scattered in the literature. Our intention has been to collect them in one place, to prove them in a coherent and direct way, and to provide complete references. Of course, the selection of topics reflects the author's personal taste. In many cases, the proofs presented here are not the original ones, and most results can be viewed from other directions also.

In the following, we fix an algebraically closed field $F$ of prime characteristic $p$. (Many of the results below have analogues for fields that are not algebraically closed, and some also make sense for fields of characteristic 0, but we do not want to change our hypotheses from theorem to theorem.) Moreover, $A$ will always denote a finite-dimensional associative unitary algebra over $F$ and $G$ a finite group.

We start by considering the *commutators* in $A$; these are the elements of the form $ab-ba$ where $a, b \in A$. The commutators themselves do not form an $F$-subspace of $A$, in general. Thus we consider the $F$-subspace $\mathbf{K}A$ of $A$ spanned by all commutators in $A$ and call $\mathbf{K}A$ the *commutator subspace* of $A$. In general, $\mathbf{K}A$ is neither a subalgebra nor an ideal, but a $\mathbf{Z}A$-submodule of $A$; this follows from the fact that $z(ab-ba) = (za)b - b(za)$ for $a, b \in A$ and $z \in \mathbf{Z}A$.

For example, if $A$ is the complete matrix algebra $\mathrm{Mat}(n, F)$ of some positive degree $n$ over $F$ then $\mathbf{K}A \subseteq \mathrm{Ker}(\mathrm{trace})$ since $\mathrm{trace}(ab) = \mathrm{trace}(ba)$ for $a, b \in A$. On the other hand, if we denote the standard basis of $A$ by $e_{ij}$ $(i, j = 1, \ldots, n)$ — this means that $e_{ij}e_{kl} = \delta_{jk}e_{il}$ for $i, j, k, l = 1, \ldots, n$ where $\delta_{jk}$ is the Kronecker symbol — then $e_{ij} = e_{ii}e_{ij} - e_{ij}e_{ii} \in \mathbf{K}A$ for $i \neq j$, and $e_{ii} - e_{11} = e_{i1}e_{1i} - e_{1i}e_{i1} \in \mathbf{K}A$

* Supported by a Heisenberg-fellowship of the DFG

for $i = 2, \ldots, n$; in particular, $\dim \mathbf{K}A \geq n^2 - 1$. Since $\dim \operatorname{Ker}(\text{trace}) = n^2 - 1$ we conclude that

$$\mathbf{K}(\operatorname{Mat}(n, F)) = \{a \in \operatorname{Mat}(n, F) : \operatorname{trace}(a) = 0\}. \tag{1}$$

Similarly, we have

$$\mathbf{K}FG = \{\textstyle\sum_{g \in G} \alpha_g g \in FG : \sum_{g \in K} \alpha_g = 0 \textit{ for } K \in \operatorname{Cl}(G)\} \tag{2}$$

where $\operatorname{Cl}(G)$ denotes the set of conjugacy classes of $G$; in order to see this observe first that $\mathbf{K}FG$ is spanned by all elements $gh - hg = gh - g^{-1}(gh)g$ where $g, h \in G$, and these are contained in the right hand side which is an $F$-subspace of $A$. On the other hand, let $a = \sum_{g \in G} \alpha_g g \in FG$ such that $\sum_{g \in K} \alpha_g = 0$ for $K \in \operatorname{Cl}(G)$. If we fix an element $g_K \in K$ then $g - g_K \in \mathbf{K}FG$ for $g \in K \in \operatorname{Cl}(G)$ since $g - g_K = h(h^{-1}g) - (h^{-1}g)h$ for some $h \in G$. Thus $a = \sum_{K \in \operatorname{Cl}(G)} \sum_{g \in K} \alpha_g (g - g_K) \in \mathbf{K}FG$.

If $I$ is an ideal in an arbitrary algebra $A$ then, as is easily checked,

$$\mathbf{K}(A/I) = \mathbf{K}A + I/I. \tag{3}$$

Similarly, if $A = A_1 \times A_2$ is the direct product of algebras $A_1, A_2$ then

$$\mathbf{K}(A_1 \times A_2) = \mathbf{K}A_1 \times \mathbf{K}A_2. \tag{4}$$

Using (3) and (4) we now show that always

(5) $\dim A/\mathbf{J}A + \mathbf{K}A$ *coincides with the number of maximal ideals of* $A$,

where $\mathbf{J}A$ denotes the (Jacobson) radical of $A$; indeed, (3) implies that $\dim A/\mathbf{J}A + \mathbf{K}A = \dim(A/\mathbf{J}A)/\mathbf{K}(A/\mathbf{J}A)$, and $A$ and $A/\mathbf{J}A$ have the same number of maximal ideals. So it suffices to prove that $\dim B/\mathbf{K}B$ coincides with the number of maximal ideals in $B := A/\mathbf{J}A$. But, by Wedderburn's theorem, $B$ is the direct product of complete matrix algebras over $F$, so by (4) we may assume that $B = \operatorname{Mat}(n, F)$ for a positive integer $n$. In this case, (1) implies that $\dim B/\mathbf{K}B = 1$. Since $B$ has exactly one maximal ideal (namely 0) this proves (5).

Essentially, this argument and the following ones go back to R. Brauer [3]. Note also that the number of maximal ideals of $A$ coincides with the number of simple $A$-modules.

The fact that $F$ has characteristic $p$ makes it natural to consider the map $A \to A$, $a \mapsto a^p$. We emphasize that this map is not necessarily additive. It does satisfy, however, the following two important properties:

(6) *If* $a, b \in A$ *then* $(a + b)^p \equiv a^p + b^p \pmod{\mathbf{K}A}$;

(7) *If* $k \in \mathbf{K}A$ *then* $k^p \in \mathbf{K}A$.

Indeed, we can write $(a + b)^p$ as a sum of $2^p$ terms $c_1 \cdots c_p$ where $c_i \in \{a, b\}$ for $i = 1, \ldots, p$. Moreover, we then have

$$c_1 \cdots c_p \equiv c_2 \cdots c_p c_1 \equiv \cdots \equiv c_p c_1 \cdots c_{p-1} \pmod{\mathbf{K}A}.$$

Thus

$$c_1 \cdots c_p + c_2 \cdots c_p c_1 + \cdots + c_p c_1 \cdots c_{p-1} \equiv p c_1 \cdots c_p \equiv 0 \pmod{\mathbf{K}A}.$$

This shows that all of the $2^p$ terms $c_1 \cdots c_p$ except $a^p$ and $b^p$ cancel mod $\mathbf{K}A$, and we have proved (6). From this we then see that

$$(xy - yx)^p \equiv (xy)^p - (yx)^p \equiv (xy)^{p-1}xy - y(xy)^{p-1}x \equiv 0 \pmod{\mathbf{K}A}$$

for $x, y \in A$. Moreover, if $k_1^p, k_2^p \in \mathbf{K}A$ then $(k_1 + k_2)^p \in \mathbf{K}A$ by (6). This implies that (7) holds.

We can restate (6) and (7) in the following convenient way:

(8) *The map $A/\mathbf{K}A \to A/\mathbf{K}A$, $a + \mathbf{K}A \mapsto a^p + \mathbf{K}A$, is well-defined and additive.*

Its $n$-th power is given by $a + \mathbf{K}A \mapsto a^{p^n} + \mathbf{K}A =: (a + \mathbf{K}A)^{p^n}$ for every non-negative integer $n$. The kernels of these maps are of interest since

(9) $\qquad \mathbf{J}A + \mathbf{K}A = \{x \in A : x^{p^n} \in \mathbf{K}A$ *for some non-negative integer* $n\}$;

indeed, if $x^{p^m}, y^{p^n} \in \mathbf{K}A$ for elements $x, y \in A$ and non-negative integers $m, n$ then $(x + y)^{p^{m+n}} \equiv x^{p^{m+n}} + y^{p^{m+n}} \equiv 0 \pmod{\mathbf{K}A}$ by (6) and (7). Thus the right hand side of (9) is an $F$-subspace $\mathbf{T}A$ of $A$ containing $\mathbf{K}A$ (by (7)) and every nilpotent element in $A$; in particular, $\mathbf{J}A + \mathbf{K}A \subseteq \mathbf{T}A$. On the other hand, if $x \in \mathbf{T}A$ then $x^{p^n} \in \mathbf{K}A$ for some non-negative integer $n$, so $(x + \mathbf{J}A)^{p^n} = x^{p^n} + \mathbf{J}A \in \mathbf{K}A + \mathbf{J}A/\mathbf{J}A = \mathbf{K}(A/\mathbf{J}A)$ by (3) which proves that $\mathbf{T}A/\mathbf{J}A \subseteq \mathbf{T}(A/\mathbf{J}A)$. Thus it suffices to show that $\mathbf{T}B = \mathbf{K}B$ for $B := A/\mathbf{J}A$. It is an easy consequence of (4) that we have $\mathbf{T}(B_1 \times B_2) = \mathbf{T}B_1 \times \mathbf{T}B_2$ for algebras $B_1, B_2$ over $F$. Since, by Wedderburn's theorem, $B$ is a direct product of complete matrix algebras over $F$ we may therefore assume that $B = \mathrm{Mat}(d, F)$ for a positive integer $d$. In this case, (1) implies that $\dim B/\mathbf{K}B = 1$. Thus, since $\mathbf{K}B \subseteq \mathbf{T}B \subseteq B$, it suffices to prove that $\mathbf{T}B \neq B$. But, if we denote the standard basis of $\mathrm{Mat}(d, F)$ by $e_{ij}$ $(i, j = 1, \ldots, d)$ then $e_{11} \notin \mathbf{K}B$ by the proof of (1). Since $e_{11}^2 = e_{11}$ this means that $e_{11} \notin \mathbf{T}B$, and (9) is proved.

It is easy to get a description of $\mathbf{J}FG + \mathbf{K}FG$ from (9). But before we state the result we would like to remind the reader of some group-theoretical facts and terminology.

An element $g \in G$ is called a *p-element* (*p′-element*) if its order is a power of $p$ (prime to $p$). We denote the set of all $p$-elements ($p'$-elements) in $G$ by $G_p$ (by $G_{p'}$). In general, neither $G_p$ nor $G_{p'}$ are subgroups of $G$. Moreover, $G$ usually contains elements which are neither $p$-elements nor $p'$-elements. However, every element $g \in G$ can be written in the form $g = g_p g_{p'}$ where $g_p \in G_p$, $g_{p'} \in G_{p'}$ and $g_p g_{p'} = g_{p'} g_p$. Moreover, in this case the elements $g_p$ and $g_{p'}$ are uniquely determined by $g$. They are called the *p-factor* and the *p′-factor*, respectively, of $g$.

For $g \in G$, the *p-section* (*p′-section*) of $g$ in $G$ consists of all elements $h \in G$ whose $p$-factor $h_p$ ($p'$-factor $h_{p'}$) is conjugate to $g_p$ (to $g_{p'}$) in $G$. Thus $G$ splits into disjoint $p$-sections ($p'$-sections), and each $p$-section ($p'$-section) of $G$ splits into disjoint conjugacy classes of $G$. Moreover, every $p'$-section $S$ of $G$ contains a unique *p′-conjugacy class*, i.e. a conjugacy class consisting of $p'$-elements; this $p'$-conjugacy class consists of the $p'$-factors of all elements in $S$. In particular, the number of $p'$-sections of $G$ coincides with the number of $p'$-conjugacy classes of $G$. We denote the set of $p$-sections ($p'$-sections) of $G$ by $\mathrm{Sec}_p(G)$ (by $\mathrm{Sec}_{p'}(G)$). Then it is routine to verify that the following properties are equivalent, for elements $g, h \in G$:

(10) *$g$ and $h$ are contained in the same $p'$-section of $G$;*

(11) *$g^{\exp_p(G)}$ and $h^{\exp_p(G)}$ are conjugate in $G$;*

(12) *$g^{p^n}$ and $h^{p^n}$ are conjugate in $G$ for every non-negative integer $n$ such that $p^n \geq \exp_p(G)$;*

(13) *$g^{p^n}$ and $h^{p^n}$ are conjugate in $G$ for some non-negative integer $n$.*

Here $\exp_p(G)$ denotes the $p$-part of the exponent $\exp(G)$ of $G$, i.e. the exponent of a Sylow $p$-subgroup of $G$. The $p'$-sections of $G$ come up in the following description of $\mathbf{J}FG + \mathbf{K}FG$:

(14) $\mathbf{J}FG + \mathbf{K}FG = \{\sum_{g\in G} \alpha_g g \in FG : \sum_{g\in S} \alpha_g = 0$ *for* $S \in \mathrm{Sec}_{p'}(G)\}$;

Indeed, by (9), an element $a = \sum_{g\in G} \alpha_g g \in FG$ is contained in $\mathbf{J}FG + \mathbf{K}FG$ if and only if $a^{p^n} \in \mathbf{K}FG$ for some non-negative integer $n$. Moreover, by (7), we may assume that $n \geq \exp_p(G)$, and (6) implies that $a^{p^n} \equiv \sum_{g\in G} \alpha_g^{p^n} g^{p^n} \pmod{\mathbf{K}FG}$. Thus, by (2), $a \in \mathbf{J}FG + \mathbf{K}FG$ if and only if

$$0 = \sum_{g\in G,\ g^{p^n}\in K} \alpha_g^{p^n} = \Big(\sum_{g\in G,\ g^{p^n}\in K} \alpha_g\Big)^{p^n}$$

for $K \in \mathrm{Cl}(G)$. But this means that $\sum_{g\in S} \alpha_g = 0$ for $S \in \mathrm{Sec}_{p'}(G)$, by the equivalence of (10), (12) and (13). Thus (14) is proved.

We note, however, that no description of $\mathbf{J}FG$ itself in terms of the group basis of $FG$ is known. Not even a group-theoretical description of $\dim \mathbf{J}FG$ exists in general. Even in special cases it may not be easy to determine $\dim \mathbf{J}FG$. For an example, see (VII.3.10) in [6]. On the other hand, (14) and (5) imply readily that

(15) $|\mathrm{Sec}_{p'}(G)|$ *is the number of maximal ideals of $FG$.*

This result and its proof are due to R. Brauer [1,3]. Let us analyze it more closely. For an arbitrary algebra $A$ over $F$ and a non-negative integer $n$, the proof of (9) shows that $\mathbf{T}_n A := \{x \in A : x^{p^n} \in \mathbf{K}A\}$ is an $F$-subspace and even a $\mathbf{Z}A$-submodule of $A$. Moreover, by (7), (9) and the finite dimension of $A$,

(16) $\mathbf{K}A = \mathbf{T}_0 A \subseteq \mathbf{T}_1 A \subseteq \mathbf{T}_2 A \subseteq \cdots$;

(17) $\sum_n \mathbf{T}_n A = \mathbf{J}A + \mathbf{K}A$;

(18) *There exists a non-negative integer $n$ such that $\mathbf{T}_n A = \mathbf{J}A + \mathbf{K}A$.*

The proof of (14) can be modified to show that, for any finite group $G$ and any non-negative integer $n$,

(19) $\mathbf{T}_n FG = \{\sum_{g\in G} \alpha_g g \in FG : \sum_{g\in C^{p^{-n}}} \alpha_g = 0$ *for* $C \in \mathrm{Cl}(G)\}$;

here, for a subset $X$ of $G$, $X^{p^{-n}}$ denotes the set of all elements $g \in G$ satisfying $g^{p^n} \in X$. It is easy to compute $\dim \mathbf{T}_n FG$ from (19); it coincides with the number of equivalence classes of the following equivalence relation on $G$: Two elements $g, h \in G$ are called *$p^n$-equivalent* if $g^{p^n}$ and $h^{p^n}$ are conjugate in $G$. It is easy to see that $\exp_p(G)$ is the smallest power of $p$ such that, for $g, h \in G$, $g$ and $h$ lie in the same $p'$-section of $G$ if and only if $g$ and $h$ are $p^n$-equivalent. Thus, by (14) and (19),

$$\min\{p^n : \mathbf{T}_n FG = \mathbf{J}FG + \mathbf{K}FG\} = \exp_p(G). \tag{20}$$

A more explicit way to describe the same fact is the following one (using (6) and (7)):

$$\exp_p(G) = \min\{p^n : x^{p^n} \in \mathbf{K}FG \text{ for } x \in \mathbf{J}FG\}. \tag{21}$$

This result was first proved by the author in [9 II] (in a stronger form; see (78) below). It was inspired by an argument of G.O. Michler in [12].

Let us give a few consequences. If $m$ is a non-negative integer such that $p^m < \exp_p(G)$ then $x^{p^m} \notin \mathbf{K}FG$ for some $x \in \mathbf{J}FG$; in particular, $x^{p^m} \neq 0$ and $(\mathbf{J}FG)^{p^m} \neq 0$. This gives a lower bound for the *Loewy length* of $FG$, i.e. the smallest non-negative integer $t$ such that $(\mathbf{J}FG)^t = 0$. Some authors call $t$ the *nilpotency index* of $FG$. For example, we have the following:

(22) *If $G$ has cyclic Sylow $p$-subgroups of order $p^n$ then* $(\mathbf{J}FG)^{p^{n-1}} \neq 0$;

(23) *If $p = 2$ and if $G$ has dihedral or quaternion Sylow 2-subgroups of order $2^n$ then* $(\mathbf{J}FG)^{2^{n-2}} \neq 0$.

The analogous results are weaker when the Sylow $p$-subgroups of $G$ have small exponent. But they are still strong enough to show:

(24) *If $p$ divides the order of $G$ then* $\mathbf{J}FG \neq 0$.

This is one half of what is usually called Maschke's theorem. We shall get to the other half below. But we would first like to recall some properties of symmetric algebras.

An algebra $A$ over $F$ is called *symmetric* if there is a non-degenerate symmetric bilinear form $A \times A \to F$, $(a, b) \mapsto (a|b)$, satisfying $(ab|c) = (a|bc)$ for $a, b, c \in A$. In this case $(.|.)$ is called a *symmetrizing* bilinear form on $A$.

For example, $A = \mathrm{Mat}(n, F)$ is symmetric where $(a|b) := \mathrm{trace}(ab)$ for $a, b \in A$; this is easily checked (and well-known). Similarly, $FG$ is symmetric where $(g|h) := \delta_{g,h^{-1}}$ for $g, h \in G$ and $\delta$ is the Kronecker symbol.

If $A_1$ and $A_2$ are symmetric algebras over $F$ then so is $A_1 \times A_2$; a symmetrizing bilinear form on $A_1 \times A_2$ is constructed in the obvious way from symmetrizing bilinear forms on $A_1$ and $A_2$. Thus it follows from Wedderburn's theorem that

(25) *Every semisimple algebra over $F$ is symmetric.*

Suppose now that $A$ and $A/I$ are both symmetric where $I$ is an ideal of $A$. Moreover, suppose that $(.|.)$ and $[.|.]$ are symmetrizing bilinear forms on $A$ and $A/I$, respectively. Then the map $A \to F$, $a \mapsto [a + I|1 + I]$, is obviously linear. Since $(.|.)$

induces a bijection between $A$ and $\mathrm{Hom}_F(A,F)$ there is an element $z \in A$ such that $[a+I|1+I] = (a|z)$ for $a \in A$. Thus

$$[a+I|b+I] = [ab+I|1+I] = (ab|z) = (a|bz)$$

for $a, b \in A$; in particular,

$$(a|bz) = [a+I|b+I] = [b+I|a+I] = (b|az) = (az|b) = (a|zb)$$

for $a, b \in A$. Hence $bz = zb$ for $b \in A$ which implies that $z \in \mathbf{Z}A$. Moreover, if $b \in I$ then $(a|bz) = [a+I|b+I] = [a+I|0] = 0$ for $a \in A$, so $bz = 0$. Conversely, if $b \in A$ and $bz = 0$ then $[a+I|b+I] = (a|bz) = 0$ for $a \in A$, so $b + I = 0$ and $b \in I$. This shows:

(26) *If $A$ and $A/I$ are symmetric algebras then $I = \{a \in A : az = 0\}$ for some $z \in \mathbf{Z}A$.*

This result and some of the following are due to Nakayama [13]. We can apply the considerations above with $I = 0$; in this case $(.|.)$ and $[.|.]$ are both symmetrizing bilinear forms on $A$. The arguments above show that there are elements $y, z \in \mathbf{Z}A$ such that $[a|b] = (a|bz)$ and $(a|b) = [a|by]$ for $a, b \in A$. Thus $(a|1) = (a|yz)$ for $a \in A$ which implies that $yz = 1$. This proves that $z$ is a unit in $A$, so any two symmetrizing bilinear forms on $A$ differ only by a unit in $\mathbf{Z}A$. This fact can be used to show that most of the results below are independent of the choice of the particular symmetrizing bilinear form on $A$. Combining (25) and (26) we obtain

(27) *If $A$ is symmetric then $\mathbf{J}A = \{a \in A : as = 0\}$ for some $s \in \mathbf{Z}A$.*

Moreover, for any $F$-subspace $V$ of $A$, $V^\perp := \{a \in A : (V|a) = 0\}$ is obviously an $F$-subspace of $A$, and the following well-known properties are satisfied, for arbitrary $F$-subspaces $V$ and $W$ of $A$:

(28) *If $V \subseteq W$ then $W^\perp \subseteq V^\perp$;*
(29) $\dim V + \dim V^\perp = \dim A$;
(30) $V^{\perp\perp} = V$;
(31) $(V+W)^\perp = V^\perp \cap W^\perp$ *and* $(V \cap W)^\perp = V^\perp + W^\perp$.

If $I$ is an ideal of $A$ then $(a|I) = (a|IA) = (aI|A)$ and $(I|a) = (AI|a) = (A|Ia)$. From this we obtain:

(32) *If $I$ is an ideal of $A$ then $I^\perp = \{a \in A : aI = 0\} = \{a \in A : Ia = 0\}$; in particular, $I^\perp$ is an ideal of $A$.*

Applying this with $I = \mathbf{J}A$ we see that

$$\mathbf{J}A^\perp = \{a \in A : a\mathbf{J}A = 0\} = \{a \in A : \mathbf{J}Aa = 0\},$$

so $\mathbf{J}A^\perp$ is the *socle* $\mathbf{S}A$ of $A$, considered as a right or left $A$-module. Choosing $s$ as in (27) we then have $\mathbf{J}A = \{a \in A : asA = 0\} = (sA)^\perp$, so (30) implies:

(33) *If $A$ is symmetric then* $\mathbf{S}A = sA$ *for some element* $s \in \mathbf{Z}A$.

In other words, the socle of any symmetric algebra is a principal ideal generated by a central element. Since $s \in \mathbf{S}A \cap \mathbf{Z}A$ a consequence is:

(34) $\mathbf{S}A = (\mathbf{S}A \cap \mathbf{Z}A)A$, *for any symmetric algebra $A$.*

For elements $a, b, c \in A$, we have the following:

$$\begin{aligned}(ab - ba|c) &= (ab|c) - (ba|c) = (a|bc) - (c|ba) \\ &= (bc|a) - (cb|a) = (bc - cb|a).\end{aligned}$$

Thus $c \in \mathbf{K}A^{\perp}$ if and only if $bc = cb$ for $b \in A$. This shows that

(35) $\mathbf{K}A^{\perp} = \mathbf{Z}A$, *for any symmetric algebra $A$.*

Moreover, if $V$ is a $\mathbf{Z}A$-submodule of $A$ then so is $V^{\perp}$ since $(v|za) = (vz|a) = 0$ for $v \in V$, $z \in \mathbf{Z}A$, $a \in V^{\perp}$; in particular, $\mathbf{T}_n A^{\perp}$ is a $\mathbf{Z}A$-submodule of $A$ for any non-negative integer $n$. Thus, by (16), we obtain the following chain of ideals of $\mathbf{Z}A$:

$$\mathbf{Z}A = \mathbf{K}A^{\perp} = \mathbf{T}_0 A^{\perp} \supseteq \mathbf{T}_1 A^{\perp} \supseteq \mathbf{T}_2 A^{\perp} \supseteq \cdots, \tag{36}$$

and, by (31), (17) and (35),

$$\bigcap_n \mathbf{T}_n A^{\perp} = (\sum_n \mathbf{T}_n A)^{\perp} = (\mathbf{J}A + \mathbf{K}A)^{\perp} = \mathbf{J}A^{\perp} \cap \mathbf{K}A^{\perp} = \mathbf{S}A \cap \mathbf{Z}A. \tag{37}$$

If $X$ is a subset of $G$ and if we define $X^+ := \sum_{x \in X} x \in FG$ then $(X^+|a) = \sum_{g \in X} \alpha_{g^{-1}}$ for $a := \sum_{g \in G} \alpha_g g \in FG$. Thus (19) means that

$$\mathbf{T}_n FG = \{a \in FG : (a|C^{p^{-n}+}) = 0 \textit{ for } C \in \mathrm{Cl}(G)\},$$

so we obtain, by (30):

$$\mathbf{T}_n FG^{\perp} = F\{C^{p^{-n}+} : C \in \mathrm{Cl}(G)\}, \textit{ for any non-negative integer } n. \tag{38}$$

Here $FX$ denotes the $F$-subspace spanned by $X$, for any subset $X$ of an arbitrary algebra $B$ over $F$. An application of (38) with $n = 0$ gives the well-known result that $\mathbf{Z}FG$ is spanned by the *class sums* $C^+$, $C \in \mathrm{Cl}(G)$. On the other hand, if $p^n \geq \exp_p(G)$ then, by (20) and (37), $\mathbf{T}_n FG^{\perp} = \mathbf{S}FG \cap \mathbf{Z}FG$ and, by (10) and (13), $F\{C^{p^{-n}+} : C \in \mathrm{Cl}(G)\} = F\{S^+ : S \in \mathrm{Sec}_{p'}(G)\}$. Thus, by (38),

$$\mathbf{S}FG \cap \mathbf{Z}FG = F\{S^+ : S \in \mathrm{Sec}_{p'}(G)\}. \tag{39}$$

The right hand side of (39) is usually called the *Reynolds ideal* of $\mathbf{Z}FG$ since, in [19], Reynolds proved that it is indeed an ideal of $\mathbf{Z}FG$. Note that this fact is obvious from (39) which, as far as we know, was first stated by the author in [9 I]; see also M.F. O'Reilly [15]. Combining (34) and (39), we obtain:

(40) $\mathbf{S}FG$ *is the ideal of $FG$ generated by all $S^+$, $S \in \mathrm{Sec}_{p'}(G)$.*

Since $\mathbf{J}FG = \mathbf{S}FG^{\perp}$ by (30) this implies, by (32), that

(41) $$\mathbf{J}FG = \{x \in FG : xS^{+} = 0 \text{ for } S \in \mathrm{Sec}_{p'}(G)\}.$$

This result is due to R. Brauer [2] and was later rediscovered by Y. Tsushima [23] (see also T. Okuyama [14]). Note that, if $p$ does not divide $|G|$ then $\{1\}$ is a $p'$-section of $G$, so (41) implies that $0 = \mathbf{J}FG \cdot 1$. Thus we have proved the other half of Maschke's theorem:

(42) *If $p$ does not divide $|G|$ then $\mathbf{J}FG = 0$.*

Let again $A$ denote an arbitrary symmetric algebra over $F$ with symmetrizing bilinear form $(.|.)$. Then, by (35), the map

$$\mathbf{Z}A \times A/\mathbf{K}A \to F, \qquad (z, a + \mathbf{K}A) \mapsto (z|a + \mathbf{K}A) := (z|a),$$

is well-defined, bilinear and non-degenerate. Moreover, it satisfies $(yz|x) = (y|zx)$ for $y, z \in \mathbf{Z}A$ and $x \in A/\mathbf{K}A$ where $A/\mathbf{K}A$ is considered as a $\mathbf{Z}A$-module. For any non-negative integer $n$, the map $F \to F$, $\alpha \mapsto \alpha^{p^n}$, is an automorphism of the field $F$. We write its inverse in the form $\alpha \mapsto \alpha^{p^{-n}}$. It is an easy consequence of (8) that, for $z \in \mathbf{Z}A$, the map $A/\mathbf{K}A \to F$, $x \mapsto (z|x^{p^n})^{p^{-n}}$, is linear. Since the non-degenerate bilinear form $(.|.)$ on $\mathbf{Z}A \times A/\mathbf{K}A$ induces a bijection between $\mathbf{Z}A$ and $\mathrm{Hom}_F(A/\mathbf{K}A, F)$ there is a unique element $\zeta_n(z) \in \mathbf{Z}A$ such that $(z|x^{p^n})^{p^{-n}} = (\zeta_n(z)|x)$ for $x \in A/\mathbf{K}A$. We have proved:

(43) *For any non-negative integer $n$ and any element $z \in \mathbf{Z}A$, there is a unique element $\zeta_n(z) \in \mathbf{Z}A$ satisfying $(z|x^{p^n}) = (\zeta_n(z)|x)^{p^n}$ for $x \in A/\mathbf{K}A$.*

We obtain a map $\zeta_n : \mathbf{Z}A \to \mathbf{Z}A$ which we consider as the adjoint of the map $A/\mathbf{K}A \to A/\mathbf{K}A$, $x \mapsto x^{p^n}$. Then $\zeta_n$ satisfies the following properties:

(44) *If $y, z \in \mathbf{Z}A$ then $\zeta_n(y + z) = \zeta_n(y) + \zeta_n(z)$ and $\zeta_n(y)z = \zeta_n(yz^{p^n})$;*
(45) *If $m$ and $n$ are non-negative integers then $\zeta_m \circ \zeta_n = \zeta_{m+n}$; in particular, $\zeta_n = \zeta_1^n$;*
(46) $\mathrm{Ker}(\zeta_n) = \mathbf{P}_n A^{\perp}$;
(47) $\mathrm{Im}(\zeta_n) = \mathbf{T}_n A^{\perp}$;

here $\mathbf{P}_n B := \{b^{p^n} : b \in B\} + \mathbf{K}B$ for any algebra $B$ over $F$. Note that, by (6), $\mathbf{P}_n B$ is an $F$-subspace of $B$. In order to prove (44), we observe that, for $y, z \in \mathbf{Z}A$, $x \in A/\mathbf{K}A$ and arbitrary non-negative integers $m$ and $n$,

$$\begin{aligned}(y + z|x^{p^n}) &= (y|x^{p^n}) + (z|x^{p^n}) = (\zeta_n(y)|x)^{p^n} + (\zeta_n(z)|x)^{p^n} \\ &= ((\zeta_n(y)|x) + (\zeta_n(z)|x))^{p^n} = (\zeta_n(y) + \zeta_n(z)|x)^{p^n},\end{aligned}$$

$$(yz^{p^n}|x^{p^n}) = (y|z^{p^n}x^{p^n}) = (\zeta_n(y)|zx)^{p^n} = (\zeta_n(y)z|x)^{p^n}$$

and

$$(z|x^{p^{m+n}}) = (\zeta_n(z)|x^{p^m})^{p^n} = (\zeta_m(\zeta_n(z))|x)^{p^{m+n}}$$

Thus (44) and (45) are consequences of (43).

For a proof of (46), note that an element $z \in \mathbf{Z}A$ is contained in $\mathrm{Ker}(\zeta_n)$ if and only if $0 = (\zeta_n(z)|x)^{p^n} = (z|x^{p^n})$ for $x \in A/\mathbf{K}A$, i.e. if and only if $0 = (z|a^{p^n})$ for $a \in A$. Thus (46) is a consequence of (35).

Finally, an element $a \in A$ is contained in $\mathrm{Im}(\zeta_n)^{\perp}$ if and only if

$$0 = (\zeta_n(\mathbf{Z}A)|a)^{p^n} = (\zeta_n(\mathbf{Z}A)|a + \mathbf{K}A)^{p^n} = (\mathbf{Z}A|a^{p^n} + \mathbf{K}A),$$

i.e. if and only if $a^{p^n} \in \mathbf{K}A$, by (35), and we have proved (47).

We now proceed to describe $\zeta_n$ in the case $A = FG$. Since the class sums form a basis of $\mathbf{Z}FG$ it suffices to compute $\zeta_n(C^+)$ for $C \in \mathrm{Cl}(G)$, by (44). For $a = \sum_{g \in G} \alpha_g g \in FG$, (6) implies that

$$\begin{aligned}(C^+|a^{p^n} + \mathbf{K}FG) &= (C^+|\sum_{g \in G} \alpha_g^{p^n} g^{p^n} + \mathbf{K}FG) \\ &= (C^+|\sum_{g \in G} \alpha_g^{p^n} g^{p^n}) = \sum_{g \in G,\, g^{p^n} \in C} \alpha_{g^{-1}}^{p^n} \\ &= (\sum_{g \in G,\, g^{p^n} \in C} \alpha_{g^{-1}})^{p^n} = (C^{p^{-n}+}|a)^{p^n} \\ &= (C^{p^{-n}+}|a + \mathbf{K}FG)^{p^n},\end{aligned}$$

so, by (43), we obtain

(48) $$\zeta_n(C^+) = C^{p^{-n}+} \textit{ for } C \in \mathrm{Cl}(G).$$

Let again $A$ be an arbitrary symmetric algebra over $F$ with symmetrizing bilinear form $(.|.)$. Then, in a way similar to (43), we have the following:

(49) *For any non-negative integer $n$ and any element $x \in A/\mathbf{K}A$, there is a unique element $\kappa_n(x) \in A/\mathbf{K}A$ such that $(z^{p^n}|x) = (z|\kappa_n(x))^{p^n}$ for $z \in \mathbf{Z}A$;*

Thus $\kappa_n$ is the adjoint of the map $\mathbf{Z}A \to \mathbf{Z}A$, $z \mapsto z^{p^n}$. The analogues of (44), (45), (46) and (47) are as follows:

(50) *If $x, y \in A/\mathbf{K}A$ and $z \in \mathbf{Z}A$ then $\kappa_n(x+y) = \kappa_n(x) + \kappa_n(y)$, $z\kappa_n(x) = \kappa_n(z^{p^n}x)$ and $\kappa_n(zx^{p^n}) = \zeta_n(z)x$;*

(51) *If $m$ and $n$ are non-negative integers then $\kappa_m \circ \kappa_n = \kappa_{m+n}$; in particular, $\kappa_n = \kappa_1^n$;*

(52) $\mathrm{Ker}(\kappa_n) = \mathbf{P}_n\mathbf{Z}A^{\perp}/\mathbf{K}A$;

(53) $\mathrm{Im}(\kappa_n) = \mathbf{T}_n\mathbf{Z}A^{\perp}/\mathbf{K}A$.

For example, if $y, z \in \mathbf{Z}A$ and $x \in A/\mathbf{K}A$ then

$$\begin{aligned}(y^{p^n}|zx^{p^n}) &= (y^{p^n}z|x^{p^n}) = (z|y^{p^n}x^{p^n}) \\ &= (\zeta_n(z)|yx)^{p^n} = (\zeta_n(z)y|x)^{p^n} = (y|\zeta_n(z)x)^{p^n},\end{aligned}$$

so (49) implies that $\kappa_n(zx^{p^n}) = \zeta_n(z)x$ for $z \in \mathbf{Z}A$ and $x \in A/\mathbf{K}A$.

The maps $\zeta_n$ and $\kappa_n$ were first defined by the author in [9 IV]. Our interest in $\kappa_n$ stems mainly from the fact that, for sufficiently large $n$, we have $\mathbf{T}_n\mathbf{Z}A = \{z \in \mathbf{Z}A : z^{p^n} = 0\} = \mathbf{JZ}A$ and

$$\mathbf{P}_n\mathbf{Z}A = F\{z^{p^n} : z \in \mathbf{Z}A\} = F\{e : e^2 = e \in \mathbf{Z}A\} =: \mathbf{E}A.$$

Note that $\mathbf{Z}A = \mathbf{JZ}A \oplus \mathbf{E}A$ and that $\mathbf{E}A$ is a subalgebra of $\mathbf{Z}A$. We denote by $\epsilon : \mathbf{Z}A \to \mathbf{E}A$ the corresponding projection. Thus $\epsilon$ is a homomorphism of algebras which, in the case $A = FG$, was first used by G.R. Robinson in [20]. Our aim is to compute the spaces $\mathbf{T}_n\mathbf{Z}FG$, $\mathbf{P}_n\mathbf{Z}FG$, $\mathbf{JZ}FG$ and $\mathbf{E}FG$, using (52) and (53). Thus we proceed to compute $\kappa_n$ in this case. This, however, seems to be more involved than the corresponding computation of $\zeta_n$ in (48). In fact, we require the so-called Brauer homomorphism which we shall introduce next.

For any subset $X$ of $G$, we first define $\pi_X : FG \to FG$ by $\pi_X(a) := \sum_{g\in X} \alpha_g g$ for $a = \sum_{g\in G} \alpha_g g \in FG$. If $Q$ is a $p$-subgroup of $G$ then $\beta_Q := \pi_{\mathbf{C}_G(Q)}$ is called the *Brauer homomorphism* with respect to $Q$. It was first introduced by R. Brauer [3] who also proved the following property which justifies the name:

(54) *For any $p$-subgroup $Q$ of $G$, the restriction of $\beta_Q$ is a homomorphism of algebras $\mathbf{Z}FG \to \mathbf{Z}F\mathbf{C}_G(Q)$.*

For a proof, it is enough to show that $\beta_Q(K^+L^+) = \beta_Q(K^+)\beta_Q(L^+)$ for $K, L \in \mathrm{Cl}(G)$. But $K^+L^+ = \sum_{g\in G} \alpha_g g$ where

$$\alpha_g = |\{(k,l) \in K \times L : kl = g\}| 1_F$$

for $g \in G$, and $\beta_Q(K^+L^+) = \sum_{g\in \mathbf{C}_G(Q)} \alpha_g g$. On the other hand, we have $\beta_Q(K^+) = (K \cap \mathbf{C}_G(Q))^+$, $\beta_Q(L^+) = (L \cap \mathbf{C}_G(Q))^+$ and $\beta_Q(K^+)\beta_Q(L^+) = \sum_{g\in\mathbf{C}_G(Q)} \alpha'_g g$ where

$$\alpha'_g = |\{(k,l) \in (K \cap \mathbf{C}_G(Q)) \times (L \cap \mathbf{C}_G(Q)) : kl = g\}| 1_F$$

for $g \in \mathbf{C}_G(Q)$. Now observe that $Q$ acts by conjugation on $\{(k,l) \in K \times L : kl = g\}$, and $\{(k,l) \in (K\cap\mathbf{C}_G(Q)) \times (L\cap\mathbf{C}_G(Q)) : kl = g\}$ is the set of fixed points under this action. Since the length of every orbit is a power of $p$ we get $\alpha_g = \alpha'_g$ for $g \in \mathbf{C}_G(Q)$, so (54) is proved.

We now proceed to compute the map $\kappa_n : FG/\mathbf{K}FG \to FG/\mathbf{K}FG$ for any non-negative integer $n$. By (50), it suffices to determine $\kappa_n(g + \mathbf{K}FG)$ for $g \in G$, but for later applications we derive a formula for $\kappa_n(K^+g + \mathbf{K}FG)$ for $g \in G$ and $K \in \mathrm{Cl}(G)$.

(55) *Let $g \in G$ with $p$-factor $u$ and $p'$-factor $s$, and let $K \in \mathrm{Cl}(G)$. Then we have $\kappa_n(K^+g + \mathbf{K}FG) = (Ku)^{p^{-n}+} s^{p^{-n}} + \mathbf{K}FG$; in particular, $\kappa_n(g + \mathbf{K}FG) = \{u\}^{p^{-n}+} s^{p^{-n}} + \mathbf{K}FG$.*

Here $s^{p^{-n}}$ denotes the unique $p'$-element $h \in G$ satisfying $h^{p^n} = s$ whereas $(Ku)^{p^{-n}}$ denotes the set of all elements $x \in G$ such that $x^{p^n} \in Ku$. In order to prove (55) we observe that, for $z \in \mathbf{Z}FG$ and $\beta := \beta_{\langle u \rangle}$,

$$\begin{aligned}(z^{p^n}|K^+g + \mathbf{K}FG) &= (z^{p^n}|K^+g) = (z^{p^n}K^+g|1) = (\beta(z^{p^n}K^+g)|1)\\ &= (\beta(z^{p^n}K^+)g|1) = (\beta(z)^{p^n}\beta(K^+)|g)\\ &= (\beta(z)^{p^n}|\beta(K^+)us) = (\beta(K^+)us|\beta(z)^{p^n})\\ &= (\beta(K^+)u|s\beta(z)^{p^n}) = (\beta(K^+)u|(s^{p^{-n}}\beta(z))^{p^n})\\ &= (\beta(K^+)u|(s^{p^{-n}}\beta(z))^{p^n} + \mathbf{K}F\mathbf{C}_G(u)).\end{aligned}$$

Now we apply (43) with $G$ replaced by $\mathbf{C}_G(u)$ to obtain

$$(z^{p^n}|K^+g + \mathbf{K}FG) = (\zeta_n(\beta(K^+)u)|s^{p^{-n}}\beta(z) + \mathbf{K}F\mathbf{C}_G(u))^{p^n}$$

where, by (48),

$$\begin{aligned}\zeta_n(\beta(K^+)u) &= \zeta_n((K\cap \mathbf{C}_G(u))^+u) = \zeta_n((Ku\cap \mathbf{C}_G(u))^+)\\ &= \{h \in \mathbf{C}_G(u) : h^{p^n} \in Ku\cap \mathbf{C}_G(u)\}^+\\ &= (\mathbf{C}_G(u)\cap (Ku)^{p^{-n}})^+ = \beta((Ku)^{p^{-n}+}).\end{aligned}$$

Thus

$$\begin{aligned}(z^{p^n}|K^+g + \mathbf{K}FG) &= (\beta((Ku)^{p^{-n}+})|s^{p^{-n}}\beta(z))^{p^n}\\ &= (s^{p^{-n}}\beta(z)|\beta((Ku)^{p^{-n}+}))^{p^n}\\ &= (1|s^{p^{-n}}\beta(z)\beta((Ku)^{p^{-n}+}))^{p^n}\\ &= (1|\beta(s^{p^{-n}}z(Ku)^{p^{-n}+}))^{p^n}\\ &= (1|s^{p^{-n}}z(Ku)^{p^{-n}+})^{p^n} = (s^{p^{-n}}|z(Ku)^{p^{-n}+})^{p^n}\\ &= (z(Ku)^{p^{-n}+}|s^{p^{-n}})^{p^n} = (z|(Ku)^{p^{-n}+}s^{p^{-n}})^{p^n}\\ &= (z|(Ku)^{p^{-n}+}s^{p^{-n}} + \mathbf{K}FG)^{p^n},\end{aligned}$$

and (55) follows from (49).

Since the result looks somewhat technical, let us illustrate it in a special situation. If $p^n \geq \exp_p(G)$ then $\{u\}^{p^{-n}} = \emptyset$ in case $u \neq 1$, and $\{1\}^{p^{-n}} = G_p$. Thus we obtain:

(56) *If* $p^n \geq \exp_p(G)$ *then* $\kappa_n(g + \mathbf{K}FG) = G_p^+ g^{p^{-n}} + \mathbf{K}FG$ *for* $g \in G_{p'}$, *and* $\kappa_n(g + \mathbf{K}FG) = 0$ *for* $g \in G \setminus G_{p'}$.

We next compute $\mathbf{T}_n\mathbf{Z}FG$. Obviously $z \in \mathbf{T}_n\mathbf{Z}FG$ if and only if

$$0 = (z|\mathbf{T}_n\mathbf{Z}FG^{\perp}) = (z|\mathbf{T}_n\mathbf{Z}FG^{\perp}/\mathbf{K}FG) = (z|\kappa_n(FG/\mathbf{K}FG)),$$

by (53). By (55),

$$\kappa_n(FG/\mathbf{K}FG) = F\{\{u\}^{p^{-n}+}s + \mathbf{K}FG : u \in G_p,\ s \in \mathbf{C}_G(u)_{p'}\}.$$

Thus $z \in \mathbf{T}_n\mathbf{Z}FG$ if and only if $0 = (z|\{u\}^{p^{-n}+}s) = (z\{u\}^{p^{-n}+}|s)$ for $u \in G_p$ and $s \in \mathbf{C}_G(u)_{p'}$. Hence we have proved:

(57) $\mathbf{T}_n\mathbf{Z}FG = \{z \in \mathbf{Z}FG : z\{u\}^{p^{-n}+} \in F[G \setminus \mathbf{C}_G(u)_{p'}]$ *for* $u \in G_p\}$.

Again this formula simplifies considerably if we take $p^n \geq \exp_p(G)$. For in this case $\{u\}^{p^{-n}} = \emptyset$ for $u \neq 1$, and $\{1\}^{p^{-n}} = G_p$. Thus $z^{p^n} = 0$ if and only if $zG_p^+ \in F[G \setminus G_{p'}]$. But now we note that $G_p \in \mathrm{Sec}_{p'}(G)$, so $G_p^+ \in \mathbf{S}FG \cap \mathbf{Z}FG$ by (39). Thus also $zG_p^+ \in \mathbf{S}FG \cap \mathbf{Z}FG$, so we can write $zG_p^+ = \sum_{S \in \mathrm{Sec}_{p'}(G)} \alpha_S S^+$ where $\alpha_S \in F$ for $S \in \mathrm{Sec}_{p'}(G)$. Since every $p'$-section contains $p'$-elements we see that $zG_p^+ \in F[G \setminus G_{p'}]$ is equivalent to $zG_p^+ = 0$. We have proved:

(58) *Let* $z \in \mathbf{Z}FG$ *and* $p^n \geq \exp_p(G)$. *Then* $z^{p^n} = 0$ *if and only if* $zG_p^+ = 0$.

This fact was obtained by the author in [9 II]. As a consequence, we have the following description of $\mathbf{JZ}FG$:

(59) $$\mathbf{JZ}FG = \{z \in \mathbf{Z}FG : zG_p^+ = 0\} = \{z \in \mathbf{Z}FG : z^{\exp_p(G)} = 0\}.$$

The first equality in (59) goes back to a paper by K. Iizuka and A. Watanabe [8]. We now turn our attention to $\mathbf{P}_n\mathbf{Z}FG$.

(60) *For* $K \in \mathrm{Cl}(G)$ *and* $z \in \mathbf{Z}FG$, $K^+z^{p^n}$ *is a linear combination of elements* $g \in G$ *satisfying* $(Ku^{-1})^{p^{-n}} \cap \mathbf{C}_G(Q) \neq \emptyset$ *for* $Q \in \mathrm{Syl}_p(\mathbf{C}_G(g))$ *where* $u$ *denotes the* $p$*-factor of* $g$.

(Here $\mathrm{Syl}_p(H)$ denotes the set of all Sylow $p$-subgroups of a finite group $H$.) Indeed, let $z \in \mathbf{Z}FG$ and write $K^+z^{p^n} = \sum_{g \in G} \alpha_g g$ where $\alpha_g \in F$ for $g \in G$. Moreover, let $g \in G$ such that $\alpha_g \neq 0$, and denote the $p$-factor of $g$ by $u$ and its $p'$-factor by $s$. Then, by (55), with $v := u^{-1}$ and $t := s^{-1}$, we have

$$\begin{aligned} 0 \neq \alpha_g &= (z^{p^n}K^+|g^{-1}) = (z^{p^n}|K^+g^{-1}) = (z^{p^n}|K^+g^{-1} + \mathbf{K}FG) \\ &= (z|\kappa_n(K^+g^{-1} + \mathbf{K}FG))^{p^n} = (z|(Kv)^{p^{-n}+}t^{p^{-n}} + \mathbf{K}FG)^{p^n}; \end{aligned}$$

in particular, $(Kv)^{p^{-n}+}t^{p^{-n}} \notin \mathbf{K}FG$. Any Sylow $p$-subgroup $Q$ of $\mathbf{C}_G(g)$ acts by conjugation on $(Kv)^{p^{-n}}t^{p^{-n}}$, and any two elements in the same orbit lie in the same coset of $\mathbf{K}FG$. Thus

$$\begin{aligned} 0 \neq (Kv)^{p^{-n}+}t^{p^{-n}} + \mathbf{K}FG &= ((Kv)^{p^{-n}}t^{p^{-n}} \cap \mathbf{C}_G(Q))^+ + \mathbf{K}FG \\ &= ((Kv)^{p^{-n}} \cap \mathbf{C}_G(Q))^+t^{p^{-n}} + \mathbf{K}FG; \end{aligned}$$

in particular, $(Kv)^{p^{-n}} \cap \mathbf{C}_G(Q) \neq \emptyset$, and (60) is proved.

Again, this looks somewhat technical, so let us consider the special case where $p^n \geq \exp_p(G)$. In this case $\mathbf{P}_n\mathbf{Z}FG = \mathbf{E}FG$ by (59). Moreover, if $h \in (Ku^{-1})^{p^{-n}} \cap \mathbf{C}_G(Q)$ then $h^{p^n} \in \mathbf{C}_G(u)_{p'}$ and $h^{p^n} \in Ku^{-1}$, so $h^{p^n}u \in K$. This shows that $K$ and $u$ are contained in the same $p$-section of $G$. We have proved:

(61) *If* $U \in \mathrm{Sec}_p(G)$ *then* $(\mathbf{Z}FG \cap FU)\mathbf{E}FG \subseteq \mathbf{Z}FG \cap FU$; *in particular,* $\mathbf{E}FG \subseteq FG_{p'}$.

The last part follows from the first by taking the $p$-section $G_{p'}$ containing 1. The results in (61) go back to K. Iizuka [7] and M. Osima [16]; see also D. Passman [18]. By (61), $\mathbf{Z}FG \cap FU$ is an $\mathbf{E}FG$-submodule of $\mathbf{Z}FG$, so the restriction of $\pi_U$ is an $\mathbf{E}FG$-homomorphism $\mathbf{Z}FG \to \mathbf{Z}FG \cap FU$ for $U \in \mathrm{Sec}_p(G)$.

We now proceed to apply these results in computing the map $\epsilon = \epsilon_G : \mathbf{Z}FG \to \mathbf{E}FG$ with kernel $\mathbf{JZ}FG$. We do this in somewhat greater generality than necessary. The following property of $\epsilon$ is an evident consequence of its definition and (54):

(62) *If* $Q$ *is a p-subgroup of* $G$ *then* $\beta_Q \circ \epsilon_G = \epsilon_{\mathbf{C}_G(Q)} \circ \beta_Q$ *on* $\mathbf{Z}FG$.

Next we observe the following group-theoretical fact.

(63) *For* $u \in G_p$, *the map* $K \mapsto K \cap u\mathbf{C}_G(u)_{p'}$ *is a bijection between the set of conjugacy classes of* $G$ *contained in the p-section* $U$ *of* $u$ *in* $G$ *and the set of conjugacy classes of* $\mathbf{C}_G(u)$ *contained in the p-section* $u\mathbf{C}_G(u)_{p'}$ *of* $u$ *in* $\mathbf{C}_G(u)$. *Moreover, if* $K \in \mathrm{Cl}(G)$ *and* $K \subseteq U$ *then* $K \cap u\mathbf{C}_G(u)_{p'} = S \cap u\mathbf{C}_G(u)_{p'}$ *where* $S$ *denotes the* $p'$*-section of* $\mathbf{C}_G(u)$ *containing* $K \cap u\mathbf{C}_G(u)_{p'}$.

Thus the map $\pi_{u\mathbf{C}_G(u)_{p'}}$ restricts to a linear bijection

$$\mathbf{Z}FG \cap FU \to \mathbf{Z}F\mathbf{C}_G(u) \cap uF\mathbf{C}_G(u)_{p'}.$$

Hence, in order to compute $\epsilon(z)K^+$ for an element $z \in \mathbf{Z}FG$ and a conjugacy class $K$ of $G$ it suffices to compute $\pi_{u\mathbf{C}_G(u)_{p'}}(\epsilon(z)K^+)$ for a $p$-element $u$ in the $p$-section of $G$ containing $K$, by (61).

(64) *Let* $z \in \mathbf{Z}FG$, *let* $K \in \mathrm{Cl}(G)$, *and let* $u$ *be a p-element in the p-section of* $G$ *containing* $K$. *Then* $\pi_{u\mathbf{C}_G(u)_{p'}}(\epsilon(z)K^+) = \pi_{u\mathbf{C}_G(u)_{p'}}(zS^+)$ *where* $S$ *denotes the* $p'$*-section of* $\mathbf{C}_G(u)$ *containing* $K \cap u\mathbf{C}_G(u)_{p'}$.

Indeed, by (54) and (62), with $\pi := \pi_{u\mathbf{C}_G(u)_{p'}}$ and $\beta := \beta_{\langle u\rangle}$, we have

$$\begin{aligned}\pi(\epsilon(z)K^+) &= \pi(\beta(\epsilon(z)K^+)) = \pi(\beta(\epsilon(z))\beta(K^+)) \\ &= \pi(\epsilon_{\mathbf{C}_G(u)}(\beta(z))\beta(K^+)).\end{aligned}$$

Since $\pi$ is a homomorphism of $\mathbf{E}F\mathbf{C}_G(u)$-modules by (61), we obtain from (63):

$$\begin{aligned}\pi(\epsilon(z)K^+) &= \epsilon_{\mathbf{C}_G(u)}(\beta(z))\pi(\beta(K^+)) \\ &= \epsilon_{\mathbf{C}_G(u)}(\beta(z))(K \cap u\mathbf{C}_G(u)_{p'})^+ \\ &= \epsilon_{\mathbf{C}_G(u)}(\beta(z))(S \cap u\mathbf{C}_G(u)_{p'})^+ \\ &= \epsilon_{\mathbf{C}_G(u)}(\beta(z))\pi(S^+) = \pi(\epsilon_{\mathbf{C}_G(u)}(\beta(z))S^+).\end{aligned}$$

But $S^+ \in \mathbf{S}F\mathbf{C}_G(u)$ by (40), so $\epsilon_{\mathbf{C}_G(u)}(\beta(z))S^+ = \beta(z)S^+$, and we conclude that $\pi(\epsilon(z)K^+) = \pi(\beta(z)S^+) = \pi(zS^+)$ which was to be proved.

The formula in (64) was obtained by the author in [9 IV]. The idea to look for such a formula came from a paper [21] by G.R. Robinson. The special case where $K = \{1\}$ already appeared in [9 III] and was also observed by M. Broué in [5]:

(65) *If* $z \in \mathbf{Z}FG$ *then* $\epsilon(z) = \pi_{G_{p'}}(zG_p^+)$.

This is clear since $\epsilon(z) = \pi_{G_{p'}}(\epsilon(z))$ by (61).

We now turn to applications to block theory. For an arbitrary algebra $A$ over $F$, $\mathbf{Z}A$ contains only finitely many primitive idempotents $e_1, \ldots, e_r$. These are called the *block idempotents* of $A$ and satisfy $e_ie_j = 0$ for $i \neq j$ and $e_1 + \cdots + e_r = 1_A$. For $i = 1, \ldots, r$, $B_i := Ae_i$ is then an ideal of $A$ and an algebra over $F$ with identity element $e_i$. Moreover, $B_1, \ldots, B_r$ are called the *blocks* of $A$ and satisfy $A = B_1 \oplus \cdots \oplus B_r$.

If $A = FG$ then the map $\nu : FG \to F$, $\sum_{g\in G} \alpha_g g \mapsto \sum_{g\in G} \alpha_g$, is a homomorphism of algebras which is called the *augmentation map* of $FG$. Since $1 = \nu(1) = \nu(e_1 + \cdots + e_r) = \nu(e_1) + \cdots + \nu(e_r)$ we have $\nu(e_i) \neq 0$ for some $i \in \{1, \ldots, r\}$. Moreover, since $\nu(e_i)\nu(e_j) = \nu(e_ie_j) = \nu(0) = 0$ for $i \neq j$ the index $i$ is uniquely determined. We have proved:

(66) $\nu(e_0(FG)) \neq 0$ *for a unique block idempotent* $e_0(FG)$ *of* $FG$.

We call $e_0(FG)$ the *principal block idempotent* and the corresponding block

$$B_0(FG) := FGe_0(FG)$$

of $FG$ the *principal block* of $FG$. It is easy to verify that $xG^+ = \nu(x)G^+$ for $x \in FG$; in particular,

$$e_0(FG)G^+ = G^+. \tag{67}$$

Since $\pi_U : \mathbf{Z}FG \to \mathbf{Z}FG \cap FU$ is an $\mathbf{E}FG$-homomorphism, we obtain $U^+ = \pi_U(G^+) = \pi_U(e_0(FG)G^+) = e_0(FG)\pi_U(G^+) = e_0(FG)U^+$ for $U \in \mathrm{Sec}_p(G)$, so we have proved (cf. M. Osima [17], for example):

(68) $U^+ \in B_0(FG) \cap \mathbf{Z}FG = \mathbf{Z}B_0(FG)$ *for* $U \in \mathrm{Sec}_p(G)$;

in particular, $G_{p'}^+ \in \mathbf{Z}B_0(FG)$. Since $\mathbf{Z}B_0(FG)$ is a local algebra over $F$ we can write $G_{p'}^+ = \alpha e_0(FG) + x$ where $\alpha \in F$ and $x \in \mathbf{J}\mathbf{Z}B_0(FG)$. Then $\alpha = \nu(\alpha e_0(FG) + x) = \nu(G_{p'}^+) = |G_{p'}|1_F \neq 0$ because of the following well-known group-theoretical fact:

(69) *For any finite group* $G$, $|G_{p'}| \not\equiv 0 \pmod p$.

Indeed, any Sylow $p$-subgroup $P$ of $G$ acts on $G_{p'}$ by conjugation, so we have $|G_{p'}| \equiv |\mathbf{C}_G(P)_{p'}| \pmod p$. But $\mathbf{C}_G(P) = \mathbf{Z}(P) \times \mathbf{O}_{p'}(\mathbf{C}_G(P))$ by the Schur-Zassenhaus theorem, so

$$|G_{p'}| \equiv |\mathbf{C}_G(P)_{p'}| \equiv |\mathbf{O}_{p'}(\mathbf{C}_G(P))_{p'}| \not\equiv 0 \pmod p,$$

and (69) is proved.

We conclude that $G_{p'}^{+} = |G_{p'}|e_0(FG) + x$ (M. Osima [17]) and

$$\epsilon(G_{p'}^{+}) = |G_{p'}|e_0(FG). \tag{70}$$

Thus (64) implies:

(71) *Let $K \in \mathrm{Cl}(G)$, let $u$ be a $p$-element in the $p$-section of $G$ containing $K$, and denote by $S$ the $p'$-section of $\mathbf{C}_G(u)$ containing $K \cap u\mathbf{C}_G(u)_{p'}$. Then*

$$\pi_{u\mathbf{C}_G(u)_{p'}}(e_0(FG)K^{+}) = (|G_{p'}|1_F)^{-1}\pi_{u\mathbf{C}_G(u)_{p'}}(G_{p'}^{+}S^{+}).$$

Of particular importance is the special case $K = 1$ where (61) leads to the following formula:

$$e_0(FG) = (|G_{p'}|1_F)^{-1}\pi_{G_{p'}}(G_{p'}^{+}G_{p}^{+}). \tag{72}$$

The preceding two results are due to the author [10]. If $Q$ is a $p$-subgroup of $G$ then $\beta_Q \circ \pi_{G_{p'}} = \pi_{\mathbf{C}_G(Q)_{p'}} \circ \beta_Q$ and, by an argument similar to the proof of (69), $|G_{p'}| \equiv |\mathbf{C}_G(Q)_{p'}| \pmod p$, so (72) implies that

$$\begin{aligned}\beta_Q(e_0(FG)) &= (|G_{p'}|1_F)^{-1}\pi_{\mathbf{C}_G(Q)_{p'}}(\beta_Q(G_{p'}^{+}G_{p}^{+}))\\ &= (|\mathbf{C}_G(Q)_{p'}|1_F)^{-1}\pi_{\mathbf{C}_G(Q)_{p'}}(\beta_Q(G_{p'}^{+})\beta_Q(G_{p}^{+}))\\ &= (|\mathbf{C}_G(Q)_{p'}|1_F)^{-1}\pi_{\mathbf{C}_G(Q)_{p'}}(\mathbf{C}_G(Q)_{p'}^{+}\mathbf{C}_G(Q)_{p}^{+})\\ &= e_0(F\mathbf{C}_G(Q)).\end{aligned}$$

Thus we have proved the following result which is known as Brauer's Third Main Theorem [4]:

(73) *For any $p$-subgroup $Q$ of $G$, $\beta_Q(e_0(FG)) = e_0(F\mathbf{C}_G(Q))$.*

Now we turn to arbitrary blocks of $FG$ and start by defining defect groups. For $K \in \mathrm{Cl}(G)$, we set $\mathrm{Def}_p(K) := \bigcup_{g\in K}\mathrm{Syl}_p(\mathbf{C}_G(g))$. Thus $\mathrm{Def}_p(K)$ is a conjugacy class of $p$-subgroups of $G$ which we call the *$p$-defect groups* of $K$. If $D$ is a $p$-defect group of $K$ and if $|D| = p^d$ then $d$ is called the *$p$-defect* of $K$.

For any $p$-subgroup $Q$ of $G$, we then have

$$\begin{aligned}\mathrm{Ker}(\beta_Q) \cap \mathbf{Z}FG &= F\{K^{+} : K \in \mathrm{Cl}(G),\ K \cap \mathbf{C}_G(Q) = \emptyset\}\\ &= F\{K^{+} : K \in \mathrm{Cl}(G),\ Q \not\subseteq \mathbf{C}_G(g) \text{ for } g \in K\}\\ &= F\{K^{+} : K \in \mathrm{Cl}(G),\ Q \not\subseteq R \text{ for } R \in \mathrm{Def}_p(K)\}.\end{aligned}$$

For subgroups $H$ and $K$ of $G$, we write $K \leq_G H$ if $gKg^{-1} \subseteq H$ for some $g \in G$. Then, for any $p$-subgroup $P$ of $G$, $I_P(FG) := \mathbf{Z}FG \cap \bigcap_Q \mathrm{Ker}(\beta_Q)$ where $Q$ ranges over all $p$-subgroups of $G$ satisfying $Q \not\leq_G P$ is an ideal of $\mathbf{Z}FG$. Moreover, for $K \in \mathrm{Cl}(G)$, we have $K^{+} \in I_P(FG)$ if and only if $Q \not\leq_G R$ for any $p$-subgroup $Q$ of $G$ satisfying $Q \not\leq_G P$ and any $p$-defect group $R$ of $K$, i.e. if and only if $R \leq_G P$ for $R \in \mathrm{Def}_p(K)$. Thus

$$I_P(FG) = F\{K^{+} : K \in \mathrm{Cl}(G),\ R \leq_G P \textit{ for } R \in \mathrm{Def}_p(K)\}, \tag{74}$$

for any $p$-subgroup $P$ of $G$. Moreover, we have $I_P(FG) = I_{gPg^{-1}}(FG)$ for $g \in G$, $I_Q(FG) \subseteq I_P(FG)$ for any subgroup $Q$ of $P$, and $I_S(FG) = \mathbf{Z}FG$ for any Sylow $p$-subgroup $S$ of $G$. For any block idempotent $e$ of $FG$, the minimal (with respect to inclusion) elements in the set of all $p$-subgroups $P$ of $G$ satisfying $e \in I_P(FG)$ are called the *defect groups* of $e$ and of the corresponding block $B := FGe$. Now Rosenberg's Lemma implies:

(75) *If $P$ is a defect group of $e$ then $e \notin \sum_Q I_Q(FG)$ where $Q$ ranges over the set of proper subgroups of $P$.*

This also shows:

(76) *For any block idempotent $e$ in $FG$, the defect groups of $e$ form a conjugacy class of $p$-subgroups of $G$.*

In particular, all defect groups of $e$ have the same order, say $p^d$. Then $d$ is called the *defect* of $e$ and of the corresponding block $B := FGe$. Since $\beta_Q(e_0(FG)) \neq 0$ for any $p$-subgroup $Q$ of $G$ the defect groups of the principal block of $FG$ are just the Sylow $p$-subgroups of $G$. We now prove:

(77) *Let $P$ be a $p$-subgroup of $G$, and let $n$ be a non-negative integer such that $p^n \geq \exp(P)$. Then $x^{p^n} \in I_P(FG)^\perp$ for $x \in \mathbf{J}FG + \mathbf{K}FG$.*

For otherwise there are $K \in \mathrm{Cl}(G)$ and $Q \in \mathrm{Def}_p(K)$ such that $Q \subseteq P$ and

$$0 \neq (K^+|x^{p^n}) = (K^+|x^{p^n} + \mathbf{K}FG) = ((\zeta_n(K^+)|x)^{p^n} = (K^{p^{-n}+}|x)^{p^n};$$

in particular, $K^{p^{-n}} \neq \emptyset$. Let $g \in K^{p^{-n}}$, so that $g^{p^n} \in K$. We may assume $Q \in \mathrm{Syl}_p(\mathbf{C}_G(g^{p^n}))$. Since $g_p \in \mathbf{C}_G(g^{p^n})$ and $p^n \geq \exp(Q)$ we have $g^{p^n} = (g_p g_{p'})^{p^n} = g_p^{p^n} g_{p'}^{p^n} = g_{p'}^{p^n} \in G_{p'}$, so $K \subseteq G_{p'}$. It follows easily that $K^{p^{-n}}$ coincides with the $p'$-section of $g_{p'}$ in $G$. Thus $K^{p^{-n}+} \in \mathbf{S}FG \cap \mathbf{Z}FG = (\mathbf{J}FG + \mathbf{K}FG)^\perp$ by (39) and (37) which contradicts the fact that $(K^{p^{-n}+}|x) \neq 0$.

We have the following characterization of the exponent of defect groups of blocks, going back to the author's paper [9 II]:

(78) *Let $B$ be a block of $FG$ with defect group $D$. Then* $\exp(D) = \min\{p^n : \mathbf{T}_n B = \mathbf{J}B + \mathbf{K}B\}$.

In order to prove this suppose first that $p^n \geq \exp(D)$, and let $x \in \mathbf{J}B + \mathbf{K}B$. Then $x^{p^n} \in I_D(FG)^\perp$ by (77). We denote the block idempotent of $B$ by $e$. Since $\mathbf{Z}FGe \subseteq I_D(FG)$ we obtain $0 = (ze|x^{p^n}) = (z|ex^{p^n}) = (z|x^{p^n})$ for $z \in \mathbf{Z}FG$, so $x^{p^n} \in \mathbf{K}FG$ by (35). Since $xe = x$ this means that $x \in \mathbf{T}_n B$. On the other hand, $\mathbf{T}_n B \subseteq \mathbf{J}B + \mathbf{K}B$, so $\mathbf{T}_n B = \mathbf{J}B + \mathbf{K}B$.

Now suppose that $p^n < \exp(D)$ and write $e = \sum_{y \in G} \alpha_y y$ where $\alpha_y \in F$ for $y \in G$. By (75), there is an element $g \in \mathbf{C}_G(D)$ such that $\alpha_g \neq 0$. Moreover, $g \in G_{p'}$ by (61). We choose an element $d \in D$ of maximal order and set $h := (g^{-1})^{p^{-n}} \in$

$C_G(D)_{p'}$. Since $d-1$ is nilpotent and commutes with $h$ we have $(d-1)he \in \mathbf{J}B+\mathbf{K}B$ by (9). On the other hand,

$$((d-1)he)^{p^n} = d^{p^n}h^{p^n}e - h^{p^n}e = d^{p^n}g^{-1}e - g^{-1}e = \sum_{y\in G} \alpha_y(d^{p^n}g^{-1}y - g^{-1}y).$$

If $\alpha_y \neq 0$ then $y \in G_{p'}$ by (61), and $d^{p^n}g^{-1} \notin G_{p'}$, so $0 = (\alpha_y d^{p^n}g^{-1}y|1)$ for $y \in G$, and $(((d-1)he)^{p^n}|1) = -\alpha_g \neq 0$. Thus $((d-1)he)^{p^n} \notin \mathbf{K}FG$ by (35), so $((d-1)he)^{p^n} \notin \mathbf{K}B$ by (4), and $(d-1)he \notin \mathbf{T}_nB$. Hence $\mathbf{T}_nB \neq \mathbf{J}B + \mathbf{K}B$.

The following consequences of (78) are the analogues of (21), (22), (23) and (24):

(79) *Let $B$ be a block of $FG$ with defect group $D$. Then* $\exp(D)$ *is the smallest power $p^n$ of $p$ such that $x^{p^n} \in \mathbf{K}B$ for $x \in \mathbf{J}B$; in particular, if $D$ is cyclic of order $p^n$ then $(\mathbf{J}B)^{p^{n-1}} \neq 0$, and if $p=2$ and $D$ is dihedral or quaternion of order $2^n$ then $(\mathbf{J}B)^{2^{n-2}} \neq 0$.*

(80) *If the block $B$ of $FG$ has positive defect then $\mathbf{J}B \neq 0$.*

This last fact is essentially due to R. Brauer, as is its converse:

(81) *If the block $B$ of $FG$ has defect 0 then $\mathbf{J}B = 0$.*

Indeed, if $B$ has defect 0 then (78) implies that $\mathbf{J}B \subseteq \mathbf{K}B \subseteq \mathbf{K}FG$; in particular, $0 = (\mathbf{J}B|1) = ((\mathbf{J}B)(FG)|1) = (\mathbf{J}B|FG)$ by (35), so $\mathbf{J}B = 0$.

We obtain the following consequence:

(82) $(\mathbf{S}FG \cap \mathbf{Z}FG)^2 = I_1(FG)^2 = \sum_B \mathbf{Z}B$ *where $B$ ranges over the set of blocks of defect 0 in $FG$.*

Indeed, if $B$ is a block of defect 0 in $FG$ with block idempotent $e$ then $\mathbf{Z}B = e\mathbf{Z}FG \subseteq I_1(FG)$. On the other hand, any conjugacy class of $p$-defect 0 in $G$ is a $p'$-section of $G$, so (39) implies that $I_1(FG) \subseteq \mathbf{S}FG \cap \mathbf{Z}FG$. Hence $\mathbf{Z}B = (\mathbf{Z}B)^2 \subseteq I_1(FG)^2 \subseteq (\mathbf{S}FG \cap \mathbf{Z}FG)^2$. Conversely, if $B$ is a block of positive defect in $FG$ with block idempotent $e$ then $e \notin \mathbf{S}FG$ by (80), so $(\mathbf{S}FG \cap \mathbf{Z}FG)e$ is a proper ideal in the local algebra $\mathbf{Z}B$. Thus $(\mathbf{S}FG \cap \mathbf{Z}FG)e \subseteq \mathbf{J}\mathbf{Z}B \subseteq \mathbf{J}B \subseteq \mathbf{J}FG$ and $(\mathbf{S}FG \cap \mathbf{Z}FG)^2e = 0$. Hence (82) is proved.

This result is due to K. Iizuka and A. Watanabe [8]. By (39), it implies that $(G_p^+)^2 \in \sum_B \mathbf{Z}B$ where $B$ ranges over the set of blocks of defect 0 in $FG$. But if $e$ is a block idempotent of defect 0 in $FG$ and if $K$ is a conjugacy class of $p$-elements in $G$ then $K^+e$ is a linear combination of class sums of conjugacy classes of defect 0 in $FG$ by (74). On the other hand, $K^+e$ is a linear combination of class sums of conjugacy classes in the $p$-section of $G$ containing $K$, by (61), so $K^+e = 0$ unless $K = 1$. Thus $e = G_p^+e = (G_p^+)^2e$, and we have proved the following result of Y. Tsushima [22]:

(83) $(G_p^+)^2$ *is the sum of all block idempotents of defect 0 in $FG$.*

The reader interested in more results of this type should consult the author's series of papers [9], the papers in the bibliography of [9] or of this paper or the standard textbooks on modular representation theory.

## REFERENCES

[1] R. Brauer, Über die Darstellungen von Gruppen in Galoisschen Feldern, Act. Sci. No. 135, Hermann, Paris 1935

[2] R. Brauer, Number theoretical investigations on groups of finite order, Proc. Intern. Symposium on Algebraic Number Theory, Tokyo and Nikko 1955, Science Council of Japan, Tokyo 1956, 55-62

[3] R. Brauer, Zur Darstellungstheorie der Gruppen endlicher Ordnung, *Math. Z.* **63** (1956), 406-444

[4] R. Brauer, Some applications of the theory of blocks of characters of finite groups I, *J. Algebra* **1** (1964), 152-167

[5] M. Broué, On a theorem of G. Robinson, *J. London Math. Soc. (2)* **29** (1984), 425-434

[6] B. Huppert and N. Blackburn, "Finite groups II," Springer-Verlag, Berlin 1982

[7] K. Iizuka, On Brauer's theorem on sections in the theory of blocks of group characters, *Math. Z.* **75** (1961), 299-304

[8] K. Iizuka and A. Watanabe, On the number of blocks of irreducible characters of a finite group with a given defect group, *Kumamoto J. Sci. (Math.)* **9** (1973), 55-61

[9] B. Külshammer, Bemerkungen über die Gruppenalgebra als symmetrische Algebra I-IV, *J. Algebra* **72** (1981), 1-7; **75** (1982), 59-69; **88** (1984), 279-291; **93** (1985), 310-323

[10] B. Külshammer, The principal block idempotent, to appear in *Arch. Math.*

[11] G.O. Michler, Blocks and centers of group algebras, in "Lectures on Rings and Modules," LNM 246, 429-563, Springer, Berlin 1972

[12] G.O. Michler, Green correspondence between blocks with cyclic defect groups I, *J. Algebra* **39** (1976), 26-51

[13] T. Nakayama, On Frobeniusean algebras I, *Ann. Math.* **40** (1939), 611-633

[14] T. Okuyama, Some studies on group algebras, *Hokkaido Math. J.* **9** (1980), 217-221

[15] M.F. O'Reilly, Ideals in the center of a group ring, in "Proc. Second Intern. Conf. Theory of Groups, Canberra 1973," LNM 372, Springer, Berlin 1974

[16] M. Osima, Note on blocks of group characters, *Math. J. Okayama Univ.* **4** (1955), 175-188

[17] M. Osima, On a theorem of Brauer, *Proc. Japan Acad.* **40** (1964), 795-798

[18] D.S. Passman, Central idempotents in group rings, *Proc. Amer. Math. Soc.* **22** (1969), 555-556

[19] W.F. Reynolds, Sections and ideals of centers of group algebras, *J. Algebra* **20** (1972), 176-181

[20] G.R. Robinson, The number of blocks with a given defect group, *J. Algebra* **84** (1983), 493-502

[21] G.R. Robinson, Counting modular irreducible characters, *J. Algebra* **90** (1984), 556-566

[22] Y. Tsushima, On the block of defect zero, *Nagoya Math. J.* **44** (1971), 57-59

[23] Y. Tsushima, On the $p'$-section sum in a finite group ring, *Math. J. Okayama Univ.* **20** (1978), 83-86

Progress in Mathematics, Vol. 95, © 1991 Birkhäuser Verlag Basel

# Darstellungstheoretische Methoden bei der Realisierung einfacher Gruppen vom Lie Typ als Galoisgruppen

GUNTER MALLE*

## Einleitung

In dieser Arbeit wird über weitere Fortschritte bei der Realisierung endlicher einfacher Gruppen als Galoisgruppen über abelschen Zahlkörpern berichtet. Der Nachweis der Existenz von solchen Galoisrealisierungen endlicher Gruppen kann durch das Rationalitätskriterium von Belyi, Matzat und Thompson auf eine rein gruppentheoretische Frage verlagert werden (siehe [15] für eine ausführliche Darstellung). Genauer läßt sich dieses Kriterium in seiner einfachsten Form wie folgt formulieren:

**Rationalitätskriterium (Matzat, Thompson).** *Es seien $G$ eine endliche Gruppe mit trivialem Zentrum und $\chi_1, \ldots, \chi_h$ die irreduziblen komplexen Charaktere von $G$. Sind $C_1$, $C_2$, $C_3$ drei Konjugiertenklassen von $G$ mit*

$$n(C_1, C_2, C_3) := \frac{|G|}{|\mathcal{C}_G(\sigma_1)||\mathcal{C}_G(\sigma_2)||\mathcal{C}_G(\sigma_3)|} \cdot \sum_{j=1}^{h} \frac{\chi_j(\sigma_1)\chi_j(\sigma_2)\chi_j(\sigma_3)}{\chi_j(\iota)} = 1$$

*($\sigma_i \in C_i$), und erzeugen die Tripel $(\sigma_1, \sigma_2, \sigma_3)$ mit $\sigma_i \in C_i$, $\sigma_1\sigma_2\sigma_3 = \iota$, die ganze Gruppe, so kommt $G$ als Galoisgruppe über dem abelschen Zahlkörper $\mathbb{Q}(\bigcup_{j=1}^{h} \chi_j(\sigma_i))$ vor.*

Die Berechnung der *normalisierten Strukturkonstanten* $n(C_1, C_2, C_3)$ erfordert dabei offensichtlich die Kenntnis der irreduziblen Charaktere der Gruppe $G$. Im Falle exzeptioneller Gruppen vom Lie Typ wurde in [14] gezeigt, daß bei geeigneter Wahl einer der drei Klassen fast alle Summanden im Ausdruck für $n(C_1, C_2, C_3)$ verschwinden. Nimmt man genauer als $\sigma_3$ ein reguläres Element eines zyklischen *t.i.* Hall Torus $T$, und ist $\sigma_2$ halbeinfach und zu keinem Element von $T$ konjugiert, so ergeben nur die unipotenten Charaktere von $G$ Beiträge zur Strukturkonstanten. In guter Charakteristik sind, im wesentlichen durch Arbeiten von Lusztig, die Werte dieser Charaktere im Prinzip bekannt, das heißt es gibt Algorithmen zur Berechnung jedes einzelnen Wertes. Damit wurde in [14] ein grosser Teil der exzeptionellen Gruppen vom Lie Typ als Galoisgruppen nachgewiesen.

Nicht behandelt werden konnten die Fälle schlechter Charakteristik, da dort die Greenfunktionen, die zur Bestimmung der Werte unipotenter Charaktere auf unipotenten Klassen notwendig sind, bisher nicht bekannt sind. Ebenso fand sich im Fall $G = E_6(q)$ (bzw. ${}^2E_6(q)$) ein geeignetes Konjugiertenklassentripel nur für $q \equiv -1 \pmod 3$ (bzw. $q \equiv 1 \pmod 3$).

---

* Der Autor dankt der Deutschen Forschungsgemeinschaft für die Förderung im Rahmen des Schwerpunktprogramms

Hier wird gezeigt, daß zumindest in einigen Fällen genügend viele Werte der Greenfunktionen auch in schlechter Charakteristik berechnet werden können, um das Rationalitätskriterium als erfüllt nachzuweisen. Ebenso kann man die fehlende Kongruenzklasse bei $E_6(q)$ durch Galoisrealisierung der Erweiterung mit dem Graphautomorphismus $\widehat{G} := E_6(q){:}\,2$ und anschließenden Abstieg zur einfachen Gruppe erhalten. Dazu muß eine Strukturkonstante für die unzusammenhängende Gruppe $\widehat{G}$ berechnet werden. Die Charaktertheorie unzusammenhängender Gruppen ist bisher auch in guter Charakteristik kaum entwickelt worden. Daher wird hier allgemein für $BN$-Paare mit einem äußeren Automorphismus, der $B$ und $N$ invariant läßt, die Struktur der Heckealgebra $\mathcal{H}(\widehat{G}, B, 1)$ des Permutationscharakters der Boreluntergruppe untersucht, die im zusammenhängenden Fall bekanntlich isomorph zur Gruppenalgebra $\mathbb{C}W$ der Weylgruppe ist. Es stellt sich heraus, daß eine Erweiterung $\widehat{W}$ der Weylgruppe definiert werden kann, so daß die Heckealgebra $\mathcal{H}$ wiederum isomorph zu $\mathbb{C}\widehat{W}$ wird. Auch die Verallgemeinerung des Howlett-Lehrerschen Vergleichssatzes [12] für Vielfachheiten in von parabolischen Untergruppen induzierten unipotenten Charakteren gilt in der allgemeineren Situation. Daraus lassen sich schließlich genügend viele Werte der Fortsetzungen der unipotenten Charaktere von $G$ auf $\widehat{G}$ berechnen, um die gesuchte Strukturkonstante zu bestimmen. Bisher läßt sich damit beweisen:

**Hauptsatz.** *Die folgenden Gruppen kommen als Galoisgruppen regulärer Körpererweiterungen über geeigneten abelschen Definitionskörpern vor:*

*(1)* $F_4(2^n)$,
*(2)* $E_6(q)$, $\quad q \neq 2^{2n}$,
*(3)* ${}^2E_6(q)$, $\quad q = p^n \not\equiv -1 \pmod 3$,
*(4)* $E_6(q)_{sc}{:}\,2$, $\quad q = p^n,\ p > 2$.

Zusammen mit den Ergebnissen aus [14] bedeutet dies, daß außer ${}^2B_2(2^{2n+1})$, ${}^2F_4(2^{2n+1})$, $E_6(2^{2n})$, ${}^2E_6(q)$ für $q \equiv -1 \pmod 3$ und $E_7(q)$, $E_8(q)$ jeweils in schlechter Charakteristik, alle exzeptionellen Gruppen vom Lie Typ als Galoisgruppen über $\mathbb{Q}^{ab}$ vorkommen. Gleichzeitig folgt aus der Klassifikation, daß die genannten Ausnahmen die einzigen endlichen einfachen Gruppen sind, die nicht bereits als Galoisgruppen über $\mathbb{Q}^{ab}$ nachgewiesen werden konnten (siehe [15] für eine Übersicht der Ergebnisse zu den anderen einfachen Gruppen).

Im ersten Paragraphen wird der Beweis des Vergleichssatzes für zyklische Erweiterungen von $BN$-Paaren skizziert. Im zweiten Paragraphen wird der gewöhnliche Howlett-Lehrersche Vergleichssatz zur Berechnung einiger Charakterwerte für $E_6(q)$, $q \not\equiv 1 \pmod 3$, angewandt und damit die Hälfte der Aussage (2) des Hauptsatzes gezeigt. Der dritte Teil demonstriert schließlich die Anwendung der Ergebnisse des ersten Paragraphen auf die unzusammenhängenden Gruppen $E_6(q)_{sc}{:}\,2$ und deutet den Beweis sowohl von (4) als auch des Rests von (2) des Hauptsatzes an. Die vollständigen Beweise aller vier Aussagen des Satzes würden den hier gegebenen Rahmen sprengen und erscheinen daher in einer gesonderten Arbeit. Der Satz 1.7 dürfte von unabhängigem Interesse in der Darstellungstheorie von $BN$-Paaren sein.

Zu den in dieser Arbeit verwendeten Notationen: Die Bezeichnung für einfache Gruppen und für Gruppenerweiterungen folgt dem Atlas [4]. Insbesondere bedeuten $A.B$ eine Erweiterung von (Kern) $A$ mit (Faktorgruppe) $B$ und $A{:}\,B$ eine zerfallende Erweiterung. Für Elemente $g \in G$ ist $o(g)$ die Ordnung von $g$, $C_G(g)$ der Zentralisator von $g$ in $G$, und $Z(G)$ bezeichnet das Zentrum von $G$. Für eine Untergruppe $T \leq G$ bedeutet $\mathcal{N}_G(T)$ den Normalisator von $T$ in $G$. Weiter ist $T$ eine *t.i.*-Untergruppe (für *trivial intersection*), falls für alle $g \in G$ entweder $T \cap T^g = T$ oder $T \cap T^g = \{\iota\}$ gilt. Mit $T^\#$ bezeichnet man in diesem Fall die Menge $T \backslash \{\iota\}$. Für die Definition und Eigenschaften von Gruppen mit einem $BN$-Paar wird auf [2] oder [6] verwiesen. Schließlich findet man die Definition einer Klassenstruktur $\mathbf{C}$ und einen Beweis des Rationalitätskriteriums in [15].

## 1. Heckealgebren zu Automorphismengruppen von $BN$-Paaren

In diesem Paragraphen wird folgende Situation betrachtet: Sei $G$ eine endliche Gruppe mit einem $BN$-Paar mit einer zyklischen Erweiterung $\widehat{G} = G.\langle\rho\rangle$ vom Grad $o(\rho) = r$. Dabei normalisiere $\rho$ sowohl $B$ als auch $N$, lasse also das $BN$-Paar fest. In dem uns später interessierenden Fall einer endlichen Gruppe vom Lie Typ $G$ ist letztere Forderung durch Abändern von $\rho$ immer erfüllbar, da dort alle Boreluntergruppen $B$ und darin alle maximalen Tori $T = B \cap N$ jeweils konjugiert sind.

Aus der $\rho$-Invarianz von $B$ und $N$ folgert man sofort, daß $\rho$ auch auf der Weylgruppe $W = N/(B \cap N)$ operiert. Seien $S = \{s_1, \ldots, s_n\}$ die involutorischen Erzeugenden von $W$ nach Definition eines $BN$-Paares. Nach [6], 65.16, ist die Menge $S$ in $W$ eindeutig bestimmt und $(W, S)$ bildet eine Coxetergruppe. Daher operiert $\rho$ auch auf $S$ und damit als (eventuell trivialer) Graphautomorphismus auf dem zu $(W, S)$ gehörenden Diagramm. Das Bild $\rho(s_i)$ ist also eine andere Involution aus $S$, und wir schreiben $\rho(s_i) =: s_{\bar{i}}$. Da $\rho$ im Gegensatz zu allen nichttrivialen Elementen aus $W$ die Boreluntergruppe $B$ festläßt, zerfällt die Erweiterung von $W$ mit $\rho$; wir bezeichnen das semidirekte Produkt als $\widehat{W} = W{:}\langle\bar{\rho}\rangle$.

Für die so definierten Gruppen $\widehat{G}$ wird nun die Zerlegung des von der Boreluntergruppe $B$ induzierten Einscharakters $1_B$ in irreduzible Charaktere untersucht. Das Ziel ist, in Erweiterung der Theorie für die Gruppen $G$ selbst, eine Beschreibung der Heckealgebra $\mathcal{H}(\widehat{G}, B)$ von $1_B^{\widehat{G}}$. Es zeigt sich, daß $\mathcal{H}(\widehat{G}, B)$ isomorph zur Gruppenalgebra $\mathbf{C}\widehat{W}$ der erweiterten Weylgruppe über $\mathbf{C}$ ist. Die Beweise hierzu verlaufen in genauer Analogie und unter teilweiser Ausnutzung der entsprechenden im Fall von $BN$-Paaren.

Die Menge $\widehat{S} = \{s_0 = \bar{\rho}, s_1, \ldots, s_n\}$ bildet ein Erzeugendensystem von $\widehat{W}$, wobei die definierenden Relationen aus denen von $W$ unter Hinzunahme der Vertauschungsregeln $s_i^{s_0} = s_{\bar{i}}$ und der Relation $s_0^r = \iota$ erhalten werden:

$$\widehat{W} = \langle s_0, \ldots, s_n \mid (s_i s_j)^{m_{ij}} = \iota,\ 1 \leq i, j \leq n;\ s_i^{s_0} = s_{\bar{i}},\ s_0^r = \iota \rangle.$$

Eine *Längenfunktion* auf $\widehat{W}$ werde wie folgt definiert: ein $w \in \widehat{W}$ hat eine eindeutige Darstellung $w = w' s_0^\nu$ mit $w' \in W$, $0 \leq \nu < r$. Dann sei $l(w) := l(w')$ (Länge in der Coxetergruppe $W$); es gilt also $l(w^{s_0}) = l(w)$.

Wie im Fall der Gruppen $G$ selbst definieren wir nun ein Erzeugendensystem für die Heckealgebra $\mathcal{H}(\widehat{G}, B)$ (siehe [5], 11.34):

*Definition 1.1.* Für $w \in \widehat{W}$ seien $\operatorname{ind} w := (B : B \cap {}^w B)$ und

$$a_w := |B|^{-1} \cdot \sum_{x \in BwB} x \quad \in \quad \mathcal{H}(\widehat{G}, B).$$

Speziell heißen $q_i := \operatorname{ind} s_i$, $0 \leq i \leq n$, die *Indexparameter*.

Insbesondere gilt also ind $s_0 = 1$. Da $\widehat{W}$ offensichtlich ein Vertretersystem doppelter Nebenklassen von $B$ in $\widehat{G}$ ist, bilden die eben definierten Elemente $a_w$ eine Basis der Endomorphismenalgebra $\mathcal{H} = \mathcal{H}(\widehat{G}, B)$.

**Bemerkung 1.2.** *$\mathcal{H} = \langle a_\iota = 1, a_{s_0}, \ldots, a_{s_n} \rangle$ mit*

$$\begin{aligned} a_{s_i}^2 &= q_i 1 + (q_i - 1) a_{s_i}, \\ (a_{s_i} a_{s_j})^{k_{ij}} &= (a_{s_j} a_{s_i})^{k_{ij}} && \textit{falls} \quad m_{ij} = 2k_{ij}, \\ (a_{s_i} a_{s_j})^{k_{ij}} a_{s_i} &= (a_{s_j} a_{s_i})^{k_{ij}} a_{s_j} && \textit{falls} \quad m_{ij} = 2k_{ij} + 1, \end{aligned}$$

*für $1 \le i, j \le n$, sowie $a_w a_{s_0} = a_{ws_0}$ und $a_{s_0} a_w = a_{s_0 w}$ für alle $w \in \widehat{W}$, ist eine Präsentation von $\mathcal{H}$.*

**Beweis.** Sei $\mathcal{A}$ eine assoziative $\mathbb{C}$-Algebra mit Erzeugenden $a_0, \ldots, a_n$, die obige Relationen erfüllen. Nach [6], 67.6, gibt es einen Algebrenepimorphismus $f$ von $\mathcal{H}' = \langle 1, a_{s_1}, \ldots, a_{s_n} \rangle$ auf $\mathcal{A}' = \langle a_1, \ldots, a_n \rangle$ mit $f(a_{s_i}) = a_i$, denn $\mathcal{H}'$ ist gerade die Heckealgebra $\mathcal{H}(G, B)$. Man erweitert nun $f$ auf $\mathcal{H}$ durch die Festsetzung $f(a_{s_0}) = a_0$ und setzt durch $f(a_{ws_0^\nu}) = f(a_w) a_0^\nu$ auf die Vektorraumbasis $\{a_w \mid w \in \widehat{W}\}$ fort. Die Linearität ist damit gesichert. Aufgrund der letzten beiden Relationen in der Bemerkung verhält sich die Multiplikation mit $a_0$ in $\mathcal{A}$ wie die mit $a_{s_0}$ in $\mathcal{H}$, und die Abbildung $f$ erhält auch die multiplikative Struktur. Die in der Bemerkung aufgeführten Erzeugenden und Relationen bilden also eine Präsentation von $\mathcal{H}$.

Q.E.D.

Um die beiden Algebren $\mathcal{H}(G, B)$ und $\mathbb{C}W$ in Beziehung zu setzen, führte Tits die generische Algebra ein. Durch Spezialisierung der darin vorkommenden Unbestimmten zu $q_i$ erhält man daraus die Heckealgebra, durch Spezialisierung zu 1 die Gruppenalgebra der Weylgruppe. Damit läßt sich die Isomorphie der beiden Algebren zeigen und so ein enger Zusammenhang zwischen den Charakteren herstellen. Es erweist sich, daß auch in unserem allgemeineren Fall eine generische Algebra exisitiert, die wahlweise zu $\mathcal{H}(\widehat{G}, B)$ oder $\mathbb{C}\widehat{W}$ spezialisiert werden kann.

**Bemerkung 1.3.** *Sei $R$ eine kommutative $\mathbb{Q}$-Algebra mit Elementen $\{u_1, \ldots, u_n\}$, so daß $u_i = u_j$ gilt, falls $s_i$ in $\widehat{W}$ zu $s_j$ konjugiert ist. Sei $A$ ein freier $R$-Modul mit Basis $\{e_w \mid w \in \widehat{W}\}$. Dann gibt es eine $R$-Algebrastruktur auf $A$, die durch*

$$e_{s_i} e_w = e_{s_i w} \qquad \textit{falls} \qquad l(s_i w) \ge l(w),$$

$$e_{s_i} e_w = u_i e_{s_i w} + (u_i - 1) e_w \qquad \textit{falls} \qquad l(s_i w) < l(w)$$

*eindeutig bestimmt ist.*

**Beweis.** Nach [6], 68.1, existiert jedenfalls eine Teilalgebra $A' = \langle e_w \mid w \in W \rangle$, die die vorgeschriebenen Relationen erfüllt. Zum Beweis der Existenz der Algebrastruktur auf $A$ geht man genau wie im Beweis zu 68.1 in [6] vor. Die Details werden daher nicht ausgeführt.

Q.E.D.

*Definition 1.4.* Die in Bemerkung 1.3 definierte Algebra $A$ heißt die *generische Algebra* zu $(\widehat{W}, \widehat{S})$.

Dies führt uns zum ersten Ergebnis dieses Abschnitts (siehe auch [9], II, Theorem 1.6, wo dieses Resultat bereits erwähnt wird):

**Satz 1.5.** *Seien $G$ eine endliche Gruppe mit einem $BN$-Paar, $\widehat{G} = G.\langle\rho\rangle$ eine endliche zyklische Erweiterung mit $B^\rho = B$ und $N^\rho = N$, weiter $\widehat{W} = \langle s_0 = \bar{\rho}, s_1, \ldots, s_n\rangle$, wobei $S = \{s_1, \ldots, s_n\}$ die kanonischen Erzeugenden der Weylgruppe $W = B/(B \cap N)$ von $G$ sind. Dann existiert ein Isomorphismus von $\mathbb{C}$-Algebren*

$$\mathcal{H} \cong \mathbb{C}\widehat{W}$$

*zwischen der Heckealgebra $\mathcal{H} = \mathcal{H}(\widehat{G}, B, 1_B)$ und der Gruppenalgebra $\mathbb{C}\widehat{W}$ der erweiterten Weylgruppe von $\widehat{G}$.*

**Beweis.** Sei $A$ gemäß Bemerkung 1.3 die generische Algebra zu $(\widehat{W}, \widehat{S})$ über $R = \mathbb{C}[u_1, \ldots, u_n]$ mit $u_i = u_j$ falls $s_i$ und $s_j$ in $\widehat{W}$ konjugiert sind. Man prüft leicht nach, daß die Erzeugenden $e_w$ von $A$ auch die in Bemerkung 1.2 für $\mathcal{H}$ angegebenen Relationen (mit $q_i$ ersetzt durch $u_i$) erfüllen, und genau wie dort beweist man, daß dies bereits ein Präsentation von $A$ liefert. Daher ist $\mathcal{H}$ die Spezialisierung $A_f$ von $A$ unter der Abbildung $f : R \to \mathbb{C}$, $u_i \mapsto \operatorname{ind} s_i$. Andererseits erhält man als Spezialisierung $A_{f'}$ zu $f' : R \to \mathbb{C}$, $u_i \mapsto 1$ die Gruppenalgebra $\mathbb{C}\widehat{W}$ der erweiterten Weylgruppe. Sowohl $\mathcal{H}$ als Endomorphismenalgebra eines halbeinfachen Moduls als auch $\mathbb{C}\widehat{W}$ als Gruppenalgebra über $\mathbb{C}$ sind halbeinfach. Mit $K = \mathbb{C}(u_1, \ldots, u_n)$ wird schließlich auch $A^K = A \otimes_R K$ eine halbeinfache $K$-Algebra, da eine ihrer Spezialisierungen halbeinfach ist. Damit ist der Deformationssatz von Tits (siehe etwa [6], 68.17, oder [2], 10.11.2) anwendbar, der unter diesen Voraussetzungen aussagt, daß die beiden Spezialisierungen $A_f$ und $A_{f'}$ isomorph sind. Q.E.D.

Der nach Satz 1.5 existierende Isomorphismus zwischen $\mathbb{C}\widehat{W}$ und $\mathcal{H}$ liefert auch eine (nicht kanonische) Bijektion $\varphi \mapsto \mu_\varphi$ zwischen den irreduziblen Charakteren $\varphi$ von $\mathbb{C}\widehat{W}$, also den $\varphi \in Irr(\widehat{W})$, und den $\mu_\varphi \in Irr(\mathcal{H})$. Im klassischen Fall $G = \widehat{G}$ heißen die irreduziblen Konstituenten von $1_B^{\widehat{G}}$ die *unipotenten Hauptseriencharaktere (HSC)*. Dabei liefert der Isomorphismus $\mathcal{H}(G, B, 1_B) \cong \mathbb{C}W$ eine Bijektion zwischen den irreduziblen Charakteren von $W$ und den unipotenten HSC von $G$. Da die beiden Algebren als Teilalgebren von $\mathcal{H}$ bzw. $\mathbb{C}\widehat{W}$ aufgefaßt werden können, ist es klar, daß die Bijektion auf der Stufe $\widehat{W}$ so gewählt werden kann, daß sie mit der Bijektion auf der Stufe $W$ kompatibel ist; das heißt, die Bijektion ist verträglich mit Einschränkung auf $G$ bzw. auf $W$.

Wir erhalten einige unmittelbare Folgerungen über die Fortsetzbarkeit der unipotenten Hauptseriencharaktere von $G$ zu $\widehat{G}$:

*Anwendungen.* Sei $G$ eine endliche Gruppe vom Lie-Typ. Dann hat $G$ ein natürliches $BN$-Paar, und jede zyklische Erweiterung von $G$ mit einem äußeren Automorphismus erfüllt die im Satz 1.5 vorausgesetzten Bedingungen.

1) Sei $\rho$ ein Körperautomorphismus von $G$ der Ordnung $r$. Dieser operiert trivial auf der Weylgruppe $W = W(G)$, also besitzen alle unipotenten Hauptseriencharaktere von $G$ jeweils $r$ verschiedene Fortsetzungen auf $\widehat{G}$.

2) Sei $G = A_n(q)$, $\rho$ der Graphautomorphismus der Ordnung zwei. Hier operiert $\rho$ zwar nichttrivial auf der Weylgruppe $W \cong S_{n+1}$, läßt aber die Klasse der erzeugenden Reflektionen, also der Transpositionen, fest. Jeder solche Automorphismus von $S_{n+1}$ ist bekanntlich ein innerer, also gilt $\widehat{W} \cong S_{n+1} \times 2$, und wieder besitzt jeder unipotente HSC von $G$ zwei Fortsetzungen auf $\widehat{G}$.

3) Sei $G = B_2(2^m)$ und $\rho$ operiere wie der Graphautomorphismus auf $G$. (Falls $m$ ungerade ist, kann $\rho$ von Ordnung zwei gewählt werden, sonst quadriert er zu einem Körperautomorphismus.) Da $\rho$ zwei Klassen der Weylgruppe $W \cong D_8$ vertauscht, operiert er als echter äußerer Automorphismus auf $W$, und man stellt fest, daß genau zwei unipotente HSC zusammenfallen, während die drei anderen Fortsetzungen auf $\widehat{G}$ besitzen.

4) Sei $G = D_n(q)$ und $\rho$ operiere wie der Graphautomorphismus der Ordnung zwei. Die erzeugenden Reflektionen von $W$ seien so numeriert, daß $s_{n-1}$ und $s_n$ zu den beiden durch $\rho$ vertauschten Punkten des Diagramms gehören. Das Bild von $\rho$ in $\widehat{W}$ werde wieder mit $s_0$ bezeichnet. Man rechnet sofort nach, daß $\{s_1, \dots, s_{n-1}, s_0\}$ die Relationen einer Coxetergruppe vom Typ $B_n$ erfüllen. Also gilt $\widehat{W} = W(D_n){:}2 \cong W(B_n)$. In [1], G-11, wird diskutiert, welche Klassen von $W(D_n)$ bei der Fusion in $W(B_n)$ zusammenfallen. Insbesondere folgt, daß für ungerades $n$ jede Klasse von $W(B_n)$ höchstens eine $W(D_n)$-Klasse enthält. In diesem Fall besitzen demnach alle unipotenten HSC Fortsetzungen auf $\widehat{G}$. Bei geradem $n$ hingegen fallen Klassen bei der Fusion in $W(B_n)$ zusammen, und zwar in wachsender Anzahl bei steigendem $n$. Dasselbe gilt daher für die unipotenten HSC.

5) Sei $G = D_4(q)$ und $\rho$ operiere wie der Graphautomorphismus der Ordnung drei. In der Weylgruppe $2^3.S_4$ werden durch $\rho$ jeweils zweimal drei Klassen zusammengeworfen. Damit besitzen nur 7 der 13 unipotenten HSC Fortsetzungen auf $\widehat{G}$. Der einzige verbleibende (kuspidale) unipotente Charakter besitzt aus Gradgründen ebenfalls eine Fortsetzung.

6) Sei $G = E_6(q)$ und $\rho$ operiere wie der Graphautomorphismus auf $G$. Die nichtriviale Operation auf der Weylgruppe muß von einem inneren Automorphismus herrühren, denn $W(E_6) \cong \mathrm{Aut}(S_4(3))$ besitzt keine äußeren Automorphismen. Daher lassen sich auch hier alle unipotenten HSC auf $\widehat{G}$ fortsetzen. Im dritten Paragraphen wird gezeigt, daß auch die beiden kuspidalen unipotenten Charaktere auf $\widehat{G}$ fortsetzbar sind.

7) Sei $G = F_4(2^m)$ und $\rho$ operiere wie der Graphautomorphismus. Die Situation ist wie im Fall 3); es zeigt sich, daß unter der Operation von $\rho$ siebenmal je zwei verschiedene Klassen von $W(F_4)$ fusionieren. Deshalb lassen sich nur elf der 25 unipotenten HSC von $G$ auf $\widehat{G}$ fortsetzen. (Für den Fall $q = 2$ vergleiche man mit den Tafeln in [4]).

8) Schließlich sei $G = G_2(3^m)$ und $\rho$ operiere wieder als Graphautomorphismus. Die Operation auf der Weylgruppe $W(G_2) \cong D_{12}$ ist nichttrivial, und genau zwei Charaktere fallen zusammen. Also besitzen nur vier der sechs unipotenten HSC von $G$ eine Fortsetzung zu $\widehat{G}$.

Die vorigen Aussagen kann man in den meisten Fällen, insbesondere wenn alle Charaktere fortsetzbar sind, bereits an den bekannten Graden der unipotenten Charaktere ablesen. (Sie sind meist alle verschieden.) Der Satz 1.5 gibt allerdings ein allgemeingültiges und strukturelles Kriterium zur Entscheidung der Fortsetzbarkeit.

Die Bijektion zwischen den irreduziblen Charakteren von $\widehat{W}$ und den Charakteren der Heckealgebra zu $1_{\widehat{B}}^{\widehat{G}}$ wird nun noch genauer untersucht. Wie im Fall von

$BN$-Paaren erhält man Aussagen über die Zerlegung der von parabolischen Untergruppen induzierten Hauptseriencharaktere in irreduzible Charaktere von $\widehat{G}$. Dazu muß zuerst der Begriff einer parabolischen Untergruppe zweckmäßig verallgemeinert werden.

*Definition 1.6.* Eine Teilmenge $J$ von $\widehat{S}$ heiße *zulässig*, falls entweder $s_0 \notin J$ oder andernfalls $J^{s_0} = J$ gilt. Die *parabolischen Untergruppen* $\widehat{W}_J$ von $\widehat{W}$ und $P_J$ von $\widehat{G}$ zu einem zulässigem $J \subseteq \widehat{S}$ sind definiert als $\widehat{W}_J = \langle s_i \mid s_i \in J \rangle$ und $P_J := B\widehat{W}_J B$.

Offensichtlich sind die $P_J$ für $s_0 \notin J$ bereits parabolische Untergruppen in $G$, andernfalls zyklische Erweiterungen von $P_{J\setminus\{s_0\}}$.

Um die irreduziblen Darstellungen der generischen Algebra $A$ zu untersuchen, arbeiten wir über dem algebraischen Abschluß $\bar{K}$ von $K = \mathbb{C}(u_1, \ldots, u_n)$, und wählen dazu feste Fortsetzungen der Spezialisierungen $f : A \to \mathcal{H}(\widehat{G}, B, 1)$, $f' : A \to \mathbb{C}\widehat{W}$ auf $A^{\bar{K}} := A \otimes_R \bar{K}$.

Wir setzen zusätzlich noch voraus, daß $G$ ein *zerfallendes $BN$-Paar der Charakteristik $p$* besitzt. Dann existiert für jede parabolische Untergruppe $P_J$ zu einer zulässigen Teilmenge $J \subseteq \widehat{S}$ eine Levizerlegung $P_J = U_J{:}L_J$, und die Leviuntergruppe ist wieder eine zyklische Erweiterung eines zerfallenden $BN$-Paares der Charakteristik $p$. Die Hauptseriencharaktere von $L_J$ sind dann nach Satz 1.5 durch die irreduziblen Charaktere von $\widehat{W}_J$ indiziert. Mittels der kanonischen Bijektion von $\mathcal{H}(L_J, B_J, 1)$ zu $\mathcal{H}(P_J, B, 1)$ entsprechen diese $\kappa$ eindeutig gelifteten irreduziblen Charakteren $\tilde{\kappa}$ von $P_J$, die den $p$-Normalteiler $U_J$ im Kern haben. Gesucht ist nun eine Formel, die angibt, wie die verallgemeinerten Induzierten $\tilde{\kappa}^{\widehat{G}}$ sich in Hauptseriencharaktere von $\widehat{G}$ zerlegen. Die späteren Rechnungen werden nämlich zeigen, daß diese Zerlegungen mehr Informationen liefern, als nur die Zerlegung der Permutationscharaktere $1_{P_J}^{\widehat{G}}$, zumindest falls $s_0$ nicht $W$ zentralisiert oder $G$ nicht den Typ $A_l$ hat.

**Satz 1.7.** *Seien $G$ eine endliche Gruppe mit zerfallendem $BN$-Paar der Charakteristik $p$ und vom Typ $(W, S)$, $\widehat{G} := G.\langle \rho \rangle$ eine zyklische Erweiterung vom Typ $(\widehat{W}, \widehat{S})$, $\zeta$ ein irreduzibler Charakter von $\widehat{G}$ mit $(1_B^{\widehat{G}}, \zeta) \neq 0$, $\varphi$ der zu $\zeta$ gehörige Charakter von $\widehat{W}$. Seien weiter $J \subset \widehat{S}$ eine zulässige Teilmenge der Indexmenge, $\kappa$ ein irreduzibler Charakter der Leviuntergruppe $L_J$ mit entsprechendem Charakter $\psi$ von $\widehat{W}_J$, $\tilde{\kappa}$ der Lift zu $P_J$. Dann gilt*

$$\left(\zeta_\varphi, \tilde{\kappa}_\psi^{\widehat{G}}\right)_{\widehat{G}} = \left(\varphi, \psi^{\widehat{W}}\right)_{\widehat{W}}.$$

Dieser Satz läßt sich auch so formulieren: Die Isometrie $\varphi \mapsto \zeta_\varphi$ von $Char(\widehat{W})$ nach $Char(\widehat{G})$ ist verträglich mit der Induktion von parabolischen Untergruppen.

Der Beweis ist eine direkte Verallgemeinerung eines eleganten Beweises von Kilmoyer (siehe [16], Theorem 3.4) auf die uns interessierende Situation.

**Beweis.** Sei $A_J$ die von $\{e_w \mid w \in \widehat{W}_J\}$ erzeugte Teilalgebra der generischen Algebra $A$. Dann gilt offensichtlich $f(A_J^{\bar{K}}) = \mathcal{H}(P_J, B, 1) \cong \mathcal{H}(L_J, B_J, 1)$, wobei der letzte Isomorphismus wie oben erwähnt kanonisch ist, und $f'(A_J^K) = \mathbb{C}\widehat{W}_J$. Weiter sei $\chi$ der eindeutig bestimmte Charakter von $A$, der zu $\zeta$ bzw. $\varphi$ spezialisiert. Die Einschränkung von $\chi$ auf die Teilalgebra $A_J$ zerlegt sich dann als

$$\chi|_{A_J} = \sum a_i \chi_i,$$

in eine positive Linearkombination irreduzibler Charaktere $\chi_i$ von $A_J$. Unter den Spezialisierungen $f$ und $f'$ wird dies zu

$$\zeta|_{\mathcal{H}(P_J,B,1)} = \sum a_i \chi_{i,f}, \qquad \varphi|_{\widehat{W}_J} = \sum a_i \chi_{i,f'}.$$

Dabei sind die $\chi_{i,f}$ irreduzible Charaktere von $\mathcal{H}(P_J, B, 1)$ und die $\chi_{i,f'}$ solche von $\widehat{W}_J$. Falls $\psi$ in der zweiten Zerlegung auftaucht, so auch $\tilde{\kappa}$ in der ersten mit derselben Vielfachheit, da sie nach Konstruktion Spezialisierungen desselben irreduziblen Charakters von $A_J$ sind. Das bedeutet aber

$$(\zeta|_{\mathcal{H}(P_J,B,1)}, \tilde{\kappa})_{P_J} = (\varphi|_{\widehat{W}_J}, \psi)_{\widehat{W}_J},$$

und daraus folgt das Ergebnis mit der Frobeniusreziprozität. Q.E.D.

In der späteren Berechnung von Charakterwerten wird es wichtig sein, daß gewisse Charaktere der generischen Algebra $A$ nicht rational sind. Wir betrachten die Erweiterung von $W(E_6)$ mit dem Graphautomorphismus. Hier seien $\varphi_{64,4}$ und $\varphi_{64,13}$ die Charaktere vom Grad 64 der Weylgruppe $W(E_6)$, weiter $\mu_{64,4}$ und $\mu_{64,13}$ die entsprechenden Charaktere der zugehörigen generischen Algebra.

**Bemerkung 1.8** [9]. *Die Fortsetzungen von $\mu_{64,4}$ und $\mu_{64,13}$ auf die generische Algebra zu $W(E_6){:}\,2$ sind nicht rational, sondern erst über $\mathbb{Q}(\sqrt{u})$ definiert.*

Für den Beweis verwendet man eine Idee von Springer, mit der für gewisse Charaktere der gewöhnlichen Heckealgebren zu $E_7$ und $E_8$ gezeigt wurde, daß sie nicht rational sind (siehe [9], II, Corollaire 3.4, wo eine vollständige Untersuchung aller Fälle durchgeführt wurde).

## 2. Gruppen vom Typ $E_6$ und $^2E_6$ als Galoisgruppen

Die einfachen Gruppen $E_6(q)$ konnten bisher nur in guter Charakteristik und für Primzahlpotenzen $q \equiv -1 \pmod 3$ als Galoisgruppen über geeigneten abelschen Erweiterungskörpern von $\mathbb{Q}$ nachgewiesen werden [14]. Die erste dieser beiden Voraussetzungen wird in diesem Paragraphen als überflüssig erkannt. Mit Hilfe der Howlett-Lehrer Induktion kann das Ergebnis aus [14] auch auf schlechte Charakteristik ausgedehnt werden, ohne daß dazu die Kenntnis der Greenfunktionen nötig wäre.

Dies ist allerdings nur unter der einschränkenden Kongruenzbedingung $q \not\equiv 1 \pmod 3$ möglich, da andernfalls die Gruppe $E_6(q)_{sc}$ ein nichttriviales Zentrum der Ordnung drei besitzt; das führt, wie Rechnungen zeigen, dazu, daß die Strukturkonstanten im allgemeinen Vielfache von drei sind, so daß das Rationalitätskriterium nicht anwendbar ist. Die Idee zur Umgehung dieses zweiten Problems liegt in der Erweiterung von $G = E_6(q)_{sc}$ durch den Graphautomorphismus der Ordung zwei zu einer Gruppe $\widehat{G}$ mit trivialem Zentrum, wie im nächsten Paragraphen gezeigt werden wird.

Sei also vorerst $q = p^n \not\equiv 1 \pmod 3$. Dann ist $G = E_6(q)_{sc}$ eine einfache Gruppe, für die wir kurz $E_6(q)$ schreiben. In [14] wurde im Fall guter Charakteristik $p$ eine Klassenstruktur $\mathbf{C}$ gefunden, für die $G$ als Galoisgruppe realisiert werden kann. Die Ausdehnung des Ergebnisses auf schlechte Charakteristik scheiterte, da für den dortigen Beweis eine explizite Kenntnis der Greenfunktionen von $E_6(q)$ nötig war; diese sind aber für $p \leq 3$ zur Zeit nicht berechnet. Es zeigt sich jedoch, daß die fehlenden Werte auf einer der unipotenten Klassen auch mit der Howlett-Lehrer Induktion berechnet werden können. Zudem liefert dies auch im Fall guter Charakteristik einen Beweis der Existenz von Galoisrealisierungen ohne Benutzung der tiefliegenden und zudem nur mit Hilfe eines Computers erhaltenen Ergebnisse von Beynon und Spaltenstein über die Greenfunktionen.

Die Klassenstruktur wird wie in [14] gewählt: Sei $C_p$ die Klasse von $G$ der in einer $p$-Sylowgruppe zentral liegenden Elemente $\rho$ mit Zentralisatorordnung

$$|\mathcal{C}_G(\rho)| = q^{36}\Phi_1^5\Phi_2^3\Phi_3^2\Phi_4\Phi_5\Phi_6,$$

$C_{q^2-1}$ die Klasse eines (halbeinfachen) Elements $\sigma$ mit Zentralisatortyp ${}^2A_2(q).A_1(q^2).(q^2-1)$ und Zentralisatorordnung

$$|\mathcal{C}_G(\sigma)| = q^5\Phi_1^3\Phi_2^4\Phi_4\Phi_6,$$

und schließlich $C_T$ die Klasse eines erzeugenden Elements $\tau$ des *t.i.*-Torus $T$ der Ordnung $|T| = |\mathcal{C}_G(\tau)| = \Phi_9(q)$. Nach [17] existieren solche Klassen für alle $q \neq 2$ (für $q = 2$ gibt es $C_{q^2-1}$ nicht).

**Bemerkung 2.1.** *Für die Klassenstruktur* $\mathbf{C} = (C_p, C_{q^2-1}, C_T)$ *von* $G = E_6(q)$ *mit* $q \not\equiv 1 \pmod 3$ *und* $q \neq 2$ *gilt* $n(\mathbf{C}) = 1$.

**Tabelle 2.2.** Auf $C_T$ nicht verschwindende Charaktere von $E_6(q)$.

| | $[\iota]$<br>$\lvert E_6(q)\rvert$ | $C_p$<br>$q^{36}\Phi_1^5\Phi_2^3\Phi_3^2\Phi_4\Phi_5\Phi_6$ | $C_{q^2-1}$<br>$q^5\Phi_1^3\Phi_2^4\Phi_4\Phi_6$ |
|---|---|---|---|
| $\phi_{1,0}$ | $1$ | $1$ | $1$ |
| $\phi_{20,2}$ | $q^2\Phi_4\Phi_5\Phi_8\Phi_{12}$ | $q^2\Phi_5(q^6+q^4+1)$ | $2q^2-q+1$ |
| $\phi_{64,4}$ | $q^4\Phi_2^3\Phi_4^2\Phi_6^2\Phi_8\Phi_{12}$ | $q^4\Phi_2^2\Phi_4\Phi_5\Phi_6(q^3-q+1)$ | . |
| $\phi_{64,13}$ | $q^{13}\Phi_2^3\Phi_4^2\Phi_6^2\Phi_8\Phi_{12}$ | $q^{13}\Phi_2^2\Phi_4\Phi_6(q^4+q^3+1)$ | . |
| $\phi_{20,20}$ | $q^{20}\Phi_4\Phi_5\Phi_8\Phi_{12}$ | $q^{20}\Phi_5$ | $-q^3(q^2-q+2)$ |
| $\phi_{1,36}$ | $q^{36}$ | . | $-q^5$ |
| $\phi_{90,8}$ | $\frac{1}{3}q^7\Phi_3^3\Phi_5\Phi_6^2\Phi_8\Phi_{12}$ | $\frac{1}{3}q^7\Phi_3^2\Phi_5\Phi_6(2q^4+1)$ | $-q\Phi_1\Phi_6$ |
| $E_6[\theta]$ | $\frac{1}{3}q^7\Phi_1^6\Phi_2^4\Phi_4^2\Phi_5\Phi_8$ | ? | . |
| $E_6[\theta^2]$ | $\frac{1}{3}q^7\Phi_1^6\Phi_2^4\Phi_4^2\Phi_5\Phi_8$ | ? | . |
| $R_{T,\theta}$ | $\Phi_1^6\Phi_2^4\Phi_3^3\Phi_4^2\Phi_5\Phi_6^2\Phi_8\Phi_{12}$ | ? | . |

Bevor wir den Beweis beginnen, sei noch an zwei Eigenschaften von Werten irreduzibler Charaktere endlicher Gruppen erinnert.

**Lemma 2.3.** *Sei $G$ eine endliche Gruppe, $\chi$ ein irreduzibler Charakter, $C$ eine Konjugiertenklasse mit Vertreter $\sigma$. Dann gelten*
*(i)* $|\chi(\sigma)| \leq \chi(\iota)$,
*(ii)* $|C| \cdot \chi(\sigma)/\chi(\iota)$ *ist eine ganze algebraische Zahl.*

**Beweis von Bemerkung 2.1.** Die einzigen auf allen drei Klassen von C nicht verschwindenden irreduziblen Charaktere von $G$ sind unter den in Tabelle 2.2 angegebenen unipotenten Charakteren, wie bereits in [14] bemerkt wurde (dies folgt leicht aus der Lusztigschen Jordanzerlegung der Charaktere von $G$). Damit ergibt sich die Behauptung also aus den in der Tabelle aufgeführten Werten und der Tatsache, daß $\chi(\tau) \equiv \chi(\iota) \pmod{\Phi_9(q)}$ und $|\chi(\tau)| \leq 1$ für $\tau \in T^\#$ gelten. Zum Beweis muß daher nur die Korrektheit der Tabelle gezeigt werden. Die Werte auf den halbeinfachen Klassen $C_T$ und $C_{q^2-1}$ können nach [7] und [13] berechnet werden, falls die Fusion der maximalen Tori der jeweiligen Zentralisatoren in die von $G$ bekannt ist. Dies wurde bereits in [14] durchgeführt. Somit bleiben nur die Werte auf $C_p$ zu bestimmen. Hier sind natürlich der Einscharakter und der Steinbergcharakter schon bekannt. Für die weitere Rechnung wird die Induktion von maximalen parabolischen Untergruppen benutzt. Diese enthalten nur Gruppen der Typen $A_n$, $D_4$ und $D_5$. Die Greenfunktionen hierfür sind entweder nach der Arbeit [11] bekannt oder können leicht bestimmt werden und daher erhält man durch Induktion ein lineares Gleichungssystem für die unbekannten Werte $\chi(\rho)$, $\rho \in C_p$. So findet man etwa

$$\phi_{20,20}(\rho) - \phi_{20,2}(\rho) = q^{20}\Phi_5 - q^2\Phi_5(q^6 + q^4 + 1).$$

Nach Lemma 2.3 gelten $|\phi_{20,2}(\rho)| < q^{20}$ und $q^{20}|\phi_{20,20}(\rho)$, woraus die Werte der Tabelle für diese beiden unipotenten Charaktere folgen. Weiter ergibt sich aus dem Gleichungssystem noch die Bedingung

$$\phi_{64,13}(\rho) - \phi_{64,4}(\rho) = q^4\Phi_2^2\Phi_4\Phi_6(q^{13} + q^{12} + q^9 - \Phi_5(q^3 - q + 1)). \qquad (*)$$

Um die fehlenden Werte zu bestimmen, wird nun die Strukturkonstante einer weiteren Klassenstruktur berechnet. Sei dazu $C_\kappa$ die Klasse eines halbeinfachen Elements $\kappa$ mit Zentralisatortyp $A_5(q).(q+1)$ (solche Elemente gibt es nach [17] für alle Primzahlpotenzen $q$). Dann sind die Werte der in der Tabelle enthaltenen Charaktere auf $C_\kappa$ wieder nach [7] und [13] berechenbar. (Sie werden hier nicht abgedruckt.) Schreibt man $\phi_{64,4}(\rho) = q^4\Phi_2^2\Phi_4\Phi_6 a$, $\phi_{64,13}(\rho) = q^{13}\Phi_2^2\Phi_4\Phi_6 b$, $\phi_{90,8}(\rho) = \frac{1}{3}q^7\Phi_3^2\Phi_5\Phi_6 c$ mit ganzen Zahlen $a, b$ und $c$ (ganz, da die genannten Vorfaktoren nach Lemma 2.3 auf jeden Fall $\chi(\rho)$ teilen), so erhält man unter Beachtung der obigen Gleichung $(*)$ zwischen $a$ und $b$

$$n(C_p, C_\kappa, C_T) = \frac{q^2\Phi_2^2\Phi_9(b - q^4 - q^3 - 1) - \Phi_5(c - 2q^4 - 1)}{q^{13}\Phi_1^3\Phi_2^3\Phi_5}. \qquad (**)$$

Aus der Gleichung $(*)$ und der Abschätzung $|\phi_{64,4}(\rho)| \leq |\phi_{64,4}(\iota)| < q^{24}$ weiß man weiter $|b| < q^5$ und wegen $|\phi_{90,8}(\rho)| < q^{30}$ folgt $|c| < q^{13}$. Da kein nichttriviales Element von $G$ sowohl ein Element aus $C_T$ als auch aus $C_p$ zentralisieren kann, muß $n(C_p, C_\kappa, C_T)$ eine ganze Zahl sein, und nach den erfolgten Abschätzungen für $b$ und $c$

kann dann nur $n(C_p, C_\kappa, C_T) = 0$ gelten. Schreibt man noch $b - q^4 - q^3 - 1 =: b_0$, $c - 2q^4 - 1 =: c_0$, so wird (**) also zu $q^2\Phi_2^2\Phi_9 b_0 = \Phi_5 c_0$.

Schließlich berechnen wir noch die gewünschte Zahl $n(\mathbf{C})$ in Abhängigkeit von $c_0$ zu

$$n(\mathbf{C}) = 1 + \frac{c_0}{q^4\Phi_1\Phi_2^3}.$$

(Da $\phi_{64,4}$ und $\phi_{64,13}$ auf $C_{q^2-1}$ verschwinden, taucht $b_0$ nicht in diesem Ausdruck auf.) Wie eben muß $n(\mathbf{C})$ eine ganze Zahl sein, also $c_0$ von $q^4\Phi_1\Phi_2^3$ geteilt werden. Für $b_0$ bedeutet das $q^2\Phi_1\Phi_2\Phi_5 | b_0$, und da wir noch $|b_0| < 2q^5$ wissen, können wir auf $b_0 = c_0 = 0$ und somit die Korrektheit der Tabelle schließen. Q.E.D.

Nach dieser Vorarbeit ist der Beweis des nächsten Ergebnisses relativ einfach, da im Wesentlichen die Überlegungen aus [14], §6, verwendet werden können.

**Satz 2.4.** *Die Gruppen $E_6(q)$, $q = p^n \not\equiv 1 \pmod 3$, $q \neq 2$, kommen zur Verzweigungsstruktur $\mathbf{C}^* = (C_p, C_{q^2-1}, C_T)^*$ als Galoisgruppen über $\mathbb{Q}^{ab}$ vor. Dabei enthalten die jeweiligen Klassen Elemente der Ordnungen $(p, q^2-1, q^6+q^3+1)$. Genauer ist das Kompositum von $\mathbb{Q}(\zeta_{q^2-1})$ mit einem Körper vom Index neun in $\mathbb{Q}(\zeta_{q^6+q^3+1})$ ein eigentlicher Definitionskörper.*

**Beweis.** Für die angegebene Klassenstruktur $\mathbf{C}$ gilt nach Bemerkung 2.1 bereits $n(\mathbf{C}) = 1$. Nach dem in der Einleitung zitierten Rationalitätskriterium ([15], II, §4.2, Folgerung 3) muß daher nur gezeigt werden, daß ein Tripel $(\rho, \sigma, \tau) \in \overline{\Sigma}(\mathbf{C})$ ganz $G$ erzeugt. In [14], Beweis zu Theorem 6.1, wurde das für diese Klassenstruktur im Falle guter Primzahlen schon bewiesen. Das dortige Vorgehen kann im Wesentlichen übernommen werden. Zuerst sieht man, daß $H := \langle \rho, \sigma, \tau \rangle$ nicht in $\mathcal{N}_G(T)$ liegt, da letztere Untergruppe für $q \neq 2$ keine durch $q^2 - 1 = o(\sigma)$ teilbare Ordnung besitzt. Statt der in [14] bewiesenen Abschätzungen für den 2- und 3-Rang von $G$ verwendet man die Schranken 2-Rang$(E_6(q)) \leq 6$ für $p \neq 2$ und 3-Rang$(E_6(q)) \leq 6$ für $p \neq 3$ aus [3]. Wegen $q \geq 3$ ist nun $|T| = o(\tau) \geq 757$, also kann das Erzeugnis $H$ des Tripels aus $\overline{\Sigma}(\mathbf{C})$ wie in [14] keine lokale Untergruppe sein. Die Möglichkeit $R \leq H \leq \mathrm{Aut}(R)$ für eine nichtabelsch einfache Untergruppe $R$ von $G$ schließt man wie in [14] aus. Nach [14], Lemma 1.3, sind damit sämtliche Möglichkeiten für $H$ außer $H = G$ untersucht und als unmöglich erkannt worden, so daß die Behauptung folgt. Q.E.D.

Wir schließen nun noch eine Lücke aus Satz 2.4, nämlich den Fall $E_6(2)$, unter Benutzung der Charaktertafel dieser Gruppe, die freundlicherweise von B. Fischer zur Verfügung gestellt wurde. Die Klassenbezeichnungen beziehen sich auf die Benennung in der CAS-Version dieser Tafel.

**Satz 2.5.** *Die Gruppe $E_6(2)$ kommt zur Verzweigungsstruktur $\mathbf{C}^* = (2a, 8b, 73a)^*$ als Galoisgruppe über dem Körper vom Index neun in $\mathbb{Q}(\zeta_{73})$ vor.*

**Beweis.** Aus der Charaktertafel errechnet man $n(2a, 8b, 73a) = 1$. Wie oben prüft man sofort, daß ein Tripel zu der vorgegebenen Klassenstruktur keine lokale Untergruppe erzeugen kann. Die einzige echte einfache Untergruppe $R$ von $G$ mit durch 73 teilbarer Ordnung ist $L_3(8)$. Von den Gruppen zwischen $L_3(8)$ und

$\mathrm{Aut}(L_3(8))$ enthalten nur $L_3(8){:}\,2$ und $L_3(8){:}\,6$ Elemente der Ordnung acht. Aber in diesen beiden ist ein Element der Ordnung 73 zu sechs bzw. achtzehn seiner primitiven Potenzen konjugiert, während es in $G$ nur zu neun Potenzen konjugiert ist. Daher kann keine der beiden Gruppen in $G$ liegen, und auch die Erzeugung ist bewiesen. Der Satz folgt damit aus dem Rationalitätskriterium. Q.E.D.

Wir wenden uns nun Galoisrealisierungen der getwisteten Gruppen ${}^2E_6(q)$ zu. Diese waren bisher nur unter der recht vagen Einschränkung '$p$ und $q$ groß genug' als Galoisgruppen nachgewiesen worden [14], wobei diese Einschränkung wieder von der ungenügenden Kenntnis der Greenfunktionen herrührte. Die Methoden des letzten Abschnitts lassen sich aber fast wortwörtlich von $E_6(q)$ auf ${}^2E_6(q)$ übertragen, und damit können hier diejenigen mit $q \not\equiv -1 \pmod 3$ als Galoisgruppen realisiert werden. Nach der verallgemeinerten Ennola-Dualität sollten die Charakterwerte von ${}^2E_6(q)$ aus denen von $E_6(q)$ einfach durch Ersetzen von $q$ durch $-q$ erhalten werden. Es stellt sich heraus, daß diese (bisher unbewiesene Vermutung) für alle beim Beweis vorkommenden Charakterwerte zutrifft. Aufgrund der großen Ähnlichkeit der Beweise wird der Fall ${}^2E_6(q)$ nicht weiter ausgeführt, sondern nur das Ergebnis angegeben: $C_p$ sei die Klasse der in einer $p$-Sylowgruppe von $G$ zentral liegenden Elemente $\rho$ mit Zentralisatorordnung

$$|\mathcal{C}_G(\rho)| = q^{36}\Phi_1^3\Phi_2^5\Phi_3\Phi_4\Phi_6^2\Phi_{10},$$

$C_{q^2-1}$ die Klasse eines (halbeinfachen) Elements $\sigma$ mit Zentralisatortyp $A_2(q).A_1(q^2).(q^2-1)$ und Zentralisatorordnung

$$|\mathcal{C}_G(\sigma)| = q^5\Phi_1^4\Phi_2^3\Phi_3\Phi_4,$$

(vergleiche Table 1 in [8]) und schließlich $C_T$ die Klasse eines erzeugenden Elements $\tau$ des $t.i$-Torus $T$ der Ordnung $|T| = |\mathcal{C}_G(\tau)| = \Phi_{18}(q)$. Sowohl die Klasse $C_p$ als auch Klassen vom Typ $C_T$ exisitieren offensichtlich für alle $q$. In $\mathcal{C}_G(\sigma)$ liegt ein maximaler Torus der Ordnung $\Phi_1\Phi_2\Phi_3\Phi_4$ mit Weylgruppe der Ordnung 12, und daraus rechnet man leicht nach, daß es Klassen $C_{q^2-1}$ für alle $q > 2$ gibt. Man beachte, daß dies nicht genau dieselbe Klassenstruktur für $G$ wie in [14], §7, ist.

**Satz 2.6.** *Die Gruppen* ${}^2E_6(q)$, $q = p^n \not\equiv -1 \pmod 3$, *kommen zur Verzweigungsstruktur* $\mathbf{C}^* = (C_p, C_{q^2-1}, C_T)^*$ *als Galoisgruppen über* $\mathbb{Q}^{ab}$ *vor. Dabei enthalten die jeweiligen Klassen Elemente der Ordnungen* $(p, q^2-1, q^6-q^3+1)$. *Genauer ist das Kompositum von* $\mathbb{Q}(\zeta_{q^2-1})$ *mit einem Körper vom Index neun in* $\mathbb{Q}(\zeta_{q^6-q^3+1})$ *ein eigentlicher Definitionskörper.*

## 3. Gruppen $E_6(q)_{sc}{:}\,2$ als Galoisgruppen

Um auch für die fehlenden Kongruenzen Galoisrealisierungen zu erhalten, wird die Erweiterung von $E_6(q)$ mit dem Graphautomorphismus der Ordnung zwei untersucht. Da diese Erweiterung unabhängig von der Kongruenzklasse von $q \pmod 3$ existiert, wird hierdurch ein einheitlicher Zugang zu der gesamten Familie erhalten. Allerdings erzeugen die Elemente der zu wählenden Klassenstruktur für $p = 2$ nicht die gesamte Gruppe. Daher kann mit dieser Methode in Charakteristik zwei keine Galoisrealisierung erhalten werden. Es sei also im weiteren $q = p^n$, $p > 2$ und $G = E_6(q)_{sc}$, die einfach zusammenhängende Gruppe vom Typ $E_6$ über dem Körper $\mathbb{F}_q$. Dann erfüllt die Erweiterung $\widehat{G}$ von $G$ mit dem Graphautomorphismus die Voraussetzungen zu den Sätzen des ersten Paragraphen.

Der Zentralisator des Graphautomorphismus $\rho$ in $\widehat{G}$ hat die Struktur $F_4(q) \times 2$, wie man aus der Operation von $\rho$ auf Steinbergerzeugenden von $G$ leicht nachrechnet. Also zerfällt die Erweiterung $G.\langle\rho\rangle$ und $\rho$ operiert nichttrivial auf dem (für $q \equiv 1 \pmod 3$) vorhandenen) Zentrum von $G$, so daß die üblichen Rationalitätskriterien [15] auf $\widehat{G}$ anwendbar sind. (Zum ersten Mal wurde dieser 'Trick' zur Umgehung des Zentrums bei dreifachen Überlagerungen sporadischer Gruppen von Feit [10] vorgestellt.)

Sei $u$ ein Element aus $C_G(\rho) = F_4(q)$ mit $|C_{F_4}(u)| = q^{24}\Phi_1^3\Phi_2^3\Phi_3\Phi_4\Phi_6$, liege also zentral in einer $p$-Sylowgruppe von $F_4(q)$ (diese Klasse ist für $p \neq 2$ eindeutig). Dann hat das Element $u\rho$ von $\widehat{G}$ die Ordnung $2p$ und die Zentralisatorordnung $2q^{24}\Phi_1^3\Phi_2^3\Phi_3\Phi_4\Phi_6$ in $\widehat{G}$. Anhand der Liste von Konjugiertenklassen von $G$ in [17] rechnet man nach, daß $u$ in $G$ die Zentralisatorordnung $q^{36}\Phi_1^5\Phi_2^3\Phi_4\Phi_5\Phi_6$ besitzt (denn einzig dafür enthält der Zentralisator eine symplektische Gruppe $C_3(q)$, siehe auch [2], S.402). Die Klasse von $u\rho$ sei mit $C_{2p}$ bezeichnet.

Sei weiter $C_T$ die Klasse eines regulären Elements $\tau$ des maximalen Torus $T$ von $G$ der Ordnung $q^6 + q^3 + 1$. Dann ist $|C_{\widehat{G}}(\tau)| = q^6 + q^3 + 1$, denn der äußere Automorphismus operiert nichttrivial auf $T$. Auf dem regulären Element $\tau \in C_T$ des Coxetertorus verschwinden fast alle irreduziblen Charaktere von $G$. Nach der Lusztig'schen Jordanzerlegung der Charaktere können nur unipotente Charaktere und die Deligne-Lusztig Charaktere $R_{T,\theta}$ zum Torus $T$ von Null verschiedene Werte auf $C_T$ annehmen (siehe dazu auch den vorigen Abschnitt, wo diese Klasse bereits benutzt wurde). Da der äußere Automorphismus fixpunktfrei auf $T\backslash Z(G)$ operiert, fallen je zwei der $R_{T,\theta}$, $\theta \neq 1$, unter der Operation von $\rho$ zusammen, so daß ihre Werte auf äußeren Klassen gleich Null sind. Es bleiben somit die unipotenten Charaktere von $G$ zu untersuchen. Zumindest die in der Hauptserie liegenden können nach den Ergebnissen im ersten Paragraphen alle auf $\widehat{G}$ fortgesetzt werden. Aus der Multiplizitätenformel in Satz 1.7 kann nun Information über die Werte der Fortsetzungen der unipotenten Charaktere erhalten werden. Dazu ist aber die Kenntnis der unipotenten Charaktere aller (maximaler) Leviuntergruppen nötig. Diese enthalten wieder jeweils ein $BN$-Paar vom Index höchstens zwei, so daß Satz 1.7 auch hier angewandt werden kann. So können die Werte aller nichtkuspidalen unipotenten Charaktere in den Levifaktoren $D_4(q){:}\,2$ und $A_5(q){:}\,2$ auf der Klasse der äußeren Involution berechnet werden.

Nach dem Ausflug zu Levifaktoren maximaler $\rho$-invarianter parabolischer Untergruppen kehren wir zur Bestimmung der uns eigentlich interessierenden Werte unipotenter Charaktere in $\widehat{G} = E_6(q)_{sc}{:}\,2$ zurück. Die Bezeichnung der unipotenten Charaktere von $G$ folgt wie schon bei Tabelle 2.2 der in [2], 13.9.

**Bemerkung 3.1.** *Die Werte der Fortsetzungen derjenigen unipotenten Charaktere von $\widehat{G} = E_6(q)_{sc}{:}\,2$, die auf $T^\#$ nicht verschwinden, auf $C_T$ und den äußeren Klassen $[\rho]$ und $C_{2p}$, sind wie in Tabelle 3.2 angegeben.*

**Tabelle 3.2.** Einige unipotente Charaktere von $E_6(q)_{sc}{:}2$.

| | $[\iota]$ | $C_T$ | $[\rho]$ | $C_{2p}$ |
|---|---|---|---|---|
| | $\lvert E_6(q)_{sc}\rvert$ | $\Phi_9$ | $\lvert F_4(q)\rvert$ | $q^{24}\Phi_1^3\Phi_2^3\Phi_3\Phi_4\Phi_6$ |
| $\phi_{1,0}$ | $1$ | $1$ | $1$ | $1$ |
| $\phi_{20,2}$ | $q^2\Phi_4\Phi_5\Phi_8\Phi_{12}$ | $-1$ | $q\Phi_4\Phi_8\Phi_{12}$ | $q(q^6+q^4+1)$ |
| $\phi_{64,4}$ | $q^4\Phi_2^3\Phi_4^2\Phi_6^2\Phi_8\Phi_{12}$ | $1$ | . | . |
| $\phi_{64,13}$ | $q^{13}\Phi_2^3\Phi_4^2\Phi_6^2\Phi_8\Phi_{12}$ | $1$ | . | . |
| $\phi_{20,20}$ | $q^{20}\Phi_4\Phi_5\Phi_8\Phi_{12}$ | $-1$ | $q^{13}\Phi_4\Phi_8\Phi_{12}$ | $q^{13}$ |
| $\phi_{1,36}$ | $q^{36}$ | $1$ | $q^{24}$ | . |
| $\phi_{90,8}$ | $\frac{1}{3}q^7\Phi_3^3\Phi_5\Phi_6^2\Phi_8\Phi_{12}$ | $-1$ | $\frac{1}{3}q^4\Phi_3^2\Phi_6^2\Phi_8\Phi_{12}$ | $\frac{1}{3}q^4\Phi_3\Phi_6(2q^4+1)$ |
| $E_6[\theta]$ | $\frac{1}{3}q^7\Phi_1^6\Phi_2^4\Phi_4^2\Phi_5\Phi_8$ | $-1$ | $\frac{1}{3}q^4\Phi_1^4\Phi_2^4\Phi_4^2\Phi_8$ | $-\frac{1}{3}q^4\Phi_1^3\Phi_2^3\Phi_4$ |
| $E_6[\theta^2]$ | $\frac{1}{3}q^7\Phi_1^6\Phi_2^4\Phi_4^2\Phi_5\Phi_8$ | $-1$ | $\frac{1}{3}q^4\Phi_1^4\Phi_2^4\Phi_4^2\Phi_8$ | $-\frac{1}{3}q^4\Phi_1^3\Phi_2^3\Phi_4$ |

Die Ausdrücke in der zweiten Zeile der Tabelle geben die Zentralisatorordnungen in $G$, nicht in $\widehat{G}$, an (Atlas-Notation).

**Beweis.** Man überlegt sich leicht, daß genau die in der Tabelle aufgeführten unipotenten Charaktere nicht auf $T^{\#}$ verschwinden (es sind gerade diejenigen, deren Grad nicht von $\Phi_9$ geteilt wird), und mit [7] lassen sich auch die Werte auf $\tau$ bestimmen (siehe dazu auch [14], §6). Nun wird Satz 1.7 angewendet. Aus dem linearen Gleichungssystem für die Werte auf $[\rho]$ und $C_{2p}$ erhält man auch die Gleichung

$$\phi_{20,20}(\rho)-\phi_{20,2}(\rho)=q\Phi_4\Phi_8\Phi_{12}(q^{12}-1).$$

Nach dem bereits wiederholt verwendeten Lemma 2.3 müssen $q^8\Phi_4\Phi_8\Phi_{12}|\phi_{20,20}(\rho)$ und $|\phi_{20,2}(\rho)|\le q^2\Phi_4\Phi_5\Phi_8\Phi_{12}$ und gelten. Dies hat als einzige Lösung die in der Tabelle angegebenen Werte.

Die Fortsetzungen der beiden Charaktere $\phi_{64,4}$ und $\phi_{64,13}$ auf $\widehat{G}$ gehören nach Bemerkung 1.8 zu nichtrationalen Charakteren der generischen Algebra $A$ zum Permutationscharakter von $\widehat{G}$ nach $B$. Unter Benutzung der Ree'schen Formel [5], 11.34, zeigt man, daß die Fortsetzungen von beiden Charakteren auf $\rho$ und $u\rho$ den Wert Null annehmen müssen.

Weitere Informationen über die fehlenden Werte können wiederum durch die Berechnung einiger Strukturkonstanten erhalten werden. Die Argumentation verläuft ähnlich wie im Beweis zu Bemerkung 2.1 und wird hier nicht ausgeführt. Man erhält damit die noch fehlenden Einträge von Tabelle 3.2. Q.E.D.

**Satz 3.3.** *Die Gruppen $E_6(q)_{sc}{:}2$, $q=p^n$, $p>2$, kommen zur Verzweigungsstruktur $\mathbf{C}^*=(C_{2p},C_{2p},C_T)^*$ als Galoisgruppen über $\mathbb{Q}^{ab}$ vor. Dabei enthalten die jeweiligen Klassen Elemente der Ordnungen $(2p,2p,q^6+q^3+1)$. Genauer ist ein Teilkörper vom Index 18 in $\mathbb{Q}(\zeta_{q^6+q^3+1})$ ein eigentlicher Definitionskörper.*

**Beweis.** Aus Tabelle 3.2 erhält man sofort $n(\mathbf{C})=1$. Zur Anwendung des Rationalitätskriteriums auf die Gruppe $\widehat{G}=E_6(q)_{sc}{:}2$ mit trivialem Zentrum muß also nur noch gezeigt werden, daß ein Tripel $(\sigma_1, \sigma_2, \tau) \in \overline{\Sigma}(\mathbf{C})$ ganz $\widehat{G}$ erzeugt. Dazu betrachten wir die Gruppe $\overline{G} := \widehat{G}/Z(E_6(q)_{sc})$ (die im Fall $q \not\equiv 1 \pmod 3$) mit $\widehat{G}$ übereinstimmt). Offensichtlich erzeugen die Bilder $\bar{\sigma}_1$, $\bar{\sigma}_2$, $\bar{\tau}$ unter der kanonischen Abbildung $\widehat{G} \to \overline{G}$ genau dann $\overline{G}$, wenn die Urbilder $\widehat{G}$ erzeugen. In $\overline{G}$ erzeugt aber $\bar{\tau}$ eine selbstzentralisierende zyklische *t.i.*-Halluntergruppe $\overline{T}$. Damit ist Lemma 1.3 aus [14] anwendbar, welches die Möglichkeiten für das Erzeugnis $\overline{H} := \langle \bar{\sigma}_1, \bar{\sigma}_2, \bar{\tau} \rangle$ stark einschränkt. In der Tat ist man wieder in der in [14], Theorem 6.1, untersuchten Situation. Die einzige Schwierigkeit ensteht beim Fall $L_3(q^3) \leq \overline{H} \leq \mathrm{Aut}(L_3(q^3))$. Offensichtlich kann nicht $\overline{H} = L_3(q^3)$ sein, denn dann wäre $\overline{H} \cap E_6(q)$ eine Untergruppe von $L_3(q^3)$ vom Index 2. Daher muß $L_3(q^3) < \overline{H}$ gelten, und $A := \langle L_3(q^3), \bar{\sigma}_1^p \rangle$ ist die Erweiterung von $L_3(q^3)$ vom Grad 2 mit dem Graphautomorphismus. Für diese kann mittels Satz 1.7 genügend Information über die Werte der Fortsetzungen der unipotenten Charaktere erhalten werden, um durch Vergleich von Strukturkonstanten in $G$ und in $A$ zu einem Widerspruch zu gelangen. Die Annahme $R = L_3(q^3)$ ist damit falsch, und es folgt $H = \widehat{G}$. Q.E.D.

Als unmittelbare Folgerung haben wir

**Satz 3.4.** *Bezeichne $E_6(q)$ die einfache Gruppe vom Typ $E_6$ und sei $q = p^n$ für $p > 2$. Dann sind $E_6(q)$, $E_6(q){:}2$, sowie für $q \equiv 1 \pmod 3$ auch die Darstellungsgruppen $3{\cdot}E_6(q)$ Galoisgruppen über $\mathbf{Q}^{ab}$.*

**Beweis.** Dies folgt durch Abstieg von $E_6(q)_{sc}{:}2$ zu $E_6(q)_{sc}$. Die Galoiserweiterung für $E_6(q)_{sc}{:}2$ ist nach Konstruktion an genau drei Punkten verzweigt. Da $E_6(q)_{sc}$ in $E_6(q)_{sc}{:}2$ den Index zwei hat, ist der Fixkörper von $E_6(q)_{sc}$ in dieser Erweiterung ein rationaler Funktionenkörper mit dem Definitionskörper der ursprünglichen Galoiserweiterung als Konstantenkörper. Die Gruppe $E_6(q)_{sc}$ ist bekanntlich genau die Darstellungsgruppe von $E_6(q)$. Die einfache Gruppe $E_6(q)$ besitzt als Faktorgruppe von $E_6(q)_{sc}$ natürlich ebenfalls eine Galoisrealisierung über demselben Definitionskörper. Dasselbe Argument gilt für $E_6(q){:}2$ als Faktorgruppe von $E_6(q)_{sc}{:}2$. Q.E.D.

Wir erhalten sogar einige Galoisrealisierungen über $\mathbf{Q}$:

**Satz 3.5.** *Die Gruppen $E_6(p)_{sc}{:}2$, $E_6(p){:}2$, $E_6(p)_{sc}$ und $E_6(p)$ kommen für Primzahlen $p \equiv 4, 5, 6, 9, 16, 17 \pmod{19}$ zur Verzweigungsstruktur $(C_{2p}, C_{2p}, C_{19})^*$ als Galoisgruppen über $\mathbf{Q}$ vor.*

**Beweis.** Für die im Satz angegebenen Kongruenzen hat $p$ die multiplikati ve Ordnung 9 modulo 19. Daher teilt in diesem Fall 19 nur $\Phi_9(p)$, nicht aber $\Phi_i(p)$ mit $i < 9$, die Primzahl 19 ist also ein primitiver Teiler von $\Phi_9(p)$ im Sinne von Zsigmondy. In dem maximalen Torus $T$ von $G := E_6(q)_{sc}$ existieren daher Elemente der Ordnung 19, die in $\widehat{G} = G.2$ wegen $|\mathcal{N}_G(T)/T| = 9$, und da $\rho$ fixpunktfrei auf $T$ operiert, rational sind. Die Werte der unipotenten Charaktere auf der zugehörigen Klasse $C_{19}$ sind wie auf $C_T$ in Tabelle 3.2, da unipotente Charaktere auf $T^{\#}$ konstante Werte annehmen. Also gilt $n(C_{2p}, C_{2p}, C_{19}) = 1$, und es bleibt nur noch die Erzeugung zu beweisen. Wegen $p \geq 5$ kann das Erzeugnis $H$ eines Tripels aus der zugehörigen Klassenstruktur nicht in $\mathcal{N}_{\widehat{G}}(T)$ liegen. Die Abschätzungen des 2- und 3-Rangs aus [14], Theorem 6.1, genügen, um auch 2- und 3-lokale Untergruppen als Kandidaten auszuschließen, denn sowohl 2 als auch 3 sind Primitivwurzeln modulo 19. Als primitiver Teiler von $\Phi_9(p)$ teilt 19 auch nicht die Ordnungen der parabolischen Untergruppen von $G$, und wegen $19 \nmid |W(E_6)|$ kann $H$ schließlich nach Lemma 1.7(2) in [14] nicht $r$-lokal für eine Primzahl $r \notin \{2,3,p\}$ sein. Mit [14], Lemma 1.3, angewandt auf $\overline{G} := \widehat{G}/Z(\widehat{G})$ gilt somit $R \leq \overline{H} \leq \mathrm{Aut}(R)$ mit $\overline{H} := H/(H \cap Z(\widehat{G}))$ für eine einfache nichtabelsche Untergruppe $R$ von $\overline{G}$.

Für ein solches $R$ kommen nur die Gruppen in Lemma 6.1 in [14] in Frage. Von diesen haben nur $L_2(19)$, $L_2(37)$, $HN$, $Th$, $B$, $M$, $J_1$, $ON$, $J_3$ und eventuell die Gruppen vom Lie-Typ in Charakteristik $p$ eine durch 19 teilbare Ordnung. Die neun Einzelfälle lassen sich leicht durch die vorgegebene Klassenstruktur und die geforderten Kongruenzen ausschließen. So kann etwa $R = J_3$ wegen $|J_3| = 2^7.3^5.5.17.19$ nur für $p \in \{5,17\}$ vorkommen. Aber $J_3{:}2$ hat keine äußeren Elemente der Ordnung 10 [4], so daß $p = 5$ unmöglich ist. Schließlich sind die $(34,34,19)$-Strukturkonstanten von $J_3{:}2$ nach [4] viel zu groß, als daß sie von $(C_{34}, C_{34}, C_{19})$ in $\widehat{G}$ stammen könnten.

Von den Gruppen vom Lie-Typ in Charakteristik $p$ bleibt wie in [14], Theorem 6.1, neben $E_6(p)$ nur noch die Möglichkeit $L_3(p^3)$. Wie oben kann auch dieser Fall ausgeschlossen werden. Damit folgt die Aussage des Satzes aus dem Rationalitätskriterium, da alle drei Klassen rational sind. Q.E.D.

## LITERATUR


[1] A.Borel *et al.*, Seminar on algebraic groups and related finite groups, *Lecture Notes in Mathematics* **131**, Springer Verlag 1970, Berlin Heidelberg New York.

[2] R. W. Carter, *Finite groups of Lie type: Conjugacy classes and complex characters*, John Wiley and Sons 1985, Chichester.

[3] A. M. Cohen, G. M. Seitz, The $r$-rank of the groups of exceptional Lie type, *Indag. Math.* **49** (1987), S.251–259.

[4] J. H. Conway, R. T. Curtis, S. P. Norton, R. A. Parker, R. A. Wilson, *Atlas of finite groups*, Clarendon Press 1985, Oxford.

[5] C. W. Curtis, I. Reiner, *Methods in representation theory I*, John Wiley 1981, New York.

[6] C. W. Curtis, I. Reiner, *Methods in representation theory II*, John Wiley 1987, New York.

[7] P. Deligne, G. Lusztig, Representations of reductive groups over finite fields, *Ann. of Math.* **103** (1976), S.103–161.

[8] D. I. Deriziotis, M. W. Liebeck, Centralizers of semisimple elements in finite twisted groups of Lie type, *J. Lond. Math. Soc. (2)* **31** (1985), S.48–54.

[9] F. Digne, J. Michel, Fonctions $L$ des variétés de Deligne-Lusztig et descente de Shintani, *Mémoires Soc. Math. France* **20** (1985).

[10] W. Feit, Some finite groups with centers which are Galois groups, in:*Group*

*Theory, Proceedings of the 1987 Singapore Conference*, W. de Gruyter 1989, Berlin - New York.

[11] J. A. Green, Characters of the finite general linear groups, *Trans. Am. Math. Soc.* **80** (1955), S.402–447.

[12] R. B. Howlett, G. I. Lehrer, Representations of generic algebras and finite groups of Lie type, *Trans. Am. Math. Soc.* **280** (1983), S.753–779.

[13] G. Lusztig, On the unipotent characters of the exceptional groups over finite fields, *Invent. Math.* **60** (1980), S.173–192.

[14] G. Malle, Exceptional groups of Lie type as Galois groups, *J. reine angew. Math.* **392** (1988), S.70–109.

[15] B. H. Matzat, *Konstruktive Galoistheorie*, Springer Lecture Notes 1284, Berlin - Heidelberg - New York 1987.

[16] K. McGovern, Multiplicities of principal series representations of finite groups with split $BN$-pairs, *J. Algebra* **77** (1982), S.419–442.

[17] K. Mizuno, The conjugate classes of Chevalley groups of type $E_6$, *J. Fac. Sci. Univ. Tokyo* **24** (1977), S.525–563.

Progress in Mathematics, Vol. 95, © 1991 Birkhäuser Verlag Basel

# Some new developments and open questions in the character theory of finite groups

OLAF MANZ

## Introduction

The following article consists of two mainly distinct parts. The first three sections deal with module actions on finite vector spaces. We start considering the critical case of quasi-primitive modules and indicate how those techniques apply to obtain bounds for the order of solvable linear groups and regular orbits of solvable primitive permutation groups on the power set. Finally, module actions with large centralizers are investigated. Sections 4–6 are concerned with several central problems in character theory, being Brauer's height-0 conjecture, the modular Ito problem, the shape of the character degree graph and Huppert's $\rho$-$\sigma$-conjecture. If the group in question is solvable, then the methods of sections 1–3 can be used to obtain at least partial results. Relying on the classification of finite simple groups, some of those can be extended to $p$-solvable or even arbitrary finite groups.

Several outlines of proofs are included. We thus hope to point out their main ideas, but we completely ignore their technical details (which often turn out to be very tedious). The reader is invited to confer [MW 2] instead. Emphasis is also put on the formulation of open questions, both for solvable and non-solvable groups. This might serve as a stimulation for the reader to go on for progress.

My contributions to sections 3–6 were obtained during the DFG-project "Representations of finite groups and finite-dimensional algebras". I am grateful to the DFG for its generous support. I would also like to thank all participants of the project I was able to cooperate with. Specials thanks to Tom Wolf, who helped familiarize me with those techniques presented in this article.

## § 1 Quasi- and pseudo-primitive modules

Let $V$ be a faithful, irreducible $G$-module. If $V$ is imprimitive, then $V$ is induced from an irreducible module $W$ of a maximal subgroup $H$ of $G$. Set $n = |G : H|$. Then

$$G \leq H/\mathbf{C}_H(W) \wr S,$$

where $S = G/\mathrm{core}_G(H)$ is a primitive permutation group on $n$ letters. In fact, $G$ as a linear group on $V$ is isomorphic to a subgroup of $H/\mathbf{C}_H(W) \wr S$ as a linear group on $W \oplus \overset{n}{\cdots} \oplus W$, where $S$ permutes the components. It should be clear that induction arguments often work in the imprimitive case. If $V$ is primitive, then Clifford's Theorem implies that $V_N$ is homogeneous for all normal subgroups $N \trianglelefteq G$. This situation turns out to be the critical one and will therefore be studied in section 1.

An irreducible $G$-module $V$ is called *quasi-primitive* if $V_N$ is homogeneous for all $N \trianglelefteq G$. At times, it is more convenient to weaken this condition to $V_N$ homogeneous for all characteristic subgroups $N \leq G$. We then call $V$ a *pseudo-primitive* $G$-module. The reason simply is that the property "characteristic" is transitive and thus induction can be applied.

Let $V$ be a faithful, irreducible $G$-module and $A$ an abelian and normal (characteristic, respectively) subgroup of $G$. If $V$ is quasi-primitive (pseudo-primitive, respectively), then $V_A$ is homogeneous and thus $A$ is cyclic. This observation limits the structure of the Fitting-subgroup $\mathbf{F}(G)$ of $G$ considerably. The following result is an immediate consequence of a Theorem of P. Hall ([Hu1;III,13.10]).

**1.1 Theorem:** *Assume that $V$ is a faithful, pseudo-primitive $G$-module. Let $p_1, \ldots, p_n$ be the distinct prime divisors of $F = \mathbf{F}(G)$ and let $Z$ be the socle of the cyclic group $\mathbf{Z}(F)$; thus $|Z| = p_1 \cdots p_n$. Then there exist subgroups $E$, $T \leq G$ such that*
*(1) $F = ET$, $Z = E \cap T$ and $T = \mathbf{C}_F(E)$.*
*(2) The Sylow-subgroups of $E$ are extra-special or cyclic of prime order.*
*(3) The exponent of $E$ divides $2 \cdot p_1 \cdot p_2 \cdots p_n$.*
*(4) The Sylow-subgroups of $T$ are cyclic, quaternion, dihedral or semi-dihedral.*
*(5) There exists $U \leq T$ of index at most 2 such that $U = \mathbf{C}_T(U)$ is cyclic, $U \trianglelefteq G$ and $EU \trianglelefteq G$.*
*(6) If $F \leq G'$, then $U = T = \mathbf{Z}(F)$.*

In the quasi-primitive case, there even exist restrictions for the Fitting-factor group of a solvable group.

**1.2 Theorem (T. Wolf [Wo 1])** *Assume that $G$ is solvable and that $V$ is a faithful and quasi-primitive $G$-module. Set $F = \mathbf{F}(G)$, let $Z$ be the socle of the cyclic group $\mathbf{Z}(F)$, and $A = \mathbf{C}_G(Z)$. Then there exist normal subgroups $E, T \trianglelefteq G$ which satisfy (1)—(6) of Theorem 1.1. Moreover, we have*
*(7) $T = \mathbf{C}_G(E)$ and $F = \mathbf{C}_A(E/Z)$.*
*(8) $E/Z \cong F/T$ is a completely reducible and faithful $A/F$-module (possibly of mixed characteristic).*
*(9) Let $P$ be an non-abelian Sylow-subgroup of $E$. Then $A$ acts symplecticly on $P/\mathbf{Z}(P)$.*

The proof of $E, T \trianglelefteq G$ investigates each Sylow-$p$-subgroup of $F$ separately. If $p$ is odd, it again relies on P. Halls's Theorem and does not even require the solvability of $G$. As far as $p = 2$ is concerned, a very ingenious induction argument is needed.

Let $\chi$ be a faithful irreducible ordinary character of an extra-special group P. Then $\chi(1) = |P/\mathbf{Z}(P)|^{1/2}$ and Theorem 1.1 has the following consequence:

**1.3 Corollary:** *Assume that $V$ is a finite, faithful and pseudo-primitive $G$-module. Let $e^2 = |F : T| = |EU : U|$ and $V_U \cong W \oplus \cdots \oplus W$ for an irreducible $U$-module $W$. Then $|V| = |W|^{et}$ for an integer $t \in \mathbf{N}$.*

Let $q$ be a prime power and $V = \mathrm{GF}(q^m)$, considered as a $\mathrm{GF}(q)$-vector space of dimension $m$. Fix $a \in V^*$, $w \in V$ and $\sigma \in \mathrm{Gal}(\mathrm{GF}(q^m))$. We define a map

$$T : V \longrightarrow V,\ T(x) = ax^\sigma + w.$$

Then $T$ is a permutation on $V$ and $T$ is trivial if and only if $a = 1$, $\sigma = 1$ and $w = 0$. Thus we have the following subgroups of the symmetric group $S_{q^m}$ :

(I) The *translations* $A(q^m) = \{x \longrightarrow x + w \mid w \in \mathrm{GF}(q^m)\}$.
(II) The *multiplications* $\Gamma_o(q^m) = \{x \longrightarrow ax \mid a \in \mathrm{GF}(q^m)^*\}$.
(III) The *semi-linear group* $\Gamma(q^m) = \{x \longrightarrow ax^\sigma \mid a \in \mathrm{GF}(q^m)^*, \sigma \in \mathrm{Gal}(\mathrm{GF}(q^m))\}$.
(IV) The *affine semi-linear group* $A\Gamma(q^m) = \{x \longrightarrow ax^\sigma + w \mid a \in GF(q^m)^*, w \in \mathrm{GF}(q^m), \sigma \in \mathrm{Gal}(\mathrm{GF}(q^m))\}$.

Note that $\Gamma_o(q^m)$ is cyclic of order $q^m - 1$, $\Gamma(q^m)/\Gamma_o(q^m) \cong \mathrm{Gal}(\mathrm{GF}(q^m))$ is cyclic of order $m$ and $A\Gamma(q^m)$ is the semi-direct product of $A(q^m)$ and $\Gamma(q^m)$. Also $V \cong A(q^m)$, where $\Gamma(q^m)$ acts on $A(q^m)$ by conjugation and on $V$ by semi-linear mappings. Thus the structure of semi-linear groups is well understood. They play an important role in the study of solvable groups.

**1.4 Theorem (B. Huppert [Hu 1;II,3.11])** *Suppose that $G$ acts faithfully on a $\mathrm{GF}(q)$-vector space $V$ of dimension $m$ ($q$ a prime power). Assume that $G$ has a normal abelian subgroup $A$ for which $V_A$ is irreducible. Then the points of $V$ may be labelled as the elements of $\mathrm{GF}(q^m)$ in such a way that $G \leq \Gamma(q^m)$, $A \leq \Gamma_o(q^m)$ and $VG \leq A\Gamma(q^m)$, where $V = A(q^m)$.*

This applies to pseudo-primitive modules as follows:

**1.5 Corollary:** *Suppose that $G$ is solvable and that $V$ is a faithful, pseudo-primitive $\mathrm{GF}(q)[G]$-module of dimension $m$. If $F = \mathbf{F}(G)$ is abelian, then $V_F$ is irreducible and $G \leq \Gamma(q^m)$.*

## § 2 Solvable linear groups and solvable primitive permutation groups

In this section, we apply the methods of § 1, and we present two results which have turned out to be crucial for several questions in the representation theory of solvable groups. The proof of the first one is very typical. General methods show that a possible counterexample has to be relatively small. What then follows is a rather tedious step-by-step analysis of those remaining cases. Sometimes, they all can be ruled out. If however exceptional cases really occur, it is even more difficult to detect them completely. A common candidate for such an exception is the group SL(2,3).

**2.1 Theorem (T. Wolf [Wo 1])** *Let $V \neq 0$ be a completely reducible, faithful and finite $G$-module for a solvable group $G$. Then $|G| \leq \frac{1}{\lambda}|V|^{9/4}$, where $\lambda = (24)^{1/3}$.*

Sketch of the proof: We proceed by induction on $|G| \cdot |V|$ and may assume that $V$ is irreducible.

Assume that $V$ is not quasi-primitive. We then choose $N \trianglelefteq G$ maximal such that $V_N$ is not homogeneous and write $V_N = U_1 \oplus \cdots \oplus V_m$ for $m > 1$ homogeneous components $U_i$ of $V_N$. Now $G/N$ faithfully and primitively permutes the $U_i$. Let $M/N$ be a chief-factor of $G$. Then $|M/N| = m$ and $M/N$ is a faithful, irreducible $G/M$-module (cf. the comments about solvable primitive permutation groups following this theorem). By induction, we have that $|N| \leq \frac{1}{\lambda}m|V|^{9/4}$ and $|G/N| \leq m \cdot \frac{1}{\lambda}m^{9/4}$. It follows that $|G| \leq \frac{m^{13/4}}{\lambda^m} \cdot \frac{1}{2}|V|^{9/4}$, and we may assume $\lambda^m < m^{13/4} < m^{10/3}$, i.e.,

$24^m < m^{10}$. Thus $2 \leq m \leq 5$, and it is easy to see that the assertion also holds in these cases.

We may assume from now on that $V$ is quasi-primitive and we adopt the notations of Theorem 1.2. Set $e^2 = |F:T| = |E:Z|$ and $V_U \cong W \oplus \cdots \oplus W$ for an irreducible $U$-module W. If $e > 1$, the inductive hypothesis implies $|A/F| \leq \frac{1}{\lambda}(e^2)^{9/4}$.

Observe that $|U| \mid |W| - 1$ and $|G/A| \cdot |T| \leq |U|^2$. Thus

$$|G| \leq \frac{1}{\lambda}(e^2)^{9/4} \cdot e^2 \cdot |W|^2 = \frac{1}{\lambda}e^{13/2}|W|^2.$$

By Corollary 1.3, $|V| = |W|^{et}$, and it remains to assume that

$$e^{13/2}|W|^2 > |V|^{9/4} = (|W|^{e \cdot t})^{9/4}.$$

Since $|W| \geq 3$, it follows that $e^{26} > 3^{9e-8}$ and $e \leq 5$.

These cases have to be considered separately. Let finally $e = 1$. Since $U = C_G(U) = F$, Lemma 1.5 implies that $G \leq \Gamma(r^n)$ ($r = \mathrm{char}(V)$ and $n = \dim V$). The assertion of the Theorem can be checked in this case.

Suppose that $G$ is a primitive solvable permutation group on a finite set $\Omega$ with point-stabilizer $G_\alpha$ ($\alpha \in \Omega$). Then $G$ has a unique minimal normal subgroup $M$, $G = M \cdot G_\alpha$, $M \cap G_\alpha = 1$, $C_G(M) = M$ and $M$ acts regularly on $\Omega$. Consequently, $|\Omega| = |M| = p^f$ is a prime power. Moreover, the mapping $m \longrightarrow \alpha m$ ($m \in M$) is a permutation isomorphism between $M$ and $\Omega$; here $G_\alpha$ acts on $M$ by conjugation. We may consider $G$ as a subgroup of the affine linear group $AGL(f,p)$, where $M$ is the normal subgroup consisting of translations (see II,2.2,II,3.2 and II,3.5 of [Hu 1]). Theorem 2.1 has the following consequence.

**2.2 Corollary.** *Let $G$ be a primitive solvable permutation group on a finite set $\Omega$. Then $|G| \leq \frac{1}{2}|\Omega|^{13/4}$.*

For a permutation group $G$ on $\Omega$ one might ask whether there exists a *regular orbit* on the power set $\mathcal{P}(\Omega)$, i.e. whether there exists a subset $\Delta \subseteq \Omega$ such that the setwise stabilizer $\mathrm{stab}_G(\Delta)$ of $\Delta$ is trivial. For primitive solvable permutation groups, D. Gluck has given a complete answer.

**2.3 Theorem (D. Gluck [Gl 1]).** *Let $G$ be a primitive solvable permutation group on a finite set $\Omega$. Then $G$ has a regular orbit on the power set $\mathcal{P}(\Omega)$ except in the following cases.*

*(1) $|\Omega| = 3$ and $G \cong D_6$, the dihedral group of order 6.*
*(2) $|\Omega| = 4$ and $G \cong A_4$, the alternating group on 4 letters, or $G \cong S_4$, the symmetric group on 4 letters.*
*(3) $|\Omega| = 5$ and $G \cong F_{10}$ or $G \cong F_{20}$, the Frobenius groups of order 10 and 20, respectively.*
*(4) $|\Omega| = 7$ and $G \cong F_{42}$, the Frobenius group of order 42.*
*(5) $|\Omega| = 8$ and $G \cong A\Gamma(2^3)$, the affine semi-linear group on $\mathrm{GF}(2^3)$.*
*(6) $|\Omega| = 9$ and $G$ is the semi-direct product of $Z_3 \times Z_3$ with a group $S$, where $S \cong D_8$, the dihedral group of order 8, or $S \cong SD_{16}$, the semi-dihedral group of order 16, or $S \cong SL(2,3)$ or $S \cong \mathrm{GL}(2,3)$.*

Using Corollary 2.2, it is very easy to show that exceptions can only occur for $|\Omega| < 81$: We use the structure of a primitive solvable permutation group introduced above. For $g \in G^*$, let $s(g)$ be the number of fixed points and $n(g)$ the number of cycles. We show that $n(g) \leq \frac{3}{4}|\Omega|$ and may thus assume that $g \in G_\alpha$. As $G_\alpha$ acts faithfully on $M$, we get $|\mathbf{C}_M(g)| \leq \frac{1}{2}|M|$, and by the permutation isomorphism, $s(g) \leq \frac{1}{2}|\Omega|$. Consequently, $n(g) \leq s(g) + \frac{1}{2}(|\Omega| - s(g)) \leq \frac{3}{4}|\Omega|$.

Since $g$ stabilizes exactly $2^{n(g)}$ subsets of $\Omega$, Corollary 2.2 yields

$$|\{(g,\Delta) | g \in G^*,\ \Delta \subseteq \Omega,\ g \in \operatorname*{stab}_G(\Delta)\}| \leq \frac{1}{2}|\Omega|^{13/4} \cdot 2^{\frac{3}{4}|\Omega|}.$$

Thus we can certainly find a regular orbit of $G$ on $\mathcal{P}(\Omega)$ provided that

$$\frac{1}{2}|\Omega|^{13/4} \cdot 2^{\frac{3}{4}|\Omega|} < 2^{|\Omega|}, \text{ or equivalently}$$

$$13 \log_2 |\Omega| < |\Omega| + 4.$$

This holds for $|\Omega| \geq 81$.

Rather easy to handle is also the case where $|\Omega|$ is a prime. Unfortunately, the proper prime powers less than 81 require a very detailed step-by-step analysis (which we skip here). We state two consequences of Theorem 2.3.

**2.4 Corollary ([MW 1])** *Let $G$ be a primitive solvable permutation group on a finite set $\Omega$. Let $q$ be a prime divisor of $|G|$, and assume that for all $\Delta \subseteq \Omega$,* $\mathrm{stab}_G(\Delta)$ *contains a Sylow-$q$ subgroup of $G$. Then one of the following cases occurs:*
*(I)* $|\Omega| = 3$, $q = 2$ *and* $G \cong S_3 \cong D_6$
*(II)* $|\Omega| = 5$, $q = 2$ *and* $G \cong F_{10} \cong D_{10}$
*(III)* $|\Omega| = 8$, $q = 3$ *and* $G \cong A\Gamma(2^3)$.

**2.5 Corollary ([GM 1])** *Let $G$ be a solvable permutation group on a finite set $\Omega$. Then there exists a subset $\Delta \subseteq \Omega$ such that* $\mathrm{stab}_G(\Delta)$ *is a $\{2,3\}$-group.*

It is natural to ask whether Gluck's classification can be extended to all primitive permutation groups. Clearly the symmetric groups $S_n$ ($n \geq 3$) and the alternating groups $A_n$ ($n \geq 4$) form an infinite series of exceptions. The following result provides some hope.

**2.6 Theorem (P. Cameron, P. Neumann, J. Saxl [CNS 1])** *Let $G$ be a primitive permutation group on a finite set $\Omega$ with $A_{|\Omega|} \not\leq G$. Then, with only finitely many exceptions, $G$ has a regular orbit on $\mathcal{P}(\Omega)$.*

The proof relies on the classification of finite simple groups. J. Key, J. Siemons and A. Wagner considered subgroups $G$ in $P\Gamma L(d+1, p^f)$ and $A\Gamma L(d, p^f)$, which act primitively on the points of the natural geometry. If no alternating group of the same degree is contained, $G$ has a regular orbit on the power set, except for 35 exceptions.

## § 3 Module actions with large centralizers

Let $V$ be a faithful and irreducible $G$-module. It should have consequences for the structure of $G$ if each $\mathbf{C}_G(v)$ ($v \in V$) contains a member of a certain canonical class of subgroups of G. This section is concerned with the case where $G$ is solvable and where the class of subgroups is $\mathrm{Syl}_q(G)$ for some prime $q$.

**3.1 Theorem (D. Gluck, T. Wolf, O. Manz [Wo 2], [GW 1], [MW 1])** *Let G be a solvable group, q a prime number and $G = O^{q'}(G) \neq 1$. Suppose that G acts faithfully and irreducibly on a finite vector space V, and that $q \nmid |G : \mathbf{C}_G(v)|$ for all $v \in V$.*

*a) If V is pseudo-primitive, then one of the following holds:*

*(1) $G \leq \Gamma(r^n)$, where $|V| = r^n$.*

*(2) $|V| = 3^2$, $q = 3$ and $G = Q_8 \cdot Z_3 \cong SL(2,3)$.*

*(3) $|V| = 2^6$, $q = 2$ and $G = H \cdot Z_2$, where H is extra-special of order $3^3$ with exponent 3, and where $Z_2$ acts fixed-point-freely on $H/\mathbf{Z}(H)$ and trivially on $\mathbf{Z}(H)$.*

*b) If V is not pseudo-primitive, we choose $C \trianglelefteq G$ maximal with respect to $V_C$ not homogeneous. Let $V_1, \ldots, V_n$ be the homogeneous components of $V_C$. Then*

*(I) $n = 3$, 5 or 8, $q = 2$, 2 or 3 and $G/C \cong D_6$, $D_{10}$ or $A(2^3)$, respectively.*

*(II) $C/\mathbf{C}_C(V_i)$ acts transitively on $V_i^*$ ($i = 1, \ldots, n$).*

*(III) C is metabelian, unless $|V_i| = 3^2$, $5^2$, $7^2$, $11^2$, $23^2$ or $3^4$.*

The proof heavily relies on the methods developed so far. Without going into details we want to emphasize its main ideas.

Outline of the proof of Theorem 3.1:

a) We apply Theorem 1.1, and set $F = \mathbf{F}(G)$ and $K = O^q(G)$.

*Step 1:* The hypothesis "pseudo-primitive" allows us to argue by induction. More precisely, if N char G and $q \mid |N|$, then $N = G$. In particular, G is q-nilpotent with q-complement K.

*Step 2:* Observe that $F \leq K \leq G'$. Therefore $U = T = \mathbf{Z}(F)$ and U is central in K. If $F = U$, then $K \leq G' \leq \mathbf{C}_G(F) \leq F$ and assertion (1) of the theorem holds. Thus $F' \neq 1$.

*Step 3:* Let $Q_o \in \mathrm{Syl}_q(G)$ and $g \in Q_o$ an element of order q. We choose $P \in \mathrm{Syl}_p(F)$ such that $g \notin \mathbf{C}_G(P)$ and consider V as an $P < g >$-module. Hall-Higman type arguments (cf. [Hu1; V,17.13] and [HB1;IX,2.6] show that

$$\dim \mathbf{C}_V(g) \leq \begin{cases} \frac{1}{2}\dim V & q \neq 2 \\ \frac{2}{3}\dim V & q = 2 \end{cases}$$

*Step 4:* Observe that $K/F$ acts faithfully and completely reducibly on $F/(U \cdot \Phi G)$. (Use Gaschütz's Theorem [Hu 1;III,4.5] and $U \leq \mathbf{Z}(K)$). By Theorem 2.1, $|K/F| \leq |F/(U \cdot \Phi(G))|^{9/4}$.

*Step 5:* We consider the set $\{(v, Q) \mid v \in V,\ Q \in \mathrm{Syl}_q(G),\ Q \leq \mathbf{C}_G(v)\}$. Since by hypothesis, each $v \in V$ is centralized by some Sylow-q subgroup, counting yields $|\mathrm{Syl}_q(G)||\mathbf{C}_V(Q_o)| \geq |V|$.

*Step 6:* Set $e^2 = |F : U|$ and $V_U \cong W \oplus \ldots \oplus$ for an irreducible U-module W. Then $|U| \mid |W| - 1$ and Corollary 1.3 yields $|V| = |W|^{et}$ ($t \in \mathbf{N}$).

Now $|\mathrm{Syl}_q(G)| \leq |K| = |U| \cdot |F/U| \cdot |K/F| \leq |W|e^2(e^2)^{9/4}$ (by Step 4) and $|\mathbf{C}_V(Q_o)| \leq |\mathbf{C}_V(g)| \leq |V|^{2/3} = (|W|^{et})^{2/3}$ (by Step 3).

Thus Step 5 implies

$$|W|e^2(e^2)^{9/4} \cdot (|W|^{et})^{2/3} \geq |W|^{et}, \text{ and therefore}$$

$$e^{13/2} \geq |W|^{\frac{1}{3}et-1} \geq 3^{e/3-1}\ .$$

This estimation shows that it remains to consider only some small values of $e$. We have to point out however that this still requires some efforts. Note that the exceptional cases (2) and (3) of the Theorem did not appear yet.

b) Observe that $G/C$ primitively and faithfully permutes the $V_i$. Let $\Delta \subseteq \{1, \ldots, n\}$, without loss of generality $\Delta = \{1, \ldots, k\}$, $k \leq n$. We choose $v_i \in V_i^*$ $(i \in \Delta)$ and set $v = v_1 + \cdots + v_k$. By our hypothesis, some Sylow-$q$-subgroup $Q$ of $G$ centralizes $v$. Therefore, $1 \neq QC/C \in \mathrm{Syl}_q(G/C)$ must leave $\Delta$ invariant and assertion (I) follows from Corollary 2.4.

Assertion (II) now is an easy consequence. What it says is that the semi-direct product of $V_i$ with $C/\mathbf{C}_C(V_i)$ is a 2-fold transitive permutation group on $V_i$. Those have been classified, namely

**3.2 Theorem (B. Huppert [HB 1;XII,7.3])** *Let $G$ be a solvable doubly transitive permutation group on a finite set $\Omega$. Then $|\Omega| = r^f$ is a prime power and unless $r^f = 3^2, 5^2, 7^2, 11^2, 23^2$ or $3^4$, the points of $\Omega$ can be labelled as the elements of $GF(r^f)$ such that $G \leq A\Gamma(r^f)$. Here, $A(r^f)$ is the unique minimal normal subgroup of $G$.*

We finish the outline of the proof of Theorem 3.1. Since $\Gamma(r^f)$ is metacyclic, in the non-exceptional cases of (III), $C/\mathbf{C}_C(V_i)$ is metacyclic for all $i$ and thus $C$ is metabelian.

Observe that $V$ in a)(3) is pseudo-primitive, but not quasi-primitive.

For Part a) of Theorem 3.1, the hypothesis $G = O^{q'}(G)$ is not really necessary, in (2) then the addition $G \cong \mathrm{GL}(2,3)$ occurs. We also mention that if $q \mid n$ but $q \nmid r^n - 1$, then the centralizer condition is met by $\Gamma(r^n)$ acting on $\mathrm{GF}(r^n)$.

It is hard to see how to proceed in Theorem 3.1 for non-solvable groups. All main ingredients of the proof namely take advantage of the solvability of $G$. We note that C. Hering has extended Huppert's classification to all 2-fold transitive permutation groups (see [HB 1;XII,7.5]).

## § 4 Brauer's height-0 conjecture and Ito's problem

The most spectacular application of Theorem 3.1 certainly was the proof of Brauer's height-0 conjecture for solvable groups. It is based on

**4.1 Fong's Reduction Theorem (cf. [Fe 1;X.1])** *Let $p$ be a prime, $B$ a $p$-block of $G$ and $\delta(B)$ its defect group. If $|G|_p = p^a$, $|\delta(B)| = p^d$ and $\chi \in \mathrm{Irr}(G) \cap B$, then the height $ht(\chi)$ of $\chi$* is defined by $\chi(1)_p = p^{a-d+ht(\chi)}$.

a) Let $N = O_{p'}(G)$ and $\beta \in \mathrm{Irr}(N)$ lying under $B$. Then induction of characters yields a height-preserving bijection between the blocks of $I_G(\beta)$ lying over $\beta$ and those of $G$ lying over $\beta$. If $b$ is the correspondent of $B$, then $\delta(B) \cong \delta(b)$. Thus induction arguments work provided that $I_G(\beta) < G$.

b) If $I_G(\beta) = G$ and $G$ is $p$-solvable, then there is a group $\tilde{G}$ (which is obtained as a central extension of $G$) with the following properties:

(1) $\tilde{G}$ has a $p$-block $\tilde{B}$ of maximal defect such that $\delta(B) \cong \delta(\tilde{B})$ and such that there exists a height-preserving bijection from $B \cap \mathrm{Irr}(G)$ onto $\tilde{B} \cap \mathrm{Irr}(\tilde{G})$.

(2) There exists $\alpha \in \mathrm{Irr}(O_{p'}(\tilde{G}))$ such that $\tilde{B} \cap \mathrm{Irr}(\tilde{G}) = \mathrm{Irr}(\tilde{G}|\alpha)$.

(3) $O_{p',p}(\tilde{G})$ has a normal Sylow-$p$-subgroup.

**4.2 Brauer's height-0-conjecture:** Let $B$ be a $p$-block of $G$. Then

$$ht(\chi) = 0 \text{ for all } \chi \in B \cap \mathrm{Irr}(G) \iff \delta(B) \text{ is abelian.}$$

This conjecture has been established for $p$-solvable groups. Before we show how this was done, we recall a well-known Theorem of Ito.

**4.3 Theorem (N. Ito; see [Is 1;6.15 and 12.34]).** *Let $p$ be a prime and $G$ $p$-solvable. Then $p \nmid \chi(1)$ for all $\chi \in \mathrm{Irr}(G)$ if and only if $G$ has a normal abelian Sylow-$p$-subgroup.*

**4.4 Theorem (P. Fong, see [Fe 1;X]).** *Let $G$ be $p$-solvable. If $\delta(B)$ is abelian, then $ht(\chi) = 0$ for all $\chi \in B \cap \mathrm{Irr}(G)$.*

*Proof:* By induction on $|G|$ we may assume that 4.1 b) occurs. Since then $\delta(\tilde{B})$ is an abelian Sylow-$p$-subgroup of $\tilde{G}$, the $p$-length of $\tilde{G}$ is 1, and (3) implies that $\tilde{G}$ has a normal abelian Sylow-$p$-subgroup. Apply Theorem 4.3.

The converse was proved by Gluck and Wolf. They first did it for solvable groups and then extended their result to $p$-solvable ones via the classification of finite simple groups ([GW 2]). We don't discuss this extension, but show how Theorem 3.1 applies in the solvable case.

**4.5 Generalized Gluck-Wolf Theorem ([GW 1]).** *Let $N \trianglelefteq G$, $\alpha \in \mathrm{Irr}(N)$ and $\pi$ a set of primes. Assume that $G/N$ is solvable and that $\chi(1)/\alpha(1)$ is a $\pi'$-number for all $\chi \in \mathrm{Irr}(G|\alpha)$. Then a Hall-$\pi$-subgroup of $G/N$ is abelian.*

*Ideas of the proof.* First observe that the hypothesis is inherited by normal subgroups $L$ with $N \leq L \leq G$. Arguing by induction on $|G/N|$, standard arguments yield

(1) $O_{\pi'}(G/N) = 1$ and $O^{\pi'}(G/N) = G/N$.
(2) If $K/N = O^{\pi}(G/N)$, then $|G/K| = q \in \pi$.
(3) If $M/N = O_{\pi}(G/N)$, then $M/N$ is a self-centralizing chief factor of $G$, and $1 \neq |K/M|$ is a $\pi'$-number.
(4) $I_G(\alpha) = G$, $I_G(\theta) = G$ for some $\theta \in \mathrm{Irr}(M|\alpha)$ and $\nmid |G : I_G(\lambda)|$ for all $\lambda \in \mathrm{Irr}(M/N)$.

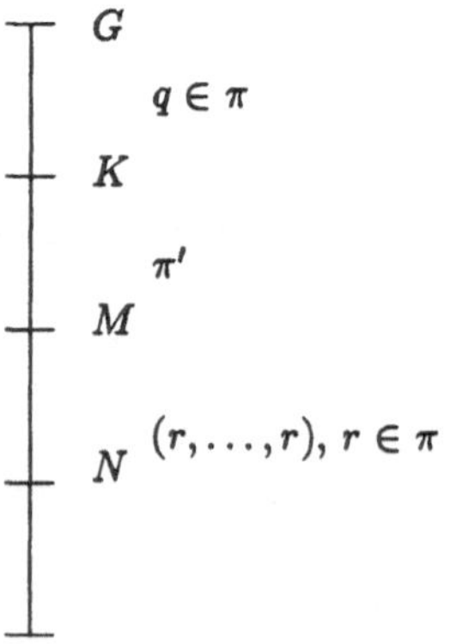


We now consider the faithful and irreducible action of $G/M$ on $\mathrm{Irr}(M/N) =: V$. If $V$ is quasi-primitive, then Theorem 3.1 a) implies that $K/M$ is cyclic or $G/M \cong$

$SL(2,3)$. In both cases $\theta$ extends to $\theta^* \in \mathrm{Irr}(G)$. Thus $\theta^*\varphi \in \mathrm{Irr}(G|\alpha)$ for all $\varphi \in \mathrm{Irr}(G/M)$, and a contradiction is reached, since we can find such a $\varphi$ with $\varphi(1) = q$.

If $V$ is not quasi-primitive, then Theorem 3.1 b) applies. Unfortunately, those three cases still require an extremely detailed analysis, which we skip.

**4.6 Corollary:** *Let $G$ be solvable. If $ht(\chi) = 0$ for all $\chi \in B \cap \mathrm{Irr}(G)$, then $\delta(B)$ is abelian.*

*Proof:* By induction on $|G|$, case b) of Fong's Reduction Theorem occurs. The hypothesis implies that $p \nmid \chi(1)/\alpha(1)$ for all $\chi \in \tilde{B} \cap \mathrm{Irr}(\tilde{G}) = \mathrm{Irr}(\tilde{G}|\alpha)$ and Theorem 4.5 yields that $\delta(B) \cong \delta(\tilde{B}) \in \mathrm{Syl}_p(\tilde{G})$ is abelian.

For several simple groups, Brauer's height-0 conjecture has been checked. T. Berger and R. Knörr [BK 1] have proved the following reduction theorem: If the assertion "$\delta(B)$ abelian $\Rightarrow ht(\chi) = 0$ for all $\chi \in B \cap \mathrm{Irr}(G)$" does not hold, then there exists a central extension of a simple group which is a counterexample. To the knowledge of the author, a similar reduction theorem for the converse assertion could not be obtained yet.

We return to Ito's Theorem 4.3. It is natural to ask whether one can remove the hypothesis "$p$-solvable" there, and whether an analogous statement holds for absolutely irreducible $p$-modular representations, which we label by their Brauer-characters $IBr_p(G)$. The answer to these questions was found by G. Michler [Mi 1]. His key result relies on the classification of finite simple groups:

If $G$ is a non-abelian simple group and $p$ an odd prime dividing $|G|$, then $G$ has a $p$-block of non-maximal defect. In fact, finite simple groups of Lie-type have a $p$-block of defect 0 ($p \neq 2$).

This fact and an easy Clifford reduction yields

**4.7 Theorem (see [Mi 2]).** *Let $p$ be a prime.*

*a) (N. Ito, G. Michler) $p|\chi(1)$ for all $\chi \in \mathrm{Irr}(G)$ $\iff$ $P \in \mathrm{Syl}_p(G)$ is normal and abelian.*

*b) (G. Michler, Y. Okuyama) $p|\beta(1)$ for all $\beta \in IBr_p(G)$ $\iff$ $P \in \mathrm{Syl}_p(G)$ is normal.*

Note that in part b), the prime in question and the characteristic of the representation coincide. What can be said if $q \nmid \beta(1)$ for all $\beta \in IBr_p(G)$ and $q \neq p$? The group $G = S_4$ in characteristic $p = 3$ shows that a Sylow-$q$-subgroup need not be normal nor abelian ($q = 2$). But again Theorem 3.1 provides an answer for solvable groups $G$.

**4.8 Theorem (T. Wolf, O. Manz [MW 1]).** *Let $G$ be solvable and $p$, $q$ be distinct primes. Assume that $q \nmid \beta(1)$ for all $\beta \in IBr_p(G)$. Then $G/O_{p,q}(G)$ has $q$-length at most 1. Moreover both $q$-factors of $G$ are abelian and $Q \in \mathrm{Syl}_q(G)$ is metabelian.*

To see how Theorem 3.1 applies we may assume that $O_p(G) = 1$ and $O^{q'}(G) = G$. By Gaschütz's Theorem (see [Hu 1;III,4.5]), $\mathbf{F}(G)/\Phi(G)$ is a direct sum of irreducible $G$-modules $V_i$. Set $\hat{V}_i = IBr_p(V_i)$ and $C_i = \mathbf{C}_G(V_i)$. Then $G/C_i$ acts faithfully and irreducibly on $\hat{V}_i$, and each $\lambda \in \hat{V}_i$ is centralized by a Sylow-$q$-subgroup of G. If each $\hat{V}_i$ is pseudo-primitive, then Theorem 3.1 a) implies that each $G/C_i$ has $q$-length at most 1 and the same is true for $G/\mathbf{F}(G)$. If however one of the $\hat{V}_i$ is not quasi-primitive, then Theorem 3.1 b) yields $q \in \{2,3\}$ and considerably restricts the structure of $G/C_i$. To handle those exceptional cases is by far not as complicated as it is in Theorem 4.5.

We add two more results in the same direction.

**4.9 Theorem [MW 1].** *We assume the hypothesis of Theorem 4.8.*

a) *If $q \nmid p-1$ and if $(p,q) \neq (2,3)$, then $G/O_p(G)$ has a normal abelian Sylow-q-subgroup.*

b) *Set $H = O^{q'}(G)$. If $(p,q) \neq (7,3)$, then $H/O_p(H)$ has p-length at most 1.*

Suppose again that $q \nmid \beta(1)$ for all $\beta \in IBr_p(G)$ $(p \neq q)$. It was shown in [MW 1] that the $p$-solvability of $G$ implies its $q$-solvability and that Theorem 4.8 and 4.9 still hold. Without the assumption "$p$-solvable" however $G$ need not be $q$-solvable. We note that

$$\beta(1) \text{ is a 2-power} \quad \text{for all } \beta \in IBr_2(SL(2,2^f)), \text{ and}$$
$$\beta(1) \text{ is odd} \quad \text{for all } \beta \in IBr_p(PSL(2,p)).$$

Hence it makes no sense to use the notion of $q$-length in general. What remains is:

**Conjecture:** Suppose that $q \nmid \beta(1)$ for all $\beta \in IBr_p(G)$ $(p \neq q)$. Then $Q \in \mathrm{Syl}_q(G)$ is metabelian.

As a first step one should be able to show that a possible counterexample to this conjecture is a simple group (or a central extension of a simple group, or its automorphism group). More insight into the list of primitive permutation groups without a regular orbit on the power set might be helpful (cf. Theorem 2.6).

## § 5 The character degree graph

Set $\rho(G) = \{q \text{ prime } | q|\chi(1) \text{ for some } \chi \in \mathrm{Irr}(G)\}$, the set of *relevant primes* for $\mathrm{Irr}(G)$. Observe that by the Ito-Michler Theorem 4.7 a), $q \notin \rho(G)$ if and only if $G$ has a normal, abelian Sylow-$q$-subgroup. We construct a graph $\Gamma(G)$ with vertex set $\rho(G)$, where we draw an edge between distinct $q, r \in \rho(G)$ if and only if $qr \mid \chi(1)$ for some $\chi \in \mathrm{Irr}(G)$. A distance function $d(q,r)$ is defined in the usual way:

$$d(q,r) = \text{lenght of the shortest path between } q \text{ and } r.$$

In particular, $d(q,q) = 0$ and $d(q,r) = \infty$ if $q$ and $r$ lie in different connected components of $\Gamma(G)$. If $\Lambda$ is a connected component of $\Gamma(G)$, then the diameter of $\Lambda$ is defined as

$$\mathrm{diam}(\Lambda) = \max\{d(q,r) | q,r \in \Lambda\}. \qquad \text{Finally}$$
$$\mathrm{diam}(\Gamma(G)) = \max\{\mathrm{diam}\,\Lambda | \Lambda \text{ a component of } \Gamma(G)\}.$$

The following problems arise:

(1) Bound the number $n(\Gamma(G))$ of connected components of $\Gamma(G)$.
(2) Bound the diameter $\mathrm{diam}(\Gamma(G))$ of $\Gamma(G)$.
(3) Has $n(\Gamma(G)) > 1$ any influence on the structure of $G$ and on the shape of the connected components of $\Gamma(G)$?

Before we state several positive results, some (solvable and non-solvable) examples will be given below.

$\Gamma(A\Gamma(2^{10}))$

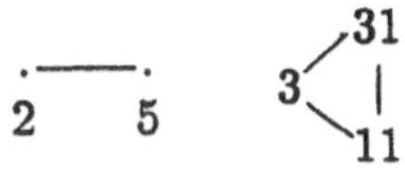


$\Gamma(A\Gamma(2^{21})$

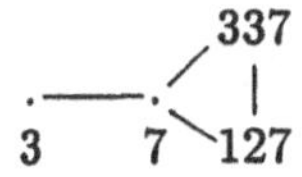


$\Gamma(A_5)$

$\Gamma(J_1)$

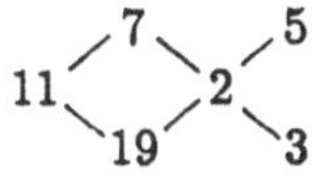


Not surprisingly, Theorem 3.1 again serves as an important tool in the solvable case. In fact the following Lemma is a consequence.

**5.1 Lemma:** *Suppose that $V$ is a faithful, irreducible and finite module for a solvable group $G$. Assume that $\pi$ is a non-empty set of prime divisors of $|G|$ and $\mathbf{C}_G(v)$ contains a Hall-$\pi$-subgroup of $G$ for all $v \in V$. Then there exists $\chi \in \mathrm{Irr}(G)$ such that $q|\chi(1)$ for all $q \in \pi$.*

**5.2 Theorem (P.P. Palfy, W. Willems, T. Wolf, O. Manz [Pa 1], [MWW 1]).** *Let $G$ be solvable and let $\pi$ be a set of primes contained in $\Gamma(G)$. Assume that $|\pi| \geq 3$. Then there exist distinct $v, w \in \pi$ such that $v \cdot w|\chi(1)$ for some $\chi \in \mathrm{Irr}(G)$.*

Using Lemma 5.1 we show how the Theorem can be proved if $|\pi| \geq 4$. We set $\pi = \{r, s, t, u\}$. Working by induction on $|G|$ it is not too hard to see that $F = \mathbf{F}(G)$ is a non-abelian Sylow-$r$-subgroup of $G$ and $F'$ is a minimal normal subgroup of $G$. Set $C = \mathbf{C}_G(F') \geq F$ and let $1 \neq \alpha \in \mathrm{Irr}(F')$. Since $r|\delta(1)$ for all $\delta \in \mathrm{Irr}(F|\alpha)$, it follows that $r|\sigma(1)$ for all $\sigma \in \mathrm{Irr}(G|\alpha)$ and $I_G(\alpha)$ contains a Hall-$\{s,t,u\}$-subgroup of $G$ for all $\alpha \in \mathrm{Irr}(F')$. As $\mathrm{Irr}(F')$ is a faithful, irreducible $G/C$-module, we may assume by Lemma 5.1 that $|G/C|$ is divisible by at most one prime in $\pi\backslash\{r\}$. Without loss of generality, $C$ contains a Hall-$\{s,t\}$-subgroup of $G$. Since $C/F$ is an $r'$-group, there exists $\beta \in \mathrm{Irr}(F|\alpha)$ that is invariant in $C$ (see [Is 1;13.3]). Thus $\beta$ extends to $\beta^* \in \mathrm{Irr}(C)$. As $\xi\beta^* \in \mathrm{Irr}(C|\beta)$ for all $\xi \in \mathrm{Irr}(C/F)$ and $r|\beta(1)$, we have that $\xi(1)$ is coprime to $s \cdot t$ for all such $\xi$. By Ito's Theorem 4.3, $C/F$ and hence $G/F$ have normal abelian Sylow-$s$ and Sylow-$t$-subgroups, say $S/F \neq 1$ and $T/F \neq 1$. Since $S/F \times T/F$ acts faithfully and completely reducibly on $F/\Phi(F)$, there exists $\Psi \in \mathrm{Irr}(ST/\Phi(F))$ with $s \cdot t|\Psi(1)$. Take $\chi \in \mathrm{Irr}(G|\Psi)$ to complete the proof.

It is now an easy exercise to deduce

**5.3 Corollary.** *Let $G$ be solvable. Then*

*a)* $n(\Gamma(G)) \leq 2$.

*b)* $\mathrm{diam}(\Gamma(G)) \leq 3$.

*c) If $n(\Gamma(G)) = 2$, then both components of $\Gamma(G)$ are complete graphs.*

We note that a) is best possible and even allows an elementary proof. It is an open problem however, whether $\operatorname{diam}(\Gamma(G)) \leq 2$ holds for solvable groups $G$. We give an answer to the first part of question (3) above, the proof of which relies on Theorems 4.5 and 5.5.

**5.4 Theorem (P. Palfy, R. Staszewski, O. Manz [Pa 1], [MS 1]).** *Let $G$ be solvable with $n(\Gamma(G)) = 2$. Then $G$ has a normal series with two abelian and one nilpotent factor, unless $G/\mathbf{F}(G) \leq GL(2,3)$.*

For non-solvable groups $G$, the graph $\Gamma(G)$ is much less understood. At any rate, for both problems (1) and (2) one cannot expect a better bound than 3. This follows from the examples $A_5$ and $J_1$ above. Partial, but rather unsatisfactory results exist for the diameter. Here, we concentrate on the number of connected components and present a reduction theorem. Its proof uses the Ito-Michler-Theorem 4.7 a) and the following remarkable result of Isaacs which does not rely on the classification of finite simple groups.

**5.5 Theorem (I.M. Isaacs [Is 2]).** *Suppose that a group $H \neq 1$ acts faithfully and coprimely on a group $G$. If $H$ fixes each non-linear $\chi \in \operatorname{Irr}(G)$, then $G$ is solvable and the nilpotency length of $G$ is at most 2.*
(Indeed, the structure of $G$ is much more restricted.)

**5.6 Proposition (R. Staszewski, W. Willems, O. Manz [MSW 1]).** *Let $G$ be non-solvable. We set*

$$k = \max\{n(\Gamma(E)) | E \textit{ a simple, non-abelian composition factor of } G\}.$$

*Then $n(\Gamma(G)) \leq k$.*

We show how the results above apply and let $N$ be a maximal normal subgroup of $G$.

Case 1: $G/N$ simple, non-abelian. Let $\ell$ be minimal such that $\Gamma(G/N)$ is covered by $\ell$ connected components $\Delta_1, \ldots, \Delta_\ell$ of $\Gamma(G)$. Then $\ell \leq k$. If $k < n(\Gamma(G))$, we choose a different component $\Delta_{\ell+1}$ and $\chi \in \operatorname{Irr}(G)$ whose prime divisors belong to $\Delta_{\ell+1}$. By Theorem 4.7 a), $|G/N|$ and $\chi(1)$ are coprime. Thus $\chi_N \in \operatorname{Irr}(N)$ and $\chi\,\varphi \in \operatorname{Irr}(G)$ for all $\varphi \in \operatorname{Irr}(G/N)$, a contradiction.

Case 2: $|G/N| = p$ for a prime $p$. By induction we have $n(\Gamma(N)) \leq k$. We may thus assume that $p \nmid \varphi(1)$ for all $\varphi \in \operatorname{Irr}(N)$ and $I_G(\varphi) = G$ for all non-linear $\varphi \in \operatorname{Irr}(N)$. By Theorem 4.7 a), $N$ has a normal abelian Sylow-$p$-subgroup $P$. Let $P_o \in \operatorname{Syl}_p(G)$. If $[N/P, P_o/P] \neq 1$, then Theorem 5.5 implies that $N/P$ and therefore $G$ is solvable, a contradiction. Consequently, $P_o \trianglelefteq G$ and we start the whole procedure again with a maximal normal subgroup $N_o$ such that $P_o \leq N_o$. This eventually yields the assertion of Proposition 5.6.

A subset $S \subseteq \operatorname{Irr}(G)$ is called a *covering-set* if for each $q \in \rho(G)$ there exists $\chi \in S$ such that $q|\chi(1)$. The *covering-number* $\operatorname{cn}(G)$ is defined as

$$\operatorname{cn}(G) = \min\{|S| | S \text{ a covering-set}\}.$$

Observe that $n(\Gamma(G)) \leq \operatorname{cn}(G)$.

**5.7 Theorem.** *We have $\operatorname{cn}(G) \leq 3$, provided that*

*(I) $G$ is simple of Lie-type ([MSW 1]), or*
*(II) $G$ is an alternating group or a sporadic simple group (D. Alvis, M. Barry [AB 1]).*

Thus Proposition 5.6 and Theorem 5.7 imply via the classification of finite simple groups:

**5.8 Corollary:** *If $G$ is non-solvable, then $n(\Gamma(G)) \leq 3$.*

What remains is the question whether even $\mathrm{cn}(G)$ admits a universal upper bound. More bravely, does even $\mathrm{cn}(G) \leq 3$ hold? A positive answer would imply Corollary 5.8. It seems clear, however, that the techniques of Proposition 5.6 do not work for $\mathrm{cn}(G)$.

Let $p$ be a prime. Then one can define a graph $\Gamma_p(G)$ with respect to $IBr_p(G)$, as it was done for $\mathrm{Irr}(G)$. Although the same questions can be asked for the modular graph $\Gamma_p(G)$, it has not been studied in great detail yet. For example, a result similar to Theorem 5.2 has not been obtained. At the end of this section, we gather some known facts about $\Gamma_p(G)$.

**5.9 Theorem (W. Willems, T. Wolf, O. Manz [MWW 1]).**
*a) If $G$ is solvable, then $n(\Gamma_p(G)) \leq 2$ and $\mathrm{diam}(\Gamma_p(G)) \leq 5$.*
*b) If $G$ is $p$-solvable, then $n(\Gamma_p(G)) \leq 3$.*
*c) $\limsup_{p\to\infty} n(\Gamma_p(SL(2,p))) = \infty$; in particular, there does not exist an upper bound for $n(\Gamma_p(G))$ which is independent of $G$ and $p$.*

The proof of part b) relies on Corollary 5.8 and thus on the classifiation of finite simple groups.

The bound for the diameter in part a) seems much too bad and it should be possible to improve it. One might also ask for a function $f(p)$ which bounds $n(\Gamma_p(G))$ for all $G$.

## § 6 Huppert's $\rho - \sigma$ conjecture

For a positive integer $n \in \mathbf{N}$, define $\sigma(n)$ to be the number of different prime divisors of $n$. Set

$$\sigma(G) = \max\{\, \sigma(\chi(1)) \mid \chi \in \mathrm{Irr}(G)\,\},$$

and recall that $\rho(G) = \{q \text{ prime} \mid q \mid \chi(1) \text{ for some } \chi \in \mathrm{Irr}(G)\}$.

Then B. Huppert conjectured the following local-global principle:

**6.1 $\rho$-$\sigma$-conjecture:**
a) $|\rho(G)| \leq 2 \cdot \sigma(G)$ if $G$ is solvable.
b) $|\rho(G)| \leq 3 \cdot \sigma(G)$ for arbitrary $G$.

If $\sigma(G) = 1$ (i.e. all character degrees are prime powers), the conjecture is true ([Ma 1], [MSW 1]). No general estimation for $|\rho(G)|$ in terms of $\sigma(G)$ is known in the non-solvable case. But we note that if $\mathrm{cn}(G) \leq 3$ was true, then part b) of the conjecture would follow.

For the rest of this section we restrict our attention to the solvable case. It is easy to see that $|\rho(G)| \leq 2 \cdot \sigma(G)$ would be best possible:

Let $p_1, \ldots, p_n, q_1, \ldots, q_n$ be distinct primes such that $q_i | p_i - 1$. Let $Z_i$ be cyclic of order $q_i$ and $E_i$ extra-special of order $p_i^3$. Consider $G = E_1 Z_1 \times \ldots \times E_n Z_n$, where $Z_i$ acts fixed-point-freely on $E_i/\mathbf{Z}(E_i)$ and trivially on $\mathbf{Z}(E_i)$. Then $\rho(G) = \{p_1, \ldots, p_n, q_1, \ldots, q_n\}$ and $\sigma(G) = n$.

The first contribution to Huppert's conjecture was due to I.M. Isaacs [Is 3], who proved an exponential bound for solvable groups, namely

$$|\rho(G)| \leq \sigma(G) \cdot 2^{\sigma(G)}.$$

D. Gluck [Gl 2] improved it to a quadratic bound, namely

$$|\rho(G)| \leq \sigma(G)^2 + 10\, \sigma(G).$$

In a recent paper [Gl 3], Gluck verified conjecture 6.1 a) provided that $\sigma(G) = 2$ or that all character degrees are squarefree. Although this sounds to be rather special, the proof is extremely technical and uses Lemma 5.1. The best general estimation known so far is the following

**6.2 Theorem (D. Gluck, O. Manz [GM 1]).** *Let G be solvable. Then*
*a)* $|\rho(G)| \leq 3\, \sigma(G) + 32$.
*b)* $|\rho(G)| \leq 2\, \sigma(G) + 32$, *provided that each normal Sylow-subgroup of G is abelian.*

It seems not so much the summand 32, which might cause problems in improving this bound. For groups of odd order for example, $|\rho(G)| \leq 3 \cdot \sigma(G)$ holds, as P. Palfy and A. Espuelas observed. The step from factor 3 to factor 2 (if possible) will certainly require entirely new ideas.

We are going to present some main ingredients of the proof of Theorem 6.2. Set $F = \mathbf{F}(G)$. By Gaschütz's Theorem, $F/\Phi(G)$ is a completely reducible and faithful $G/F$-module, and it is an enormous technical advantage to work with the group $G/\Phi(G)$ instead of $G$. The price we have to pay is that we loose exactly those $q \in \rho(G)$ which belong to a normal, non-abelian Sylow-$q$-subgroup of G. It happens exactly at this point where the factor 2 deteriorates to 3. We now assume that $\Phi(G) = 1$, pick an irreducible constituent $W$ of $F$, set $C = \mathbf{C}_G(W)$ and consider the faithful and irreducible action of $G/C =: H$ on $\mathrm{Irr}(W) =: V$.

The reduction to the case where $V$ is a quasi-primitive $H$-module is rather easy and uses Corollary 2.5. Hence we may assume that $H$ has the structure given in Theorem 1.2. We set $F = \mathbf{F}(H)$, $e^2 = |F : T|$, and distinguish between three cases.

(I) $F' = 1$ : Then Corollary 1.5 shows that $H$ is a semi-linear group.
(II) $F' \neq 1$ and $e > 131$ : Then $H$ has a regular orbit on $V$, i.e. there exists $v \in V$ such that $\mathbf{C}_H(v) = 1$.
(III) $F' \neq 1$ and $e \leq 131$ : Then there exists $v \in V$ such that all prime divisors of $|\mathbf{C}_H(v)|$ are smaller or equal to 131.

We do not give more details of the proof of Theorem 6.2. But we note that

$$|\{q \text{ prime } \mid q \leq 131\}| = 32.$$

This is the reason for the summand 32.

Working with solvable groups it is natural to ask whether the *derived length* $\mathrm{dl}(G)$ can be bounded in terms of $\sigma(G)$. This question has a negative answer. Within the class of $\{p,q\}$-groups namely there is no universal bound for $\mathrm{dl}(G)$, but clearly $\sigma(G) \le 2$. The following result might serve as a substitute and is a consequence of Theorem 4.5.

**6.3 Theorem (R. Staszewski, O. Manz [MS 1]).** *Let G be solvable. Then G has a characteristic series* $1 \le N_o \le N_1 \le N_2 \le \ldots \le N_{2k-1} \le N_{2k} = G$ *with the following properties:*
*(1)* $N_o$ *is abelian and contains the normal abelian* $\rho(G)'$*-Hall-subgroup A of G.*
*(2)* $N_{2i+1}/N_{2i}$ *is abelian* $(i = 0, \ldots, k-1)$.
*(3)* $\sigma(|N_{2i}/N_{2i-1}|) \le \sigma(G)$ $(i = 1, \ldots, k)$.
*(4)* $k \le 2 \cdot \sigma(G)$.

*Proof:* We argue by induction on $s = \sigma(G)$, and choose iterated commutator subgroups $N/A = (G/A)^{(j)}$ and $M/A = (G/A)^{(j+1)}$ such that $\sigma(N) = s$, but $\sigma(M) < s$. By induction, $M$ has a characteristic series

$$1 \le N_o \le \ldots \le N_{2\ell-1} \le N_{2\ell} = M$$

such that (1)—(3) hold and $\ell \le 2(s-1)$. We choose $\alpha \in \mathrm{Irr}(N)$ with $\sigma(\alpha(1)) = s$, and denote the set of prime divisors of $\alpha(1)$ by $\pi$. Since $\sigma(G) = s$, Clifford's Theorem implies $\chi(1)/\alpha(1)$ is a $\pi$-number for all $\chi \in \mathrm{Irr}(G|\alpha)$. Thus Theorem 4.5 yields that $G/N$ has an abelian Hall-$\pi'$-subgroup and consequently that the $\pi'$-length of $G/N$ is at most 1. The assertion of the Theorem now follows easily.

There are several problems left where the reader is invited to work on:
— Prove Huppert's conjecture 6.1 a) for solvable groups.
— Try to find bounds for $|\rho(G)|$ in terms of $\sigma(G)$ for arbitrary groups $G$.
— Find estimations for the covering-number $\mathrm{cn}(G)$.
— For a prime $p$, define $\rho_p(G) = \{q \text{ prime} |\ q|\beta(1) \text{ for some } \beta \in IBr_p(G)\}$ and $\sigma_p(G) = \max\{\sigma(\beta(1))|\beta \in IBr_p(G)\}$. To the knowledge of the author, estimations of $|\rho_p(G)|$ in terms of $\sigma_p(G)$ have not been found yet.

## REFERENCES

[AB1] D. Alvis, M. Barry, "Charachter degrees of simple groups", to appear in J. Algebra.

[BK1] T. Berger, R. Knörr, "On Brauer's height-0 conjecture", Nagoya J. Math. 109 (1988), 109-116.

[CNS1] P. Cameron, P. Neumann, J. Saxl, "On groups with no regular orbits on the set of subsets", Arch. Math. 43 (1984), 295-296.

[Fe 1] W. Feit, "The representation theory of finite groups", North Holland, Amsterdam, 1982.

[Gl 1] D. Gluck, "Trivial set-stabilizers in finite permutation groups", Can. J. Math. 35 (1983), 59-67.

[Gl 2] D. Gluck, "Primes dividing character degrees and character orbit sizes", Proc. Amer. Math. Soc. 101 (1987), 219-225.

[Gl 3] D. Gluck, "A conjecture about character degrees of solvable groups", to appear.

[GM 1] D. Gluck, O. Manz, "Prime factors of character degrees of solvable groups", Bull. London Math. Soc. 19 (1987), 431-437.

[GW 1] D. Gluck, T. Wolf, "Defect groups and character heights in blocks of solvable groups II", J. Algebra 87 (1984), 222-246.

[GW 2] D. Gluck, T. Wolf, "Brauer's height conjecture for $p$-solvable groups", Trans. Amer. Math. Soc. 282 (1984), 137-152.

[Hu 1] B. Huppert, "Endliche Gruppen I", Springer Verlag, Berlin, 1979.

[HB 1] B. Huppert, N. Blackburn, "Finite groups II,III", Springer Verlag, Berlin, 1982.

[Is 1] I.M. Isaacs, "Character theory of finite groups", Academic Press, New York, 1976.

[Is 2] I.M. Isaacs, "Coprime group actions fixing all non-linear irreducible characters", Can. J. Math. 41 (1989), 419-433.

[Is 3] I.M. Isaacs, "Solvable groups character degrees and sets of primes", J. Algebra 104 (1986), 209-219.

[Ma 1] O. Manz, "Endliche auflösbare Gruppen, deren sämtliche Charaktergrade Primzahlpotenzen sind", J. Algebra 94 (1985), 211-255.

[MS 1] O. Manz, R. Staszewski, "Some applications of a fundamental theorem by Gluck and Wolf in the character theory of finite groups", Math. Z. 192 (1986), 383-389.

[MSW1] O. Manz, R. Staszewski, W. Willems, "On the number of components of a graph related to character degrees", Proc. Amer. Math. Soc. 103 (1988), 31-37.

[MW 1] O. Manz, T. Wolf, "Brauer characters of $q'$-degree in $p$-solvable groups", J. Algebra 115 (1988), 75-91.

[MW 2] O. Manz, T. Wolf, "Representations of solvable groups", to appear in Cambridge University Press (Lecture Note Series).

[MWW1] O. Manz, W. Willems, T. Wolf, "The diameter of the character degree graph", J. reine angew. Math. 402 (1989),181-198.

[Mi 1] G. Michler, "A finite simple group of Lie-type has $p$-blocks with different defects, $p \neq 2$", J. Algebra 104 (1986), 220-230.

[Mi 2] G. Michler, "Brauer's conjectures and the classification of finite simple groups", Representation Theory II. Groups and orders, Springer Verlag, Berlin, 129-142.

[Pa 1] P.P. Pálfy, private communication.

[Wo 1] T. Wolf, "Solvable and nilpotent subgroups of $GL(n, q^m)$", Can. J. Math. 34 (1982), 1097-1111.

[Wo 2] T. Wolf, "Defect groups and character heights in blocks of solvable groups", J. Algebra 72 (1981), 183-209.

Progress in Mathematics, Vol. 95, © 1991 Birkhäuser Verlag Basel

# Some applications of representation theory

W. PLESKEN

## I. Introduction

Group representation theory has grown out of group theory. Therefore one way of measuring progress is by looking at the applications of representation theory to group theory and theories where groups play a role. In this note I discuss two areas of applications, which I have been involved in. A third area, which is left out here, is that of extensions. As pointed out in [HoP 89] representation theory not only provides abelian normal subgroups for extensions with given factor group, but also nonabelian (e.g. pro-$p$) normal subgroups. A fourth area, which I also only mention in passing, is that of applications to computational group theory, more precisely to the investigation of finite presentations of groups, cf. [Ple 87], [HoP 89] Chapter 7, [HoP 90].

The two areas of application in this note are maximal finite rational (and integral) matrix groups in Chapter II, and torsion free space groups in Chapter III. Whereas the first application is more or less motivated from within group theory the second comes from differential topology or geometry. Torsion free space groups are exactly the fundamental groups of compact connected flat Riemannian manifolds. Because the point group of a torsion free space group can geometrically be interpreted as holonomy group some nice group theoretical problems arise naturally. Chapter III is a survey of recent progress, whereas Chapter II also contains new results. Open problems are listed at the end of both chapters.

## II. Finite integral and rational matrix groups

### a) Integral matrix groups

The original Jordan version [Jor 80] of the Jordan-Zassenhaus Theorem [Zas 38] states that $GL_n(\mathbb{Z})$ has only finitely many conjugacy classes of finite subgroups. However, it seems that this number increases rapidly with $n$, e.g. 2, 13, 73, 710 for $n = 1, 2, 3, 4$, cf. [BBNWZ 78]. Therefore one might want to restrict attention to special classes of subgroups, e.g. the Bravais groups which can be defined as full automorphism groups of lattices with a positive definite form. But again the increase is quite big; there are 1,5,14,64,189,826 conjugacy classes of Bravais groups in dimensions 1,2,3,4,5,6, cf. [BBNWZ 78], [Ple 81], [PlH 84]. One might restrict oneself further to maximal finite subgroups of $GL_n(\mathbb{Z})$, where the number of classes are 1,2,4,9,17 for $n = 1, 2, 3, 4, 5$, cf. [Dad 65] for $n = 4$, [Rys 72] and [Bül 73] for $n = 5$. The methods in the last three papers were geometric, and it seems that they get harder to apply as the dimension increases, cf. [RyL 80] where the proof for the statements in [Rys 72] are given. Finally representation theoretic methods yield 7*,17,7*,26,20,9*,17,24,9*,28 conjugacy classes of absolutely irreducible maximal finite subgroups of $GL_n(\mathbb{Z})$ for $n = 5^*, 6, 7^*, 8, 9, 11^*, 13, 17, 19^*, 23$, where a $*$ indicates that all groups are essentially reflection groups, more precisely full automorphism groups of root systems, cf. [PlP 77-80] and [Ple 85]. For a geometric discussion of the lattices arising cf. [CoS 88].

## b) The complex of maximal finite rational matrix groups

For composite dimensions such as 6 and 8 the classifications of irreducible maximal finite subgroups of $GL_n(\mathbb{Z})$ is already rather involved. Therefore I suggest a coarser classification which might still be manageable in higher dimensions and yield some insight at the same time.

**(II.1) Definition** *The simplicial complex $M_n(\mathbb{Q})$ of maximal finite subgroups of $GL_n(\mathbb{Q})$ has the $GL_n(\mathbb{Q})$-conjugacy classes of maximal finite subgroups of $GL_n(\mathbb{Q})$ as vertices. The $s+1$ vertices $P_0, \ldots, P_s$ represented by $G_0, \ldots, G_s \leq GL_n(\mathbb{Q})$ form an $s$-simplex, if there exists $H \leq GL_n(\mathbb{Q})$ which is conjugate to subgroups of $G_0, \ldots, G_s$ and $\dim_{\mathbb{Q}} C_{\mathbb{Q}^{n\times n}}(H) = \dim_{\mathbb{Q}} C_{\mathbb{Q}^{n\times n}}(G_i)$ for $i = 0, \ldots, s$, where $C_{\mathbb{Q}^{n\times n}}(X)$ denotes the algebra of rational matrices commuting the elements of $X$ for $X \leq GL_n(\mathbb{Q})$.*

It should be noted that one might also consider a similarly defined simplicial complex $M_n^F(\mathbb{Q})$ where the commuting algebras $C_{\mathbb{Q}^{n\times n}}(X)$ are replaced by the space of invariant forms

$$F(X) = \{F \in \mathbb{Q}^{n\times n} \mid F^{tr} = F,\ gFg^{tr} = F \text{ for all } g \in X\}.$$

Clearly a simplex in $M_n(\mathbb{Q})$ is also a simplex in $M_n^F(\mathbb{Q})$, but the converse is not necessarily the case.

Immediately from the definition a semisimple subalgebra $C$ of $\mathbb{Q}^{n\times n}$ unique up to inner automorphisms of $\mathbb{Q}^{n\times n}$ is attached to every connected component of $M_n(\mathbb{Q})$, e.g. $\mathbb{Q}I_n$ if the groups represented by the vertices of the component are absolutely irreducible. In fact the representatives of the groups can be chosen in such a way that they all lie in the commuting algebra of $C$ in $\mathbb{Q}^{n\times n}$.

To distinguish maximal finiteness as a subgroup of $GL_n(\mathbb{Z})$ and as a subgroup of $GL_n(\mathbb{Q})$, define for each finite $G \leq GL_n(\mathbb{Q})$ the set $\mathcal{Z}(G) = \{L \mid L \text{ full } \mathbb{Z}G\text{-sublattice of } V\}$ where $V = \mathbb{Q}^{n\times 1}$ is the natural $\mathbb{Q}G$-module, and $\mathcal{F}_{>0}(G) = \{F \in \mathcal{F}(G) \mid F \text{ positive definite}\}$. Clearly $N_{GL_n(\mathbb{Q})}(G)$ operates on $\mathcal{Z}(G)$; the finitely many orbits are in bijection to the $GL_n(\mathbb{Z})$-conjugacy classes within the $\mathbb{Q}$-class of $G$, which is the set of subgroups of $GL_n(\mathbb{Z})$ conjugate to $G$ under $GL_n(\mathbb{Q})$. For $L \in \mathcal{Z}(G)$ and $F \in \mathcal{F}_{>0}(G)$ let $Aut(L, F) = \{g \in GL_n(\mathbb{Q}) \mid gL = L, g^{tr}Fg = F\}$. One clearly has the following.

**(II.2) Remark** *Let $G \leq GL_n(\mathbb{Q})$ be finite.*

*(i) The action of $G$ on $L \in \mathcal{Z}(G)$ gives rise to a maximal finite subgroup of $GL_n(\mathbb{Z})$ if and only if $G = Aut(L, F)$ for all $F \in F_{>0}(G)$.*

*(ii) $G$ is a maximal finite subgroup of $GL_n(\mathbb{Q})$ if and only if $G = Aut(L, F)$ for all $(L, F) \in \mathcal{Z}(G) \times F_{>0}(G)$.*

### c) Reducible and imprimitive maximal finite subgroups of $GL_n(\mathbb{Q})$

Reducible maximal finite subgroups of $GL_n(\mathbb{Z})$ can be quite difficult to handle, e.g. their constituent groups – no matter how they are defined – need not be maximal finite in $GL_m(\mathbb{Z})$ for their degrees $m < n$, cf. [PlH 84]. These difficulties disappear when one takes maximal finite subgroups of $GL_n(\mathbb{Q})$ rather than of $GL_n(\mathbb{Z})$.

**(II.3) Definition** *For $G_1 \leq GL_{n_1}(\mathbb{Q}), \ldots, G_k \leq GL_{n_k}(\mathbb{Q})$ let $Diag(G_1, \ldots, G_k) = \{diag(g_1, \ldots, g_k) \mid g_i \in G_i \text{ for } i = 1, \ldots, k\} \leq GL_n(\mathbb{Q})$ with $n = n_1 + \cdots + n_k$.*

Clearly $Diag(G_1, \ldots, G_k) \cong G_1 \times \cdots \times G_k$ and one obviously has the following.

**(II.4) Remark** *(i) Let $G \leq GL_n(\mathbb{Q})$ be a reducible maximal finite subgroup of $GL_n(\mathbb{Q})$. Then there are $k, n_1, \ldots, n_k \in \mathbb{N}$ with $k > 1$ and $n_1 + \cdots + n_k = n$ such that $G$ is conjugate to $Diag(G_1, \ldots, G_k)$ where $G_i$ are maximal finite irreducible subgroups of $GL_{n_i}(\mathbb{Q})$ for $i = 1, \ldots, k$.*
*(ii) Let $G_0, \ldots, G_s \leq GL_n(\mathbb{Q})$ be maximal finite representing the vertices of an $s$-simplex of $M_n(\mathbb{Q})$. If one $G_i$ is reducible, each $G_i$ is conjugate to $Diag(G_{i1}, \ldots, G_{ik})$ with $G_{ij} \leq GL_{n_j}(\mathbb{Q})$ maximal finite and irreducible and for each $j = 1, \ldots, k$ the groups $G_{0j}, \ldots, G_{sj} \leq GL_{n_j}(\mathbb{Q})$ represent the vertices of an $s(j)$-simplex in $M_{n_j}(\mathbb{Q})$ for some $s(j) \leq s$.*

For $G_i \leq GL_{n_i}(\mathbb{Q})$ maximal finite for $i = 1, \ldots, k$, $Diag(G_1, \ldots, G_k)$ need not always be maximal finite in $GL_n(\mathbb{Q})$ for $n = n_1 + \cdots + n_k$.

**(II.5) Definition** *(i) $H \leq GL_m(\mathbb{Q})$ and $k \in \mathbb{N}$. The wreath product $H \wr S_k$ of $H$ with the symmetric group $S_k$ is defined as*
*$H \wr S_k := \langle diag(h_1, \ldots, h_k), P \otimes I_m \mid h_i \in H, P$ a $k \times k$-permutation matrix $\rangle$*
*$\leq GL_{km}(\mathbb{Q})$, where $I_m$ denotes the $m \times m$-matrix.*
*(ii) An irreducible subgroup $G$ of $GL_n(\mathbb{Q})$ is called imprimitive if it is conjugate to a subgroup of $H \wr S_k$ for some $H \leq GL_m(\mathbb{Q})$ with $k \neq 1$ and $mk = n$.*
*(iii) Two irreducible finite subgroups $G_i \leq GL_{n_i}(\mathbb{Q})$ for $i = 1, 2$ are called primitively related if there exists an irreducible finite $H \leq GL_m(\mathbb{Q})$ such that $G_i$ is conjugate to a*

*subgroup of* $H \wr S_{k_i}$ *where* $k_i m = n_i$ *for* $i = 1, 2$.

Clearly if $G_i \leq GL_m(\mathbb{Q})$ are maximal finite for $i = 1, \ldots, k > 1$ and all are conjugate, then $Diag(G_1, \ldots, G_k)$ is conjugate to the base group of $G_1 \wr S_k$ (and hence not maximal finite in $GL_{km}(\mathbb{Q})$.)

Note, that primitive maximal finite subgroups of $GL_{n_i}(\mathbb{Q})$ which are primitively related are necessarily of the same degree and conjugate. The desired criterion now is as follows.

**(II.6) Proposition** *Let* $G_i$ *be maximal finite irreducible subgroups of* $GL_{n_i}(\mathbb{Q})$ *for* $i = 1, \ldots, k > 1$.
*(i) If* $Diag(G_1, \ldots, G_k)$ *is maximal finite in* $GL_n(\mathbb{Q})$ *for* $n = n_1 + \cdots + n_k$*, then no two of the* $G_i$ *are primitively related.*
*(ii) Conversely assume that at most one of the* $G_i$ *has the trivial Brauer character* 1 *for* $p = 2$ *in the restriction of its natural character to* 2*-regular classes. Then, if no two of the* $G_i$ *are primitively related,* $Diag(G_1, \ldots, G_k)$ *is maximal finite in* $GL_n(\mathbb{Q})$ *for* $n = n_1 + \cdots + n_k$.

Proof. (i) Say $G_1$ and $G_2$ be primitively related. Then $G_i$ is conjugate to $H \wr S_{k_i}$ for $i = 1, 2$ and $Diag(G_1, G_2)$ is conjugate to a proper subgroup of $H \wr S_{k_1+k_2}$.
(ii) Let $L_i \in \mathcal{Z}(G_i)$ and $G = Diag(G_1, \ldots, G_k)$ and view $L_i$ as $\mathbb{Z}G$-lattice. Since $-I_{n_i} \in G_i$ and because of the hypothesis about Brauer characters for $p = 2$, no two of the $L_i/pL_i$ have common $\mathsf{F}_pG$-constituents for any prime $p$. Therefore, cf. [Ple 78],

$$\mathcal{Z}(G) = \{\oplus_{i=1}^{k} M_i \mid M_i \in \mathcal{Z}(G_i) \text{ for } i = 1, \ldots, k\}.$$

Also

$$F(G) = \{diag(F_1, \ldots, F_k) \mid F_i \in F(G_i) \text{ for } i = 1, \ldots, k\}.$$

Now the claim follows from (II.2) and Eichler's unique decomposition theorem for lattices in Euclidean space, cf. [Kne 54], also [Dad 64].

The example $G_1 = \langle -1 \rangle$ and $G_2 = W(E_7) \leq GL_7(\mathbb{Z})$ (Weyl group of root system $E_7$) shows that the technically looking assumption about the Brauer characters are necessary in (II.6)(ii), since $Diag(G_1, G_2)$ is conjugate to $W(E_8)$ in this example. (However I do not know of examples not involving $\langle -1 \rangle$ and $W(E_7)$.) In the same spirit as (II.6) is the following.

**(II.7) Proposition** *Let* $H \leq GL_m(\mathbb{Q})$ *be irreducible maximal finite and primitive. If the trivial Brauer character* 1 *for* $p = 2$ *does not occur as a constituent of the natural character of* $H$ *restricted to* 2*-regular classes, then* $H \wr S_k \leq GL_{mk}(\mathbb{Q})$ *is maximal finite.*

Proof. Similarly as in the proof of (II.6)(ii) one has $\mathcal{Z}(H \wr S_k) = \{\oplus_{i=1}^{k} L \mid L \in \mathcal{Z}(H)\}$ and $F(H \wr S_k) = \{I_k \otimes F \mid F \in F(H)\}$. The claim follows in the same way.

$\langle -1 \rangle \wr S_4$, which is contained in the Weyl group for the root system $F_4$, shows that the hypothesis on the 2-Brauer characters is needed. Again it is the only example I know.

## d) Some irreducible groups

Already Burnside, cf. [Bur 12], dealt with the question which finite subgroups of $GL_n(\mathbf{Q})$ contained all permutation matrices, cf. also [BaB 73], [Ple 85]. Some of these groups give examples of irreducible maximal finite subgroups of $GL_n(\mathbf{Q})$ for every $n$, namely the following, where $W(X)$ and $Aut(X)$ denote the Weyl- resp. automorphism group of a root system $X \in \{A_n, B_n, F_4, E_6, E_7, E_8\}$.

**(II.8) Proposition** *(i) For $n \neq 4$, $\langle -1 \rangle \wr S_n = W(B_n) = Aut(B_n)$ is a maximal finite subgroup of $GL_n(\mathbf{Q})$. For $n = 4$, $W(B_n)$ is contained in $W(F_4)$, which is maximal finite in $GL_4(\mathbf{Q})$.*
*(ii) For $n \neq 7, 8$, $Aut(A_n) = \langle -I_n \rangle \times W(A_n) (\cong C_2 \times S_{n+1})$ is a maximal finite subgroup of $GL_n(\mathbf{Q})$. For $n = 7, 8$, $Aut(A_n)$ is contained in $Aut(E_n) = W(E_n)$, which is maximal finite in $GL_n(\mathbf{Q})$.*

Two questions arise: When are $Aut(B_n)$ and $Aut(A_n)$ in the same connected component of $M_n(\mathbf{Q})$ and, in case they are, when do they form 2-simplex in $M_n(\mathbf{Q})$.

**(II.9) Remark** *Let $G_1, G_2 \leq GL_n(\mathbf{Q})$ be maximal finite. If they represent vertices of the same connected component of $M_n(\mathbf{Q})$, then there exists an $X \in GL_n(\mathbf{Q})$ with $X^{tr} F(G_1) X = F(G_2)$, i.e. the forms in $F(G_1)$ are rationally equivalent to those in $F(G_2)$.*

To apply this rather obvious remark, we note that $F(\langle -1 \rangle \wr S_n)) = \mathbf{Q} I_n$ and for suitable choice of basis $F(Aut(A_n)) = \mathbf{Q}(I_n + J_n)$ where $J_n \in \mathbf{Q}^{n \times n}$ has all entries equal to 1.

**(II.10) Lemma** *$I_n + J_n$ is rationally equivalent to a multiple of $I_n$ in exactly the following cases:*

*(i) $n \equiv 0 \pmod 4$ and $n + 1$ is a square.*

*(ii) $n \equiv 3 \pmod 4$.*

*(iii)* $n \equiv 1 \pmod 4$ *and if* $p^\alpha$ *is the biggest power of a prime* $p \equiv 3 \pmod 4$ *dividing* $n+1$, *then* $\alpha$ *is even.*

Proof. According to the description of the Witt group $W(\mathbb{Q})$ given in [Sch 85], we only have to compare $\partial_p(I_n + J_n)$ with $\partial_p(\alpha I_n)$ for all primes $p$. Now for finite $p$ one has

$$0 = \partial_p(I_{n+1}) = \partial_p((I_n + J_n) \oplus (n+1)) = \partial_p(I_n + J_n) + \partial_p((n+1)),$$

because $I_n + J_n$ can be obtained as the Gram matrix of $\langle e_0 + \cdots + e_n\rangle^\perp$ with respect to the basis $(e_0 - e_2, \ldots, e_0 - e_n)$, where $(e_0, \ldots, e_n)$ is an orthonormal basis of $\mathbb{Q}^{n+1}$. Hence $\partial_p(I_n + J_n) = -\partial_p((n+1))$ for all primes $p$. Since the torsion part of $W(\mathbb{Q})$ has exponent 4, $\partial_p(\alpha I_n)$ is 0 for $n \equiv 0 \pmod 4$. Hence for $n \equiv 0 \pmod 4$ $\alpha I_n \sim_{\mathbb{Q}} I_n$ for all $\alpha \in \mathbb{Q}^x$ and $\alpha I_n \sim_{\mathbb{Q}} I_n + J_n$ iff $n+1$ is a square. If $n \equiv 2 \pmod 4$, $\partial_p(\alpha I_n) = \partial_p(\alpha I_2) = \partial_p(\alpha) + \partial_p(\alpha)$ which can never be $-\partial_p(n+1)$ (since $n+1$ can also not be a square). Hence $\alpha I_n \not\sim_{\mathbb{Q}} I_n + J_n$ in this case. For $n \equiv 3 \pmod 4$ one has $(n+1)I_n \sim_{\mathbb{Q}} I_n + J_n$; the case $n \equiv 1 \pmod 4$ can be handled similarly.

For instance the case $n = 9$ shows, that the converse of (II.9) does not hold: $10I_9 \sim_{\mathbb{Q}} I_9 + J_9$, but from [PlP 77,80] Part III it is clear that $\langle -1\rangle \wr S_9$ and $Aut(A_9)$ do not lie in the same component of $M_9(\mathbb{Q})$.

**(II.11) Corollary** *If* $n$ $(\neq 4)$ *is even and* $n+1$ *is not a square or if* $n \equiv 1 \pmod 4$ *and* $n+1$ *has a prime divisor* $p \equiv 3 \pmod 4$ *to an odd power in its primary decomposition, then* $Aut(B_n)$ *and* $Aut(A_n)$ *represent vertices in different connected components of* $M_n(\mathbb{Q})$.

This answer is far from being complete. As for the second problem, the classification of finite simple groups helps. The definitions yield immediately the following lemma.

**(II.12) Lemma** *Let* $n \neq 3, 4, 7, 8$. $Aut(A_n)$ *and* $Aut(B_n)$ *represent vertices of the same 1-simplex if and only if there exists a 2-transitive permutation group* $G$ *on* $n+1$ *letters with permutation character* $1 + \mu$, *where* $\mu$ *is induced up from a rational linear character of a subgroup of index* $n$.

**(II.13) Theorem** *Let* $n \neq 3, 4, 7, 8$. $Aut(A_n)(\cong C_2 \times S_{n+1})$ *and* $Aut(B_n) = \langle -1\rangle \wr S_n$ *represent the vertices of a 1-simplex of* $M_n(\mathbb{Q})$ *if and only if* $n = 2^\alpha - 1$ *for* $\alpha \geq 4$ *or* $n = 11$.

Proof. For instance the 1-dimensional affine group $Aff_1(2^\alpha) = \mathbb{F}_{2^\alpha} \rtimes \mathbb{F}_{2^\alpha}^*$ of the field $\mathbb{F}_{2^\alpha}$ on $n+1 = 2^\alpha$ points and the Mathieu group $M_{11}$ on $n+1 = 12$ points yields the desired 2-transitive groups according to (II.12). We have to show that no other degrees turn up. Let $G$ be 2-transitive on $n+1$ letters as in (II.12).

(a) $G$ has no fixpoint free element $g$ of odd order. Because otherwise $\mu(g) = -1$, which is impossible for an element of odd order in a (rational) monomial representation ($g \mapsto -1$ yields a homomorphism $\langle g \rangle \longrightarrow \langle -1 \rangle$).
(b) $\mu$ is in the principal 2-block. Because the restriction of $\mu$ to 2-regular classes hat the 2-Brauer character 1 as constituent, with $\mu$ being monomial.
(c) $n$ satifies the restrictions of (II.11).

Now (a) implies immediately that a regular normal abelian subgroup of $G$ must be of order $2^\alpha$. Hence by [Cam 81] Proposition 5.3 one may assume that $G$ has a unique minimal simple normal subgroup. Going through the list in [Cam 81] (a),(b),(c) or if neither of the three works simply the absence of subgroups of index $n$ in those groups rules out all possibilities. E.g. (b) rules out the Suzuki groups and $PSU(3, 2^\alpha)$, (a) works usually, but breaks down for instance for $PSL_2(31)$, where one of course knows that there are no subgroups of index $n = 31$.

From (II.6) and [Ple 85] Lemma III.2 it follows that $Aut(A_n) \wr S_m$ is a maximal finite subgroup of $GL_{nm}(\mathbb{Q})$ for $n$ even and $n \neq 8$. $Aut(A_8) \wr S_m$ is conjugate to a subgroup of the maximal finite subgroup $W(E_8) \wr S_m$ of $GL_{8m}(\mathbb{Q})$. If $m$ is a multiple of 4, the forms $F(Aut(A_n) \wr S_m)$ $(= \langle I_m \otimes (I_n + J_n) \rangle_{\mathbb{Q}})$ are rationally equivalent to the ones in $F(\langle -1 \rangle \wr S_{nm})$ $(= \mathbb{Q} I_{nm})$. The fact that $\langle -1 \rangle \wr S_4$ is not maximal finite in $GL_4(\mathbb{Q})$ enriches $M_n(\mathbb{Q})$ for $4|n$. Those components of $M_n(\mathbb{Q})$ whose vertices represent reducible groups in $GL_n(\mathbb{Q})$, for short the reducible components of $M_n(\mathbb{Q})$, are more or less dealt with by lower dimensional irreducible groups via (II.6). We now only concentrate on $M_n^{irr}(\mathbb{Q})$, which is the full subcomplex of $M_n(\mathbb{Q})$ containing only the irreducible components of $M_n(\mathbb{Q})$, i.e. the vertices representing irreducible maximal finite subgroups of $GL_n(\mathbb{Q})$. As a consequence of [Dad 65], [PlP 77,80] we get the description for $M_n^{irr}(\mathbb{Q})$ for some small $n$ in (II.14) below. Note that every irreducible maximal finite subgroup of $GL_n(\mathbb{Q})$ (even $GL_n(\mathbb{Z})$) for the dimensions $n$ treated below is already absolutely irreducible. For $n \neq 6, 8, 10$ this is clear from the quoted papers, for $n = 6$ it is proved in [Ple 81], for $n = 8$, which is a rather involved case, in [Sou 91]. The maximal finite irreducible subgroups of $GL_{10}(\mathbb{Q})$ will be derived in this paper.

**(II.14) Theorem** *$M_n^{irr}(\mathbb{Q})$ is given as follows.*

*n=2:* $A_2\bullet$ $\quad\bullet B_2$

*n=3:* $v$ $\quad\bullet A_3(B_3)$

*n=4:* $A_4\bullet$ $\quad\bullet 2A_2$ $\quad\bullet F_4$

*n=5:* $A_5\bullet$ $\quad\bullet B_5$

*n=6:* $A_6\bullet$ $\quad\bullet E_6$ $\quad\bullet B_6$ $\quad\bullet G_2(6) \cong C_2 \times S_5$

$\qquad|$ $\qquad\qquad|$

$\bullet$ $\qquad\qquad\bullet 3A_2$

$G_1(6)$
$\cong C_2 \times PGL_2(7)$

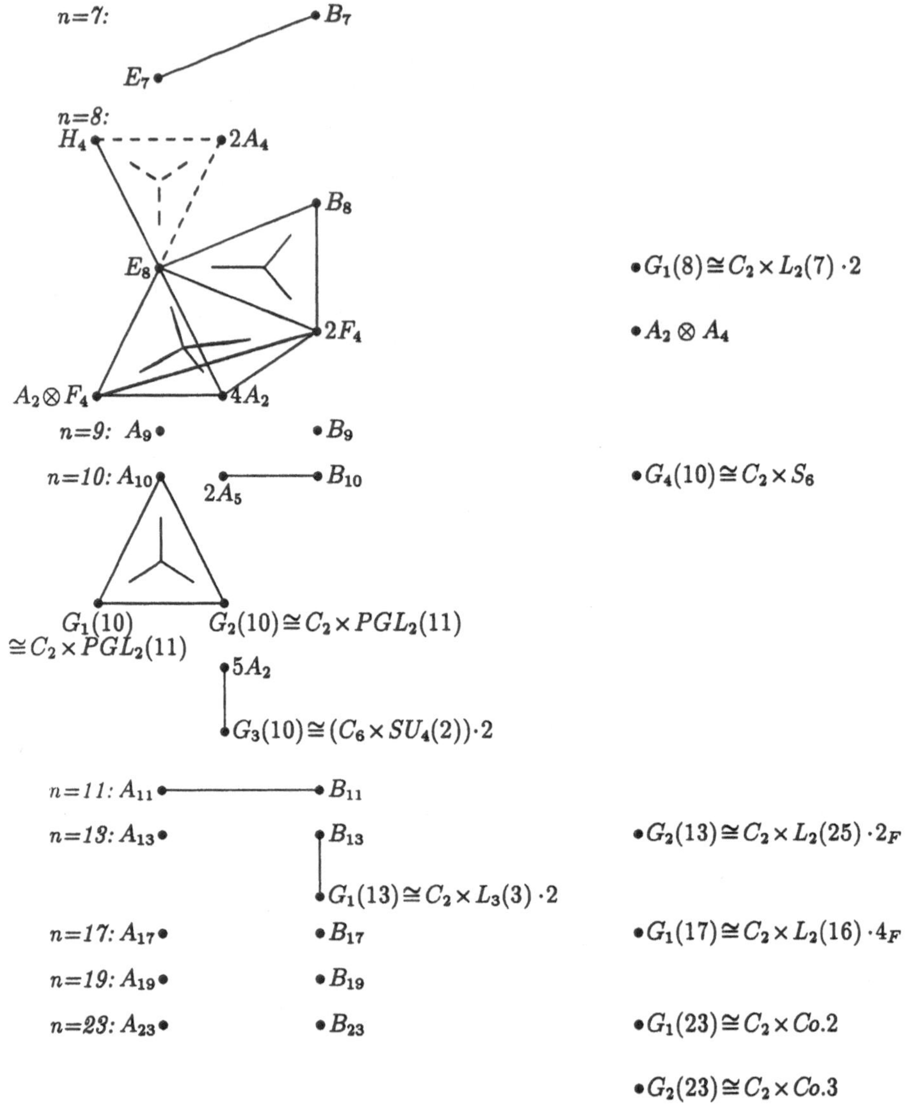


*where the vertices are marked by the names of the root systems X (or in two cases tensor products of root lattices), whenever the corresponding vertex represents groups conjugate to Aut(X). For $n = 8$, $H_4$ denotes a class of groups isomorphic to an extension of the central product of $SL_2(5)$ by itself by a group of order 4. Simplices belonging to $M_n^F(\mathbb{Q})$ rather than to $M_n(\mathbb{Q})$ are indicated by broken lines. The rest of the notation is selfexplanatory, e.g. $L_2(16)\cdot 4_F$ means $PSL_2(16)$ extended by a field automorphism group of order 4.*

A comment might be appropriate for dimension 8. A subgroup yielding the 3-dimensional simplex is isomorphic to $D_6 \times GL_2(3)$. The 2-simplex in $M_8^F(\mathbb{Q})$ with vertices $H_4$, $2A_4$, and $E_8$ comes from a group isomorphic to $C_4 \times (\mathbf{F}_5 \rtimes \mathbf{F}_5^*)$, the 1-simplex $H_4, E_8$ in $M_8(\mathbb{Q})$ from a non-split extension of this group by $C_2$, i.e. a non-split extension of $C_{20}$ by $Aut(C_{20})$. Note, in [PlP 77,80] part V the automorphism group of the form $F_{15}$ there, which corresponds to our vertex $H_4$, is given incorrectly: only a subgroup of index 2 is described.

## e) Irreducible, not absolutely irreducible maximal finite matrix groups

Because of (II.14) one might wonder whether there are maximal finite irreducible but not absolutely irreducible subgroups of $GL_n(\mathbf{Q})$. In fact H. Zassenhaus asked me this question for $GL_n(\mathbf{Z})$ already long before (II.14) was known some 15 years ago. Using the classification of the finite simple groups one can actually construct examples, even for $GL_n(\mathbf{Q})$. The smallest such $n$ satisfies $n \leq 36$, It might be so that the smallest $n$ for the $GL_n(\mathbf{Z})$-case is smaller than the one for $GL_n(\mathbf{Q})$.

**(II.15) Theorem** *$GL_{36}(\mathbf{Q})$ has an irreducible but not absolutely irreducible subgroup $G \cong C_2 \times PGL_2(13)$.*

Proof. From the character table of $PGL_2(13)$ and the computation of the local Schur indices of the character it is clear that an irreducible representation $\Delta : PGL_2(13) \hookrightarrow GL_{36}(\mathbf{Q})$ exists with character $\chi_\Delta = \chi_{12_1} + \chi_{12_2} + \chi_{12_3}$ where $\chi_{12_i}$ are absolutely irreducible real characters of $PGL_2(13)$ taking values in $\mathbf{Q}[\zeta_7 + \zeta_7^{-1}]$, where $\zeta_7$ is a primitive 7-th root of 1. Let $G = \langle -I_{36}, \Delta(PGL_2(13)\rangle$. Claim: $G$ is maximal finite. Assume $M$ with $G < M \leq GL_{36}(\mathbf{Q})$ is a finite subgroup of $GL_{36}(\mathbf{Q})$ with $G$ maximal in $M$. We have to find a contradiction. Clearly for sufficiently big $n$ the $n$-th derived subgroup $H := M^{(n)}$ of $M$ is perfect and contains $G' \cong PSL_2(13)$. Since $\Delta_{|PSL_2(13)}$ is primitive as a rational representation, $H$ contains at most $Z(H)$ as abelian normal subgroup or any reducible normal subgroup for that matter, as one sees easily from Clifford theory. Since $Out(PSL_2(13)) \cong C_2$ and $PSL_2(13)$ cannot be the outer automorphism group of a simple or nonabelian charateristically simple group, one sees that $M$ is simple or quasisimple with $Z(M) \cong C_2$, the latter because $C_{\mathbf{Q}^{36\times36}}(G') \cong \mathbf{Q}[\zeta_7+\zeta_7^{-1}]$ has only $-1$ as roots of unity. A quick inspection of character tables shows that $H/Z(H)$ cannot be a sporadic simple group or an alternating group. Finite groups of Lie type are defined as linear groups acting on finite vector spaces $\mathbf{F}_{p^\alpha}^n$. $SL_2(13)$ (rather than $PSL(13)$ since we have to allow for the Schur multiplier of $PSL_2(13)$) acts faithfully irreducibly on the following vector spaces $(\mathbf{F}_{p^\beta})^m$ as one sees by computing the decomposition numbers and looking at the values of Brauer characters mod $p$:

$$
\begin{array}{ll}
(m,\beta) \in \{(6,2),(12,3)\} & \text{for } p=2 \\
(m,\beta) \in \{(7,1),(13,1),(14,1)\} & \text{for } p=3 \\
(m,\beta) \in \{(6,2),(12,1),(14,1),(14,2)\} & \text{for } p=7 \\
(m,\beta) \in \{(d,1) \mid d=2,3,\ldots,13\} & \text{for } p=13
\end{array}
$$

and the $m$'s for $p \nmid |PSL_2(13)|$ are the degrees of the irreducible Frobenius characters. All that remains to check is that the simple groups of Lie type defined by an action on $\mathbf{F}_{p^\alpha}^n$ or their 2-fold covers do not have an irreducible rational character of degree

36 (or $\mathbb{Q}[\zeta_7 + \zeta_7^{-1}]$-valued character of degree 12) in $(n, 2)$ satisfies $n \geq m$ and $\alpha \geq \beta$ for some $(n, \beta)$ given above. E.g. for $p = 3$ the pair $(7, 1)$ yields $G_2(3)$ as a possible isomorphism type for $M$, but the character table of $G_2(3)$ does not have a character of degree 36. In most cases such as $Sp_6(4)$ for $p = 2$, $PSL_2(13^2)$ for $p = 13$ etc. that need to be checked the table in [LaS 74] is sufficient. Sometimes one has to look at specific character tables. q.e.d.

## f) The maximal finite irreducible subgroups of $GL_{10}(\mathbb{Q})$.

It will become clear how much easier it is to find the conjugacy classes of irreducible maximal finite subgroups of $GL_n(\mathbb{Q})$ than the ones of $GL_n(\mathbb{Z})$, mainly because other classification results are more directly applicable.

**(II.16) Lemma** *Let $G \leq GL_{10}(\mathbb{Q})$ be finite irreducible, but not absolutely irreducible. Then the $\mathbb{Q}$-algebra $\overline{\mathbb{Q}G}$ spanned by the matrices of $G$ is one of the following:*
*(i) $\overline{\mathbb{Q}G} \cong \mathbb{Q}[\zeta_{11}]$ ($\zeta_{11}$ primitive 11-th root of 1) and $G \cong C_{11}$ or $G \cong C_{22}$.*
*(ii) $\overline{\mathbb{Q}G} \cong \mathbb{Q}[\zeta_{11} + \zeta_{11}^{-1}]^{2\times 2}$ and $G \cong D_{22}$ or $G \cong D_{44}$.*
*(iii) $\overline{\mathbb{Q}G} \cong K^{5\times 5}$ where $K : \mathbb{Q} = 2$. In this case $K = \mathbb{Q}[\sqrt{-11}]$ and $G$ is isomorphic to $PSL_2(11)$, $PSL_2(11) \times C_2$ or $C_n \rtimes C_5$ with $n = 11$ or $22$;*
*or $K = \mathbb{Q}[\zeta_3]$ and $G$ is isomorphic to a subgroup of $C_6 \times SU_4(2)$ or of $C_6 \wr S_5$; or $K = \mathbb{Q}[\zeta_4]$ and $G$ is isomorphic to a subgroup of $C_4 \times S_6$ or of $C_4 \wr S_5$.*

Proof. $\overline{\mathbb{Q}G} = D^{k\times k}$ where $D$ is a division algebra with centre $K$ and index $i$. Then since $\mathbb{Q}^{10\times 1}$ is isomorphic to the unique irreducible module $D^{k\times 1}$, one has $10 = dim_{\mathbb{Q}}K \cdot i^2 \cdot k$. Hence $i = 1$ and $K = D$.
(i) $dim_{\mathbb{Q}}K = 10$, $k = 1$. In this case $\mathbb{Q}G$ is a field and $G$ is cyclic of order $n$ with $\varphi(n) = 10$, where $\varphi$ denotes the Euler $\varphi$-function. Hence $n = 11$ or $22$.
(ii) $dim_{\mathbb{Q}}K = 5$ and $k = 2$. In this case $G$ must contain elements of order $x_1, \ldots, x_l$ such that $K \subseteq \mathbb{Q}[\zeta_{x_1}, \ldots, \zeta_{x_l}]$. Again an $x_i \in \{11, 22\}$ must show up and $K = \mathbb{Q}[\zeta_{11} + \zeta_{11}^{-1}]$. In particular $G$ has an element $g_{11}$ of order 11 which is conjugate to its inverse by an element $h$. Since $h^2$ commutes with $g_{11}$ it can only have order 1 or 2. Looking at the 11-adic Schur indices leaves $|\, h \,| = 2$ as only possibility. Either directly or by appealing to [Bli 17] one gets now $G = \langle g_{11}, h\rangle$ or $G = \langle -g_{11}, h\rangle$.
(iii) $dim_{\mathbb{Q}}K = 2$ and $k = 5$. This follows from the classification of finite primitive subgroups of $GL_5(\mathbb{C})$ in [Bra 67] and the list of maximal subgroups of $SU_4(2)$ in [CC-NPW 85]. q.e.d.

**(II.17) Corollary** *No irreducible but not absolutely irreducible subgroup of $GL_{10}(\mathbb{Q})$ is maximal finite.*

Proof. The groups in (II.16) are contained in groups $G \leq GL_{10}(\mathbb{Q})$ with $G \cong C_{22} \rtimes C_{10}$ ((i) and ((ii)), $G \cong C_2 \times PGL_2(11)$, $G \cong (C_6 \times SU_4(2)) \cdot 2$, $G \cong C_2 \wr S_{10}$, $G \cong D_{12} \wr S_5$, $(C_2 \times S_6) \wr S_2$. q.e.d.

**(II.18) Lemma** *Let $G \leq GL_n(\mathbb{Q})$ be irreducible finite. If $G$ has an abelian noncyclic normal subgroup $N$, then $G$ is imprimitive.*

Proof. Abelian noncyclic groups have no faithful irreducible representations. Hence by Clifford theory, the restriction of the natural representation of $G$ to $N$ must have more than one isotypic component. q.e.d.

**(II.19) Lemma** *Let $G \leq GL_{10}(\mathbb{Q})$ be irreducible, maximal finite, and primitive. Then any soluble normal subgroup of $G$ is cyclic of order 2 or 6. In the last case $G \cong (C_6 \times SU_4(2)) \cdot 2$, where the outer automorphism acts nontrivially on $C_6$ and on $SU_4(2)$.*

Proof. Let $N \trianglelefteq G$ minimal with $\langle -I_{10}\rangle < N$ $\langle -I_{10}\rangle \neq N$. Assume $N$ is soluble but not cyclic. By (II.18) $N$ is not abelian. $N/\langle -I_{10}\rangle$ is therefore an elementary abelian 2-group and $N' = \langle -I_{10}\rangle$. By Clifford this leaves $N \cong D_8$. But this contradicts the minimality of $N$, since $D_8$ has a characteristic subgroup of order 4. Hence $N$ must be cyclic. Similarly as in (II.16) one gets $N \cong C_4$ or $N \cong C_6$. In both cases the centralizer of $N$ in $G$, which is of index at most 2 in $G$, is not absolutely irreducible and (II.17) implies $G : C_G(N) = 2$. Going through the candidates for $C_G(N)$ leaves the above mentioned possibility for $G$. q.e.d.

**(II.20) Theorem** *Let $G \leq GL_{10}(\mathbb{Q})$ be irreducible and maximal finite.*
*(i) If $G$ is imprimitive; then $G$ is conjugate to $Aut(B_{10})$, $Aut(5A_2)$ or $Aut(2A_5)$.*
*(ii) If $G$ is primitive and $11 \mid |G|$, then $G$ contains an irreducible subgroup isomorphic to $C_{11} \rtimes C_{10} (\cong Aff_1(\mathbb{F}_{11}))$ and up to conjugacy there are three possibilities for $G$; namely $Aut(A_{10}) \cong C_2 \times S_{11}$, $G_1(10) \cong G_2(10) \cong C_2 \times PGL_2(11)$.*
*(iii) If $G$ is primitive and $11 \nmid |G|$, then one has up to conjugacy two possibilities, namely $G_3(10) \cong (C_6 \times SU_4(2)) \cdot 2$ and $G_4(10) \cong C_2 \times S_6$.*

Proof. (i) is clear from the lower dimensional classifications and (II.7).
Now let $G$ be primitive.
(ii) If $11 \mid |G|$, (II.16) and (II.17) imply that $G$ contains a subgroup $H \cong C_{11} \rtimes C_5$. Clearly $\mathcal{Z}(H)$ contains 5 isomorphism classes of lattices and $dim F(H) = 1$. By computing the automorphism groups, one arrives at the above result. (Note $C_{11} \rtimes C_{10}$ also acts on these lattices; the lattices come in 2 pairs of dual lattices and one lattice which is dual to a multiple of itself. Note also that [Fei 74] could be quoted.)
(iii) Let $11 \nmid |G|$. The case that $G$ has a soluble normal subgroup bigger than $\langle -I_{10}\rangle$ has been dealt with in (II.19) and leads to $G_3(10)$. Therefore the terminating term $G^{(\infty)}$ of the derived series of $G$ is quasisimple with centre of order at most 2. If $G^{(\infty)}$ is imprimitive or reducible over $\mathbb{C}$, one is lead again to $G_3(10)$ or via the primitive permutation groups of degree 5 or 10 (more precisely their 2-fold covering folds) to the

alternating group $A_6 \cong L_2(9)$ on 10 letters. This leads to $G_4(10) \cong C_2 \times S_6$ since the irreducible character of degree 10 does not extend rationally to $M_{10}$ or $PGL_2(9)$, cf. [CCNPW 85]. What is left is the possibility of $G^{(\infty)}$ to be complex primitive. Either from [Fei 76] or by using the classification of finite simple groups (and the character tables of their covering groups) one sees that there are no further possibilities. (Note the Minkowski bound says $|G|$ divides $2^{18} \cdot 3^6 \cdot 5^2 \cdot 7 \cdot 11$, but we are in the case $11 \nmid |G|$.) q.e.d.

**(II.21) Theorem** *$M_{10}^{irr}(\mathbf{Q})$ is as described in (II.14), i.e. its components consist of a 2-simplex, two 1-simplices, and one 0-simplex.*

Proof. The three groups $G$ with $11 \mid |G|$ all contain an absolutely irreducible subgroup isomorphic to $C_{11} \rtimes C_{10}$; hence one gets the 2-simplex with vertices $A_{10}$, $G_1(10)$, and $G_2(10)$. This is a full component of $M_{10}^{irr}(\mathbf{Q})$, since no other groups admit an invariant bilinear form $F$ with $\partial_{11}(F) \neq 0$.
$G_4(10)$ forms a component by itself, because all its proper subgroups $U$ either contain $A_6$ and satisfy $\mathcal{Z}(U) = \mathcal{Z}(G_4(10))$ in this case or they are reducible, cf. [HoP 89] pg. 309.
$B_{10}$ and $2A_5$ form a 2-simplex an irreducible group $H \cong M_{10}$ (Mathieu group of degree 10) conjugates into $Aut(B_{10})$ and $Aut(2A_5)$. This simplex is a full component of $M_{10}^{irr}(\mathbf{Q})$, since no other groups except $G_4(10)$ have invariant forms $F$ with $\partial_p(F) = 0$ for all $p = 2, 3, \ldots$.
$5A_2$ and $G_3(10)$ form a 2-simplex because an absolutely irreducible group $H \cong C_2^4 \rtimes S_5$ occurs as a subgroup of both, $Aut(5A_2)$ and $G_3(10)$. Since $\partial_3(F) \neq 0$ for all forms involved, they form a component by itself, cf also [HoP 89] pg. 280 and pg. 337. q.e.d.

## g) Some open problems

The following problems might vary in difficulty. They more or less grow naturally out of the preceding discussion.

(P1) Which rational quadratic forms $F$ allow finite irreducible subgroups for the orthogonal group $O(F)$ over the rationals?

Comment: Clearly, if one restricts oneself to absolutely irreducible finite subgroups, a certain multiple of $F$ must have $\partial_p(F) = 0$ for all finite primes $p > n+1$. Conversely if all $\partial_p F = 0$ for all finite primes, then $F \sim_{\mathbf{Q}} I_n$, where $n$ denotes the dimension. Note also that with any two "admissible" forms, their tensor product is also admissible.

(P2) Decide when $A_n$ and $B_n$ lie in the same component of $M_n(\mathbf{Q})$.

(P3) Give reasonable upper and lower bounds for
a) the number of components of $M_n^{irr}(\mathbf{Q})$,

b) the dimensions of the simplices in $M_n(\mathbb{Q})$.

(P4) Are there shortcuts for the derivation of the conjugacy classes of maximal finite subgroups of $GL_n(\mathbb{Z})$ once $M_n(\mathbb{Q})$ is known?

Comment. Clearly if $G \leq GL_n(\mathbb{Z})$ is maximal finite, it is conjugate under $GL_n(\mathbb{Q})$ to at least one maximal finite subgroup $H$ of $GL_n(\mathbb{Q})$. We may call this $G$ weakly belongs to $H$. If moreover $dimC_{\mathbb{Q}^{n\times n}}(G) = dimC_{\mathbb{Q}^{n\times n}}(H)$ (or $dimF(G) = dimF(H)$ in case we deal with $M_n^F(\mathbb{Q})$), then we say $G$ strongly belongs to $H$. In the later sense, $G$ either belongs to no vertex of $M_n(\mathbb{Q})$ or to a $k$-simplex for some $k \geq 0$. The last is quite common, e.g. always occurs for $\mathbb{C}$-irreducible groups and $M_n^{irr}(\mathbb{Q})$ provides valuable information.

(P5) Find the minimal dimensions $n$ for which one has maximal finite irreducible but not absolutely irreducible subgroups of $GL_n(\mathbb{Q})$ resp. of $GL_n(\mathbb{Z})$.

## III. Torsion free space groups

### a) Connection with Euclidean space forms

Crystallographic space groups are discrete subgroups of the Euclidean group $Iso(\mathbb{R}^n)$ of all isometries of Euclidean $n$-space $R^n$ which contain $n$ linearly independent translations. Group theoretically they are characterized as extensions of free abelian groups (= translation lattice $L$) by a faithfully acting finite group called the point group $P$ of $R$, $P \leq Aut(L)$, cf. [BBNWZ 78] or [HoP 89]. A space group $R \leq Iso(\mathbb{R}^n)$ acts fixed point freely on $\mathbb{R}^n$ if and only if $R$ is torsion free. In this situation the orbit space $R \setminus \mathbb{R}^n$ forms a compact connected flat $n$-dimensional Riemannian manifold, for short, an $n$-dimensional flat manifold. Conversely an $n$-dimensional flat manifold has the Euclidean $n$-space $\mathbb{R}^n$ as its universal covering space and an $n$-dimensional crystallographic space group with (fixed point) free action on $\mathbb{R}^n$ as fundamental group. The geometric interpretation of the point group in this context is that of the (linear) holonomy group of the flat manifold, cf. [Wol 67], [Cha 86]. So talking about a flat manifold with a given holonomy group amounts to talk about a torsion-free space group with a point group specified up to isomorphism. Two sort of problems arise.

(1) Classify all torsion-free space groups in a given dimension.

(2) For a given finite group $G$ classify all torsion-free space groups with point group isomorphic to $G$ up to isomorphism.

(1) can and has been done in small dimensions, there are 2,10, resp. 74, isomorphism types in dimensions 2,3, resp.4, all listed in [BBNWZ 78]. There are also partial results in [Szc 90] for dimension 5.

(2) is usually far too difficult since it would as a partial result also involve a classification of all indecomposable $\mathbb{Z}G$-lattices $L$. Because for a torsion free space group $R$ with point group isomorphic to $G$, the semidirect product $L \rtimes R$ would be another such space group. However, it had been remarked in [Mey 84] that the problem of extensions and that of lattices can be dealt with almost independently for general space groups. Namely for a given finite group $G$ there are finitely many extensions $E_i$ of $\mathbb{Z}G$-lattices $L_i$ by $G$ such that any extension $E$ of any $\mathbb{Z}G$-lattice $L$ by $G$ can be obtained as a factor group of $L \rtimes E_i$ for one of the finitely many $E_i$ modulo a pure $\mathbb{Z}G$-sublattice of $L \oplus L_i$ isomorphic to $L_i$. Applied to the special case of torsion free space groups this shows that the following two numbers are well defined, cf. also [AuK 57],[HSa 86].

**(III.1) Definition** *Let $G$ be a finite group.*
*(i) $m(G)$ denotes the smallest integer $m$ such that exists an $m$-dimensional torsion free space group with point group isomorphic to $G$.*
*(ii) $n(G)$ denotes the smallest natural number $n$ such that any torsion free space group $R$ with point group isomorphic to $G$ has a normal subgroup $N$ contained in the translation lattice $T(R)$ such that $R/N$ is still torsion free and $dim T(R)/N \leq n$.*

## b) Minimal dimensions: the invariant m(G)

The determination of $m(G)$ for a finite group $G$ can be quite a challenging task. In [Ple 89] $p$-adic integral representation theory has been introduced as a tool for this as follows. If $R$ is a space group with point group $G$, it is embedded in its $p$-adic completion $R_p$ where the translation subgroup $T(R)$ has been enlarged to $F(R)_p := \mathbb{Z}_p \otimes T(R) = T(R_p)$. Clearly $R$ is $p$-torsion free if and only if the $p$-adic space group $R_p$ is $p$-torsion free. The key observation is as follows.

**(III.2) Remark** *Let $R$ be a $p$-torsion free $p$-adic space group with point group $G$. Then $R$ contains a (closed) subgroup $S$ which is a Frattini extension of $T(S) = T(R) \cup S$ by $G$.*

Cf. e.g. [PlH 89] for $p$-adic space groups, Frattine extension etc.. So above $S$ is the extension of a $\mathbb{Z}G$-lattice $T(S)$ by $G$ such that the Frattini subgroup $\phi(S/pT(S))$ contains $T(S)/pT(S)$. Hence $T(S)$ is an image of the (universal) Frattini lattice $\Omega^2(\mathbb{Z}_p)$, where $G$ acts trivially on $\mathbb{Z}_p$ and $\Omega$ is the Heller shift with respect $\mathbb{Z}_p(G)$. So the problem becomes: Which factor groups $R_n/N$ of the universal Frattini extension $R_n$ of $\Omega^2(\mathbb{Z}_p)$ by $G$, $N \subseteq T(R_n) = \Omega^2(\mathbb{Z}_p)$, are still $p$-torsion free? In case the Sylow $p$-subgroups of $G$ are cyclic, one obtains the following result.

**(III.3) Theorem ([Ple 89])** *Let $G$ be a finite group with a cyclic Sylow $p$-subgroup $\neq 1$.*

*(1) For any (p)-torision free space group $R$ with point group $G$, the character of $G$ afforded by the translation lattice $T(R)$ contains a certain irreducible complex character $\chi$, which is attached to the second neighbour of 1 in the Brauer tree of principal p-block of $G$.*

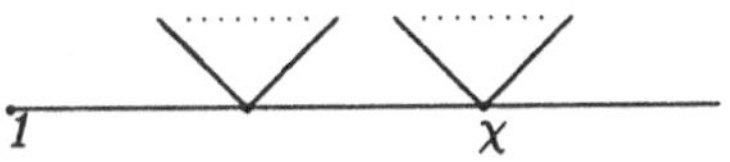


*(If this vertex in the Brauer tree is exceptional $\chi$ is any one of the exceptional characters with $\mathbb{Q}[\chi] : \mathbb{Q}$ maximal).*

*(2) Conversely, if $V$ is any $\mathbb{Q}G$-module having the above $\chi$ as constituent in its character, then there exists a p-torsion free extension of a full $\mathbb{Z}G$-sublattice of $V$ by $G$.*

The question had been asked whether the point group $G$ of a torsion free space group $R$ can act irreducibly on $T(R)$. Note the character of $T(R)$ would then belong to the principle $p$-block of $G$ for any prime $p||G|$. Based on this, a generalization of (III.3)(1), and the classification of finite simple groups, the following has been proved.

**(III.4) Theorem [HiS 90]** *Let $G \neq 1$ be a finite group. There exists no torsion free space group $R$ with point group $G$ such that $G$ acts irreducibly on its translation lattice $T(R)$.*

(III.2) has also been used in [Ple 89] to compute $m(PSL_2(p))$ for all prime numbers $p$. For instance $m(PSL(7)) = 15$ and the character of the translation lattice is the sum of the irreducible characters of degree 7 and 8, which comes close to a counterexample to (III.4).

As a variant the number $m'(G)$ = minimal dimension of a torsion free space group $R$ with $R/T(R) \cong G$ and $R : R'$ finite (or equivalently $Z(R) = 1$ or 1 not contained in the $G$-character afforded by $T(R)$.) $m'(G)$ is well defined if and only if $G$ does not have a normal complement for any cyclic Sylow subgroup, cf. [HSa 86] or [ClW 89]. (Part of this can also be obtained from (III.2)). Some recent references for this can be found in [Ple 89].

### c) Maximal minimal dimensions: The invariant $n(G)$

The definition of $n(G)$ was more closely investigated in [ClW 89]. the prove the following two resuls.

**(III.5) Theorem [ClW 89]** *Let $G$ be a finite group, $\mathcal{X}$ a set of representatives* *of the conjugacy classes of subgroups of prime order of G. Let*

$$L = \oplus_{H \in \mathcal{X}} (\mathbb{Z}_H)^G$$

*where $\mathbb{Z}_H$ is the $\mathbb{Z}H$-lattice $\mathbb{Z}$ with trivial $H$-action. Then any torsion free space group $R$ with point group $G$ has an epimorphic image in a certain torsion free extension of $L$ by $G$.*

As a consequence one has

$$n(G) \leq \sum_{H \in \mathcal{X}} |G : H|$$

**(III.6) Theorem [ClW 89]** *Let $G$ and $\mathcal{X}$ as in (III.5). Then*

$$n(G) = \sum_{H \in \mathcal{X}} |G : H|$$

*holds if and only if $G$ is a $p$-group.*

## d) Open problems

(P1) Develop further theory for the determination of $m(G)$ and $m'(G)$ for finite groups $G$.

Comment. The methods used so far are either restricted to small (soluble) groups $G$, where one constructs some torsion free extensions by ad hoc methods and then proves that the dimension cannot be decreased or the $p$-adic approach using the Frattini module in [Ple 89]. In the latter approach one sometimes has to deal with the following problem.

(P2) Given a character $\chi$ of a finite group $G$ and a subgroup $U \leq G$ of prime order $p$. Decide whether or not a $\mathbb{Z}_pG$- lattice $L$ with character $\chi$ exists such that the $\mathbb{Z}_pU$-lattice $\mathbb{Z}_p$ with trivial $H$-action is a direct summand of the restriction $L_{\mathbb{Z}_pU}$ of $L$ to $\mathbb{Z}_pU$.

As far as the second invariant $n(G)$ goes one of course has the following problem.

(P3) Determine $n(G)$ for finite groups $G$ which are not of prime power order.

In view of the approach taken in [Mey 84], the following problem is of interest.

(P4) Investigate the torsion free extensions $R$ of free abelian groups by finite groups $G$, such that no proper epimorphic image of $R$ is torsion free.

Finally there is a variant of the whole set up for finite groups rather than space groups: Consider extensions $E$ of $\mathbb{F}_pG$- modules $M$ by $G$, such that each $x \in E$ with order $|xM| = p$ in $E/M$ has order $p^2$. Again one can define minimal dimensions etc..

## REFERENCES


[Ban 73] Ei. Bannai, Et. Bannai, *On some finite subgroups of* $GL(n, \mathbb{Q})$. J. Fac. Sci. Univ. Tokyo, Sect. IA 20,(3) (1973), 319-340.

[Bli 17] H. F. Blichfeldt, *Finite Collineation Groups.* Chicago: University of Chicago Press 1917.

[Bra 67] R. Brauer, *Über endliche lineare Gruppen von Primzahlgrad.* Math. Annalen 169, 73-96 (1967).

[BBNWZ 78] H. Brown, R. Bülow, J. Neubüser, H. Wondratschek, H. Zassenhaus, *Crystallographic Groups of Four-Dimensional Space.* Wiley 1977.

[Bül 73] R. Bülow, *Über Dadegruppen in* $GL(5, \mathbb{Z})$. Dissertation RWTH Aachen 1973.

[Bur 12] W. Burnside, *The determination of all groups of rational linear substitutions of finite order which contain the symmetric group in the variables.* Proc. London Math. Cos. (2) 10 (1912), 284-308.

[Cam 81] P.J. Cameron, *Finite Permutation Groups and Finite Simple Groups.* Bull. London Math. Soc. 13 (1981), 1-22.

[CoS 88] J.H. Conway, N.J.A. Sloane, *Low-dimensional lattices. II. Subgroups of* $GL(n, \mathbb{Z})$. Proc. R. Soc. London A 419, 29-68 (1988).

[CCNPW 85] J. H. Conway, R. T. Curtis, S. P. Norton, R. A. Parker, R. A. Wilson, *Atlas of finite groups.* Oxford University Press 1985.

[CuR 62] C.W. Curtis, I. Reiner, *Representation theory of finite groups and associative algebras.* Interscience, New York, 1962.

[Dad 64] E.C. Dade, *Integral Systems of Imprimitivity*. Math. Annalen 154, 383-386 (1964).

[Dad 65] E.C. Dade, *The maximal finite groups of* $4 \times 4$ *matrices*. Ill.J.Math. 9 (1965), 99-122.

[Fei 74] W. Feit, *On integral representations of finite groups*. Proc. London Math Soc. (3) 29 (1974), 633-683.

[Fei 76] W. Feit, *On finite linear groups in dimension at most 10*. 397-407 in *Proceedings Conference on Finite Groups* (ed W. R. Scott and F. Gross), New York Academic Press 1976.

[HoP 89] D. F. Holt, W. Plesken, *Perfect Groups*. Oxford University Press 1989.

[HoP 90] D. F. Holt, W. Plesken, *A cohomological criterion for a finitely presented group to be infinite*. submitted 1990.

[Jor 80] C. Jordan, *Mémoire sur l'équivalence des formes*. J. Ecole Polytech. 48 (1880), 112-150. Oeuvres de C. Jordan, Vol.III, Gauthier-Villars, Paris, 1962, 421-460.

[Kne 54] M. Kneser, *Zur Theorie der Kristallgitter*. Math. Ann. 127, 105 - 106 (1954).

[LaS 74] V. Landazuri, G.M. Seitz, *On the minimal degrees of projective representations of the finite Chevalley groups*. Journal of Algebra 32, 418-443 (1974).

[PlP 77,80] W. Plesken, M. Pohst, *On maximal finite irreducible subgroups of* $GL(n, \mathbb{Z})$. *I. The five- and seven-dimensional case. II. The six-dimensional case*, Math. Comp. 31 (138)(1977), 536-577; *III. The nine-dimensional case. IV. Remarks on even dimensions with applications to* $n = 8$. *V. The eight-dimensional case and a complete description of dimensions less than ten*, Math. Comp. 34 (149) (1980), 245-301.

[Ple 78] W. Plesken *On reducible and decomposable representations of orders*. J. reine angew. Math. 297 (1978), 188-210.

[Ple 81] W. Plesken, *Bravais groups in low dimensions.* Proceedings of a Conference on Crystallographic Groups, Bielefeld 1979. match 10 (1981), 97-119.

[Ple 85] W. Plesken, *Finite unimodular groups of prime degree and circulants.* J. of Algebra vol. 97, no. 1 (1985), 286-312.

[Ple 87] W. Plesken, *Towards a Soluble Qutient Algorithm.* J. Symbolic Computation (1987) 4, 111-122.

[PlH 84] W. Plesken, W. Hanrath, *The lattices of six-dimensional euclidean space.* Math. of Comput. vol. 43, no. 168 (1984), 573-587.

[Rys 72a] S.S. Ryškov, *On maximal finite groups of integer* $n \times n$*-matrices.* Dokl. Akad. Nauk SSSR 204 (1972), 561-564. Sov. Math. Dokl. 13 (1972), 720-724.

[Rys 72b] S.S. Ryškov, *Maximal finite groups of integral* $n \times n$ *matrices and full groups of integral automorphisms of positive quadratic forms (Bravais models).* Tr. Mat. Inst. Steklov 128 (1972), 183-211. Proc. Steklov Inst. Math. 128 (1972), 217-250.

[RyL 80] S.S. Ryškov, Z.D. Lomakina, *Proof of a theorem on maximal finite groups of integral* $5 \times 5$*-matrices.* Proc. Steklov Inst. Math. (1982) Issue 1, 225-235.

[Scha 85] W. Scharlau, *Quadratic and Hermitian forms.* Springer-Verlag 1985.

[Sou 91] B. Souvignier, *Irreduzible Bravaisgruppen.* Diplomarbeit Aachen 1991.

[Zas 38] H. Zassenhaus, *Neuer Beweis der Endlichkeit der Klassenzahl bei unimodularer Äquivalenz endlicher ganzzahliger Substitutionsgruppen.* Hamb. Abh. 12 (1938), 276-288.

## Additional References for Chapter III

[Auk 57] L. Auslander, M. Kuranishi, *On the holonomy group of locally Euclidean spaces.* Ann. Math. 65, 411-415 (1957).

[Cha 86] L.S. Charlap, *Bieberbach groups and flat manifolds.* Berlin Heidelberg New Zork: Springer 1986.

[ClW 89] G. Cliff, A. Weiss, *Torsion free space groups and permutation lattices for finite groups.* Contemporary Mathem. vol 93 (Representation Theory, Group Rings, and Coding Thery) AMS (1989), 123-132.

[HiS 90] G. Hiss, A. Szczepanski, *On torsion free crystallographic groups.* submitted 1990.

[HSa 86] H. Hiller, C.-H. Sah, *Holonomy of flat manifolds with* $b_1 = 0$. Q.J. Math., Oxf.II Ser. 37, 177-187 (1986).

[Mey 84] J. Meyer, *Minimal extensions of finite groups by free abelian groups.* Arch. Math. 42, 16-31 (1984).

[Ple 89] W. Plesken, *Minimal dimensions for flat manifolds with prescribed holonomy.* Math. Ann. 284, 477-486 (1989).

[Szc 90] A. Szczepanski, *Five dimensional Bieberbach groups with trivial centre.* submitted 1990.

[Wol 67] J.A. Wolf, *Spaces of constant curvature.* New York: McGraw-Hill 1967.

Progress in Mathematics, Vol. 95, © 1991 Birkhäuser Verlag Basel

# A construction of orders of finite global dimension

ALFRED WIEDEMANN

## Introduction

Let $R$ be a complete Dedekind domain with field of quotients $K$. We consider in this paper classical $R$-orders. Such an order is a subring of a finite-dimensional semisimple $K$-algebra containing a $K$-basis of this algebra and is finitely generated as a module over $R$. In particular, we are interested in the global dimension of classical $R$-orders.

Many orders arising naturally in other contexts, turn out to be of finite global dimension, and thus the concept of the global dimension is of interest for these orders. Let us mention first the Auslander orders, that are the endomorphism rings of additive generators for the category of lattices over orders of finite lattice type, which have global dimension at most two [3]. This is the main reason for the general interest in orders of global dimension two.

More recently many efforts have been made to transfer the concept of a quasi-hereditary algebra into the theory of orders. First there is the notion of a quasi-hereditary $k$-algebra for an arbitrary commutative ring $k$ introduced and studied by Cline, Parshall and Scott [4]. A different concept of a quasi-hereditary $R$-order was introduced and developed by König [12]. However, both concepts lead to $R$-orders of finite global dimension.

Starting with classical investigations by Jategaonkar and Tarsy, initiated by Kaplansky [9, 10, 20, 21], an extensive attempt has been made in papers by Kirkman and Kuzmanovich [11], Fujita [5] and by Roggenkamp and the author [23] to understand the global dimension of so-called tiled orders. By a tiled order we mean here an $R$-order $\Lambda$ in a $K$-algebra $A$, such that $\Lambda$ contains a complete set of pairwise orthogonal primitive idempotents of $A$, and moreover, the endomorphism ring of each indecomposable projective $\Lambda$-lattice is the maximal order in a finite-dimensional skewfield over $K$.

Overviewing all these concepts, it is interesting to note that all $R$-orders $\Lambda$ which occur in the above contexts, including arbitrary $R$-orders of global dimension 2 [13] and for which the global dimension was extensively studied so far, have the common property that they contain at least one primitive idempotent $e$ such that $e\Lambda e$ is of finite global dimension, or equivalently, $e\Lambda e$ is the maximal $R$-order in a skewfield [13].

This observation led to the question whether there exist at all any orders of finite global dimension which do not have this property. Examples of artin algebras with the analog property were first considered by Green [7] and recently rediscovered by Happel [8]. So we were interested in finding examples of classical orders of finite global dimension having the property that the endomorphism ring of all the indecomposable projective lattices has infinite global dimension.

The purpose of this note is therefore to show that endomorphism rings of certain well chosen lattices over Bäckström orders [14] have finite global dimension; that is done in the first section. This result then provides a handy construction for interesting examples of orders of finite global dimension which need not have an indecomposable projective lattice with endomorphism ring the maximal order in a skewfield.

As a first example of such an order we present the following $R$-order $\Lambda$ in $(R)_5$, the $5 \times 5$-matrix ring over an arbitrary complete Dedekind domain $R$ with radical $\pi$, and where $R \equiv R$ denotes the set of pairs $(x, y) \in R^{(2)}$ with $x - y \in \pi$:

$$\Lambda = \begin{bmatrix} R & \pi & \pi & R & \pi \\ \pi & R & \pi & R & R \\ \pi & \pi & R & \pi & R \\ \pi & \pi & \pi & R & \pi \\ \pi & \pi & \pi & \pi & R \end{bmatrix}.$$

Obviously, $\Lambda$ has two indecomposable projective lattices whose endomorphi sm rings are nonmaximal orders, and one easily checks that $\Lambda$ has global dimension three. Moreover, it is easy to recognize $\Lambda$ as the endomorphism ring of a right lattice over the Bäckström order

$$B = \begin{bmatrix} R & \pi \\ \pi & R \end{bmatrix}.$$

This observation was made by K.W. Roggenkamp in a joint discussion, and it led to the general result stated in Theorem 1 below.

Further examples of this kind are presented in the third section, where we also present an order in a full $3 \times 3$-matrix ring over a field which has global dimension three. Although this order is not tiled, in a wider sense, it also can be regarded as a counterexample to an old conjecture of Tarsy, saying that the global dimension of an order in a full $n \times n$-matrix ring is at most $n - 1$ or infinite [20]. Recently, this conjecture was disproved by Fujita, who gave a series of examples of tiled orders of global dimension $n$ in full $n \times n$-matrix rings for $n \geq 6$ [5].

We freely shall use the Auslander-Reiten theory for orders as developed in [1, 2, 15, 19].

I am grateful to K.W. Roggenkamp and Th. Weichert for several helpful and stimulating discussions on the topic of this paper.

## 1. Certain endomorphism rings of finite global dimension

Our result is mainly based on the description of the Auslander-Reiten quiver of Bäckström orders as described in [17]. We recall briefly the main concepts and notations from there in order to be able to formulate Theorem 1. For more details the reader should consult directly that paper.

Let $R$ be a commutative complete discrete valuation ring with field of quotients $K$ and residue class field $k$. Recall that an $R$-order $B$ in a semisimple $K$-algebra $A$ is called a Bäckström order, if there exists a hereditary order $\Gamma$ in $A$ such that

$$Rad\Gamma = RadB.$$

Then $\Gamma$ is uniquely determined by $B$. As in [17], we denote by $I$ the radical $RadB$ and associate to $B$ the $k$-algebra

$$\mathcal{D} = \begin{bmatrix} \Gamma/I & \Gamma/I \\ 0 & \Lambda/I \end{bmatrix}.$$

Then the species $S = S(B, \Gamma)$ of $B$ is defined as $k$-modulated species of $\mathcal{D}$ in the sense of Gabriel [6]. In particular, $\mathcal{D}$ is the $k$-tensor algebra of the dual of $S$. Moreover note that since $(Rad\mathcal{D})^2 = 0$, the vertices of $S$ either are sinks or sources.

The main result of [17] describes how one gets the Auslander-Reiten quiver $\mathcal{A}(B)$ of $B$ from the Auslander-Reiten quiver $\mathcal{A}(S)$ of the representations of $S$. First note that the sinks of $\mathcal{A}(S)$ correspond to the simple injective representations of $S$, whereas the sources of $\mathcal{A}(S)$ correspond to the simple projective representations of $S$. In order to construct $\mathcal{A}(B)$ from $\mathcal{A}(S)$ one first has to delete the sinks of $\mathcal{A}(S)$. Then the sinks in the resulting new quiver correspond to the nonsimple injective representations of $S$. In order to get now the connected components of $\mathcal{A}(B)$ which contain the projective $B$-lattices, one suitably has to identify the set of the new sinks with the set of sources of $\mathcal{A}(S)$.

Because of this construction we can speak of the preprojective part of $\mathcal{A}(B)$ which by definition means the full subquiver of $\mathcal{A}(B)$ whose vertices correspond through the above construction to the noninjective preprojective representations of $S$. If $S$ is a disjoint union of connected species $S_i, i = 1, ..., m$, then the preprojective part of $\mathcal{A}(B)$ decomposes accordingly into a finite disjoint union of connected components $Z_i, i = 1, ..., m$.

The following figure wants to illustrate the above procedure how $\mathcal{A}(B)$ is obtained from $\mathcal{A}(S)$:

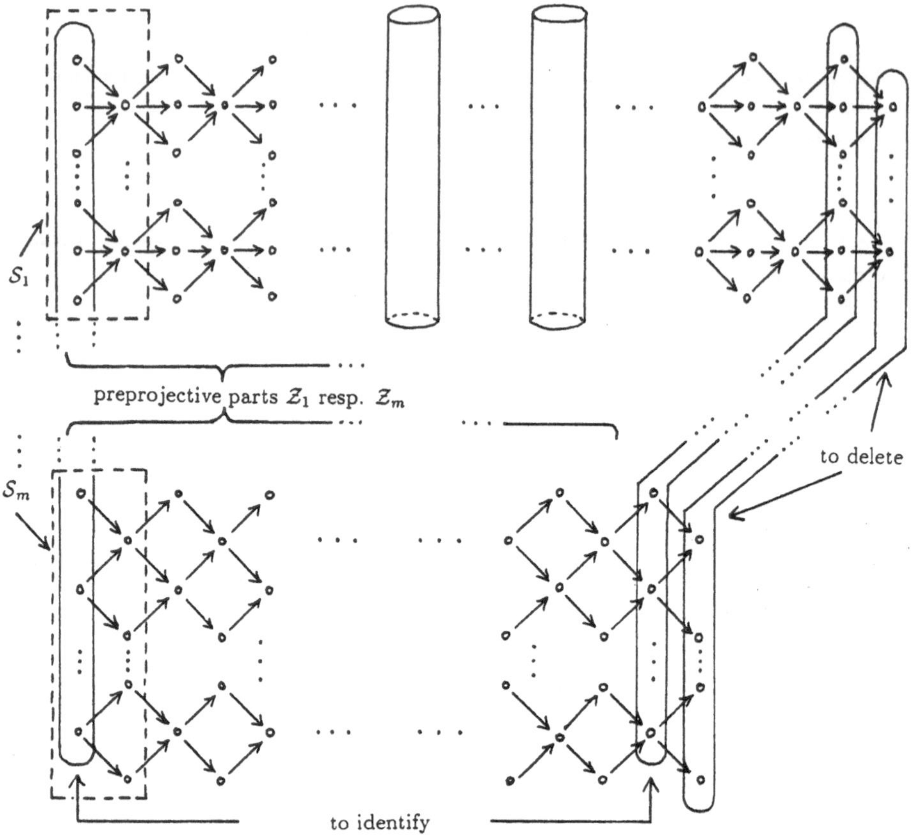


We shall freely identify the set of vertices of $\mathcal{A}(B)$ with the indecomposable $B$-lattices. Now let $\mathcal{M}$ be a finite set of nonisomorphic indecomposable $B$-lattices. Recall that a section of $Z_i$ consists of a full connected subquiver of $Z_i$ with the property that it hits each orbit under the Auslander-Reiten translation on $Z_i$ exactly once. We say that $\mathcal{M}$ contains a section of $Z_i$, if there exists a section of $Z_i$ whose vertices all represent lattices in $\mathcal{M}$.

We are now interested in the endomorphism ring of the direct sum of the finitely many indecomposable $B$-lattices in $\mathcal{M}$, and therefore, we put

$$G = \bigoplus_{M \in \mathcal{M}} M$$

and

$$\Lambda = End_B(G).$$

With this notation, we can now formulate our result.

**Theorem 1.** *Assume that $\mathcal{M}$ only contains $B$-lattices belonging to the preprojective part of $\mathcal{A}(B)$, and $\mathcal{M}$ contains no $\Gamma$-lattices. If for each $i = 1, \cdots, m$ for which $\mathcal{M} \cap$*

*$Z_i$ is nonempty, $\mathcal{M}$ also contains a section of $Z_i$, then the endomorphism ring $\Lambda = End_B(G)$ is an R-order of finite global dimension.*

PROOF. The indecomposable projective $\Lambda$-lattices are of the form $Hom_B(U,G)$ for an indecomposable $B$-lattice $U$ belonging to $\mathcal{M}$.

We have to show that $rad_\Lambda Hom_B(U,G)$ has finite homological dimension over $\Lambda$. We write $rad_\Lambda(U,G)$ for $rad_\Lambda Hom_B(U,G)$ and note that this is the set of all $B$-homomorphisms from $U$ to $G$ which are not split mono.

First we claim that $Hom_B(H,G)$ has finite homological dimension for each indecomposable $\Gamma$-lattice $H$. Assume that this $H$ belongs to the left border of a component $Z = Z_i$ of the preprojective part of $\mathcal{A}(B)$ as defined above. If $Z \cap \mathcal{M} = \emptyset$, then it is easy to see that there exists an indecomposable $\Gamma$-lattice $H'$ with

$$Hom_B(H',G) \cong Hom_B(H,G),$$

and $H'$ belongs to the left border of another component $Z'$ with $Z' \cap \mathcal{M} \neq \emptyset$.

We therefore may assume from the beginning that $Z \cap \mathcal{M} \neq \emptyset$, and by hypothesis, we may fix a section $\mathcal{T}$ of $Z$ belonging to $\mathcal{M}$. In particular, this implies that $H$ is not an injective $B$-lattice. Since $H \notin \mathcal{M}$, and if

$$0 \to H \to W \to V \to 0$$

is the almost split sequence with kernel $H$, then the induced sequence

$$0 \to Hom_B(V,G) \to Hom_B(W,G) \to Hom_B(H,G) \to 0$$

is exact. We now want to show that both $Hom_B(V,G)$ and $Hom_B(W,G)$ have finite homological dimension over $\Lambda$, implying that $Hom_B(H,G)$ also has.

For this let $Z'$ be the set of vertices of $Z$ which lie between the sources of $Z$ and the chosen section $\mathcal{T}$ of $Z$. Note furthermore that the length of any oriented path in $Z$, starting in a vertex of $Z'$ and avoiding vertices in $\mathcal{M}$, is bounded from above. Moreover, if the successors of a vertex $X \in Z' \setminus \mathcal{M}$ all lie in $\mathcal{T}$, then all indecomposable direct summands of the middle term and the cokernel of the almost split sequence with kernel $X$ lie in $\mathcal{M}$. Applying $Hom_B(-,G)$ to that sequence, shows that $Hom_B(X,G)$ has homological dimension 1 over $\Lambda$. We may therefore use induction on the length of longest oriented paths in $Z'$ avoiding $\mathcal{M}$, to conclude as desired that both $Hom_B(V,G)$ and $Hom_B(W,G)$ have finite homological dimension over $\Lambda$.

Now let $U$ be arbitrary in $\mathcal{M}$. If $U$ is not an injective $B$-lattice, then the almost split sequence

$$0 \to U \to W \to V \to 0$$

provides an exact sequence

$$0 \to Hom_B(V,G) \to Hom_B(W,G) \to rad_\Lambda(U,G) \to 0.$$

If $V$ or one of the indecomposable summands of $W$ does not belong to $\mathcal{M}$, call such a lattice $X$, then if $X$ is not injective, the almost split sequence with kernel $X$ provides a similar exact sequence, and it is therefore enough to show that a $\Lambda$-lattice of the form $Hom_B(X,G)$ has finite homological dimension whenever $X$ lies in a component $Z$ of the preprojective part of $\mathcal{A}(B)$ such that there is within $Z$ no oriented path from $X$ to any vertex of $\mathcal{M} \cap Z$. But then by the factorization property of almost split sequences, each homomorphism from $X$ to $G$ factors through a $\Gamma$-lattice because all components $Z_i$ of the preprojective part of $\mathcal{A}(B)$ only have $\Gamma$-lattices as sources. Then

$$Hom_B(X,G) \cong Hom_B(\Gamma X, G),$$

and this $\Lambda$-lattice has finite homological dimension by the first part of this proof.

In the case where the $U$ or the $X$ from above is an injective $B$-lattice, all its successors already are $\Gamma$-lattices, and one has that $rad_\Lambda(U,G)$ or $rad_\Lambda(X,G)$ respectively is of the form $Hom_B(H,G)$ for a $\Gamma$-lattice $H$. Therefore it also has finite homological dimension.

REMARK. In view of the examples we have in mind, we have here restricted to the particularly nice situation of a Bäckström order $B$. However, it is clear that the above proof also works in more general situations as for generalized Bäckström orders [16] or even for subhereditary orders in the simply connected case [18], when the notions of the preprojective part and of a section suitably are adjusted.

## 2. A necessary condition for finite global dimension

Recall from [18] that a classical $R$-order $\Lambda$ is called subhereditary if there exists a hereditary overorder $\Gamma$ of $\Lambda$ with $Rad\Gamma \subset \Lambda$. From the theory of subhereditary orders developed in [18, 22], it is easy to derive a necessary and sufficient condition for a subhereditary order $\Lambda$ to have finite global dimension in terms of the associated artin algebra $\mathcal{D}$, analogously defined as in section 1. In this explicit form we state it here, it was first remarked by Th. Weichert.

We start with an $R$-order $\Lambda$ in a hereditary order $\Gamma$ such that

$$Rad\Gamma = I \subset \Lambda \subset \Gamma.$$

As in section 1, to the order $\Lambda$ one associates the algebra

$$\mathcal{D} = \begin{bmatrix} \Gamma/I & \Gamma/I \\ 0 & \Lambda/I \end{bmatrix}$$

which is socle-projective. Then it is known that the category of $\Lambda$-lattices is representation equivalent to the category of socle-projective $\mathcal{D}$-modules without simple direct summands [18]. Furthermore, this representation equivalence is provided by associating to a $\Lambda$-lattice $M$ the $\mathcal{D}$-module

$$\tilde{M} = \begin{bmatrix} \Gamma M/IM \\ M/IM \end{bmatrix},$$

on which $\mathcal{D}$ acts by usual matrix multiplication from the left. If $M$ is an indecomposable nonprojective $\Lambda$-lattice with projective cover sequence

$$0 \to \Omega M \to P \to M \to 0,$$

we may decompose $\Omega M = M' \oplus G$, where $G$ is the maximal direct summand of $\Omega M$ which is a $\Gamma$-lattice. From [18, 22] it is then easy to derive that the sequence

$$0 \to \tilde{M}' \oplus \begin{bmatrix} I^{-1}G/G \\ 0 \end{bmatrix} \to \tilde{P} \to \tilde{M} \to 0$$

is an exact sequence of $\mathcal{D}$-modules, and moreover, it is a projective cover sequence for $\tilde{M}$.

As a consequence, one can derive now a necessary and sufficient condition for $\Lambda$ having finite global dimension in terms of the occurance of the simple projective $\mathcal{D}$-modules in a minimal projective resolutions of the indecomposable injective $\mathcal{D}$-modules. In view of our application where $\Lambda$ is subhereditary with respect to a maximal order $\Gamma$ in a simple $K$-algebra, we restrict here to this simple case in order to avoid complicated notations. So let $H$ denote the up to isomorphism unique indecomposable $\Gamma$-lattice.

**Theorem 2.** *The order $\Lambda$ as above has finite global dimension if and only if $\mathcal{D}$ has finite global dimension and the simple projective $\mathcal{D}$-module does not occur in a minimal projective resolution of the $\mathcal{D}$-module $\tilde{H}$.*

PROOF. If $\Lambda$ has finite global dimension, then the minimal projective resolution of each $\Lambda$-lattice $M$ provides a projective resolution of $\tilde{M}$ over $\mathcal{D}$ with the same length. It is also easy to see that each nonprojective direct summand of $Rad\mathcal{D}$ occurs as a direct summand of $R\tilde{a}d\Lambda$. This implies that $gldim\mathcal{D} < \infty$. Furthermore, $H$ cannot occur as a syzygy in a minimal projective resolution of itself, and from the consideration above, it is clear that the simple projective $\mathcal{D}$-module $S = \begin{bmatrix} I^{-1}H/H \\ 0 \end{bmatrix}$ cannot occur in a minimal projective resolution of $\tilde{H}$.

Now assume conversely that $\mathcal{D}$ has finite global dimension. This just means that each $\Lambda$-lattice has a finite resolution involving only projective $\Lambda$-lattices or $\Gamma$-lattices. If $S$ does not occur in a minimal finite projective resolution of $\tilde{H}$, then this resolution directly is induced by a projective resolution of $H$, forcing that $H$ has finite homological dimension over $\Lambda$. Together with the first remark, this implies that $gldim\Lambda < \infty$.

## 3. Examples of untiled and thick orders of finite global dimension

We propose to call a classical order $\Lambda$ untiled if for each primitive idempotent $e \in \Lambda$, the order $e\Lambda e$ has infinite global dimension, and moreover, we call $\Lambda$ a thick order, if $\Lambda$ has no irreducible projective lattice. Note that "untiled" is not the same as "not tiled". Clearly, each thick order is untiled, because a local order has finite global dimension if and only if it is a maximal order in a skewfield [13].

We now use Theorem 1 to present explicitly some thick and untiled orders of finite global dimension.

EXAMPLE 1. As in the introduction, let

$$B = \begin{bmatrix} R & \pi \\ \pi & R \end{bmatrix}.$$

Then $B$ is a Bäckström order with associated species $\mathcal{S}(B,(R)_2)$ in the sense of section 1 of type

$$\tilde{A}_{1,2} : \bullet \xrightarrow{(2,2)} \bullet$$

[14, 17], and thus, $B$ is of infinite lattice type. The preprojective part of $\mathcal{A}(B)$ is made up by the right $B$-lattices $M_i, i \geq 2$ of the form

$$M_i = \begin{bmatrix} R & \pi \\ R & R \\ & R \\ & \vdots \\ R & \\ R & R \\ \pi & R \end{bmatrix}_{i\times 2}$$

with $i$ rows, and each pair $(M_j, M_{j+1}), j \geq 2$ makes up a section in the preprojective part of $\mathcal{A}(B)$. If we put $G = B \oplus M_3$, we get the order $\Lambda$ from the introduction as $\Lambda = End_B(G)$. If we more generally choose an ascending chain of natural numbers $2 \leq \alpha_1 < \alpha_2 < \cdots < \alpha_s, s \geq 2$, and put $G = \bigoplus_{i=1}^{s} M_{\alpha_i}$, then by Theorem 1, $\Lambda = End_B(G)$ has finite global dimension if $\alpha_{j+1} = \alpha_j + 1$ for some $j \geq 1$:

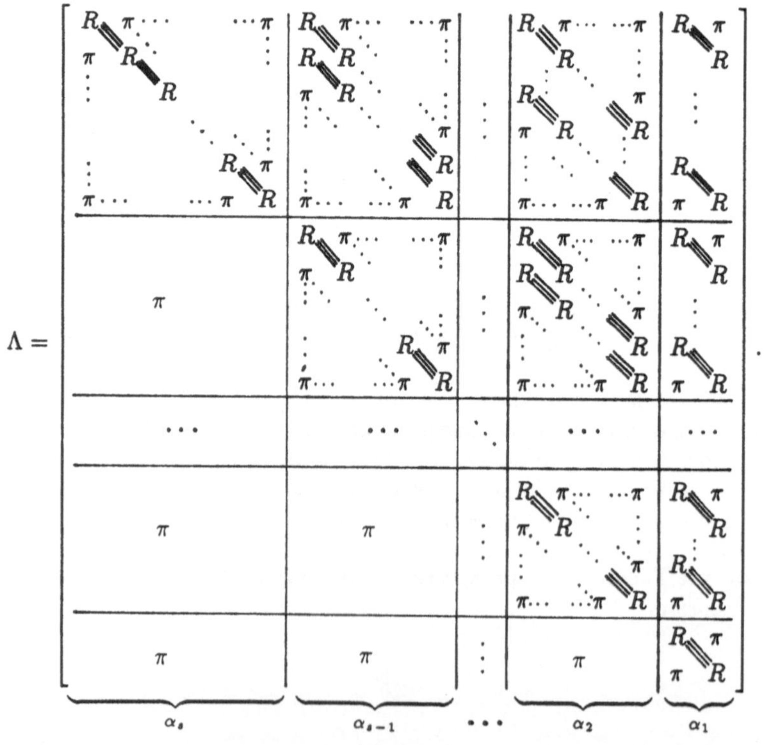

Furthermore, it easily can be checked that an order of the shape of $\Lambda$ as above still has finite global dimension if the $\alpha_i$ satisfy $\alpha_1 \leq \cdots \leq \alpha_s$ and $\alpha_{j+1} = \alpha_j + 1$ for at least one $j$. Note that if $\alpha_k = \alpha_{k+1}$ for some $k$, then $\Lambda$ is not any more the endomorphism ring of a $B$-lattice as above.

Now we want to show the converse, namely that an order $\Lambda$ as displayed above has infinite global dimension if for all $j$, $\alpha_{j+1} \neq \alpha_j + 1$ holds.

For this we use Theorem 2. Clearly, $\Lambda$ is subhereditary with respect to $\Gamma$ being the full matrix ring over $R$ of the same size as $\Lambda$. Let $\mathcal{D}$ be the algebra as defined in section 2. In a matrix $C$, the Cartan matrix of $\mathcal{D}$, we collect the multiplicities with which the simple $\mathcal{D}$-modules occur as a composition factor in the indecomposable projective $\mathcal{D}$-modules, and we obtain

$$C = \begin{bmatrix} 1 & \alpha_s & \alpha_{s-1} & \cdots & & \cdots & \alpha_1 \\ 0 & 1 & \alpha_s - \alpha_{s-1} + 1 & \ddots & & & \alpha_s - \alpha_1 + 1 \\ \vdots & \ddots & \ddots & \ddots & & & \vdots \\ & & & \ddots & \ddots & \ddots & \\ & & & \ddots & 1 & \alpha_3 - \alpha_2 + 1 & \alpha_3 - \alpha_1 + 1 \\ \vdots & & & & & 1 & \alpha_2 - \alpha_1 + 1 \\ 0 & \cdots & & & \cdots & 0 & 1 \end{bmatrix}.$$

The condition of Theorem 2 now means that the column in which we collect the corresponding multiplicities of the simple $\mathcal{D}$-modules in $\tilde{H}$, namely

$$\begin{bmatrix} 1 \\ \alpha_s \\ \cdot \\ \cdot \\ \cdot \\ \alpha_1 \end{bmatrix},$$

must be an integral linear combination of the second up to the last column of $C$. This implies that the determinant of the matrix

$$\begin{bmatrix} \alpha_s & \alpha_{s-1} & \cdots & & \cdots & \alpha_1 & 1 \\ 1 & \alpha_s - \alpha_{s-1} + 1 & & & & \alpha_s - \alpha_1 + 1 & \alpha_s \\ 0 & \ddots & \ddots & & & \vdots & \vdots \\ & & \ddots & \ddots & & \vdots & \vdots \\ \vdots & & & \ddots & 1 & \alpha_2 - \alpha_1 + 1 & \alpha_2 \\ 0 & \cdots & & \cdots & 0 & 1 & \alpha_1 \end{bmatrix}$$

must vanish.

On the other hand, this determinant is

$$(\alpha_s + 1)(\alpha_1 - 1) \prod_{j=1}^{s-1} (\alpha_{j+1} - \alpha_j - 1).$$

Because we assumed $\alpha_1 \geq 2$, this implies indeed that $\alpha_{j+1} - \alpha_j = 1$ for at least one $j$.

EXAMPLE 2. Whereas the orders discussed in Example 1 were found by ad hoc methods before Theorem 1 was established, the following thick order $\Lambda$ of finite global dimension only could be found after the knowledge of Theorem 1. We start from the Bäckström order

$$B = \begin{bmatrix} R & R & R \\ \pi & R & R \\ \pi & \pi & R \end{bmatrix}$$

having associated species $\mathcal{S}(B,(R)_3)$ of type

$D_4 :$ 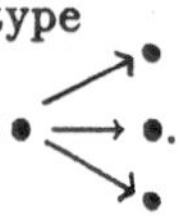

The successors of $B$ in $\mathcal{A}(B)$ are

$$M_1 = \begin{bmatrix} \pi & R & R \\ \pi & \pi & R \end{bmatrix},$$

$$M_2 = \begin{bmatrix} R & R & R \\ \pi & \pi & R \end{bmatrix},$$

$$M_3 = \begin{bmatrix} R & R & R \\ \pi & R & R \end{bmatrix}.$$

Together with $B$ they make up a section of $\mathcal{A}(B)$ in the sense of section 1. Then the $B$-endomorphisms of $G = B \oplus M_1 \oplus M_2 \oplus M_3$ are given by the order

$$\Lambda = \begin{bmatrix} R & R & R & R & R & \pi & R & \pi & R \\ \pi & R & R & \pi & R & \pi & R & \pi & \pi \\ \pi & \pi & R & \pi & \pi & \pi & \pi & \pi & \pi \\ \pi & R & R & R & R & \pi & R & \pi & \pi \\ \pi & \pi & R & \pi & R & \pi & \pi & \pi & \pi \\ R & R & R & R & R & R & R & \pi & R \\ \pi & \pi & R & \pi & \pi & \pi & R & \pi & \pi \\ R & R & R & R & R & \pi & R & R & R \\ \pi & R & R & \pi & R & \pi & R & \pi & R \end{bmatrix}$$

which has global dimension three.

EXAMPLE 3. Let $k$ be a field having an extension field $f$ with $| f : k | = 3$. We put $R = k[[T]]$ the power series ring in one variable $T$ over $k$, and let $\Omega = f[[T]]$. Then $B = k + T\Omega$ is a Bäckström order in the quotient field of $\Omega$ with associated species $\mathcal{S}(B, \Omega)$ of type

$$G_2 : \bullet \xrightarrow{(1,3)} \bullet.$$

In $\mathcal{A}(B)$, $B$ has one successor $M$, being an indecomposable $B$-lattice in $\Omega^{(2)}$. Then $B$ and $M$ together make up a section in $\mathcal{A}(B)$, and hence $\Lambda = End_B(B \oplus M)$ has finite global dimension. Clearly, $\Lambda$ is an untiled $R$-order in $(\Omega)_3$, and going through the proof of Theorem 1, it follows that $\Lambda$ has global dimension three. Thus $\Lambda$ can be regarded as a counterexample to Tarsy's first conjecture in a $3 \times 3$-matrix ring over a field [20].

## REFERENCES

[1] M. Auslander, Existence theorems for almost split sequences, in "Ring Theory II", Proceedings Second Oklahoma Conference, p. 1-44, Dekker, New York, 1977.

[2] M. Auslander, Functors and morphisms determined by objects, in "Representation Theory of Algebras", Lect. Notes in Pure and Appl. Math., Vol. 37, p. 1-244, Dekker, 1979.

[3] M. Auslander and K.W. Roggenkamp, A characterization of orders of finite lattice type, Invent. Math. 17 (1972), 79-84.

[4] E. Cline, B. Parshall and L.L.Scott, Integral and quasi-hereditary algebras, Preprint 1989.

[5] H. Fujita, Tiled orders of finite global dimension, Trans. Amer. Math. Soc., to appear.

[6] P. Gabriel, Indecomposable representations II, Symp. Math. Ist. Naz. Alta Mat. 11 (1973), 81-104.

[7] E.L. Green, Remarks on projective resolutions, in "Representation Theory II" Lect. Notes in Math. 832, pp. 255-279, Springer, Berlin 1980.

[8] D. Happel, A family of algebras with two simple modules, Preprint 1989.

[9] V.A. Jategaonkar, Global dimension of triangular orders over a discrete valuation ring, Proc. Amer. Math. Soc. 38 (1973), 8-14.

[10] V.A. Jategaonkar, Global dimension of tiled orders over a discrete valuation ring, Trans. Amer. Math. Soc. 196 (1974), 313-330.

[11] E. Kirkman and J. Kuzmanovich, Global dimensions of a class of tiled orders, J. Algebra 127 (1989), 57-72.

[12] S. König, Quasi-hereditary orders, Manuscr. Math. 68 (1990), 417-433.

[13] S. König and A. Wiedemann, Global dimension two orders are quasi-hereditary, Manuscr. Math. 66 (1989), 17-23.

[14] C.M. Ringel and K.W. Roggenkamp, Diagrammatic methods in the representation theory of orders, J. Algebra 60 (1979), 11-42.

[15] K.W. Roggenkamp, The lattice type of orders II, in "Integral Representations and Applications", Lect. Notes in Math. 882, pp. 430-470, Springer, Berlin 1981.

[16] K.W. Roggenkamp, Auslander-Reiten species for socle determined categories of hereditary algebras and for generalized Bäckström orders, Mitt. Math. Sem. Giessen 159 (1983).

[17] K.W. Roggenkamp, Auslander-Reiten species of Bäckström orders, J. Algebra 85 (1983), 449-476.

[18] K.W. Roggenkamp, Lattices over subhereditary orders and socle-projective modules, J. Algebra 121 (1989), 40-67.

[19] K.W. Roggenkamp and J. Schmidt, Almost split sequences for integral group rings and orders, Comm. Algebra 4 (1976), 893-917.

[20] R.B. Tarsy, Global dimension of orders, Trans. Amer. Math. Soc. 151 (1970), 335-340.

[21] R.B. Tarsy, Global dimension of triangular orders, Proc. Amer. Math. Soc. 28 (1971), 423-426.

[22] A. Wiedemann, Projective resolutions and the global dimension of subhereditary orders, Arch. Math. 53 (1989), 461-468.

[23] A. Wiedemann and K.W. Roggenkamp, Path orders of global dimension two, J. Algebra 80 (1983), 113-133.

Progress in Mathematics, Vol. 95, © 1991 Birkhäuser Verlag Basel

# Duality and forms in representation theory

WOLFGANG WILLEMS

## Introduction

Duality and the use of the geometry caused by duality are powerful tools in representation theory. A very nice example for this is Okuyama's proof that $G$ has a normal Sylow 2-subgroup if 2 does not divide the dimension of any absolutely simple module in characteristic 2 (see [12]).

The aim of this paper is to present a survey on recent advances where emphasis is put mainly on the case of characteristic 2.

The cooperation with R. Gow and P. Sin on problems in that subtile situation made the progress possible. However, in some cases the answer is not as satisfactory as we would like. For instance: the question about a Frobenius-Schur indicator in characteristic 2 is still open; the Witt index can be determined easily only for solvable groups.

Thus the paper may be seen as a survey and a stimulation for further study.

All groups are assumed to be finite, all modules of finite dimension over the field $k$.

## 1. Duality and Forms

Let $M$ be a $kG$-module, $M^*$ its dual and let $M^* \otimes_k M^*$ denote the space of all bilinear forms on $M$. Via the $kG$-isomorphism

$$\begin{aligned} M^* \otimes_k M^* &\longrightarrow \mathrm{Hom}_k(M, M^*) \\ b &\longrightarrow (m \to b(m, \cdot)) \end{aligned}$$

the space $(M^* \otimes_k M^*)^G$ of $G$-invariant bilinear forms corresponds to $\mathrm{Hom}_{kG}(M, M^*)$, and the set of non-degenerate $G$-invariant forms to the $G$-isomorphisms from $M$ onto $M^*$. Moreover, if $b$ is a non-degenerate $G$-invariant form on $M$, then

$$(M^* \otimes_k M^*)^G = \{b_\alpha := b(\cdot\alpha, \cdot) | \alpha \in \mathrm{End}_{kG}(M)\} \qquad 1.1$$

and $(M^* \otimes_k M^*)^G$ carries an algebra structure via $b_\alpha b_\beta = b_{\alpha\beta}$ for $\alpha, \beta \in \mathrm{End}_{kG}(M)$.

Note that $S^2(M^*)$, $\Lambda^2(M^*)$ describes the space of quadratic, respectively symplectic forms on $M$ and $S^2(M^*)^G$, $\Lambda^2(M^*)^G$ those forms which are $G$-invariant. Furthermore, the $kG$-homomorphism "evaluation on the diagonal".

$$\begin{aligned} \zeta : M^* \otimes_k M^* &\longrightarrow S^2(M^*) \\ b &\longrightarrow q \end{aligned} \qquad 1.2$$

is defined by $q(m) := b(m,m)$ for $m \in M$. Obviously, $\operatorname{Ker}\zeta = \Lambda^2(M^*)$. Comparing now dimensions yields the exact sequence:

$$0 \longrightarrow \Lambda^2(M^*) \longrightarrow M^* \otimes_k M^* \xrightarrow{\zeta} S^2(M^*) \longrightarrow 0\ . \qquad 1.3$$

Observe that for $\operatorname{Char} k \neq 2$ the map

$$\eta : S^2(M^*) \longrightarrow M^* \otimes_k M^*$$
$$q \longrightarrow b$$

defined by
$$b(m_1,m_2) = \frac{1}{2}(q(m_1+m_2) - q(m_1) - q(m_2))$$

for $m_i \in M$ naturally splits the sequence 1.3. Hence for a selfdual module in characteristic different from 2,

$$\Lambda^2(M^*)^G \oplus S^2(M^*)^G = End_{kG}(M)\ . \qquad 1.4$$

In the following, we are mainly interested in forms on simple modules and their projective covers. Note here that a projective module is selfdual if and only if its head is selfdual. Furthermore, a form on a simple module is non-degenerate if it is non-zero. In order to be brief, we call a module $M$ of quadratic, respectively symplectic type if $M$ carries a non-degenerate $G$-invariant quadratic respectively symplectic form.

## 2. Forms in characteristic $\neq 2$

Throughout this section the characteristic of the field $k$ is supposed to be different from 2. Thus we have 1.4 for all self-dual modules. Remember that a non-degenerate $G$-invariant form $b$ on $M$ defines an involution $\alpha \to \alpha^{ad}$ on $\operatorname{End}_{kG}(M)$ via

$$b(m_1\alpha^{ad}, m_2) = b(m_1, m_2\alpha) \quad \text{for} \quad m_i \in M\ .$$

Thus, if $b$ is symplectic, $c$ symmetric and both are non-degenerate and $G$-invariant, then by 1.1

$$b(\cdot,\cdot) = c(\cdot\alpha,\cdot)\ ,$$

hence $-\alpha = \alpha^{ad}$ w.r.t. $c$ as an easy calculation shows. Moreover, by 1.1 and 1.4 we obtain the well-known fact:

**Proposition 2.1.** *Let $M \cong M^*$.*
(a) *If $M$ is indecomposable, then $M$ is of symplectic or quadratic type.*
(b) *If $M$ is absolutely indecomposable, then $M$ cannot be of both types.*

According to this result it is very natural to ask: Is there any connection between the geometric type of an indecomposable projective module $P \cong P^*$ and the type of its simple (selfdual) head $\overline{P} = P/PJ(kG)$. Indeed, there is one as we shall see.

If $\beta$ denotes the $k$-character on $\overline{P}$, then it is well-known that

$$\lambda = \sum_{g\in G} \beta(g^{-1})g \in Z(kG)$$

and $\lambda$ maps $P$ onto $\mathrm{Soc}(P)$ by right multiplication. Let $b \in (P^* \otimes_k P^*)^G$. Since $b(PJ(kG), \mathrm{Soc}(P)) = 0$ we obtain a $\overline{b} \in (\overline{P}^* \otimes_k \overline{P}^*)^G$ by putting

$$\overline{b}(\overline{x}, \overline{y}) = b(x, y\lambda) \ \text{ for } x, y \in P .$$

Note that $\overline{b}$ is non-degenerate, if $b$ is non-degenerate. Furthermore $\overline{\phantom{x}} : \mathrm{End}_{kG}(P) \to \mathrm{End}_{kG}(\overline{P})$ defined by $\overline{x}\,\overline{\alpha} = \overline{x\alpha}$ is a $k$-algebra epimorphism with kernel $J(\mathrm{End}_{kG}(P))$ the radical of $\mathrm{End}_{kG}(P)$. Thus by 1.1

$$\begin{aligned} \overline{\phantom{x}} : (P^* \otimes_k P^*)^G &\longrightarrow (\overline{P}^* \otimes_k \overline{P}^*)^G \\ b &\longrightarrow \overline{b} \end{aligned}$$

defines also a $k$-algebra epimorphism. Since $\lambda = \lambda^{ad}$ w.r.t. any non-degenerate $G$-invariant form $b$ on $P$ we obtain by 1.4

**Proposition 2.2.** *Let $M$ be a simple $kG$-module. Then $M$ is of quadratic, respectively symplectic type if and only if its projective cover $P_G(M)$ is of quadratic, respectively symplectic type.*

Let $\widehat{\phantom{x}}$ denote the involution $g \to g^{-1}$ for $g \in G$ extended to a $k$-antialgebra automorphism of $kG$. Let $P = ekG$ with a primitive idempotent $e$ of $kG$. Then there is a $G$-invariant non-degenerate pairing

$$b : ekG \times \hat{e}kG \longrightarrow k \qquad 2.3$$

defined by $b(ea, \hat{e}b) = \alpha(ea\widehat{\hat{e}b}) = \alpha(ea\hat{b}e)$ where $\alpha$ is a functional giving ekGe the structure of a symmetric algebra. ($\alpha$ may be chosen as the restriction of the trace map $\sum_g a_g g \to a_1$ which makes kG to a symmetric algebra.) Thus, $P^* \cong \hat{e}kG$. Moreover,

$$P \text{ is of quadratic type if } e = \hat{e} . \qquad 2.4$$

Next consider the $k$-algebra homomorphism

$$\hat{e}kGe \longrightarrow ekGe \qquad 2.5$$

given by left multiplication with $e$. Since $\hat{e}kG$ and $ekG$ are projective, this is an automorphism whenever $e\hat{e}$ is not contained in the Jacobson radical $J(kG)$ of $kG$. Suppose $b \neq 0$ is a $G$-invariant symmetric form on the head $M = \overline{P}$ of $P$. Choose $m \in M$ with $b(m, m) \neq 0$. Via the $kG$-epimorphism

$$\begin{aligned} kG &\longrightarrow M \\ a &\longrightarrow ma \end{aligned}$$

we may find a primitive idempotent $e'$ such that $m = me'$ and $\overline{e'}\,\overline{kG} \cong M$.

To avoid notation, let $e$ be $e'$. We obtain

$$0 \neq b(m, m) = b(me, me) = b(m, me\hat{e}) .$$

In particular, $e\hat{e} \notin J(kG)$. Thus by the isomorphism 2.5, there exists an $a \in kG$ such that

$$e\hat{e}ae = e .$$

If we put $f = \hat{e}ae$, then $f$ and $\hat{f}$ are non-zero idempotents in the local ring $\hat{e}kGe$. Hence $f = \hat{f}$ and $\overline{f}\,\overline{kG} \cong M$. This fact has been observed by Landrock and Manz [11]. Together with 2.2 we have proved

**Proposition 2.6.** *Let $e$ be a primitive idempotent of $kG$. Then $ekG$ is of quadratic type if and only if there is an idempotent $f = \hat{f}$ with $fkG \cong ekG$.*

However, this observation is not appropriate in determining the type of a given simple module $M \cong M^*$ or equivalently its projective cover. In the classical case, this may be done by use of the Frobenius-Schur indicator provided the squaring map on $G$ and the character values on $M$ are known, *i.e.*

**Theorem 2.7** (Frobenius, Schur [7]). *If $M$ is a simple $\mathbf{C}G$-module with character $\chi$, then*

$$\nu_2(\chi) := \frac{1}{|G|} \sum_{g \in G} \chi(g^2) \in \{1, -1, 0\}$$

*with value 1, -1, 0 respectively if and only if $M$ is of quadratic, symplectic, not selfdual type respectively.*

With that beautiful result in mind one is attempted to look for a Frobenius-Schur indicator in characteristic $p$. Indeed, for $p \neq 2$ as always assumed throughout this section there is an analogue. In order to state the result, let $d_{\chi,\beta}$ denote the decomposition number with respect to $\chi \in \mathrm{Irr}(G)$ and $\beta \in \mathrm{IBr}(G)$. Note that a simple $kG$-module $M$ is selfdual if and only if its Brauer character is real valued.

The $p$-analogue of 2.7 is now as follows.

**Theorem 2.8** (Feit, Thompson, Willems, [15]). *Let $\beta$ be the Brauer character of a simple selfdual $kG$-module $M$. Then we have*
(a) *There exists $\chi = \overline{\chi} \in \mathrm{Irr}(G)$ with $d_{\chi\beta} \equiv 1 (\mathrm{mod}\, 2)$.*
(b) *$\chi$ is of quadratic, respectively symplectic type if and only if $\beta$ is of quadratic, respectively symplectic type.*
*Thus $\nu_2(\chi)$ may serve as a Frobenius-Schur indicator for $\beta$.*

**Proof.** Part (a) of 2.8 may be easily seen as follows. Consider the generalized character $\psi$ defined by

$$\psi(g) = \beta(g_{p'})$$

where $g_{p'}$ is the $p'$-part of the decomposition $g = g_{p'}g_p = g_pg_{p'}$ into $p-$ and $p'$-parts. Clearly

$$\psi = \sum_{\substack{(\chi,\overline{\chi})\\ \nu_2(\chi)=0}} a_\chi(\chi + \overline{\chi}) + \sum_{\substack{\chi\\ \nu_2(\chi)\neq 0}} a_\chi \chi \text{ with } a_\chi \in \mathbf{Z}$$

and therefore

$$\beta = \sum_{\substack{(\chi,\overline{\chi})\\ \nu_2(\chi)=0}} \sum_{\alpha \in \mathrm{IBr}(G)} a_\chi(d_{\chi\alpha} + d_{\overline{\chi}\alpha})\alpha + \sum_{\substack{\chi\\ \nu_2(\chi)\neq 0}} \sum_{\alpha \in \mathrm{IBr}(G)} a_\chi d_{\chi\alpha}\alpha\,.$$

Comparing the coefficients of $\beta$ yields

$$1 = \sum_{\substack{(\chi,\overline{\chi})\\ \nu_2(\chi)=0}} 2a_\chi d_{\chi\beta} + \sum_{\substack{\chi\\ \nu_2(\chi)\neq 0}} a_\chi d_{\chi\beta}\,,$$

proving (a). Observe that (a) holds true for $p = 2$.

To prove part (b), first observe the following general fact. Given a $kG$-module $V$ over an arbitrary field and a simple $kG$-module $M \cong M^*$ with odd multiplicity as a composition factor of $V$. Then, if $V$ is of symmetric, respectively symplectic type, then $M$ is of symmetric, respectively symplectic type. For, if $X$ is a simple submodule of $V$, then with respect to the given bilinear form
either (1) $V = X \perp X^{\perp}$,
or (2) $X^{\perp}/X$ is of the same type as $V$ and $V/X^{\perp} \cong X^*$.
In case (1) with $X \cong M$ we already have the assertion. Otherwise repeat the argument with $V = X^{\perp}$ or $V = X^{\perp}/X$.

Thus according to (a) it is sufficient to find a lattice $L$ affording $\chi$ such that its reduction $\widetilde{L}$ has a filtration

$$0 \subset X_1 \subset \ldots \subset X_t = \widetilde{L}$$

with $kG$-modules $X_i$ and $X_{i+1}/X_i$ of the same type as $\chi$. Indeed, this can be done with $t = 2$. The interested reader is referred to [15] or ([13], §5).

## 3. Forms in characteristic 2

In this section let $k$ be a perfect field of characteristic 2. For a finite dimensional $k$-space $M$ let $M^{(2)}$ denote the Frobenius twist of $M$, *i.e.* if $\sigma : x \to x^2$ denotes the Frobenius automorphism on k, then the scalars are acting on $M$ by

$$\lambda \circ m := \sigma^{-1}(\lambda)m \quad (m \in M, \lambda \in k) .$$

Furthermore, let

$$\begin{aligned} \Theta : S^2(M^*) &\longrightarrow \Lambda^2(M^*) \\ q &\longrightarrow b \end{aligned}$$

denote "the polarization" which is defined by

$$b(m_1, m_2) = q(m_1 + m_2) - q(m_1) - q(m_2) \quad (m_i \in M) .$$

Note that $M^{(2)*}$ is just the kernel of $\Theta$. With this notation and the diagonal map $\zeta$ of 1.2 we have the following diagram

$$\begin{array}{c} 0 \\ \uparrow \\ 0 \to M^{(2)*} \to S^2(M^*) \overset{\Theta}{\to} \Lambda^2(M^*) \to 0 \\ \zeta \uparrow \\ M^* \otimes_k M^* \\ \uparrow \\ \Lambda^2(M^*) \\ \uparrow \\ 0 \end{array}$$

where both, the horizontal and vertical sequences are exact. If $M$ is endowed with a $kG$-structure, then the diagram may be considered in the category of $kG$-modules. Taking then fixed points we obtain the diagram

$$
\begin{array}{ccccccccc}
 & & & & \vdots & & & & \\
 & & & & \uparrow & & & & \\
 & & & & H^1(G,\Lambda^2(M^*)) & & & & \\
 & & & & \uparrow & & & & \\
0 \to & M^{(2)*G} & \to & S^2(M^*)^G & \overset{\Theta}{\to} & \Lambda^2(M^*)^G & \to & H^1(G,M^{(2)*}) & \to \dots \\
 & & & & \uparrow & & & & \\
 & & & & (M^* \otimes_k M^*)^G & & & & \\
 & & & & \uparrow & & & & \\
 & & & & \Lambda^2(M^*)^G & & & & \\
 & & & & \uparrow & & & & \\
 & & & & 0 & & & &
\end{array}
\qquad (D)
$$

Suppose that $M$ is a selfdual $kG$-module without $G$-fixed points. Hence $\Theta$ is a monomorphism and by the diagram above, $\Lambda^2(M^*)^G \neq 0$ since $(M^* \otimes_k M^*)^G \cong \mathrm{End}_{kG}(M^*) \neq 0$. This is just Fong's Lemma (see [6]).

At this point it is natural to ask:

1. What conditions on $G$ and $M$ imply that $S^2(M^*)^G \neq 0$, or even stronger, that $S^2(M^*)^G \cong \Lambda^2(M^*)^G$ via the polarization map?
2. Is there a Frobenius-Schur indicator for simple modules, projective indecomposable modules, *i.e.* a formula easy to handle which determines the existence or non-existence of a non-degenerate $G$-invariant quadratic form?

The latter question is quite challenging and has not yet found an answer. The situation is very "elusive" as Thompson already remarked in [15]. However, question 1 can be answered in a wide range of groups and modules using diagram $(D)$ and results on vanishing cohomology in degree 1. This is one of the main subjects in paper [14]. To see how the methods work we like to mention

**Proposition 3.1** ([14], Prop. 2.6). *Let $M \cong M^*$ be a faithful indecomposable $kG$-module. If $O_{2'}(G) \neq 1$, then $S^2(M^*)^G \cong \Lambda^2(M^*)^G$ via the polarization.*

**Proof.** By a well-known result of R. Brauer, $O_{2'}(G)$ acts trivially on the principal 2-block. Thus $H^1(G, M^{(2)*}) = 0$ since $M^{(2)*}$ is faithful. Now the horizontal sequence in the diagram $(D)$ yields the assertion.

Dealing with problem 1 we may always assume that $M$ is faithful. If we specialize to simple modules we may also assume that $O_{2',2}(G) = 1$ by 3.1. For such groups non-vanishing of the first cohomology group for a faithful simple module heavily restricts the structure of $G$. Because of its own interest we state more generally as we need for our purpose.

**Theorem 3.2** ([16]). *Let $1 = O_{p',p}(G) < G$ and let $k$ be a field of characteristic $p$. Then the number of minimal normal subgroups of $G$ is equal to*

$$\mathrm{Min}\,\{r \,|\, r \in \mathbf{N}, \text{ there exists a faithful simple } kG\text{-module } M \text{ with}$$

$$H^r(G,M) \neq 0\} \ .$$

This result again in combination with diagram $(D)$ leads immediately to

**Proposition 3.3.** *Let $M \cong M^*$ be a faithful simple module for $G \neq 1$. If $\Theta : S^2(M^*)^G \to \Lambda^2(M^*)^G$ is not an isomorphism, i.e. not onto, then the socle of $G$ is a minimal normal non-abelian subgroup of $G$.*

*Remarks 3.4.* (a) Let the situation be as in 3.3, *i.e.* $\Theta : S^2(M^*)^G \to \Lambda^2(M^*)^G$ is not onto for a faithful simple selfdual module $M$. Let $S$ denote the socle of $G \neq 1$. Then by Clifford's theorem

$$M_{|S} = e\left(\bigoplus_g (N \otimes g)\right) \ ,$$

where $N$ is a simple kS-module and $e \in \mathbf{N}$.

If $N \ncong N^*$, then it is easy to see that $S^2(M^*)^S \cong \Lambda^2(M^*)^S$ via $\Theta$, from which $S^2(M^*)^G \cong \Lambda^2(M^*)^G$ follows. Thus $N \cong N^*$ and $M \cong L \otimes_{kI} kG$, where $L \cong L^*$ is a simple $kI$-module and $I$ is the inertial group of $N$ in $G$. Furthermore,

$$\Theta : S^2(L^*)^I \longrightarrow \Lambda^2(L^*)^I \qquad (*)$$

is not onto, otherwise

$$S^2(L^*)^G \cong \Lambda^2(L^*)^G$$

via inducing forms.

Let $K = \mathrm{Ker}(L) \leq I$ and let $S = S_1 \times \ldots \times S_r$ with simple groups $S_i$. By $(*)$ and the diagram $(D)$, we have $H^1(I,L) \neq 0$. Hence $H^1(S,L) \neq 0$ and the use of the Künneth formula shows that only one of the $S_i$, say $S_1$, acts non-trivially on $L$.

Therefore $I \subseteq N_G(S_1)$. Putting $\overline{I} = I/K$ we have $C_{\overline{I}}(\overline{S}_1) \trianglelefteq \overline{I}$ and $\overline{S}_1 \cap C_{\overline{I}}(\overline{S}_1) = 1$.

If $C_{\overline{I}}(\overline{S}_1) \neq 1$, then $\overline{I}$ has at least two minimal normal subgroups. This is a contradiction to the fact that

$$\Theta : S^2(L^*)^{\overline{I}} \longrightarrow \Lambda^2(L^*)^{\overline{I}}$$

is not onto by $(*)$ and 3.3. Hence $C_{\overline{I}}(\overline{S}_1) = 1$ and $\overline{S}_1 \trianglelefteq N_{\overline{I}}(\overline{S}_1)$, which proves that $\overline{I}$ is an almost simple group. Thus for simple faithful selfdual modules question 1 allows a reduction to almost simple groups.

(b) In the special case of Lie-type groups information about the first cohomology group may be found in [4], [10].

*Remarks 3.5.*

(a) As in section 2 we may ask about the connection between the type of a simple selfdual module $M$ and its projective cover $P = P_G(M)$. Unfortunately, there is not one. It may happen — even in the class of solvable groups — that $M$ is of quadratic type, but $P$ is not and vice versa (see [14]). In view of this fact, Theorem 3.9 below is very surprising.

(b) Note that a projective module $P \cong P^*$ is of quadratic type if and only if $P$ is of symplectic type. This follows from diagram $(D)$ since $H^1(G,P) = 0$. Suppose $2 \mid |G|$ and let $t$ be an involution of $G$. Then

$$b(g,h) = \begin{cases} 1 & \text{if } tg = h \\ 0 & \text{otherwise} \end{cases}$$

defines by $k$-bilinear extension a non-degenerate $G$-invariant symplectic form on the regular module $kG$. The restriction of $b$ on the projective cover of the trivial module remains non-degenerate. This is a consequence of the next Lemma which we also need in the sequel of this section.

**Lemma 3.6.** *Let $k$ be any field and let $G$ be a finite group. Suppose $P$ is a projective $kG$-module with a non-degenerate $G$-invariant symmetric or symplectic form $b$. If $P$ is orthogonal indecomposable w.r.t. $b$, then either $P$ is an indecomposable $kG$-module or $P \cong P_0 \oplus P_0^*$ with an indecomposable $kG$-module $P_0$.*

**Proof.** Let $P = P_1 \oplus \ldots \oplus P_s$ with indecomposable $kG$-modules $P_i$. Suppose that $s \geq 2$. Clearly, $b(\mathrm{Soc}(P_i), P_i) = 0$ for all $i$. After renumbering the $P_i$ if necessary we may assume that

$$b(\mathrm{Soc}(P_1), P_2) \neq 0 .$$

We claim that $b$ is non-degenerate on $P_1 \oplus P_2$. Suppose that $M$ is a simple submodule of the radical of $b$ restricted to $P_1 \oplus P_2$. Since $M \neq \mathrm{Soc}(P_1)$, there exists an $\alpha \in \mathrm{Hom}_{kG}(\mathrm{Soc}(P_2), \mathrm{Soc}(P_1))$ such that

$$M = \{s_2\alpha + s_2 | s_2 \in Soc(P_2)\} .$$

Note that for all $s_2 \in \mathrm{Soc}(P_2)$, we have

$$0 = b(s_2\alpha + s_2, P_2) = b(s_2\alpha, P_2) ,$$

hence $\alpha = 0$ and $M = Soc(P_2)$.

But $b(P_1, \mathrm{Soc}(P_2)) = 0$ contradicts $b(Soc(P_1), P_2) \neq 0$. Thus $b$ is non-degenerate on $P_1 \oplus P_2$. Furthermore, $Soc(P_1)^*$ is isomorphic to the head of $P_2$. Hence $P_2 \cong P_1^*$ as required.

**Lemma 3.7.** *Let $G$ be a finite group of Lie type defined over a field of characteristic 2. Let $k$ be a splitting field for $G$ of characteristic 2 and let $St$ denote the Steinberg module for $G$. If $P \cong P^*$ is an indecomposable projective $kG$-module, then there exists a simple $kG$-module $M \cong M^*$ such that the multiplicity of $P$ in $M \otimes_k St$ is odd.*

**Proof.** To be precise we may suppose that the following situation of fields is given. Let $K_0$ be an algebraic number field, $\rho$ a prime ideal in the ring of integers of $K_0$ containing 2. Let $R$ be the ring of integers in the $\rho$-adic completion of $K_0$ and let $(\pi)$ be the maximal ideal of $R$. Finally put $K = \mathrm{Quot}(R)$ and $k = \bar{R} = R/(\pi)$. Suppose that $K$ and $k$ are splitting fields for $G$.

Let $\mathrm{IBr}(G) = \{\varphi_1, \ldots, \varphi_h\}$ be the set of irreducible Brauer characters with values in $K$ where the numbering is chosen such that $\varphi_1, \ldots, \varphi_s$ are exactly the real valued characters. Let $\Phi_1, \ldots, \Phi_h$ denote the corresponding projective characters over $K$ and let $\varphi_s = \Phi_s = \Phi$ be the Steinberg character. Finally, let $g_1, \ldots, g_h$ be a set of representatives of the 2'-conjugacy classes of $G$, the numbering again chosen such that $g_1, ..., g_s$ are real.

Consider now the character equation

$$\varphi_i\Phi = \sum_{j=1}^{h} a_{ij}\Phi_j \text{ with } a_{ij} \in \mathbf{N} \cup \{0\} \ (i = 1, \ldots, h) \ .$$

Since for $i = 1, \ldots, s$ the left hand side is real valued, we have

$$\varphi_i\Phi = \sum_{j=1}^{s} a_{ij}\Phi_j + \sum_{j=s+1}^{h} a_{ij}(\Phi_j + \overline{\Phi}_j) \ (i = 1, \ldots, s) \ .$$

(Here $\overline{\phantom{x}}$ denotes complex conjugation.)

By ([5], Chap.IV, 2.5), we have

$$\Phi_j(g_k) = |C_G(g_k)|_2 \, \alpha_j(g_k) \text{ with } a_j(g_k) \in R \text{ for all } j \text{ and } k$$

and by ([3], 6.4.7),

$$\Phi(g_k) = \pm |C_G(g_k)|_2 \ \text{ for all } k \ .$$

For $k = 1, \ldots, s$ it follows

$$\pm\varphi_i(g_k) \, |C_G(g_k)|_2 = \sum_{j=1}^{s} a_{ij} \, |C_G(g_k)|_2 \, \alpha_j(g_k) + \sum_{j=s+1}^{h} 2a_{ij} \, |C_G(g_k)|_2 \, \alpha_j(g_k) \ .$$

Hence

$$\widetilde{\varphi_i(g_k)} = \sum_{j=1}^{s} \widetilde{a_{ij}} \widetilde{\alpha_j(g_k)} \text{ for } i, k = 1, \ldots, s \ .$$

By a result [8] of Gow we have

$$\det(\widetilde{\varphi_i(g_k)})_{i,k=1,\ldots,s} \neq 0 \ .$$

Thus, to a given $j \in \{1, \ldots, s\}$ there exists an $i \in \{1, \ldots, s\}$ such that $2 \not| a_{ij}$.Q.E.D.

Finally we need the known fact that in characteristic 2 the tensor product of two symplectic spaces is the polarization of a quadratic space, more precisely

**Proposition 3.8** ([14], [1]). *Let $M$ and $N$ be finite dimensional $k$-spaces where the field $k$ is perfect and of characteristic 2. Then in the pullback diagram*

$$\begin{array}{ccccccc}
0 \to (M \otimes_k N)^{(2)*} \to & S^2(M^* \otimes_k N^*) & \overset{\Theta}{\to} & \Lambda^2(M^* \otimes_k N^*) & \to 0 \\
\| & \cup\!\mathsf{I} & & \cup\!\mathsf{I} & \\
0 \to (M \otimes_k N)^{(2)*} \to & E(M,N) & \overset{\Theta}{\to} & \Lambda^2(M^*) \otimes_k \Lambda^2(N^*) & \to 0
\end{array}$$

*the bottom row splits naturally. The splitting is given by the map*

$$\begin{aligned} \Phi : E(M,N) &\longrightarrow (M \otimes_k N)^{(2)*} \\ q &\longrightarrow \Phi_q \end{aligned} ,$$

where $\Phi_q$ is defined by $\Phi_q(m \otimes n) = q(m \otimes n)$ $(m \in M, n \in N)$.

**Theorem 3.9** (Gow, Willems). *Let $G$ be a finite group of Lie type defined over a field of characteristic 2. Let $k$ be a splitting field for $G$, perfect and also of characteristic 2. If $P \cong P^*$ is a projective indecomposable module of $kG$, then $P$ is of quadratic type.*

**Proof.** By 3.7 there exists a simple selfdual $kG$-module $M$ such that the multiplicity of $P$ in $M \otimes_k St$ is odd. Recall that the Steinberg module is projective, simple and selfdual, hence of quadratic type by Fong's Lemma and 3.5 (b). Thus we may assume that $M$ is not the trivial module. Again by Fong's lemma, $M$ is of symplectic type. Thus by 3.8, $M \otimes_k St$ is of quadratic type and the assertion of the theorem follows by 3.6.

*Remark 3.10.* Let $\overline{G}$ be a simply connected simple algebraic group defined over an algebraically closed field of characteristic 2 with Frobenius map $\sigma$ such that the group $G = \overline{G}^{\sigma}$ of fixed points of $\sigma$ is finite. By ([3], 3.7.3), two $\sigma$-stable semisimple elements are conjugate in $G^{\sigma}$ if and only if they are conjugate in $G$. If $g$ is a semisimple $\sigma$-stable element of $G$, then $g$ is contained in a $\sigma$-stable maximal torus $T$ of $G$. If furthermore $-1$ belongs to the Weyl group $W(T)$ of $T$, then $g$ is conjugate to $g^{-1}$ in $G$, hence in $G^{\sigma}$. Thus in case $-1 \in W(T)$, all $\sigma$-stable semisimple elements of $G$ are real. This is fullfilled if $G$ is of type

$$B_l, C_l, D_l \ (l \text{ even}), E_7, E_8, F_4 \text{ and } G_2 .$$

(see [2], Planche I-IX). Hence for such groups all simple $kG$-modules are selfdual and of symplectic type. Those which have more than two non-trivial factors in Steinberg's tensor product decomposition are of quadratic type by 3.8. According to 3.9, the projective covers are all of quadratic type.

## 4. Problems and Remarks

**a) Find a Frobenius-Schur indicator for simple modules in characteristic 2.**

This problem was already mentioned in section 2. For $p$ odd, the indicator relies on the connection between the geometries in characteristic 0 and $p$ via constant reduction (see 2.8). This is far away to be understood if $p = 2$.

*Examples.*

(i) Let $G = \langle x, y | o(x) = 4, o(y) = 3, y^x = y^{-1} \rangle$. Then there is a $\beta \in \mathrm{IBr}_2(G)$ of quadratic type which has two lifts $\chi_1, \chi_2 \in \mathrm{Irr}_{\mathbb{C}}(G)$, one with Frobenius-Schur indicator $+1$, the other with indicator $-1$.

(ii) The Tits simple group ${}^2F_4(2)'$ has two simple, non real-valued complex characters of degree 26. In characteristic 2 both reduce to the same simple Brauer character which is of quadratic type.

(iii) Let $2|n$ and let $\chi = \pi - 1 \in \mathrm{Irr}_{\mathbb{C}}(A_n)$ where $\pi$ is the permutation character of the alternating group $A_n$. Clearly, $v_2(\chi) = 1$ and $\chi = 1 + \beta$ on 2'-elements with $\beta \in \mathrm{IBr}_2(A_n)$. $\beta$ is of quadratic type if and only if $4|n$.

In this context, even the answer to the following problem seems to be unknown. Let $G$ be solvable and $\beta = \overline{\beta} \in \mathrm{IBr}_2(G)$. Suppose that $\beta$ is not the trivial character. Thus $\beta$ is of quadratic type by Fong's Lemma and 3.3. Does there exist a lift $\chi \in \mathrm{Irr}_{\mathbb{C}}(G)$ of $\beta$ with $v_2(\chi) = 1$? (According to the work of M. Isaacs on liftability there is always a real-valued lift, see [9].)

**b) Let $k$ be finite. Find a handy criterion to determine the Witt index of a simple even dimensional module of quadratic type.**

Recall that the Witt-index of a quadratic non-degenerate $k$-space $(M, q)$ is the dimension of a maximal totally isotropic subspace of $M$. If $2 \not| \dim M$, then the Witt index is $\frac{\dim M - 1}{2}$. In case $2 | \dim M$, there is an ambiguity; the index is either $\frac{\dim M}{2}$ or $\frac{\dim M}{2} - 1$ and usually it is a tedious task to find the value. The situation is much better if $M$ is endowed with a $kG$-structure since we have

**Theorem** [14]. *Let $M$ be an even-dimensional $kG$-module with non-degenerate $G$-invariant quadratic form $q$. Suppose that $G$ has an elementary abelian $r$-subgroup $A$ for some odd prime $r$ and $M^A = 0$. Let $s$ denote the multiplicative order of $|k| \pmod r$. Then $(M, q)$ has Witt index $\frac{\dim M}{2}$ if and only if $s \big| \frac{\dim M}{2}$.*

Unfortunately, even for simple modules this is not applicable in any case since the assumption $M^A = 0$ is rather restrictive. For instance, if $k$ is a field of characteristic 2, then the alternating group $A_7$ has two simple modules of quadratic type, a 6- and a 14-dimensional one. For the 6-dimensional module, the theorem applies with $|A| = 7$, but for the other one it does not.

We would like to mention here, that for $G$ solvable and $M$ simple and faithful over a field of characteristic 2, the theorem provides a complete answer to **b)**. (Note that in this case we can find $A \trianglelefteq G$, hence $M^A = 0$ by Clifford's theorem.)

This means: A solvable group $G$ cannot be embedded as an irreducible group in both $O^+(2n, 2^f)$ and $O^-(2n, 2^f)$ for any $n$. Does this hold true if we drop the assumption on solvability?

Because of the important role of classical groups and their representations in many areas, further development would be very desirable.

## REFERENCES


[1] M. Aschbacher, "On the maximal subgroups of classical groups", Inv. Math. 76 (1984), 469–514.

[2] N. Bourbaki, "Groupes et algèbres de Lie", Chap. 4, 5 et 6, Herman, Paris (1968).

[3] R.W. Carter, "Finite groups of Lie type, conjugacy classes and complex characters", Wiley (1985).

[4] E. Cline, B. Parshall, L. Scott, "Cohomology of finite groups of Lie type I, II", Publ. Math. IHES 45 (1975), 169–191 and J. Algebra 45 (1977), 182–198.

[5] W. Feit, "The representation theory of finite groups", North-Holland (1982).

[6] P. Fong, "On decomposition numbers of $J_1$ and $R(q)$", in: Symposia Mathematica 13 (1974), 414–422.

[7] G. Frobenius, I. Schur, "Über die reellen Darstellungen der endlichen Gruppen", Collected works of Frobenius 3, 355–377.

[8] R. Gow, "Real valued characters and the Schur index", J. Alg. 40 (1979), 258–270.

[9] I.M. Isaacs, "Lifting Brauer characters for solvable groups II", J. Alg. 51 (1978), 476–490.

[10] W. James, B. Parshall, "On the 1-cohomology of finite groups of Lie type", in: Proceed. of the Conference on finite groups, Utah (1975), 313–328.

[11] P. Landrock, O. Manz, "Symmetric forms and idempotents", Preprint.

[12] T. Okuyama, "On a problem of Wallace", Preprint.

[13] D. Quillen, "The Adams conjecture", Topology 10 (1971), 53–65.

[14] P. Sin, W. Willems, "$G$-invariant quadratic forms", submitted to J. Reine Angew. Math.

[15] J. G. Thompson, "Bilinear forms in characteristic $p$ and the Frobenius-Schur indicator", in: Group Theory, Beijing (1984), SLN 1184, 221–230.

[16] W.Willems, "On the irreducible faithful modules and their cohomology", to appear Bull. London Math. Soc.

## Progress in Mathematics

*Edited by:*

J. Oesterlé
Département de Mathématiques
Université de Paris VI
4, Place Jussieu
75230 Paris Cedex 05
France

A. Weinstein
Department of Mathematics
University of California
Berkeley, CA 94720
U.S.A.

*Progress in Mathematics* is a series of books intended for professional mathematicians and scientists, encompassing all areas of pure mathematics. This distinguished series, which began in 1979, includes authored monographs and edited collections of papers on important research developments as well as expositions of particular subject areas.

We encourage preparation of manuscripts in such forms as LaTeX or AMS TeX for delivery in camera-ready copy which leads to rapid publication, or in electronic form for interfacing with laser printers or typesetters.

Proposals should be sent directly to the editors or to: Birkhäuser Boston, 675 Massachusetts Avenue, Cambridge, MA 02139, U.S.A.

A complete list of titles in this series is available from the publisher.

49 SCHLICHTKRULL. Hyperfunctions and Harmonic Analysis on Symmetric Spaces
50 BOREL, ET AL. Intersection Cohomology
51 BERTIN/GOLDSTEIN. Séminaire de Théorie des Nombres. Paris 1982-83
52 GASQUI/GOLDSCHMIDT. Déformations Infinitésimales des Structures Conformes Plates
53 LAURENT. Théorie de la Deuxième Microlocalisation dans le Domaine Complexe
54 VERDIER/LE POTIER. Module des Fibres Stables sur les Courbes Algébriques: Notes de l'Ecole Normale Supérieure, Printemps, 1983
55 EICHLER/ZAGIER. The Theory of Jacobi Forms
56 SHIFFMANN/SOMMESE. Vanishing Theorems on Complex Manifolds
57 RIESEL. Prime Numbers and Computer Methods for Factorization
58 HELFFER/NOURRIGAT. Hypoellipticité Maximale pour des Opérateurs Polynomes de Champs de Vecteurs
59 GOLDSTEIN. Séminaire de Théorie des Nombres, Paris 1983-84
60 PROCESI. Geometry Today: Giornate Di Geometria, Roma, 1984
61 BALLMANN/GROMOV/SCHROEDER. Manifolds of Nonpositve Curvature
62 GUILLOU/MARIN. A la Recherche de la Topologie Perdue
63 GOLDSTEIN. Séminaire de Théorie des Nombres, Paris 1984-85
64 MYUNG. Malcev-Admissible Algebras
65 GRUBB. Functional Calculus of Pseudo-Differential Boundary Problems

66 CASSOU-NOGUÈS/TAYLOR. Elliptic Functions and Rings and Integers
67 HOWE. Discrete Groups in Geometry and Analysis: Papers in Honor of G.D. Mostow on His Sixtieth Birthday
68 ROBERT. Autour de L'Approximation Semi-Classique
69 FARAUT/HARZALLAH. Deux Cours d'Analyse Harmonique
70 ADOLPHSON/CONREY/GHOSH/YAGER. Number Theory and Diophantine Problems: Proceedings of a Conference at Oklahoma State University
71 GOLDSTEIN. Séminaire de Théorie des Nombres, Paris 1985-1986
72 VAISMAN. Symplectic Geometry and Secondary Characteristic Classes
73 MOLINO. Riemannian Foliations
74 HENKIN/LEITERER. Andreotti-Grauert Theory by Integral Formulas
75 GOLDSTEIN. Séminaire de Théorie des Nombres, Paris 1986-1987
76 COSSEC/DOLGACHEV. Enriques Surfaces
77 REYSSAT. Quelques Aspects des Surfaces de Riemann
78 BORHO/BRYLINSKI/MACPHERSON. Nilpotent Orbits, Primitive Ideals, and Characteristic Classes
79 MCKENZIE/VALERIOTE. The Structure of Decidable Locally Finite Varieties
80 KRAFT/PETRIE/SCHWARZ. Topological Methods in Algebraic Transformation Groups
81 GOLDSTEIN. Séminaire de Théorie des Nombres, Paris 1987-1988
82 DUFLO/PEDERSEN/VERGNE. The Orbit Method in Representation Theory: Proceedings of a Conference Held in Copenhagen, August to September 1988
83 GHYS/DE LA HARPE. Sur les Groupes Hyperboliques d'après Mikhael Gromov
84 ARAKI/KADISON. Mappings of Operator Algebras: Proceedings of the Japan-U.S. Joint Seminar, University of Pennsylvania, Philadelphia, Pennsylvania, 1988
85 BERNDT/DIAMOND/HALBERSTAM/HILDEBRAND. Analytic Number Theory: Proceedings of a Conference in Honor of Paul T. Bateman
86 CARTIER/ILLUSIE/KATZ/LAUMON/MANIN/RIBET. The Grothendieck Festschrift: A Collection of Articles Written in Honor of the 60th Birthday of Alexander Grothendieck, Volume I
87 CARTIER/ILLUSIE/KATZ/LAUMON/MANIN/RIBET. The Grothendieck Festschrift: A Collection of Articles Written in Honor of the 60th Birthday of Alexander Grothendieck, Volume II
88 CARTIER/ILLUSIE/KATZ/LAUMON/MANIN/RIBET. The Grothendieck Festschrift: A Collection of Articles Written in Honor of the 60th Birthday of Alexander Grothendieck, Volume III
89 VAN DER GEER/OORT/STEENBRINK. Arithmetic Algebraic Geometry
90 SRINIVAS. Algebraic K-Theory
91 GOLDSTEIN. Séminaire de Théorie des Nombres, Paris 1988-1989
92 CONNES/DUFLO/JOSEPH/RENTSCHLER. Operator Algebras, Unitary Representations, and Invariant Theory
93 AUDIN. Hamiltonian Torus Actions
94 MORA/TRAVERSO. Effective Methods in Algebraic Geometry
95 MICHLER/RINGEL. Representation Theory of Finite Groups and Finite-Dimensional Algebras